Mathematics

with Applications to Business, Economics, and Social Sciences

Richard Bouldin
University of Georgia, Athens

SAUNDERS COLLEGE PUBLISHING

Philadelphia   New York   Chicago
San Francisco   Montreal   Toronto
London   Sydney   Tokyo   Mexico City
Rio de Janeiro   Madrid

Address orders to:
383 Madison Avenue
New York, NY 10017

Address editorial correspondence to:
West Washington Square
Philadelphia, PA 19105

Text Typeface: Times Roman
Compositor: Progressive Typographers
Acquisitions Editor: Leslie Hawke
Developmental Editor: Jay Freedman
Project Editor: Sally Kusch
Copy Editor: Charlotte Nelson
Art Director: Carol Bleistine
Art/Design Assistant: Virginia A. Bollard
Text Design: Caliber Design Planning, Inc.
Cover Design: Lawrence R. Didona
Text Artwork: J & R. Technical Services, Inc.
Production Manager: Tim Frelick
Assistant Production Manager: Maureen Iannuzzi

Cover credit: D. Hamilton/THE IMAGE BANK

Library of Congress Cataloging in Publication Data

Bouldin, Richard.
   Mathematics, with applications to business,
economics, and social sciences.

   Includes index.
   1. Mathematics—1961-   I. Title.
QA37.2.B66   1984   510   84-10562
ISBN 0-03-062164-X

Mathematics with Applications to Business,
Economics and Social Sciences

ISBN 0-03-062164-X

3456   32   987654321

CBS COLLEGE PUBLISHING
Saunders College Publishing
Holt, Rinehart and Winston
The Dryden Press

Preface
for Instructor

The purpose of this book is to present the material frequently taught in a two-quarter or two-semester mathematics course offered to business students and others. The goal of the book is to be clear and persuasive. I believe that the ideas are offered in a way that would convince anyone that this material is natural, accessible and powerful. The book takes the points of view of the small-business person, investor and consumer. Most of the students taking such courses are not destined to be corporate managers but rather sales people, clerks, assistant managers, and small-business managers. The book also considers macroeconomics from the point of view of a business person and voting citizen.

Exercises There are more exercises than an instructor will ordinarily assign, and the number assigned will probably vary from one class to another. One instructor might assign every fourth exercise to a well-prepared class with good aptitude, while another instructor might assign every other exercise to a class that needs a lot of practice. Since the odd-numbered exercises have answers given in the back of the text and the even-numbered exercises do not, an instructor can assign only problems with answers given, or only problems without answers given, or any combination desired.

The exercises increase in algebraic complication as the numbers increase. The beginning exercises require very few steps, and the answers are frequently integers. Thus, the instructor can assign only the easier early exercises if desired, or the instructor can construct homework assignments using predominantly the later more difficult exercises. Word problems are provided at the end of almost every section. Exercises not assigned can be used for classroom examples, quizzes, and review work.

The section at the end of each chapter entitled *Review Problems* is intended as a comprehensive test on the problem-solving methods of the preceding chapter. Since students will ordinarily use this section as a study device, it is substantially longer than an in-class exam on the same material. The Review Problems give the student the opportunity to practice at associating a technique of solution with a problem that is not identified according to section. The answers to all Review Problems are given in the back of the text.

Applications   Applications to business, economics, and personal finance are integrated into the body of the text. Most of these applications are introduced through examples given shortly after the relevant mathematics is presented. If new terms are required, they are covered immediately prior to the examples. The personal-finance applications should interest all students. Understanding concepts like "rate of return," "profit margin," "inflation," and "mortgage" is important for any educated consumer.

The section at the end of each chapter entitled *Social Science Applications* provides applications which can be used to motivate the material for social science students not involved with business or economics. Since each application is clearly labeled by discipline, the instructor can choose those applications most appropriate to a particular class. Occasionally some non-mathematical material will be presented in the first paragraphs of the section to facilitate the subsequent applications.

Sequence of Chapters   The chapters of this text have been constructed in a way that permits flexibility in the order of presentation. Chapter 13, Sets and Counting, is independent of all other chapters, and all earlier chapters are independent of it. Chapter 14, Probability, depends only on Chapter 13, except for the optional Section 14.5, which also depends on Chapter 3, Systems of Equations and Matrices. In a program that emphasizes calculus the chapters might be covered in the order that they appear in the book. In another program with less emphasis on calculus, Chapters 13 and 14 might be taught immediately after Chapter 6, Mathematics of Money, or immediately after Chapter 5, Linear Programming, if Chapter 6 is not covered.

In any program that leaves the teaching of linear programming to the business school, Chapters 13 and 14 might be inserted in place of Chapter 5. It is even possible to teach Chapter 13, Sets and Counting, before Chapter 1, Elementary Algebra; in this case the instructor can use the language of sets while teaching subsequent chapters. Below we symbolically indicate some of the possible sequences for teaching the chapters.

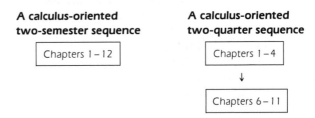

A calculus-oriented two-semester sequence	A calculus-oriented two-quarter sequence
Chapters 1 – 12	Chapters 1 – 4
	↓
	Chapters 6 – 11

A two-semester sequence that only introduces calculus

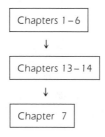

Chapters 1 – 6

↓

Chapters 13 – 14

↓

Chapter 7

A two-quarter sequence with no linear programming that only introduces calculus

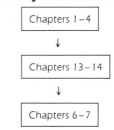

Chapters 1 – 4

↓

Chapters 13 – 14

↓

Chapters 6 – 7

Career Profiles   Following each chapter a Career Profile, accompanied by a photograph, gives a brief insight into an occupation. The Career Profiles are not intended to be definitive, or even thorough; they are quick glimpses intended to stimulate the student's imagination. Staffs specializing in career counselling can provide the interested student with more extensive sources such as *The Occupational Outlook Handbook* issued by the U.S. Department of Labor.

Characteristics of Quantitative Education   This text does not ordinarily present ideas merely for the sake of exposing the student to them. Only material that is needed is presented, and the topics introduced are developed and applied. This approach and the informal style aimed at developing the student's intuition permit the text to concentrate on basic skills and central concepts. In contrast, the survey technique is frequently used in the humanities and is sometimes employed in texts such as this one. Some breadth is sacrificed for the sake of developing reliable skills and a sound grasp of central concepts.

Calculator Use   A student does not need a calculator in order to use this book. However, the text does try to exploit the fact that many students do have a calculator. An inexpensive calculator with keys to evaluate exponential and logarithm functions can be used to make "function," "limit," and other concepts more concrete. Almost every chapter contains Calculator Examples, which are problems that would be unreasonable without the use of a calculator. Many sections have Calculator Problems, which require the use of a calculator, at the end of the regular exercises. Many of the regular exercises can be solved more quickly and easily with a calculator, but they can also be solved by the customary use of a pencil and paper.

vi

Acknowledgments I am indebted to my following colleagues at the University of Georgia who provided me with helpful reactions, suggestions and advice: C.H. Edwards, Jr., Thomas C. Gard, D. Kannan, Frank G. Lether, Carol W. Penney, and David E. Penney.

I am grateful to the following reviewers for their ideas, opinions and suggestions: Stephen Andrilli (LaSalle College), James J. Buckley (University of Alabama in Birmingham), Paul Deland (California State University at Fullerton), Albert Fadell (SUNY at Buffalo), Matt Hassett (Arizona State University), Evan G. Houston (University of North Carolina at Charlotte), Arnold W. Miller (University of Texas at Austin), Robert Moreland (Texas Tech. University), Robert Packard (Northern Arizona University), Richard Porter (Northeastern University), Michael Racine (University of Ottawa), Donald J. Sparks (Kennesaw College), Manfred Stoll (University of South Carolina), David Wend (University of Montana at Bozeman).

I want to thank my developmental editor, Jay Freedman, for many helpful suggestions, Carol Brown for solving each problem in the last 13 chapters, and Robert Prince for valuable assistance in checking the accuracy of many items. The Career Profiles were written by JoAnne Simpson Growney.

Richard Bouldin

Contents

Appendix   669

Answers   703

Index   I-1

List of Applications

Business

Consumer/Investor Topics

Economics

Social Science

General Interest

1

Elementary Algebra

1.1 Real Numbers

A basic tool for the mathematics that we shall consider is the system of **real numbers**. The real numbers are usually described by means of a picture called a **number line**. On the horizontal straight line below, we first label a point with 0, the symbol for zero. Next, we choose some unit of length, and we mark off multiples of that length to the right, labeling them 1, 2, 3, ... (In the preceding sentence the three dots indicate that the list continues indefinitely in the pattern shown.) We mark off multiples of that length to the left, away from 0, and we label them $-1, -2, -3, ...$

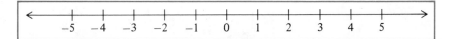

Every point on the number line represents a real number, and every real number is represented by some point on the number line. The following numbers are called the **counting numbers** or **natural numbers.**

1, 2, 3, . . .

The **integers** are the following numbers, which include the counting numbers.

. . . , $-3, -2, -1, 0, 1, 2, 3, . . .$

We have indicated the integers on our number line.

Between any two integers, such as 3 and 4, are many real numbers. Some of those numbers are

$$3.1, 3.11, 3.2, 3.3333, \frac{7}{2}, \frac{15}{4}.$$

Each of these numbers is called a **rational number** because each can be written as the ratio of two integers, as we show.

$$3.1 = \frac{31}{10}, \qquad 3.11 = \frac{311}{100}, \qquad 3.2 = \frac{32}{10}, \qquad 3.3333 = \frac{33,333}{10,000}$$

There are other numbers between the integers besides the rational numbers. Any number that is not a rational number is said to be an **irrational number.** Examples of irrational numbers are

$$\pi, \sqrt{2}, \sqrt{3}, \sqrt{7}, \sqrt{13}.$$

In fact, for any counting number n, either $\sqrt{n}$ is an integer or it is an irrational number. For any irrational number the decimal representation neither terminates nor repeats in a pattern. Consequently, when we want to indicate the value of π (this is the Greek letter pi) we write

$$\pi \approx 3.1416.$$

This means π is approximately equal to the rational number 3.1416. The symbol $\approx$ stands for "is approximately equal to."

EXAMPLE 1

Locate each of the following numbers on a number line: 2, 2.5, $\frac{17}{8}$, $\sqrt{7}$, 3. Identify which of these numbers are counting numbers, which are integers, and which are rational numbers.

Solution
Before locating the numbers on the following number line, we simplify $\frac{17}{8}$ to $2\frac{1}{8}$ and we use a calculator to get $\sqrt{7} \approx 2.646$.

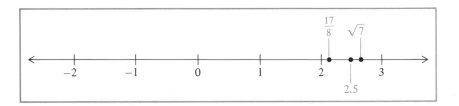

The only counting numbers among the given numbers are 2 and 3; these are also the only integers. The rational numbers are 2, 2.5, $\frac{17}{8}$, and 3. Since $\sqrt{7}$ is not an integer, it must be an irrational number. ∎

One way that we express the relation between two real numbers is to write

$$a < b,$$

which means that a is left of b on the number line. This is read "a is less than b." Sometimes we write

$$a \le b,$$

which means either a is to the left of b or else it coincides with b on the number line. This is read "a is less than or equal to b." Sometimes it is convenient to reverse the symbols and write

$$a > b,$$

which means a is to the right of b. This is read "a is greater than b." We may also write

$$a \geq b,$$

which means either a is right of b or else it coincides with b. This is read "a is greater than or equal to b." All of the symbols $<, \leq, >$, and $\geq$ are referred to as **inequality signs,** and each of the preceding statements about a and b is called an **inequality.** Often two inequalities like

$$a \leq c \quad \text{and} \quad c < b$$

will be written together as

$$a \leq c < b.$$

On a number line, we can picture the numbers described by given inequalities, as the next example shows.

EXAMPLE 2

Locate on a number line every real number b satisfying the inequalities $-1 \leq b < 2$. Do the same for the inequalities $0 < b \leq 3$.

Solution
On a number line we locate -1 and 2; the heavy line indicates all possible choices of b that satisfy the first set of inequalities.

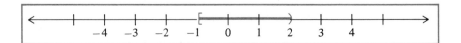

The bracket symbol, [, indicates that -1 is included as a possible choice for b. The parenthesis symbol,), indicates that 2 is not included as a possible choice.

On the next number line, we locate 0 and 3. Again, the heavy line indicates all possible choices of b that satisfy the second set of inequalities.

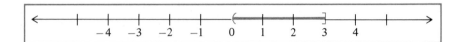

The symbol] indicates that 3 is a possible choice; the symbol (indicates that 0 is not a possible choice. ■

Inequalities are useful in describing some practical situations.

EXAMPLE 3

The price p of a certain textbook must be at least \$16 for the publisher to avoid losing money. If the price of the text is more than \$25, it will not be competitive with other available texts. Write inequalities that describe the possible choices of p.

Solution

The inequality below shows that p must be at least \$16.

$$\$16 \le p$$

We now indicate that p cannot be more than \$25.

$$p \le \$25$$

Thus, the inequalities that describe the possible choices of p are

$$\$16 \le p \le \$25. \quad \blacksquare$$

Another way that we describe the position of a real number on the number line is by giving its absolute value, which is the distance between 0 and the number. We write $|a|$ for the **absolute value** of the real number a. We define $|a|$ to equal a provided $a \ge 0$, and if $a < 0$ then $|a|$ equals $-a$. A real number a is said to be **nonnegative** provided $a \ge 0$; we say that a is **positive** provided $a > 0$. Thus, any nonnegative number equals its absolute value. A real number a is said to be **nonpositive** provided $a \le 0$; we say that a is **negative** provided $a < 0$. Thus, for any negative number a the absolute value $|a|$ equals $-a$, which is a positive number. The absolute value of any real number is nonnegative. Indeed, the absolute value of any real number is the distance along the number line between that number and 0.

For any real number a, we shall refer to $|a|$ as the **magnitude** of the number. To say that a number is small in magnitude means that the number is close to 0. If a number is large in magnitude, then it is far from 0. Note that -5 is less than 1,

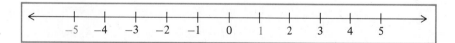

but the magnitude of -5, that is, $|-5| = -(-5) = 5$, is greater than the magnitude of 1, because $|1| = 1$.

EXAMPLE 4

Simplify each of the following expressions so that no absolute value sign appears.

$$|2(3-5)|, \quad |5| - |-5|, \quad 3\left|\frac{5-3}{2}\right|$$

Solution

$$|2(3-5)| = |2(-2)| = |-4| = 4$$
$$|5| - |-5| = 5 - 5 = 0$$

$$3\left|\frac{5-3}{2}\right| = 3\left|\frac{2}{2}\right| = 3|1| = 3(1) = 3 \; \blacksquare$$

There are two familiar operations defined for the real numbers—addition and multiplication. The properties of these operations are the basis for most algebraic manipulations.

Properties of Addition

Let a, b, and c be any real numbers.

1. $a + b = b + a.$ **Commutative property**
2. $(a + b) + c = a + (b + c).$ **Associative property**
3. $a + 0 = a = 0 + a.$ 0 is the **identity** for addition
4. $a + (-a) = 0 = -a + a.$ $-a$ is the **additive inverse** of a

Frequently, we do not write $a + (-b)$; instead we write $a - b$, which means the same thing. Do not be fooled into thinking that $-a$ is a negative number because of the minus sign in front. If a is a negative number like -2, for example, then $-a$ is a positive number.

$$-a = -(-2) = 2$$

The properties of multiplication of real numbers are similar to the properties of addition.

Properties of Multiplication

Let a, b, and c be any real numbers.

1. $ab = ba.$ **Commutative property**
2. $(ab)c = a(bc).$ **Associative property**
3. $a \cdot 1 = a = 1 \cdot a.$ 1 is the **identity** for multiplication
4. provided $a \neq 0$,

$$a\left(\frac{1}{a}\right) = 1 = \left(\frac{1}{a}\right)a. \qquad \frac{1}{a} \text{ is the } \textbf{multiplicative inverse} \text{ of } a$$

Frequently, we write a/b instead of $a(1/b)$. It is important to note that $\frac{1}{0}$ is not defined; consequently, $a/0$ is not defined. **Division by 0 is not defined.**

An important property that relates addition and multiplication is given next.

The Distributive Property

For any real numbers a, b, and c we have

$$a(b + c) = ab + ac$$

and

$$a(b - c) = ab - ac.$$

We use the preceding properties of real numbers to simplify many algebraic expressions, as indicated in the next example.

EXAMPLE 5

Let a, b, and c denote real numbers. Identify which properties and definitions have been used in the following simplification.

$$\frac{a + b}{c} = (a + b)\frac{1}{c} = \frac{1}{c}(a + b) = \left(\frac{1}{c}\right)a + \left(\frac{1}{c}\right)b = \frac{a}{c} + \frac{b}{c}$$

Solution

Beside each step we cite the definition or property used.

$$\frac{a + b}{c} = (a + b)\frac{1}{c}$$ Definition of $\dfrac{a + b}{c}$

$$= \frac{1}{c}(a + b)$$ Commutative property of multiplication

$$= \left(\frac{1}{c}\right)a + \left(\frac{1}{c}\right)b$$ Distributive property

$$= \frac{a}{c} + \frac{b}{c}$$ Definitions of $\dfrac{a}{c}$ and $\dfrac{b}{c}$ and the commutative property of multiplication ■

Exercises   1.1

For each number listed below, indicate whether it is a counting number, an integer, a rational number, or an irrational number.

1	-115	2	39
3	51.7635	4	-119.2332
5	$\dfrac{11}{3}$	6	$\dfrac{9}{2}$
7	$-\sqrt{6}$	8	$\sqrt{10}$
9	$\dfrac{1}{\sqrt{5}}$	10	$\dfrac{1}{\pi}$
11	$\sqrt{36}$	12	$\sqrt{81}$

Sketch a number line and locate each of the following numbers on the number line.

13 1.3

14 $-.5$

15 $\dfrac{11}{2}$

16 $\dfrac{9}{-4}$

17 $\sqrt{49}$

18 $\sqrt{25}$

In problems 19 through 30, sketch a number line and locate every real number b satisfying the given inequalities.

19 $4 \le b < 5.2$

20 $-2 \le b < -1.5$

21 $-3 < b \le 2.4$

22 $-5.3 < b \le -1$

23 $0 < b < 3.6$

24 $-5 < b < -2.2$

25 $-4 \le b \le -2$

26 $-1 \le b \le 4$

27 $3.8 < b$

28 $-2 \le b$

29 $b < -4.5$

30 $b < 2.8$

Simplify each of the following expressions so that no absolute value sign appears.

31 $|\sqrt{11}|$

32 $|\sqrt{19}|$

33 $\dfrac{5}{|-2|}$

34 $\dfrac{10}{|-3|}$

35 $|3(5-4)|$

36 $|4(5-9)|$

37 $|7|+|-3|$

38 $|-4|+|-2|$

39 $|8|-|-5|$

40 $|-10|-|2|$

41 $6|5-2|$

42 $5|7-3|$

43 $\left|\dfrac{13-(-2)}{5}\right|$

44 $\left|\dfrac{4-(-16)}{2}\right|$

45 $|-5||2|$

46 $|3.5||-4|$

47 $|2||9-14|$

48 $\left|\dfrac{22}{-7}\right||3+4|$

In problems 49 through 66, identify each property and definition that has been used in the indicated simplification. Let a, b, and c denote real numbers.

49 $a+b=b+a$

50 $a+(b+c)=(b+c)+a$

51 $a(1-1)=a-a$

52 $(3-2)b=b(3-2)=3b-2b$

53 $a+(b+c)=(a+b)+c$

54 $(1+a)+(b+c)=1+[a+(b+c)]$

55 $3(5a)=15a$

56 $2+(7+a)=9+a$

57 $a+(b-b)=a$

58 $[a+(-a)]b=0$

59    $13 + 0 = 13$

60    $0 + 101 = 101$

61    $(ab)\dfrac{1}{b} = a\left[b\left(\dfrac{1}{b}\right)\right] = a \cdot 1 = a$

62    $\dfrac{1}{a}(ab) = \left[\left(\dfrac{1}{a}\right)a\right]b = 1 \cdot b = b$

63    $a(b - 5) = ab - 5a$

64    $(a + 3)b = ab + 3b$

65    $\dfrac{a}{a} = a\left(\dfrac{1}{a}\right) = 1$

66    $b - b = b + (-b) = 0$

67   **Calculator Problem**   Determine the rational number with four digits after the decimal point that best approximates $\sqrt{2}$.

68   **Calculator Problem**   Determine the rational number with four digits after the decimal point that best approximates $\sqrt{3}$.

Solve each of the following.

69   The price p of a ticket to a certain movie theatre must be at least $2 or the theatre will lose money; the price cannot be more than $5 or the theatre will not be competitive. Write inequalities that describe the possible choices of p.

70   The price p of a new soft drink must be at least 25¢, or the manufacturer will lose money; the price must be less than 50¢ for the drink to be competitive. Write inequalities that describe the possible choices of p.

71   The number of people, say n, who can simultaneously use a certain suspended walkway in a hotel lobby must not exceed 100. Write two inequalities describing n.

72   The number of people, say n, who can simultaneously ride a certain cable car at a certain ski resort must not exceed 20. Write two inequalities describing n.

73   The number of subscribers to a certain magazine, say n, will vary according to the price of the magazine. At the lowest possible price the number is 100,000; at the highest price that the magazine would ever charge the number is 100. Write two inequalities that describe the possible values of n.

1.2   Linear Equations and Inequalities

An **equation** is a statement that two mathematical quantities are equal. For example, $a = 3$ is a simple equation. The symbol $=$ means "is equal to"; it is called the **equal sign.** Many equations ($2x + 1 = 5$, for example) involve a variable like x. A **variable** is a symbol that can be replaced by any one of a variety of numbers. We **solve** a given equation by finding each number that makes the equation true when it is substituted for the variable; a number that makes the equation true is called a **solution.** For example, the number 2 is a solution of the equation $2x + 1 = 5$, since $2(2) + 1 = 4 + 1 = 5$.

In solving equations we shall use the properties of addition and multiplication given in Section 1.1. We shall also use the basic properties given next.

Basic Properties of an Equation

Let a, b, and c be any real numbers.

1.   If $a = b$, then $a + c = b + c$.   Addition property

2.   If $a = b$, then $a - c = b - c$.   Subtraction property

3.   If $a = b$, then $ac = bc$.   Multiplication property

4.   If $a = b$ and $c \neq 0$, then $\dfrac{a}{c} = \dfrac{b}{c}$.   Division property

Of course, division by c and multiplication by $1/c$ are the same. The next example will demonstrate how these properties are used in solving a given equation.

EXAMPLE 1

Solve the equation $2x + 1 = 5$.

Solution

Since this is our first example, we shall present the steps in greater detail than we would normally. Beside each step we indicate the property of an equation that justifies the step.

$$2x + 1 = 5$$
$$2x + 1 - 1 = 5 - 1 \qquad \text{Subtraction property}$$
$$2x + 0 = 4$$
$$2x = 4$$
$$\frac{1}{2}(2x) = \frac{1}{2}(4) \qquad \text{Multiplication property}$$
$$\left[\left(\frac{1}{2}\right)2\right]x = \frac{1}{2}(4)$$
$$(1)x = 2$$
$$x = 2$$

Thus, 2 is the only solution of the given equation. ■

Later we shall use steps analogous to the preceding to solve "matrix" equations. Now let us solve another simple equation.

EXAMPLE 2

Solve the equation $\dfrac{s}{2} - 1 = \dfrac{s}{3}$.

Solution

In order to eliminate the fractions, we begin by multiplying both sides of the equation by the product of the two denominators. Again we cite each property of an equation used.

$$6\left(\frac{s}{2} - 1\right) = 6\left(\frac{s}{3}\right) \qquad \text{Multiplication property}$$

$$6\left(\frac{s}{2}\right) - 6 = 6\left(\frac{s}{3}\right)$$

$$6\left(\frac{1}{2}\right)s - 6 = 6\left(\frac{1}{3}\right)s$$

$$\left[6\left(\frac{1}{2}\right)\right]s - 6 = \left[6\left(\frac{1}{3}\right)\right]s$$

$$3s - 6 = 2s$$

$$3s - 6 - 2s = 2s - 2s \qquad \text{Subtraction property}$$

$$(3s - 2s) - 6 = 2s - 2s$$

$$s - 6 = 0$$

$$s - 6 + 6 = 0 + 6 \qquad \text{Addition property}$$

$$s + 0 = 0 + 6$$

$$s = 6$$

Thus, 6 is the only solution of the given equation. ■

Any equation that can be written in the form $ax + b = 0$ where a and b are real numbers, with $a \neq 0$, is called a **linear equation.** For example, $2x - 4 = 0$ is a linear equation. Although the equation $2x + 1 = 5$ does not appear to have the correct form to be a linear equation, it can be written in the correct form. We perform the necessary steps.

$$2x + 1 = 5$$

$$2x + 1 - 5 = 5 - 5$$

$$2x - 4 = 0$$

We say that two equations are **equivalent** if one equation can be transformed into the other by using the elementary rules of algebra. For example, the equation $2x - 4 = 0$ is equivalent to the equation $2x + 1 = 5$. **Two equivalent equations have the same solutions.**

The next example demonstrates a procedure often used to solve linear equations.

EXAMPLE 3

Solve the equation

$$.5y + 1.3y - 2 = .9y + .4y + 3.$$

Solution

We get all of the addends that involve the variable y to appear on the left side of the equation. We get the other addends to appear on the right side.

$$.5y + 1.3y - 2 = .9y + .4y + 3$$

$$(.5 + 1.3)y - 2 = (.9 + .4)y + 3$$

$$1.8y - 2 = 1.3y + 3$$

$$1.8y - 2 + (2 - 1.3y) = 1.3y + 3 + (2 - 1.3y)$$

$$1.8y - 1.3y = 3 + 2$$

$$(1.8 - 1.3)y = 5$$

$$.5y = 5$$

Now we divide both sides by the number that multiplies y.

$$\frac{.5y}{.5} = \frac{5}{.5}$$

$$\frac{.5}{.5}y = \frac{5.0}{.5}$$

$$y = 10$$

This last equation is equivalent to the equation given in the problem; thus, the two equations have the same solutions. Obviously, 10 is the only solution of this last equation, and so it is the only solution of the given equation. ∎

Some linear equations do not look like linear equations. The next example involves such an equation.

EXAMPLE 4

Solve the equation

$$\frac{3}{2(x + 1)} + \frac{5}{x + 1} = 4 \qquad x \neq -1.$$

Solution

To eliminate the fractions we could multiply both sides of the equation by the product of the denominators. However, we observe that if we multiply both sides by $2(x + 1)$, which is simpler than the product of the denominators, it will get rid of the fractions.

$$2(x + 1)\left[\frac{3}{2(x + 1)} + \frac{5}{x + 1}\right] = 2(x + 1)[4]$$

$$3 + 5(2) = 8(x + 1)$$

$$13 = 8x + 8$$

$$13 - 13 = 8x + 8 - 13$$

$$0 = 8x - 5$$

$$-8x = 8x - 5 - 8x$$

$$-8x = -5$$

$$\frac{1}{-8}(-8x) = \frac{1}{-8}(-5)$$

$$x = \frac{-5}{-8} = \frac{5}{8}$$

At the fifth step, when we have $0 = 8x - 5$, it is clear that the given equation can be written in the form of a linear equation. The only solution of the given equation is the number $\frac{5}{8}$. ■

In Section 1.1 we introduced simple inequalities. Now we want to consider inequalities that involve a variable ($2x + 1 \leq 5$, for example). We say that we have **solved** a given inequality when we find all values for the variable that make the inequality true. The collection of real numbers that make the inequality true is said to be the **solution**. The process of solving an inequality is very similar to the process for solving an equation. We shall use the following basic properties.

Basic Properties of an Inequality

Let a, b, and c be any real numbers.

1. If $a < b$, then $a + c < b + c$.   Addition property

2. If $a < b$, then $a - c < b - c$.   Subtraction property

3. If $a < b$ and c is positive,   Multiplication by a positive
    then $ac < bc$.   number

4. If $a < b$ and c is negative,   Multiplication by a nega-
    then $bc < ac$.   tive number

Each of these statements remains true if the symbol $<$ is replaced by any one of the symbols $\leq$, $>$, $\geq$.

The next example demonstrates how the preceding properties are used.

EXAMPLE 5

Solve the inequality $2x + 1 < 5$.

Solution

Each time we use a basic property of inequalities we cite the property.

$$2x + 1 < 5$$

$$2x + 1 - 1 < 5 - 1 \qquad \text{Subtraction property}$$

$$2x < 4$$

$$\frac{1}{2}(2x) < \frac{1}{2}(4) \qquad \text{Multiplication by a positive}$$

$$x < 2$$

The last inequality is equivalent to the given inequality. Thus, the solution of the given inequality consists of all real numbers less than 2, which is depicted on the number line below.

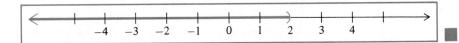

Remember that when each side of an inequality is multiplied by a negative number, then the sides are reversed. This is illustrated by the next example.

EXAMPLE 6

Solve the inequality $4 - \frac{1}{5}x \geq 7$.

Solution

$$4 - \frac{1}{5}x \geq 7$$

$$4 - \frac{1}{5}x - 4 \geq 7 - 4 \qquad \text{Subtraction property}$$

$$-\frac{1}{5}x \geq 3$$

$$-5(3) \geq -5\left(-\frac{1}{5}x\right) \qquad \text{Multiplication by a negative}$$

$$-15 \geq x$$

The solution of the given inequality consists of all real numbers less than or equal to -15, which is depicted on the number line below.

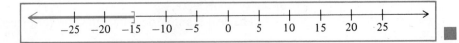

Our four inequality signs permit us to write the same fact in more than one way. For example, $2 \leq 5$ can be written as $5 \geq 2$, and both express the same fact.

Any inequality that can be written in one of the forms $ax + b < 0$ or $ax + b \leq 0$, with $a \neq 0$, is called a **linear inequality.** The inequality given in the statement of Example 6 appears to not be a linear inequality; below we rewrite that inequality to show that it can be written in the correct form.

$$4 - \frac{1}{5}x \geq 7$$

$$7 \leq 4 - \frac{1}{5}x$$

$$\frac{1}{5}x + 7 \leq \frac{1}{5}x + 4 - \frac{1}{5}x$$

$$\frac{1}{5}x + 7 \leq 4$$

$$\frac{1}{5}x + 7 - 4 \leq 4 - 4$$

$$\frac{1}{5}x + 3 \leq 0$$

This last inequality has the form of a linear inequality.

We shall now apply linear equations and inequalities to various practical situations. Recall that **percent,** which is "by the hundred" in Latin, means "hundredths." Thus, 40% is the same as .40. (The symbol % stands for "percent.") When a product is sold below its usual listed price, we say that the product is **discounted,** and the amount of the discount is often stated as a percent. A $12 item that is discounted 10% sells for

$$\$12 - .10(\$12) = \$12 - \$1.20 = \$10.80.$$

EXAMPLE 7

University Bookstore buys books from most publishers at a 20% discount off the price marked on the book. What is the price to a student of a book that costs the bookstore $16?

Solution
We let x stand for the unknown price of the book to a student. Since the price to the bookstore is discounted 20%, the bookstore pays $(x - .20x)$ for the book, and we have the following equation

$$x - .20x = \$16 \quad \text{or} \quad .80x = \$16.$$

This linear equation is easily solved.

$$\left(\frac{1}{.80}\right).80x = \left(\frac{1}{.80}\right)\$16$$

$$x = \$20 \quad \blacksquare$$

The **rate of return** on an investment is the amount that the investment pays the investor in a year divided by the amount of the investment. This is frequently stated as a percent. For example, if a $100 investment pays $18 per year then the rate of return on the investment is $\frac{18}{100} = .18 = 18\%$.

EXAMPLE 8

Suppose part of $10,000 is invested in stocks and the rest is invested in bonds. The rate of return on the stocks is 25% and the rate of return on the bonds is 16%. If the total amount earned in one year is $2140, how much is invested in stocks and how much is in bonds?

Solution
We let x stand for the unknown amount invested in stocks. Since the total amount invested is $10,000, it must be that ($10,000 − x$) is invested in bonds. Since the rate of return on the stocks is 25%, the amount that the stocks pay in one year is $(25\%)x = .25x$. Since the rate of return on the bonds is 16%, the amount that the bonds pay in one year is $(16\%)(\$10,000 − x) = .16(\$10,000 − x)$. Since the total amount earned in one year is $2140, we have the equation

$$.25x + .16(\$10,000 − x) = \$2140.$$

We simplify and solve this linear equation.

$$.25x + \$1600 − .16x = \$2140$$
$$.09x + \$1600 = \$2140$$
$$.09x = \$2140 − \$1600$$
$$.09x = \$540$$
$$x = \frac{\$540}{.09} = \$6000$$
$$\$10,000 − x = \$10,000 − \$6000 = \$4000$$

Thus, $6000 is invested in stocks and $4000 is invested in bonds. ∎

The **net income** of a company is the income that the company reports on its income tax return. The **profit margin** of a company is the net income divided by the total sales; the profit margin is usually stated as a percent. A store that does a lot of business, like a grocery store, may have a profit margin of 1% or 2%. Another store, like a jewelry store, may have a profit margin of 50%.

EXAMPLE 9

Suppose the owner of a small business must have a net income of $20,000 to maintain his life style. The owner anticipates that his total sales will be $260,000 for the next year. What inequality must his profit margin satisfy to provide him with the income that he needs?

Solution

Let x be the unknown profit margin and note that

$$\frac{\text{net income}}{\$260,000} = x$$

or

net income = $\$260,000x$.

Since the owner's net income must be at least \$20,000, the following inequality must hold.

net income = $\$260,000x \geq \$20,000$

We solve this inequality for x.

$$x \geq \frac{\$20,000}{\$260,000} = .0769 = 7.69\%$$

The owner's profit margin must be at least 7.69%. ■

Exercises   1.2

Solve each of the following equations.

1 $5x + 7 = 22$

2 $3 - 2x = -11$

3 $5 - 6x = 0$

4 $7x - 9 = 0$

5 $-2(s + 3) = 8$

6 $5(s - 1) + 20 = 0$

7 $2(s + 5) = 3(s - 2) + 7$

8 $4(s - 3) - 2(s + 4) = 0$

9 $y + 2(3y - 5) = 11$

10 $4(3 - 2y) + 3y = -8$

11 $2(y + 11) - 3(5 - y) = 7$

12 $5(3 - y) + 2(3y - 1) = 15$

13 $\frac{x}{2} + \frac{x}{3} = 1$

14 $\frac{x}{2} - \frac{x}{5} = 3$

15 $\frac{3x}{4} + 2x = 3$

16 $5 - \frac{2x}{3} = x$

17 $\frac{x+1}{5} + \frac{x-2}{3} = -1$

18 $\frac{1}{2}(x + 3) + 6 = \frac{1}{3}(x - 1)$

19 $.2s + .9s - 3 = .6s + 2 + .4s$

20 $1.1s + .7s - .5s = .3s + 15$

21 $.5s + 4(s - 3) = 6$

22 $1.2(2 - s) + .2(s + 5) = 0$

23 $\frac{1}{s} + \frac{4}{s} = 2$

24 $\frac{5}{s} + 3 = \frac{3}{2s}$

25 $\frac{3}{x+1} + \frac{1}{2(x+1)} = 4$

26 $\frac{2}{x+3} + 5 = \frac{3}{x+3}$

27 $\frac{5}{x+1} - \frac{3}{x+1} = 2$

28 $\frac{1}{5x} + \frac{3}{5x} = 1$

29 $\frac{2}{x-3} + \frac{4}{5} = \frac{1}{x-3}$

30 $\frac{2}{3} - \frac{3}{x+1} = \frac{4}{x+1}$

Solve each of the following inequalities.

31 $3x + 2 < 29$

32 $1 + 5x < 26$

33 $6 + 2x \geq 30$

34 $4x + 3 \geq 27$

35 $5 - 2x \leq -21$

36 $5 - 11x \leq -28$

37 $\frac{1}{3}y + 2 > 6$

38 $9 + \frac{1}{2}y > 7$

39 $-\frac{1}{5}y + 3 < -\frac{3}{5}$

40 $4 - \frac{1}{9}y < \frac{2}{9}$

41 $5y - 2 \geq 3y + 4$

42 $7y + 5 \geq y - 19$

43 $3(w + 2) - 3 \leq 2(3w - 5)$

44 $7(w - 1) + 3(w + 4) \leq 10$

45 $\frac{w}{3} - \frac{w}{5} > 4 + \frac{w}{2}$

46 $\frac{2w}{5} + 8 > \frac{3w}{4} + 2$

47 $\frac{w + 3}{2} - \frac{3w + 5}{3} < 1$

48 $\frac{2w + 1}{3} + 5 < \frac{w - 5}{2}$

49 $3x + 2(5x - 1) \geq 24$

50 $2(4 - 3x) + 2x \geq 12$

Solve each of the following.

51 A merchant advertises that he will sell all merchandise at a 10% discount from the prices indicated on the price tags provided cash, rather than credit, is used to make the purchase. What is the cash price of an item marked $25.50?

52 If cash purchases receive a 15% discount from the amounts shown on the price tags, what is the cash price of an item marked $13.60?

53 All members of United Buyers' Service receive a 20% discount on items purchased from participating retailers. How much does a member pay for an item that costs a nonmember $81.20 at a participating retailer?

54 At a participating retailer, what is the price that a member of United Buyers' Service (see problem 53) pays for an item marked $9.80 for nonmembers?

55 Economy Printers has a system of graduated discounts. For the first 1000 copies the price is $20.60. For the second 1000 copies the buyer gets a 5% discount from $20.60; each additional 1000 copies, up to a total of 10,000 copies, receives an additional 5% discount from $20.60. What is the price of 5000 copies?

56 What is the price of 8500 copies from Economy Printers? (See problem 55.)

57 Suppose an investment in A gives an 8% rate of return and an investment in B, a riskier enterprise, gives a 12% rate of return. If an investor wants an 11% rate of return on the investment of $5000, how much should go into A and how much into B?

58 Suppose the investor in problem 57 only requires a 9% rate of return. How should the $5000 be divided between investments A and B?

59 Suppose $20,000 is divided between investment A with a 22% rate of return and investment B with a 15% rate of return. If the total annual

return is $4015, then how is the $20,000 divided between the two investments?

60 Suppose the total annual return on the $20,000 described in problem 59 is $3770. How must the $20,000 be divided between the two investments?

61 Suppose $12,000 is put into investment A, which has a rate of return of 18%, and $8000 is put into investment B. If the total annual return is $3280, what is the rate of return for investment B?

62 Suppose the total annual return in problem 61 is $2960. What is the rate of return for investment B?

63 A certain business has total annual sales of $760,000, and the owner of the business estimates that he needs at least $125,000 in income to meet his obligations. How large must his profit margin be in order to provide him with the necessary income?

64 Suppose the owner in problem 63 needs an income of $152,000 or more. How large must his profit margin be in order to provide him with the necessary income?

65 Fred Jackson owns a business with a profit margin of 12%. If Fred requires an annual income of $18,000 or more, how large must the total sales of Fred's business be?

66 Suppose Fred Jackson, from problem 65, increases his profit margin to 16%. How large must total sales be to provide Fred with the necessary income?

1.3 Formal Algebraic Expressions

When numbers, represented by symbols, are combined by the operations of addition, subtraction, multiplication, division, or extraction of roots, the resulting expression is called an **algebraic expression.** Each addend in an algebraic expression is called a **term,** and **like terms** are addends that involve the same power of the variable. For example, in the expression $x^4 + 6x^3 + 5x^2 + 3x - 9 - 11x^2$ the only like terms are $5x^2$ and $-11x^2$. The number that multiplies the power of the variable is called a **coefficient.** The next example illustrates how an algebraic expression is simplified by combining the coefficients of like terms.

EXAMPLE 1

Simplify the algebraic expression

$$(2x^5 - 5x^3 + 11x^2 - 7) + (9x^4 + x^3 - 6x + 9).$$

Solution

First, we rearrange the terms so that the powers of the variable decrease as we read left to right.

$$(2x^5 - 5x^3 + 11x^2 - 7) + (9x^4 + x^3 - 6x + 9) = 2x^5 + 9x^4 + x^3 - 5x^3 + 11x^2 - 6x + 9 - 7$$

Now we combine the coefficients of like terms.

$$2x^5 + 9x^4 + (x^3 - 5x^3) + 11x^2 - 6x + (9 - 7) = 2x^5 + 9x^4 - 4x^3 + 11x^2 - 6x + 2 \quad \blacksquare$$

The like terms may involve roots and negative powers of the variable, as the next example illustrates.

EXAMPLE 2

Simplify the algebraic expression

$$\frac{5}{x} + 2\sqrt{x} - x + 3\sqrt{x} - \frac{2}{x} + 7x.$$

Solution

We get like terms together and then combine the coefficients.

$$\frac{5}{x} + 2\sqrt{x} - x + 3\sqrt{x} - \frac{2}{x} + 7x = \left(\frac{5}{x} - \frac{2}{x}\right) + (2\sqrt{x} + 3\sqrt{x}) + (7x - x)$$

$$= \frac{3}{x} + 5\sqrt{x} + 6x \quad \blacksquare$$

Sometimes we must compute the product of two algebraic expressions before we can simplify, as in the next example. Recall these three important rules of exponents.

$$x^n x^m = x^{n+m}, \qquad x^{-n} = \frac{1}{x^n}, \qquad \text{and} \qquad \frac{x^n}{x^m} = x^{n-m}.$$

All of the rules of exponents will be reviewed in Section 2.4.

EXAMPLE 3

Simplify the algebraic expression

$$(7x + 2)(x^2 - 5x + 1) + 3x - 6.$$

Solution

First, we compute the product. We multiply $7x$ times each term in the second factor, and then we multiply 2 times each term in the second factor.

$$(7x + 2)(x^2 - 5x + 1) + 3x - 6 = 7x^3 - 35x^2 + 7x + 2x^2 - 10x + 2 + 3x - 6$$

Then we group like terms and combine the coefficients.

$$7x^3 - 35x^2 + 7x + 2x^2 - 10x + 2 + 3x - 6 = 7x^3 + (2x^2 - 35x^2) + (7x - 10x + 3x) + (2 - 6)$$
$$= 7x^3 - 33x^2 - 4 \quad \blacksquare$$

In many computations, we get different results according to the order we follow in performing the indicated operations. For example, the algebraic expression $3x + 1$ is not equal to $3(x + 1)$. In the first expression, x is multiplied by 3 and then 1 is added, whereas in the second expression, 1 is added to x before multiplying by 3.

Sometimes we indicate the order to use in performing operations by means of parentheses. **The operation inside all but the innermost set of** parentheses, brackets, and braces is performed first, and the innermost parentheses (or brackets or braces) are removed. The process continues until all grouping symbols are removed, as illustrated in the next example.

EXAMPLE 4

Simplify the algebraic expression

$$x^2[4(x^3 - 2x^2 + 9) + 7] + 3.$$

Solution
Here there is one set of brackets and one set of parentheses. The first operation to be performed is to multiply 4 times $(x^3 - 2x^2 + 9)$; then we add 7 and simplify.

$$x^2[4(x^3 - 2x^2 + 9) + 7] + 3 = x^2[4x^3 - 8x^2 + 36 + 7] + 3$$
$$= x^2[4x^3 - 8x^2 + 43] + 3$$

Now we multiply x^2 times $[4x^3 - 8x^2 + 43]$.

$$x^2[4x^3 - 8x^2 + 43] + 3 = 4x^5 - 8x^4 + 43x^2 + 3 \quad ■$$

Sometimes we need to add fractions that have algebraic expressions in the denominators. This requires that we find the common denominator, as indicated in the next example.

EXAMPLE 5

Write the indicated sum as a single fraction.

$$\frac{-2}{x-3} + \frac{2}{x+3}$$

Solution
The common denominator is the product of the two denominators. In order to rewrite each fraction over the common denominator, we multiply the first one by $\frac{x+3}{x+3}$ and the second one by $\frac{x-3}{x-3}$. Then we sum and simplify.

$$\frac{-2}{x-3} + \frac{2}{x+3} = \frac{(-2)}{(x-3)} \frac{(x+3)}{(x+3)} + \frac{(2)}{(x+3)} \frac{(x-3)}{(x-3)}$$
$$= \frac{-2(x+3)}{x^2-9} + \frac{2(x-3)}{x^2-9}$$
$$= \frac{-2x-6}{x^2-9} + \frac{2x-6}{x^2-9}$$
$$= \frac{-2x-6+2x-6}{x^2-9}$$
$$= \frac{-12}{x^2-9} \quad ■$$

Some fractions can be simplified by eliminating factors that the numerator and denominator have in common. **Factoring** is just the reverse of the distributive property mentioned in Section 1.1. The next example illustrates a simple case.

EXAMPLE 6

Simplify the fraction

$$\frac{x^5 + 3x^4 - 2x^3 + 7x^2}{5x^3 - 6x^2}$$

Solution

We recognize that x^2 is a factor of both the numerator and the denominator. After factoring the numerator and denominator, we remove the factor common to both, provided that it is not zero.

$$\frac{x^5 + 3x^4 - 2x^3 + 7x^2}{5x^3 - 6x^2} = \frac{x^2(x^3 + 3x^2 - 2x + 7)}{x^2(5x - 6)}$$

$$= \frac{x^2}{x^2} \cdot \frac{x^3 + 3x^2 - 2x + 7}{5x - 6}$$

$$= \frac{x^3 + 3x^2 - 2x + 7}{5x - 6}, \qquad \text{provided } x^2 \neq 0 \quad \blacksquare$$

In simplifying fractions, the following factorizations are often useful.

$$x^2 - a^2 = (x - a)(x + a)$$

$$x^3 - a^3 = (x - a)(x^2 + ax + a^2)$$

$$x^3 + a^3 = (x + a)(x^2 - ax + a^2)$$

$$x^4 - a^4 = (x^2 - a^2)(x^2 + a^2) = (x - a)(x + a)(x^2 + a^2)$$

In each of the above equations, the symbol a stands for a fixed real number. Each equation is verified by completing the multiplication indicated on the right-hand side.

The next example illustrates the use of the preceding factorization formulas.

EXAMPLE 7

Simplify the fraction $\dfrac{2x^3 - 54}{2x - 6}$.

Solution

First, we observe that 2 is a factor that is common to the numerator and the denominator. We eliminate the factor of 2, and then we factor the numerator according to the formula for the difference of cubes.

$$\frac{2x^3 - 54}{2x - 6} = \frac{2(x^3 - 27)}{2(x - 3)} = \frac{x^3 - 27}{x - 3}$$

$$= \frac{(x - 3)(x^2 + 3x + 9)}{x - 3}$$

Now we can remove the factor $x - 3$, which appears in the numerator and denominator.

$$\frac{(x-3)(x^2+3x+9)}{x-3} = x^2 + 3x + 9, \qquad \text{provided } (x-3) \neq 0$$

Thus, we have

$$\frac{2x^3 - 54}{2x - 6} = x^2 + 3x + 9, \qquad \text{provided } (x-3) \neq 0. \; \blacksquare$$

Exercises 1.3

Simplify each of the following algebraic expressions.

1 $(5x^2 - 7x + 11) + (12x - 10)$
2 $(9x^2 + 3x - 5) - (8x + 6)$
3 $(\sqrt{2}x^2 + \sqrt{5}x + 1) - (3\sqrt{2}x^2 + 9)$
4 $(\sqrt{3}x^2 - 7x) + (2\sqrt{3}x^2 + 2x + 3)$
5 $(9x^3 + 6x - 3) + (4x^3 + 2x^2 + 5)$
6 $(x^3 + x^2 + x + 1) - (x^3 - x^2 + x - 1)$
7 $(4\sqrt{x} + 2x + 1) + (3 - \sqrt{x})$
8 $(x - 3\sqrt{x} + 2) - (x^2 - \sqrt{x} + 7)$
9 $(\sqrt[4]{x} + \sqrt[3]{x} + \sqrt{x}) - (7\sqrt[3]{x} - 3\sqrt{x})$
10 $(5x - \sqrt{x} + 2\sqrt[3]{x}) + (\sqrt{x} - 9x + 1)$
11 $\left(\frac{9}{x} + 2 + 5x\right) + \left(\frac{1}{x^2} - \frac{1}{x} + 3\right)$
12 $\left(6 - \frac{3}{x}\right) + \left(4x - 2 + \frac{2}{x}\right)$
13 $\left(\sqrt{x} + \frac{1}{x}\right) + \left(x - \frac{2}{x}\right) + (4\sqrt{x} - 5x)$
14 $(3x^2 - \sqrt{x} + 1) - \left(2\sqrt{x} + \frac{3}{x}\right) + \left(\frac{2}{x} - 4x^2 - 2\right)$
15 $(x + 1)(x - 1) + 2 + x - x^2$
16 $(x + 2)(x + 2) - 4 - 4x$
17 $x(5x^2 - 6x + 3) - x^2 - x + 1$
18 $x(x^3 + 7x + 1) + x^2 - 3$
19 $(3x + 2)(4x - 5) + 10$
20 $(x - 5)(7x + 2) + 3x - 5$
21 $(x + 9)(x^2 - 2x + 3) + x^2 - 4$
22 $(2x - 1)(x^2 + x + 11) - 15x + 10$
23 $(x^2 + 9)(x^2 + 4x + 1) - 4x^3 - 9$
24 $(3x^2 + 4)(x^2 - x + 2) + x + 6$

25 $10(x^2 + x - 2) - 5x(x + 3)$

26 $x(3x - 5) + 4(x^2 - 2x + 1)$

27 $(x + 1)(x^2 + 3x + 2) - (x^2 + 5)(6x - 7)$

28 $(x^2 + x + 1)(3x - 2) + (x + 4)(x^2 - 9)$

29 $\left(\dfrac{1}{x} + x\right)(2x + 3)$

30 $(3x^2 - 5x + 6)\left(5 - \dfrac{2}{x}\right)$

31 $\left(\sqrt{x} + 12 - \dfrac{1}{x}\right)(x^2 + 7x - 1)$

32 $\left(\dfrac{3}{x} + 5 - x\right)\left(4\sqrt{x} + \dfrac{2}{\sqrt{x}}\right)$

33 $x^3\left(\dfrac{1}{x} - \dfrac{2}{x^2} + \dfrac{5}{x^3}\right) + (7x^2 + 8)$

34 $x^{-2}(x^2 + 8x + 6) - \left(\dfrac{4}{x} + 2 + 3x\right)$

35 $(\sqrt{x} + 2)(\sqrt{x} - 2) + 4$

36 $(\sqrt{x} - 1)(\sqrt{x} + 1) + 1$

37 $\dfrac{5}{x + 1} - \dfrac{5}{x - 1}$

38 $\dfrac{11}{x - 2} - \dfrac{11}{x + 2}$

39 $\dfrac{3}{x - 5} + \dfrac{7x}{x + 5}$

40 $\dfrac{4x}{x - 1} + \dfrac{1}{x + 1}$

41 $\dfrac{5}{x + 1} + \dfrac{3}{x - 2}$

42 $\dfrac{9}{x - 3} - \dfrac{2}{x + 5}$

43 $\dfrac{x - 3}{x + 4} + \dfrac{4}{x - 2}$

44 $\dfrac{x + 1}{x - 1} - \dfrac{3}{x + 6}$

45 $\dfrac{x + 3}{x - 5} - \dfrac{x - 1}{x + 2}$

46 $\dfrac{x + 4}{x + 1} + \dfrac{x + 5}{x - 3}$

47 $\dfrac{x}{x - 1} + \dfrac{7}{x + 3} - \dfrac{3x + 1}{x - 5}$

48 $\dfrac{1}{x + 3} - \dfrac{x}{x - 7} + \dfrac{x - 3}{x + 2}$

49 $\dfrac{x}{x - 5} + \dfrac{2x - 1}{x + 6} - \dfrac{4}{x + 1}$

50 $\dfrac{11}{x + 1} - \dfrac{3x + 1}{x - 5} + \dfrac{x}{x + 3}$

51 $\dfrac{4x^3 + 3x^2}{x^2}$

52 $\dfrac{5x^4 - x^3 + 2x^2}{x^3 + x^2}$

53 $\dfrac{x^3 - 2x^2 + 5x}{x^2 - 2x}$

54 $\dfrac{9x^3 + 11x^2 - 7x}{x}$

55 $\dfrac{x^3}{x^6 + x^5 + x^4}$

56 $\dfrac{x^2}{5x^4 - 3x^3 + x^2}$

57 $\dfrac{9x + x^2}{3x^4 - x^2 + 5x}$

58 $\dfrac{5x^2 - 6x^3}{9x^5 - 7x^4 + 4x^3}$

$$59 \quad \frac{x^2 - 25}{2x + 10}$$

$$60 \quad \frac{3x^2 - 3}{2x + 2}$$

$$61 \quad \frac{x^3 - 8}{3x - 6}$$

$$62 \quad \frac{x^3 - 27}{5x - 15}$$

$$63 \quad \frac{2x + 12}{x^2 - 36}$$

$$64 \quad \frac{3x + 3}{4x^2 - 4}$$

$$65 \quad \frac{x^2 + 2x + 1}{x^2 - 1}$$

$$66 \quad \frac{x^2 + 4x + 4}{x^2 - 4}$$

$$67 \quad \frac{x^2 - 25}{x^3 - 125}$$

$$68 \quad \frac{x^2 - 16}{x^3 - 64}$$

1.4 Quadratic Equations and Inequalities

Any equation that can be written in the form $ax^2 + bx + c = 0$ where a, b, and c are real numbers, with $a \neq 0$, is called a **quadratic equation.** For example, $x^2 - 2x + 1 = 0$ is a quadratic equation. Although the equation $x^2 + 5 = 2x + 4$ does not appear to have the correct form to be a quadratic equation, it can be written in the correct form, as follows.

$$x^2 + 5 = 2x + 4$$

$$x^2 + 5 - 2x - 4 = 0$$

$$x^2 - 2x + 1 = 0$$

These steps show that the equations $x^2 + 5 = 2x + 4$ and $x^2 - 2x + 1 = 0$ are equivalent. Since any two equivalent equations have the same solutions, these two equations have the same solutions. **A quadratic equation may have no real solution, one real solution, or two real solutions.** Any solution of a quadratic equation is called a **root** of the equation.

One of the basic methods for solving a quadratic equation $ax^2 + bx + c = 0$ with $a = 1$ is to factor the left side. This method is based on the following principle.

Principle of Zero Products

For any real numbers a and b, if $ab = 0$ then either $a = 0$ or $b = 0$ or both a and b equal zero.

In general, if the left side of the quadratic equation $x^2 + bx + c = 0$ can be factored, then the form of the factorization is $x^2 + bx + c = (x - k)(x - m)$ for some values of k and m. We multiply the factors on the right of the above equation and compare the resulting coefficients to the coefficients on the left side of the equation.

$$\begin{aligned} x^2 + bx + c &= (x - k)(x - m) \\ &= x^2 - mx - kx + km \\ &= x^2 - (k + m)x + km \end{aligned}$$

Comparing the coefficients we see that

$$km = c \quad \text{and} \quad -(k+m) = b$$

or

$$\boxed{km = c \quad \text{and} \quad k+m = -b.}$$

The last equation enables us to find the factors $(x-k)$ and $(x-m)$, as illustrated below in Example 1. **It follows from the Principle of Zero Products that k and m are the roots of the equation**

$$x^2 + bx + c = 0.$$

We now illustrate this method of factorization in the following examples.

EXAMPLE 1

Solve $x^2 - x - 6 = 0$ by factoring.

Solution

Here the coefficients are $b = -1$ and $c = -6$; the factorizations of $c = -6$, using only integers, are

$$-6 = (3)(-2), \qquad -6 = (-3)(2), \qquad -6 = (6)(-1), \qquad -6 = (-6)(1).$$

We find the factorization with the property that the sum of the factors is $-b = -(-1) = 1$. The sum of the factors in each factorization is

$$3 + (-2) = 1, \qquad -3 + 2 = -1, \qquad 6 + (-1) = 5, \qquad -6 + 1 = -5.$$

The factorization of -6 that we need is $(3)(-2) = -6$. The factorization of $x^2 - x - 6$ is then

$$x^2 - x - 6 = (x-3)(x-[-2]) = (x-3)(x+2).$$

This factorization is easily verified by computing the product $(x-3)(x+2)$ to see that it is $x^2 - x - 6$.

We write the quadratic equation to be solved as

$$(x-3)(x+2) = 0.$$

According to the Principle of Zero Products, in order for the equation above to be true at least one of the equations

$$x - 3 = 0, \qquad x + 2 = 0$$

must be true. The solutions of these equations are

$$x = 3, \qquad x = -2.$$

Thus, the roots of the quadratic equation to be solved are 3 and -2. ∎

We illustrate this method of factorization again, with less explanation.

EXAMPLE 2

Solve $x^2 + 2x + 1 = 0$ by factoring.

Solution

The relevant factorizations of $c = 1$ are

$$1 = (1)(1), \qquad 1 = (-1)(-1).$$

We select the factorization such that the sum of the factors is $-b = -2$. The sums of the factors are

$$1 + 1 = 2, \qquad -1 + (-1) = -2.$$

The factorization of 1 that we need is $(-1)(-1) = 1$. Therefore, the factorization of $x^2 + 2x + 1$ is

$$x^2 + 2x + 1 = (x - [-1])(x - [-1]) = (x + 1)(x + 1).$$

According to the Principle of Zero Products, the roots of the quadratic equation to be solved are -1 and -1. There is only one root. ■

Sometimes a quadratic expression $ax^2 + bx + c$, with $a \neq 1$, can be factored. The product of the coefficients of x in the factors must equal a, and the product of the constants in the factors must equal c. We illustrate the method.

EXAMPLE 3

Solve $2x^2 + 5x - 3 = 0$ by factoring.

Solution

The relevant factorizations of $c = -3$ are

$$-3 = (-3)(1), \qquad -3 = (3)(-1).$$

The relevant factorizations of $a = 2$ are

$$2 = (2)(1), \qquad 2 = (-2)(-1).$$

With trial and error we find the following factorization.

$$2x^2 + 5x - 3 = (2x - 1)(x + 3)$$

Various possibilities exist for the factors, such as

$$(2x + 1)(x - 3), \qquad (-2x - 1)(-x + 3), \qquad \text{and} \quad (2x - 1)(x + 3).$$

We choose the one that results in the proper middle term, $5x$.

According to the Principle of Zero Products, the expression $2x^2 + 5x - 3$ equals 0 precisely when one of the equations

$$2x - 1 = 0, \qquad x + 3 = 0$$

holds. The solutions of these equations are easily found to be

$$x = \frac{1}{2}, \qquad x = -3.$$

Thus, the roots of the quadratic equation to be solved are $\frac{1}{2}$ and -3. ■

Not all quadratic equations can be solved by the preceding methods of factoring. However, all quadratic equations can be solved by the quadratic formula, which we now give.

Quadratic Formula

The solutions of the quadratic equation $ax^2 + bx + c = 0$, with $a \neq 0$, are given by

$$x = \frac{-b \pm \sqrt{b^2 - 4ac}}{2a}.$$

We illustrate the use of the Quadratic Formula with the next two examples.

EXAMPLE 4

Solve $x^2 - 2x - 1 = 0$ with the Quadratic Formula.

Solution
Since the quadratic expression is on the left side of the equation and 0 is on the right side, we can easily identify a, b, and c. We have $a = 1$, $b = -2$, $c = -1$, and we substitute these values into the Quadratic Formula.

$$x = \frac{-(-2) \pm \sqrt{(-2)^2 - 4(1)(-1)}}{2(1)}$$

$$x = \frac{2 \pm \sqrt{4 + 4}}{2} = \frac{2 \pm \sqrt{4(2)}}{2}$$

$$x = \frac{2 \pm 2\sqrt{2}}{2} = 1 \pm \sqrt{2}$$

The roots of the quadratic equation to be solved are $1 + \sqrt{2}$ and $1 - \sqrt{2}$. ■

Example 2 showed that a given quadratic equation might have only one root. Using the Quadratic Formula, we show that a quadratic equation may have no real roots. This is illustrated in the next example.

EXAMPLE 5

Solve $x^2 + 1 = 0$ with the Quadratic Formula.

Solution
First, we check that the quadratic expression is on the left of the equation and 0 is on the right side. We recognize that

$$a = 1, \quad b = 0, \quad c = 1,$$

and we substitute these values into the Quadratic Formula.

$$x = \frac{-0 \pm \sqrt{(0)^2 - 4(1)(1)}}{2(1)}$$

$$x = \frac{\pm \sqrt{-4}}{2}$$

Since the square root of a negative number, like -4, is not a real number, there are no real roots. In Section 2.1, we shall carefully consider the properties of square roots. ■

CALCULATOR EXAMPLE

Solve $x^2 + 7x + 2 = 0$ with the Quadratic Formula.

Solution

We recognize that

$$a = 1, \qquad b = 7, \qquad c = 2.$$

We substitute these values into the Quadratic Formula and use a calculator to compute the appropriate square root.

$$x = \frac{-7 \pm \sqrt{49 - 4(1)(2)}}{2(1)}$$

$$x = \frac{-7 \pm \sqrt{41}}{2} \approx \frac{-7 \pm 6.403}{2}$$

$$x \approx -.298 \quad \text{or} \quad -6.702 \quad ■$$

A **quadratic inequality** is an inequality with one of the following forms:

$$ax^2 + bx + c > 0 \quad \text{or} \quad ax^2 + bx + c \geq 0$$

or

$$ax^2 + bx + c < 0 \quad \text{or} \quad ax^2 + bx + c \leq 0 \qquad \text{with } a \neq 0.$$

Examples of quadratic inequalities include

$$2x^2 + x + 3 > 0, \qquad y^2 - 5 \geq 0, \qquad w^2 + 7w < 0, \qquad \text{and} \quad 6x^2 \leq 0.$$

The basic method for solving quadratic inequalities is indicated in the next example.

EXAMPLE 6

Solve the quadratic inequality $x^2 - 4x + 3 > 0$.

Solution

First, we factor the quadratic expression. Since the only relevant factorizations of 3 are $3 = (-3)(-1)$ and $3 = (3)(1)$, it is easy to see that

$$x^2 - 4x + 3 = (x - 3)(x - 1).$$

Thus, the quadratic inequality to be solved is equivalent to

$$(x - 3)(x - 1) > 0.$$

The product of $(x - 3)$ and $(x - 1)$ is positive precisely when $(x - 3)$ and $(x - 1)$ have the same sign. Either both are positive or else both are negative. The inequalities

$$x - 3 > 0 \quad \text{and} \quad x - 1 > 0$$

can be simplified to

$$x > 3 \quad \text{and} \quad x > 1.$$

The values of x satisfying these two inequalities are shown on the next number lines.

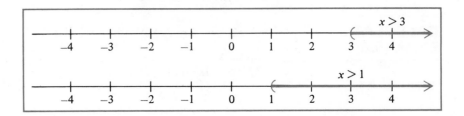

Both $(x - 3)$ and $(x - 1)$ are positive precisely when $x > 3$; thus, the solution includes all x greater than 3.

The product of $(x - 3)$ and $(x - 1)$ is also positive when both factors are negative. The inequalities

$$x - 3 < 0 \quad \text{and} \quad x - 1 < 0$$

can be simplified to

$$x < 3 \quad \text{and} \quad x < 1.$$

The values of x satisfying these two inequalities are shown on the next number lines.

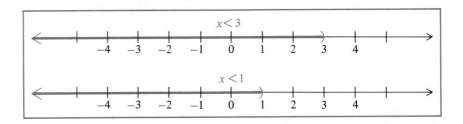

Both $(x - 3)$ and $(x - 1)$ are negative precisely when $x < 1$; thus, the solution includes all x less than 1.

The solution is all x less than 1 or greater than 3, which is pictured below.

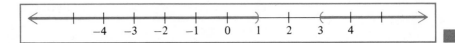

The next example is solved by the same method as Example 6. However, the reasoning is more complicated because the factors must have opposite signs.

EXAMPLE 7

Solve the quadratic inequality $x^2 - x - 12 \leq 0$.

Solution

First, we factor the quadratic expression. The relevant factorizations of $c = -12$ are

$$-12 = (-12)(1), \quad -12 = 12(-1), \quad -12 = 2(-6),$$
$$-12 = 6(-2), \quad -12 = 4(-3), \quad -12 = 3(-4).$$

Since the sum of factors should be $-b = 1$, we see that the factorization of the quadratic expression is

$$x^2 - x - 12 = (x - 4)(x + 3).$$

The inequality to be solved is equivalent to

$$(x - 4)(x + 3) \leq 0.$$

A solution occurs when one of the factors is 0, which happens for $x = 4$, -3, or when the factors have opposite signs. If the first factor were positive and the second factor were negative, we would have the inequalities

$$x - 4 > 0 \quad \text{and} \quad x + 3 < 0.$$

These inequalities simplify to

$$x > 4 \quad \text{and} \quad x < -3.$$

Since -3 is left of 4, it is not possible for x to be both right of 4 and left of -3. Thus, it is not possible for the first factor to be positive and the second factor to be negative.

If the first factor were negative and the second factor were positive, we would have the inequalities

$$x - 4 < 0 \quad \text{and} \quad x + 3 > 0.$$

These inequalities simplify to

$$x < 4 \quad \text{and} \quad x > -3.$$

The values of x satisfying these two inequalities are shown on these number lines.

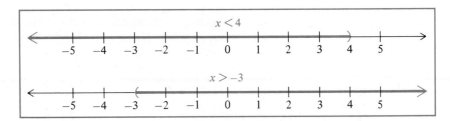

The values of x satisfying both inequalities are shown below.

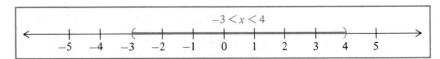

Recalling that $x = 4, -3$ are also solutions, we can now describe all solutions of the given quadratic inequality. The solution is all x satisfying the inequalities

$$-3 \leq x \leq 4,$$

which is pictured below.

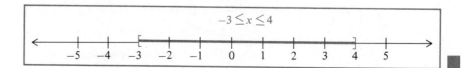

Some of the techniques given in this section are used to solve problems that look different from the preceding examples. In the next example we solve an equation that does not look like a quadratic equation.

EXAMPLE 8

Solve $\dfrac{5}{x - 3} - \dfrac{4}{x + 2} = 1$.

Solution
We multiply both sides of the equation by the product of the two denominators, that is, $(x - 3)(x + 2)$, and we simplify.

$$5(x + 2) - 4(x - 3) = (x - 3)(x + 2)$$
$$5x + 10 - 4x + 12 = x^2 - x - 6$$
$$x^2 - 2x - 28 = 0$$

This quadratic equation is equivalent to the given equation provided $x - 3 \neq 0$ and $x + 2 \neq 0$.

Using the Quadratic Formula, we find that the roots of the last equation are $x = 1 + \sqrt{29}, 1 - \sqrt{29}$. Since each of our steps above is reversible, these two roots make the given equation true provided neither denominator is 0. Thus, the solution is $x = 1 \pm \sqrt{29}$. ■

Sometimes we can solve an equation that involves higher powers of x than x^2 by factoring it and setting each factor equal to 0. The next example illustrates the process.

EXAMPLE 9

Solve $x^4 - 16 = 0$.

Solution
From Section 1.3 we recall the factorization

$$w^2 - a^2 = (w - a)(w + a).$$

Let $w = x^2$ and $a^2 = 16$; then, we have

$$x^4 - 16 = (x^2 - 4)(x^2 + 4).$$

The equation to be solved can be rewritten as

$$(x^2 - 4)(x^2 + 4) = 0.$$

According to the Principle of Zero Products, $x^4 - 16 = 0$ is true for precisely those values of x satisfying one of the equations

$$x^2 - 4 = 0, \qquad x^2 + 4 = 0.$$

The first equation is solved by factoring.

$$(x - 2)(x + 2) = x^2 - 4 = 0$$
$$x = 2, -2$$

The Quadratic Formula shows that there are no real roots for $x^2 + 4 = 0$. Thus, the solution of the equation $x^4 - 16 = 0$ is $x = 2, -2$. ■

Factoring is sometimes required in order to simplify an algebraic expression. Some relevant examples were given in Section 1.3; the next example requires the factorization techniques given in this section.

EXAMPLE 10

Simplify the fraction $\dfrac{x^2 + 2x - 15}{x^2 - 2x - 3}$.

Solution
We factor the numerator and denominator using methods given earlier in this section.

$$\frac{x^2 + 2x - 15}{x^2 - 2x - 3} = \frac{(x + 5)(x - 3)}{(x + 1)(x - 3)} = \frac{x + 5}{x + 1}, \qquad \text{provided } x - 3 \neq 0. \blacksquare$$

Sometimes it is useful to factor a quadratic expression even though the factorization involves irrational numbers. The Quadratic Formula can be used to determine the roots of the associated quadratic equation, and then the expression can be factored as indicated at the beginning of this section. We illustrate the process.

EXAMPLE 11

Factor the quadratic expression $x^2 + 2x - 1$.

Solution
Using the Quadratic Formula, we find that the roots of the equation

$$x^2 + 2x - 1 = 0$$

are

$$x = \frac{-2 \pm \sqrt{4 - 4(1)(-1)}}{2(1)}$$
$$x = \frac{-2 \pm \sqrt{8}}{2} = \frac{-2 \pm 2\sqrt{2}}{2}$$
$$x = -1 \pm \sqrt{2}.$$

Now we see from the discussion of factoring at the beginning of this section that k and m are $-1 + \sqrt{2}$ and $-1 - \sqrt{2}$. We have the factorization

$$x^2 + 2x - 1 = (x + 1 - \sqrt{2})(x + 1 + \sqrt{2}). \quad \blacksquare$$

Exercises 1.4

Solve the following equations by factoring.

1	$x^2 - 1 = 0$	2	$x^2 - 9 = 0$
3	$x^2 + 3x = 0$	4	$x^2 - 5x = 0$
5	$3y^2 - 4y = 0$	6	$-2y^2 + y = 0$
7	$x^2 + 2x - 3 = 0$	8	$x^2 - 3x - 10 = 0$
9	$w^2 - 6w - 7 = 0$	10	$w^2 - 2w - 48 = 0$
11	$p^2 + 6p - 40 = 0$	12	$p^2 - 6p - 27 = 0$
13	$x^2 + 3x + 2 = 0$	14	$x^2 + 7x + 12 = 0$
15	$y^2 + 8y + 12 = 0$	16	$y^2 + 15y + 56 = 0$
17	$x^2 + 15x + 50 = 0$	18	$x^2 + 23x + 132 = 0$
19	$t^2 - 4t + 3 = 0$	20	$t^2 - 6t + 8 = 0$
21	$x^2 - 13x + 40 = 0$	22	$x^2 - 13x + 42 = 0$
23	$p^2 - 22p + 120 = 0$	24	$p^2 - 25p + 150 = 0$
25	$2x^2 - 5x - 3 = 0$	26	$3x^2 + 5x - 2 = 0$
27	$2x^2 + 13x + 20 = 0$	28	$3x^2 + 2x - 5 = 0$
29	$6r^2 + r - 2 = 0$	30	$10r^2 - 11r - 6 = 0$
31	$18x^2 - 30x - 12 = 0$	32	$24x^2 - 20x - 16 = 0$

Find all real roots for each of the following equations by means of the Quadratic Formula.

33	$x^2 + 2 = 0$	34	$x^2 + 5 = 0$
35	$w^2 - 7 = 0$	36	$w^2 - 11 = 0$
37	$x^2 - 2x - 2 = 0$	38	$x^2 - 4x - 1 = 0$
39	$p^2 + 6p + 2 = 0$	40	$p^2 + 10p + 23 = 0$
41	$x^2 - 2x + 3 = 0$	42	$x^2 + 4x + 9 = 0$
43	$4x^2 + 12x + 10 = 0$	44	$9x^2 - 6x + 4 = 0$
45	$25r^2 + 10r - 9 = 0$	46	$16r^2 - 24r + 1 = 0$

Solve each of the following equations by any appropriate means.

47	$x^2 = 8 - 2x$	48	$x^2 = 18 + 3x$
49	$x(x + 1) = 30$	50	$x(x + 2) = 80$
51	$\dfrac{2}{x + 2} + \dfrac{8}{x - 12} = 3$	52	$\dfrac{3}{x - 5} + \dfrac{1}{x + 4} = 2$
53	$\dfrac{x}{x - 6} - \dfrac{2}{x + 1} = 4$	54	$\dfrac{5}{x + 3} + \dfrac{2x}{x - 1} = -1$
55	$\dfrac{1}{x - 1} = 3x + 2$	56	$\dfrac{2}{x + 3} = 2x - 1$

57 $5x + 1 - \dfrac{2}{x} = 0$

58 $3x - 7 + \dfrac{3}{x} = -6$

59 $\dfrac{2x}{x+3} = 5x$

60 $\dfrac{x}{x-5} = 3x$

61 $9 + \dfrac{3x}{x+1} = 2x - 1$

62 $5 - \dfrac{x}{2x+3} = 3 - x$

63 $x^4 - 81 = 0$

64 $x^4 - 625 = 0$

65 $x^3 - 2x^2 + x = 0$

66 $x^3 + 6x^2 + 9x = 0$

67 $x^4 + x^2 - 2 = 0$

68 $x^4 + 7x^2 + 12 = 0$

69 $\sqrt{x+1} = x - 3$

70 $\sqrt{2x+3} = x + 1$

71 $\sqrt{x-5} + 3x = 2$

72 $\sqrt{x+4} - 2x = 5$

Simplify each of the following expressions.

73 $\dfrac{x^2 + 3x}{x^3 + 2x^2 + x}$

74 $\dfrac{x^2 - 2x}{x^3 + x^2 - 2x}$

75 $\dfrac{x^2 - 1}{x + 1}$

76 $\dfrac{x^2 - 4}{3x - 6}$

77 $\dfrac{x - 2}{x^2 + x - 6}$

78 $\dfrac{x - 1}{x^2 - 2x + 1}$

79 $\dfrac{x^2 - x - 20}{x^2 + 2x - 8}$

80 $\dfrac{x^2 - 3x - 18}{x^2 + 2x - 48}$

81 $\dfrac{2x^2 - x - 3}{2x^2 + 9x - 18}$

82 $\dfrac{5x^2 - 6x - 8}{5x^2 - 11x - 12}$

Solve each inequality.

83 $x^2 + 3x - 10 > 0$

84 $x^2 - 2x - 3 > 0$

85 $x^2 + 2x - 24 \geq 0$

86 $x^2 + 2x - 80 \geq 0$

87 $4x^2 + 5x > 0$

88 $3x^2 - x > 0$

89 $2x^2 + 3 \geq 0$

90 $x^2 + 5 \geq 0$

91 $x^2 - 5x - 14 < 0$

92 $x^2 - x - 12 < 0$

93 $x^2 - 2x - 80 \leq 0$

94 $x^2 - 4x - 45 \leq 0$

95 $3x^2 + 7x < 0$

96 $9x^2 - 2x < 0$

97 $x^2 + 17 \leq 0$

98 $5x^2 + 4 \leq 0$

99 $x^2 + 6x + 4 \leq 0$

100 $x^2 - 10x + 23 \leq 0$

101 ***Calculator Problem*** Solve $3x^2 - 9x + 2 = 0$ with the Quadratic Formula.

102 ***Calculator Problem*** Solve $4x^2 - 11x + 3 = 0$ with the Quadratic Formula.

103 ***Calculator Problem*** Solve $2.01x^2 + 5.03x + 1.66 = 0$ with the Quadratic Formula.

104 ***Calculator Problem*** Solve $.87x^2 + 4.26x + 2.33 = 0$ with the Quadratic Formula.

1.5   Applications of Quadratic Equations

We shall now apply quadratic equations to various situations. The first example illustrates how quadratic equations may be used.

EXAMPLE 1

It costs a certain manufacturer $(.2x^2 + x + 20)$ dollars to make x units of its product. How many units of its product can it make with an expenditure of $120?

Solution
Since the costs equal $120, we solve the equation below.

$$.2x^2 + x + 20 = 120$$
$$.2x^2 + x - 100 = 0$$
$$x^2 + 5x - 500 = 0$$
$$(x + 25)(x - 20) = 0$$
$$x = -25, 20$$

The roots of the given equation are -25 and 20; we discard the root -25, since the number of units manufactured must be a positive number. Thus, the answer $x = 20$. ■

The next example shows how easily a quadratic equation can arise in a geometric problem.

EXAMPLE 2

A rectangular advertising poster has an area of 200 square inches and the length is 10 inches more than the width. Determine the dimensions of the poster.

Solution
If x represents the unknown width, then $(x + 10)$ represents the unknown length. Since the area of a rectangle is the product of the width and length, we must solve the following equation.

$$x(x + 10) = 200$$
$$x^2 + 10x - 200 = 0$$
$$(x + 20)(x - 10) = 0$$
$$x = -20, 10$$

Since x is a positive dimension, -20 makes no sense in the problem. The poster must have a width of 10 inches and a length of $(10 + 10) = 20$ inches. ■

Another applied geometric problem is solved in Example 3.

EXAMPLE 3

A restaurant owner wants to enclose a rectangular patio near his restaurant for customers to use during nice weather. The patio should cover 120 square feet. To screen the patio from a road on one side and a parking lot on the opposite side, the owner will use a high wall that costs $10 per linear foot; on the other two sides will be a low wall that costs $5 per linear foot. If the owner plans to spend exactly $320 on the walls, what are the dimensions of the patio?

Solution
If x is the length of the patio along the road and y is the length of the patio perpendicular to the road, then $xy = 120$. We use $y = 120/x$ and we solve the equation that sets the costs of the wall equal to $320.

$$10(2x) + 5(2)\left(\frac{120}{x}\right) = 320$$

$$20x + \frac{1200}{x} = 320$$

$$x + \frac{60}{x} = 16$$

$$x^2 + 60 = 16x$$

$$x^2 - 16x + 60 = 0$$

$$(x - 10)(x - 6) = 0$$

$$x = 10, 6$$

Our cost equation has two solutions. Consequently, the owner could construct a patio that is 10 feet by $\frac{120}{10} = 12$ feet or else 6 feet by $\frac{120}{6} = 20$ feet. ■

The **Pythagorean Theorem** in plane geometry says that the square of the length of the hypotenuse of a right triangle equals the sum of the squares of the lengths of the other two legs. Below is a right triangle with the hypotenuse and legs labeled. This theorem is used in the next example.

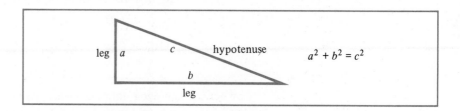

EXAMPLE 4

A wading pool for children is constructed in the shape of a right triangle. The hypotenuse is 50 feet long, and the sum of the two legs is 70 feet. Determine the dimensions of the pool.

Solution

We use the Pythagorean Theorem, which was explained in the paragraph preceding this example; we solve the equation dictated by that theorem. If x represents the length of one leg then the other leg is $70 - x$, and the following equation applies the Pythagorean Theorem to our problem.

$$x^2 + (70 - x)^2 = (50)^2$$
$$x^2 + 4900 - 140x + x^2 = 2500$$
$$2x^2 - 140x + 2400 = 0$$
$$x^2 - 70x + 1200 = 0$$
$$(x - 40)(x - 30) = 0$$
$$x = 30, 40$$

Either the first leg of the right triangle is 30 feet and the second leg is $70 - 30 = 40$ feet, or else the first leg is 40 feet and the second leg is $70 - 40 = 30$ feet. ■

The next example shows how a quadratic equation arises in a practical business decision.

EXAMPLE 5

A movie theatre owner is going to reduce the price of a ticket from $5.50 in order to get more customers. He decides that the average number of customers per day will be $(100 + 100x)$ where x is the reduction in the ticket price in dollars. How much should x be in order for the average daily ticket sales to total $750?

Solution

After the price reduction, a ticket will cost $(5.50 - x)$ dollars and the number of tickets sold will be $(100 + 100x)$. We set the product of these two numbers equal to $750, the total ticket sales, and we solve the resulting equation.

$$(5.50 - x)(100 + 100x) = 750$$
$$550 + 550x - 100x - 100x^2 = 750$$
$$-100x^2 + 450x - 200 = 0$$
$$10x^2 - 45x + 20 = 0$$
$$(10x - 5)(x - 4) = 0$$
$$x = .5, 4$$

The owner can achieve total ticket sales of $750 by reducing the ticket price $.50 or $4.00. Because a reduction of 50¢ results in fewer customers, there would be less wear and tear on the carpets, seats, and such. There would also be less revenue from refreshments. ■

We conclude these applications with an illustration of how quadratic inequalities arise in practical problems.

EXAMPLE 6

The profit for a certain business is $(150x - 10x^2 - 200)$ dollars where x is the number of customers served. Determine how many customers should be served in order for the profit to exceed $300.

Solution

We solve the inequality that says profit is greater than $300.

$$150x - 10x^2 - 200 > 300$$

$$-10x^2 + 150x - 200 - 300 > 0$$

$$-x^2 + 15x - 50 > 0$$

$$x^2 - 15x + 50 < 0$$

$$(x - 10)(x - 5) < 0$$

We consider the two ways that the product of $(x - 10)$ and $(x - 5)$ can be negative. Either $(x - 10) < 0$ and $(x - 5) > 0$ or else $(x - 10) > 0$ and $(x - 5) < 0$. We consider the first possibility.

$$x - 10 < 0 \quad \text{and} \quad x - 5 > 0$$

$$x < 10 \quad \text{and} \quad x > 5$$

These inequalities are satisfied for x chosen such that $10 > x > 5$. In other words the profit exceeds $300 when the number of customers served is 6, 7, 8, or 9. Now we consider the case $(x - 10) > 0$ and $(x - 5) < 0$.

$$x - 10 > 0 \quad \text{and} \quad x - 5 < 0$$

$$x > 10 \quad \text{and} \quad x < 5$$

Clearly there is no choice of x that is to the right of 10 and to the left of 5. Thus, the solution already given is the only solution. ■

Exercises 1.5

1 It costs a certain business $(.1x^2 + x + 50)$ dollars to serve x customers. How many customers can be served with an expenditure of $250?

2 The cost of serving x customers in a certain business is $(x^2 + 10x + 100)$ dollars. If $1300 is spent serving customers, how many customers are served?

3 The suppliers of a certain product will produce x units of the product when the price of the product is $(x^2 + 25x)$ dollars per unit. How many units will be produced when the price is $350?

4 When the price of a ton of a certain commodity is $(.01x^2 + 5x)$ then suppliers of the commodity will produce x tons of it. How many tons will be produced when the price is $5000 per ton?

5 A rectangular advertising poster must have an area of 112 square inches and the sum of the width and the length must be 22 inches. Determine the dimensions of the poster.

6   A rectangular poster must have an area of 150 square inches and the perimeter must be 50 inches. Determine the dimensions of the poster.

7   A rectangular poster must have an area of 288 square inches and the length must be twice the width. Determine the dimensions of the poster.

8   A rectangular poster must have an area of 216 square inches and the width must be two thirds of the length. Determine the dimensions of the poster.

9   A rectangular patio with 192 square feet of area is to be enclosed with a wall costing $10 per linear foot. If $560 is to be spent on the wall, what are the dimensions of the patio?

10 A rectangular patio with 224 square feet of area is to be enclosed. Along the two short sides of the patio will be walls costing $20 per linear foot; along the long sides will be walls costing $10 per linear foot. If $880 is to be spent on the walls, what are the dimensions of the patio?

11 A wading pool is constructed in the shape of a right triangle. The hypotenuse is 100 feet long and the short leg is three fourths of the long leg. Determine the dimensions of the pool.

12 A playground in the shape of a right triangle is enclosed by a fence 36 feet long. If the longest side of the fence is 15 feet, what are the lengths of the other two sides?

13 A circular advertising poster has an area of 100π square inches. Determine the radius of the circular poster.

14 A circular garden is constructed with an area of 400π square feet. Determine the radius of the circular garden.

15 A cable television service reduces its subscription rate from $20 per month in order to get more subscribers. After the monthly subscription rate is reduced x dollars there will be $(10{,}000 + 500x)$ subscribers. How much should x be in order for the total monthly dues to be $192,000?

16 How much should x be in problem 15 in order for the total monthly dues to be $168,000?

17 A fence that is 100 feet long is cut into two pieces. One piece is shaped into a square and the other piece is shaped into a rectangle that is twice as long as it is wide. If the sum of the areas of the square and the rectangle is 300 square feet, what are the dimensions of each?

18 In problem 17, suppose the fence is 96 feet long and the sum of the two areas is 277 square feet. Determine the dimensions of the square and the rectangle.

19 Suppose the sum of two positive numbers is 8 and the product of the two is 15.75. Determine each of the numbers.

20 Suppose the sum of two positive numbers is 8.5 and the product of the two is 15. Determine each of the numbers.

21 The profit of a certain manufacturer is $(175x - x^2 - 800)$ dollars, where x is the number of units made. Determine how many units should be made in order for the profit to be more than $6700.

22 The profit of a certain small business is $(20x - x^2 - 40)$ dollars, where x is the number of customers. Determine how many customers should be served in order for the profit to exceed $44.

23 The costs of a certain business are $(x^2 + 40)$ dollars, where x is the number of customers. Determine how many customers should be served in order for costs not to exceed \$121.

24 The costs of a certain manufacturer are $(x^2 + 1200)$ dollars, where x is the number of units manufactured. Determine how many units should be made in order for costs not to exceed \$1600.

25 A certain business estimates that the number of its customers is $(x^2 + 4x + 60)$, where x is the number of television ads per week. How many ads are required for the number of customers to equal or exceed 200?

26 In problem 25, what should x be in order for the number of customers to equal or exceed 540?

1.6   Absolute Value and Inequalities (Optional)

In this section, we shall consider the properties of absolute value and solve some inequalities that involve absolute value. As indicated in Section 1.1, the absolute value of the number a is denoted $|a|$ and defined by

$$|a| = \begin{cases} a & \text{if} \quad a \geq 0 \\ -a & \text{if} \quad a < 0. \end{cases}$$

Thus, $|1.5| = 1.5$ and $|0| = 0$ since $1.5 \geq 0$ and $0 \geq 0$; on the other hand, $|-2.2| = -(-2.2) = 2.2$ since $-2.2 < 0$.

Now we state the basic properties of absolute value.

Properties of Absolute Value

Let a and b be real numbers.

$$|ab| = |a||b| \tag{1}$$

$$\left| \frac{a}{b} \right| = \frac{|a|}{|b|}, \qquad b \neq 0 \tag{2}$$

$$|a + b| \leq |a| + |b| \tag{3}$$

Each of these properties follows from the definition of absolute value by considering all the possible signs of a and b. The verifications are somewhat tedious. We shall verify property (1) to show what is involved. If a and b are both nonnegative, then $|a| = a$, $|b| = b$, and (ab) is nonnegative. Thus, we have

$$|ab| = ab = |a||b|.$$

If a and b are both negative then $|a| = -a$, $|b| = -b$, and (ab) is positive. Thus, we have

$$|ab| = ab = (-a)(-b) = |a||b|.$$

If either a or b—say b, for example—is negative and the other is nonnegative, then $|a| = a$, $|b| = -b$, and $(ab) \leq 0$. Thus, we have

$$|ab| = -(ab) = a(-b) = |a||b|.$$

We have verified property (1) for all choices of signs for a and b.

The next example illustrates how the properties of absolute value can be used.

EXAMPLE 1

Show that $|3x + 5| \leq 11$ is true when $|x| \leq 2$.

Solution

$$
\begin{aligned}
|3x + 5| &\leq |3x| + |5| && \text{By property (3)} \\
&\leq |3||x| + |5| && \text{By property (1)} \\
&\leq 3|x| + 5 \\
&\leq 3(2) + 5 && \text{Because } |x| \leq 2 \\
&\leq 11 \qquad \blacksquare
\end{aligned}
$$

The next example illustrates how to use the properties and definition of absolute value to solve an equation.

EXAMPLE 2

Solve the equation $|x/2| = 6$.

Solution
First we simplify the equation.

$$\left|\frac{x}{2}\right| = \frac{|x|}{2} = 6 \qquad \text{By property (2)}$$

$$|x| = 12$$

Now we use the definition to discover all choices of x that make this last equation true. If $x \geq 0$, then the equation becomes

$$|x| = x = 12,$$

and we discover that $x = 12$ is a solution. If $x < 0$, then the equation becomes

$$|x| = -x = 12 \quad \text{or} \quad x = -12,$$

and we see that $x = -12$ is a solution. The two answers are $x = 12$ and $x = -12$. $\blacksquare$

The following rule is useful in solving some inequalities involving absolute value.

Rule 1   For $a \geq 0$, the solution of the inequality

$$|w| \leq a$$

is described by the inequalities

$$-a \leq w \leq a.$$

On the number line below we illustrate the values of w satisfying the inequalities $-a \leq w \leq a$.

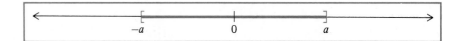

The next example shows how Rule 1 can be used.

EXAMPLE 3

Solve the inequality $|x - 3| \leq 4$.

Solution
Let w in Rule 1 be $(x - 3)$. The values of $(x - 3)$ that make the given inequality true are

$$-4 \leq x - 3 \leq 4.$$

Add 3 to each side of each inequality to get a description of the choices of x that make the given inequality true

$$-1 \leq x \leq 7.$$

This is the solution. ■

The next rule is very similar to Rule 1.

Rule 2   For $a > 0$, the solution of the inequality

$$|w| < a$$

is described by the inequalities

$$-a < w < a.$$

Below we illustrate the w satisfying the inequalities $-a < w < a$.

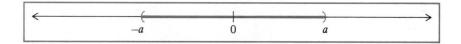

The next example illustrates the use of Rule 2.

EXAMPLE 4

Solve the inequality $|5x + 2| < 1$.

Solution
Let w in Rule 2 be $(5x + 2)$. The values of $(5x + 2)$ that make the given inequality true are

$$-1 < 5x + 2 < 1.$$

Using the elementary properties of inequalities, we simplify the above inequalities.

$$-1 < 5x + 2 < 1$$

$$-3 < 5x < -1$$

$$-\frac{3}{5} < x < -\frac{1}{5}$$

These last inequalities describe the solution. ■

The next rule indicates how to proceed when the absolute value is the larger quantity in the inequality.

Rule 3   For $a \geq 0$, the solution of the inequality

$$|w| \geq a$$

is described by the inequalities

either $w \geq a$ or $w \leq -a$.

On the number line below we show all the w satisfying one of the inequalities $w \geq a$ or $w \leq -a$.

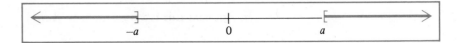

We apply Rule 3 in the next example.

EXAMPLE 5

Solve the inequality $|7 - 3x| \geq 2$.

Solution

Let w in Rule 3 be $(7 - 3x)$. The values of $(7 - 3x)$ that make the given inequality true are

either $7 - 3x \geq 2$ or $7 - 3x \leq -2$.

Using the elementary properties of inequalities, we simplify the above inequalities.

$$7 - 3x \geq 2 \quad \text{or} \quad 7 - 3x \leq -2$$

$$-3x \geq -5 \quad \text{or} \quad -3x \leq -9$$

$$x \leq \frac{5}{3} \quad \text{or} \quad x \geq 3$$

The last inequalities describe the solution. ■

It should be emphasized that x makes the inequality in Example 5 true provided $x \le \frac{5}{3}$. Alternatively, the inequality is true provided $x \ge 3$. There is no choice of x that satisfies both of the inequalities

$$x \le \frac{5}{3}, \qquad x \ge 3.$$

Either one of the inequalities $x \le \frac{5}{3}$ or $x \ge 3$ suffices to have

$$|7 - 3x| \ge 2.$$

The next rule is very similar to Rule 3.

Rule 4   For $a \ge 0$, the solution of the inequality

$$|w| > a$$

is described by the inequalities

$$\text{either} \quad w > a \quad \text{or} \quad w < -a.$$

On the number line below we show all the w satisfying one of the inequalities $w > a$ or $w < -a$.

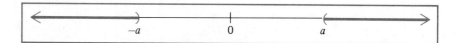

We apply Rule 4 in the next example.

EXAMPLE 6

Solve the inequality $\left| \dfrac{5}{x} + 1 \right| > 3$.

Solution
Let w in Rule 4 be $((5/x) + 1)$. The values of $((5/x) + 1)$ that make the given inequality true are

$$\text{either} \quad \frac{5}{x} + 1 > 3 \quad \text{or} \quad \frac{5}{x} + 1 < -3.$$

We simplify these inequalities somewhat.

$$\frac{5}{x} + 1 > 3 \quad \text{or} \quad \frac{5}{x} + 1 < -3$$

$$\frac{5}{x} > 2 \quad \text{or} \quad \frac{5}{x} < -4$$

$$\frac{1}{x} > \frac{2}{5} \quad \text{or} \quad \frac{1}{x} < -\frac{4}{5}.$$

Each of these inequalities requires careful attention because inequalities are dramatically changed when we multiply both sides by a negative quantity. For the inequality

$$\frac{1}{x} > \frac{2}{5}$$

to be true, $1/x$ must be positive and so x must be positive. Now we can solve this inequality.

$$\left(\frac{5}{2}x\right)\frac{1}{x} > \left(\frac{5}{2}x\right)\frac{2}{5}, \qquad x > 0$$

$$\frac{5}{2} > x > 0$$

For the inequality

$$\frac{1}{x} < -\frac{4}{5}$$

to be true, $1/x$ must be negative, and so x must be negative. Thus, the product $\left(-\frac{5}{4}x\right)$ is positive and we have

$$\left(-\frac{5}{4}x\right)\frac{1}{x} < \left(-\frac{5}{4}x\right)\left(-\frac{4}{5}\right), \qquad x < 0$$

$$-\frac{5}{4} < x < 0.$$

The values of x that make the given inequality true are described by the inequalities

$$\frac{5}{2} > x > 0 \quad \text{or} \quad -\frac{5}{4} < x < 0. \quad \blacksquare$$

Our last rule indicates how to solve equations involving absolute value.

Rule 5   For $a \geq 0$, the solution of the equation

$$|w| = a$$

is described by the equations

either $w = a$ or $w = -a.$

The next example illustrates the use of Rule 5.

EXAMPLE 7

Solve the equation $|5x - 10| = 4.$

Solution
Let w in Rule 5 be $(5x - 10)$. The values of $(5x - 10)$ that make the given equation true are

either $5x - 10 = 4$ or $5x - 10 = -4.$

We solve these equations routinely.

$$5x - 10 = 4 \quad \text{or} \quad 5x - 10 = -4$$
$$5x = 14 \quad \text{or} \quad 5x = 6$$
$$x = \frac{14}{5} \quad \text{or} \quad x = \frac{6}{5}$$

The two numbers that solve the given equation are $\frac{14}{5}$ and $\frac{6}{5}$. ■

Each of the five rules involved a number a that was required to be nonnegative. Because the absolute value of any number is nonnegative, the inequalities

$$|w| \leq a, \qquad |w| < a$$

have no solutions for negative a. The inequalities

$$|w| \geq a, \qquad |w| > a$$

are true for all w when a is a negative number. Thus, solving the inequalities with negative a is not difficult. For negative a the equation $|w| = a$ has no solution.

Exercises   1.6

Solve each of the following equations.

1 $|2x + 3| = 7$

2 $|3x - 5| = 2$

3 $|9x + 4| = 0$

4 $\left|\dfrac{x}{3} + 1\right| = 0$

5 $|-3x + 2| = -1$

6 $|-7x + 9| = -3$

7 $\left|\dfrac{3}{x} + 4\right| = 5$

8 $\left|6 - \dfrac{2}{x}\right| = 8$

9 $|9x| + 3 = 6$

10 $|-5x| - 2 = 7$

11 $3|4x - 5| - 1 = 8$

12 $2|3 - 4x| + 3 = 11$

13 $|7x| + |5| = |-19|$

14 $|-2x| - |3| = |7|$

15 $2\left|\dfrac{3}{x} + 1\right| + 5 = |-11|$

16 $3\left|8 - \dfrac{4}{x}\right| - 4 = |11|$

Solve each of the following inequalities.

17 $|2x - 12| \leq 20$

18 $|3x + 9| \leq 21$

19 $|4 - 5x| < 6$

20 $|9 + 7x| < 10$

21 $\left|\dfrac{x}{2} + 3\right| \leq 1$

22 $\left|5 - \dfrac{x}{3}\right| \leq 2$

23 $\left|\dfrac{1}{x}+7\right|<9$

24 $\left|10-\dfrac{3}{x}\right|<12$

25 $|9x+1|>15$

26 $|7-4x|>20$

27 $|3-5x|\geq 1$

28 $|2x+10|\geq 2$

29 $\left|\dfrac{5}{x}+1\right|>-2$

30 $\left|8-\dfrac{3}{x}\right|>-9$

31 $\left|11-\dfrac{5}{x}\right|<-15$

32 $\left|\dfrac{13}{x}+11\right|<-2$

33 $\left|\dfrac{2}{x}+1\right|\geq 3$

34 $\left|7-\dfrac{3}{x}\right|\geq 9$

35 $5|10x+3|>|20|$

36 $3|5-6x|>|-27|$

37 $4|-7x|+|5|\leq|-19|$

38 $10|5x|-|-7|\leq|18|$

39 $\left|\dfrac{3}{x}\right|+2<9$

40 $7-\left|\dfrac{2}{x}\right|<10$

Review of Terms

Important Mathematical Terms
real numbers, *p. 2*
number line, *p. 2*
counting numbers, *p. 2*
natural numbers, *p. 2*
integers, *p. 2*
rational number, *p. 3*
irrational number, *p. 3*
inequality signs, *p. 4*
inequality, *p. 4*
absolute value, *p. 5*
nonnegative, *p. 5*
positive, *p. 5*
nonpositive, *p. 5*
negative, *p. 5*
magnitude, *p. 5*
commutative property, *p. 6*
associative property, *p. 6*
identity, *p. 6*
inverse, *p. 6*
distributive property, *p. 7*
equation, *p. 9*

equal sign, *p. 9*
variable, *p. 9*
solve, *p. 10*
solution, *p. 10*
linear equation, *p. 11*
equivalent, *p. 12*
linear inequality, *p. 15*
percent, *p. 15*
algebraic expression, *p. 19*
term, *p. 19*
like terms, *p. 19*
coefficient, *p. 19*
common denominator, *p. 21*
factoring, *p. 21*
quadratic equation, *p. 25*
root, *p. 25*
quadratic expression, *p. 27*
Quadratic Formula, *p. 28*
quadratic inequality, *p. 29*
Pythagorean Theorem, *p. 37*

Important Terms from the Applications
discount, *p. 15*
rate of return, *p. 15*
net income, *p. 16*

profit margin, *p. 16*
cost, *p. 36*
profit, *p. 39*

Review Problems

For each of the following numbers, indicate whether it is a counting number, an integer, a rational number, or an irrational number.

1 -3.51 2 $\dfrac{22}{7}$ 3 111 4 $\sqrt{10}$ 5 -13

On a number line locate every value of x that satisfies the given inequalities.

6 $x < 2$ 7 $-3 \leq x \leq 5$
8 $6 < x \leq 10$ 9 $-9 < x < 8$
10 $10 \leq x < 11$

Identify each property of the real numbers that has been used in the indicated simplification. Let a, b, and c denote real numbers.

11 $a(b + c) = ab + ac$ 12 $3 + a = a + 3$
13 $(2 + 1) + 4 = 2 + 5$ 14 $3(ab) = (3a)b$
15 $9 + 0 = 9$

Solve each of the following equations.

16 $9x - 12 = 5$ 17 $2(x - 1) + 3(x + 2) = 6$

18 $-8x = 3 + 5(x - 2)$ 19 $\dfrac{x}{3} + \dfrac{4x}{5} = 1$

20 $\dfrac{5}{x} + 3 = \dfrac{2}{x}$ 21 $\dfrac{3}{x - 2} + \dfrac{7}{x - 2} = 4$

22 $\dfrac{5}{2(x + 1)} - 2 = \dfrac{3}{x + 1}$

Solve each of the following inequalities.

23 $5x - 7 < 20$ 24 $9 - 2x \geq 11$

25 $6 - \dfrac{x}{3} \leq 8$ 26 $3(2x + 5) > 2(x + 3)$

27 $\dfrac{x}{2} + 3 < 5 + \dfrac{x}{3}$ 28 $\dfrac{2x + 1}{5} + 6 > \dfrac{x - 3}{2}$

Simplify each algebraic expression that follows.

29 $(x^2 + 2x - 3) + (4x^2 - 5x + 7)$
30 $(9x + 7) + (6x^2 - 10)$
31 $(x^3 + x + \sqrt{x}) - (x^3 + 3\sqrt{x})$
32 $x\left(3x^2 + 9 - \dfrac{2}{x}\right) + \left(5x - 1 + \dfrac{5}{x}\right)$
33 $(x + 1)(3x - 2) + x^2 + 3x + 5$
34 $(x - 3)(x^2 + 4x + 6) - x(x^2 + 3x - 1)$

35 $\dfrac{7}{x-2} - \dfrac{7}{x+2}$

36 $\dfrac{2}{x-3} + \dfrac{5}{x+4}$

37 $\dfrac{x}{x+1} + \dfrac{x+5}{x-1}$

38 $\dfrac{2x}{x-3} + \dfrac{5}{x+2} - \dfrac{x+1}{x-1}$

39 $\dfrac{5x^3 - 6x^2 + 10x}{x}$

40 $\dfrac{x^2}{9x^4 - 11x^3}$

41 $\dfrac{x^2 - 2x}{x^3 + 3x^2 + 4x}$

42 $\dfrac{x^2 - 16}{x+4}$

43 $\dfrac{x^2 + 4x + 4}{x^2 - 4}$

Solve each of the following equations by factoring.

44 $x^2 - 25 = 0$

45 $x^2 + 2x + 1 = 0$

46 $x^2 + x - 2 = 0$

47 $7x^2 - 11x = 0$

48 $x^2 - 7x + 12 = 0$

49 $2x^2 - 5x - 3 = 0$

50 $8x^2 - 14x + 3 = 0$

Find all real roots for each of the equations that follow by means of the Quadratic Formula.

51 $x^2 - 2x - 1 = 0$

52 $x^2 - 4x + 5 = 0$

53 $x^2 + 9 = 0$

54 $x^2 - 5x = 0$

Solve each of the following equations by any appropriate means.

55 $x^2 - 2 = x$

56 $x(x+1) = 12$

57 $\dfrac{3}{x-2} + \dfrac{5}{x+1} = 4$

58 $\dfrac{3}{x-1} + 2 = x+1$

59 $x^3 - x^2 - 6x = 0$

60 $x^4 - 1 = 0$

61 $\sqrt{x+2} = x-3$

62 $x^4 + 3x^2 - 4 = 0$

Solve each of the following inequalities.

63 $x^2 - 13x + 42 < 0$

64 $x^2 - 2x \le 15$

65 $x^2 + 2x > 8$

66 $x^2 - x - 6 \ge 0$

Solve the following equations and inequalities involving absolute value.

67 $|7x + 3| = 5$

68 $|8 - 3x| = 10$

69 $\left| 1 + \dfrac{2}{x} \right| = 4$

70 $|-5x| - |-2| = |-3|$

71 $|4x + 6| \le 12$

72 $|8 - 2x| < 6$

73 $\left| \dfrac{x}{3} + 5 \right| > 8$

74 $\left| \dfrac{6}{x} + 5 \right| \ge 9$

75 $4|2 - 3x| - 1 \le 11$

Solve each of the following problems.

76 A merchant advertises that all merchandise will be sold at a 15% discount off the prices indicated on the price tags. What is the selling price of an item marked $31?

77 A magazine publisher has a system of graduated discounts. For the first 100 copies the price is $5 each. For the second 100 copies the buyer gets a 10% discount; the buyer gets an additional discount of 10% on all copies beyond the first 200. What is the price of 400 copies?

78 Suppose $30,000 is divided between investment A with a 25% rate of return and investment B with a 20% rate of return. If the total annual return is $6600 then how is the $30,000 divided between the two investments?

79 A certain business has total annual sales of $900,000, and the owners of the business need $198,000 in income. What should the profit margin be to provide the necessary income?

80 It costs a certain business $(.3x^2 + 15x + 180)$ dollars to serve x customers. How many customers can be served with an expenditure of $600?

81 A rectangular advertising poster must have an area of 162 square inches and the length must be twice the width. Determine the dimensions of the poster.

82 A playground in the shape of a right triangle is enclosed by a fence 48 feet long. If the longest side of the fence is 20 feet, what are the lengths of the other two sides?

83 A magazine reduces its subscription rate from the regular $60 per year in order to get more subscribers. After the subscription rate is reduced x dollars there will be $(100,000 + 1000x)$ subscribers. How much should x be in order for the total annual payments from subscribers to be $5,500,000?

84 The profit of a certain manufacturer is $(100x - x^2 - 600)$ dollars, where x is the number of units made. Determine how many units should be made in order for the profit to exceed $1800.

85 Suppose the sum of two positive numbers is 15.5 and the product of the two is 55. Determine each of the numbers.

Social Science Applications

Metric System
We list three of the basic units of length in the metric system along with the equivalent length in an appropriate English unit.

1 centimeter $\approx$ .3937 inch
1 meter $\quad \approx$ 1.094 yards
1 kilometer $\approx$ 0.621 mile

1 Write five inequalities that express the relative length of a centimeter, inch, meter, yard, kilometer, and mile.

2 Write equations showing how to convert from centimeters to inches, from meters to yards, and from kilometers to miles.

3   Write equations showing how to convert from inches to centimeters, from yards to meters, and from miles to kilometers.

4   *Home Economics*   One liter is equivalent to 1.057 liquid quarts, and 4 quarts equal 1 gallon. Which is the wiser purchase, a 4-liter bottle of wine for $8.50 or a gallon bottle of the same wine priced at $8.30?

5   *Physical Geography*   To convert a temperature reading from a Celsius thermometer to an equivalent Fahrenheit reading, multiply by $\frac{9}{5}$ and then add 32. Write an equation that shows how to convert a Celsius temperature to an equivalent Fahrenheit temperature.

6   *Psychology*   A person's intelligence quotient (or I.Q.) is the person's mental age, as determined by intelligence tests, multiplied by 100 and divided by the person's chronological age. What is the mental age of a 20-year old with an I.Q. of 140?

7   *Sociology*   The average per capita income in Connecticut in 1980, which was $11,720, is what percent of the average per capita income in Mississippi in 1980, which was $6,580?

8   *Law Enforcement*   One method for determining the speed of a car involved in an accident is by measuring the skid marks left by the tires on the road. An automobile traveling at a speed of s (in miles per hour) will have a skid approximately b feet long, according to the formula

$$b = .041s^2.$$

What was the speed of a car that skidded 35 feet?

9   *Political Science*   The amount of an ad valorem tax is, by definition, a percentage of the value of the item taxed. Taxes on real estate are typically ad valorem taxes. If a property tax equals .6% times the value of the home taxed and the home is worth $100,000, how much is the tax?

CAREER PROFILE
Accountant

Accounting has been called "the language of business." For this reason, completion of an undergraduate major in accounting not only prepares an individual for a career as an accountant but also is an important stepping-stone toward business ownership or upper-echelon management positions.

Because accountants work with numerical information, it is frequently supposed that mathematics courses are an important part of an accountant's training. Rather than needing knowledge of advanced topics in mathematics, an accountant needs a mathematics background primarily as a means of learning to think logically and systematically and to do careful and well-organized work.

Mathematical aptitude is essential for an accountant, but there are few specific mathematical requirements. In fact, accountants and other professionals who like mathematics often study it partly as an avocation. Mathematical study stimulates them beyond routine thinking; it offers a challenge and a recreational diversion. Off the job, many professionals whose work activity demands systematic and regulated thinking turn to mathematical puzzles and games for satisfying relaxation.

To be a good accountant, an individual must pay attention to small details, often concentrating for long periods of time and striving for correctness. He or she must be able to analyze and interpret facts and figures and to draw sound conclusions from them. To be successful, an accountant must be personable and able to communicate recommendations to clients effectively.

The services of accountants are in heavy demand in times of recession as well as in times of prosperity. The ability to assess financial relationships between projected costs and anticipated sales and to ascertain the tax advantages and disadvantages of certain business procedures is always in demand. About 900,000 individuals in the United States were employed as accountants (and auditors) in 1980. Employment for accountants is expected to grow faster than that for most other occupations because of increasing pressures on both businesses and governments to improve financial management procedures.

Most entry-level positions in accounting require a bachelor's degree in accounting. Experience with computers is becoming a necessity as well. Permanent employment and advancement generally require the Certified Public Accountant (CPA) certificate. For persons with a bachelor's degree in accounting, this certificate requires passage of a four-part, $2\frac{1}{2}$-day examination (administered by the American Institute of Certified Public Accountants and taken during a student's senior year) together with two years' experience in public accounting. The requirement that CPA candidates have training beyond the bachelor's degree and, in some cases, have a master's degree, is being considered in some states; in addition, some states already require that a certain number of hours of continuing education must be completed before certification is renewed.

The Institute of Internal Auditors, Inc., confers the Certified Internal Auditor (CIA) title to accounting graduates of accredited colleges and universities who have completed three years' experience in internal auditing and who have passed a four-part examination. Of the 900,000 people who worked as accountants and auditors in 1980, the number of CPAs was about 200,000 and the number of CIAs was about 10,000.

In 1980, individuals holding a bachelor's degree in accounting received starting salaries averaging around $16,000. An accountant can expect to advance to a position with an average salary in excess of $40,000. Also, because accounting is the language of business, an accountant is in a good position to advance to higher level management or consulting positions.

Because opportunities for part-time work are plentiful, accounting combines well with careers in teaching, with raising a family, or when less than full-time employment is desired.

Sources of Additional Information
American Institute of Certified Public Accountants. 1211 Avenue of the Americas, New York, NY 10036. (The institute offers information about careers in accounting and about aptitude tests administered in high schools, colleges, and public accounting firms.)

Occupational Outlook Handbook. Bureau of Labor Statistics, U.S. Department of Labor, Washington, DC 20212. (Revised every 2 years, this handbook provides information about job duties, working conditions, level and places of employment, education and training requirements, advancement possibilities, job outlook, and earnings for about 250 occupations. Some of the information for this Career Profile was obtained from this source.)

2 | Functions and Graphs

2.1 Functions

Because a student's scores on tests determine his average, we say that a student's average is a function of his test scores. The success of most businesses is determined by the volume of sales, so business success is a function of sales. Any time we choose a value for x in the formula $y = x^2$, then y is determined and so y is a function of x. In general, a **function** is a rule that associates with each number x in a certain set, called the domain, precisely one number y from another set called the range. More will be said about domain and range later. A student's average is associated with his test scores; if the scores on the first two tests are 90 and 100 and the score on the third test is the yet undetermined number x, then the associated average y is $(90 + 100 + x)/3$. Given the test scores, there is a unique number that is the average. If the test scores change, then the average changes, but as soon as the scores are listed, there is a unique corresponding average.

EXAMPLE 1

Demo has a hot dog stand in Dallas Mall where he sells his hot dogs for 60¢ each. Give a formula that determines Demo's monthly revenue.

Solution
Let x be the number of hot dogs that Demo sells in one month; then his revenue is

$$R = .60x.$$

Demo's revenue is a function of his sales. ∎

Most businesses have some costs before they sell anything; these costs are called the **fixed costs.** For example, Demo rented the space in Dallas Mall where he sells his hot dogs, and before he could fix any hot dogs he had to buy some restaurant equipment and pay to have it installed. In addition to fixed costs, a business has to pay for the materials and labor necessary to make its product. These are called the **variable costs** because they vary according to how much of the product is made.

EXAMPLE 2

Demo paid for all of the equipment in his hot dog stand with two loans. Each month Demo pays $350 on his loans and $400 in rent. If it costs Demo 15¢ to prepare and sell one hot dog, what is a formula for Demo's costs?

Solution
In this case, the monthly fixed costs are $750, and the monthly variable costs are determined by the number of hot dogs sold. If x is the number of hot dogs sold per month, then the variable costs are $.15x$. Thus, Demo's total monthly costs C are given by the formula

$$C = 750 + .15x \quad \blacksquare$$

A function that invariably interests business students, is the profit function. **Profit** is defined to be revenues minus costs. In the case of Demo's hot dog stand in Examples 1 and 2, the profit P is given by the formula

$$P = .60x - (750 + .15x)$$

$$P = .45x - 750.$$

Frequently, we use the notation $P(x)$ to indicate that P is a function of x. This is also indicated by saying that x is the **independent variable** and P is the **dependent variable.** Since x goes first in taking a value, it is independent; once x takes on a value, P is determined and so P is dependent on x.

EXAMPLE 3

Let x be the independent variable and let y be the dependent variable. Does the equation $y^2 = 9x$ determine a function?

Solution
When x is chosen to be 1, then the values of y are 3 and -3. Since there is not a unique y associated with each x, this is not a function. $\blacksquare$

Most functions in elementary work are given by explicit formulas, and certain kinds of formulas arise so often that we give them names. The function $f(x)$ is **linear** provided it can be written in the form $f(x) = ax + b$, where a and b are some fixed constants and a is not zero. Here, the value of x varies and the values of a and b do not change. The function $g(x)$ is **quadratic** provided there are constants a, b, and c such that

$$g(x) = ax^2 + bx + c \quad \text{with } a \neq 0.$$

The area A of a square with an edge that is x inches long is given by $A = x^2$; this is a practical example of a quadratic function. The function $h(x)$ is **cubic** provided there are constants a, b, c, and d such that

$$h(x) = ax^3 + bx^2 + cx + d \quad \text{with } a \neq 0.$$

The volume of a cube is a cubic function of the length of the edge of the cube.

Clearly, if we continue the pattern of the above definitions we shall run out of names for the constants involved. A frequently used mathematical device for creating lots of names is to choose one letter and give it different subscripts for different names. For example, a cubic function has the form

$$h(x) = a_3x^3 + a_2x^2 + a_1x + a_0 \qquad \text{with } a_3 \neq 0$$

where a_3, a_2, a_1, and a_0 are fixed constants. Now we can describe one kind of function that is very useful in mathematics. The function $p(x)$ is a **polynomial** provided there are constants $a_n, a_{n-1}, \ldots, a_1, a_0$ such that

$$p(x) = a_nx^n + a_{n-1}x^{n-1} + \cdots + a_1x + a_0 \qquad \text{with } a_n \neq 0.$$

Here, n must be a nonnegative integer and it is said to be the **degree** of the polynomial. For example, the degree of $x^2 + 1$ is 2, the degree of $-2x + 1$ is 1, and the degree of 5 is 0.

The **domain** of a function $f(x)$ is defined to be the collection of all x for which $f(x)$ can be evaluated. Of course, the definition of the function $f(x)$ may include restrictions on which x should be used and which x should not be used.

EXAMPLE 4

Each nonnegative number has two square roots: for example, the roots of 4 are 2 and -2. So that taking square roots will be a function, we agree that $\sqrt{x}$ always denotes the nonnegative square root of x. Give the domain of the function $f(x) = \sqrt{x}$ and determine whether it is a polynomial.

Solution

If y equals $\sqrt{x}$, then $y^2 = x$ according to the meaning of square root. Since y^2 is the product of two factors with the same sign, it must be nonnegative. Thus, x must be nonnegative for $\sqrt{x}$ to make sense, so the domain of $f(x)$ is all nonnegative numbers. Although $f(x)$ can be written as $f(x) = x^{1/2}$, it is not a polynomial because $\frac{1}{2}$ is not a nonnegative integer. ■

EXAMPLE 5

Give the domain of the function $g(x) = 1/x$ and determine whether it is a polynomial.

Solution

Since division by 0 is not defined, the formula does not make sense for $x = 0$; otherwise, the formula does make sense. The domain is all nonzero numbers. Although $g(x)$ can be written as $g(x) = x^{-1}$, it is not a polynomial because -1 is not a nonnegative integer. ■

EXAMPLE 6

Compute $f(5)$, where $f(t) = t^3 - 2t^2 - t + 8$.

Solution

$$\begin{aligned} f(5) &= 5^3 - 2(5^2) - 5 + 8 \\ &= 125 - 50 - 5 + 8 = 78. \end{aligned}$$ ■

EXAMPLE 7

Compute $f(s + 2)$, where $f(s) = s^2 + s + 1$.

Solution

$$f(s+2) = (s+2)^2 + (s+2) + 1$$
$$= s^2 + 4s + 4 + s + 2 + 1$$
$$= s^2 + 5s + 7 \quad \blacksquare$$

The next example involves a computation that will arise often in Chapter 7.

EXAMPLE 8

Compute $\dfrac{f(x+h) - f(x)}{h}$ where $f(x) = x^2 + x + 1$ and h is a nonzero constant.

Solution

$$\frac{f(x+h) - f(x)}{h} = \frac{[(x+h)^2 + (x+h) + 1] - [x^2 + x + 1]}{h}$$
$$= \frac{x^2 + 2xh + h^2 + x + h + 1 - x^2 - x - 1}{h}$$
$$= \frac{2xh + h^2 + h}{h}$$
$$= 2x + h + 1 \quad \blacksquare$$

EXAMPLE 9

Specify the domain of $f(x) = \sqrt{2x - 1}$.

Solution
Since $\sqrt{y}$ is defined only for $y \geq 0$, as shown in Example 4, $f(x) = \sqrt{2x - 1}$ is defined only when $2x - 1 \geq 0$. By solving this last inequality we determine the domain.

$$2x - 1 \geq 0$$
$$2x \geq 1$$
$$x \geq \frac{1}{2}$$

The domain of $f(x)$ is all x greater than or equal to $\frac{1}{2}$. $\quad \blacksquare$

For some functions, the rule that associates y with each x is not given by the usual kind of formula.

EXAMPLE 10

Specify the domain for the function below and evaluate the function at -1, 1, 5, and 10.

$$f(u) = \begin{cases} 0 & \text{if } u \neq 5 \\ 1 & \text{if } u = 5 \end{cases}$$

Solution
The rule indicates how to evaluate the function for any real number x, and so

the domain is the set of all real numbers. Since none of -1, 1, and 10 equals 5, we have

$$f(-1) = 0, \qquad f(1) = 0, \qquad f(10) = 0$$

and, of course, we have

$$f(5) = 1. \quad \blacksquare$$

CALCULATOR EXAMPLE

Show that $f(x) = \sqrt{x}$ gets larger and larger as x takes on each of the values 1, 2, . . . , 9, 10 in that order.

Solution
The table below was constructed using the square root key of a calculator.

x	1	2	3	4	5	6	7	8	9	10
$\sqrt{x}$	1	1.41	1.73	2	2.24	2.45	2.65	2.83	3	3.16 $\blacksquare$

Exercises 2.1

In problems 1 through 20, find $f(0)$, $f(1)$, $f(-1)$, and $f(\frac{1}{2})$ for the indicated function $f(x)$.

1 $f(x) = 5$
2 $f(x) = 12$
3 $f(x) = 3x + 2$
4 $f(x) = 5x - 4$
5 $f(x) = -7x + 1$
6 $f(x) = .2x + .5$
7 $f(x) = x^2 + 1$
8 $f(x) = 3x^2 - x$
9 $f(x) = 2x^2 + x + 1$
10 $f(x) = x^2 + 2x + 1$
11 $f(x) = (x - 2)(x + 1)$
12 $f(x) = x(3x + 4)$

13 $f(x) = \dfrac{1}{x + 1}$
14 $f(x) = \dfrac{x}{x^2 + 2}$

15 $f(x) = \dfrac{x + 1}{x - 3}$
16 $f(x) = \sqrt{x^2 + 1}$

17 $f(x) = \sqrt{x + 5}$
18 $f(x) = x^3 + 1$
19 $f(x) = x^3 + x^2 + 1$
20 $f(x) = x(x^2 - 3x + 2)$

In problems 21 through 30, give the domain of the indicated function.

21 $f(x) = \sqrt{x^2 + 1}$
22 $g(x) = \sqrt{x + 5}$

23 $h(t) = \sqrt{3t - 6}$
24 $r(x) = \dfrac{1}{\sqrt{x - 3}}$

25 $p(s) = \sqrt{4 - s^2}$
26 $c(v) = \dfrac{1}{v - 5}$

27 $f(x) = \dfrac{9}{x^2 - 1}$

28 $g(x) = \dfrac{3}{x^2 + x - 2}$

29 $h(u) = \dfrac{2u}{4u^2 + 3}$

30 $f(s) = \dfrac{s^2 - 9}{s - 3}$ Caution: Do not change the function.

In problems 31 through 38, indicate whether the given function is a polynomial. If it is, specify the degree.

31 $f(x) = 11x^{23}$

32 $g(x) = \dfrac{x + 1}{x - 1}$

33 $h(x) = x^{1/3}$

34 $r(t) = t^2 - t + 5 - t^{-1}$

35 $m(s) = 16$

36 $c(x) = \dfrac{x^2 - 1}{x - 1}$

37 $r(t) = \sqrt{7}t^2 + .11t - \sqrt{5}$

38 $p(v) = \dfrac{2}{v^2 - v + 2}$

In problems 39 through 47, let x be the independent variable and y be the dependent variable. Indicate whether the equation determines a function.

39 $y^2 = x$
40 $y^3 = x$
41 $y^4 = x$
42 $y = \sqrt{x^2}$
43 $1 = x^2 + y^2$
44 $36 = 9x^2 + 4y^2$
45 $y = \begin{cases} 1 & \text{if } x \geq 0 \\ -1 & \text{if } x < 0 \end{cases}$
46 $y =$ the largest integer not larger than x
47 $xy = 0$

In problems 48 through 55, compute $\dfrac{f(x + h) - f(x)}{h}$ for each of the indicated functions.

48 $f(x) = x^2 + 5$

49 $f(x) = 4x - 2$

50 $f(x) = \dfrac{x}{3}$

51 $f(x) = 3x^2 - 6x + 7$

52 $f(x) = \sqrt{x}$
54 $f(x) = x^3$

53 $f(x) = \sqrt{3x + 1}$
55 $f(x) = 2x^3 + 1$

Solve the following:

56 For Demo's hot dog stand described in the first two examples of this section, determine how many hot dogs must be sold for revenues to equal costs.
57 Reliable Brick Company has fixed costs of $100,000; it costs 2¢ for them to make one of their bricks, which they sell for 10¢. Find a formula for

the profit of the brick company as a function of the number of bricks sold. Assume that they sell every brick that they make.

58 Speedy Tax Service charges an average fee of $35 for each tax return that it prepares. Write revenues as a function of the number of tax returns prepared.

59 Give an example of a polynomial with degree 23.

60 Write a formula for the area of a circle as a function of the radius of the circle.

61 Write a formula for the circumference of a circle as a function of the radius.

62 Write a formula for the perimeter of a square as a function of the length of one edge.

63 Write a formula for the area of a triangle with equal sides as a function of the length of one side.

64 *Calculator Problem* Show that $f(x) = x^8$ gets larger and larger as x takes on each of the values 1, 2, . . . , 9, 10, in that order.

65 *Calculator Problem* Show that $f(x) = x^9$ gets larger and larger as x takes on each of the values 1, 2, . . . , 9, 10 in that order.

66 *Calculator Problem* Show that $f(x) = 1/\sqrt{x}$ gets smaller and smaller as x takes on each of the values 1, 2, . . . , 9, 10 in that order.

2.2 Coordinate Systems

The idea of using numbers to describe such geometric ideas as length and area is very natural and very elementary. Early in the seventeenth century René Descartes had the idea of using geometry to describe numbers and to understand functions. If a picture is worth a thousand words, what could be better than a picture of a function! Most functions are described in such a way that we must calculate the values of the dependent variable one at a time. Thus, it would be very enlightening to see all of the values of the dependent variable exhibited at one time.

In honor of Descartes, what we are about to describe is called the **Cartesian coordinate system.** A straight line becomes a number line when we pick some arbitrary point on the line to be 0 and in each of the two directions away from 0 we mark off multiples of some basic unit of length. Here is a horizontal number line.

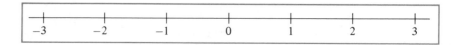

Since any function requires considering a pair of numbers, namely, some choice for the independent variable and the associated value of the dependent variable, we need two number lines. The usual arrangement is pictured next.

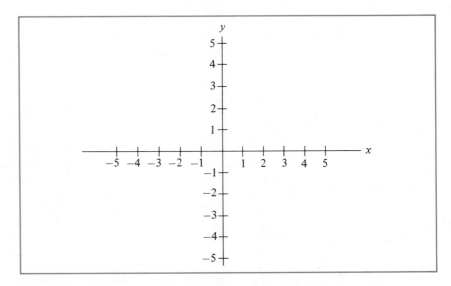

We agree to let values for the independent variable be pictured on the horizontal number line, which we call the **x-axis,** and let the values for the dependent variable be pictured by the vertical number line, which we call the **y-axis.** A quick way to describe some of the values of a function $f(x)$ is to list ordered pairs (x, y) where $y = f(x)$. For example, if $y = f(x) = x^2$ then $(0, 0), (1, 1), (-1, 1), (2, 4),$ and $(-2, 4)$ are ordered pairs given by $f(x)$, since $0 = f(0), 1 = f(1), 1 = f(-1), 4 = f(2),$ and $4 = f(-2)$. Now we describe how to picture these ordered pairs on the Cartesian coordinate system. For example, to picture $(2, 4)$, find 2 on the x-axis and move up vertically until you are right beside 4 on the y-axis. More precisely, the point picturing $(2, 4)$ is the intersection of the vertical line that crosses the x-axis at 2 and the horizontal line that crosses the y-axis at 4. Thus, each ordered pair of numbers has a corresponding point in the Cartesian coordinate plane.

Furthermore, the preceding process is reversible. Corresponding to any point in the Cartesian plane is an ordered pair that describes that point. Given a point, draw a line through the point perpendicular to the x-axis; where that line crosses the x-axis is the first number in the corresponding ordered pair. The number just obtained is called the **x-coordinate** of the point. To get the second number in the ordered pair, called the **y-coordinate,** draw a line through the point perpendicular to the y-axis; where that line crosses the y-axis is the y-coordinate of the point.

We say that we have **graphed the function** $f(x)$ when we have pictured in the coordinate plane all the ordered pairs given by the function, and the picture is called the **graph** of f.

EXAMPLE 1

Graph $y = f(x) = x + 2$.

Solution

A convenient way to list some of the ordered pairs given by the function is to use a table such as

x	y
2	4
1	3
0	2
−1	1
−2	0

We locate these ordered pairs in the next coordinate plane.

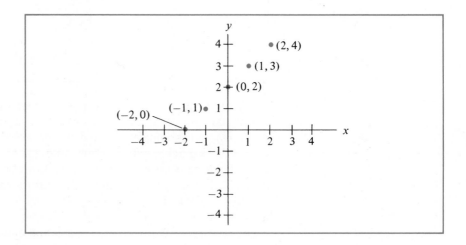

Then we draw through the points pictured the simplest curve that passes through each point. In this case, the simplest curve is a straight line, as pictured next.

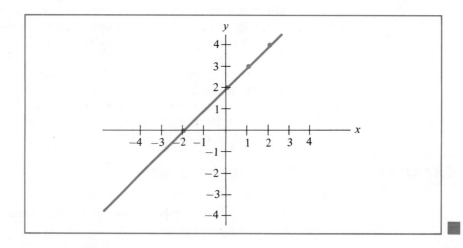

EXAMPLE 2

Graph the function that gives the profit for Demo's hot dog stand, $y = P(x) = .45x - 750$.

Solution

Here it is convenient to use a much smaller unit on the x-axis and the y-axis. Following is a table of ordered pairs and the graph. Demo has to sell a lot of hot dogs to make much money!

x	y
2000	150
1000	-300
0	-750
-1000	-1200
-2000	-1650

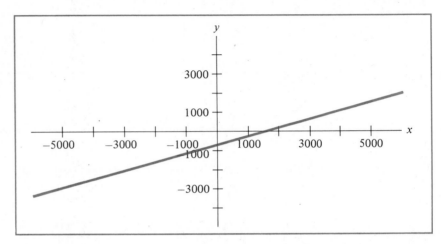

Of course, for Demo's hot dog stand, only nonnegative values of x make sense. ■

EXAMPLE 3

Graph $y = x^2$.

Solution

A table of ordered pairs and graph are given below.

x	y
-2	4
-1	1
0	0
1	1
2	4

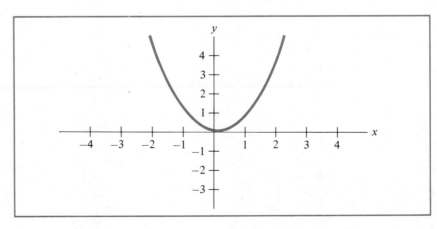

In the preceding example, it is helpful to note that if $x > 1$, then $x^2 > x$ (multiply both sides of $x > 1$ by x) and if $0 \le x < 1$, then $x^2 < x$; also, $(-x)^2 = x^2$.

To obtain accurate graphs for most functions requires more than just plotting points; in Chapter 9, many functions that could not be graphed easily by plotting points are graphed very accurately using differential calculus.

EXAMPLE 4

Graph the following function

$$f(x) = \begin{cases} 2 & \text{if } x \ne 3 \\ 1 & \text{if } x = 3 \end{cases}$$

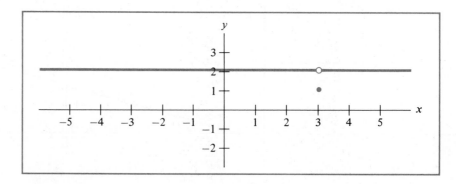

Solution

Here the hollow circle indicates that the center of the circle is omitted from the graph; the heavy dot indicates that the value of y at $x = 3$ is 1. ■

We can **graph an equation** even if it does not define a function by picturing all of the ordered pairs (x, y) where x and y make the equation true. For example, the graph of the equation $x = 2$ is the vertical line pictured next.

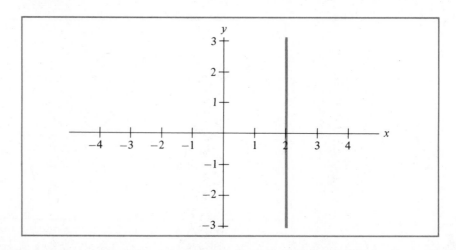

The ordered pair (x, y) satisfies the given equation provided x is 2 and y is anything; that is just what has been pictured.

EXAMPLE 5

Graph $1 = x^2 + y^2$.

Solution

We use the same procedure as if we were graphing a function.

x	y
2	no value
1	0
0	$+1, -1$
-1	0
-2	no value

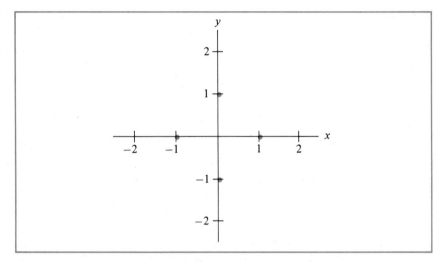

Since the preceding table resulted in only the four points pictured previously, we make a second effort at choosing some values for x. Since x^2 and y^2 are both nonnegative and their sum is 1, each must not exceed 1.

x	y
$\dfrac{1}{2}$	$\dfrac{\sqrt{3}}{2}, -\dfrac{\sqrt{3}}{2}$
$\dfrac{1}{4}$	$\dfrac{\sqrt{15}}{4}, -\dfrac{\sqrt{15}}{4}$
$-\dfrac{1}{4}$	$\dfrac{\sqrt{15}}{4}, -\dfrac{\sqrt{15}}{4}$
$-\dfrac{1}{2}$	$\dfrac{\sqrt{3}}{2}, -\dfrac{\sqrt{3}}{2}$

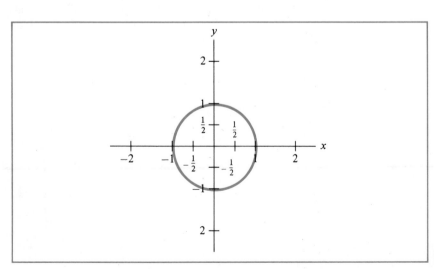

It can be proved easily from elementary plane geometry that the graph of $a = x^2 + y^2$ where $a > 0$ is a circle with center at the intersection of the *x*-axis and the *y*-axis. That point of intersection is called the **origin.** It is clear that the equation $1 = x^2 + y^2$ does not define a function since $+1$ and -1 are both *y*-values associated with an *x*-value of 0. Consequently there fails to be a unique *y* associated with $x = 0$. Looking at the graph we note that the vertical line crossing the *x*-axis at 0 intersects the graph at $(0, 1)$ and $(0, -1)$. **More generally, the graph of an equation is the graph of a function provided any vertical line that intersects the graph intersects it in only one point.**

EXAMPLE 6

Is the following graph the graph of a function?

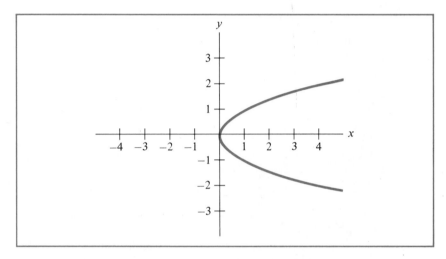

Solution

A vertical line that crosses the *x*-axis at $x = 2$ (or any other positive number) intersects the graph at two places as the next picture shows. Thus, there are two *y*-values associated with $x = 2$ and the given graph is not the graph of a function.

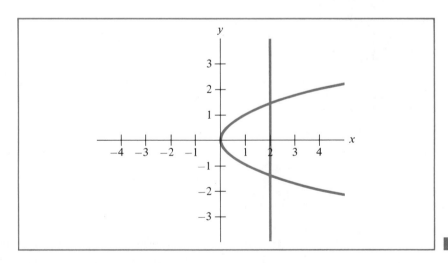

The **range** of any function $f(x)$ is defined to be the collection of values assumed by the function.

EXAMPLE 7

Find the range of the function $y = f(x) = x^2$ by examining the graph pictured in Example 3.

Solution

Recall that the range of the function $f(x)$ is the collection of all the y-values for that function. Looking at the graph, we determine that there is a point on the graph beside each $y \geq 0$ and there is no point on the graph beside any $y < 0$. Thus, the range is exactly the collection of $y \geq 0$. ■

EXAMPLE 8

Specify the domain of $f(x) = |x|$ and use the graph of the function to determine the range.

Solution

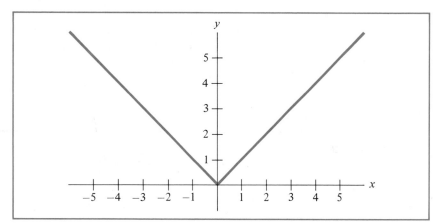

x	$|x|$
-2	2
-1	1
0	0
1	1
2	2

The definition of $|x|$ makes sense for all x and so the domain consists of all real numbers. From the definition it follows that $y = |x| \geq 0$; the graph shows that there is a point on the graph corresponding to every nonnegative y. Thus, the range consists of all nonnegative numbers. ■

It is convenient to regard the Cartesian coordinate system as dividing the plane into four parts. The upper right part, where $x \geq 0$ and $y \geq 0$, is called the **first quadrant**. (The word *quadrans* means "one fourth" in Latin.) The upper left part, where $x \leq 0$ and $y \geq 0$, is the **second quadrant**; the lower left part, where $x \leq 0$ and $y \leq 0$, is the **third quadrant**; the lower right part, where $x \geq 0$ and $y \leq 0$, is the **fourth quadrant**. In the following figure, each quadrant is labeled.

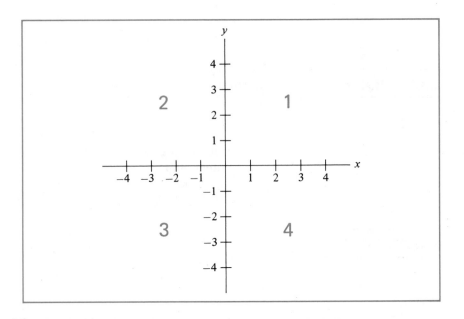

We can describe the location of the graph in Example 3 by saying that it is in the first and second quadrants, whereas the graph in Example 6 is in the first and fourth quadrants.

Exercises 2.2

In problems 1 through 20, sketch a coordinate system and place a dot at the point corresponding to the given ordered pair.

1 (1, 1)	2 (2, 1)	3 (1, 2)
4 (5, 10)	5 $(\frac{1}{2}, 6)$	6 (−1, 1)
7 (−2, 5)	8 (−5, 2)	9 $(-\frac{1}{4}, 8)$
10 $(-8, \frac{1}{4})$	11 (−1, −1)	12 (−2, −8)
13 (−8, −2)	14 $(-\frac{1}{2}, -10)$	15 (−10, −1)
16 (1, −1)	17 (2, −10)	18 (10, −2)
19 $(\frac{1}{3}, -6)$	20 $(6, -\frac{1}{3})$	

In problems 21 through 30, sketch the graph of the indicated function.

21	$y = 3x - 6$	22	$y = -x + 2$
23	$y = 5$	24	$y = \frac{1}{2}x$
25	$y = -x^2$	26	$y = x^2 + 3$
27	$y = 4 - x^2$	28	$y = 2x^2 - 8$
29	$y = \sqrt{x}$	30	$y = 1/x$

In problems 31 through 40, sketch the graph of the given equation and indicate whether it is the graph of a function.

31 $y^2 = x$

32 $y^3 = x$

33 $y^4 = x$

34 $y = \sqrt{x^2}$

35 $x = 4$

36 $4 = x^2 + y^2$

37 $y = $ the greatest integer not larger than x

38 $y = \begin{cases} 1 & \text{if } x \geq 0 \\ -1 & \text{if } x < 0 \end{cases}$

39 $0 = 2y + 4x$

40 $5 = 2y - x$

In problems 41 through 50, determine how many times the given number occurs as a y-value for the given function.

41 1 for $y = x^2$

42 -1 for $y = x^2$

43 2 for $y = \sqrt{x}$

44 0 for $y = 1/x$

45 -1 for $y = 1/x$

46 1 for $y = x^3$

47 -1 for $y = x^3$

48 0 for $y = x^2 + 2x + 1$

49 3 for $y = 2$

50 2 for $y = 2$

In problems 51 through 60, find the graph of the function and determine the range by examining the graph.

51 $y = 2x - 1$

52 $y = -x$

53 $y = -3x + 2$

54 $y = 5$

55 $y = -x^2 + 1$

56 $y = 4x^2 + 2$

57 $y = -3x^2 - 3$

58 $y = x^3$

59 $y = \begin{cases} 1 & \text{if } x \geq 0 \\ -1 & \text{if } x < 0 \end{cases}$

60 $y = $ the largest integer not larger than x

2.3 Straight Line

The graph of any linear function $y = f(x) = ax + b$ is a line, and every nonvertical line is the graph of such an equation. Consequently, the graphs in Example 1 and Example 2 in the previous section were lines. Knowing that the graph of a function is a line before starting to calculate a table of ordered pairs is a powerful bit of knowledge! Since there is only one line that passes through any given pair of points, our table of ordered pairs requires only two choices for x.

EXAMPLE 1

Graph $y = 2x + 3$.

Solution

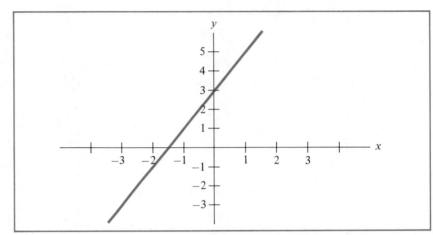

x	y
0	3
1	5

Everybody knows what a line is. However, if you ask someone to give you a definition of a line without drawing a picture, that person will probably be stumped. Although we understand "line" as a fundamental geometric idea, it is difficult to give a characterization except by picture, and any picture that we draw can only give examples of lines. The key idea to characterizing "line" is slope. The **slope between any pair of points,** say (a, b) and (c, d), is defined to be $\dfrac{d - b}{c - a}$.

This ratio is the difference in the y-coordinates of the points divided by the difference in the x-coordinates, where the points are used in the same order in calculating each difference. **An unbroken graph that is unending to the left and to the right is a nonvertical line provided there is one number that is the slope between any pair of points on the graph.** That number, which is the slope between any pair of points on the graph, is called the **slope of the line.**

EXAMPLE 2

Find the slope of the graph of $y = -x + 2$.

Solution
Since this is a linear function, the graph is a line. Since the slope between any pair of points on the line is the slope of the line, we can simply make two choices of values for x. Two points on the line are $(0, 2)$ and $(2, 0)$; thus, the slope is

$$\frac{0 - 2}{2 - 0} = -1. \quad \blacksquare$$

Note that two ordered pairs given by the linear function $y = ax + b$ are $(0, b)$ and $(1, a + b)$. Since these two points lie on the line that is the function's graph, the slope of that line is

$$\frac{(a + b) - b}{1 - 0} = a.$$

Since the letter m is usually used to denote the slope of a line, we rewrite the equation as $y = mx + b$. This is said to be the **slope-intercept equation** for the line that is its graph. This is a natural name for the equation, since m is the slope of the line and the line intercepts the y-axis at b. In general, the value where a graph crosses the y-axis is called the **y-intercept** and the value where it crosses the x-axis is called the **x-intercept**.

EXAMPLE 3

Put the equation $2x + 3y + 6 = 0$ into slope-intercept form.

Solution

We must solve for y.

$$3y = -2x - 6$$

$$y = \left(-\frac{2}{3}\right)x - 2$$

The slope of the line that is the graph of this equation is $-\frac{2}{3}$ and the y-intercept is -2. ■

EXAMPLE 4

Find the slope-intercept equation for the line that passes through $(1, 0)$ and $(3, 4)$.

Solution

Since any pair of points on the line determines the slope of the line, the slope of the line must be

$$\frac{4 - 0}{3 - 1} = \frac{4}{2} = 2.$$

A point (x, y) lies on the line if and only if the slope between (x, y) and a distinct point on the line, say $(1, 0)$, is the slope of the line.

$$\frac{y - 0}{x - 1} = 2$$

so

$$y - 0 = 2(x - 1)$$

This last equation is an equation for the line. We solve for y in order to get the slope-intercept equation

$$y = 2(x - 1)$$

$$y = 2x - 2 \quad ■$$

EXAMPLE 5

Find an equation for the line passing through $(-1, 2)$ with slope 1.

Solution
A point (x, y) lies on the line passing through $(-1, 2)$ with slope 1 if and only if the slope between (x, y) and $(-1, 2)$ is 1 or $(x, y) = (-1, 2)$.

$$\frac{y - 2}{x - (-1)} = 1$$

$$\frac{y - 2}{x + 1} = 1$$

$$y - 2 = x + 1$$

$$y = x + 3 \quad \blacksquare$$

Vertical lines are a peculiar case. Every vertical line is the graph of the equation $x = a$, where a is some fixed constant. Conversely, for every choice of the constant a the graph of $x = a$ is a vertical line. Since any two points on a vertical line have the same x-coordinate and division by 0 is undefined, a vertical line does not have a slope. For example, take $(3, 1)$ and $(3, 2)$ on the graph of $x = 3$ and try to compute the slope.

$$\frac{2 - 1}{3 - 3} = \frac{1}{0} \quad \text{undefined}$$

EXAMPLE 6

Graph $x = 3$.

Solution

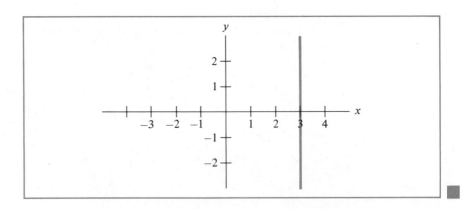

As you move your finger along a nonvertical line left to right, the line goes up or down according to the slope. A large positive slope means that the line is going up fast, and a small positive slope means that the line is going up slowly. A negative slope means that the line is going down. The next graph shows six lines with the slope of each indicated next to the line.

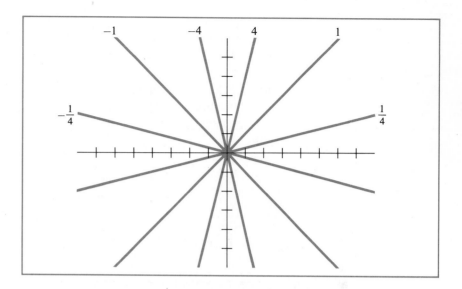

Frequently, in finding an equation whose graph is a particular line, we shall have the slope of the line m and one point (x_1, y_1) on the line, as in Example 5. Then we reason that a variable point (x, y) lies on the line provided

$$y - y_1 = m(x - x_1).$$

Thus, this equation is called a **point-slope equation** for the particular line.

EXAMPLE 7

Find a point-slope equation for the line passing through $(-2, 3)$ with slope $\frac{1}{2}$.

Solution

The point (x, y) lies on the line passing through $(-2, 3)$ with slope $\frac{1}{2}$ if and only if the slope between (x, y) and $(-2, 3)$ is $\frac{1}{2}$, or $(x, y) = (-2, 3)$.

$$\frac{y - 3}{x - (-2)} = \frac{1}{2}$$

$$\frac{y - 3}{x + 2} = \frac{1}{2}$$

$$y - 3 = \frac{1}{2}(x + 2)$$

This is a point-slope equation; we are done. ■

Business property like an auto or a building that has a useful life of more than one year can be depreciated on the owner's income tax. This means that the owner can claim the "wearing out" of the property as a cost of doing business and he can reduce his reported profit accordingly. The most basic scheme for claiming depreciation is the **straight line method.** In this method, the cost of the property to the owner is divided by the number of years in its useful life; the resulting quotient is the **depreciation** that can be claimed yearly. The **undepreciated part of the cost** is the cost minus the total depreciation already claimed.

EXAMPLE 8

Mr. McGee owns a single family house that he rents to families. The useful life of the house is 30 years and the cost was $60,000. Find a formula for the undepreciated part of the cost of the house as a function of time. Graph this function.

Solution

The annual depreciation is $60,000/30 = $2,000. If t denotes the time measured in years and U denotes the undepreciated part of cost, then

$$U(t) = 60{,}000 - 2{,}000t.$$

Since this is a linear function, the graph is easy to draw.

t	$U(t)$
0	60,000
30	0

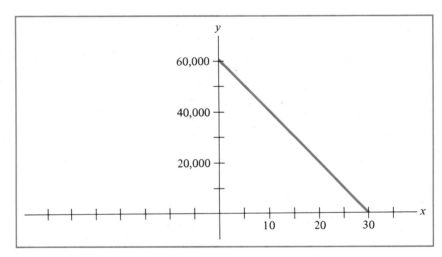

In the preceding graph, the same unit of length represents different numbers on the two axes. This is necessary in order to show the large numbers required for the y-axis. It does, however, change the apparent slope of the line. ■

Many daily newspapers list the price of a share of common stock for each corporation listed on the New York Stock Exchange. Also given is the price/earnings ratio, or p/e ratio, for that corporation. The total earnings divided by the total number of shares sold is the **earnings per share,** and the **p/e ratio** is the price of one share divided by the earnings per share. Companies with good prospects for growth sell at higher p/e ratios than companies with poor prospects. The reasoning is that the earnings, dividends, and price of the good company will go up while earnings, dividends, and price of the poor company might go down.

EXAMPLE 9

If IBM has a p/e ratio of 12, find the graph of the price of one share as a function of earnings per share.

Solution
If p denotes price and x denotes earnings per share, then

$$\frac{p}{x} = 12 \quad \text{or} \quad p = 12x$$

and the graph is given next.

x	p
0	0
3	36

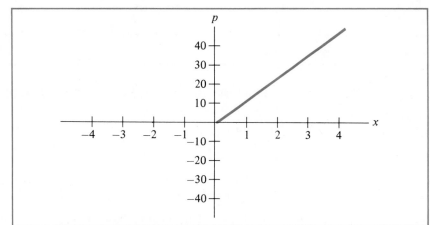

A standard method for calculating the price of a business for sale uses a **capitalization ratio.** The prospective buyer decides that his investment m must produce an income i at a certain rate of return r, that is $i/m = r$; otherwise, he would rather have the money than the business. This r is the **capitalization ratio.** Then the prospective buyer studies the business ledgers and decides what would be the likely yearly income produced by the business if he purchased it. If i denotes that income, then the price that the prospective buyer is willing to pay is $m = i/r$.

EXAMPLE 10

Sandy Gun wants to sell the maid service that she developed in her spare time. Find a formula for the selling price as a function of the yearly income that has fluctuated from year to year. Assume a prospective buyer wants a capitalization ratio of 15%. Graph that function.

Solution
If p denotes selling price and x denotes the yearly income, then

$$p = \frac{x}{.15} \approx 6.67x$$

The graph is next.

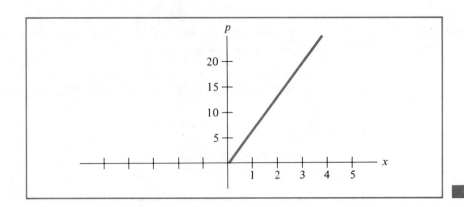

Exercises 2.3

In problems 1 through 10, find the slope-intercept equation for the line that passes through the given pair of points.

1   (1, 0), (2, 1)   2   (1, 1), (−1, 1)
3   (2, 3), (4, 0)   4   (2, 2), (4, 6)
5   (−4, 4), (4, −4)   6   (3, −2), (−2, −2)
7   (0, 0), (3, 1)   8   (−10, 2), (5, 2)
9   (0, 0), (5, −2)   10   (−2.3, 0), (7.2, 0)

In problems 11 through 20, sketch the graph of the given equation.

11   $y = x - 3$ 12 $0 = 2x + 4y - 6$
13   $0 = 3x - 9$ 14 $y = -2x + 2$
15   $0 = 10x - 5y$ 16 $0 = 3y - 12$
17   $0 = 5x + 7y - 2$ 18 $0 = x$

19   $y = \dfrac{1}{2}x + 2$ 20 $0 = y$

In problems 21 through 30, find a point-slope equation for the line passing through the given point with the given slope, if that is possible.

21   (1, 1), $m = 1$
22   (2, 1), $m = -1$
23   (0, 0), $m = 5$
24   (3, 3),   the line has no slope
25   (0, 3), $m = 2$

26   (6, 0), $m = -\dfrac{1}{3}$

27   (−2, −1),   the line has no slope
28   (−2, −2), $m = -10$
29   (0, −2), $m = -2$
30   (0, 0),   the line is vertical

In problems 31 through 40, find the slope of the line that is the graph of the given equation.

31 $0 = x + y + 1$ 32 $0 = 2x - 4y + 8$
33 $0 = 6x + 2y - 7$ 34 $y = 3x - 6$
35 $0 = 2x - 4$ 36 $0 = 5y + 10$
37 $y = -3x + 3$ 38 $0 = 2y + 9$

39 $y = \dfrac{x}{5} - 5$ 40 $0 = 5x + 1$

Solve the following

41 If Bargain Corporation always sells at a p/e ratio of 4, how much will the price increase when the earnings increase by $10?

42 If Glamour Corporation always sells at a p/e ratio of 30, how much will the price increase provided earnings increase by $5?

43 Richard paid $25,000 for a duplex and $5000 for the land under the duplex. If he uses straight line depreciation with a useful life of 25 years, what is the undepreciated part of the cost of the duplex after 5 years? Only the duplex, not the land, can be depreciated.

44 John bought a quadriplex and its lot for $90,000, and he figures the lot is worth $10,000. If he declares the quadriplex to have a useful life of 30 years and uses straight-line depreciation, how much depreciation has he claimed after 10 years? Only the quadriplex, not the land, can be depreciated.

45 A fast food restaurant in a good location produces a yearly income of $100,000. If Frank wants a capitalization ratio of .22, how much should he pay for the restaurant?

46 If Jane wants to purchase the restaurant in problem 45 and she is satisfied with the capitalization ratio of .12, how much would she pay for it?

2.4 Exponential Functions and Rules for Exponents

Linear functions are not the only important functions; in fact, many useful functions are not even polynomials. The exponential functions discussed in this section are essential in solving many diverse practical problems. Some of those problems involve "differential equations," which will be treated later. The graphs of exponential functions generally go up and down much more rapidly than the graphs of polynomials.

First we must review the elementary rules of exponents. Since 2^3 means use 2 as a factor three times and 2^5 means use 2 as a factor five times, it is clear that the product $2^3 \cdot 2^5$ equals 2^8. The general rule is that $a^b a^c = a^{b+c}$ where a is some positive number and b and c are any numbers. Since $(2^3)^5$ means use 2^3 as a factor five times and 2^3 means use 2 as a factor three times, it is clear that $(2^3)^5$ amounts to using 2 as a factor fifteen times. The general rule is that $(a^b)^c = a^{bc}$. Each of the general rules is natural, and the form of the rule is dictated by common sense observations.

Rules for Exponents   For a, b positive numbers and x, y any numbers, the following rules hold:

1. $a^x a^y = a^{x+y}$

2. $\dfrac{a^x}{a^y} = a^{x-y}$

3. $(a^x)^y = a^{xy}$
4. $(ab)^x = a^x b^x$
5. $a^0 = 1$
6. $a^x = a^y$ precisely when $x = y$

EXAMPLE 1

Simplify and calculate the quantity $\left(\dfrac{3^{100}}{3^{90}}\right)^{1/2}$.

Solution

$$\left(\frac{3^{100}}{3^{90}}\right)^{1/2} = (3^{100-90})^{1/2} = (3^{10})^{1/2} \qquad \text{By Rule 2}$$

$$(3^{10})^{1/2} = 3^5 \qquad \text{By Rule 3}$$

$$3^5 = 243 \qquad \text{By computation} \ \blacksquare$$

Recall that 3^{-1} means the reciprocal of 3, that is, $3^{-1} = \frac{1}{3}$. In general, a^{-b} means $1/a^b$. A convenient notation for the square root of 3 is $3^{1/2}$, and the square root of any nonnegative number a is written $a^{1/2}$. In general, an nth root of a is a number b with the property that $b^n = a$ and we denote b by $a^{1/n}$. Note that $(a^{1/n})^n = a^{n/n} = a$.

EXAMPLE 2

Compute $64^{5/2}$.

Solution

Using Rule 3 we can think of $64^{5/2}$ as $(64^5)^{1/2}$ or as $(64^{1/2})^5$. Although we get the same number either way, it is probably easier to use the fact that 8 is the square root of 64 and then raise 8 to the fifth power. Thus, $64^{5/2} = 8^5 = 32{,}768$. $\blacksquare$

For each positive number a, the **exponential function with base a** is represented by

$$y = f(x) = a^x.$$

The preceding remarks define a^x for $x = m/n$ where m and n are integers and $n \neq 0$; for other choices of x, a^x is defined so that the graph will have no breaks or gaps.

The following are graphs for $a = 2$ and $a = 3$.

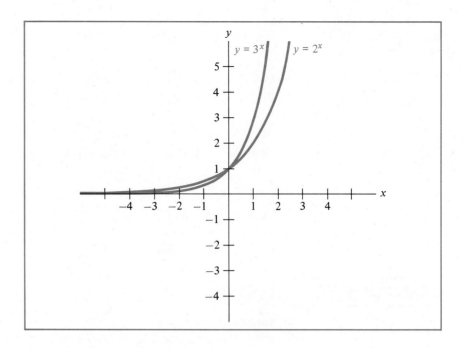

The preceding graphs are typical of the graph of a^x where $a > 1$, and the following graph is typical of the graph of a^x where $0 < a < 1$.

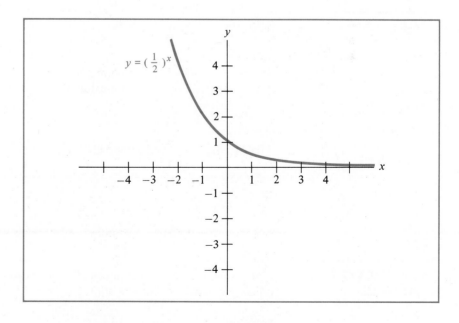

It is important to note that a^x is always positive.

EXAMPLE 3

Solve the equation $(a^x a^3)^2 = a^{10}$ for x.

Solution
Using the rules of exponents, we simplify the left side of the above equation to $a^{(x+3)(2)} = a^{2x+6}$. Thus the equation becomes

$$a^{2x+6} = a^{10}.$$

By Rule 6, this equation is true exactly when

$$2x + 6 = 10$$

or

$$2x = 4$$

or

$$x = 2. \blacksquare$$

EXAMPLE 4

Factor $a^{2x} + a^x - 2$.

Solution
$a^{2x} + a^x - 2 = (a^x)^2 + a^x - 2$. Since this is a quadratic expression with a^x in the role of the unknown, we can factor it by the methods of Section 1.4.

$$(a^x)^2 + a^x - 2 = (a^x + 2)(a^x - 1) \blacksquare$$

EXAMPLE 5

Solve the equation $2^{3-x} = 8$.

Solution
Since the left side of the equation involves the exponential function with base 2, we write the right side as a value of that function.

$$2^{3-x} = 8 = 2^3$$

The equation is solved provided

$$3 - x = 3$$

$$-x = 0$$

$$x = 0. \blacksquare$$

CALCULATOR EXAMPLE

Make a table to show the values of the functions $.2^x$, $.5^x$, 3^x, and 5^x at the following values of x: $x = -10, -8, -6, -4, -2, 0, 2, 4, 6, 8, 10$.

Solution
The following table was constructed on a calculator that displays 10 digits.

x	$.2^x$	$.5^x$	3^x	5^x
-10	9,765,625	1024	.000,016,935	.000,000,102,4
-8	390,625	256	.000,152,415,7	.000,002,56
-6	15,625	64	.001,371,742,1	.000,064
-4	625	16	.012,345,679	.0016
-2	25	4.0	.111,111,111,1	.04
0	1.0	1.0	1.0	1.0
2	.04	.25	9.0	25
4	.0016	.0625	81	625
6	.000,064	.015,625	729	15,625
8	.000,002,56	.003,906,25	6561	390,625
10	.000,000,102,4	.000,976,562,5	59,049	9,765,625

Why do the same numbers that appear in the second column also appear in the fifth column in reverse order? ■

EXAMPLE 6

The price of a barrel of oil produced by the OPEC countries was $10.46 in 1974 and it was $31 at the beginning of 1981. Assuming that the price of oil doubles every 4.5 years, find a formula for the price as a function of time and graph that function.

Solution
To say that a price has doubled means that if you multiply the old price by 2 you get the new price. If t is the number of years since 1974, then $t/4.5$ is the number of times the price has doubled since 1974. Thus, if we multiply $10.46 by $2^{t/4.5}$ then we get the price t years after 1974. The answer is

$$P = (10.46)2^{t/4.5}$$

t	p
0	10.46
4.5	20.92
8	35.87
-4.5	5.23
-8	3.05

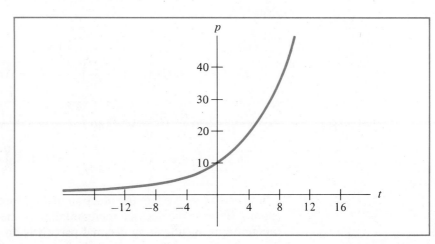

The preceding example illustrates how we **model** real-life situations mathematically. No one really believes that the price of OPEC oil will be given precisely by the function in Example 6. However, that function is reasonably consistent with past OPEC prices, so it is reasonable to consider what prices it would give in future years. In Section 2.6 we explain other aspects of modeling.

In Section 2.3 we explained the straight line method of depreciating property on income tax returns. An alternative method of depreciation is the **double declining balance** method that applies to newly constructed rental property. Under this method, the annual depreciation expense is double that computed by the straight line method. Thus, for the first year of the life of a building with a useful life of n years and a cost of c, the depreciation is $2\,\dfrac{c}{n}$. In the second year the undepreciated part of the cost is $c - 2\,\dfrac{c}{n} = c\left(1 - \dfrac{2}{n}\right)$ and the depreciation expense is $\dfrac{2}{n}\,c\left(1 - \dfrac{2}{n}\right)$. The depreciation expense D for the tth year is

$$D(t) = \frac{2}{n}\,c\left(1 - \frac{2}{n}\right)^{t-1}.$$

Since depreciation is relevant only to the computation of income tax, the domain of $D(t)$ is taken to be the set of positive integers from 1 to n.

EXAMPLE 7

Sam Smith is a contractor specializing in building duplexes. Sam built a duplex that cost him \$60,000 exclusive of the cost of the land, but he was unable to sell the duplex for 4 years. Since Sam wanted to make his taxable income as small as he could, he chose the double declining balance method of depreciation. If the useful life of the duplex is 40 years, what are the depreciation expenses that Sam claimed on his income tax for the first 3 years?

Solution

$$D(1) = \frac{2}{40}\,(60,000)\left(1 - \frac{2}{40}\right)^{1-1} = \frac{1}{20}\,(60,000) = 3,000$$

$$D(2) = \frac{2}{40}\,(60,000)\left(1 - \frac{2}{40}\right)^{2-1} = (3,000)\left(1 - \frac{1}{20}\right) = 2,850$$

$$D(3) = \frac{2}{40}\,(60,000)\left(1 - \frac{2}{40}\right)^{3-1} = (3,000)\left(\frac{19}{20}\right)^2 = 2707.50 \ \blacksquare$$

There is an exponential function that is especially important to applications. Because the decimal representation of the base of that exponential function has an unending decimal part (it is an irrational number), it is designated by e; it is 2.718 to three decimal places. The letter e was chosen in honor of the mathematician Leonhard Euler. After we have studied "deriva-tive," we shall see that e^x is the essential tool in solving many practical problems that lead to "differential equations."

Exercises 2.4

Using the fact that $2^{1/2} \approx 1.414$ to three decimal places, calculate the follow-ing numbers.

1	$2^{3/2}$	2	$(2^7)^{1/2}$
3	$2^{3/6}$	4	$2^{-1/2}$
5	$2^{-3/2}$	6	2^0
7	$2^{-5/2}$	8	$8^{1/2}$
9	$8^{1/6}$	10	$32^{1/2}$

Solve each of the following equations for x.

11	$3^{3x} = 3$	12	$5^{-x} = 5^2$
13	$10^{x-2} = 100$	14	$3^{2-x} = 27$
15	$e^{2x+1} = \dfrac{1}{e}$	16	$(2^x \cdot 2^{-5})^2 = 1$
17	$3^{x+1} \cdot 3^{x-1} = 9$	18	$5^x/5^2 = 5^3$
19	$2(3^{2x}) = 6$	20	$(3^{x+1})^{x-1} = 3^8$

Factor each of the following expressions.

21	$e^x + (e^x)^2$	22	$e^x - e^{2x}$
23	$e^{2x} - 1$	24	$e^{2x} + 2e^x + 1$
25	$e^{2x} - e^x - 6$	26	$e^{3x} + e^{2x}$
27	$9 - 3^{2x}$	28	$3^x + 3^{3x}$
29	$2^{x+3} + 2^{x-1}$	30	$5(2^x) - 2^{x+2}$

Sketch the graph of each of the following functions.

31	$f(x) = 5^x$	32	$f(x) = 3(2^x)$
33	$f(x) = 2^{x+1}$	34	$f(x) = 2^x - 1$
35	$f(x) = 2^{3x}$	36	$f(x) = \left(\dfrac{1}{3}\right)^x$
37	$g(x) = 9\left(\dfrac{1}{3}\right)^x$	38	$g(x) = 3^{2-x}$
39	$g(x) = \left(\dfrac{1}{3}\right)^x - 1$	40	$g(x) = \left(\dfrac{1}{3}\right)^{2x}$

Solve the following.

41 Use Rule 1 to show that Rule 5 must hold. (**Hint:** $a^y = a^{0+y} = a^0 a^y$.)
42 Use Rules 2 and 5 to show that $1/a^b = a^{-b}$.

43 If an average house cost \$63,000 in 1980 and if the price of an average house doubles every 8 years, what is a formula for the price as a function of time?

44 If the price of a widget was \$10 in 1960 and the price triples every 10 years, find a formula for the price as a function of time.

45 If the population of Paradise Island was 100,000 in 1980 and the population doubles every 12 years, what is a formula for the population as a function of time?

46 Pete inherits some money and he buys two new duplexes as an investment as well as a place to live. The cost of the one not occupied by Pete is \$40,000 excluding the land, and Pete estimates the useful life to be 40 years. Find Pete's depreciation on that duplex for the first 3 years by the double declining balance method.

47 A new rental house costing \$90,000 excluding the cost of the land is depreciated by the double declining balance method. If the useful life is 45 years, what is the first year's depreciation?

48 If a \$200,000 building with a useful life of 25 years is depreciated by the double declining balance method, what is the depreciation for each of the first 2 years?

49 *Calculator Problem* Show that $f(x) = 2^x$ gets larger and larger as x takes on each of the values 1, 2, . . . , 9, 10 in that order.

50 *Calculator Problem* Show that $f(x) = e^x$ gets larger and larger as x takes on each of the values 1, 2, . . . , 9, 10 in that order. (Use $e = 2.718$.)

51 *Calculator Problem* Show that $f(x) = 2^x/e^x$ gets smaller and smaller as x takes on each of the values 1, 2, . . . , 9, 10 in that order.

2.5 Logarithmic Functions and Rules for Logarithms

The logarithm function probably causes students of elementary mathematics more discomfort than any other function. The definition seems abstract and backwards! This criticism is somewhat deserved but unavoidable. Here is the definition.

Definition Let a be a fixed positive number other than 1. The **logarithm of x to the base a** is that number y with the property that $a^y = x$.
In symbols we write

$$\log_a x = y \tag{1}$$

provided

$$a^y = x. \tag{2}$$

This definition is somewhat backwards because the logarithm to the base a is defined in terms of the values of the exponential function with base a. The definition is somewhat abstract because we have not made a concrete choice for a.

Now we shall obtain two very useful equations that will make the "backwards" aspect of the previous definition more explicit and, consequently, easier to understand. Substitute equation (2) into equation (1) and get

$$\log_a a^y = y, \tag{3}$$

and substitute equation (1) into equation (2) and get

$$a^{\log_a x} = x. \tag{4}$$

Equations (3) and (4) show that $\log_a x$ and a^y are inverse functions; that is, if one of these functions is followed by the other one, then we get back to the variable we started with. Since we shall not need any inverse functions other than $\log_a x$ and a^y, we shall not study the general concept. We mention the idea here because it helps to understand equations (3) and (4). There are other important examples of inverse functions in the general study of calculus.

EXAMPLE 1

Find the value $\log_2 8$.

Solution
From the definition, the value is $y = \log_2 8$ provided $2^y = 8$. Sooner or later we notice that $8 = 2^3$, so the value to be calculated is $y = 3$. Also notice that the original expression can be written as

$$\log_2 2^3.$$

In view of equation (3) the answer is 3. ■

EXAMPLE 2

Find $\log_a 1$.

Solution
The fifth property of exponential functions listed in the previous section is $a^0 = 1$. So the preceding value can be written as

$$\log_a a^0.$$

In view of equation (3) the answer is 0, regardless of the value of a. ■

EXAMPLE 3

Find $\log_{10} 10$.

Solution
When no exponent is explicitly given, then an exponent of 1 is understood; that is, $10 = 10^1$. So the preceding value is

$$\log_{10} 10^1.$$

As usual, equation (3) gives the answer, which is 1. ■

EXAMPLE 4

Solve $\log_3 x = 10$ for x.

Solution

From property (4) it follows that

$$3^{\log_3 x} = x.$$

Using the given equation produces

$$x = 3^{\log_3 x} = 3^{10} = 59{,}049,$$

which is the answer. ■

EXAMPLE 5

Solve $\log_a x = b$ for x.

Solution

Using the given equation and equation (4) produces

$$x = a^{\log_a x} = a^b,$$

which is the answer. ■

In the definition of $\log_a x$ there are two choices of a that have practical significance. When $a = 10$, the resulting logarithm function is called the **common logarithm,** and it is usually written as $\log x$ rather than $\log_{10} x$. This is the only circumstance in which the subscript is omitted from $\log_a x$. When $a = e$ (the number, approximated by 2.718, mentioned in the previous section), the resulting logarithm function is called the **natural logarithm;** we usually write $\ln x$ rather than $\log_e x$.

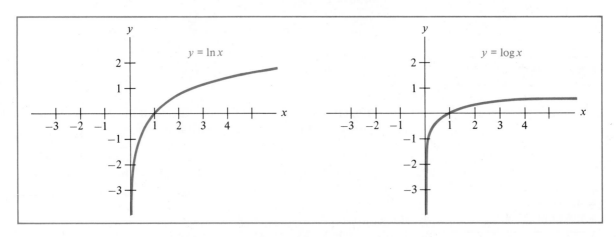

Note that $\ln x$ and $\log x$ are defined only for positive x; in general the domain of $\log_a x$ is all positive numbers, and the range is the set of all real numbers.

The following properties make $\log_a x$ an important function. Each of the properties follows from properties of a^x.

Properties of $\log_a x$

$$\log_a(xy) = \log_a x + \log_a y \tag{1}$$

$$\log_a(1/x) = -\log_a x \tag{2}$$

$$\log_a(x/y) = \log_a x - \log_a y \tag{3}$$

$$\log_a(x^b) = b \log_a x \tag{4}$$

EXAMPLE 6

According to the Table of Natural Logarithms in the back of this book,

$$\ln 2 = .693147, \quad \ln 3 = 1.098612, \quad \ln 5 = 1.609438.$$

Using these and the rules of logarithms, find each of the following:

a. ln 24   b. ln 100
c. ln .9   d. ln 180.

Solution

a. $\ln 24 = \ln(3 \cdot 8) = \ln 3 + \ln 8 = \ln 3 + \ln(2^3)$
$= \ln 3 + 3 \ln 2$
$= 1.098612 + 3(.693147)$
$= 3.178053$

b. $\ln 100 = \ln(4 \cdot 25) = \ln 4 + \ln 25 = \ln(2^2) + \ln(5^2)$
$= 2 \ln 2 + 2 \ln 5$
$= 2(.693147) + 2(1.609438)$
$= 4.605170$

c. $\ln .9 = \ln \dfrac{9}{10} = \ln 9 - \ln 10 = \ln(3^2) - \ln(2 \cdot 5)$

$= 2 \ln 3 - (\ln 2 + \ln 5)$
$= 2(1.098612) - .693147 - 1.609438$
$= -.105361$

d. $\ln 180 = \ln(5 \cdot 4 \cdot 9) = \ln 5 + \ln 4 + \ln 9$
$= \ln 5 + \ln(2^2) + \ln(3^2)$
$= \ln 5 + 2 \ln 2 + 2 \ln 3$
$= 1.609438 + 2(.693147) + 2(1.098612)$
$= 5.192956$ ■

EXAMPLE 7

Using logarithms, compute the quantity $(1.12)^{40}$ to two decimal places.

Solution
First we compute the natural logarithm of the indicated quantity as $\ln(1.12^{40}) = 40 \ln 1.12 = 40(.11329) = 4.533160$. Rounding off the above number so that we can use the Table of Exponential Values, we employ equation (4).

$$1.12^{40} = e^{\ln(1.12^{40})} \approx e^{4.53} \approx 92.76 \quad ■$$

EXAMPLE 8

If log 2.73 = .436163, what is log 2730?

Solution

$$\log 2730 = \log[(2.73)(10^3)]$$
$$= \log 2.73 + \log 10^3$$
$$= .436163 + 3$$
$$= 3.436163. \quad \blacksquare$$

EXAMPLE 9

Solve $e^{2x} = 12$ for x.

Solution

Since both sides of the equation represent the same number, the natural logarithms of both sides must be equal.
Thus,

$$\ln(e^{2x}) = \ln 12$$

and by equation (3),

$$\ln(e^{2x}) = 2x.$$

Thus we have

$$2x = \ln 12 \quad \text{or} \quad x = \frac{1}{2} \ln 12.$$

Using the Table of Natural Logarithms in the back of the book or a calculator, we get

$$x = 1.242453. \quad \blacksquare$$

EXAMPLE 10

Solve $e^{3\ln x} = 8$ for x.

Solution

Using Rule 4 of the properties of logarithms, we write the given equation as

$$e^{\ln(x^3)} = 8.$$

Then using equation (4) we have

$$e^{\ln(x^3)} = x^3 = 8$$

and so $x = 2$. $\blacksquare$

EXAMPLE 11

Write $2 \log x + \log(x + 1) - \log(x - 1)$ as the logarithm of a single expression.

Solution

$$2 \log x + \log(x + 1) - \log(x - 1) = \log(x^2) + \log(x + 1) - \log(x - 1)$$
$$= \log[x^2(x + 1)] - \log(x - 1)$$
$$= \log\left[\frac{x^2(x + 1)}{x - 1}\right] \quad \blacksquare$$

CALCULATOR EXAMPLE

Make a table to show the values of the functions $\log x$ and $\ln x$ at the following values of x: $x = 5000, 500, 50, 5, .5, .05, .005, .0005$.

Solution
The following table was constructed on a calculator that displays ten digits.

x	$\log x$	$\ln x$
5000	3.698970004	8.517193191
500	2.698970004	6.214608098
50	1.698970004	3.912023005
5	.6989700043	1.609437912
.5	−.3010299957	−.6931471806
.05	−1.301029996	−2.995732274
.005	−2.301029996	−5.298317367
.0005	−3.301029996	−7.60090246

Can you explain the similarity in the digits of the numbers in the second column? ■

EXAMPLE 12

Assume that the price of a barrel of oil doubles every 4.5 years and the price was $10.46 on January 1, 1974. (See Example 6 in Section 2.4.) Find when the price will be $1000.

Solution
Recall from the earlier example that the price p is given as a function of time t by the formula

$$p = 10.46 \cdot 2^{t/4.5}$$

Now we substitute 1000 for p and solve the resulting equation.

$$1000 = 10.46 \cdot 2^{t/4.5}$$

$$95.60 = 2^{t/4.5}$$

$$\ln 95.60 = \ln 2^{t/4.5} = (t/4.5) \cdot \ln 2$$

$$\frac{t}{4.5} = \frac{\ln 95.60}{\ln 2} = \frac{4.560}{.693} = 6.58$$

$$t = (4.5)(6.58) = 29.61$$

Thus, about $29\frac{1}{2}$ years after January 1, 1974, or in the middle of 2003, the price of oil would reach $1000 per barrel. ■

The rules for the logarithm functions give a useful equation that relates values of $\log x$ to those of $\ln x$. By equation (4) we have

$$10^{\log x} = x.$$

Using Rule 4 we get

$$\ln(10^{\log x}) = \log x \ln 10 = \ln x$$

or

$$\log x = \frac{\ln x}{\ln 10}.$$

A table in the back of the book gives the values of $\ln x$, and those values along with the last equation give the values of $\log x$. This same idea allows us to calculate the values of a^x given the table of values for e^x in the back of the book.

$$a^x = (e^{\ln a})^x = e^{x \ln a}$$

Exercises   2.5

In problems 1 through 12, find the value of the indicated quantity without using tables or calculator.

1   $\log 100$ 2 $\log .1$
3   $\log_2 32$ 4 $\log 1$
5   $\log_3 81$ 6 $\log .01$
7   $\ln e^{11}$ 8 $\log_3 \sqrt{3}$
9   $\log_2 \sqrt[5]{2}$ 10 $\ln 1/e$
11 $\log 10^e$ 12 $\ln e^{10}$

In problems 13 through 20, find the indicated value using the Table of Natural Logarithms in the back of the book or a calculator.

13 $\ln 3$ 14 $\ln 300$
15 $\ln 15$ 16 $\ln .002$
17 $\log 7$ 18 $\log 500$
19 $\log .05$ 20 $\ln 1$

In problems 21 through 40, solve the equation for x.

21 $e^{3x} = 8$ 22 $e^x e^{5x} = e^{12}$
23 $(e^{2x})^4 = e^{32}$ 24 $e^{\ln x} = 1$
25 $e^{3x}/e^x = e$ 26 $(e^5)^x = 3$
27 $e^3 e^x = e^7$ 28 $\ln(e^x) = 10$
29 $e^{2\ln x} = 16$ 30 $e^{-\ln x} = 2$
31 $10^x = 100$ 32 $10^x = 5$
33 $10^{3x} 10^{7x} = 1000$ 34 $10^{3x} 10^{-2x} = 4$
35 $10^{5x}/100 = e$ 36 $\log(10^{3x}) = 3$
37 $\log(10x) = 9$ 38 $10^{2\log x} = 25$
39 $(.01)(10^x) = 10^8$ 40 $\log(10^{5x} 10^{-2})^3 = 15$

In problems 41 through 50, write each of the following as the logarithm of a single expression.

41 $\log 2 - \log x$ 42 $\log x + \log 5$
43 $\log(x + 1) + \log x + \log(x + 2)$ 44 $\log(x + 1) - \log x + \log(x + 2)$

45 $\dfrac{1}{2} \log x + \dfrac{1}{3} \log(x + 1)$ 46 $2 \log(x + 2) - 3 \log x$

47 $\dfrac{1}{5} \log(x + 1) - 2 \log x$ 48 $\log 5 + \log x - \log 100$

49 $3 \log(x + 2) - 2 \log .1$ 50 $(\log x) + 1$

Solve the following.

51 Using the information in Example 12, find the year that the price of oil will reach $500 per barrel.
52 Using the information in Example 12, find the price of a barrel of oil in 1996.
53 The predicted population of Metropolis is given by $p(t) = 200,000\, e^{.1t}$, where t is the number of years since January 1, 1976, when the population was 200,000. Find when the population will reach one million.
54 Using the information in problem 53, find when the population of Metropolis will reach 2 million.
55 Trusty Electricity had one pound of strontium-90 at its nuclear generating plant at the beginning of 1980, and the amount remaining t years later is $p(t) = e^{-.02t}$. Strontium-90 is a radioactive element that disintegrates. Find when $\frac{1}{2}$ pound will remain.
56 Using the information in problem 55, find how much strontium-90 remains in 1992.
57 *Calculator Problem*   Show that $\log x$ gets larger and larger as x takes on each of the values 1, 2, . . . , 9, 10 in that order.
58 *Calculator Problem*   Show that $\ln x$ gets larger and larger as x takes on each of the values 1, 2, . . . , 9, 10 in that order.

2.6 Curve Fitting and Projections*

Sometimes functions in the real world can be described with explicit formulas, as we have seen in the previous sections. However, many functions that are important in the real world are not given by any explicit formulas. For example, **money stock,** which is defined as the currency in circulation plus the public's demand deposits at commercial banks, is a function of time. At the close of the banking day, it is possible to determine the size of the money stock. Plotted next are some points for the graph of M, the amount that money stock exceeds $260 billion, as a function of t, the number of years past January 1, 1976.

* This topic is included in this chapter in order to give some insights as to the origins of the functions used in many applications.

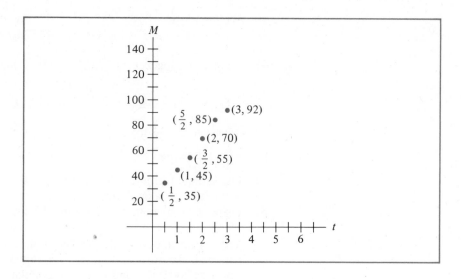

Since we do not have a formula for M as a function of t, we shall try to "fit" a straight line to the data points plotted. The customary way of measuring how well the line fits the plotted points is to use the **least squares method.** Let $M = f(t) = mt + b$ be the linear function whose graph is supposed to fit the plotted points. In the least squares method we make

$$\left(35 - f\left(\frac{1}{2}\right)\right)^2 + (45 - f(1))^2 + \left(55 - f\left(\frac{3}{2}\right)\right)^2 + (70 - f(2))^2 + \left(85 - f\left(\frac{5}{2}\right)\right)^2 + (92 - f(3))^2$$

as small as possible by a clever choice of m and b. By methods covered later in this book it can be shown that the correct choices for m and b satisfy the following equations:

$$6b + \frac{21}{2}m = 382$$

$$\frac{21}{2}b + \frac{91}{4}m = 773.5$$

Solving this pair of equations simultaneously,* we get

$$m = 24.0$$

$$b = 21.7$$

We plot the graph of $M = f(t) = 24t + 21.7$ on the same coordinate system as the previous six points for the sake of comparison.

---

* The general theory of systems of equations, which is covered in the next chapter, is not required for the simple pairs of equations occurring in this section.

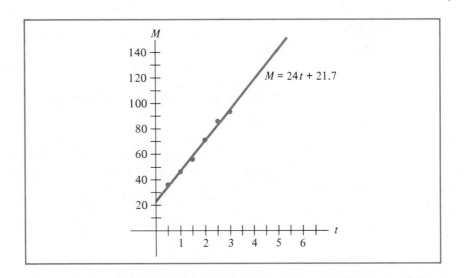

The primary reason for finding this line is so that we can look at the line beyond the six points that we have plotted and find the values of M at later times. For example,

$$M = 24(4) + 21.7 = 117.7$$

is the value that the linear function takes at $t = 4$. This means money stock would be $117.7 + 260 = \$377.7$ billion at the beginning of $4 + 1976 = 1980$ if the points plotted for 1976, 1977, and 1978 are representative of what was likely to happen in 1979. When a predicted value is obtained by fitting a curve to some plotted points and using a later value from the curve as the anticipated value of the unknown function, we say the predicted value is a **projection of the given data.**

Of course, a projection of some given data does not necessarily have to come true; it is merely a statement of what would happen if trends of the past are continued. In the case of money stock, our projection was not an excellent prediction. At the beginning of 1980 the money stock was $370 billion, which is somewhat less than the $377.7 billion that we projected. Perhaps the Federal Reserve, which controls the size of the money stock, made the same projection that we did and then changed its policy because $377.7 billion was too much money. Some economists believe that increases in money stock cause increases in the prices of consumer products.

In general, if the plotted points that are used to find the best fitting line, which is called **the least squares line,** are

$$(x_1, y_1), (x_2, y_2), \ldots, (x_n, y_n)$$

then the slope and y-intercept of the least squares line $y = mx + b$ are the solutions to the equations

$$bn + m(x_1 + x_2 + \cdots + x_n) = y_1 + y_2 + \cdots + y_n \tag{1}$$

$$b(x_1 + x_2 + \cdots + x_n) + m(x_1^2 + x_2^2 + \cdots + x_n^2) = x_1y_1 + x_2y_2 + \cdots + x_ny_n. \tag{2}$$

EXAMPLE 1

The owner of Cycles Unlimited notices that his bicycle sales have been going up rapidly over the last 4 months; the sales were 100, 108, 118, and 130. He is considering increasing the number of his salesmen and he would like to project next month's sales by using a least squares line. Find the equation of the least squares line and determine next month's projected sales.

Solution

Let S be monthly sales as a function of time t measured in months; the sales figures given report the first 4 months of the period when we are considering sales. Thus, the ordered pairs to be plotted are

$$(1, 100), \quad (2, 108), \quad (3,118), \quad (4,130)$$

and in this case equations (1) and (2) become

$$4b + 10m = 456 \tag{1'}$$

$$10b + 30m = 1190. \tag{2'}$$

Multiply equation (1') by 10 and equation (2') by -4 and then we have

$$40b + 100m = 4560 \tag{3}$$

$$-40b - 120m = -4760. \tag{4}$$

Adding equations (3) and (4) we get

$$-20m = -200$$

$$m = 10.$$

Using this solution for m in equation (1'), $b = 89$ and, consequently, the least squares line is the graph of $S = 10t + 89$. Sales for the fifth month are projected to be

$$S(5) = 10(5) + 89 = 139. \quad \blacksquare$$

EXAMPLE 2

A property developer in Sun City is trying to decide whether or not to build another subdivision of houses. The primary consideration is the population growth of Sun City, and the most recent available figures are the following: 90,000 in 1976, 100,000 in 1977, 115,000 in 1978, 125,000 in 1979, and 145,000 in 1980. Find the equation for the least squares line and determine the projected population for 1981.

Solution

Let p be the population of Sun City as a function of time t measured in years; the population figures given report the first 5 years of the period when we are considering the size of the population. Thus, the ordered pairs to be plotted are

$$(1, 90{,}000); \quad (2, 100{,}000); \quad (3, 115{,}000); \quad (4, 125{,}000); \quad (5, 145{,}000)$$

and in this case equations (1) and (2) become

$$5b + 15m = 575,000 \tag{1'}$$

$$15b + 55m = 1,860,000. \tag{2'}$$

Multiply equation (1') by 3 and equation (2') by -1,

$$15b + 45m = 1,725,000 \tag{3}$$

$$-15b - 55m = -1,860,000. \tag{4}$$

Adding equations (3) and (4), we get

$$-10m = -135,000$$

$$m = 13,500.$$

Using this solution for m in equation (1'), one finds that $b = 74,500$ and, consequently, the least squares line is the graph of $p = 13,500t + 74,500$. Population for the sixth year is projected to be

$$p(6) = 13,500(6) + 74,500 = 155,500. \ \blacksquare$$

We want to caution you that in most uses of the least squares line, many ordered pairs are used. In this presentation we used six or fewer ordered pairs in order to simplify the computations so that the ideas would not get lost in the arithmetic. In general, greater accuracy results from using more ordered pairs. Sometimes the pattern of the plotted points corresponding to the given ordered pairs suggests that the graph of a quadratic function should be fitted to the points rather than a linear function. On occasion, the graph of a cubic function will be fitted to the plotted points. Such instances of least squares curve fitting should be covered in a statistics course.

Exercises   2.6

In problems 1 through 20, find the equation of the least squares line $y = mx + b$ for the given ordered pairs.

1 (1, 1), (2, 5), (3, 8)
2 (1, 10), (2, 18), (3, 30)
3 (1, -8), (2, -3), (3, 2)
4 (1, -1), (2, -6), (3, -14)
5 (5, 20), (9, 30),(13, 38)
6 (1, 3), (2, 7), (3, 10), (4, 12)
7 (1, 12), (2, 10), (3, 9), (4, 8)
8 (1, -20), (2, -16), (3, -12), (4, -6)
9 (1, -6), (2, -1), (3, 4), (4, 9)
10 (-3, -7), (0, 0), (3, 8), (6, 14)
11 (1, 2), (2, 4), (3, 7), (4, 11), (5, 16)

12   (1, 20), (2, 16), (3, 10), (4, 4), (5, −4)
13   (1, 10), (2, 6), (3, 0), (4, −6), (5, −10)
14   (1, −20), (2, −10), (3, −2), (4, 8), (5, 20)
15   (4, 0), (6, 8), (10, 13), (16, 18), (20, 23)
16   (1, 25), (2, 40), (3, 50), (4, 60), (5, 70), (6, 75)
17   (1, 44), (2, 37), (3, 30), (4, 23), (5,16), (6, 9)
18   (1, −15), (2, −6), (3, 3), (4, 12), (5, 21), (6, 30)
19   (1, −10), (2, −8), (3, −4), (4, 2), (5, 10), (6, 22)
20   (−4, 2), (−2, 4), (0, 6), (5, 10), (10, 14), (12, 16)

In problems 21 through 30, get the answer by making a projection of the given data using the least squares line.

21   The manager of Economy Stoves observes that his sales of wood burning stoves over the last 3 months have been 25, 36, and 45, respectively. How many stoves should he expect to sell next month?

22   The number of customers coming in Tom's new clothing store has been 300, 325, and 360 per month in each of the first 3 months that the store is open. How many customers should he expect next month?

23   The total value of Jane's stock portfolio for the past 3 years has been $10,000, $11,800, and $14,000. If the trend continues, what will her stock be worth after another year?

24   As a result of improvements in the manufacture of certain computer parts, the price of a Handy Pocket Calculator has gone from $75 to $55 and then to $38 over the past 3 years. If the trend continues, what will the calculator cost after another year?

25   Joe observes that his electric bill for the summer months has been $300, $345, and $400 for the past three years. What should Joe project as his bill for the next summer?

26   The yearly revenues for Hank's clothing store since he bought it 4 years ago have been $100,000, $105,000, $111,000, and $116,000 in that order. What should Hank project as his revenues for the next year?

27   Sandra owns a piece of land just outside of town, and each year the value of the land has increased. If the value for the past 4 years are $20,000, $22,000, $24,500, and $27,000, what should Sandra expect the value to be in another year?

28   A small town is trying to decide whether to improve its water system. The number of gallons of water consumed in each of the past 4 years were 500,000, 580,000, 640,000, and 700,000. How many gallons should the town expect to be consumed during the next year?

29   The profits made by a local newspaper in each of the last 4 months are $9000, $9100, $9220, and $9310. If the profit trend continues, what will be the profits during the next month?

30   As a result of the declining number of college age people and other trends, the enrollment at State University has been declining. Project next year's enrollment given that enrollments for the past 4 years have been 21,000, 20,750, 20,500, and 20,200.

Review of Terms

Review Problems

In problems 1 through 4, give the domain of the indicated function.

1 $f(x) = \sqrt{x+1}$

2 $g(x) = \sqrt{9x^2 + 4}$

3 $f(x) = \dfrac{5}{x-3}$

4 $g(x) = \dfrac{2x+2}{x^2+x-2}$

In problems 5 through 8 indicate whether the given function is a polynomial.

5 $f(x) = \sqrt{x}$

6 $g(x) = x^{-1}$

7 $f(x) = \pi$

8 $g(x) = \dfrac{7x^2 + 11x}{3x}$

In problems 9 through 12, let x be the independent variable and y be the dependent variable. Indicate if the equation defines a function.

9   $xy = 9$

10   $x^2 + y^2 = 25$

11   $y = \sqrt{x^2}$

12   $\sqrt{y^2} = x$

In problems 13 through 16, sketch the graph of the function and determine the range by examining the graph.

13   $y = 2 - x^2$

14   $y = x^3 + 1$

15   $y = -3x + 8$

16   $y = \begin{cases} 2 & \text{if } x > 0 \\ 0 & \text{if } x = 0 \\ -2 & \text{if } x < 0 \end{cases}$

In problems 17 through 19, find the slope-intercept equation for the line that passes through the given pair of points.

17   $(0, 5), (7, 5)$

18   $(-3, 4), (4, -3)$

19   $(1, -1), (2, 1)$

In problems 20 through 23, sketch the graph of the given equation.

20   $2y = x + 2$

21   $x + 3 = 0$

22   $y - 5 = 0$

23   $x + 3y - 9 = 0$

In problems 24 through 26, find an equation for a line passing through the given point with the given slope.

24   $(8, 9), m = 0$

25   $(-2, 2), m = 4$

26   $(0, 9), m = -9$

Solve each of the following equations for x.

27   $e^{5x}e^2 = e^{12}$

28   $(2^x/2)^3 = 1$

29   $(5^{3x+1}5^2)/5^x = 125$

30   $10^{x^2-8} = .0001$

31   $\log(10^{5x}) = 10$

32   $10^{-x} = .001$

33   $\log(100x) = 2$

34   $e^{-2\ln x} = 27$

Sketch the graph of each of the following functions.

35   $f(x) = 5(3^x)$

36   $g(x) = 3(5^{-x})$

Find the value of the indicated quantity without using tables or calculator.

37   $\log .001$

38   $\ln 1$

39   $\log_2 .5$

40   $\log_3(27)^2$

In problems 41 through 43, find an equation for the least squares line for the given ordered pairs.

41   $(2, 3), (3, 5), (4, 6)$

42   $(-1, 4), (0, 2), (1, 1), (2, -1)$

43   $(-3, 4), (-2, 5), (-1, 3), (0, 4), (1, 5)$

Solve each of the following.

44   A civic group sells soft drinks in paper cups at football games. Each cup costs 2¢, and the drink and ice that goes in the cup costs 3¢; there are no

fixed costs. Write a formula for the cost as a function of the number of soft drinks sold.

45 The civic group mentioned in problem 44 sells each soft drink for 25¢. Find formulas for revenue and profit as functions of the number of drinks sold.

46 Give an example of a polynomial with degree 7.

47 The graph of $y = x$ extends into which quadrants?

48 The annual report of a company with earnings of $2 per share and p/e ratio of 10 is so impressive that the p/e ratio increases to 12. What is the new price of a share?

49 A rental house that costs $50,000 is depreciated by the straight line method over a period of 30 years. What is the undepreciated part of the cost after 10 years?

50 A business that produces a profit of $30,000 per year is for sale. A prospective buyer who wants a capitalization ratio of .25 should offer how much?

51 A new rental house that cost $100,000 is depreciated by the double declining balance method over a period of 40 years. What is the amount of the depreciation in the tenth year?

52 Assume that the price of food doubles every 7 years and find a formula for the future price of a market basket of food now costing $100.

53 Using the formula obtained in problem 52, find how many years will pass before the $100 market basket of food will cost $250.

54 The value of an investment portfolio for the past 4 years has been $18,000, $20,000, $23,000, and $24,000. If the trend continues, what will the portfolio be worth after another year according to a least squares line projection?

Social Science Applications

Carbon Dating

All living things have a relatively constant amount of carbon-14 (a radioactive form of carbon). This results from the dependence of the food chain on vegetation and the dependence of vegetation on atmospheric carbon dioxide. After death, the carbon-14 present continues to decay and no new carbon-14 is introduced. The time that has passed since the organism died can be estimated according to the amount of carbon-14 present. The formula

$$P(t) = P_0 e^{-.00012t}$$

indicates the amount $P(t)$ of carbon-14 present after t years, assuming there was P_0 present at the time of death.

1 *Archeology* Suppose some fossils are discovered that contain .9 of the amount of carbon-14 present at the time of death. What is the age of the fossils?

2 *Psychology (Learning Curves)* A computer programmer learns to type instructions on the display screen of a computer terminal. At first the

programmer learns quickly, but as she gains a mastery of the process her progress becomes slower and slower. The number of words that she can input per minute is given by

$$n(t) = 85(1 - e^{-.004t}),$$

where t indicates the number of hours of practice at the terminal. How many words per minute can she input after 5 weeks of practice for 40 hours per week?

3 **Public Administration**   Contagious diseases sometimes reach epidemic proportions. It is important to have a function that predicts the approximate number $n(t)$ of infected people t weeks after the epidemic begins, if no steps are taken to stop the spread of the disease. A typical function for a town of 200,000 is

$$n(t) = \frac{200,000}{1 + 1,999\ e^{-.8t}}.$$

If no action is taken to stop the spread of this disease, approximately how many people will be infected after 8 weeks?

4 **Sociology**   Rumors that are spread by exchanges from one person to another person are much like contagious diseases. In a city of 800,000 the approximate number $n(t)$ of people spreading the rumor t weeks after the rumor starts is typically given by

$$n(t) = \frac{800,000}{1 + 7,999\ e^{-t}}.$$

Determine the approximate number of people spreading the rumor after 4 weeks.

5 **Sociology**   Important news events, such as an attempt to assassinate the President, are not communicated in the same way as rumors. Exchanges from person to person are secondary to the impact of the mass media, such as television and newspapers. The number $n(t)$ of people in a city of 1,000,000 who have been informed of the event, t hours after it occurred, is typically given by

$$n(t) = 1,000,000(1 - e^{-.03t}).$$

Determine approximately how many people have been informed of the event 24 hours after it occurs.

6 **Psychology**   Let x indicate the level of some stimulus, such as the volume of a radio measured in decibels, and let y indicate the smallest amount of change in that stimulus that would be noticed by a listener. Typically, we have

$$y = cx,$$

where c is a constant depending on the particular listener. Suppose that $c = .05$ for a particular listener, and determine the amount that the volume of the radio at 60 decibels must be changed for the listener to notice.

7 ***Sociology***   Recent studies indicate that the average family now spends approximately 40% of its food budget on meals prepared at restaurants. If y is the amount spent on restaurant food and x is the amount of the food budget, then we have

$$y = .4x.$$

Determine the amount spent monthly on "eating out" by a family with a monthly food budget of $500.

CAREER PROFILE
Stockbroker

Becoming a stockbroker (also known as a registered representative or an account executive) is a career choice that requires more than a college degree. An engaging personality and the ability to attract customers are sought-after assets when a securities firm considers a job applicant.

Beginning brokers spend a large fraction of their time seeking clients. Part of their sales effort may involve offering seminars in which they explain the terminology and opportunities of investing to potential customers. If a potential investor has well-thought-out financial goals, then a good broker will be able to identify and recommend investments that are the best way to achieve those goals.

Although personality is an important entry qualification, in the long run, success as a broker depends on providing continuing good advice to clients. Various college courses provide important basic information enabling the broker to do this. Courses in economics are an asset in understanding the economic cycles that influence the values of investments. A background in statistics (especially data analysis) is of value in analyzing trends in prices of securities and in applying that analysis to determine when to buy and sell assets.

Concepts in probability are useful in assessing the risks associated with certain types of investments and in estimating expected profit. Using techniques from calculus, securities analysts have developed complex formulas to help them decide whether certain stock options are advisable purchases.

The study of accounting provides a broker with the background that is essential to the assessment of the financial condition of a corporation whose stock might be purchased. Courses in business administration are helpful to the broker who seeks to advance to a management position within a firm.

Most employers provide on-the-job training for individuals whose personality traits indicate potential for success in selling investments. The training period lasts at least four months and may be considerably longer. Through classroom instruction or correspondence courses, trainees master the terminology and regulations of security sales and learn the advantages of the various types of investments. Certification as a registered representative requires passing standardized examinations.

Trainees usually are paid a salary (often around $1000 per month) until they meet licensing and registration requirements. After this, earnings depend largely on commissions from sales. In prosperous times, when individuals and institutions have extra funds to invest, salaries climb. Salaries average around $40,000, but urban brokers with the ability to attract and keep many customers readily earn as much as $100,000 when the economy is prospering.

Sources of Additional Information
Securities Industry Association, 20 Broad St., New York, NY 10005. (For a $1 charge, this association will provide material concerning a career in securities sales.)
Occupational Outlook Handbook. Bureau of Labor Statistics, U.S. Department of Labor, Washington, DC 20212. (Revised every 2 years, this handbook provides information about job duties, working conditions, level and places of employment, education and training requirements, advancement possibilities, job outlook, and earnings for about 250 occupations. Some of the information for this Career Profile was obtained from this source.)

3

Systems of Equations and Matrices

3.1 Small Systems and Equilibrium Price

A **system of equations** is a set of equations to be considered simultaneously. For example,

$$x + y = 2$$

$$x - y = 4$$

is a system of equations; this system has two equations and each equation has two unknowns represented by the symbols x and y. A **solution to a system of equations** is a list of values for the unknowns with the property that each equation is true when each unknown is replaced by its corresponding value. For example, $x = 3$, $y = -1$ is a solution to the system of equations above, since

$$3 + (-1) = 2$$

$$3 - (-1) = 4.$$

We say that we have **solved** a given system of equations when we find all solutions.

It is easy to solve such a simple system of equations as the one given above. This is indicated in Example 1.

EXAMPLE 1

Solve the following system of equations.

$$x + y = 2$$

$$x - y = 4$$

Solution

By adding the left sides of the two equations and the right sides of the two equations, we get an equation involving x only.

$$x + y = 2$$
$$\underline{x - y = 4}$$
$$2x + 0y = 6$$
$$2x = 6$$

Since this last equation is linear with only one unknown, it is easy to solve.

$$\left(\frac{1}{2}\right) 2x = \left(\frac{1}{2}\right) 6$$
$$x = 3$$

Substituting this value for x in the original first equation, we get an equation involving y only.

$$3 + y = 2$$
$$y = 2 - 3 = -1$$

The only solution to the given system is $x = 3$, $y = -1$. ■

A system of equations with the form

$$a_1 x + b_1 y = c_1$$
$$a_2 x + b_2 y = c_2,$$

where a_1, b_1, c_1, a_2, b_2, and c_2 are constants, is an example of a **linear system;** that is, each equation is a linear equation. The general definition of a linear system is given in the next section.

There is a simple geometric interpretation for the process of solving a linear system. On the next axes we have graphed each of the linear equations in Example 1.

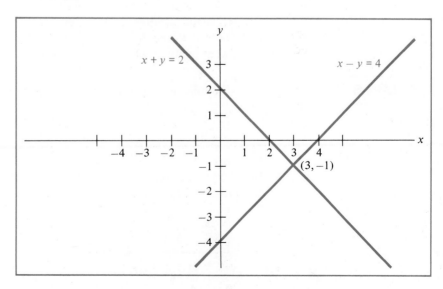

The coordinates of every point on the graph of $x + y = 2$ make this equation true, and the coordinates of every point on the graph of $x - y = 4$ make that equation true. The coordinates of the point where the two lines intersect must satisfy both equations. **If the graphs of the two equations in a linear system intersect, then the coordinates of the point of intersection give the only solution for the system.**

There are two other possibilities for the graphs of the equations in a linear system with two equations in two unknowns. One possibility is that the two graphs coincide, because both equations give the same line. The final possibility is that the graphs are parallel lines, as indicated on the following axes.

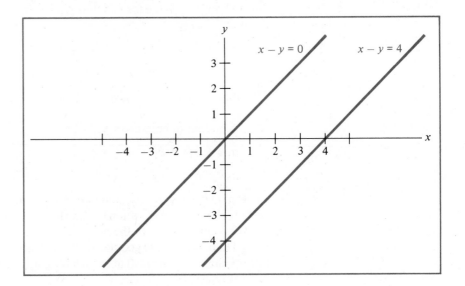

In this case the two lines do not intersect; there are no values for x and y that make both of the equations $x - y = 0$ and $x - y = 4$ true. Example 2 illustrates what happens when you try to solve such a system algebraically.

EXAMPLE 2

Solve the following system.

$$x - y = 0$$
$$x - y = 4$$

Solution
In order to eliminate one of the unknowns, we subtract the second equation from the first equation.

$$\begin{array}{r} x - y = 0 \\ x - y = 4 \\ \hline 0x + 0y = -4 \\ 0 = -4 \end{array}$$

This last equation is obviously false! If there were values for x and y that made both equations true, then the difference in the left sides would equal the difference in the right sides. Thus, it must be that there are no values for x and y that make both equations true. This system has no solutions. ∎

As already mentioned, there is one more possibility for the solution of a linear system, and it is illustrated by the next example.

EXAMPLE 3

Solve the following system.

$$.5x - .5y = 2$$
$$x - y = 4$$

Solution
In order to eliminate one unknown we multiply the first equation by -2. (Because of the multiplication property given in Section 1.2, we know that the resulting equation has the same solutions as the original one.)

$$-x + y = -4$$
$$x - y = 4$$

Now we add the two equations.

$$-x + y = -4$$
$$\underline{x - y = 4}$$
$$0x + 0y = 0$$
$$0 = 0$$

This last equation is obviously true! It indicates that $-x + y = -4$ is just the equation $x - y = 4$ multiplied by -1. Any values for x and y that makes one of these equations true make the other equation true also. Thus, the solution to the given linear system is all choices of x and y that satisfy the equation $x - y = 4$. Customarily, we solve this equation for one unknown in terms of the other.

$$x - y = 4$$
$$x = 4 + y$$

No matter how the value for y is chosen, if the value for x is given by the equation

$$x = 4 + y,$$

then we have a solution. The usual description of the solutions is the following.

$$x = 4 + y, \qquad y \text{ is arbitrary}$$

Since there are infinitely many choices of values for y, each with a corresponding value for x, there are infinitely many solutions to the original system. We could have obtained a slightly different description of the same solutions by solving the equation $x - y = 4$ for y instead of x. ■

As indicated by the preceding examples, **a linear system of equations may have exactly one solution, no solutions, or an infinite number of solutions.** When a linear system with two unknowns has infinitely many solutions, the graphs of the equations in the given system coincide with a single line. The graph of $.5x - .5y = 2$ is identical with the graph of $x - y = 4$.

Systems of equations arise in many natural ways. The next example indicates one way.

EXAMPLE 4

Suppose a sum of money is divided between investment A that has a rate of return of 22% and investment B with a rate of return of 16%. Together the investments pay $9800. The next year, the rates of return are 20% for A and 18% for B; the total return of the two is $9600. How much was put into each investment?

Solution

Let x be the amount of money put into investment A and let y be the amount put into B. For the first year, the return from A is $.22x$ and the return from B is $.16y$; the sum of these two is the total return of $9800. This gives us the first equation below; the second equation is obtained similarly.

$$.22x + .16y = \$9800$$

$$.20x + .18y = \$9600$$

In order to solve this system, we multiply both sides of the first equation by .20 and both sides of the second equation by .22. Thus, we have

$$.044x + .032y = \$1960$$

$$.044x + .0396y = \$2112.$$

By subtracting the second equation from the first, we get an equation involving y only.

$$-.0076y = -\$152$$

$$y = \quad \$20,000$$

Substituting this value for y in the original first equation, we get an equation involving x only.

$$.22x + .16(\$20,000) = \$\ 9,800$$

$$.22x + 3200 = \$\ 9,800$$

$$.22x = \$\ 6,600$$

$$x = \$30,000$$

The amount put into investment A was $30,000 and the amount put into investment B was $20,000. ■

The method that we have used to solve the first four examples is called the **elimination method.** In this method, each equation is multiplied by an appropriate number so that the difference of the resulting equations has one or more variables missing. Of course, either equation might be multiplied by 1, if we do not wish to change the coefficients.

For these small systems, there is another possible approach to finding a solution. We can solve one of the equations for one of the unknowns; then we substitute for that unknown in the other given equation. This is called the **substitution method.** It is demonstrated in the next example.

EXAMPLE 5

A manufacturer requires 40 hours to make one unit of product A and 60 hours to make a unit of product B. The manufacturer spends 1400 hours making 30 units of the two products. How many units of each product were made?

Solution

Let x and y be the number of units of product A and product B, respectively, that are produced. Since a total of 30 units are produced, we have the equation

$$x + y = 30.$$

Each unit of product A requires 40 hours, so x units require $40x$ hours. The y units of product B require $60y$ hours. Since the total hours spent on all units is 1400, we have the equation

$$40x + 60y = 1400.$$

We must solve the following system.

$$x + y = 30$$
$$40x + 60y = 1400$$

We solve the first equation for y:

$$y = 30 - x.$$

Then we substitute for y in the second equation and solve for x:

$$40x + 60(30 - x) = 1400$$
$$-20x + 1800 = 1400$$
$$-20x = -400$$
$$x = 20.$$

Using this value in the formula for y we get

$$y = 30 - 20 = 10.$$

The only solution to the system is $x = 20$, $y = 10$.

(In this example, it would have been just as easy to solve the first equation for x, and substitute for x in the second equation.) ∎

Now we consider how supply and demand determine the market price for a particular product. In economics the **demand function** for a product, denoted $p = D(x)$, gives the price p that consumers will pay when total consumer demand is x. The **supply function** for the product, denoted $p = S(x)$, gives the price that a producer of the product will charge when the total supply is x. We assume that the total supply equals the total consumer demand. The point of intersection of the graphs of these two functions is the **equilibrium point;** the coordinates of that point are the **market demand,** x_0, and the **market price,** p_0. These are the demand and price that will occur in a free market.

EXAMPLE 6

For a certain product the demand function is $p = D(x) = 7 - x$ and the supply function is $p = S(x) = 2x + 1$. Determine the market demand and the market price for this product.

Solution
We must solve the following linear system.

$$p = 7 - x$$
$$p = 2x + 1$$

The form of the system suggests the use of the substitution method.

$$7 - x = 2x + 1$$
$$6 = 3x$$
$$2 = x$$

Using this value of x in the formula for p we get

$$p = 7 - 2 = 5.$$

The market demand is 2, and the market price is 5. ■

On the next axes, at the top of the following page, we graph the demand function and the supply function from Example 6.

Note that the line that is the graph of the demand function has negative slope. Increasing consumer demand is associated with decreasing prices. In contrast, the line that is the graph of the supply function has positive slope. When prices go up the suppliers of the product are eager to supply more of the product. Consequently, as prices go up the supply goes up. Even when the demand function is not linear, it falls as we read left to right; even when the supply function is not linear, it rises as we read left to right.

Once the supply and demand functions for a commodity have been determined, they can be used to predict the effect of new taxes or price supports for that commodity. This is illustrated in the next example.

EXAMPLE 7

The demand and supply functions for a certain commodity are $p = D(x) = 7.4 - .1x$ and $p = S(x) = .2x + 2$. Suppose the government decides to place

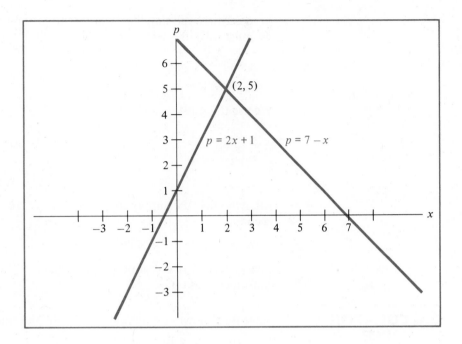

a $1.20 tax on the sale of each unit of this commodity. Determine the change in the price to the consumer and the supply of the product.

Solution
To determine the market price and market demand before the tax, we solve the system

$$p = 7.4 - .1x$$
$$p = .2x + 2.$$

We solve it by the substitution method.

$$7.4 - .1x = .2x + 2$$
$$5.4 = .3x$$
$$x = 18$$
$$p = 7.4 - .1(18) = 5.60$$

The price that suppliers receive is not changed by the tax. The proceeds of the tax go to the government. What changes is the demand function. When the suppliers' price is p, the price that the consumers pay is $(p + 1.20)$ and the new demand function is

$$p + 1.20 = D(x) = 7.4 - .1x$$

or

$$p = 6.2 - .1x.$$

To determine the suppliers' price p and the amount of supply after the tax, we must solve the system

$$p = 6.2 - .1x$$
$$p = .2x + 2.$$

We use the substitution method.

$$6.2 - .1x = .2x + 2$$
$$4.2 = .3x$$
$$x = 14$$

After the tax, the supply drops from 18 to 14. The price to the supplier becomes

$$p = .2(14) + 2 = 4.80.$$

The price to the consumer is $4.80 + 1.20 = 6.00$; thus, the price to the consumer goes up .40. ■

CALCULATOR EXAMPLE

Solve the following system by the elimination method.

$$10.23x - 8.14y = 27.19$$
$$3.36x + 9.07y = 35.42$$

Solution
Subtract 10.23 times the second equation from 3.36 times the first equation. The x term is eliminated, and we get

$$[3.36(-8.14) - 10.23(9.07)]y = 3.36(27.19) - 10.23(35.42)$$

or

$$-120.1365y = -270.9882$$

or

$$y \approx 2.2557.$$

Using this value in the first equation, we determine x.

$$10.23x - 8.14(2.2557) \approx 27.19$$
$$10.23x \approx 8.14(2.2557) + 27.19$$
$$10.23x \approx 45.5514$$
$$x \approx 4.4527 \text{ ■}$$

Exercises 3.1

Solve each system below by elimination. Graph each equation in the system and use the graphs to check the solution you have obtained.

1 $-2x + y = -2$ 2 $-x + y = 1$
    $x + y = 2$ $y = 3$

3   $x - 2y = 0$
    $.5x - y = -2$

4   $\frac{1}{3}x + y = 3$

    $x + 3y = -2$

5   $15x - 3y = 3$
    $-5x + y = -1$

6   $12x + 4y = 8$
    $3x + y = 2$

Solve each system below by elimination.

7   $x - 3y = 0$
    $4x + 6y = 1$

8   $2x + y = -1$
    $3x + 2y = 2$

9   $2x + 3y = 1$

    $\frac{1}{3}x + \frac{1}{2}y = \frac{1}{6}$

10      $x - .2y = .1$
    $-10x + 2y = -1$

11      $5x - 3y = 2$
    $-30x + 18y = 6$

12   $2x + 7y = 0$
    $8x + 28y = 4$

13   $y = 5x + 2$
    $5x - y = 2$

14   $x = -.5y - 3$
    $2x + y = 5$

15   $10x - 15y + 12 = 0$
    $x + 9y - 15 = 0$

16   $3v - 5w = 0$
    $6v = 10w$

17   $v = 7w - 3$
    $7w - v = 3$

18      $.24x + .13y = .12$
    $1.44x + .78y = .72$

19   $-.02x + .81y = .15$
    $.1x - 4.05y = .75$

20   $.14r - .08s = .24$
    $.06r + .22s = .12$

21   $.32r + .24s = .56$
    $.27r - .09s = .18$

22   $.10x + .35 - 2.5y = 0$
    $.16x - .26 + .12y = 0$

23   $\frac{3}{4}m - \frac{5}{4}n = 2$

    $1.5m - 2.5n = -4$

24   $2m + 2.4n = 7$

    $\frac{1}{2}m + \frac{3}{5}n = 3$

Solve each system below by substitution.

25   $y = \frac{12}{15}x + 2$

    $4x - 5y = -10$

26   $y = -\frac{10}{16}x + 1$

    $5x + 8y = 8$

27   $x = .8y + 3$
    $10x - 8y = 9$

28   $x = .24y + .52$
    $25x - 6y = 4$

29   $y = 5x - 3$
    $y = -3x + 5$

30   $y = -2x + 4$
    $y = -x - 3$

Solve each of the following problems.

31   Suppose that business A had a rate of return of 12% in 1981 and business
     B had a rate of return of 20%. In 1981, the total return on the two
     businesses was $24,000. In 1982, each business has a rate of return of

18% and the total return was $27,000. How much was invested in each business?

32 A certain investment portfolio consists of stocks and bonds. In 1 year the total return on the portfolio was $10,000; the rates of return on the stocks and the bonds were 10% and 15%, respectively. The next year the total return was $9600; the rates of return on the stocks and the bonds were 8% and 16%, respectively. Determine the amount invested in stocks.

33 A manufacturer requires 70 hours to make a light chair and 90 hours to make a heavy chair. If 180 chairs were made in 14,200 hours, how many of those chairs were heavy chairs?

34 A manufacturer uses 20 pounds of steel in making a small bicycle and 30 pounds of steel in a large bicycle. If 2200 pounds of steel are used in the production of 90 bicycles, how many are small and how many are large?

35 The demand function for a certain product is $p = D(x) = 50 - .5x$ and the supply function is $p = S(x) = .8x + 11$. Determine the market demand and the market price for this product.

36 For a certain product, the demand function is $p = D(x) = 96 - 1.6x$ and the supply function is $p = S(x) = 1.4x + 12$. Determine the equilibrium point for this product.

37 A vending machine's coin receptacle is filled with dimes and quarters. If there are 320 coins and their worth is $50, how many of the coins are dimes?

38 If four hamburgers and two soft drinks cost $4.40 and two hamburgers and four soft drinks cost $3.40, what is the price of a hamburger?

39 A maintenance worker bought 90 screws. The wood screws cost 12¢ each and the sheet metal screws cost 20¢ each. If the total cost was $14, how many of each kind of screw was purchased?

40 A shipment of jars costing $69 contains two sizes. The large jars cost 80¢ each and the small jars cost 60¢ each. If the total number of jars is 95, determine how many small jars are included.

41 *Calculator Problem* Solve the following system by the elimination method.

$$3.25x + 5.06y = 1.15$$
$$6.12x - 8.87y = 9.44$$

42 *Calculator Problem* Solve the following system by the elimination method.

$$9.02x + 3.67y = 14.54$$
$$4.27x + 8.75y = 19.43$$

43 *Calculator Problem* Solve the following system by the elimination method.

$$33.24x - 16.78y = 4.58$$
$$4.39x - 28.63y = 8.54$$

3.2 Gaussian Elimination

Often it is necessary to solve a system of equations with more than two equations or more than two unknowns. A **linear system** is any system that can be written in the form

$$a_{11}x_1 + a_{12}x_2 + \cdots + a_{1n}x_n = b_1$$
$$a_{21}x_1 + a_{22}x_2 + \cdots + a_{2n}x_n = b_2$$
$$\cdots \cdots \cdots \cdots \cdots \cdots \cdots$$
$$a_{m1}x_1 + a_{m2}x_2 + \cdots + a_{mn}x_n = b_m.$$

Each of the symbols a_{ij} and b_i stands for a constant; each symbol x_i stands for an unknown. The symbols a_{ij} are called **coefficients.** By counting the right sides of the equations, we see that there are m equations; by counting the number of variables in each equation, we see that there are n unknowns. **We say that such a linear system is $m \times n$, which is read "m by n."**

As in Section 3.1, a **solution to a linear system,** like the one above, is a list of values for each unknown such that each equation is true when each unknown is replaced by its corresponding value. **A linear system of equations may have one solution, no solutions, or an infinite number of solutions.**

In this section, we shall solve some linear systems that are more complicated than the 2×2 systems that we solved in Section 3.1. Our approach to solving linear systems is justified by the following theorem, which follows from the properties of equations given in Section 1.2.

Theorem The solutions to a linear system are unchanged when the system is transformed by one of the following operations.

1. Two of the equations are interchanged.
2. One equation is multiplied by a nonzero constant.
3. One equation is replaced by that equation plus a constant multiple of another equation.

We illustrate this process in Example 1.

EXAMPLE 1

Solve the given system.

$$x_1 + x_2 + x_3 = 1$$
$$2x_1 + 3x_2 + 3x_3 = 4$$
$$3x_1 + 2x_2 + 5x_3 = 6$$

Solution

Replace the second equation with the second equation minus 2 times the first equation. Replace the third equation with the third equation minus 3 times the first equation.

$$x_1 + x_2 + x_3 = 1$$
$$x_2 + x_3 = 2$$
$$-x_2 + 2x_3 = 3$$

Replace the third equation with the third equation plus the second equation.

$$x_1 + x_2 + x_3 = 1$$
$$x_2 + x_3 = 2$$
$$3x_3 = 5$$

From the third equation, we get $x_3 = \frac{5}{3}$. We substitute this into the second equation and solve for x_2.

$$x_2 + \frac{5}{3} = 2$$

$$x_2 = \frac{1}{3}$$

We substitute for x_3 and x_2 in the first equation and we solve for x_1.

$$x_1 + \frac{1}{3} + \frac{5}{3} = 1$$

$$x_1 = -1$$

Thus, the only solution is $x_1 = -1$, $x_2 = \frac{1}{3}$, $x_3 = \frac{5}{3}$. ∎

The method that was used in Example 1 is called the **Gaussian elimination method.** This method was invented by the mathematician Carl F. Gauss (1777–1855). In this method, we choose a variable and use one equation to eliminate that variable from the remaining equations. Then we repeat the process with the equations having a reduced number of variables. When this process is completed, then the solution is found by working from the equation with the fewest variables to the equation with the most variables.

Now we indicate how the method works when the number of equations is greater than the number of unknowns.

EXAMPLE 2

Solve the given system.

$$x_1 - x_2 + 2x_3 = 3$$
$$x_1 + 2x_2 - x_3 = 1$$
$$2x_1 - x_2 + 3x_3 = 4$$
$$2x_1 + x_2 - 2x_3 = 0$$

Solution
Replace the second equation with that equation minus the first equation.

Replace each of the last two equations with that equation minus 2 times the first equation.

$$x_1 - x_2 + 2x_3 = 3$$
$$3x_2 - 3x_3 = -2$$
$$x_2 - x_3 = -2$$
$$3x_2 - 6x_3 = -6$$

Multiply the second and fourth equations by $\frac{1}{3}$.

$$x_1 - x_2 + 2x_3 = 3$$
$$x_2 - x_3 = -\frac{2}{3}$$
$$x_2 - x_3 = -2$$
$$x_2 - 2x_3 = -2$$

Replace each of the last two equations with that equation minus the second equation.

$$x_1 - x_2 + 2x_3 = 3$$
$$x_2 - x_3 = -\frac{2}{3}$$
$$0 = -\frac{4}{3}$$
$$- x_3 = -\frac{4}{3}$$

The third equation is obviously false; this indicates that the system does not have any solutions. ■

Frequently, when the number of equations exceeds the number of variables, the system will have no solutions. The next example shows what can happen when the number of equations is smaller than the number of variables.

EXAMPLE 3

Solve the given system.

$$x_1 - x_2 + 3x_3 - x_4 = 2$$
$$2x_1 + x_2 - 3x_3 + x_4 = 0$$
$$3x_1 + x_2 + x_3 + x_4 = -2$$

Solution

We replace the second equation with that equation minus 2 times the first equation. We replace the third equation with that equation minus 3 times the first equation.

$$x_1 - x_2 + 3x_3 - x_4 = 2$$

$$3x_2 - 9x_3 + 3x_4 = -4$$

$$4x_2 - 8x_3 + 4x_4 = -8$$

Multiply the second equation by $\frac{1}{3}$ and the third equation by $\frac{1}{4}$.

$$x_1 - x_2 + 3x_3 - x_4 = 2$$

$$x_2 - 3x_3 + x_4 = -\frac{4}{3}$$

$$x_2 - 2x_3 + x_4 = -2$$

We replace the third equation with that equation minus the second equation.

$$x_1 - x_2 + 3x_3 - x_4 = 2$$

$$x_2 - 3x_3 + x_4 = -\frac{4}{3}$$

$$x_3 = -\frac{2}{3}$$

Using the third equation we substitute for x_3 in the second equation, which we solve for x_2.

$$x_2 - 3\left(-\frac{2}{3}\right) + x_4 = -\frac{4}{3}$$

$$x_2 = -\frac{10}{3} - x_4$$

Now we substitute for x_3 and x_2 in the first equation.

$$x_1 - \left(-\frac{10}{3} - x_4\right) + 3\left(-\frac{2}{3}\right) - x_4 = 2$$

$$x_1 = \frac{2}{3}$$

The solutions are $x_1 = \frac{2}{3}$, $x_2 = -\frac{10}{3} - x_4$, $x_3 = -\frac{2}{3}$, $x_4 =$ an arbitrary number. There are an infinite number of solutions to the system, one for each choice of a value for x_4. ■

The Gaussian elimination process can be speeded up by using a mathematical device called a matrix. A **matrix** is a rectangular array of numbers, called **entries,** written within brackets. Below are examples of matrices.

$$\begin{bmatrix} 2 & 1 \\ -1 & 3 \end{bmatrix} \quad [1 \quad 5 \quad 3] \quad \begin{bmatrix} 7 \\ 0 \\ 9 \\ 5 \end{bmatrix} \quad \begin{bmatrix} 8 & 1 & 2 \\ 0 & 5 & 4 \\ 1 & -3 & 0 \end{bmatrix}$$

Associated with each linear system is a matrix called the **coefficient matrix.** Below is a 3×3 linear system and the coefficient matrix associated with the linear system. Each row consists of the coefficients of the corresponding equation, with the symbols for the variables omitted. Also, each column contains all of the coefficients of one variable; for instance, the first column consists of the coefficients of x_1 in the three equations.

$$\begin{aligned} 2x_1 - x_2 + 5x_3 &= 1 \\ 3x_1 + 2x_2 - x_3 &= 3 \\ x_1 + x_2 - 2x_3 &= -1 \end{aligned} \qquad \begin{bmatrix} 2 & -1 & 5 \\ 3 & 2 & -1 \\ 1 & 1 & -2 \end{bmatrix}$$

If another column is added to the right of the coefficient matrix to display the constants from the right-hand sides of the above equations, that matrix is called the **augmented matrix.** For the linear system given previously the augmented matrix is

$$\begin{bmatrix} 2 & -1 & 5 & | & 1 \\ 3 & 2 & -1 & | & 3 \\ 1 & 1 & -2 & | & -1 \end{bmatrix}.$$

A matrix with m rows and n columns is said to have **order** $m \times n$, which is read "m by n." This definition is consistent with the comparable definition for systems given at the beginning of the section. Using matrices we can execute the Gaussian elimination process, which was given in the preceding three examples, more quickly. In this procedure, we get zeroes in the augmented matrix by some operations called the **elementary row operations.**

Definition   We refer to the following as the elementary row operations.

1. Two rows of the matrix are interchanged.
2. One row is multiplied by a nonzero constant.
3. One row is replaced by that row plus a constant multiple of another row.

Compare these with the theorem preceding Example 1, and you will see that they correspond exactly to the operations on a system of equations that leave the solutions unchanged. We now demonstrate how to perform the process of Gaussian elimination using an augmented matrix.

EXAMPLE 4   Solve the given system.

$$2x_1 + x_2 - x_3 = 1$$
$$3x_1 - x_2 + 2x_3 = 3$$
$$x_1 - x_2 - x_3 = 0$$

Solution

This system is 3×3 and the corresponding coefficient matrix has order 3×3; the augmented matrix has order 3×4. The augmented matrix is

$$\begin{bmatrix} 2 & 1 & -1 & | & 1 \\ 3 & -1 & 2 & | & 3 \\ 1 & -1 & -1 & | & 0 \end{bmatrix}.$$

Our first goal is to get 1 in the upper left corner by means of elementary row operations. This can be accomplished easily by interchanging the first and third rows.

$$\begin{bmatrix} 1 & -1 & -1 & | & 0 \\ 3 & -1 & 2 & | & 3 \\ 2 & 1 & -1 & | & 1 \end{bmatrix}$$

Now we get zeros below the 1 in the first column. Replace the second row with that row minus 3 times the first row. Replace the third row with that row minus 2 times the first row.

$$\begin{bmatrix} 1 & -1 & -1 & | & 0 \\ 0 & 2 & 5 & | & 3 \\ 0 & 3 & 1 & | & 1 \end{bmatrix}$$

Now we get a 1 as the second entry in the second row. Multiply the second row by $\frac{1}{2}$.

$$\begin{bmatrix} 1 & -1 & -1 & | & 0 \\ 0 & 1 & \frac{5}{2} & | & \frac{3}{2} \\ 0 & 3 & 1 & | & 1 \end{bmatrix}$$

Next we get a 0 below the second entry in the second row. We replace the third row with that row minus 3 times the second row.

$$\begin{bmatrix} 1 & -1 & -1 & | & 0 \\ 0 & 1 & \frac{5}{2} & | & \frac{3}{2} \\ 0 & 0 & -\frac{13}{2} & | & -\frac{7}{2} \end{bmatrix}$$

Now we write the linear system corresponding to this augmented matrix.

$$x_1 - x_2 - x_3 = 0$$
$$x_2 + \frac{5}{2}x_3 = \frac{3}{2}$$
$$-\frac{13}{2}x_3 = -\frac{7}{2}$$

From the last equation we obtain the value of x_3.

$$x_3 = \frac{7}{13}$$

Substituting for x_3 in the second equation, we obtain the value of x_2.

$$x_2 + \frac{5}{2}\left(\frac{7}{13}\right) = \frac{3}{2}$$

$$x_2 = \frac{2}{13}$$

Substituting the x_2 and x_3 in the first equation, we solve for x_1

$$x_1 - \left(\frac{2}{13}\right) - \left(\frac{7}{13}\right) = 0$$

$$x_1 = \frac{9}{13}$$

The only solution to the original linear system is $x_1 = \frac{9}{13}, x_2 = \frac{2}{13}, x_3 = \frac{7}{13}$. ■

It is not difficult to see that this is the same process of Gaussian elimination that was used in the first three examples. The use of the augmented matrix makes it easier to focus on the important numbers. When the explanation for the steps is abbreviated, then the procedure with the augmented matrix is much quicker than Gaussian elimination.

We can stop using the elementary row operations when the first non-zero entry in each row has only zeroes below it. In the final matrix in Example 4, the first entry in the first row is nonzero, and the first entry in each of the other rows is zero. The first nonzero entry in the second row (the 1 in the second column) has a zero below it. Of course, the third row has no entries below it, so we don't worry about it. In terms of Gaussian elimination, for each equation given we have removed one variable from all but one equation.

In the next example, we demonstrate the same process with a scheme for giving abbreviated explanations as to the steps taken. We denote the ith row by R_i; thus, when we write $R_2 - 3R_1$ we are indicating that the second row is replaced by that row minus 3 times the first row.

EXAMPLE 5

Solve the given system.

$$x_1 + x_2 + x_3 + x_4 = 1$$
$$x_1 + x_2 - x_3 - x_4 = 0$$
$$x_1 + x_2 - x_3 + x_4 = 2$$

Solution

We write the augmented matrix and perform the appropriate steps with the abbreviations indicated in the paragraph preceding this example. The linear system is 3×4 and the augmented matrix, given below, has order 3×5.

$$\begin{bmatrix} 1 & 1 & 1 & 1 & 1 \\ 1 & 1 & -1 & -1 & 0 \\ 1 & 1 & -1 & 1 & 2 \end{bmatrix}$$

$$\begin{matrix} R_2 - R_1 \\ R_3 - R_1 \end{matrix} \begin{bmatrix} 1 & 1 & 1 & 1 & 1 \\ 0 & 0 & -2 & -2 & -1 \\ 0 & 0 & -2 & 0 & 1 \end{bmatrix}$$

$$\begin{matrix} \\ \\ R_3 - R_2 \end{matrix} \begin{bmatrix} 1 & 1 & 1 & 1 & 1 \\ 0 & 0 & -2 & -2 & -1 \\ 0 & 0 & 0 & 2 & 2 \end{bmatrix}$$

Since we have all the zeroes necessary we write down the corresponding system and find the solution.

$$x_1 + x_2 + x_3 + x_4 = 1$$
$$-2x_3 - 2x_4 = -1$$
$$2x_4 = 2$$

$$x_4 = 1$$
$$-2x_3 - 2(1) = -1$$
$$x_3 = -\frac{1}{2}$$

$$x_1 + x_2 + \left(-\frac{1}{2}\right) + 1 = 1$$

$$x_1 = \frac{1}{2} - x_2$$

The solutions are $x_1 = \frac{1}{2} - x_2$, $x_2 =$ an arbitrary number, $x_3 = -\frac{1}{2}$, $x_4 = 1$. ■

When there are infinitely many choices for the numbers that solve a given linear system, as in Example 5, then there may be more than one way to describe the solutions. Another description of the solutions for Example 5 is $x_1 =$ an arbitrary number, $x_2 = \frac{1}{2} - x_1$, $x_3 = -\frac{1}{2}$, $x_4 = 1$.

The next example shows that an $n \times n$ system may have an infinite number of solutions.

EXAMPLE 6

Solve the given system

$$x_1 - x_2 \quad = 1$$
$$x_2 + x_3 = 2$$
$$-x_1 + x_2 \quad = -1.$$

Solution

We write the augmented matrix, and we perform the appropriate steps using the abbreviations used in Example 5.

$$\begin{bmatrix} 1 & -1 & 0 & | & 1 \\ 0 & 1 & 1 & | & 2 \\ -1 & 1 & 0 & | & -1 \end{bmatrix}$$

$$R_3 + R_1 \quad \begin{bmatrix} 1 & -1 & 0 & | & 1 \\ 0 & 1 & 1 & | & 2 \\ 0 & 0 & 0 & | & 0 \end{bmatrix}$$

We have all the zeroes necessary, so we solve the corresponding system.

$$\begin{aligned} x_1 - x_2 \quad &= 1 \\ x_2 + x_3 &= 2 \end{aligned}$$

or

$$\begin{aligned} x_3 &= 2 - x_2 \\ x_1 &= 1 + x_2 \end{aligned}$$

The solutions are $x_1 = 1 + x_2$, $x_2 =$ an arbitrary number, $x_3 = 2 - x_2$. There are an infinite number of solutions. ◼

The final example shows that an $n \times n$ system may have no solutions at all.

EXAMPLE 7

Solve the given system

$$\begin{aligned} x_1 \quad\quad - x_3 &= 2 \\ - x_2 + x_3 &= 0 \\ -x_1 \quad\quad + x_3 &= 1 \end{aligned}$$

Solution

We continue to use the abbreviations from Example 5.

$$\begin{bmatrix} 1 & 0 & -1 & | & 2 \\ 0 & -1 & 1 & | & 0 \\ -1 & 0 & 1 & | & 1 \end{bmatrix}$$

$$R_3 + R_1 \quad \begin{bmatrix} 1 & 0 & -1 & | & 2 \\ 0 & -1 & 1 & | & 0 \\ 0 & 0 & 0 & | & 3 \end{bmatrix}$$

The corresponding system is

$$\begin{aligned} x_1 \quad - x_3 &= 2 \\ - x_2 + x_3 &= 0 \\ 0 &= 3. \end{aligned}$$

The last equation can never be true regardless of how x_1, x_2, and x_3 are chosen. Thus, the system has no solutions. ■

Exercises 3.2

Solve each system by using the elementary row operations on the augmented matrix.

1. $x_1 + x_2 - x_3 = -1$
   $-x_1 + x_2 \qquad = 2$
   $2x_1 \qquad + 3x_3 = 5$

2. $x_1 \qquad + 2x_3 = 3$
   $3x_2 - x_3 = -2$
   $3x_1 + x_2 + x_3 = 4$

3. $x + 2y + 3z = 1$
   $2x + y - z = 2$
   $x + 3y - 4z = 3$

4. $x + 3y - 2z = 0$
   $-x - 2y + z = 2$
   $x + y \qquad = -3$

5. $2w - x - y = 4$
   $w + 2x + y = 2$
   $w - 3x - 2y = 2$

6. $3w - 2x + y = 0$
   $2w + x - 3y = 1$
   $w - 3x + 4y = -1$

7. $x_1 - x_2 - x_3 + x_4 = -3$
   $x_1 + x_2 \qquad + 3x_4 = 2$
   $x_2 + 2x_3 - x_4 = -1$

8. $x_1 + 3x_2 - x_3 \qquad = 0$
   $x_1 - x_2 + x_3 + x_4 = 1$
   $x_1 \qquad - 2x_3 + x_4 = 2$

9. $w + 3x + 2y + z = 1$
   $2w \qquad + 3y - z = -3$
   $2x - y + z = 3$

10. $2x - y - z = 2$
    $w \qquad + y + 3z = 1$
    $2w + x \qquad - z = 0$

11. $2x_1 - x_2 + 3x_3 - 2x_4 = 0$
    $x_1 + 2x_2 - x_3 + x_4 = 1$
    $3x_1 + x_2 + 2x_3 - x_4 = -1$

12. $x_1 - 3x_2 + 2x_3 - 4x_4 = -1$
    $2x_1 + x_2 - x_3 + 2x_4 = 2$
    $3x_1 - 2x_2 + x_3 - 2x_4 = 0$

13. $x_1 + x_2 - x_3 = 0$
    $x_1 - x_2 + 2x_3 = 1$
    $2x_1 \qquad + x_3 = 1$
    $2x_1 + 2x_2 - x_3 = 3$

14. $x_1 - 2x_2 + x_3 = 4$
    $2x_1 - x_2 - x_3 = 4$
    $x_1 + x_2 - 2x_3 = 0$
    $2x_1 + x_2 + x_3 = -2$

15. $r + s + t = -1$
    $r - s + t = 1$
    $r + s - t = 0$
    $r - s - t = 2$

16. $r + s - 2t = 3$
    $2r + 2s + t = 4$
    $r - s - t = -1$
    $3r + s + t = -2$

17. $-r + 2s - 3t = 1$
    $2r - s + t = 0$
    $r + s - 2t = 1$
    $3r - 3s + 4t = -1$

18. $2r - 3s + 4t = 0$
    $r + 2s - 2t = 3$
    $r - 5s + 6t = -3$
    $3r - s + 2t = 3$

19. $x_1 - x_2 - x_3 + x_4 = 1$
    $x_1 + 2x_2 + x_3 \qquad = -1$
    $-x_1 + x_2 - x_3 + x_4 = 0$
    $2x_1 \qquad + x_3 - x_4 = 3$

20. $x_1 + 3x_2 - 2x_3 + x_4 = 0$
    $x_1 \qquad + 3x_3 - 2x_4 = 2$
    $x_1 + x_2 \qquad + 3x_4 = -1$
    $x_1 - x_2 + x_3 \qquad = -2$

21. $w + 2x - y + 3z = 0$
    $w - x + 2y - z = 1$
    $2w + x + y + 2z = 2$
    $w + x - y - z = -1$

22. $w - 3x + 2y - z = 4$
    $-4x + 3y - 3z = 6$
    $w + x - y + 2z = -2$
    $2w - 2x + y + z = 2$

3.3 **Matrix Multiplication and Matrix Equations**

Any linear system of equations can be written as one single equation through the use of matrices. This approach to linear systems has important theoretical and practical consequences. In Section 3.4 we consider some of the theoretical consequences, and in Section 4.2 we develop some significant practical applications.

Definition Let A be a matrix with order $m \times n$ and denote the entry in the ith row and jth column by a_{ij}. Let B be a matrix with order $n \times p$ and denote the entry in the ith row and the jth column by b_{ij}. The **product matrix** **AB** has order $m \times p$ and the entry in the ith row and jth column is

$$a_{i1}b_{1j} + a_{i2}b_{2j} + a_{i3}b_{3j} + \cdots + a_{in}b_{nj}.$$

Matrix multiplication is usually learned by studying examples like Example 1.

EXAMPLE 1

Find the product AB given

$$A = \begin{bmatrix} 2 & 1 \\ 4 & 3 \end{bmatrix} \quad \text{and} \quad B = \begin{bmatrix} 5 \\ 6 \end{bmatrix}.$$

Solution
Begin by noting that the order of A is 2×2 and the order of B is 2×1. Thus, the order of AB is 2×1; AB consists of a single column. Multiply the elements of the first row of A and the corresponding elements of the column of B.

$$\begin{bmatrix} 2 & 1 \\ 4 & 3 \end{bmatrix}\begin{bmatrix} 5 \\ 6 \end{bmatrix} \qquad 2(5) + 1(6) = 16$$

Thus, 16 is the first entry in the only column of AB. Now, multiply the elements of the second row of A with the corresponding elements of B.

$$\begin{bmatrix} 2 & 1 \\ 4 & 3 \end{bmatrix}\begin{bmatrix} 5 \\ 6 \end{bmatrix} \qquad 4(5) + 3(6) = 38$$

Thus, 38 is the second entry in the only column of AB. This product is usually written

$$AB = \begin{bmatrix} 2 & 1 \\ 4 & 3 \end{bmatrix}\begin{bmatrix} 5 \\ 6 \end{bmatrix} = \begin{bmatrix} 16 \\ 38 \end{bmatrix}. \quad \blacksquare$$

In the next example, we perform the correct steps in the correct order without so much explanation.

EXAMPLE 2

Find the product CD where

$$C = \begin{bmatrix} 3 & 0 & 1 \\ 0 & 2 & -1 \end{bmatrix} \quad \text{and} \quad D = \begin{bmatrix} 1 & 5 \\ 4 & 2 \\ 6 & 0 \end{bmatrix}$$

Solution
C is 2×3 and D is 3×2, so CD is 2×2.

$$\begin{bmatrix} 3 & 0 & 1 \\ 0 & 2 & -1 \end{bmatrix} \begin{bmatrix} 1 & 5 \\ 4 & 2 \\ 6 & 0 \end{bmatrix}$$
$3(1) + 0(4) + 1(6) = 9$
in first row, first column of product

$$\begin{bmatrix} 3 & 0 & 1 \\ 0 & 2 & -1 \end{bmatrix} \begin{bmatrix} 1 & 5 \\ 4 & 2 \\ 6 & 0 \end{bmatrix}$$
$3(5) + 0(2) + 1(0) = 15$
in first row, second column of product

$$\begin{bmatrix} 3 & 0 & 1 \\ 0 & 2 & -1 \end{bmatrix} \begin{bmatrix} 1 & 5 \\ 4 & 2 \\ 6 & 0 \end{bmatrix}$$
$0(1) + 2(4) - 1(6) = 2$
in second row, first column of product

$$\begin{bmatrix} 3 & 0 & 1 \\ 0 & 2 & -1 \end{bmatrix} \begin{bmatrix} 1 & 5 \\ 4 & 2 \\ 6 & 0 \end{bmatrix}$$
$0(5) + 2(2) - 1(0) = 4$
in second row, second column of product

We have determined that the product is

$$CD = \begin{bmatrix} 3 & 0 & 1 \\ 0 & 2 & -1 \end{bmatrix} \begin{bmatrix} 1 & 5 \\ 4 & 2 \\ 6 & 0 \end{bmatrix} = \begin{bmatrix} 9 & 15 \\ 2 & 4 \end{bmatrix}. \blacksquare$$

It is important to note when two matrices can be multiplied. **To compute the product AB where the order of A is $m \times n$ and the order of B is $p \times q$, it must be that $n = p$. Then the product matrix has order $m \times q$.**

A matrix that consists of one column is called a **column matrix** and a matrix that has only one row is called a **row matrix**. If B is a column matrix, then so is the product AB; if A is a row matrix, then so is the product AB.

Most of the algebraic properties of matrices will be developed systematically in Section 4.1. We should observe immediately that matrix multiplication has some properties similar to multiplication of real numbers and some properties that are quite different. If A, B, and C and three matrices for which the products below are defined, then

$$(AB)C = A(BC).$$

This is the **associative property** of matrix multiplication. This is analogous to the equation

$$(ab)c = a(bc)$$

for multiplication of real numbers.

For real numbers a and b, the two products ab and ba are equal. This is called the **commutative property** of multiplication. For matrices A and B, however, the two products AB and BA can be quite different. For example, let A and B be given by

$$A = [3 \quad 0 \quad 2], \qquad B = \begin{bmatrix} 1 \\ -1 \\ 0 \end{bmatrix}.$$

Then the two products AB and BA are given by

$$AB = [3 \quad 0 \quad 2] \cdot \begin{bmatrix} 1 \\ -1 \\ 0 \end{bmatrix} = [3]$$

and

$$BA = \begin{bmatrix} 1 \\ -1 \\ 0 \end{bmatrix} \cdot [3 \quad 0 \quad 2] = \begin{bmatrix} 3 & 0 & 2 \\ -3 & 0 & -2 \\ 0 & 0 & 0 \end{bmatrix}$$

Since A is 1×3 and B is 3×1, the product AB is 1×1, whereas the product BA is 3×3. Thus, **matrix multiplication is not commutative.**

We say that **matrix A equals matrix C** provided A and C have the same order and each entry in A equals the corresponding entry in C. For example, the first entry in the first row of A must equal the first entry in the first row of C. To indicate that A equals C, we write $A = C$. The next example uses the idea of equal matrices.

EXAMPLE 3

Verify that the values of x_1, x_2, and x_3 that make the following matrix equation true

$$\begin{bmatrix} 2 & 3 & 1 \\ 0 & -1 & 1 \\ 1 & 1 & 2 \end{bmatrix} \begin{bmatrix} x_1 \\ x_2 \\ x_3 \end{bmatrix} = \begin{bmatrix} 1 \\ 0 \\ 4 \end{bmatrix}$$

solve the linear system

$$\begin{aligned} 2x_1 + 3x_2 + x_3 &= 1 \\ -x_2 + x_3 &= 0 \\ x_1 + x_2 + 2x_3 &= 4. \end{aligned}$$

Solution

We simplify the matrix equation by computing the product on the left.

$$\begin{bmatrix} 2 & 3 & 1 \\ 0 & -1 & 1 \\ 1 & 1 & 2 \end{bmatrix} \begin{bmatrix} x_1 \\ x_2 \\ x_3 \end{bmatrix} = \begin{bmatrix} 2x_1 + 3x_2 + x_3 \\ -x_2 + x_3 \\ x_1 + x_2 + 2x_3 \end{bmatrix} = \begin{bmatrix} 1 \\ 0 \\ 4 \end{bmatrix}$$

For the last two matrices to be equal, it must be that the entries of one equal the corresponding entries of the other.

$$\begin{aligned} 2x_1 + 3x_2 + x_3 &= 1 \\ -x_2 + x_3 &= 0 \\ x_1 + x_2 + 2x_3 &= 4 \end{aligned}$$

Since this is the linear system given in the example, we see that solving the matrix equation is equivalent to solving the linear system. ∎

Any linear system can be quickly written as a matrix equation. By definition, any linear system can be written in the form

$$a_{11}x_1 + a_{12}x_2 + \cdots + a_{1n}x_n = b_1$$
$$a_{21}x_1 + a_{22}x_2 + \cdots + a_{2n}x_n = b_2$$
$$\cdots\cdots\cdots\cdots\cdots\cdots$$
$$a_{m1}x_1 + a_{m2}x_2 + \cdots + a_{mn}x_n = b_m,$$

where each a_{ij} and b_j stands for a constant and each symbol x_i stands for an unknown. The single matrix equation with the same solution is

$$\begin{bmatrix} a_{11} & a_{12} & \cdots & a_{1n} \\ a_{21} & a_{22} & \cdots & a_{2n} \\ \cdots & \cdots & \cdots & \cdots \\ a_{m1} & a_{m2} & \cdots & a_{mn} \end{bmatrix} \begin{bmatrix} x_1 \\ x_2 \\ \vdots \\ x_n \end{bmatrix} = \begin{bmatrix} b_1 \\ b_2 \\ \vdots \\ b_m \end{bmatrix}.$$

Sometimes a matrix equation like this is written in the briefer notation

$$AX = B,$$

where A stands for the coefficient matrix with entries of the form a_{ij}, X stands for the column matrix of unknowns, and B stands for the column matrix of constants on the right side of the equation.

In certain problems, the easiest way to give the data for the linear system is to use a matrix. This situation is illustrated in the next example.

EXAMPLE 4

A company produces three products, denoted a, b, and c; each product involves three production lines, denoted I, II, and III. The number of hours that each product spends on each production line is indicated in the matrix below.

$$\text{production lines} \quad \begin{array}{c} \\ \text{I} \\ \text{II} \\ \text{III} \end{array} \overset{\displaystyle \overset{\text{products}}{\begin{array}{ccc} a & b & c \end{array}}}{\begin{bmatrix} 1 & 2 & 3 \\ 3 & 1 & 1 \\ 2 & 3 & 2 \end{bmatrix}}$$

This matrix shows, for example, that product c requires 3 hours on line I, since under c and beside I is the numeral 3. Suppose the numbers of hours that the three lines are available are 160, 140, and 200, respectively. Determine how many of each product can be made.

Solution

Let x_1, x_2, and x_3 denote the number of units of a, b, and c, respectively, that are produced. The linear system given by the preceding data can be written as the matrix equation

$$\begin{bmatrix} 1 & 2 & 3 \\ 3 & 1 & 1 \\ 2 & 3 & 2 \end{bmatrix} \begin{bmatrix} x_1 \\ x_2 \\ x_3 \end{bmatrix} = \begin{bmatrix} 160 \\ 140 \\ 200 \end{bmatrix}.$$

We solve the linear system by using the method of elimination on the augmented matrix. We continue to abbreviate our steps as we did in Example 5 of Section 3.2.

$$\begin{bmatrix} 1 & 2 & 3 & | & 160 \\ 3 & 1 & 1 & | & 140 \\ 2 & 3 & 2 & | & 200 \end{bmatrix}$$

$$\begin{matrix} R_2 - 3R_1 \\ R_3 - 2R_1 \end{matrix} \quad \begin{bmatrix} 1 & 2 & 3 & | & 160 \\ 0 & -5 & -8 & | & -340 \\ 0 & -1 & -4 & | & -120 \end{bmatrix}$$

$$\begin{matrix} \text{interchange} \\ R_2 \text{ and } R_3 \end{matrix} \quad \begin{bmatrix} 1 & 2 & 3 & | & 160 \\ 0 & -1 & -4 & | & -120 \\ 0 & -5 & -8 & | & -340 \end{bmatrix}$$

$$\begin{matrix} \\ R_3 - 5R_2 \end{matrix} \quad \begin{bmatrix} 1 & 2 & 3 & | & 160 \\ 0 & -1 & -4 & | & -120 \\ 0 & 0 & 12 & | & 260 \end{bmatrix}$$

Writing the linear system from the last matrix, we have

$$x_1 + 2x_2 + 3x_3 = 160$$
$$-x_2 - 4x_3 = -120$$
$$12x_3 = 260$$

From the third equation, we get

$$x_3 = 21\frac{2}{3}.$$

We substitute this into the second equation and solve for x_2.

$$-x_2 - 4\left(21\frac{2}{3}\right) = -120$$

$$x_2 = 33\frac{1}{3}$$

Now we substitute for x_2 and x_3 in the first equation.

$$x_1 + 2\left(33\frac{1}{3}\right) + 3\left(21\frac{2}{3}\right) = 160$$

$$x_1 = 28\frac{1}{3}$$

In actual practice, it is not possible to make a fraction of a unit, so the answer is

$$x_1 = 28, \qquad x_2 = 33, \qquad x_3 = 21. \quad \blacksquare$$

The next example describes another situation in which the use of a matrix equation is helpful.

EXAMPLE 5

A fast food restaurant offers three choices of menu for breakfast. In the matrix below the breakfasts are denoted by the symbols I, II, and III; the menu for each is listed in the column below the symbol. At the end of the day the restaurant manager observes that 120 eggs, 200 sausage patties, and 180 pancakes were served that day. Determine how many times each breakfast was ordered.

$$
\begin{matrix}
 & \text{breakfasts} \\
 & \begin{matrix} \text{I} & \text{II} & \text{III} \end{matrix} \\
\begin{matrix} \text{eggs} \\ \text{sausage patties} \\ \text{pancakes} \end{matrix} &
\begin{bmatrix} 2 & 1 & 0 \\ 2 & 1 & 2 \\ 0 & 3 & 3 \end{bmatrix}
\end{matrix}
$$

Solution
Let the number of times the breakfasts I, II, and III were ordered be denoted by x_1, x_2, and x_3, respectively. The matrix equation to be solved is

$$
\begin{bmatrix} 2 & 1 & 0 \\ 2 & 1 & 2 \\ 0 & 3 & 3 \end{bmatrix}
\begin{bmatrix} x_1 \\ x_2 \\ x_3 \end{bmatrix} =
\begin{bmatrix} 120 \\ 200 \\ 180 \end{bmatrix}.
$$

The augmented matrix is

$$
\left[\begin{array}{ccc|c} 2 & 1 & 0 & 120 \\ 2 & 1 & 2 & 200 \\ 0 & 3 & 3 & 180 \end{array} \right].
$$

The easiest way to get the necessary zeroes in this matrix is the following.

$$
R_2 - R_1 \qquad
\left[\begin{array}{ccc|c} 2 & 1 & 0 & 120 \\ 0 & 0 & 2 & 80 \\ 0 & 3 & 3 & 180 \end{array} \right]
$$

We interchange R_2 and R_3.

$$
\left[\begin{array}{ccc|c} 2 & 1 & 0 & 120 \\ 0 & 3 & 3 & 180 \\ 0 & 0 & 2 & 80 \end{array} \right]
$$

Now we solve the corresponding system.

$$2x_1 + x_2 = 120$$
$$3x_2 + 3x_3 = 180$$
$$2x_3 = 80$$

Thus, $x_3 = 40$; we substitute this into the second equation and solve for x_2.

$$3x_2 + 3(40) = 180$$

$$x_2 = 20$$

We substitute for x_2 and x_3 in the first equation, and we solve for x_1.

$$2x_1 + (20) = 120$$

$$x_1 = 50$$

The only solution is $x_1 = 50$, $x_2 = 20$, $x_3 = 40$. ∎

Exercises 3.3

In the following problems, the order of each of the matrices A and B is given. Determine whether the products AB and BA exist; determine the order of each product when it exists.

1   A is 2×3, B is 3×3

2   A is 2×3, B is 2×2

3   A is 3×3, B is 3×3

4   A is 2×2, B is 2×2

5   A is 3×1, B is 1×3

6   A is 3×4, B is 2×3

7   A is 3×4, B is 4×2

8   A is 9×5, B is 5×3

9   A is 1×10, B is 10×4

10   A is 4×4, B is 4×1

Compute each of the following products when that is possible.

11   $\begin{bmatrix} 1 & 0 \\ -1 & 2 \end{bmatrix}\begin{bmatrix} 3 & -2 \\ 1 & 5 \end{bmatrix}$

12   $\begin{bmatrix} 8 & 3 \\ 1 & 5 \end{bmatrix}\begin{bmatrix} 2 \\ -3 \end{bmatrix}$

13   $[1 \ \ 0]\begin{bmatrix} 5 \\ 7 \end{bmatrix}$

14   $\begin{bmatrix} 1 \\ 0 \end{bmatrix}[5 \ \ 7]$

15   $\begin{bmatrix} -1 & 1 \\ 3 & 2 \end{bmatrix}[5 \ \ 9]$

16   $\begin{bmatrix} 1 & 0 \\ 0 & 1 \end{bmatrix}\begin{bmatrix} 4 & 3 \\ 2 & 1 \end{bmatrix}$

17   $\begin{bmatrix} 7 & 3 \\ 5 & 2 \end{bmatrix}\begin{bmatrix} 1 & 0 \\ 0 & 1 \end{bmatrix}$

18   $\begin{bmatrix} 1 & 0 & -3 \\ 2 & -1 & 0 \end{bmatrix}\begin{bmatrix} -2 \\ 0 \\ -1 \end{bmatrix}$

19   $\begin{bmatrix} .5 & .5 \\ .5 & .5 \end{bmatrix}\begin{bmatrix} .5 & .5 \\ .5 & .5 \end{bmatrix}$

20   $\begin{bmatrix} 0 & 0 \\ 1 & 0 \end{bmatrix}\begin{bmatrix} 0 & 0 \\ 1 & 0 \end{bmatrix}$

21   $\begin{bmatrix} 1 & 0 & 0 \\ 0 & 1 & 0 \\ 0 & 0 & 1 \end{bmatrix}\begin{bmatrix} 5 \\ 3 \\ 6 \end{bmatrix}$

22   $\begin{bmatrix} 9 & -1 & 2 \\ 0 & 3 & -5 \end{bmatrix}\begin{bmatrix} 1 & 0 & 0 \\ 0 & 1 & 0 \\ 0 & 0 & 1 \end{bmatrix}$

23   $[4 \ \ 5 \ \ 1]\begin{bmatrix} 0 \\ -2 \\ 3 \end{bmatrix}$

24   $\begin{bmatrix} 1 \\ -1 \\ 1 \end{bmatrix}[5 \ \ 7 \ \ 9]$

25 $\begin{bmatrix} 2 & 3 & 5 \\ -1 & 1 & 0 \\ 7 & -2 & 1 \end{bmatrix} \begin{bmatrix} -3 & 1 & 6 \\ 4 & 9 & -2 \\ 10 & 0 & 1 \end{bmatrix}$

26 $\begin{bmatrix} 5 & 2 & 1 \\ 0 & 1 & 3 \\ 9 & 8 & 7 \end{bmatrix} \begin{bmatrix} 4 \\ 3 \end{bmatrix}$

27 $\begin{bmatrix} 9 & 1 & 1 \\ 0 & 2 & 7 \\ 5 & 8 & 3 \end{bmatrix} \begin{bmatrix} 10 & 0 & 9 \end{bmatrix}$

28 $\begin{bmatrix} 3 & 2 & 9 \\ 8 & 0 & 1 \\ 5 & 7 & 1 \end{bmatrix} \begin{bmatrix} 6 \\ 3 \\ 1 \end{bmatrix}$

29 $\begin{bmatrix} 1 & 0 & 1 & 2 \\ 3 & -1 & 0 & 1 \end{bmatrix} \begin{bmatrix} 4 \\ 1 \\ -2 \\ 0 \end{bmatrix}$

30 $\begin{bmatrix} -3 & 1 \\ 4 & 8 \\ 5 & -2 \\ 0 & 2 \end{bmatrix} \begin{bmatrix} 1 & 0 & 2 \\ 3 & 2 & -1 \end{bmatrix}$

For each system below, write a matrix equation that has the same solution.

31 $3x_1 + 2x_2 = 1$
$x_1 - 5x_2 = 3$

32 $x - y = 4$
$4x + 3y = 1$

33 $2x + y - z = 3$
$x - 3y + z = -1$

34 $5x_1 + x_2 - 3x_3 = 1$
$x_1 - 3x_2 + x_3 = 2$

35 $x_1 + 3x_2 = 4$
$x_1 - x_2 = 1$
$3x_1 + 4x_2 = 0$

36 $x - y + 2z = 1$
$x + 3y - z = 5$
$2x - y + 3z = 0$

37 $x_1 - 3x_2 + x_3 - 4 = 0$
$x_1 + x_2 + 5x_3 - 1 = 3$
$x_1 - x_2 - x_3 + 5 = 2$

38 $x - 5 + y = z$
$x + 2y - 3 = z$
$3x - y + z - 4 = 0$

Solve each of the following.

39 Find the solution in Example 4 supposing that the number of hours that the production lines I, II, and III are available are 150, 150, and 180, respectively.

40 Find the solution in Example 4 supposing that the number of hours that the production lines I, II, and III are available are 170, 150, and 190, respectively.

41 Find the solution in Example 5 supposing that 115 eggs, 195 sausage patties, and 195 pancakes were served.

42 Find the solution in Example 5 supposing that 130 eggs, 220 sausage patties, and 165 pancakes were served.

43 A financial advisor has constructed investment portfolios for each of three clients. We refer to the clients as Adams, Brown, and Cain. Each portfolio consists of bonds, stocks, and cash reserves. The matrix below indicates the decimal fraction of each client's portfolio that is invested in each of the three areas.

	Adams	Brown	Cain
Bonds	.4	.3	.6
Stocks	.4	.6	.3
Cash	.2	.1	.1

Determine the total amount in each portfolio assuming that the total amounts in bonds, stocks, and cash are $166,000, $142,000, and $62,000, respectively.

44 Find the solution in problem 43 supposing that the total amounts in bonds, stocks, and cash are $100,000, $115,000, and $35,000.

3.4 Inverse of a Matrix

Matrices that have the same number of columns as rows are said to be **square.** One of the square matrices with order $n \times n$ that is particularly important is

$$I_n = \begin{bmatrix} 1 & 0 & \cdots & 0 & 0 \\ 0 & 1 & \cdots & 0 & 0 \\ \cdot & \cdot & & \cdot & \cdot \\ 0 & 0 & \cdots & 1 & 0 \\ 0 & 0 & \cdots & 0 & 1 \end{bmatrix}.$$

The matrix I_n has 1's down the **main diagonal** (which consists of the entries from the upper left corner to the lower right corner) and 0's everywhere else. This matrix is called the **identity matrix.** If A is any $n \times n$ matrix, then both of the matrix equations

$$AI_n = A, \qquad I_n A = A$$

are true. This is analogous to the role that 1 plays in the multiplication of real numbers; if a is any real number, then we have

$$a \cdot 1 = a, \qquad 1 \cdot a = a.$$

EXAMPLE 1

Compute the products AI_2, I_2A, BI_3, and I_3B where

$$a = \begin{bmatrix} 3 & 2 \\ -1 & 5 \end{bmatrix}, \qquad B = \begin{bmatrix} 2 & -1 & 1 \\ -5 & 2 & 3 \\ 0 & 4 & 0 \end{bmatrix}$$

Solution

Computing the products of the matrices is made easy by the fact that most of the products of entries are 0.

$$\begin{bmatrix} 3 & 2 \\ -1 & 5 \end{bmatrix}\begin{bmatrix} 1 & 0 \\ 0 & 1 \end{bmatrix} = \begin{bmatrix} 3 & 2 \\ -1 & 5 \end{bmatrix}, \qquad \begin{bmatrix} 1 & 0 \\ 0 & 1 \end{bmatrix}\begin{bmatrix} 3 & 2 \\ -1 & 5 \end{bmatrix} = \begin{bmatrix} 3 & 2 \\ -1 & 5 \end{bmatrix}$$

$$\begin{bmatrix} 2 & -1 & 1 \\ -5 & 2 & 3 \\ 0 & 4 & 0 \end{bmatrix}\begin{bmatrix} 1 & 0 & 0 \\ 0 & 1 & 0 \\ 0 & 0 & 1 \end{bmatrix} = \begin{bmatrix} 2 & -1 & 1 \\ -5 & 2 & 3 \\ 0 & 4 & 0 \end{bmatrix}$$

$$\begin{bmatrix} 1 & 0 & 0 \\ 0 & 1 & 0 \\ 0 & 0 & 1 \end{bmatrix}\begin{bmatrix} 2 & -1 & 1 \\ -5 & 2 & 3 \\ 0 & 4 & 0 \end{bmatrix} = \begin{bmatrix} 2 & -1 & 1 \\ -5 & 2 & 3 \\ 0 & 4 & 0 \end{bmatrix} \qquad \blacksquare$$

In fact, if A is any $m \times n$ matrix and B is any $n \times p$ matrix, then the following equations hold.

$$AI_n = A, \qquad I_n B = B.$$

For an $n \times n$ matrix D, there is sometimes an $n \times n$ matrix C with the property that the equations

$$CD = I_n, \qquad DC = I_n$$

hold. Such a matrix C is said to be the **inverse of D,** and C is usually denoted by D^{-1}. The preceding equations and notation are analogous to properties and notation for the real numbers. If a is any nonzero real number, then we have

$$a^{-1} \cdot a = 1, \qquad a \cdot a^{-1} = 1$$

where a^{-1} is the reciprocal of a. For

$$A = \begin{bmatrix} 2 & 3 \\ -1 & 4 \end{bmatrix}$$

we verify that the inverse is

$$A^{-1} = \begin{bmatrix} \frac{4}{11} & \frac{-3}{11} \\ \frac{1}{11} & \frac{2}{11} \end{bmatrix}.$$

We compute $A^{-1}A$ and AA^{-1}.

$$A^{-1}A = \begin{bmatrix} \frac{4}{11} & \frac{-3}{11} \\ \frac{1}{11} & \frac{2}{11} \end{bmatrix}\begin{bmatrix} 2 & 3 \\ -1 & 4 \end{bmatrix} = \begin{bmatrix} 1 & 0 \\ 0 & 1 \end{bmatrix}$$

$$AA^{-1} = \begin{bmatrix} 2 & 3 \\ -1 & 4 \end{bmatrix}\begin{bmatrix} \frac{4}{11} & \frac{-3}{11} \\ \frac{1}{11} & \frac{2}{11} \end{bmatrix} = \begin{bmatrix} 1 & 0 \\ 0 & 1 \end{bmatrix}$$

We emphasize that **not all square matrices have an inverse. Furthermore, any matrix that is not square does not have an inverse.** A matrix that has an inverse is said to be **invertible.** An example of a square matrix that is not invertible is

$$A = \begin{bmatrix} 0 & 0 \\ 1 & 0 \end{bmatrix}.$$

Note that

$$A^2 = AA = \begin{bmatrix} 0 & 0 \\ 1 & 0 \end{bmatrix} \begin{bmatrix} 0 & 0 \\ 1 & 0 \end{bmatrix} = \begin{bmatrix} 0 & 0 \\ 0 & 0 \end{bmatrix}.$$

If A had an inverse, then the following false equation would hold.

$$\begin{bmatrix} 0 & 0 \\ 0 & 0 \end{bmatrix} = A^{-1} \begin{bmatrix} 0 & 0 \\ 0 & 0 \end{bmatrix} = A^{-1}(AA) = (A^{-1}A)A = I_n A = \begin{bmatrix} 0 & 0 \\ 1 & 0 \end{bmatrix}$$

The remainder of this section will be spent explaining a technique for finding the inverse of a given matrix. The technique uses the elementary row operations defined in Section 3.2. When we are solving a linear system by using the elementary row operations on the augmented matrix, we continue until we have zeroes below the main diagonal. A matrix with zeroes below the main diagonal is said to be **upper triangular.** If all the entries in a matrix above the main diagonal are zeroes, then the matrix is said to be **lower triangular.** A matrix that is both upper triangular and lower triangular is said to be **diagonal.** Clearly I_n is a diagonal matrix.

In order to discover the method for finding an inverse matrix, we make the following observation. If B is the inverse matrix for a given matrix A, then multiplying A by the first column of B must give the first column of I_n. We have

$$\begin{bmatrix} a_{11} & \cdots & a_{1n} \\ \vdots & & \vdots \\ a_{n1} & \cdots & a_{nn} \end{bmatrix} \begin{bmatrix} b_{11} \\ \vdots \\ b_{n1} \end{bmatrix} = \begin{bmatrix} 1 \\ 0 \\ \vdots \\ 0 \end{bmatrix}.$$

For a given matrix A, we can use elementary row operations, as in Gaussian elimination, to transform the augmented matrix associated with the preceding matrix equation (or the equivalent system) into

$$\begin{bmatrix} 1 & \cdots & 0 & | & b_{11} \\ \vdots & & \vdots & | & \vdots \\ 0 & \cdots & 1 & | & b_{n1} \end{bmatrix}.$$

In the same way, multiplying A by the second column of B must give the second column of I_n, and so on for each of the n columns. It can be verified that in each case **we use the same sequence of elementary row operations.** This means that, instead of augmenting the matrix A by one column of I_n each time and doing n Gaussian eliminations, we can take a shortcut. We augment A with all of I_n at once, and perform only one elimination sequence. At the end of the sequence, A will have been reduced to I_n, and the inverse matrix B will appear on the right.

EXAMPLE 2

Find the inverse of $A = \begin{bmatrix} 2 & 3 \\ -1 & 4 \end{bmatrix}$.

Solution

First, let's see how each of the columns of the inverse matrix would be found separately.

To find first column	To find second column
$\begin{bmatrix} 2 & 3 & \vert & 1 \\ -1 & 4 & \vert & 0 \end{bmatrix}$	$\begin{bmatrix} 2 & 3 & \vert & 0 \\ -1 & 4 & \vert & 1 \end{bmatrix}$
$R_1 + R_2 \quad \begin{bmatrix} 1 & 7 & \vert & 1 \\ -1 & 4 & \vert & 0 \end{bmatrix}$	$R_1 + R_2 \quad \begin{bmatrix} 1 & 7 & \vert & 1 \\ -1 & 4 & \vert & 1 \end{bmatrix}$
$R_2 + R_1 \quad \begin{bmatrix} 1 & 7 & \vert & 1 \\ 0 & 11 & \vert & 1 \end{bmatrix}$	$R_2 + R_1 \quad \begin{bmatrix} 1 & 7 & \vert & 1 \\ 0 & 11 & \vert & 2 \end{bmatrix}$
$\frac{1}{11} R_2 \quad \begin{bmatrix} 1 & 7 & \vert & 1 \\ 0 & 1 & \vert & \frac{1}{11} \end{bmatrix}$	$\frac{1}{11} R_2 \quad \begin{bmatrix} 1 & 7 & \vert & 1 \\ 0 & 1 & \vert & \frac{2}{11} \end{bmatrix}$
$R_1 - 7R_2 \quad \begin{bmatrix} 1 & 0 & \vert & \frac{4}{11} \\ 0 & 1 & \vert & \frac{1}{11} \end{bmatrix}$	$R_1 - 7R_2 \quad \begin{bmatrix} 1 & 0 & \vert & \frac{-3}{11} \\ 0 & 1 & \vert & \frac{2}{11} \end{bmatrix}$

Using the shortcut, we augment A with both columns of I_2 and perform the same sequence of elementary row operations:

$$\overset{A}{\overbrace{\begin{bmatrix} 2 & 3 \\ -1 & 4 \end{bmatrix}}} \overset{I_2}{\overbrace{\begin{bmatrix} 1 & 0 \\ 0 & 1 \end{bmatrix}}} \rightarrow \begin{bmatrix} 1 & 7 & \vert & 1 & 1 \\ -1 & 4 & \vert & 0 & 1 \end{bmatrix} \rightarrow \begin{bmatrix} 1 & 7 & \vert & 1 & 1 \\ 0 & 11 & \vert & 1 & 2 \end{bmatrix} \rightarrow \begin{bmatrix} 1 & 7 & \vert & 1 & 1 \\ 0 & 1 & \vert & \frac{1}{11} & \frac{2}{11} \end{bmatrix} \rightarrow \begin{bmatrix} 1 & 0 & \vert & \frac{4}{11} & \frac{-3}{11} \\ 0 & 1 & \vert & \frac{1}{11} & \frac{2}{11} \end{bmatrix}$$

$$\underbrace{}_{I_2} \quad \underbrace{}_{B}$$

Since the left half of the augmented matrix is I_2, we are done. The right half is B, the inverse of A, as we verified earlier in this section. ∎

Now we demonstrate the process on a 3×3 matrix.

EXAMPLE 3

Determine A^{-1} where

$$A = \begin{bmatrix} 1 & -1 & 0 \\ 0 & 1 & 2 \\ 2 & 1 & -1 \end{bmatrix}$$

Solution

The first part of the process is the same as if we were transforming the augmented matrix for a linear system. We get a 1 as the first entry in the first row, and we get 0's as the other entries in the first column.

$$\begin{bmatrix} 1 & -1 & 0 & | & 1 & 0 & 0 \\ 0 & 1 & 2 & | & 0 & 1 & 0 \\ 2 & 1 & -1 & | & 0 & 0 & 1 \end{bmatrix}$$

$$R_3 - 2R_1 \quad \begin{bmatrix} 1 & -1 & 0 & | & 1 & 0 & 0 \\ 0 & 1 & 2 & | & 0 & 1 & 0 \\ 0 & 3 & -1 & | & -2 & 0 & 1 \end{bmatrix}$$

Then we get 1 as the second entry in the second row and 0's as the other entries in the second column.

$$\begin{matrix} R_1 + R_2 \\ \\ R_3 - 3R_2 \end{matrix} \quad \begin{bmatrix} 1 & 0 & 2 & | & 1 & 1 & 0 \\ 0 & 1 & 2 & | & 0 & 1 & 0 \\ 0 & 0 & -7 & | & -2 & -3 & 1 \end{bmatrix}$$

Now we get 1 as the third entry in the third row and 0's as the other entries in the third column.

$$-\frac{1}{7}R_3 \quad \begin{bmatrix} 1 & 0 & 2 & | & 1 & 1 & 0 \\ 0 & 1 & 2 & | & 0 & 1 & 0 \\ 0 & 0 & 1 & | & \frac{2}{7} & \frac{3}{7} & -\frac{1}{7} \end{bmatrix}$$

$$\begin{matrix} R_1 - 2R_3 \\ R_2 - 2R_3 \end{matrix} \quad \begin{bmatrix} 1 & 0 & 0 & | & \frac{3}{7} & \frac{1}{7} & \frac{2}{7} \\ 0 & 1 & 0 & | & -\frac{4}{7} & \frac{1}{7} & \frac{2}{7} \\ 0 & 0 & 1 & | & \frac{2}{7} & \frac{3}{7} & -\frac{1}{7} \end{bmatrix}$$

Since the last matrix on the left is I_3, the process is complete and the desired inverse is the last matrix on the right. For brevity, we write

$$\begin{bmatrix} 1 & -1 & 0 \\ 0 & 1 & 2 \\ 2 & 1 & -1 \end{bmatrix}^{-1} = \begin{bmatrix} \frac{3}{7} & \frac{1}{7} & \frac{2}{7} \\ -\frac{4}{7} & \frac{1}{7} & \frac{2}{7} \\ \frac{2}{7} & \frac{3}{7} & -\frac{1}{7} \end{bmatrix} \quad \blacksquare$$

In some problems there are very few ways that the elementary row operations can be used. The following example demonstrates this.

EXAMPLE 4

Determine A^{-1} where

$$A = \begin{bmatrix} 0 & 0 & -1 \\ 0 & -1 & 0 \\ 1 & 0 & 0 \end{bmatrix}.$$

Solution

There is only one elementary row operation that results in a 1 as the first entry in the first row.

$$\begin{bmatrix} 0 & 0 & -1 & | & 1 & 0 & 0 \\ 0 & -1 & 0 & | & 0 & 1 & 0 \\ 1 & 0 & 0 & | & 0 & 0 & 1 \end{bmatrix}$$

Interchange R_1 and R_3.

$$\left[\begin{array}{ccc|ccc} 1 & 0 & 0 & 0 & 0 & 1 \\ 0 & -1 & 0 & 0 & 1 & 0 \\ 0 & 0 & -1 & 1 & 0 & 0 \end{array}\right]$$

$$\begin{array}{c} \\ (-1)R_2 \\ (-1)R_3 \end{array} \left[\begin{array}{ccc|ccc} 1 & 0 & 0 & 0 & 0 & 1 \\ 0 & 1 & 0 & 0 & -1 & 0 \\ 0 & 0 & 1 & -1 & 0 & 0 \end{array}\right]$$

We are done, and

$$\left[\begin{array}{ccc} 0 & 0 & -1 \\ 0 & -1 & 0 \\ 1 & 0 & 0 \end{array}\right]^{-1} = \left[\begin{array}{ccc} 0 & 0 & 1 \\ 0 & -1 & 0 \\ -1 & 0 & 0 \end{array}\right]. \quad \blacksquare$$

It is curious to see how the process works when it is applied to a matrix that does not have an inverse. We demonstrate this phenomenon in the next example.

EXAMPLE 5

Try to determine A^{-1} where

$$A = \left[\begin{array}{cc} 0 & 0 \\ 1 & 0 \end{array}\right].$$

Solution

$$\left[\begin{array}{cc|cc} 0 & 0 & 1 & 0 \\ 1 & 0 & 0 & 1 \end{array}\right]$$

Interchange R_1 and R_2.

$$\left[\begin{array}{cc|cc} 1 & 0 & 0 & 1 \\ 0 & 0 & 1 & 0 \end{array}\right]$$

Now the process comes to a halt, because there is no elementary row operation that results in the second entry in the second row becoming 1. The conclusion is that

$$\left[\begin{array}{cc} 0 & 0 \\ 1 & 0 \end{array}\right]$$

does not have an inverse. $\quad\blacksquare$

EXAMPLE 6

Determine if A is invertible where

$$A = \left[\begin{array}{ccc} 1 & 2 & 4 \\ 0 & 1 & 2 \\ 1 & 1 & 2 \end{array}\right].$$

Solution

We carry out the process indicated in detail in Example 3. We continue to use our standard abbreviations for the elementary row operations.

$$\left[\begin{array}{ccc|ccc} 1 & 2 & 4 & 1 & 0 & 0 \\ 0 & 1 & 2 & 0 & 1 & 0 \\ 1 & 1 & 2 & 0 & 0 & 1 \end{array}\right]$$

$$\begin{array}{c} \\ \\ R_3 - R_1 \end{array} \left[\begin{array}{ccc|ccc} 1 & 2 & 4 & 1 & 0 & 0 \\ 0 & 1 & 2 & 0 & 1 & 0 \\ 0 & -1 & -2 & -1 & 0 & 1 \end{array}\right]$$

$$\begin{array}{c} R_1 - 2R_2 \\ \\ R_3 + R_2 \end{array} \left[\begin{array}{ccc|ccc} 1 & 0 & 0 & 1 & -2 & 0 \\ 0 & 1 & 2 & 0 & 1 & 0 \\ 0 & 0 & 0 & -1 & 1 & 1 \end{array}\right]$$

There is no elementary row operation that results in having 1 as the third entry in the third row without changing one of the 0's in the second column to a nonzero entry. Thus, the process has ended. It is not possible to transform the given matrix into I_3 by elementary row operations. Thus, the given matrix is not invertible. ∎

One method for solving a matrix equation involves finding the inverse of the coefficient matrix, as we indicate in the next example.

EXAMPLE 7

Solve the matrix equation below by finding the inverse of the coefficient matrix.

$$\left[\begin{array}{ccc} 1 & 2 & 3 \\ 3 & 1 & 1 \\ 2 & 3 & 2 \end{array}\right] \left[\begin{array}{c} x_1 \\ x_2 \\ x_3 \end{array}\right] = \left[\begin{array}{c} 160 \\ 140 \\ 200 \end{array}\right]$$

Solution

We carry out the process described in detail in Example 3.

$$\left[\begin{array}{ccc|ccc} 1 & 2 & 3 & 1 & 0 & 0 \\ 3 & 1 & 1 & 0 & 1 & 0 \\ 2 & 3 & 2 & 0 & 0 & 1 \end{array}\right]$$

$$\begin{array}{c} R_2 - 3R_1 \\ R_3 - 2R_1 \end{array} \left[\begin{array}{ccc|ccc} 1 & 2 & 3 & 1 & 0 & 0 \\ 0 & -5 & -8 & -3 & 1 & 0 \\ 0 & -1 & -4 & -2 & 0 & 1 \end{array}\right]$$

$$\begin{array}{c} \text{Interchange} \\ R_2 \text{ and } R_3; \\ \text{then compute} \\ (-1)R_2 \end{array} \left[\begin{array}{ccc|ccc} 1 & 2 & 3 & 1 & 0 & 0 \\ 0 & 1 & 4 & 2 & 0 & -1 \\ 0 & -5 & -8 & -3 & 1 & 0 \end{array}\right]$$

$$R_1 - 2R_2 \qquad \begin{bmatrix} 1 & 0 & -5 & -3 & 0 & 2 \\ 0 & 1 & 4 & 2 & 0 & -1 \\ 0 & 0 & 12 & 7 & 1 & -5 \end{bmatrix}$$

$$R_3 + 5R_2$$

$$\tfrac{1}{12} R_3 \qquad \begin{bmatrix} 1 & 0 & -5 & -3 & 0 & 2 \\ 0 & 1 & 4 & 2 & 0 & -1 \\ 0 & 0 & 1 & \frac{7}{12} & \frac{1}{12} & -\frac{5}{12} \end{bmatrix}$$

$$R_1 + 5R_3 \qquad \begin{bmatrix} 1 & 0 & 0 & -\frac{1}{12} & \frac{5}{12} & -\frac{1}{12} \\ 0 & 1 & 0 & -\frac{1}{3} & -\frac{1}{3} & \frac{2}{3} \\ 0 & 0 & 1 & \frac{7}{12} & \frac{1}{12} & -\frac{5}{12} \end{bmatrix}$$

$$R_2 - 4R_3$$

Since the last matrix on the left is I_3, the last matrix on the right is the desired inverse. We have

$$\begin{bmatrix} 1 & 2 & 3 \\ 3 & 1 & 1 \\ 2 & 3 & 2 \end{bmatrix}^{-1} = \begin{bmatrix} -\frac{1}{12} & \frac{5}{12} & -\frac{1}{12} \\ -\frac{1}{3} & -\frac{1}{3} & \frac{2}{3} \\ \frac{7}{12} & \frac{1}{12} & -\frac{5}{12} \end{bmatrix}$$

The column matrix of unknowns equals the product given below; we explain and justify this in the first paragraph after this example.

$$\begin{bmatrix} x_1 \\ x_2 \\ x_3 \end{bmatrix} = \begin{bmatrix} -\frac{1}{12} & \frac{5}{12} & -\frac{1}{12} \\ -\frac{1}{3} & -\frac{1}{3} & \frac{2}{3} \\ \frac{7}{12} & \frac{1}{12} & -\frac{5}{12} \end{bmatrix} \begin{bmatrix} 160 \\ 140 \\ 200 \end{bmatrix} = \begin{bmatrix} 28\frac{1}{3} \\ 33\frac{1}{3} \\ 21\frac{1}{3} \end{bmatrix}.$$

This is the solution of the matrix equation. ■

In general, if we want to solve the matrix equation

$$AX = B$$

then we apply A^{-1} to both sides of the equation.

$$X = I_n X = (A^{-1}A)X = A^{-1}(AX) = A^{-1}B$$

This gives the column matrix of unknowns as the product $A^{-1}B$. This is precisely the product computed in Example 7.

The matrix equation in Example 7 arose in the **production scheduling** problem given in Example 4 of Section 3.3. The production lines described in Example 4 would probably be available different numbers of hours each month. One month the lines are available for 160, 140, and 200 hours, respectively, and we have determined how many units of products A, B, and C can be produced. The next month the lines might be available for 180, 160, and 210 hours, respectively. If we had to solve the preceding month's scheduling problem by the augmented matrix approach of Example 4, Section 3.3, then we must begin again with the new matrix equation

$$\begin{bmatrix} 1 & 2 & 3 \\ 3 & 1 & 1 \\ 2 & 3 & 2 \end{bmatrix} \begin{bmatrix} x_1 \\ x_2 \\ x_3 \end{bmatrix} = \begin{bmatrix} 180 \\ 160 \\ 210 \end{bmatrix}.$$

If we solved the preceding month's scheduling problem by finding the inverse of the coefficient matrix as in Example 6 of this section, however, we can easily obtain the solution of this new matrix equation with the computation that follows.

$$\begin{bmatrix} x_1 \\ x_2 \\ x_3 \end{bmatrix} = \begin{bmatrix} 1 & 2 & 3 \\ 3 & 1 & 1 \\ 2 & 3 & 2 \end{bmatrix}^{-1} \begin{bmatrix} 180 \\ 160 \\ 210 \end{bmatrix} = \begin{bmatrix} -\frac{1}{12} & \frac{5}{12} & -\frac{1}{12} \\ -\frac{1}{3} & -\frac{1}{3} & \frac{2}{3} \\ \frac{7}{12} & \frac{1}{12} & -\frac{5}{12} \end{bmatrix} \begin{bmatrix} 180 \\ 160 \\ 210 \end{bmatrix} = \begin{bmatrix} 33\frac{1}{3} \\ 26\frac{2}{3} \\ 30\frac{5}{6} \end{bmatrix}.$$

In many situations, the coefficient matrix is exactly the same for many different linear systems that must be solved. Examples 4 and 5 of Section 3.3 give two such situations. The systems are different because the constants on the right sides of the equations change. If we solve the matrix equation by finding the inverse of the coefficient matrix, we can avoid starting from the beginning to solve each new linear system.

In Section 4.3, we shall discover a convenient way to determine whether a given square matrix has an inverse. In that section we define the determinant of a square matrix. If the determinant is not zero, then the matrix is invertible, and if the determinant is zero, then the matrix is not invertible. By computing the determinant of a matrix, we can decide whether it has an inverse before starting to use the elementary row operations.

Exercises 3.4

In each of the problems that follow, determine if the two matrices given are inverses for each other.

1 $\begin{bmatrix} 1 & 0 \\ -1 & 1 \end{bmatrix}$ and $\begin{bmatrix} 1 & 0 \\ 1 & 1 \end{bmatrix}$

2 $\begin{bmatrix} 0 & 1 \\ -1 & 1 \end{bmatrix}$ and $\begin{bmatrix} 1 & -1 \\ 1 & 0 \end{bmatrix}$

3 $\begin{bmatrix} 2 & 1 \\ -2 & 3 \end{bmatrix}$ and $\begin{bmatrix} \frac{3}{8} & -\frac{1}{8} \\ \frac{1}{4} & \frac{1}{4} \end{bmatrix}$

4 $\begin{bmatrix} 3 & 2 \\ -1 & 4 \end{bmatrix}$ and $\begin{bmatrix} \frac{2}{7} & -\frac{1}{7} \\ \frac{1}{14} & \frac{3}{14} \end{bmatrix}$

5 $\begin{bmatrix} 1 & -2 \\ -1 & 3 \end{bmatrix}$ and $\begin{bmatrix} 3 & 2 \\ 1 & -1 \end{bmatrix}$

6 $\begin{bmatrix} 2 & -2 \\ 2 & 2 \end{bmatrix}$ and $\begin{bmatrix} \frac{1}{4} & -\frac{1}{4} \\ \frac{1}{4} & \frac{1}{4} \end{bmatrix}$

7 $\begin{bmatrix} 1 & 0 & 1 \\ 0 & 1 & 0 \\ 1 & 0 & 1 \end{bmatrix}$ and $\begin{bmatrix} 1 & 0 & -1 \\ 0 & 1 & 0 \\ 0 & 0 & 1 \end{bmatrix}$

8 $\begin{bmatrix} 1 & -1 & 1 \\ 0 & 1 & 0 \\ -1 & 1 & -1 \end{bmatrix}$ and $\begin{bmatrix} 1 & 1 & 1 \\ 1 & 1 & 1 \\ 1 & 1 & 1 \end{bmatrix}$

9 $\begin{bmatrix} 1 & 2 & -2 \\ 1 & 3 & -1 \\ 3 & 2 & 0 \end{bmatrix}$ and $\begin{bmatrix} .2 & -.4 & .4 \\ -.3 & .6 & -.1 \\ -.7 & .4 & .1 \end{bmatrix}$

10 $\begin{bmatrix} 2 & 1 & -1 \\ 0 & 2 & 0 \\ -1 & 1 & 0 \end{bmatrix}$ and $\begin{bmatrix} 0 & .5 & -1 \\ 0 & .5 & 0 \\ -1 & 1.5 & -2 \end{bmatrix}$

In each of the following problems, determine A^{-1}, if A is invertible.

11 $\begin{bmatrix} 2 & 0 \\ 0 & 3 \end{bmatrix}$ 12 $\begin{bmatrix} 5 & 0 \\ 0 & 4 \end{bmatrix}$

13 $\begin{bmatrix} 2 & 1 \\ 0 & 0 \end{bmatrix}$ 14 $\begin{bmatrix} 0 & 0 \\ 3 & 1 \end{bmatrix}$

15 $\begin{bmatrix} 1 & 0 \\ 1 & 1 \end{bmatrix}$ 16 $\begin{bmatrix} 1 & 2 \\ 0 & 3 \end{bmatrix}$

17 $\begin{bmatrix} 1 & 1 \\ 2 & 2 \end{bmatrix}$ 18 $\begin{bmatrix} 3 & 3 \\ -1 & -1 \end{bmatrix}$

19 $\begin{bmatrix} 1 & 0 \\ 0 & 1 \end{bmatrix}$ 20 $\begin{bmatrix} -1 & 0 \\ 0 & -1 \end{bmatrix}$

21 $\begin{bmatrix} 2 & -1 \\ 1 & 3 \end{bmatrix}$ 22 $\begin{bmatrix} 1 & 5 \\ 0 & 2 \end{bmatrix}$

23 $\begin{bmatrix} 1 & 0 & 0 \\ 0 & 2 & 0 \\ 0 & 0 & 3 \end{bmatrix}$ 24 $\begin{bmatrix} 9 & 0 & 0 \\ 0 & 5 & 0 \\ 0 & 0 & 2 \end{bmatrix}$

25 $\begin{bmatrix} 1 & -1 & 2 \\ 2 & 1 & 3 \\ 0 & 0 & 0 \end{bmatrix}$ 26 $\begin{bmatrix} 4 & 1 & 3 \\ 0 & 0 & 0 \\ 2 & -2 & 5 \end{bmatrix}$

27 $\begin{bmatrix} 1 & 2 & -1 \\ 0 & 3 & -2 \\ 0 & 0 & -1 \end{bmatrix}$ 28 $\begin{bmatrix} 3 & 0 & 0 \\ 1 & 5 & 0 \\ 4 & -1 & 2 \end{bmatrix}$

29 $\begin{bmatrix} 1 & 1 & 1 \\ 1 & -1 & 1 \\ 1 & 0 & 1 \end{bmatrix}$ 30 $\begin{bmatrix} 2 & 1 & 2 \\ -1 & 3 & -1 \\ 1 & 4 & 1 \end{bmatrix}$

Solve each matrix equation by finding the inverse of the coefficient matrix.

31 $\begin{bmatrix} 1 & 2 \\ 0 & 3 \end{bmatrix} \begin{bmatrix} x_1 \\ x_2 \end{bmatrix} = \begin{bmatrix} 4 \\ 6 \end{bmatrix}$

32 $\begin{bmatrix} 1 & 4 \\ 2 & 2 \end{bmatrix} \begin{bmatrix} x_1 \\ x_2 \end{bmatrix} = \begin{bmatrix} 9 \\ 7 \end{bmatrix}$

33 $\begin{bmatrix} 5 & 8 \\ 9 & 3 \end{bmatrix} \begin{bmatrix} x \\ y \end{bmatrix} = \begin{bmatrix} 20 \\ 30 \end{bmatrix}$

34 $\begin{bmatrix} 3 & 10 \\ 8 & 7 \end{bmatrix} \begin{bmatrix} x \\ y \end{bmatrix} = \begin{bmatrix} 32 \\ 26 \end{bmatrix}$

35 $\begin{bmatrix} 1 & 5 & 4 \\ 3 & 2 & 2 \\ 5 & 4 & 2 \end{bmatrix} \begin{bmatrix} x_1 \\ x_2 \\ x_3 \end{bmatrix} = \begin{bmatrix} 200 \\ 140 \\ 180 \end{bmatrix}$

36 $\begin{bmatrix} 4 & 3 & 0 \\ 0 & 6 & 4 \\ 1 & 2 & 2 \end{bmatrix} \begin{bmatrix} x_1 \\ x_2 \\ x_3 \end{bmatrix} = \begin{bmatrix} 60 \\ 90 \\ 40 \end{bmatrix}$

Review of Terms

Important Mathematical Terms

system of equations, *p. 110*	elementary row operations, *p. 125*
solution to a system, *p. 110*	product matrix, *p. 131*
solved, *p. 110*	column matrix, *p. 132*
linear system, *p. 121*	row matrix, *p. 132*
elimination method, *p. 115*	associative property, *p. 132*
substitution method, *p. 115*	commutative property, *p. 132*
coefficients, *p. 121*	equal matrices, *p. 133*
$m \times n$, *p. 121*	main diagonal, *p. 139*
Gaussian elimination	identity matrix, *p. 139*
method, *p. 122*	inverse matrix, *p. 140*
matrix, *p. 125*	invertible, *p. 140*
entries, *p. 125*	upper triangular, *p. 141*
coefficient matrix, *p. 125*	lower triangular, *p. 141*
augmented matrix, *p. 125*	diagonal, *p. 141*
order, *p. 125*	

Important Terms from the Applications

demand function, *p. 116*	market demand, *p. 116*
supply function, *p. 116*	market price, *p. 116*
equilibrium point, *p. 116*	production scheduling, *p. 146*

Review Problems

Solve each system by substitution. In the first two problems, graph each equation and use the graphs to check your solution.

1 $y = 2x - 1$
$y = -x + 1$

2 $y = \dfrac{3}{5}x + \dfrac{2}{5}$
$y = 5x - 6$

$$3 \quad y = -3x + 4 \qquad\qquad 4 \quad 3x - 9y = 11$$
$$5x - 7y + 8 = 0 \qquad\qquad\qquad x = 2y - 1$$

Solve each system by using the elementary row operations on the augmented matrix.

$$5 \quad x + y = 3 \qquad\qquad 6 \quad .5x + .1y = 4$$
$$x - y = 2 \qquad\qquad\qquad .1x - .3y = -2$$

$$7 \quad 3v - 5w = 2 \qquad\qquad 8 \quad \frac{1}{3}m - \frac{1}{2}n = 1$$
$$2v + 4w = -3$$
$$\frac{5}{6}m + \frac{2}{3}n = 3$$

$$9 \quad x_1 + x_2 - 4x_3 = 0 \qquad\qquad 10 \quad x + 3y - 4z = 5$$
$$x_1 - x_2 - x_3 = -3 \qquad\qquad\qquad 5y - 7z = 2$$
$$x_1 - x_2 + 2x_3 = 2 \qquad\qquad\qquad x + z = -1$$
$$\qquad\qquad\qquad\qquad\qquad\qquad x - 2y = 3$$

$$11 \quad 3r - s + 2t + u = 0 \qquad 12 \quad w + x - 3y + 2z = -5$$
$$r + 4s - 5t - u = 2 \qquad\qquad w - 2x + y + 5z = 6$$
$$r + s + t + 2u = 5 \qquad\qquad w + x - y - z = 1$$
$$\qquad\qquad\qquad\qquad\qquad\qquad x + 6y + 3z = 4$$

Compute each of the following products when it is possible.

$$13 \quad \begin{bmatrix} 1 & 0 & 2 \\ 3 & 1 & -1 \end{bmatrix} \begin{bmatrix} 5 & 3 \\ -2 & 1 \end{bmatrix}$$

$$14 \quad \begin{bmatrix} 4 \\ 6 \end{bmatrix} \begin{bmatrix} 1 & 5 & 7 & 9 \end{bmatrix}$$

$$15 \quad \begin{bmatrix} 0 & 3 & 2 \end{bmatrix} \begin{bmatrix} 1 \\ -1 \\ 4 \end{bmatrix}$$

$$16 \quad \begin{bmatrix} 1 & -1 \\ -1 & 2 \end{bmatrix} \begin{bmatrix} 3 & 0 \\ 5 & 2 \end{bmatrix}$$

$$17 \quad \begin{bmatrix} 0 & 1 & 1 \\ 1 & 3 & -3 \end{bmatrix} \begin{bmatrix} 2 & 1 & -1 \\ 4 & 3 & 0 \\ 0 & -1 & 5 \end{bmatrix}$$

$$18 \quad \begin{bmatrix} -3 & -4 & 2 \\ 1 & -1 & 5 \\ 0 & 3 & 1 \end{bmatrix} \begin{bmatrix} 1 \\ 2 \\ -1 \end{bmatrix}$$

$$19 \quad \begin{bmatrix} 0 & 0 & 0 & -1 \\ 0 & 0 & -1 & 0 \\ 0 & -1 & 0 & 0 \\ -1 & 0 & 0 & 0 \end{bmatrix} \begin{bmatrix} 0 & 0 & 0 & -1 \\ 0 & 0 & -1 & 0 \\ 0 & -1 & 0 & 0 \\ -1 & 0 & 0 & 0 \end{bmatrix}$$

For each system, write a matrix equation that has the same solution.

20 $4x_1 - 5x_2 = 9$
    $13x_1 + 17x_2 = 23$

21 $9x - 3y + 2z = 1$
    $2x + 25y - 7z = 35$
    $12x - 13y + 43z = 81$

In each of the following problems, determine A^{-1}, if A is invertible.

22 $\begin{bmatrix} 33 & 0 \\ 0 & 22 \end{bmatrix}$

23 $\begin{bmatrix} 18 & 0 & 0 \\ 0 & 5 & 0 \\ 0 & 0 & 43 \end{bmatrix}$

24 $\begin{bmatrix} 5 & 0 & 3 \\ 2 & 0 & 1 \\ -1 & 0 & 2 \end{bmatrix}$

25 $\begin{bmatrix} 8 & 3 \\ 1 & 2 \end{bmatrix}$

26 $\begin{bmatrix} 1 & 1 & 0 \\ 0 & 1 & 1 \\ 1 & 0 & 1 \end{bmatrix}$

27 $\begin{bmatrix} 1 & 3 & -2 \\ 2 & -1 & 1 \\ -1 & 3 & 4 \end{bmatrix}$

Solve each of the matrix equations that follow by finding the inverse of the coefficient matrix.

28 $\begin{bmatrix} 1 & 4 \\ 5 & 2 \end{bmatrix} \begin{bmatrix} x \\ y \end{bmatrix} = \begin{bmatrix} 34 \\ 22 \end{bmatrix}$

29 $\begin{bmatrix} 3 & 5 & 2 \\ 1 & 3 & 6 \\ 2 & 1 & 5 \end{bmatrix} \begin{bmatrix} x_1 \\ x_2 \\ x_3 \end{bmatrix} = \begin{bmatrix} 100 \\ 50 \\ 200 \end{bmatrix}$

Solve each of the following.

30 Suppose that investment A had a rate of return of 16% in 1982 and investment B had a rate of return of 18%. In 1982, the total return on the two investments was $4300. In 1983, each investment had a rate of return of 14% and the total return was $3500. Determine the amount put into each investment.

31 The demand function for a certain product is $p = D(x) = 12 - .2x$ and the supply function is $p = S(x) = .5x + 4$. Determine the market demand and the market price for this product.

32 A vending machine's coin receptacle is filled with nickels and quarters. If there are 440 coins and their worth is $57.20, how many of the coins are quarters?

33 The workers in a certain office ordered hot dogs, hamburgers, and soft drinks from a restaurant. Hot dogs cost 75¢ each, hamburgers cost $1 each, and soft drinks cost 50¢ each. There were 15 items ordered. If twice as many hamburgers as hot dogs were ordered and the total bill was $11.25, how many hamburgers were ordered?

Social Science Applications

Cryptography (The Science of Codes)

We describe a simple process for constructing a code. First, replace each letter of the alphabet with a corresponding number from 1 through 26. For example, assign a number to each letter moving through the alphabet in alphabetical order.

A	B	C	D	E	F	G	H	I	J	K	L	M
↕	↕	↕	↕	↕	↕	↕	↕	↕	↕	↕	↕	↕
1	2	3	4	5	6	7	8	9	10	11	12	13

N	O	P	Q	R	S	T	U	V	W	X	Y	Z
↕	↕	↕	↕	↕	↕	↕	↕	↕	↕	↕	↕	↕
14	15	16	17	18	19	20	21	22	23	24	25	26

Take a message, such as "attack," and group the letters in bunches of three, working from left to right.

att ack

Replace each letter with the corresponding number.

1 20 20   1 3 11

Choose some invertible 3×3 matrix, such as

$$\begin{bmatrix} 1 & 1 & 0 \\ 1 & -1 & 1 \\ 0 & 1 & 1 \end{bmatrix}$$

Apply the matrix to each bunch of three numbers regarded as a row vector.

$$(1\ 20\ 20) \begin{bmatrix} 1 & 1 & 0 \\ 1 & -1 & 1 \\ 0 & 1 & 1 \end{bmatrix} = (21\ 1\ 40)$$

$$(1\ 3\ 11) \begin{bmatrix} 1 & 1 & 0 \\ 1 & -1 & 1 \\ 0 & 1 & 1 \end{bmatrix} = (4\ 9\ 14)$$

The coded message is "21 1 40 4 9 14."

To decode the message, apply the inverse of the given matrix to each bunch of three numbers.

$$(21\ 1\ 40) \begin{bmatrix} \frac{2}{3} & \frac{1}{3} & -\frac{1}{3} \\ \frac{1}{3} & -\frac{1}{3} & \frac{1}{3} \\ -\frac{1}{3} & \frac{1}{3} & \frac{2}{3} \end{bmatrix} = (1\ 20\ 20)$$

$$(4\ 9\ 14) \begin{bmatrix} \frac{2}{3} & \frac{1}{3} & -\frac{1}{3} \\ \frac{1}{3} & -\frac{1}{3} & \frac{1}{3} \\ -\frac{1}{3} & \frac{1}{3} & \frac{2}{3} \end{bmatrix} = (1\ 3\ 11)$$

Reassemble the bunches of three, and replace each number with the corresponding letter.

1 20 20 1 3 11
a t t a c k

1   Use the preceding process to decode the message "39 20 41 29 −13 33 19 −13 22."

2   Use the preceding process to decode the message "14 32 21 17 18 28 27 −1 35."

3   Write the message "all is not lost" in the preceding code.

4   Write the message "now is the time" in the preceding code.

5   *Political Science*   Voting districts 1 and 2 are side-by-side, and together they contain 200,000 black voters and 240,000 white voters. A judge orders that the boundaries be redrawn so that 50% of the voters in district 1 are black and 60% of the voters in district 2 are white. Determine the number of voters that each district must contain.

6   *Public Administration*   A county government has $2.1 million for the pay raises for its 1250 school teachers. The government decides to give its less senior teachers a $1500 raise and its more senior teachers a $1800 raise. How many teachers will get each raise?

7   *Sociology*   A sociologist distributes two questionnaires, A and B, to 1000 respondents. There are 12 questions on questionnaire A and 15 questions on B. The results are tabulated by a computer, and the computer indicates that it graded 13,320 questions. How many questionnaires of each type were tabulated?

8   *Psychology*   Three personality types, known as A, B, and C, have been studied relative to heart disease. A type A person tends to be aggressive, impatient, and hostile; a type B person tends to be dynamic but calm and fond of people; a type C person tends to be easygoing, retiring, and noncompetitive. A psychologist administers tests for patience and for hostility to each of 200 subjects. Experience indicates that the percentage of each personality type failing each test is given by the table

	Personality Type		
	A	B	C
Failing patience test	98%	25%	2%
Failing hostility test	94%	30%	6%

If 90 of the subjects failed the patience test and 92 failed the hostility test, how many of the subjects possess each personality type?

9   *Sociology*   A sociologist uses a questionnaire to study the impact of family status on job satisfaction. He classifies the completed questionnaires according to whether the respondent is (a) single, (b) married with no children, or (c) married with children. The following three questions are asked: (1) Do you enjoy your job? (2) Can you easily meet your essential needs with your income? (3) If you were starting over, would you pursue the same career? The percentages of positive responses to the three questions by group (a) are 80%, 90%, and 85%; for group (b), the percentages of positive responses are 70%, 95%, and 90%; for group (c), the percentages are 50%, 70%, and 80%. Construct a matrix like the ones in Examples 4 and 5 of Section 3.3 to exhibit the preceding data.

CAREER PROFILE
Programmer/Systems Analyst

Do you want to work with computers? Then, regardless of your major, take at least three or four computer science courses. If you seek employment in computer-related work, plan to start as a programmer; nearly everyone starts there! It is common for those who work as programmers to have completed a college major in computer science, but mathematics and science majors and also economics, business, and language majors (English or foreign) with several additional computer science courses also find jobs in this high-demand field.

The largest employers of programmers are manufacturing firms, banks, insurance companies, data processing service organizations, government agencies, and software firms. Since employers who offer responsible positions prefer programmers with experience, work as a student intern or a part-time job with a small company is a good complement to courses of study.

Programmers are assigned to write the detailed step-by-step computer instructions to accomplish tasks described in a more general way by systems analysts who have carefully studied the overall task the computer is to perform. If, for example, a business organization decides that a new inventory system is desired, the organization's managers first will discuss with their systems analysts the exact nature of the system to be designed. The analyst will advise the managers about what is feasible and about the costs and benefits of different options. Once general guidelines for a system have been agreed upon, the systems analyst must determine what data and equipment are needed for computation and must identify the steps to follow in the development of the system.

The individual who writes and corrects programs that accomplish the steps specified by the analyst often holds the title of applications programmer. Assignments range from simple programs that can be written in a few hours to programs that use complex mathematical formulas or many data files and require more than a year of work.

Applications programmers generally have a specialty: they are business-oriented, engineering-oriented, or science-oriented. Another type of specialist, the systems programmer, maintains the general instructions (that is, software) that control the operation of the entire computer system.

Most programmers are college graduates. Employers who use computers for scientific or engineering applications seek programmers with degrees in computer or information science, mathematics, engineering, or the physical sciences. Mathematics courses essential to this preparation include calculus, linear algebra, discrete mathematics (including graph theory), numerical analysis, and simulation.

Some employers who use computers for business applications do not require college degrees. Still, they prefer applications programmers who have had some college courses in data processing and accounting.

Some small organizations do not employ systems analysts to direct the work of their programmers. Instead, workers called programmer-analysts are responsible for both systems analysis and programming.

Almost half of all persons employed as systems analysts have transferred into this occupation from another, primarily from computer programmer. In addition to having the background specified for programmers, analysts need to have familiarity with several programming languages. Courses in computer concepts, systems analysis, and data base management systems are helpful. Systems analysts need to be able to think logically and to organize a large number of tasks simultaneously.

Salaries for programmers and analysts vary considerably from employer to employer, but few in this field start at salaries less than $15,000. Experienced, good programmers and analysts earn annual salaries of over $30,000. The number of jobs for programmers and analysts is expected to grow throughout the 1980s but not as rapidly as in the past. Because good software has become available for purchase, in the future many computer users will decide to buy it rather than to hire programming personnel.

Sources of Additional Information

American Federation of Information Processing Societies, 1815 North Lynn Street, Arlington, VA 22209. (Information about the occupations of programmer and systems analyst are available from this federation.)

The Institute for Certification of Computer Professionals, 35 E. Wacker Drive, Suite 2828, Chicago, Il. 60601. (Information about professional certification of programmers and analysts is available from the institute.)

Occupational Outlook Handbook. Bureau of Labor Statistics, U.S. Department of Labor, Washington, D.C. 20212. (Revised every 2 years, this handbook provides information about job duties, working conditions, level and places of employment, education and training requirements, advancement possibilities, job outlook, and earnings for about 250 occupations. Some of the information for this Career Profile was obtained from this source.)

4

More on Matrices

4.1 Algebraic Operations with Matrices

In Chapter 3, we introduced and developed the theory of matrices in a manner similar to the way Arthur Cayley (1821–1895) presented the original theory. Matrices and matrix multiplication were introduced to facilitate the handling of systems of linear equations. After the usefulness of matrices was established, then matrices were studied as an example of an interesting algebraic system.

Like the real numbers, matrices can be combined by using certain **operations.** Let us define three elementary matrix operations.

Definition Let A and B be matrices with order $m \times n$. Let a_{ij} and b_{ij} denote the entries in the ith row and jth column of A and B, respectively. The **sum $A + B$** is defined to be the $m \times n$ matrix with $a_{ij} + b_{ij}$ in the ith row and jth column. The **difference $A - B$** is defined to be the $m \times n$ matrix with $a_{ij} - b_{ij}$ in the ith row and jth column. For any real number k the **product kA** is defined to be the $m \times n$ matrix with ka_{ij} in the ith row and the jth column.

Note that it is possible to add or subtract two matrices only when each has the same order.

EXAMPLE 1

Determine the sum $A + B$, the difference $A - B$, and the product $5A$ where

$$A = \begin{bmatrix} 1 & -1 & 0 \\ 3 & 2 & -4 \end{bmatrix}, \qquad B = \begin{bmatrix} 0 & 1 & 4 \\ -3 & 5 & 1 \end{bmatrix}$$

Solution

To compute $A + B$, we add corresponding entries in each of the two matrices.

$$A + B = \begin{bmatrix} 1 & -1 & 0 \\ 3 & 2 & -4 \end{bmatrix} + \begin{bmatrix} 0 & 1 & 4 \\ -3 & 5 & 1 \end{bmatrix} = \begin{bmatrix} 1+0 & -1+1 & 0+4 \\ 3-3 & 2+5 & -4+1 \end{bmatrix} = \begin{bmatrix} 1 & 0 & 4 \\ 0 & 7 & -3 \end{bmatrix}$$

To compute $A - B$, we subtract corresponding entries in each of the two matrices.

$$A - B = \begin{bmatrix} 1 & -1 & 0 \\ 3 & 2 & -4 \end{bmatrix} - \begin{bmatrix} 0 & 1 & 4 \\ -3 & 5 & 1 \end{bmatrix} = \begin{bmatrix} 1-0 & -1-1 & 0-4 \\ 3-(-3) & 2-5 & -4-1 \end{bmatrix} = \begin{bmatrix} 1 & -2 & -4 \\ 6 & -3 & -5 \end{bmatrix}$$

To compute $5A$, we multiply each entry in A by 5.

$$5A = 5 \begin{bmatrix} 1 & -1 & 0 \\ 3 & 2 & -4 \end{bmatrix} = \begin{bmatrix} 5 & -5 & 0 \\ 15 & 10 & -20 \end{bmatrix} \blacksquare$$

Multiplication of a matrix by a real number is called **scalar multiplication;** the multiplication of two matrices, as presented in Chapter 3, is called **matrix multiplication.** The basic properties for scalar multiplication, matrix multiplication, addition of matrices, and subtraction of matrices are given below.

Properties of Matrix Operations

In the equations below, A, B, and C are matrices. The order of each matrix (the number of rows and columns) must be such that the indicated operations are defined. The symbol I denotes the identity matrix with an order appropriate for the indicated multiplication.

1. $A(BC) = (AB)C$ Associative property for multiplication

2. $AI = A, \quad IA = A$ Multiplicative identity

3. $A + (B + C) = (A + B) + C$ Associative property for addition

4. $A + B = B + A$ Commutative property

5. $k(A + B) = kA + kB$ First distributive property

6. $A(B + C) = AB + AC$ Second distributive property

7. $(B + C)A = BA + CA$ Third distributive property

8. $A - B = A + (-1)B$

EXAMPLE 2

Verify the associative property of matrix multiplication for the following matrices:

$$A = [-1, \quad 2], \qquad B = \begin{bmatrix} 0 & 1 & 3 \\ -2 & 0 & -1 \end{bmatrix}, \qquad C = \begin{bmatrix} 5 \\ -2 \\ 0 \end{bmatrix}.$$

Solution
Since

$$BC = \begin{bmatrix} 0 & 1 & 3 \\ -2 & 0 & -1 \end{bmatrix} \begin{bmatrix} 5 \\ -2 \\ 0 \end{bmatrix} = \begin{bmatrix} -2 \\ -10 \end{bmatrix},$$

we have

$$A(BC) = [-1 \quad 2] \begin{bmatrix} -2 \\ -10 \end{bmatrix} = [-18].$$

On the other hand, since

$$AB = [-1 \quad 2] \begin{bmatrix} 0 & 1 & 3 \\ -2 & 0 & -1 \end{bmatrix} = [-4 \quad -1 \quad -5],$$

we have

$$(AB)C = [-4 \quad -1 \quad -5] \begin{bmatrix} 5 \\ -2 \\ 0 \end{bmatrix} = [-18].$$

In this case, we have verified that

$$A(BC) = (AB)C. \quad \blacksquare$$

In Section 3.2, we gave examples to show that matrix multiplication is not commutative. It is very easy to verify that matrix addition is commutative. If the entries in the ith row and jth column for A and B are a_{ij} and b_{ij}, respectively, then $a_{ij} + b_{ij}$ is in the ith row and jth column of $A + B$. The entry in the ith row and jth column of $B + A$ is $b_{ij} + a_{ij}$. Since $a_{ij} + b_{ij} = b_{ij} + a_{ij}$, we see that $A + B = B + A$. Most of the eight properties listed earlier have similar verifications.

Just as for real numbers, several operations can be combined in a single expression, as the next example shows.

EXAMPLE 3

Determine the matrix $3A - 2B$ where

$$A = \begin{bmatrix} -1 & 1 & 0 \\ 0 & 2 & -2 \\ 1 & 0 & -1 \end{bmatrix}, \quad B = \begin{bmatrix} 2 & 1 & 3 \\ -2 & 0 & -1 \\ -1 & 1 & 0 \end{bmatrix}.$$

Solution

$$3A - 2B = 3 \begin{bmatrix} -1 & 1 & 0 \\ 0 & 2 & -2 \\ 1 & 0 & -1 \end{bmatrix} - 2 \begin{bmatrix} 2 & 1 & 3 \\ -2 & 0 & -1 \\ -1 & 1 & 0 \end{bmatrix}$$

$$= \begin{bmatrix} -3 & 3 & 0 \\ 0 & 6 & -6 \\ 3 & 0 & -3 \end{bmatrix} - \begin{bmatrix} 4 & 2 & 6 \\ -4 & 0 & -2 \\ -2 & 2 & 0 \end{bmatrix}$$

$$= \begin{bmatrix} -7 & 1 & -6 \\ 4 & 6 & -4 \\ 5 & -2 & -3 \end{bmatrix} \quad \blacksquare$$

The properties of matrix operations allow us to solve many matrix equations with steps analogous to those used to solve elementary equations. We illustrate this in the next two examples.

EXAMPLE 4

Determine the matrix X by solving the matrix equation

$$A + 3X = B + C.$$

Solution

We subtract A from both sides of the equation.

$$3X = B + C - A$$

We do not indicate any order for performing the operations on the right side, since the associative property causes us to get the same answer whether the addition or subtraction is completed first. We multiply both sides by $\frac{1}{3}$, to get the answer

$$X = \frac{1}{3}(B + C - A). \; \blacksquare$$

The next example demonstrates a process that we shall use in the next section.

EXAMPLE 5

Assume that the matrix $A + I$ is invertible and determine X by solving the given matrix equation.

$$AX + X = B$$

Solution

First we introduce the identity matrix, and then we use the distributive property of matrix multiplication.

$$AX + X = B$$
$$AX + IX = B$$
$$(A + I)X = B$$

Now we apply the inverse of $A + I$ to both sides.

$$(A + I)^{-1}(A + I)X = (A + I)^{-1}B$$
$$IX = (A + I)^{-1}B$$
$$X = (A + I)^{-1}B \; \blacksquare$$

In Examples 4 and 5, we solved for the matrix X in terms of the matrices A, B, C, and I, which we assume to be given. When all the entries are given for all of the matrices other than X, then we can find the entries of X. We illustrate this process in Example 6.

EXAMPLE 6

Determine the matrix X by solving the matrix equation below.

$$\begin{bmatrix} 0 & 1 \\ -1 & 0 \end{bmatrix} X + \begin{bmatrix} 2 \\ 3 \end{bmatrix} = \begin{bmatrix} 5 \\ 1 \end{bmatrix}$$

Solution
We get X on one side of the equation with a matrix of numbers on the other side.

$$\begin{bmatrix} 0 & 1 \\ -1 & 0 \end{bmatrix} X + \begin{bmatrix} 2 \\ 3 \end{bmatrix} = \begin{bmatrix} 5 \\ 1 \end{bmatrix}$$

$$\begin{bmatrix} 0 & 1 \\ -1 & 0 \end{bmatrix} X = \begin{bmatrix} 5 \\ 1 \end{bmatrix} - \begin{bmatrix} 2 \\ 3 \end{bmatrix} = \begin{bmatrix} 3 \\ -2 \end{bmatrix}$$

We find the inverse of $\begin{bmatrix} 0 & 1 \\ -1 & 0 \end{bmatrix}$ to be $\begin{bmatrix} 0 & -1 \\ 1 & 0 \end{bmatrix}$ by using the methods of

Section 3.4; then, we apply the inverse to both sides of the matrix equation.

$$\begin{bmatrix} 0 & -1 \\ 1 & 0 \end{bmatrix} \begin{bmatrix} 0 & 1 \\ -1 & 0 \end{bmatrix} X = \begin{bmatrix} 0 & -1 \\ 1 & 0 \end{bmatrix} \begin{bmatrix} 3 \\ -2 \end{bmatrix}$$

$$IX = \begin{bmatrix} 2 \\ 3 \end{bmatrix}$$

$$X = \begin{bmatrix} 2 \\ 3 \end{bmatrix}$$

This is the solution sought. ■

Exercises 4.1

In problems 1 through 10, perform the indicated operations, if possible.

1 $\begin{bmatrix} 3 & 0 \\ 1 & -2 \end{bmatrix} + \begin{bmatrix} 1 & 4 \\ -5 & 0 \end{bmatrix}$

2 $\begin{bmatrix} .5 & .2 \\ 1.4 & 2.1 \end{bmatrix} - \begin{bmatrix} .8 & .6 \\ .7 & 1.2 \end{bmatrix}$

3 $\begin{bmatrix} -7 & 2 & 9 \\ 4 & -3 & 1 \end{bmatrix} + \begin{bmatrix} 0 & 0 & 0 \\ 0 & 0 & 0 \end{bmatrix}$

4 $\begin{bmatrix} 0 & 0 & 0 \\ 0 & 0 & 0 \\ 0 & 0 & 0 \end{bmatrix} + \begin{bmatrix} 1 & 0 & 8 \\ 2 & 3 & -1 \\ 5 & 4 & 9 \end{bmatrix}$

5 $2\begin{bmatrix} 1 \\ -1 \\ 4 \end{bmatrix} - \begin{bmatrix} 5 \\ 6 \\ -3 \end{bmatrix}$

6 $\begin{bmatrix} \frac{1}{2} & \frac{1}{3} & \frac{1}{5} \end{bmatrix} + \begin{bmatrix} \frac{2}{5} & \frac{3}{4} & \frac{-2}{3} \end{bmatrix}$

7 $\begin{bmatrix} 8 & -5 \\ 4 & 3 \end{bmatrix} - 3\begin{bmatrix} 1 & 0 \\ 0 & 1 \end{bmatrix}$

8 $5\begin{bmatrix} 1 & 0 & 0 \\ 0 & 1 & 0 \\ 0 & 0 & 1 \end{bmatrix} - \begin{bmatrix} 2 & 0 & 1 \\ 0 & 3 & 0 \\ -1 & 0 & 4 \end{bmatrix}$

9 $\begin{bmatrix} 3 & 2 \\ 1 & 0 \\ -5 & 4 \end{bmatrix} + \begin{bmatrix} -6 & 0 & 4 \\ 1 & 3 & -2 \end{bmatrix}$

10 $\begin{bmatrix} 4 \\ -3 \end{bmatrix} + \begin{bmatrix} -1 & 0 \\ 1 & 0 \end{bmatrix}$

In problems 11 through 20, perform the indicated operations, if possible, where

$$A = \begin{bmatrix} 2 & -1 & 0 \\ 5 & 3 & -2 \\ 0 & 6 & 7 \end{bmatrix}, \quad B = \begin{bmatrix} 0 & 6 & 4 \\ 1 & -2 & 2 \\ 3 & 0 & 0 \end{bmatrix}, \quad C = \begin{bmatrix} -1 & 0 & 2 \\ 0 & 0 & 3 \\ 1 & 4 & 7 \end{bmatrix}.$$

11 $A + B + C$ 12 $A - B + C$
13 $4A - 5B$ 14 $3A + 2C$

15 $A + \dfrac{1}{2}(B + C)$ 16 $5(A - B) + C$

17 $2(A + C) - (B + C) - C$ 18 $3(A - B) + 3(B - C) + 3C$
19 $AB + C$ 20 $A + BC$

In problems 21 through 30, solve the given equation for X in terms of A, B, C, and I. Assume $A + I$ and $B + I$ are invertible.

21 $2X - A = B$ 22 $3X + 4A = B$
23 $X - A = B - C$ 24 $5A - X - B = C$
25 $AX + B = C - X$ 26 $A + BX = C - X$
27 $B - 2XA = 2X + C$ 28 $C - 5XB - X = 4X + A$
29 $A - 2BX - 2X + 3A = C$ 30 $5C + 3XA = -3X - 2B$

In problems 31 through 40, solve the given equation for X.

31 $2X = \begin{bmatrix} 10 & 8 \\ 4 & 6 \end{bmatrix}$

32 $3X + \begin{bmatrix} 1 & 1 \\ -1 & -1 \end{bmatrix} = \begin{bmatrix} 10 & 7 \\ 2 & 8 \end{bmatrix}$

33 $5X - 10 \begin{bmatrix} 1 & 0 \\ 0 & 1 \end{bmatrix} = \begin{bmatrix} 0 & 0 \\ 0 & 0 \end{bmatrix}$

34 $2 \begin{bmatrix} 0 & 0 \\ 0 & 0 \end{bmatrix} + 4X = \begin{bmatrix} 0 & 1 \\ 1 & 0 \end{bmatrix}$

35 $\begin{bmatrix} 2 & 1 \\ 0 & 3 \end{bmatrix} + X = \begin{bmatrix} 5 & 6 \\ 3 & 4 \end{bmatrix}$

36 $2X - 3 \begin{bmatrix} 0 & 1 \\ 0 & -1 \end{bmatrix} + 4 \begin{bmatrix} 1 & 1 \\ 0 & 0 \end{bmatrix} = \begin{bmatrix} 0 & 0 \\ 0 & 0 \end{bmatrix}$

37 $\begin{bmatrix} 1 & 0 \\ 0 & 1 \end{bmatrix} X = \begin{bmatrix} 3 & 2 \\ 0 & -4 \end{bmatrix}$

38 $X \begin{bmatrix} 0 & 1 \\ 1 & 0 \end{bmatrix} = \begin{bmatrix} 2 & 2 \\ 3 & 3 \end{bmatrix}$

39 $\begin{bmatrix} 1 & 0 \\ 2 & 3 \end{bmatrix} X = \begin{bmatrix} 3 & -2 \\ 1 & 5 \end{bmatrix}$

40 $X \begin{bmatrix} 2 & -1 \\ 0 & 5 \end{bmatrix} - \begin{bmatrix} 1 & 1 \\ 1 & 1 \end{bmatrix} = \begin{bmatrix} 0 & 0 \\ 0 & 0 \end{bmatrix}$

41 Verify the associative property for matrix addition.
42 Verify the distributive property for scalar multiplication.
43 Verify property 8 of properties of matrix operations.
44 Verify property 6 of properties of matrix operations for 2×2 matrices.
45 Verify property 7 of properties of matrix operations for 2×2 matrices.

4.2   Input-Output Matrices

Wassily Leontief received the 1973 Nobel prize in economics for his development of input-output matrices. The purpose of an input-output matrix is to understand one industry's activities in a broad economic context. Such a matrix indicates the interdependence of several (or many) industries in the production of final products for consumers. This approach is used by companies to predict market trends and to prepare for anticipated consumer demand.

Our examples are oversimplified for the sake of easy computation. With high speed computers, realistic problems involving hundreds of industries can be made manageable.

Table 4.1 shows the interaction of two hypothetical industries. Each entry indicates millions of dollars of product. The row beside an industry shows how that industry's output is used: the industry uses some of its own output, the other industry uses some, and the remainder goes to consumers (represented by "final demand"). For example, Industry 1 uses $220 million of its own output, it sells $700 million of its output to Industry 2, and it sells $180 million of output to consumers, for a total of $1100 million.

TABLE 4.1   A Hypothetical Two-Industry Economy

	Purchases			
	Industry 1	Industry 2	Final Demand	Total Output
Industry 1	220	700	180	1100
Industry 2	330	1400	1070	2800
Other items	550	700		
Totals	1100	2800		

Each column gives the values of that industry's purchases from itself and from the other industry, and the total of other items on its balance sheet. For example, costs of raw materials, total wages paid to labor, and profit are other items that the industry "purchases." By agreement, the amount of "other items" is chosen so that an industry's total purchases equal its total output.

The data in the table show the transactions that take place in this economy for the specific amounts of final (consumer) demand indicated. In order to use the method of the input-output matrix, we must make the following assumption: **the amount of each purchase of each industry divided by the total output of the industry is a constant ratio despite changes in the final demand.** In particular, if consumer demand increases then each of the purchases of Industry 1 must increase; however, each purchase, as a fraction of its total output, is unchanged.

In the case of Table 4.1, this means that the amount of Product 1 purchased by Industry 1 is the constant fraction

$$\frac{220}{1100} = 0.2$$

times the output of Industry 1. Thus, if the total output of Industry 1 is denoted by X_1 then that industry will have to purchase $0.2X_1$ of Product 1. The same is assumed to be true for both industries and for both products. The other fractions are

$$\frac{330}{1100} = 0.3 \qquad \text{for Product 2 bought by Industry 1}$$

$$\frac{700}{2800} = 0.25 \qquad \text{for Product 1 bought by Industry 2}$$

$$\frac{1400}{2800} = 0.5 \qquad \text{for Product 2 bought by Industry 2}$$

It is easiest to remember that each entry in the "Purchases" columns is divided by the total at the bottom of that column (which is equal to the total output for that industry).

Now let us consider what happens when the final (consumer) demand changes. If the final demand for an industry's product increases, the industry must increase its total output to meet that demand. At the same time, it will have to purchase more of its own output and the output of the other industry in order to achieve the new total output, and this will require a further increase in the total output. We need a method for predicting the new total outputs of both industries that will meet the new final demand. This method consists of representing the economy by a matrix equation that we can solve.

The matrix equation involves three matrices. The first is a column matrix that contains the new final demands, which will be given:

$$D = \begin{bmatrix} D_1 \\ D_2 \end{bmatrix} \qquad \text{where} \quad \begin{cases} D_1 = \text{new final demand for Product 1} \\ D_2 = \text{new final demand for Product 2.} \end{cases}$$

The second is a column matrix of the new total outputs, which gives the solution to the problem that we are trying to solve:

$$X = \begin{bmatrix} X_1 \\ X_2 \end{bmatrix} \qquad \text{where} \quad \begin{cases} X_1 = \text{new total output of Industry 1} \\ X_2 = \text{new total output of Industry 2.} \end{cases}$$

The third matrix is a column matrix that gives the amount of each product purchased by the industries (that is, the amount of output that does **not** go to consumers):

$$P = \begin{bmatrix} P_1 \\ P_2 \end{bmatrix} \quad \text{where} \quad \begin{cases} P_1 = \text{industrial purchases of Product 1} \\ P_2 = \text{industrial purchases of Product 2.} \end{cases}$$

Since the total output of each industry is exactly the sum of industrial purchases plus final demand, we can write the matrix equation as

$$X = P + D.$$

(Remember that the matrix equation represents two ordinary equations.)

Because we are assuming that all purchases are constant fractions of total output, the new industrial purchases (for the data in Table 4.1) are

$$P_1 = 0.2\, X_1 + 0.25\, X_2$$
$$P_2 = 0.3\, X_1 + 0.5\, X_2.$$

Notice that these two equations can be written as the single matrix equation

$$P = \begin{bmatrix} 0.2 & 0.25 \\ 0.3 & 0.5 \end{bmatrix} \begin{bmatrix} X_1 \\ X_2 \end{bmatrix}.$$

If the square matrix of fractional coefficients is denoted by A, then this equation is simply

$$P = AX.$$

The matrix A, which describes the interaction of the two industries, is called the **input-output matrix** for the two-industry economy.

Now we can substitute $P = AX$ into the matrix equation for the total outputs.

$$X = P + D$$
$$X = AX + D$$

We can solve the last equation for X, assuming that $(I - A)$ is invertible.

$$X = AX + D$$
$$X - AX = D$$
$$IX - AX = D$$
$$(I - A)X = D$$
$$(I - A)^{-1}(I - A)X = (I - A)^{-1}D$$
$$IX = (I - A)^{-1}D$$
$$X = (I - A)^{-1}D$$

The last matrix equation allows us to calculate the new total outputs from the known matrices A and D.

As an illustration, assume that the final demands for Product 1 and

Product 2 change to $200 million and $1000 million, respectively. What new total outputs will be required to meet these demands?

The result of our analysis shows that the matrix of new total outputs will be

$$X = (I - A)^{-1}D$$

$$= \left(\begin{bmatrix} 1 & 0 \\ 0 & 1 \end{bmatrix} - \begin{bmatrix} 0.2 & 0.25 \\ 0.3 & 0.5 \end{bmatrix} \right)^{-1} \begin{bmatrix} 200 \\ 1000 \end{bmatrix}$$

$$= \begin{bmatrix} 0.8 & -0.25 \\ -0.3 & 0.5 \end{bmatrix}^{-1} \begin{bmatrix} 200 \\ 1000 \end{bmatrix}.$$

Using the method of Section 3.4, we compute the indicated inverse matrix. Substituting that inverse in the last equation and performing the multiplication, we get the answer to the problem.

$$X = \begin{bmatrix} 1.538 & 0.769 \\ 0.923 & 2.462 \end{bmatrix} \begin{bmatrix} 200 \\ 1000 \end{bmatrix} = \begin{bmatrix} 1077 \\ 2647 \end{bmatrix}.$$

We have determined that the new total outputs of Industry 1 and Industry 2 should be $1077 and $2647 million, respectively. Note that, under the assumption that each purchase is a constant fraction of total output, production and demand balance for each product.

$$0.2(1077) + 0.25(2647) + 200 = 1077$$

$$0.3(1077) + 0.5(2647) + 1000 = 2647$$

Let us now summarize the general process for using input-output matrices. We begin with the matrix equation

$$X = AX + D,$$

which quantifies the equation

total demand = industry demand + final demand.

Proceeding from top to bottom, the column matrix D lists the final demand for each of the industries involved. Proceeding from top to bottom, the column matrix X lists the total output of each industry that is required for the final demands listed in D. The matrix A is the input-output matrix that describes the interdependence of the industries involved; this matrix is obtained from a table as previously indicated in this section. Using the steps given on page 168, we solve the matrix equation by evaluating

$$X = (I - A)^{-1}D.$$

We demonstrate the process with further examples.

EXAMPLE 1

The next table shows the interaction of two industries. Assume that the interdependence of the industries remains unchanged when the final demands for the outputs of Industries 1 and 2 become 1400 and 1200, respectively. All numbers indicate millions of dollars. Find the new total output for each industry.

| | Purchases | | | |
	Industry 1	Industry 2	Final Demand	Total Output
Industry 1	150	350	1500	2000
Industry 2	400	200	1000	1600
Other items	1450	1050		
Totals	2000	1600		

Solution
By dividing each of the first two entries in each of the first two columns by the total of the column, we obtain the entries for the input-output matrix A. This matrix describes the purchases made by each industry as fractions of the total output of that industry.

$$A = \begin{bmatrix} \frac{150}{2000} & \frac{350}{1600} \\ \frac{400}{2000} & \frac{200}{1600} \end{bmatrix} = \begin{bmatrix} .0750 & .2188 \\ .2000 & .1250 \end{bmatrix}.$$

Now we compute the inverse of $I - A$ using the notation employed in Section 3.4.

$$I - A = \begin{bmatrix} 1 & 0 \\ 0 & 1 \end{bmatrix} - \begin{bmatrix} .075 & .2188 \\ .200 & .125 \end{bmatrix} = \begin{bmatrix} .925 & -.2188 \\ -.200 & .875 \end{bmatrix}$$

$$\begin{bmatrix} .925 & -.2188 & | & 1 & 0 \\ -.200 & .875 & | & 0 & 1 \end{bmatrix}$$

$\frac{1}{.925} R_1$
$$\begin{bmatrix} 1 & -.2365 & | & 1.0811 & 0 \\ -.200 & .875 & | & 0 & 1 \end{bmatrix}$$

$.200 R_1 + R_2$
$$\begin{bmatrix} 1 & -.2365 & | & 1.0811 & 0 \\ 0 & .8277 & | & .2162 & 1 \end{bmatrix}$$

$\frac{1}{.8277} R_2$
$$\begin{bmatrix} 1 & -.2365 & | & 1.0811 & 0 \\ 0 & 1 & | & .2612 & 1.2082 \end{bmatrix}$$

$R_1 + .2365 R_2$
$$\begin{bmatrix} 1 & 0 & | & 1.1429 & .2857 \\ 0 & 1 & | & .2612 & 1.2082 \end{bmatrix}$$

The last matrix on the right is the desired inverse. The above calculations were performed on a hand calculator and rounded off to the fourth decimal place.

Now we compute the product of $(I - A)^{-1}$ and D, the column matrix giving the new final demands for the two industries.

$$X = (I - A)^{-1}D = \begin{bmatrix} 1.1429 & .2857 \\ .2612 & 1.2082 \end{bmatrix}\begin{bmatrix} 1400 \\ 1200 \end{bmatrix}$$
$$= \begin{bmatrix} 1942.90 \\ 1815.52 \end{bmatrix}.$$

Thus, the new total outputs for Industries 1 and 2 are $1942.90 and $1815.52 million, respectively. ◼

In a truly realistic example there might be 20 industries. Our energies are fully taxed by the following example with three industries.

EXAMPLE 2

The table below shows the interaction of three industries. Assume that the interdependence of the industries remains unchanged when the final demands for the outputs of Industries 1, 2, and 3 become 100, 50, and 50, respectively. All numbers indicate millions of dollars. Find the new total output for each industry.

	Purchases			Final Demand	Total Output
	Industry 1	Industry 2	Industry 3		
Industry 1	20	45	45	90	200
Industry 2	30	10	30	30	100
Industry 3	30	30	10	30	100
Other items	120	15	15		
Totals	200	100	100		

Solution

By dividing each of the first three entries in each of the first three columns by the total of the column, we obtain the entries for the input-output matrix A.

$$A = \begin{bmatrix} \frac{20}{200} & \frac{45}{100} & \frac{45}{100} \\ \frac{30}{200} & \frac{10}{100} & \frac{30}{100} \\ \frac{30}{200} & \frac{30}{100} & \frac{10}{100} \end{bmatrix} = \begin{bmatrix} .10 & .45 & .45 \\ .15 & .10 & .30 \\ .15 & .30 & .10 \end{bmatrix}$$

Now we compute the inverse of $I - A$ using the notation employed in Section 3.4. All calculations are rounded off to the second decimal place.

$$I - A = \begin{bmatrix} 1 & 0 & 0 \\ 0 & 1 & 0 \\ 0 & 0 & 1 \end{bmatrix} - \begin{bmatrix} .10 & .45 & .45 \\ .15 & .10 & .30 \\ .15 & .30 & .10 \end{bmatrix} = \begin{bmatrix} .90 & -.45 & -.45 \\ -.15 & .90 & -.30 \\ -.15 & -.30 & .90 \end{bmatrix}$$

$$\begin{bmatrix} .90 & -.45 & -.45 & | & 1 & 0 & 0 \\ -.15 & .90 & -.30 & | & 0 & 1 & 0 \\ -.15 & -.30 & .90 & | & 0 & 0 & 1 \end{bmatrix}$$

$$\frac{1}{.90} R_1 \qquad \begin{bmatrix} 1.00 & -.50 & -.50 & | & 1.11 & 0 & 0 \\ -.15 & .90 & -.30 & | & 0 & 1 & 0 \\ -.15 & -.30 & .90 & | & 0 & 0 & 1 \end{bmatrix}$$

$$\begin{matrix} R_2 + .15\, R_1 \\ R_3 + .15\, R_1 \end{matrix} \qquad \begin{bmatrix} 1.00 & -.50 & -.50 & | & 1.11 & 0 & 0 \\ 0 & .83 & -.38 & | & .17 & 1 & 0 \\ 0 & -.38 & .83 & | & .17 & 0 & 1 \end{bmatrix}$$

$$\frac{1}{.83} R_2 \quad \begin{bmatrix} 1.00 & -.50 & -.50 & 1.11 & 0 & 0 \\ 0 & 1.00 & -.46 & .20 & 1.20 & 0 \\ 0 & -.38 & .83 & .17 & 0 & 1 \end{bmatrix}$$

$$\begin{array}{c} R_1 + .5R_2 \\ \\ R_3 + .38R_2 \end{array} \quad \begin{bmatrix} 1.00 & 0 & -.73 & 1.21 & .60 & 0 \\ 0 & 1.00 & -.46 & .20 & 1.20 & 0 \\ 0 & 0 & .66 & .25 & .46 & 1.00 \end{bmatrix}$$

$$\frac{1}{.66} R_3 \quad \begin{bmatrix} 1.00 & 0 & -.73 & 1.21 & .60 & 0 \\ 0 & 1.00 & -.46 & .20 & 1.20 & 0 \\ 0 & 0 & 1.00 & .38 & .70 & 1.52 \end{bmatrix}$$

$$\begin{array}{c} R_1 + .73R_3 \\ R_2 + .46R_3 \end{array} \quad \begin{bmatrix} 1.00 & 0 & 0 & 1.49 & 1.11 & 1.11 \\ 0 & 1.00 & 0 & .37 & 1.52 & .70 \\ 0 & 0 & 1.00 & .38 & .70 & 1.52 \end{bmatrix}$$

The last matrix on the right is the desired inverse.

Now we compute the product of $(I - A)^{-1}$ and D, the column matrix giving the new final demands for the three industries.

$$X = (I - A)^{-1}D = \begin{bmatrix} 1.49 & 1.11 & 1.11 \\ .37 & 1.52 & .70 \\ .38 & .70 & 1.52 \end{bmatrix} \begin{bmatrix} 100 \\ 50 \\ 50 \end{bmatrix}$$

$$= \begin{bmatrix} 260 \\ 148 \\ 149 \end{bmatrix}$$

Thus, the new total outputs for Industries 1, 2, and 3 are $260, $148, and $149 million, respectively. ■

Exercises 4.2

1 Construct the input-output matrix corresponding to the table below.

	Purchases			
	Industry 1	Industry 2	Final Demand	Total Output
Industry 1	10	50	40	100
Industry 2	30	20	50	100
Other items	60	30		
Totals	100	100		

2 Assume that the interdependence of the industries in problem 1 remains unchanged when the final demands for the outputs of Industries 1 and 2

become 30 and 60, respectively. All numbers indicate millions of dollars. Find the new total output for each industry.

3 Making the same assumptions as in problem 2, find the new total output for each industry when the final demands for the outputs of Industries 1 and 2 become 50 and 60, respectively.

4 Construct the input-output matrix corresponding to the table below.

	Purchases		Final Demand	Total Output
	Industry 1	Industry 2		
Industry 1	200	300	300	800
Industry 2	400	200	400	1000
Other items	200	500		
Totals	800	1000		

5 Assume that the interdependence of the industries in problem 4 remains unchanged when the final demands for the outputs of Industries 1 and 2 become 400 and 500, respectively. All numbers indicate millions of dollars. Find the new total output for each industry.

6 Making the same assumptions as in problem 5, find the new total output for each industry when the final demands for the outputs of Industries 1 and 2 become 250 and 450, respectively.

7 Construct the input-output matrix corresponding to the table below.

	Purchases			Final Demand	Total Output
	Industry 1	Industry 2	Industry 3		
Industry 1	40	150	100	110	400
Industry 2	120	50	100	230	500
Industry 3	80	200	0	120	400
Other items	160	100	200		
Totals	400	500	400		

8 Assume that the interdependence of the industries in problem 7 remains unchanged when the final demands for the outputs of Industries 1, 2, and 3 become 140, 250, and 150, respectively.

9 Making the same assumptions as in problem 8, find the new total output for each industry when the final demands for the outputs of Industries 1, 2, and 3 become 100, 200, and 150, respectively.

10 Construct the input-output matrix corresponding to the following table.

	Purchases			Final Demand	Total Output
	Industry 1	Industry 2	Industry 3		
Industry 1	100	200	500	200	1000
Industry 2	250	50	400	100	800
Industry 3	350	300	100	750	1500
Other items	300	250	500		
Totals	1000	800	1500		

11 Assume that the interdependence of the industries in problem 10 remains unchanged when the final demands for the outputs of Industries 1, 2, and 3 become 200, 200, and 700, respectively. All numbers indicate millions of dollars. Find the new total output for each industry.

12 Making the same assumptions as in problem 11, find the new total output for each industry when the final demands for the outputs of Industries 1, 2, and 3 become 300, 200, and 600, respectively.

13 Assume that $I - A$ is invertible and solve the matrix equation below for X.

$$X = XA + B$$

14 Assume that $(A + B)$ is invertible and solve the matrix equation below for X.

$$XA = C - XB$$

15 Assume that A is invertible and solve the matrix equation below for X.

$$AX - B = I$$

4.3 Determinants

Systems of linear equations were solved by methods based on determinants before matrices, as we know them today, were introduced. A determinant is a single number associated with a given square matrix. How we calculate the number depends on the order of the matrix. For any 2×2 matrix, the calculation of the determinant is quick and easy. For a 3×3 matrix, the determinant is more difficult to calculate. For matrices with order $n \times n$ and $n \geq 4$, the calculation of the determinant is much more trouble.

The determinant has many applications. The most fundamental applications are based on the fact that a square matrix is invertible precisely when the determinant of that matrix is not equal to zero. Thus, before trying to calculate the inverse of a given matrix, it is sometimes wise to calculate the determinant of the matrix to discover whether the matrix has an inverse. In the study of functions with more than one variable, the determinant is an important device for notation. In advanced calculus there is an important idea called the Jacobian, and it is a determinant. In the study of differential equations, there is an important determinant called the Wronskian.

We begin with the simplest ways to calculate a determinant.

Definition   The determinant of the matrix A is denoted $|A|$. With a specified matrix, we simply replace the brackets with vertical bars to indicate the determinant. The determinant of a 2×2 matrix is given by the formula

$$\Delta = \begin{vmatrix} a_{11} & a_{12} \\ a_{21} & a_{22} \end{vmatrix} = a_{11}a_{22} - a_{12}a_{21}$$

The determinant of a 1×1 matrix is the single entry in the matrix.

Note that the determinant of a 2×2 matrix is the product down the main diagonal minus the product down the other diagonal. We illustrate the process with some specific matrices.

EXAMPLE 1

Find each of the determinants below.

$$\begin{vmatrix} 2 & 3 \\ -1 & 5 \end{vmatrix}, \quad \begin{vmatrix} 1 & 1 \\ 1 & 1 \end{vmatrix}, \quad \begin{vmatrix} a & b \\ 0 & c \end{vmatrix}$$

Solution

$$\begin{vmatrix} 2 & 3 \\ -1 & 5 \end{vmatrix} = (2)(5) - (3)(-1) = 10 + 3 = 13$$

$$\begin{vmatrix} 1 & 1 \\ 1 & 1 \end{vmatrix} = (1)(1) - (1)(1) = 1 - 1 = 0$$

$$\begin{vmatrix} a & b \\ 0 & c \end{vmatrix} = ac - b(0) = ac \quad ■$$

Definition The determinant of a 3×3 matrix is given by the formula

$$\Delta = \begin{vmatrix} a_{11} & a_{12} & a_{13} \\ a_{21} & a_{22} & a_{23} \\ a_{31} & a_{32} & a_{33} \end{vmatrix} = \begin{aligned} & a_{11}a_{22}a_{33} + a_{12}a_{23}a_{31} + a_{13}a_{21}a_{32} \\ & - a_{13}a_{22}a_{31} - a_{11}a_{23}a_{32} - a_{12}a_{21}a_{33} \end{aligned}$$

In practice, we do not use this formula but rather the following scheme for producing the formula. Copy the first two columns just to the right of the given array. Compute the product of the terms along each diagonal line slanted down and put a plus sign in front of the product. Compute the product of the terms along each diagonal line slanted up and put a minus sign in front of the product. The determinant is the sum of the six products.

$$\begin{vmatrix} a_{11} & a_{12} & a_{13} \\ a_{21} & a_{22} & a_{23} \\ a_{31} & a_{32} & a_{33} \end{vmatrix} \begin{matrix} a_{11} & a_{12} \\ a_{21} & a_{22} \\ a_{31} & a_{32} \end{matrix}$$

We illustrate the process with specific matrices.

EXAMPLE 2

Compute the indicated determinants.

$$\begin{vmatrix} 2 & 0 & 1 \\ -1 & 3 & 5 \\ 4 & 0 & 3 \end{vmatrix}, \quad \begin{vmatrix} 1 & 1 & -1 \\ 1 & -1 & 1 \\ 0 & -2 & 2 \end{vmatrix}$$

Solution

$$\begin{vmatrix} 2 & 0 & 1 \\ -1 & 3 & 5 \\ 4 & 0 & 3 \end{vmatrix} \begin{matrix} 2 & 0 \\ -1 & 3 \\ 4 & 0 \end{matrix} = (2)(3)(3) + 0 + 0 - (1)(3)(4) - 0 - 0$$

$$= 18 - 12$$

$$= 6$$

Notice that every diagonal line that contains a zero has a product of zero.

$$\begin{vmatrix} 1 & 1 & -1 & 1 & 1 \\ 1 & 1 & 1 & 1 & -1 \\ 0 & -2 & 2 & 0 & 2 \end{vmatrix} = (1)(-1)(2) + 0 + (-1)(1)(-2) + 2 - (1)(1)(2)$$
$$= -2 + 2 + 2 - 2$$
$$= 0 \quad \blacksquare$$

We begin to describe the elaborate process by which we calculate the determinants for matrices that are $n \times n$ with $n \geq 4$. The technique that we are going to describe is called the **expansion-in-minors technique**. First, we define a minor.

Definition   With any entry in any matrix, such as a_{11} in the matrix below, we associate a determinant that we call the **minor** for that entry. The minor for the entry in the ith row and jth column is denoted M_{ij}. The minor is the determinant of the matrix that remains after we delete the row and the column in which the entry appears.

In the 3×3 matrix below, we have drawn a line through the row and the column to be deleted to get the minor for a_{11}.

$$\begin{vmatrix} a_{11} & a_{12} & a_{13} \\ a_{21} & a_{22} & a_{23} \\ a_{31} & a_{32} & a_{33} \end{vmatrix}$$

The minor for a_{11}, denoted M_{11}, is $\begin{vmatrix} a_{22} & a_{23} \\ a_{32} & a_{33} \end{vmatrix} = a_{22}a_{33} - a_{23}a_{32}$.

EXAMPLE 3

Determine the minor for the entry in the third row and the third column of the matrix below.

$$\begin{bmatrix} 1 & 1 & -1 \\ 1 & -1 & 1 \\ 0 & -2 & 2 \end{bmatrix}$$

Solution
We draw lines through the row and the column to be deleted; then we compute the minor indicated.

$$\begin{bmatrix} 1 & 1 & -1 \\ 1 & -1 & 1 \\ 0 & -2 & 2 \end{bmatrix} \qquad \begin{vmatrix} 1 & 1 \\ 1 & -1 \end{vmatrix} = (1)(-1) - (1)(1)$$
$$= -1 - 1 = -2$$

The minor is -2. $\quad \blacksquare$

Definition The determinant Δ of a 4×4 matrix is given by the expansion-in-minors formula below. The analogous formula defines the determinant of any $n \times n$ matrix with $n \geq 5$.

$$\Delta = \begin{vmatrix} a_{11} & a_{12} & a_{13} & a_{14} \\ a_{21} & a_{22} & a_{23} & a_{24} \\ a_{31} & a_{32} & a_{33} & a_{34} \\ a_{41} & a_{42} & a_{43} & a_{44} \end{vmatrix} = (-1)^{1+1}a_{11}M_{11} + (-1)^{1+2}a_{12}M_{12} + (-1)^{1+3}a_{13}M_{13} + (-1)^{1+4}a_{14}M_{14}$$

Note that the factor just before a_{ij} is $(-1)^{i+j}$.

EXAMPLE 4

Compute the determinant

$$\Delta = \begin{vmatrix} 1 & 0 & 2 & 3 \\ -2 & 1 & 4 & 0 \\ 3 & 2 & -1 & 1 \\ 0 & 5 & 1 & 2 \end{vmatrix}.$$

Solution
The minor for the first entry in the first row is

$$\begin{vmatrix} 1 & 4 & 0 \\ 2 & 1 & 1 \\ 5 & 1 & 2 \end{vmatrix} = -2 + 20 + 0 - 0 - 1 - 16 = 1.$$

Since the second entry in the first row is 0, the product $(-1)^{1+2}a_{12}M_{12}$ must be zero, so we do not bother to compute the minor for this entry. The minor for the third entry in the first row is.

$$\begin{vmatrix} -2 & 1 & 0 \\ 3 & 2 & 1 \\ 0 & 5 & 2 \end{vmatrix} = -8 + 0 + 0 - 0 - (-10) - 6 = -4.$$

The minor for the fourth entry in the first row is

$$\begin{vmatrix} -2 & 1 & 4 \\ 3 & 2 & 1 \\ 0 & 5 & 1 \end{vmatrix} = -4 + 0 + 60 - 0 - 10 - 3 = 43.$$

The determinant Δ is given by the formula

$$\Delta = (-1)^2(1)(1) + (-1)^3(0) + (-1)^4(2)(-4) + (-1)^5(3)(43)$$
$$= 1 - 8 - 129$$
$$= -136. \blacksquare$$

We offer without verification some facts that make determinants easier to calculate.

Fact 1 The value of the determinant is not altered regardless of which row or column is chosen for the expansion-in-minors formula. In our definition we used the first row; we could have used any row or any column and we would have obtained the same value.

The next example will show how Fact 1 can be used to advantage.

EXAMPLE 5

Compute the determinant

$$\Delta = \begin{vmatrix} 3 & -2 & 0 & 1 \\ 4 & 5 & 0 & 6 \\ -3 & 7 & 1 & -2 \\ 1 & 2 & 0 & 4 \end{vmatrix}.$$

Solution

We can compute Δ with much less work if we use the expansion-in-minors formula applied to the third column. Since the first, second, and fourth entries in the third column are zeroes, the product of each of those entries and its minor is 0. Thus, we do not need to determine those minors.

The minor of the third entry in the third column is

$$\begin{vmatrix} 3 & -2 & 1 \\ 4 & 5 & 6 \\ 1 & 2 & 4 \end{vmatrix} = 60 - 12 + 8 - 5 - 36 + 32 = 47.$$

Thus,

$$\Delta = (-1)^{1+3}(0) + (-1)^{2+3}(0) + (-1)^{3+3}(1)(47) + (-1)^{3+4}(0) = 47. \quad \blacksquare$$

Fact 2 If each entry in some row, or column, of the matrix A is 0, then the determinant $|A|$ equals 0. If two rows, or columns, of the matrix A are identical, then the determinant $|A|$ equals 0.

In view of Fact 2, each determinant below can be seen to be 0 without making any calculations.

$$\begin{vmatrix} 3 & 5 & 0 \\ 4 & 9 & 0 \\ -2 & 2 & 0 \end{vmatrix}, \quad \begin{vmatrix} 5 & 4 & 2 & 3 \\ -1 & 1 & 0 & 6 \\ 1 & -1 & 4 & 2 \\ -1 & 1 & 0 & 6 \end{vmatrix}, \quad \begin{vmatrix} 1 & 3 & 1 & 0 \\ 2 & 0 & 2 & 5 \\ 1 & 4 & 1 & 2 \\ -1 & 6 & -1 & 3 \end{vmatrix}$$

Fact 3 If the matrix A is upper triangular or lower triangular, then the determinant $|A|$ equals the product of the entries in the main diagonal.

EXAMPLE 6

Evaluate each of the determinants below.

$$\begin{vmatrix} 1 & 3 & -2 \\ 0 & 4 & -1 \\ 0 & 0 & 5 \end{vmatrix}, \quad \begin{vmatrix} 2 & 0 & 0 & 0 \\ 0 & -1 & 0 & 0 \\ 0 & 0 & 3 & 0 \\ 0 & 0 & 0 & 5 \end{vmatrix}, \quad \begin{vmatrix} 3 & 0 & 0 & 0 \\ 5 & 2 & 0 & 0 \\ 7 & 3 & 10 & 0 \\ -4 & 9 & 2 & -1 \end{vmatrix}$$

Solution
The first matrix, for which we are calculating the determinant, is upper triangular and so the determinant is the product of the entries in the main diagonal. Its determinant is

$$(1)(4)(5) = 20.$$

The second matrix, for which we are computing the determinant, is both upper triangular and lower triangular. Its determinant is

$$(2)(-1)(3)(5) = -30.$$

The third matrix above is lower triangular, and its determinant is

$$(3)(2)(10)(-1) = -60. \quad\blacksquare$$

Fact 4 If each entry in some row of the matrix A is multiplied by a real number c, then the determinant is multiplied by c. If each entry in some column of A is multiplied by c, then the determinant is multiplied by c.

EXAMPLE 7

Evaluate the determinant below.

$$\begin{vmatrix} 2 & 4 & 6 \\ 6 & 0 & 1 \\ 3 & 6 & 9 \end{vmatrix}$$

Solution
The following equations follow from Fact 4.

$$\begin{vmatrix} 2 & 4 & 6 \\ 6 & 0 & 1 \\ 3 & 6 & 9 \end{vmatrix} = 2\begin{vmatrix} 1 & 2 & 3 \\ 6 & 0 & 1 \\ 3 & 6 & 9 \end{vmatrix} = (2)(3)\begin{vmatrix} 1 & 2 & 3 \\ 6 & 0 & 1 \\ 1 & 2 & 3 \end{vmatrix}$$

Since the first row of the matrix on the extreme right of the above equations is identical with the third row of that matrix, the determinant of the matrix is 0, by Fact 2. Thus, the value of the given determinant is 0. ∎

Fact 5 The matrix A is invertible if and only if its determinant $|A|$ is not 0.

EXAMPLE 8

Determine which of the matrices below are invertible.

$$\begin{bmatrix} 1 & 2 \\ 3 & -1 \end{bmatrix}, \quad \begin{bmatrix} 2 & -1 & 1 \\ -1 & 3 & 0 \\ 1 & 2 & 1 \end{bmatrix}, \quad \begin{bmatrix} 2 & 6 & 2 & 0 \\ -1 & 0 & -1 & 0 \\ 7 & 3 & 5 & 1 \\ 4 & 0 & 4 & 0 \end{bmatrix}$$

Solution
We determine whether each matrix is invertible by computing its determinant and using Fact 5.

$$\begin{vmatrix} 1 & 2 \\ 3 & -1 \end{vmatrix} = -1 - 6 = -7$$

Thus, the first matrix is invertible.

$$\begin{vmatrix} 2 & -1 & 1 \\ -1 & 3 & 0 \\ 1 & 2 & 1 \end{vmatrix} \begin{matrix} 2 & -1 \\ -1 & 1 \\ 1 & 2 \end{matrix} = 6 + 0 - 2 - 3 - 0 - 1 = 0.$$

Thus, the second matrix is not invertible.

To calculate the determinant of the last matrix we use the expansion in minors starting with the fourth column.

$$\begin{vmatrix} 2 & 6 & 2 & 0 \\ -1 & 0 & -1 & 0 \\ 7 & 3 & 5 & 1 \\ 4 & 0 & 4 & 0 \end{vmatrix} = 0 + 0 + (-1)^{3+4}(1) \begin{vmatrix} 2 & 6 & 2 \\ -1 & 0 & -1 \\ 4 & 0 & 4 \end{vmatrix} + 0$$

The determinant of the 3 × 3 matrix given on the right of the equation must be 0 because the first and third columns are identical. Thus, the determinant of the third matrix is 0, and the matrix is not invertible. ■

When the determinant of a large matrix is computed without the use of any simplifying facts, the process is extremely tedious. It requires a vast number of computations. For a large matrix, it is easier to compute the inverse than it is to compute the determinant by the expansion-in-minors formula. Fortunately, there are standard computer programs that compute the inverse and determinant for such large matrices.

Exercises 4.3

Evaluate each of the following determinants. Use the properties of determinants when that is possible.

1 $\begin{vmatrix} 3 & -2 \\ 1 & 5 \end{vmatrix}$

2 $\begin{vmatrix} 8 & -1 \\ 5 & -3 \end{vmatrix}$

3 $\begin{vmatrix} .2 & -.1 \\ .8 & .7 \end{vmatrix}$

4 $\begin{vmatrix} -.3 & .6 \\ .2 & 1.5 \end{vmatrix}$

5 $\begin{vmatrix} 7 & 7 \\ 7 & 7 \end{vmatrix}$

6 $\begin{vmatrix} 5 & 0 \\ 6 & 0 \end{vmatrix}$

7 $\begin{vmatrix} 2 & 113 \\ 0 & -5 \end{vmatrix}$

8 $\begin{vmatrix} 4 & 0 \\ e & -7 \end{vmatrix}$

9 $\begin{vmatrix} 7 & 0 & 3 \\ 2 & 5 & 1 \\ -1 & 4 & 0 \end{vmatrix}$

10 $\begin{vmatrix} 1 & 5 & 6 \\ 0 & -2 & -1 \\ 4 & 0 & 3 \end{vmatrix}$

11 $\begin{vmatrix} 3 & -4 & 9 \\ 1 & 3 & 4 \\ 3 & -4 & 9 \end{vmatrix}$

12 $\begin{vmatrix} 5 & 11 & 5 \\ 7 & 3 & 7 \\ 0 & -1 & 0 \end{vmatrix}$

13 $\begin{vmatrix} 1 & 54 & 129 \\ 0 & 2 & 78 \\ 0 & 0 & 3 \end{vmatrix}$

14 $\begin{vmatrix} 4 & 0 & 0 \\ 210 & -2 & 0 \\ 6.3 & 95 & 3 \end{vmatrix}$

15 $\begin{vmatrix} .6 & .9 & -.2 \\ 0 & 0 & 0 \\ 1.3 & .5 & .8 \end{vmatrix}$

16 $\begin{vmatrix} .7 & .3 & 0 \\ .5 & -.1 & 0 \\ 2.2 & .4 & 0 \end{vmatrix}$

17 $\begin{vmatrix} 4 & 0 & -4 \\ 3 & 5 & 4 \\ 5 & 0 & -5 \end{vmatrix}$

18 $\begin{vmatrix} 1 & -2 & -3 \\ 3 & 0 & 0 \\ -2 & 2 & 3 \end{vmatrix}$

19 $\begin{vmatrix} 3 & 4 & 0 & -1 \\ 2 & 0 & 1 & 0 \\ 0 & 1 & -1 & 2 \\ 3 & 2 & 1 & -2 \end{vmatrix}$

20 $\begin{vmatrix} 1 & 2 & 0 & 5 \\ 0 & 5 & 3 & 2 \\ 0 & -1 & 2 & 4 \\ 3 & 0 & 1 & 0 \end{vmatrix}$

21 $\begin{vmatrix} 1 & 0 & 0 & 0 \\ 8 & 5 & 0 & 0 \\ 9 & -4 & 3 & 0 \\ 7 & 8 & 11 & 4 \end{vmatrix}$

22 $\begin{vmatrix} 6 & -5 & 7 & 4 \\ 0 & 2 & 9 & 6 \\ 0 & 0 & 1 & 5 \\ 0 & 0 & 0 & 0 \end{vmatrix}$

23 $\begin{vmatrix} 3 & 2 & -1 & 1 \\ 4 & 0 & 2 & 0 \\ 0 & 5 & -3 & 1 \\ 3 & 2 & -1 & 1 \end{vmatrix}$

24 $\begin{vmatrix} 0 & 1 & 2 & 1 \\ -1 & 0 & 1 & 0 \\ 1 & -1 & 0 & -1 \\ 2 & 0 & -1 & 0 \end{vmatrix}$

Determine which of the following matrices are invertible.

25 $\begin{bmatrix} 1 & 1 \\ 1 & 1 \end{bmatrix}$

26 $\begin{bmatrix} 1 & 1 \\ 0 & 1 \end{bmatrix}$

27 $\begin{bmatrix} 3 & 5 \\ 4 & 0 \end{bmatrix}$

28 $\begin{bmatrix} 2 & 3 \\ 4 & 6 \end{bmatrix}$

29 $\begin{bmatrix} 5 & 2 & 4 \\ 0 & 0 & 0 \\ -5 & -2 & -4 \end{bmatrix}$

30 $\begin{bmatrix} 1 & 1 & 1 \\ 1 & 1 & 0 \\ 1 & 0 & 0 \end{bmatrix}$

31 $\begin{bmatrix} 1 & 2 & 3 \\ 2 & 3 & 0 \\ 3 & 0 & 0 \end{bmatrix}$

32 $\begin{bmatrix} 0 & 2 & 4 & 0 \\ 5 & 3 & -1 & 0 \\ 0 & 6 & 2 & 1 \\ 0 & 3 & 6 & 0 \end{bmatrix}$

33 $\begin{bmatrix} 1 & 1 & 1 & 1 \\ 0 & 1 & 1 & 1 \\ 0 & 0 & 1 & 1 \\ 0 & 0 & 0 & 1 \end{bmatrix}$

34 $\begin{bmatrix} 5 & 4 & 3 & 2 \\ 4 & 3 & 2 & 0 \\ 3 & 2 & 0 & 0 \\ 2 & 0 & 0 & 0 \end{bmatrix}$

4.4   Cramer's Rule

In this section, we present a method for solving systems of linear equations that was discovered in the first half of the eighteenth century. This method was used by Gabriel Cramer (1704–1752) approximately 100 years before the term "matrix" was used. The method is called Cramer's Rule.

Although Cramer's Rule is usually the least efficient method for solving a given specific system, there are good reasons for studying it. One reason is that the method is very widely used, due to its simplicity. Another reason is that we can solve for one particular unknown that interests us, without solving for all of the unknowns. In theoretical work, especially advanced calculus, Cramer's Rule is significant because it allows us to write a simple looking formula for each unknown. Although the method of Gaussian elimination is extremely efficient for solving a given specific system, it does not produce reasonable formulas for the unknowns in an abstract system.

We begin the explanation of Cramer's Rule by obtaining the relevant formulas for 2×2 systems through Gaussian elimination. Suppose that we want to solve the system

$$a_{11}x + a_{12}y = b_1$$
$$a_{21}x + a_{22}y = b_2.$$

In order to eliminate x, we multiply the first equation by a_{21} and the second equation by a_{11}.

$$a_{11}a_{21}x + a_{12}a_{21}y = a_{21}b_1$$
$$a_{11}a_{21}x + a_{11}a_{22}y = a_{11}b_2$$

Subtracting the first equation from the second, we get

$$(a_{11}a_{22} - a_{12}a_{21})y = a_{11}b_2 - a_{21}b_1.$$

Thus, we get a formula for y.

$$y = \frac{a_{11}b_2 - a_{21}b_1}{a_{11}a_{22} - a_{12}a_{21}}$$

Note that

$$a_{11}a_{22} - a_{12}a_{21} = \begin{vmatrix} a_{11} & a_{12} \\ a_{21} & a_{22} \end{vmatrix}$$

$$a_{11}b_2 - a_{21}b_1 = \begin{vmatrix} a_{11} & b_1 \\ a_{21} & b_2 \end{vmatrix}$$

and so

$$y = \frac{\begin{vmatrix} a_{11} & b_1 \\ a_{21} & b_2 \end{vmatrix}}{\begin{vmatrix} a_{11} & a_{12} \\ a_{21} & a_{22} \end{vmatrix}}.$$

Similarly we can show that

$$x = \frac{\begin{vmatrix} b_1 & a_{12} \\ b_2 & a_{22} \end{vmatrix}}{\begin{vmatrix} a_{11} & a_{12} \\ a_{21} & a_{22} \end{vmatrix}}.$$

Of course, these formulas assume that the determinant of the matrix of coefficients is not zero, since it appears in the denominator.

EXAMPLE 1

Solve the system below by Cramer's Rule.

$$2x + 5y = 1$$
$$3x - y = 2$$

Solution

First, we check that the determinant of the matrix of coefficients is not zero.

$$\begin{vmatrix} 2 & 5 \\ 3 & -1 \end{vmatrix} = -2 - 15 = -17$$

We can use this value in computing x and y.

$$x = \frac{\begin{vmatrix} 1 & 5 \\ 2 & -1 \end{vmatrix}}{\begin{vmatrix} 2 & 5 \\ 3 & -1 \end{vmatrix}} = \frac{-1 - 10}{-17} = \frac{-11}{-17} = \frac{11}{17}$$

$$y = \frac{\begin{vmatrix} 2 & 1 \\ 3 & 2 \end{vmatrix}}{\begin{vmatrix} 2 & 5 \\ 3 & -1 \end{vmatrix}} = \frac{4 - 3}{-17} = -\frac{1}{17} \quad \blacksquare$$

For the abstract 3×3 system below, we indicate the solution as determined by Cramer's Rule. For

$$a_{11}x_1 + a_{12}x_2 + a_{13}x_3 = b_1$$
$$a_{21}x_1 + a_{22}x_2 + a_{23}x_3 = b_2$$
$$a_{31}x_1 + a_{32}x_2 + a_{33}x_3 = b_3$$

the solution is

$$x_1 = \frac{\begin{vmatrix} b_1 & a_{12} & a_{13} \\ b_2 & a_{22} & a_{23} \\ b_3 & a_{32} & a_{33} \end{vmatrix}}{\begin{vmatrix} a_{11} & a_{12} & a_{13} \\ a_{21} & a_{22} & a_{23} \\ a_{31} & a_{32} & a_{33} \end{vmatrix}}, \quad x_2 = \frac{\begin{vmatrix} a_{11} & b_1 & a_{13} \\ a_{21} & b_2 & a_{23} \\ a_{31} & b_3 & a_{33} \end{vmatrix}}{\begin{vmatrix} a_{11} & a_{12} & a_{13} \\ a_{21} & a_{22} & a_{23} \\ a_{31} & a_{32} & a_{33} \end{vmatrix}}, \quad x_3 = \frac{\begin{vmatrix} a_{11} & a_{12} & b_1 \\ a_{21} & a_{22} & b_2 \\ a_{31} & a_{32} & b_3 \end{vmatrix}}{\begin{vmatrix} a_{11} & a_{12} & a_{13} \\ a_{21} & a_{22} & a_{23} \\ a_{31} & a_{32} & a_{33} \end{vmatrix}}$$

As in the case of a 2×2 system, the determinant of the coefficient matrix appears in each denominator. So we must assume that this determinant is not 0.

EXAMPLE 2

Solve the system below by Cramer's Rule.

$$2x_1 - 3x_2 + x_3 = 0$$
$$x_1 + x_2 - x_3 = 1$$
$$5x_1 - x_2 + x_3 = 0$$

Solution

We compute the determinant of the matrix of coefficients to make sure that we can use Cramer's Rule.

$$\begin{vmatrix} 2 & -3 & 1 \\ 1 & 1 & 1 \\ 5 & -1 & 1 \end{vmatrix} \begin{matrix} 2 & -3 \\ 1 & 1 \\ 5 & -1 \end{matrix} = 2 + 15 - 1 - 5 - 2 + 3 = 12$$

Since the determinant is not 0, we can use Cramer's Rule. We have not wasted any time with the preceding computation since we can use the value of the determinant just computed in the formulas for x_1, x_2, and x_3.

$$x_1 = \frac{\begin{vmatrix} 0 & -3 & 1 \\ 1 & 1 & 1 \\ 0 & -1 & 1 \end{vmatrix} \begin{matrix} 0 & -3 \\ 1 & 1 \\ 0 & -1 \end{matrix}}{\begin{vmatrix} 2 & -3 & 1 \\ 1 & 1 & -1 \\ 5 & -1 & 1 \end{vmatrix}} = \frac{0 + 0 - 1 - 0 - 0 + 3}{12} = \frac{2}{12} = \frac{1}{6}$$

$$x_2 = \frac{\begin{vmatrix} 2 & 0 & 1 \\ 1 & 1 & 1 \\ 5 & 0 & 1 \end{vmatrix} \begin{matrix} 2 & 0 \\ 1 & 1 \\ 5 & 0 \end{matrix}}{\begin{vmatrix} 2 & -3 & 1 \\ 1 & 1 & -1 \\ 5 & -1 & 1 \end{vmatrix}} = \frac{2 + 0 + 0 - 5 - 0 - 0}{12} = \frac{-3}{12} = -\frac{1}{4}$$

$$x_3 = \frac{\begin{vmatrix} 2 & -3 & 0 \\ 1 & 1 & 1 \\ 5 & -1 & 0 \end{vmatrix} \begin{matrix} 2 & -3 \\ 1 & 1 \\ 5 & -1 \end{matrix}}{\begin{vmatrix} 2 & -3 & 1 \\ 1 & 1 & -1 \\ 5 & -1 & 1 \end{vmatrix}} = \frac{0 - 15 + 0 - 0 + 2 - 0}{12} = -\frac{13}{12}$$

Observe that the determinants appearing in the numerators in Cramer's Rule have a predictable form. In the matrix of coefficients the first column enumerates, from top to bottom, the coefficients of x_1, the first unknown, in the three equations. The second column enumerates the coefficients of x_2, and the third column enumerates the coefficients of x_3. The determinant in the numerator of the **formula for x_1** has the column of **coefficients of x_1** replaced by a column consisting of the constants appearing on the right sides of the equations. The determinant in the **formula for x_2** has the column of **coefficients of x_2** replaced by the column of constants from the right sides of the equations. Similarly, we obtain the formula for x_3.

We have demonstrated Cramer's Rule for 2×2 systems and 3×3 systems. Now we state the general theorem and formulas.

Cramer's Rule   Let a system of n linear equations in n unknowns be given by

$$a_{11}x_1 + a_{12}x_2 + \cdots + a_{1n}x_n = b_1$$
$$a_{21}x_1 + a_{22}x_2 + \cdots + a_{2n}x_n = b_2$$
$$\cdots \cdots \cdots \cdots \cdots \cdots \cdots$$
$$a_{n1}x_1 + a_{n2}x_2 + \cdots + a_{nn}x_n = b_n.$$

Denote the matrix of coefficients by A and let Δ be its determinant. Provided Δ is not 0, the system has a unique solution. The solution is given by

$$x_1 = \frac{\Delta_1}{\Delta}, x_2 = \frac{\Delta_2}{\Delta}, \ldots, x_n = \frac{\Delta_n}{\Delta}$$

where Δ_k is the determinant of the matrix obtained by replacing the kth column of A by the column of constants from the right sides of the given equations (for $k = 1, 2, \ldots, n$).

If $\Delta = 0$, then the given system has no solutions for some choices of $b_1, \ldots, b_n$ and infinitely many solutions for other choices of $b_1, \ldots, b_n$.

EXAMPLE 3

Find the value of x_1 in a solution for the system

$$x_1 + x_3 - x_4 = 0$$
$$x_2 + x_3 - 2x_4 = 1$$
$$2x_1 - 3x_3 + x_4 = 0$$
$$x_1 + 4x_3 + x_4 = 0.$$

Solution

To compute the determinant of the matrix of coefficients, we use the expansion-in-minors procedure given in Section 4.3. We use the second column in the expansion.

$$\begin{vmatrix} 1 & 0 & 1 & -1 \\ 0 & 1 & 1 & -2 \\ 2 & 0 & -3 & 1 \\ 1 & 0 & 4 & 1 \end{vmatrix} = (-1)^{2+2}(1) \begin{vmatrix} 1 & 1 & -1 \\ 2 & -3 & 1 \\ 1 & 4 & 1 \end{vmatrix}$$

$$= -3 + 1 - 8 - 3 - 2 - 4$$
$$= -19$$

To solve for x_1, we replace the first column of coefficients with the column of constants from the right sides of the equations.

$$x_1 = \frac{\begin{vmatrix} 0 & 0 & 1 & -1 \\ 1 & 1 & 1 & -2 \\ 0 & 0 & -3 & 1 \\ 0 & 0 & 4 & 1 \end{vmatrix}}{\begin{vmatrix} 1 & 0 & 1 & -1 \\ 0 & 1 & 1 & -2 \\ 2 & 0 & -3 & 1 \\ 1 & 0 & 4 & 1 \end{vmatrix}} = \frac{0}{-19} = 0$$

The calculation of the determinant was made easy by the observation that the first two columns are identical. ■

CALCULATOR EXAMPLE

Use Cramer's Rule to determine the value of x in a solution for the following system.

$$5.31x + 8.01y = 12.44$$

$$6.03x + 2.37y = 9.68$$

Solution

$$x = \frac{\begin{vmatrix} 12.44 & 8.01 \\ 9.68 & 2.37 \end{vmatrix}}{\begin{vmatrix} 5.31 & 8.01 \\ 6.03 & 2.37 \end{vmatrix}} = \frac{-48.054}{-35.7156} \approx 1.3455 \quad ■$$

Exercises 4.4

Solve each of the following systems by Cramer's Rule or indicate why Cramer's Rule does not apply.

1. $x + 2y = 5$
    $3x - y = -1$

2. $.1x - .1y = 1$
    $.2x + .3y = -1$

3. $x - y = -1$
    $-x + y = 1$

4. $.5x + .4y = 10$
    $1.5x + 1.2y = 30$

5. $y = 3x + 1$
    $x = 4y - 2$

6. $9 = 2x - 3y$
    $y = -x + 2$

7. $2r - 7s = 4$
    $6r + 3s = 1$

8. $u + 5v = -1$
    $3u - 2v = 5$

9. $x - 2y + 2z = 0$
    $x + y - z = 1$
    $2x - y + z = 2$

10. $3x - 4y + 2z = 1$
    $x + 3y - 3z = 0$
    $4x - y - z = 5$

11. $x_1 - x_2 = 1$
    $x_2 + x_3 = 0$
    $x_1 - x_3 = -1$

12. $x_1 + x_2 + x_3 = 3$
    $2x_1 - x_2 = 1$
    $-x_1 + 2x_3 = 0$

13 $\quad .1x - .1y + .3z = 0$
$\quad\quad .2y - .4z = 1$
$\quad\quad .2x - .3y = -1$

14 $\quad .5x + .1y - .2z = 1$
$\quad\quad .2x + .3z = -1$
$\quad\quad .3y - .1z = 0$

For each system, find the value of x in a solution.

15 $\quad x + 3y + z = 2$
$\quad -x - y + 3z = 1$
$\quad 2x + y - z = -1$

16 $\quad u - v + x = 0$
$\quad u + 2v - x = 1$
$\quad 3u - 4v + x = 0$

17 $\quad 5x + 3y - 2z = 0$
$\quad 6x + 7y + 8z = 0$
$\quad 9x - 10y + 12z = 0$

18 $\quad .2x - .3y + .5z = 0$
$\quad .4x + .7y - .9z = 0$
$\quad .6x + .8y - .1z = 0$

19 $\quad w - x + y - z = 1$
$\quad 2w + x + 3y = 2$
$\quad w + 3x - y + 4z = 0$
$\quad -w + x - y = -1$

20 $\quad x + 2y + z = 0$
$\quad w + x - z = 1$
$\quad -w + 2x - y + z = 0$
$\quad w + 3x + 3z = -1$

21 $\quad w - x + y + z = 0$
$\quad w + x - y + z = 0$
$\quad w + x + y - z = 0$
$\quad -w + x + y + z = 0$

22 $\quad 3w + x + y + z = 0$
$\quad w + 3x + y + z = 0$
$\quad w + x + 3y + z = 0$
$\quad w + x + y + 3z = 0$

23 *Calculator Problem* Use Cramer's Rule to determine the value of x in a solution to the system

$$2.36x - 5.23y = 8.61$$

$$9.04x + 6.01y = 20.57.$$

24 *Calculator Problem* Use Cramer's Rule to determine the value of x in a solution to the system

$$8.76x + 2.34y = 18.05$$

$$5.44x + 6.58y = 16.02.$$

25 *Calculator Problem* Use Cramer's Rule to determine the value of y in a solution to the system

$$4.68x + 5.06y = 20.48$$

$$9.23x - .83y = 4.78.$$

26 *Calculator Problem* Use Cramer's Rule to determine the value of y in a solution to the system

$$.98x - .64y = 3.85$$

$$.32x - .83y = 6.04.$$

Review of Terms

Important Mathematical Terms
matrix operations, *p. 160*
matrix sum, *p. 160*
determinant, *p. 174*
expansion in minors, *p. 176*

matrix difference, *p. 160*   Cramer's Rule, *p. 185*
scalar product, *p. 161*
Important Terms from the Applications
final demand, *p. 167*   purchases, *p. 166*
total output, *p. 167*   other items, *p. 166*
input-output matrix, *p. 168*

Review Problems

Perform the indicated operations, if possible.

1 $\begin{bmatrix} 2 & -1 \\ 3 & 0 \end{bmatrix} + \begin{bmatrix} 1 & 4 \\ -2 & 3 \end{bmatrix}$

2 $\begin{bmatrix} 5 \\ -2 \\ 1 \end{bmatrix} - \begin{bmatrix} -1 \\ 3 \\ 4 \end{bmatrix}$

3 $\begin{bmatrix} 1 \\ -2 \end{bmatrix} + 3 \begin{bmatrix} 0 & 2 \\ 0 & 3 \end{bmatrix}$

4 $\begin{bmatrix} 1 & 2 & -1 \\ 0 & 3 & 2 \\ 5 & 0 & 1 \end{bmatrix} \begin{bmatrix} 3 & 1 & 0 \\ 0 & -1 & 2 \\ 4 & 0 & 1 \end{bmatrix} - \begin{bmatrix} 2 & 3 & 1 \\ 0 & 2 & -1 \\ 4 & 0 & 2 \end{bmatrix}$

5 $6[2 \ 0 \ 1] \begin{bmatrix} 3 \\ 5 \\ -1 \end{bmatrix} - [4]$

Solve for X in each equation given below.

6 $X \begin{bmatrix} 0 & 1 \\ 1 & 0 \end{bmatrix} = \begin{bmatrix} 2 & 0 \\ 1 & 3 \end{bmatrix}$

7 $X + \begin{bmatrix} 5 \\ 2 \end{bmatrix} = \begin{bmatrix} 3 \\ -1 \end{bmatrix}$

8 $\begin{bmatrix} 2 & 1 \\ 1 & 2 \end{bmatrix} X + I = \begin{bmatrix} 5 & 2 \\ 1 & 3 \end{bmatrix}$

9 $\begin{bmatrix} 1 & 3 \\ 2 & 1 \end{bmatrix} X - \begin{bmatrix} 0 & 3 \\ 2 & 0 \end{bmatrix} X = \begin{bmatrix} 4 & 2 \\ -2 & 1 \end{bmatrix}$

10 Construct the input-output matrix corresponding to the table below.

| | Purchases | | | |
	Industry 1	Industry 2	Final Demand	Total Output
Industry 1	10	70	120	200
Industry 2	90	20	40	150
Other items	100	60		
Totals	200	150		

11 Assume that the interdependence of the industries in problem 10 remains unchanged when the final demands for outputs of Industries 1 and 2 become 100 and 50, respectively. All numbers indicate millions of dollars. Find the new total output for each industry.

12 Making the same assumptions as in problem 11, find the new total output for each industry when the final demands for the outputs of Industries 1 and 2 become 140 and 80, respectively.

Evaluate each determinant below. Use the properties of determinants when that is possible.

13 $\begin{vmatrix} 8 & 4 \\ 4 & 2 \end{vmatrix}$

14 $\begin{vmatrix} 1 & 1 & 1 \\ 1 & 1 & 1 \\ 1 & 1 & 1 \end{vmatrix}$

15 $\begin{vmatrix} 3 & 2 & 5 \\ 0 & 1 & 0 \\ -1 & -2 & 1 \end{vmatrix}$

16 $\begin{vmatrix} 7 & 9 & 0 \\ 4 & 2 & 0 \\ 3 & 1 & 0 \end{vmatrix}$

17 $\begin{vmatrix} 2 & 1 & -1 & 3 \\ -1 & 4 & 2 & 5 \\ 0 & 1 & -1 & 9 \\ 0 & 0 & 0 & 1 \end{vmatrix}$

18 $\begin{vmatrix} 1 & -1 & 1 & -1 \\ 2 & 0 & -1 & 1 \\ 3 & -3 & 3 & -3 \\ 4 & 0 & -2 & 2 \end{vmatrix}$

Determine which of the following matrices are invertible.

19 $\begin{bmatrix} 1 & -2 \\ 2 & 1 \end{bmatrix}$

20 $\begin{bmatrix} 3 & 5 & -2 \\ -1 & 1 & 0 \\ 2 & 6 & -2 \end{bmatrix}$

21 $\begin{bmatrix} 2 & 7 & -6 & 5 \\ 0 & 3 & 8 & 4 \\ 0 & 0 & 1 & 9 \\ 0 & 0 & 0 & -1 \end{bmatrix}$

Solve each system by Cramer's Rule or indicate why Cramer's Rule does not apply.

22 $.4x + .5y = 0$
    $.3x - .2y = 1$

23 $r + 3s = 1$
    $2r - s = 2$

24 $x_1 + 2x_2 - x_3 = 3$
    $x_1 - x_2 + x_3 = -1$
    $2x_1 + x_2 = 2$

25 $x = 1 - 2y$
    $3x = 2 + z$
    $2z = -1 - 3y$

For each of the following systems, find the value of x in a solution.

26 $x + 2y - 3z = 1$
    $3x - y + z = 0$
    $5x + 2z = 0$

27 $w + x + y = 0$
    $3w - x - y = 0$
    $w + 2x + y - z = 0$
    $x - 2y = 0$

Social Science Applications

Graph Theory*

Graph theory uses matrix computations to understand the possible communication links between stations. The stations might be individuals, committees, or even different departments in a government agency. In the picture, the arcs show the communication links among three stations, represented by the points P_1, P_2, and P_3.

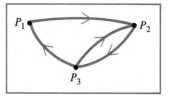

The arrows indicate the directions of possible communications. For example, two-way communication is possible between P_2 and P_3 only.

The possible communication links among the three stations are reported by the 3×3 matrix.

$$A = \begin{bmatrix} a_{11} & a_{12} & a_{13} \\ a_{21} & a_{22} & a_{23} \\ a_{31} & a_{32} & a_{33} \end{bmatrix} = \begin{bmatrix} 0 & 1 & 0 \\ 0 & 0 & 1 \\ 1 & 1 & 0 \end{bmatrix}.$$

Here, we set a_{ij} equal to 1 provided $i \neq j$ and P_i has a direct communication link with P_j; otherwise, a_{ij} equals 0, for $1 \le i, j \le 3$. For example, a_{12} equals 1 since P_1 can communicate directly with P_2, but a_{21} equals 0 since P_2 cannot communicate directly with P_1.

The following remarkable fact is proved in graph theory. The entry of A^m in the ith row and jth column equals the number of different ways to link P_i to P_j, using exactly m arcs. For example, the computation below shows that the entry of A^2 in the first row and second column is 0; thus, there is no link between P_1 and P_2 that uses exactly two arcs.

$$A^2 = \begin{bmatrix} 0 & 1 & 0 \\ 0 & 0 & 1 \\ 1 & 1 & 0 \end{bmatrix}\begin{bmatrix} 0 & 1 & 0 \\ 0 & 0 & 1 \\ 1 & 1 & 0 \end{bmatrix} = \begin{bmatrix} 0 & 0 & 1 \\ 1 & 1 & 0 \\ 0 & 1 & 1 \end{bmatrix}$$

1   Determine the number of different ways to link P_1 to P_2 using exactly three arcs.

2   Determine the number of different ways to link P_2 to P_3 using exactly three arcs.

3   Which point can be linked to another point in more than one way using exactly four arcs.

---

* The term "graph" is used here to mean a path built from simple arcs, not graph of a function.

Solve problems 4 through 7 by Cramer's Rule.

4   ***Political Science***   Voting districts 3 and 4 are side-by-side, and together they contain 220,000 black voters and 180,000 white voters. A judge orders that the boundaries be redrawn so that 60% of the voters in district 3 are black and 50% of the voters in district 4 are white. Determine the number of voters that each district must contain.

5   ***Public Administration***   A county government has $1.51 million for the pay raises for its 800 employees. The government decides to give its most valuable employees a $2000 raise and its less valuable employees a $1700 raise. How many employees will get each raise?

6   ***Sociology***   A sociologist distributes two questionnaires, A and B, to 500 respondents. There are ten questions on questionnaire A and eight questions on B. The results are tabulated by a computer, and the computer indicates that it graded 4560 questions. How many questionnaires of each type were tabulated?

7   ***Psychology***   An experimental psychologist has previously determined that 85% of her trained mice can get through the maze test in the required time. Only 5% of untrained mice can get through the test in the required time. If 100 mice are given the maze test and 37 get through it in the required time, how many of the mice are trained?

CAREER PROFILE
Economist

The good economist is at once a mathematician, a historian, and a forecaster. He or she must be able to make sense of the present, in light of the past, for the purpose of the future. Trained economists are valued contributors to many organizations, including private businesses, public utilities, governments, and universities.

In the Depression of the 1930s, the federal government began to lean heavily on economic principles to make decisions. After World War II, private businesses began to realize the extent to which economic theory could be applied to solve business problems and to develop sound policies.

193

At the present time, approximately 75% of the individuals (about 50,000) employed as economists work in the private business sector. Employers include large manufacturing firms, transportation companies, utilities, banks, insurance firms, investment brokerages, and others. Firms not large enough to warrant hiring their own economists engage economic consultants as needed. Regardless of their size, businesses are aware that government policies and economic trends and cycles have an effect on their opportunities and profits.

The areas of expertise in which an economist may specialize are many and varied. A **bank economist** may specialize in short-term and long-term forecasts of the markets for money and credit. The main function of an **industrial economist** could be to develop and apply economic methods for forecasting company sales. An **investment economist** will be concerned with relating the economic outlook to investment analysis and portfolio decisions. Each analyst applies principles of economics to forecast how external events will affect his or her organization and how economic factors should influence its internal operations.

A bachelor's degree in economics is sufficient educational preparation for many beginning jobs. However, because colleges and universities are presently graduating many economics majors, job competition is becoming keen. As we move through the 1980s, although the number of positions for economists is expected to increase, graduate training will become a more frequent requirement for advancement to responsible positions.

Working economists consistently emphasize the importance of their college courses in applied statistics and accounting. Learning to program and use computers is another essential facet of the educational preparation of an economist. Facility with formulas and graphs is vital. A thorough understanding of important aspects of economic theory depends on a good background in both calculus and linear algebra.

Although the training of an economist must place heavy emphasis on the development of quantitative skills, success in this career field also depends on the development of effective skills in oral and written communication. Economists must be able to make the results of their work understandable to a wide range of people. Those whose decisions are based on the economist's analyses and forecasts may lack knowledge of sophisticated economic ideas, and economists must be able to translate their knowledge and conclusions into language that is clear to these decision makers.

College graduates with majors in economics earn more than those who major in any other social science; in fact, they consistently receive better starting salaries than general business majors. (Accounting majors, however, generally start higher.) Beginning economists with good academic records can expect salaries in excess of $15,000. In 1980, the median salary of business economists was $38,000. Highest average incomes have occurred for those economists working abroad and for those working in the securities and investment industries. Economists' salaries that exceed $50,000 are not unusual.

Sources of Additional Information
American Economic Association, 1313 21st Avenue South, Nashville, TN 37212. (This organization will supply information on schools offering graduate training in economics.)

National Association of Business Economists, 28349 Chagrin Boulevard, Suite 201, Cleveland, OH 44122. (NABE will provide additional information about careers in business economics.)

Occupational Outlook Handbook. Bureau of Labor Statistics, U.S. Department of Labor, Washington, DC 20212. (Revised every 2 years, this handbook provides information about job duties, working conditions, level and places of employment, education and training requirements, advancement possibilities, job outlook, and earnings for about 250 occupations. Some of the information for this Career Profile was obtained from this source.)

5

Linear Programming

5.1 Solving Linear Inequalities

In Section 1.2 we developed the elementary algebra necessary to solve a single linear inequality in one variable such as

$$ax + b < 0 \quad \text{or} \quad ax + b \le 0$$

where a and b represent constants. In order to solve an inequality of the form

$$cx + d > 0 \quad \text{or} \quad cx + d \ge 0$$

we first multiply both sides of the inequality by -1. Then the resulting inequality

$$-cx - d < 0 \quad \text{or} \quad -cx - d \le 0$$

can be solved by the methods already developed.

Now we consider an inequality in two variables, such as

$$ax + by + c < 0,$$

and those inequalities that are obtained by replacing $<$ in this inequality with $\le$, $>$, or $\ge$. Here a, b, and c represent constants. Any inequality in one of these forms will be called a **linear inequality** in two unknowns.

Suppose we want to solve the inequality

$$x + y - 2 < 0.$$

This means that we want to find all ordered pairs of numbers, say (k, l), such that the given inequality is true when x is replaced by k and y is replaced by l. The first step in solving the inequality is to consider the equation

$$x + y - 2 = 0,$$

which results from replacing the inequality sign with an equal sign. The graph of this equation is a straight line that can be determined by finding two ordered pairs that solve the equation. We have sketched the graph by using the pairs $(2, 0)$ and $(0, 2)$.

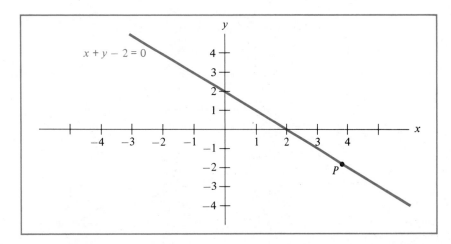

Each point (x, y) on the line makes the equation $x + y - 2 = 0$ true. If we select any point on the line, say P, and we move downward (decreasing the value of y) or we move to the left (decreasing the value of x), we obtain a point (x, y) such that $x + y - 2 < 0$. In fact, the set of solutions to the inequality

$$x + y - 2 < 0$$

is the set of all points in the shaded area of the next graph.

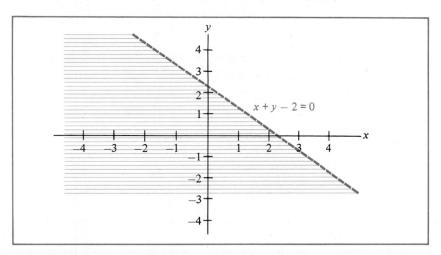

The solution set is the set of points below or to the left of the dotted line $x + y - 2 = 0$. The line is dotted to show that the points on the line are not solutions.

We illustrate this method further with another example.

EXAMPLE 1

Show graphically all of the ordered pairs (x, y) that solve the inequality

$$x + 2y \geq 0.$$

Solution

First we graph the equation

$$x + 2y = 0.$$

We note that $(0, 0)$ and $(2, -1)$ are two points on the line $x + 2y = 0$.

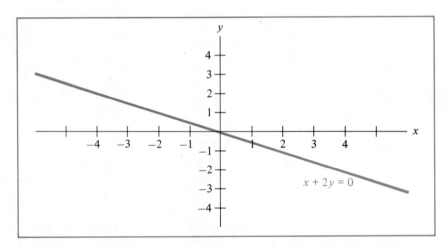

If (x, y) is some point on the line, then $x + 2y = 0$; if we move up, we increase y and, if we move right, we increase x. In either case, we have $x + 2y > 0$. Thus, the set of solutions to

$$x + 2y \geq 0$$

includes the line itself and everything to the right or above the line. The solution set is shaded on the next graph.

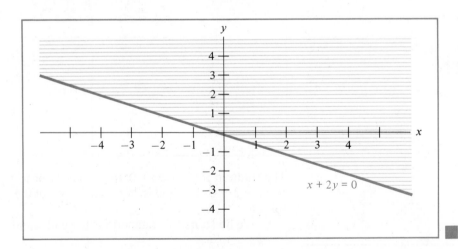

We can now give a simple procedure for graphing the solution set of any linear inequality in two unknowns. First, we sketch the line that is the graph of the equation obtained by changing the inequality sign to an equal sign. We note that the graph of the solution set includes all of the half plane on one side of that line. We can determine which side of the line is part of the solution set by choosing a point on one side and testing its coordinates in the given linear inequality. If the coordinates make the inequality true, then that is the half plane included in the solution set. If the coordinates of our chosen point do not make the inequality true, then the other side of the line is the half plane included in the solution set. The line itself is included in the solution set if and only if the inequality sign is $\leq$ or $\geq$.

EXAMPLE 2

Show graphically all of the points (x, y) that solve the inequality

$$x - 3y - 3 \geq 0.$$

Solution
First, we graph the equation $x - 3y - 3 = 0$.

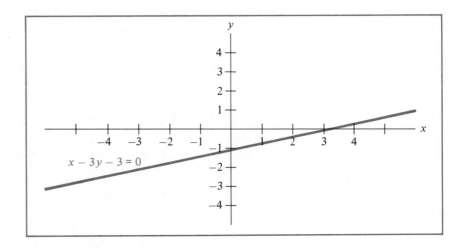

Second, we choose a point, for example $(0, 0)$, on one side of the line and test its coordinates in the given linear inequality. Note that

$$0 - 3(0) - 3 < 0,$$

so the coordinates of $(0, 0)$ do not make the inequality true. Thus, it must be the half plane below or to the right of the line that is included in the solution set. On the next page we shade the relevant half plane.

Since the inequality sign in the given inequality is $\geq$, the line is included in the solution set. ■

Now we are ready to consider a system of linear inequalities. **A system of linear inequalities** is a collection of two or more linear inequalities that are to be considered simultaneously. For inequalities in two unknowns, a **solution** is a point (x, y) whose coordinates make each inequality true. The **solution set** is the collection of all solutions.

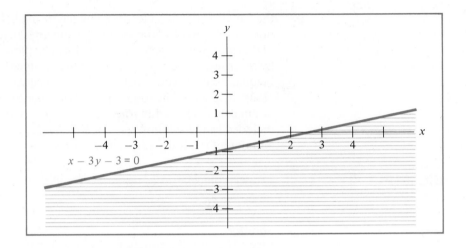

The next example demonstrates an appropriate procedure for solving a system of linear inequalities.

EXAMPLE 3

Show graphically all of the points (x, y) that solve the system

$$y + 2x - 3 < 0,$$
$$y - x + 2 > 0.$$

Solution
We use horizontal shading to identify the solution set for the first inequality and vertical shading to identify the solution set for the second inequality.

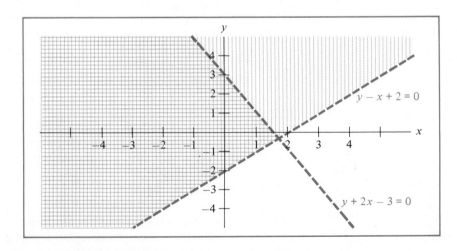

The region that has both horizontal and vertical shading indicates the solution set for the given system of linear inequalities. ▮

It is possible for a system of linear inequalities to have no solutions, as indicated in the next example.

EXAMPLE 4

Show graphically the solution set for the system

$$y - x - 2 > 0,$$

$$y - x + 3 < 0.$$

Solution

We use horizontal shading to identify the solution set for the first inequality and vertical shading to identify the solution set for the second inequality.

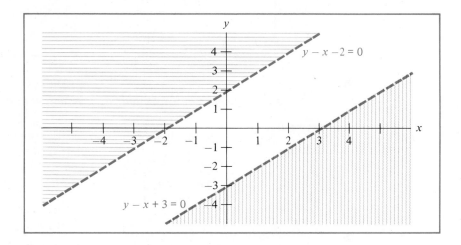

(By the methods of Section 7.1, we see that the two lines are parallel.) Since the solution sets of the two inequalities do not overlap, there is no point (x, y) satisfying both inequalities. The solution set is empty. ■

The next example displays another possibility for the solution set of a system of linear inequalities.

EXAMPLE 5

Show graphically the solution set for the system

$$2x + y - 1 \geq 0,$$

$$2x + y + 2 \geq 0.$$

Solution

We use horizontal shading to identify the solution set for the first inequality and vertical shading to identify the solution set for the second inequality. The region that has both horizontal and vertical shading indicates the solution set for the given system. Again, the two lines are parallel. In this case, the half plane, including the line, that solves the first inequality is the solution to the system. The picture is on the next page.

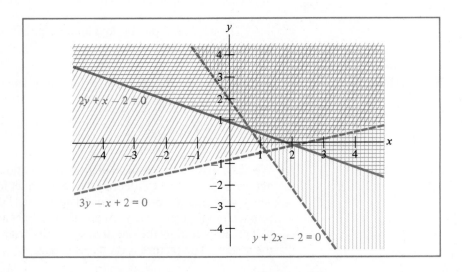

Our last example involves a system with three linear inequalities.

EXAMPLE 6

Show graphically the solution set for the system

$$2y + x - 2 \geq 0,$$
$$y + 2x - 2 > 0,$$
$$3y - x + 2 > 0.$$

Solution
We use horizontal, vertical, and diagonal shadings to identify the solution sets for the first, second, and third inequalities, respectively.

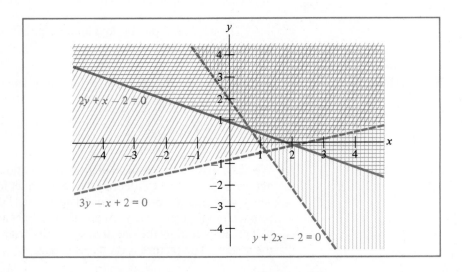

The preceding picture is so complicated that we redraw the picture with only the solution set for the given system shaded.

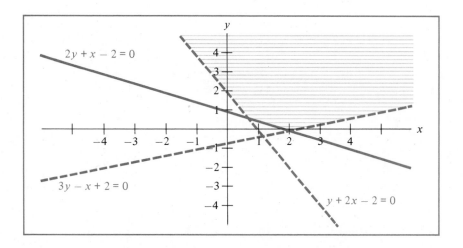

We have drawn dotted line segments for the parts of lines that bound the solution set but are not included in the solution set. The part of the line $2y + x - 2 = 0$ that bounds the solution set is included in the solution set. ■

Exercises 5.1

Show graphically in the plane the set of solutions for each inequality.

1 $x < 3$

2 $x \geq -1$

3 $y \leq -2$

4 $y > 5$

5 $x < y$

6 $y \leq x$

7 $y < 3x$

8 $x \leq -2y$

9 $y < -x + 2$

10 $2x + y < 4$

11 $3x + 4y \leq -1$

12 $x \leq 3 - y$

13 $y < 5x + 2$

14 $2y - 4x < 5$

15 $-3x + y \leq -2$

16 $0 \leq 6x - 3y + 9$

17 $2x + 4y - 10 \leq 0$

18 $5x - 10y - 3 < 2$

19 $\frac{1}{2}x + \frac{1}{3}y \leq \frac{1}{5}$

20 $.2x - .3y + .6 < .1$

21 $3x + .1y - \frac{1}{2} > \frac{1}{3}$

22 $.4x - 5y + \frac{1}{5} \geq \frac{1}{4}$

Show graphically in the plane the set of solutions for each system.

23 $y - x > 0$
   $y + x < 0$

24 $y - x < 0$
   $y + x < 0$

25 $y - 2 \le 0$
   $2y - x \ge 0$

26 $y - 2 \ge 0$
   $2y - x \le 0$

27 $x - 4 < 0$
   $x + y - 2 > 0$

28 $-x + y + 1 < 0$
   $2x + y - 2 > 0$

29 $-x + y + 1 \le 0$
   $2x + y - 2 \le 0$

30 $y - x - 2 < 0$
   $y - 3x - 2 \ge 0$

31 $y - x - 2 \le 0$
   $y - 3x - 2 > 0$

32 $x + y \ge -3$
   $5x + y \ge -15$

33 $y + x \ge 0$
   $y - x \ge 0$
   $y - 2 \ge 0$

34 $y - 5x \le 0$
   $5y - x \ge 0$
   $y + x - 1 \ge 0$

35 $x + y + 3 < 0$
   $-x + y + 3 > 0$
   $x - 2 > 0$

36 $2x + y + 8 > 0$
   $-x + y - 4 > 0$
   $y - 5 < 0$

37 $5y - x + 5 \le 0$
   $y - 4x + 20 < 0$
   $y + 3x - 31 < 0$

38 $2y - x - 2 < 0$
   $y - 2x + 1 > 0$
   $2y + x + 2 \le 0$

39 $2x + y - 1 < 0$
   $2x + y - 2 > 0$

5.2   Solving Linear Programming Problems by Graphing

A **linear programming problem in two variables** consists of an **objective function** of the form

$$z = ax + by,$$

where a and b are constants and z is a quantity to be maximized (or minimized) together with a set of constraints that can be written as linear inequalities in x and y. If the coordinates of the point (x, y) satisfy the inequalities that express the constraints, then (x, y) is said to be a **feasible solution.** It can be proved that the set of feasible solutions for a given linear programming problem has the following property: whenever (x_1, y_1) and (x_2, y_2) are feasible solutions, then every point (x, y) on the line segment connecting (x_1, y_1) and (x_2, y_2) is a feasible solution. More generally, any set of points (x, y) with this property is said to be **convex.** A triangle with its interior, a rectangle with its interior, and a circle with its interior are examples of convex sets. We illustrate this with the pictures on the next page.

The point of intersection of two straight line segments on the boundary of a convex set is called a **vertex.** It can be proved that the **set of feasible solutions for a given linear programming problem is a convex set whose boundary consists of straight line segments.** Thus, the set of feasible solutions might be a triangle with its interior or a rectangle with its interior. The set of feasible solutions could not be a circle with its interior, despite the fact that a circle with its interior is convex.

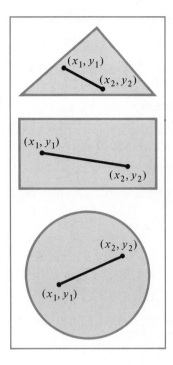

We say that (x, y) is a **solution** to the given linear programming problem provided (1) it is a feasible solution and (2) the coordinates of (x, y) make the objective function a maximum or a minimum, according to which is desired. In the graphing method for solving linear programming problems we use the following theorem.

Theorem If the objective function has a maximum (or minimum), then it must occur as a value at a vertex.

We demonstrate this method in the first example. The example represents a practical problem that has been oversimplified somewhat for the sake of clarity.

EXAMPLE 1

An accountant accumulates $20,000 in her Individual Retirement Account at a local bank. She decides to use the money to make an investment in bonds; she plans to put some of the money into AAA bonds paying 13% and some into BBB bonds paying 15%. Since AAA bonds are safer than BBB bonds, she wants at least $5000 more in AAA bonds. Since BBB bonds have a higher rate of return than AAA bonds, she wants at least $6000 invested in BBB bonds. How much should be invested in each type of bond to achieve a maximum return on the investment?

Solution
Let x and y represent the two unknowns.

x = amount invested in AAA bonds

y = amount invested in BBB bonds

The condition that the amount invested in bonds cannot exceed the available $20,000 can be written

$x + y \le 20,000.$

The condition that the amount of AAA bonds must exceed the amount in BBB bonds by at least $5000 can be written

$x - y \ge 5,000.$

The condition that the amount in BBB bonds must be at least $6000 can be written

$y \ge 6,000.$

The return on the investment R is the sum of the return on the AAA bonds, which is $.13x$, and the return on the BBB bonds, which $.15y$. Thus, we have

$R = .13x + .15y.$

Tersely stated, the problem is to maximize the objective function $R = .13x + .15y$ while satisfying the constraints

$$x + y \le 20,000$$
$$x - y \ge \ 5,000$$
$$y \ge \ 6,000$$
$$x \ge 0, \qquad y \ge 0.$$

The last two inequalities are required by the sense of the problem.

Next we have graphed the inequalities and shaded the region that represents the feasible solutions.

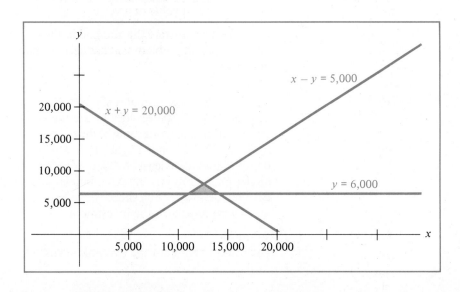

The basic fact that we use to solve this problem is that the maximum value of R must occur at one of the vertices of the shaded region, which represents the feasible solutions. The lower left corner of the shaded region is the point of intersection of the graphs of $x - y = 5,000$ and $y = 6,000$. We easily find the coordinates by substitution.

$$x - 6,000 = 5,000$$

$$x = 11,000$$

$$(x, y) = (11,000, 6,000)$$

The lower right corner is the point of intersection of the graphs of $x + y = 20,000$ and $y = 6,000$. Again we obtain the coordinates by substitution.

$$x + 6,000 = 20,000$$

$$x = 14,000$$

$$(x, y) = (14,000, 6,000)$$

At the upper corner, the graphs of $x - y = 5,000$ and $x + y = 20,000$ intersect; we solve this pair of equations simultaneously.

$$x - y = 5,000$$

$$x + y = 20,000$$

Add the two equations to get

$$2x = 25,000$$

$$x = 12,500.$$

Substitute this value for x in the first equation to find y.

$$12,500 - y = 5,000$$

$$y = 7,500$$

$$(x, y) = (12,500, 7,500)$$

Having found the coordinates of each vertex, we compute the value of R at each vertex.

Vertex	Return on Investment
(11,000, 6,000)	$.13(11,000) + .15(6,000) = 2,330$
(14,000, 6,000)	$.13(14,000) + .15(6,000) = 2,720$
(12,500, 7,500)	$.13(12,500) + .15(7,500) = 2,750$

Now we conclude that the maximum return on investment occurs when $12,500 is invested in AAA bonds and $7,500 is invested in BBB bonds. ∎

In the next example we use the graphing method to find a minimum for a cost function.

EXAMPLE 2

Mr. McDonald uses two fertilizers. Fertilizer no. 1 has 10 units of nitrogen, 20 units of phosphorus, and 5 units of potash in each pound; it costs $9 per pound. Fertilizer no. 2 has 10 units of nitrogen, 5 units of phosphorus, and 10 units of potash in each pound; it costs $7 per pound. Mr. McDonald needs a mix of these two fertilizers with at least 20 units of phosphorus, at least 10 units of potash, and not more than 20 units of nitrogen. How should he combine the two fertilizers to minimize his costs?

Solution

Let x and y represent the two unknowns.

$x =$ amount of fertilizer no. 1 in the mix

$y =$ amount of fertilizer no. 2 in the mix

The amount of nitrogen in the mix is $10x + 10y$ and the constraint on nitrogen can be written

$$10x + 10y \leq 20.$$

The constraints on the phosphorus and potash can be written

$$20x + 5y \geq 20$$
$$5x + 10y \geq 10.$$

The cost of 1 pound of the mix is $C = 9x + 7y$. The problem is to minimize

$$C = 9x + 7y$$

while satisfying the constraints

$$10x + 10y \leq 20$$
$$20x + 5y \geq 20$$
$$5x + 10y \geq 10$$
$$x \geq 0, \quad y \geq 0.$$

The last two inequalities are required by the sense of the problem.

At the top of the next page we have graphed the inequalities and shaded the region that represents the feasible solutions.

The uppermost vertex is the point of intersection of the graphs of $20x + 5y = 20$ and $10x + 10y = 20$. We solve these equations simultaneously by subtracting 2 times the second equation from the first equation.

$$20x + 5y = 20$$
$$10x + 10y = 20$$
$$-15y = -20$$

$$y = \frac{4}{3}$$

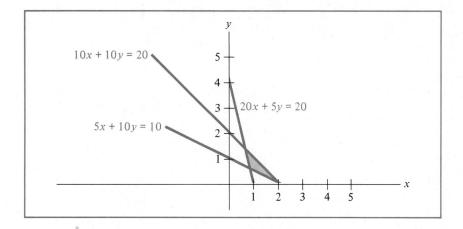

We substitute for y in the second equation and solve for x.

$$10x + 10\left(\frac{4}{3}\right) = 20$$

$$10x = \frac{20}{3}$$

$$x = \frac{2}{3}$$

$$(x, y) = \left(\frac{2}{3}, \frac{4}{3}\right)$$

The lowest vertex is the point of intersection of the graphs of $5x + 10y = 10$ and $10x + 10y = 20$. We solve these equations simultaneously by multiplying the first one by (-2) and adding it to the second one.

$$5x + 10y = 10$$
$$10x + 10y = 20$$
$$-10y = 0$$
$$y = 0$$

Substituting this into the first equation, we get x.

$$5x = 10$$
$$x = 2$$
$$(x, y) = (2, 0)$$

The remaining vertex is the point of intersection of the graphs of $20x + 5y = 20$ and $5x + 10y = 10$. We solve these equations simultaneously by multiplying the second one by (-4) and adding it to the first one.

$$20x + 5y = 20$$

$$5x + 10y = 10$$

$$-35y = -20$$

$$y = \frac{4}{7}$$

Substituting this into the second equation, we get x.

$$5x + 10\left(\frac{4}{7}\right) = 10$$

$$5x = \frac{30}{7}$$

$$x = \frac{6}{7}$$

$$(x, y) = \left(\frac{6}{7}, \frac{4}{7}\right)$$

Having found the coordinates of each vertex, we compute the value of C at each vertex.

Vertex	Cost of Indicated Mix of Fertilizers
$(\frac{2}{3}, \frac{4}{3})$	$9(\frac{2}{3}) + 7(\frac{4}{3}) = 15\frac{1}{3}$
$(2,0)$	$9(2) + 7(0) = 18$
$(\frac{6}{7}, \frac{4}{7})$	$9(\frac{6}{7}) + 7(\frac{4}{7}) = 11\frac{5}{7}$

We conclude that the minimum cost occurs when $\frac{6}{7}$ pound of fertilizer no. 1 is mixed with $\frac{4}{7}$ pound of fertilizer no. 2. ■

In the next example, the constraints $x \geq 0$, $y \geq 0$ play a particularly important role.

EXAMPLE 3

Bergere Company manufactures two kinds of wood chairs. The parts for each chair are purchased from a lumber yard already cut. The simple chair requires 2 hours on the assembly line and another 2 hours in the finishing room, where the wood is stained and varnished. The fancy chair requires 4 hours on the assembly line and 2 hours in the finishing room. The assembly line is not available for more than 100 hours per week, and the finishing room is not available for more than 80 hours per week. The profit on the simple chair is $5, and the profit on the fancy chair is $10. Determine how many chairs of each type should be manufactured in order to maximize profit.

Solution
Let x and y represent the two unknowns.

x = number of simple chairs that are manufactured

y = number of fancy chairs that are manufactured

The constraints can be written

$$2x + 4y \leq 100$$

$$2x + 2y \leq 80$$

$$x \geq 0, \qquad y \geq 0.$$

The profit $P = 5x + 10y$ is to be maximized.

Next we have graphed the inequalities and shaded the region that represents the feasible solutions.

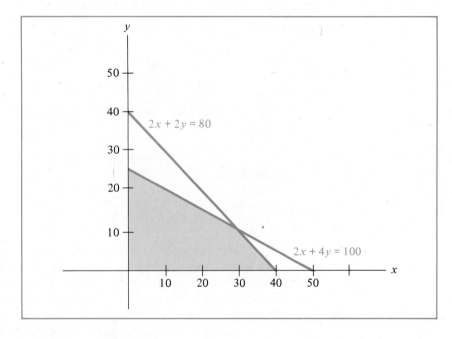

First, we find the vertex on the y-axis.

$$2x + 4y = 100$$

$$x = 0$$

$$4y = 100$$

$$y = 25$$

$$(x, y) = (0, 25)$$

Now, we find the vertex on the x-axis.

$$2x + 2y = 80$$

$$y = 0$$

$$2x = 80$$

$$x = 40$$

$$(x, y) = (40, 0)$$

Finally, we determine the vertex where the two lines intersect. We subtract the second equation from the first equation.

$$2x + 4y = 100$$

$$2x + 2y = 80$$

$$2y = 20$$

$$y = 10$$

$$2x + 4(10) = 100$$

$$2x = 60$$

$$x = 30$$

$$(x, y) = (30, 10)$$

Having found the coordinates of each vertex, we compute the value of P at each vertex.

Vertex	Profit from the Indicated Production
(0, 25)	$5(0) + 10(25) = 250$
(40, 0)	$5(40) + 10(0) = 200$
(30, 10)	$5(30) + 10(10) = 250$

We conclude that the maximum profit occurs at two different vertices. Bergere Company can achieve maximum profit by manufacturing 25 fancy chairs and no simple chairs or by manufacturing 10 fancy chairs and 30 simple chairs. ◼

In the next three examples, we illustrate other phenomena that can arise when solving linear programming problems by the graphing method.

EXAMPLE 4

Minimize the objective function

$$z = 10x + 8y$$

subject to the constraints

$$5x + y \geq 5$$

$$x + 4y \geq 4$$

$$x \geq 0, \qquad y \geq 0.$$

Solution

First, we graph the inequalities and shade the region that represents the feasible solutions to our problem.

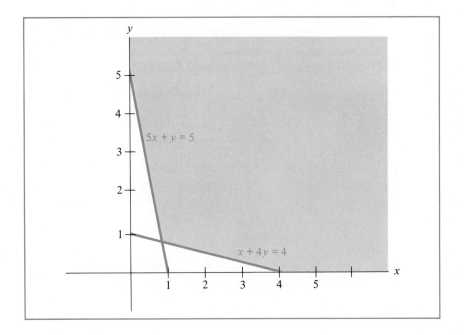

Second, we locate the vertices. To find the vertex on the y-axis, we substitute $x = 0$ into $5x + y = 5$, and we get $y = 5$. To find the vertex on the x-axis, we substitute $y = 0$ into $x + 4y = 4$, and we get $x = 4$. To find the vertex where the two lines cross, we solve the system

$$5x + y = 5$$
$$x + 4y = 4.$$

Multiply the second equation by (-5) and add that to the first equation.

$$-19y = -15$$

$$y = \frac{15}{19}$$

Substituting in the second equation of the system, we get x.

$$x + 4\left(\frac{15}{19}\right) = 4$$

$$x = \frac{16}{19}$$

Third, we evaluate the objective function at each vertex.

Vertex	Value of Objective Function
$(0, 5)$	$10(0) + 8(5) = 40$
$(4, 0)$	$10(4) + 8(0) = 40$
$(\frac{16}{19}, \frac{15}{19})$	$10(\frac{16}{19}) + 8(\frac{15}{19}) = 14\frac{14}{19} \approx 14.74$

We conclude that the minimum value of the objective function is $14\frac{14}{19}$ and the minimum occurs when $x = \frac{16}{19}$ and $y = \frac{15}{19}$. ■

The next example illustrates a possible difficulty when a maximum is sought.

EXAMPLE 5

Try to maximize the objective function

$$z = 8x + 5y$$

subject to the constraints

$$x + 3y \geq 3$$

$$x \geq 0, \qquad y \geq 0.$$

Solution

Next is the graph of the inequalities; the region representing the feasible solutions is shaded.

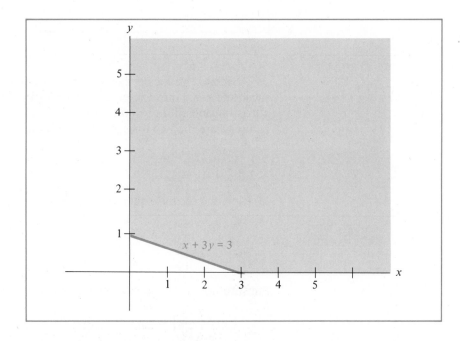

We do not bother to find the vertices of the region of feasible solutions, because it is easy to see that the objective function does not have a maximum. Along the y-axis the region of feasible solutions includes points (x, y) with y arbitrarily large. Consequently, $z = 8x + 5y$ can be made arbitrarily large. There is no maximum for the objective function. ■

Our basic theorem says that the maximum value of the objective function occurs at a vertex **provided that it has a maximum. It can be proved that** the objective function has a maximum value provided the region of feasible solutions is bounded in all directions.

The last example illustrates that the region that represents the feasible solutions may be odd looking.

EXAMPLE 6

Maximize and minimize the objective function

$$z = x + 2y$$

subject to the constraints

$$x \le 5, \quad y \le 4$$
$$3x + 2y \ge 6$$
$$x + 2y \le 9$$
$$x \ge 0, \quad y \ge 0.$$

Solution

Next is the graph of the inequalities; the region representing the feasible solutions is shaded.

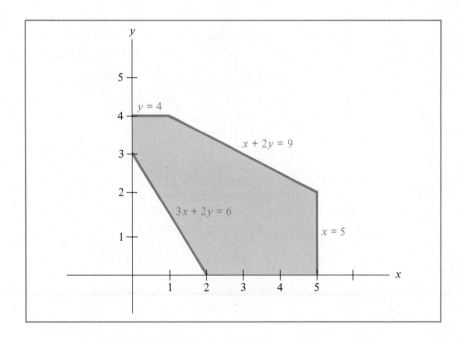

The vertices on the y-axis are $(0, 3)$ and $(0, 4)$; the vertices on the x-axis are $(2, 0)$ and $(5, 0)$. The intersection of $y = 4$ and $x + 2y = 9$ is the vertex $(1, 4)$. The intersection of $x = 5$ and $x + 2y = 9$ is the vertex $(5, 2)$. Below we compute the value of the objective function at each vertex.

Vertex	Value of Objective Function
(0, 4)	$0 + 2(4) = 8$
(0, 3)	$0 + 2(3) = 6$
(2, 0)	$2 + 2(0) = 2$
(5, 0)	$5 + 2(0) = 5$
(5, 2)	$5 + 2(2) = 9$
(1, 4)	$1 + 2(4) = 9$

The minimum occurs when $x = 2$ and $y = 0$. The maximum occurs when $x = 5$ and $y = 2$ and, also, when $x = 1$ and $y = 4$. ■

Exercises 5.2

Use the graphing method to solve each of the following linear programming problems.

1   Maximize $z = 2x + y$
subject to $x + y \leq 1$, $x \geq 0$, $y \geq 0$.

2   Maximize $z = x + 3y$
subject to $x + 2y \leq 4$, $x \geq 0$, $y \geq 0$.

3   Minimize $z = 2x + y$
subject to $x + y \geq 1$, $x \geq 0$, $y \geq 0$.

4   Minimize $z = x + 3y$
subject to $x + 2y \geq 4$, $x \geq 0$, $y \geq 0$.

5   Maximize $z = 3x + 2y$
subject to $x + 2y \leq 8$, $2x + y \leq 7$, $x \geq 0$, $y \geq 0$.

6   Maximize $z = 4x + 5y$
subject to $x + 3y \leq 15$, $3x + y \leq 13$, $x \geq 0$, $y \geq 0$.

7   Minimize $z = 3x + 2y$
subject to $x + 2y \geq 8$, $2x + y \geq 7$, $x \geq 0$, $y \geq 0$.

8   Minimize $z = 4x + 5y$
subject to $x + 3y \geq 15$, $3x + y \geq 13$, $x \geq 0$, $y \geq 0$.

9   Maximize $z = 4x + 4y$
subject to $y \leq 3$, $x \leq 2$, $x \geq 0$, $y \geq 0$.

10   Maximize $z = 2x + 2y$
subject to $y \leq 4$, $x \leq 5$, $x \geq 0$, $y \geq 0$.

11   Minimize $z = 4x + 4y$
subject to $y \geq 3$, $x \geq 2$.

12   Minimize $z = 2x + 2y$
subject to $y \geq 4$, $x \geq 5$.

13   Maximize $z = 3x + 5y$
subject to $x - y \geq 1$, $4x - y \leq 12$, $x \geq 0$, $y \geq 0$.

14   Maximize $z = 4x + y$
subject to $y - 3x \geq 1$, $y - x \leq 4$, $x \geq 0$, $y \geq 0$.

15 Minimize $z = 3x + 5y$
    subject to $x - y \le 1$, $4x - y \ge 12$, $x \ge 0$, $y \ge 0$.
16 Minimize $z = 4x + y$
    subject to $y - 3x \le 1$, $y - x \ge 4$, $x \ge 0$, $y \ge 0$.
17 Maximize $z = 5x + 4y$
    subject to $x + y \le 4$, $2x + y \le 6$, $2x - y \le 4$, $x \ge 0$, $y \ge 0$.
18 Maximize $z = 2x + 3y$
    subject to $y \le 3$, $x + y \le 5$, $x - y \le 1$, $x \ge 0$, $y \ge 0$.
19 Minimize $z = 5x + 4y$
    subject to $x + y \ge 4$, $2x + y \ge 6$, $2x - y \le 4$, $x \ge 0$, $y \ge 0$.
20 Minimize $z = 2x + 3y$
    subject to $y \le 3$, $x + y \le 5$, $x - y \le 1$, $x \ge 0$, $y \ge 0$.
21 Try to maximize and minimize $z = 10x - 8y$
    subject to $4x - y \ge 3$, $x - y \le 0$, $4x + y \le 15$, $x \ge 0$, $y \ge 0$.
22 Try to maximize and minimize $z = 3x + 4y$
    subject to $y \le 5$, $2x + y \ge 7$, $2x - y \le 5$, $x \ge 0$, $y \ge 0$.
23 Try to maximize and minimize $z = x + y$
    subject to $x + y \ge 1$, $x - y \ge -1$, $x - y \le 1$, $x \ge 0$, $y \ge 0$.
24 Try to maximize and minimize $z = 2x - y$
    subject to $2x - y \ge 0$, $x - y \le 0$, $x + y \ge 0$, $x \ge 0$, $y \ge 0$.
25 Try to maximize and minimize $z = 2x + 10y$
    subject to $x \ge 2$, $y \ge 0$, $x - y \ge -2$, $x + y \le 10$, $5x - y \le 20$.

Solve each of the following problems.

26 The sum of $10,000 is to be divided between AAA bonds paying 14% and BB bonds paying 18%. If the investment in BB bonds must not exceed the investment in the safer AAA bonds and the total return on the investment is to be maximized, how should the funds be divided?
27 The sum of $15,000 is to be divided between A bonds paying 15% and BBB bonds paying 16%. If the amount in A bonds must exceed the amount in BBB bonds by at least $3000 and the total return is to be maximized, how should the funds be divided?
28 A farmer's weekly application of fertilizer for a certain crop must contain at least 80 units of nitrogen, 120 units of phosphorus and 120 units of potash. Fifty pounds of fertilizer no. 1, which costs $6, contains 5, 15, and 5 units of these three ingredients, respectively. Fifty pounds of fertilizer no. 2, which costs $8, contains 10, 10, and 30 units of these three ingredients, respectively. Plan the farmer's weekly purchases of the two fertilizers so that his costs are minimized.
29 Suppose the farmer mentioned in problem 28 has another crop that must receive no more than 100 units of nitrogen and no less than 120 units of each of the other two ingredients. How much of each of the two fertilizers should the farmer purchase if he wants to minimize his costs?
30 Blaque Company makes a simple table and a fancy table. Each simple table requires 4 hours of assembly and 4 hours of finishing and earns $5 of profit. Each fancy table requires 5 hours of assembly and 10 hours of

finishing and earns $8 profit. If the company can spend up to 100 hours per week on assembly and up to 140 hours per week on finishing, how many tables of each type should be made for maximum profit?

31 A company uses three production crews to make two products. Product A requires 6, 3, and 5 hours, respectively, with crews no. 1, no. 2, and no. 3; its sale brings $16 profit. Product B requires 3, 6, and 4 hours, respectively, with crews no. 1, no. 2, and no. 3; it brings $14 profit. If the total hours that the crews are available are 90, 75, and 80, respectively, determine the number of each product that leads to maximum profit.

32 One ounce of food A provides 10 units of niacin, 5 units of thiamine, and 15 units of riboflavin; it costs 12¢. One ounce of food B provides 10 units of niacin, 15 units of thiamine, and 5 units of riboflavin; it costs 10¢. These two foods should be mixed so that the resulting diet contains at least 100 units of niacin and 120 units of each of the others. What mixture of the foods meets these nutrition requirements and costs the least possible?

33 Suppose that one ounce of food A, in problem 32, supplies 12 units of niacin, 8 units of thiamine, and 10 units of riboflavin; suppose it costs 10¢. Suppose one ounce of food B supplies 8 units of niacin, 12 units of thiamine, and 10 units of riboflavin; it costs 11¢. If the diet must contain 110 units of riboflavin and 120 units of each of the others, how should the foods be mixed for minimum costs?

34 A grocery store has 200 pounds of steak, 300 pounds of pork roasts, and 400 pounds of beef roasts that have not been sold and are getting old. The manager of the meat department decides to grind as much of this as possible to make hot dogs and hamburger patties. The hot dogs are 40% beef roasts, 40% pork roasts, and 20% cereal fillers; the store has plenty of cereal filler. The hamburger patties are 40% steak and 60% beef roasts. How should the meat be used?

35 A trucker can haul as much freight as he wants for a toy company or a tool company. The toy company uses a box that is 3 cubic feet and weighs 2 pounds, for which the trucker charges $1. The tool company uses a box that is 1 cubic foot and weighs 4 pounds, for which the trucker charges 80¢. The cargo capacity of the truck is 2100 cubic feet and 8000 pounds. How many boxes should the trucker take from each company to maximize his revenue?

5.3   The Simplex Method for Finding a Maximum

The graphical method presented in Section 5.2 is limited to linear programming problems that require only two variables. Many practical problems require more than two variables. Thus, we need a more general method, and the simplex method meets that need.

A linear programming problem in three variables is said to have **standard form** provided there is an objective function

$$z = c_1 x_1 + c_2 x_2 + c_3 x_3 \quad \text{with} \quad x_1 \geq 0, \; x_2 \geq 0, \; x_3 \geq 0$$

to be maximized and some constraints that can be written as linear inequalities like the ones below.

$$a_{11} x_1 + a_{12} x_2 + a_{13} x_3 \leq b_1$$
$$a_{21} x_1 + a_{22} x_2 + a_{23} x_3 \leq b_2$$
$$\cdots \cdots \cdots \cdots \cdots$$
$$a_{m1} x_1 + a_{m2} x_2 + a_{m3} x_3 \leq b_m$$

Here $b_1, b_2, \ldots, b_m$ must be nonnegative and the inequality signs in the constraints must be $\leq$. The simplex method only works for linear programming problems in standard form.

In order to present the simplest and clearest possible explanation of the simplex method, we shall begin with a problem that requires only two variables. Suppose we want to maximize the objective function

$$z = 5x + 10y$$

subject to the constraints $x \geq 0$, $y \geq 0$ and

$$x + 2y \leq 50,$$

$$x + y \leq 40.$$

The first step is to convert each inequality into an equation by introducing additional variables. In the first constraint above we see that $x + 2y$ does not exceed 50; thus, we may add a nonnegative quantity u to $x + 2y$ and take up the slack so that

$$x + 2y + u = 50.$$

Similarly, we add the nonnegative variable v to $x + y$ in order to convert the second constraint into the equation

$$x + y + v = 40.$$

The variables u and v that we have introduced are called **slack variables.**

Note that (x, y) is a feasible solution to the given linear programming problem provided x, y, u, v are nonnegative numbers that solve the system of linear equations

$$(*) \quad \begin{aligned} x + 2y + u &= 50 \\ x + y + v &= 40. \end{aligned}$$

In order to obtain a procedure similar to Gaussian elimination (described in Section 3.2), we exhibit all of the numbers involved in our linear programming problem in the **simple tableau** that follows.

$$\begin{bmatrix} x & y & u & v & z & b \\ 1 & 2 & 1 & 0 & 0 & 50 \\ 1 & 1 & 0 & 1 & 0 & 40 \\ \cdots & \cdots & \cdots & \cdots & \cdots & \cdots \\ -5 & -10 & 0 & 0 & 1 & 0 \end{bmatrix}$$

In the first two rows, we have listed the coefficients of each variable in each of two equations given in the system (*); the numbers on the right sides of those equations are listed in the b-column. To get the entries in the last row, we write the equation that defines the objective function in the form

$$z - 5x - 10y = 0.$$

Then we list the coefficients and right side of this equation in the last row.

The linear system corresponding to the augmented matrix given in the tableau has a solution so easy to obtain that we give it a special name. Each of the variables u, v, and z appears in only one equation; this is indicated in the tableau by the appearance of only one nonzero entry in the column below each of u, v, and z. These variables are called **basic variables,** and the remaining variables, that is, x and y, are called **nonbasic variables.** By choosing the nonbasic variables to be 0, we find that the values of the basic variables resulting in a solution are the numbers listed in the b-column. In our example, we get

$$x = 0, y = 0 \quad \text{and} \quad u = 50, v = 40, z = 0.$$

This solution is called the **basic solution** corresponding to the given tableau.

A tableau with n rows is called a simple tableau provided all of the columns of the $n \times n$ identity matrix appear as columns in the tableau. In our example, each of the columns of the 3×3 identity matrix appears in the tableau. Thus, it is a simple tableau. Recall from Chapter 3 that elementary row operations do not alter the solutions to a given linear system. We are looking for the solution that gives the largest possible value for z.

The simplex method uses a procedure called the **pivot operation,** which is similar to Gaussian elimination. To use a nonzero entry in the ith row and the jth column of the tableau, represented by a_{ij}, as the **pivot element,** do the following: multiply the ith row by the reciprocal of a_{ij}, and add appropriate multiples of the new ith row to the other rows so that each of the other entries in the jth column is 0. The pivot operation produces a new simple tableau with a new set of basic variables. We try to increase the value of z in the basic solution corresponding to the tableau by a wise choice of the pivot element.

The choice of the pivot element is made as follows. Examine the equation corresponding to the last row of the tableau.

$$-5x - 10y + z = 0$$
or
$$z = 5x + 10y$$

Since $x = 0$ and $y = 0$ in the basic solution corresponding to the first simple tableau, and since the coefficient for y in the above equation for z is larger than the coefficient of x, we can increase z the most by increasing the value of y. For this reason we choose the y-column of the tableau as the column containing the pivot element; we shall call this the **pivot column. For any simple tableau, the correct choice of the pivot column is the one that has the negative number with the largest magnitude in the bottom row.**

The choice of the row of the pivot element, called the **pivot row,** is dictated by our desire to keep the constants in the b-column (and, hence, the values in the corresponding basic solution) nonnegative. To find the pivot row that achieves this, divide each **positive** entry in the pivot column into the b-column entry in that same row. **The entry in the pivot column that results in the smallest nonnegative quotient is chosen as the pivot element.** Now the pivot operation produces a new simple tableau, and the value of z in the corresponding basic solution is at least as large as the value obtained from the previous tableau. We continue to choose a new pivot element and perform the pivot operation until there are no negative numbers in the bottom row (except possibly in the b-column), at which time the value of z in the corresponding basic solution will be maximized.

In our sample problem, we previously determined that the pivot column should be the second column. The quotients used to determine the pivot row are

$$\frac{50}{2} = 25, \qquad \frac{40}{1} = 40.$$

Thus, the first row is the pivot row and the pivot element is the first entry in the second column. We apply the pivot operation to the simple tableau that follows, in which the pivot element is circled.

$$\begin{array}{cccccc} x & y & u & v & z & b \\ \left[\begin{array}{cccccc} 1 & ② & 1 & 0 & 0 & 50 \\ 1 & 1 & 0 & 1 & 0 & 40 \\ \hdashline -5 & -10 & 0 & 0 & 1 & 0 \end{array}\right] \end{array}$$

We use the following elementary row operations: multiply the first row by $\frac{1}{2}$, subtract the new first row from the second row, and add 10 times the first row to the third row. The new simple tableau is

$$\begin{array}{cccccc} x & y & u & v & z & b \\ \left[\begin{array}{cccccc} \frac{1}{2} & 1 & \frac{1}{2} & 0 & 0 & 25 \\ \frac{1}{2} & 0 & -\frac{1}{2} & 1 & 0 & 15 \\ \hdashline 0 & 0 & 5 & 0 & 1 & 250 \end{array}\right] \end{array}.$$

The simplex method is completed when no entry in the last row, except possibly the b-column entry, is negative. Thus, the process is completed in our example. To read the basic solution corresponding to the final tableau, set all nonbasic variables (the ones whose columns do not consist of all 0's except for one 1) equal to 0, and read the values of the basic variables from the b-column. In this case, the basic solution is

$$x = 0, \, u = 0 \quad \text{and} \quad y = 25, \, v = 15, \, z = 250.$$

Note that $x = 0$, $y = 25$ is a feasible solution and the corresponding value of the objective function is $z = 250$. The simplex method indicates that this is the maximum value. Writing the equation corresponding to the last row, we have

or

$$5u + z = 250$$

$$z = 250 - 5u.$$

Note that we cannot increase the value of z by increasing the value of any variable. The only variable in the last equation is u, and any increase in u decreases z.

The example that we have just solved by the simplex method was solved by the graphical method in Example 3 of Section 5.2. It is interesting to locate on the graph for Example 3 each feasible solution produced as part of the basic solution corresponding to a simple tableau. The first feasible solution obtained is $(0, 0)$, which is the vertex at the origin; the second feasible solution obtained is $(0, 25)$, which is the vertex on the y-axis. The simplex method moves from one vertex to another, choosing a direction that increases the value of z.

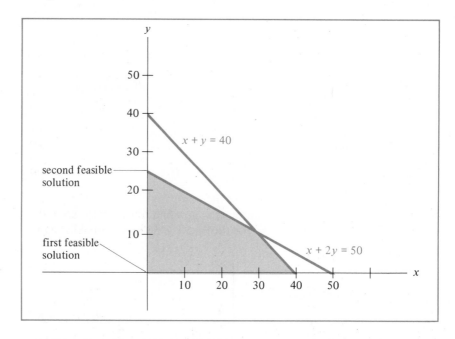

Now we solve another problem that requires only two variables. We perform the steps without explaining why they work.

EXAMPLE 1

Use the simplex method to maximize $z = 5x + 3y$ subject to the constraints $x \geq 0$, $y \geq 0$ and

$$x + 4y \leq 16$$

$$x - 3y \leq 3.$$

Solution

We introduce slack variables u and v, and we rewrite the equation that defines the objective function.

$$x + 4y + u = 16$$

$$x - 3y + v = 3$$

$$-5x - 3y + z = 0$$

Now we write the simple tableau for this problem.

$$
\begin{array}{c}
\begin{array}{cccccc} x & y & u & v & z & b \end{array} \\
\left[
\begin{array}{cccccc}
1 & 4 & 1 & 0 & 0 & 16 \\
1 & -3 & 0 & 1 & 0 & 3 \\
\cdots & \cdots & \cdots & \cdots & \cdots & \cdots \\
-5 & -3 & 0 & 0 & 1 & 0
\end{array}
\right]
\end{array}
$$

The negative number with the greatest magnitude in the last row is -5; thus, the first column is the pivot column. Since $\frac{3}{1}$ is less than $\frac{16}{1}$, the second row is the pivot row. To perform the pivot operation, we subtract the second row from the first row, and we add 5 times the second row to the third row. The result is

$$
\begin{array}{c}
\begin{array}{cccccc} x & y & u & v & z & b \end{array} \\
\left[
\begin{array}{cccccc}
0 & 7 & 1 & -1 & 0 & 13 \\
1 & -3 & 0 & 1 & 0 & 3 \\
\cdots & \cdots & \cdots & \cdots & \cdots & \cdots \\
0 & -18 & 0 & 5 & 1 & 15
\end{array}
\right]
\end{array}.
$$

The negative number with the greatest magnitude in the last row is -18; thus, the second column is the pivot column. Since 7 is the only positive entry in the second column, the first row is the pivot row. To perform the pivot operation, we multiply the first row by $\frac{1}{7}$, we add 3 times the new first row to the second row, and we add 18 times the new first row to the third row. The result is

$$
\begin{array}{c}
\begin{array}{cccccc} x & y & u & v & z & b \end{array} \\
\left[
\begin{array}{cccccc}
0 & 1 & \frac{1}{7} & -\frac{1}{7} & 0 & \frac{13}{7} \\
1 & 0 & \frac{3}{7} & \frac{4}{7} & 0 & \frac{60}{7} \\
\cdots & \cdots & \cdots & \cdots & \cdots & \cdots \\
0 & 0 & \frac{18}{7} & \frac{17}{7} & 1 & \frac{339}{7}
\end{array}
\right]
\end{array}.
$$

Since there are no negative numbers in the last row, we are done. The corresponding basic solution is

$$u = 0, \, v = 0 \quad \text{and} \quad x = \frac{60}{7}, y = \frac{13}{7}, z = \frac{339}{7}.$$

The maximum value for the objective function is $z = \frac{339}{7}$, and it occurs when $x = \frac{60}{7}, y = \frac{13}{7}$. ∎

Next we summarize the simplex method for problems like the ones we have been considering.

Simplex Method for Maxima

1. Check that the linear programming problem is in standard form.
2. Convert each constraint into an equation by introducing a slack variable. Rewrite the equation defining the objective function so that z appears on the left side and only 0 appears on the right side.
3. Write the simple tableau for the given problem.
4. The pivot column is the column, other than the b-column, that contains the negative entry with the largest magnitude in the last row. If two or more entries in the last row are negative with the largest magnitude, then choose one of them arbitrarily.
5. The pivot row is a row containing a positive entry in the pivot column such that the b-column entry of that row divided by the positive entry is the smallest such nonnegative quotient.
6. Perform the pivot operation using the indicated pivot element.
7. Repeat the process of selecting a pivot element until there are no negative entries in the last row, except possibly in the b-column.
8. The maximum value of z is the value of z in the basic solution corresponding to the final simple tableau.

Now we use the simplex method on a problem that we could not solve by the graphical method, because it involves more than two independent variables.

EXAMPLE 2

Find the maximum of the function

$$z = 6x_1 + x_2 + 10x_3$$

subject to the constraints $x_1 \geq 0$, $x_2 \geq 0$, $x_3 \geq 0$ and

$$x_1 + 3x_2 + 2x_3 \leq 10,$$

$$4x_1 + x_2 + x_3 \leq 8,$$

$$3x_1 + 5x_2 + x_3 \leq 12.$$

Solution

The problem is given in standard form. We convert the constraints into equations, and we rewrite the equation for the objective function.

$$x_1 + 3x_2 + 2x_3 + u_1 = 10$$

$$4x_1 + x_2 + x_3 + u_2 = 8$$

$$3x_1 + 5x_2 + x_3 + u_3 = 12$$

$$-6x_1 - x_2 - 10x_3 + z = 0$$

We exhibit all of the relevant numbers in a simple tableau.

$$
\begin{array}{ccccccc}
x_1 & x_2 & x_3 & u_1 & u_2 & u_3 & z & b \\
\left[\begin{array}{ccccccc|c}
1 & 3 & 2 & 1 & 0 & 0 & 0 & 10 \\
4 & 1 & 1 & 0 & 1 & 0 & 0 & 8 \\
3 & 5 & 1 & 0 & 0 & 1 & 0 & 12 \\
\hdashline
-6 & -1 & -10 & 0 & 0 & 0 & 1 & 0
\end{array}\right]
\end{array}
$$

The negative entry with the greatest magnitude in the last row is -10, so the pivot column is the x_3-column. The quotients associated with the positive entries of the pivot column are $\frac{10}{2}=5, \frac{8}{1}=8$, and $\frac{12}{1}=12$; thus, the first row is the pivot row. The pivot element is the first entry in the x_3-column. The result of the pivot operation is given below.

$$
\begin{array}{ccccccc}
x_1 & x_2 & x_3 & u_1 & u_2 & u_3 & z & b \\
\left[\begin{array}{ccccccc|c}
\frac{1}{2} & \frac{3}{2} & 1 & \frac{1}{2} & 0 & 0 & 0 & 5 \\
\frac{7}{2} & -\frac{1}{2} & 0 & -\frac{1}{2} & 1 & 0 & 0 & 3 \\
\frac{5}{2} & \frac{7}{2} & 0 & -\frac{1}{2} & 0 & 1 & 0 & 7 \\
\hdashline
-1 & 14 & 0 & 5 & 0 & 0 & 1 & 50
\end{array}\right]
\end{array}
$$

The only negative entry in the last row is -1, so the new pivot column is the x_1-column. The quotients associated with the positive entries of the pivot column are $10, \frac{6}{7}$, and $\frac{14}{5}$; thus, the pivot row is the second row. The pivot element is the second entry in the x_1-column. The result of the pivot operation is given below.

$$
\begin{array}{ccccccc}
x_1 & x_2 & x_3 & u_1 & u_2 & u_3 & z & b \\
\left[\begin{array}{ccccccc|c}
0 & \frac{11}{7} & 1 & \frac{4}{7} & -\frac{1}{7} & 0 & 0 & \frac{32}{7} \\
1 & -\frac{1}{7} & 0 & -\frac{1}{7} & \frac{2}{7} & 0 & 0 & \frac{6}{7} \\
0 & \frac{27}{7} & 0 & -\frac{1}{7} & -\frac{5}{7} & 1 & 0 & \frac{34}{7} \\
\hdashline
0 & \frac{97}{7} & 0 & \frac{34}{7} & \frac{2}{7} & 0 & 1 & \frac{356}{7}
\end{array}\right]
\end{array}
$$

Since there are no negative entries in the last row, the procedure is completed. The basic solution corresponding to the final simple tableau is $x_2=0, u_1=0, u_2=0$, and $x_3=\frac{32}{7}, x_1=\frac{6}{7}, u_3=\frac{34}{7}, z=\frac{356}{7}$. Thus, the maximum is $\frac{356}{7}$ and it occurs at the feasible solution $x_1=\frac{6}{7}, x_2=0, x_3=\frac{32}{7}$. ∎

We conclude this section by applying the simplex method to a practical problem.

EXAMPLE 3

Weir Company makes three kinds of tables: simple, fancy, and extra fancy. The next chart shows the hours required for cutting the parts, assembling the parts, and finishing each kind of table as well as the profit resulting from the sale of each type.

The cutting, assembly, and finishing lines are available for 84, 80, and 100 hours per week, respectively. Determine how many tables of each type should be made in order to maximize profit.

	Cutting	Assembly	Finishing	Profit
Simple	2	2	2	10
Fancy	4	2	4	18
Extra fancy	4	4	4	22

Solution

Let x, y, and z be the numbers of simple, fancy, and extra fancy tables made per week. Then the week's profit is

$$P = 10x + 18y + 22z.$$

The constraints are $x \geq 0$, $y \geq 0$, $z \geq 0$ and

$$2x + 4y + 4z \leq 84,$$
$$2x + 2y + 4z \leq 80,$$
$$2x + 4y + 4z \leq 100.$$

We simplify these slightly by multiplying each inequality by $\frac{1}{2}$. The result is

$$x + 2y + 2z \leq 42,$$
$$x + y + 2z \leq 40,$$
$$x + 2y + 2z \leq 50.$$

We rewrite the constraints and equation for the objective function, which is the profit function, in the usual way.

$$x + 2y + 2z + u = 42$$
$$x + y + 2z + v = 40$$
$$x + 2y + 2z + w = 50$$
$$-10x - 18y - 22z + P = 0$$

We exhibit these numbers in a simple tableau.

$$
\begin{array}{cccccccc}
x & y & z & u & v & w & P & b \\
\end{array}
$$

$$
\left[
\begin{array}{ccccccc|c}
1 & 2 & 2 & 1 & 0 & 0 & 0 & 42 \\
1 & 1 & 2 & 0 & 1 & 0 & 0 & 40 \\
1 & 2 & 2 & 0 & 0 & 1 & 0 & 50 \\
\hdashline
-10 & -18 & -22 & 0 & 0 & 0 & 1 & 0 \\
\end{array}
\right]
$$

The negative number with the greatest magnitude in the last row is -22. Thus, the z-column is the pivot column. By considering the appropriate quotients, we choose the second row as the pivot row. The result of the pivot operation follows.

$$\begin{array}{cccccccc} x & y & z & u & v & w & P & b \\ \left[\begin{array}{ccccccc|c} 0 & 1 & 0 & 1 & -1 & 0 & 0 & 2 \\ \frac{1}{2} & \frac{1}{2} & 1 & 0 & \frac{1}{2} & 0 & 0 & 20 \\ 0 & 1 & 0 & 0 & -1 & 1 & 0 & 10 \\ \hdashline 1 & -7 & 0 & 0 & 11 & 0 & 1 & 440 \end{array}\right] \end{array}$$

Since -7 is the only negative entry in the last row, the pivot column is the y-column. By considering the appropriate quotients, we choose the first row as the pivot row. The result of the pivot operation follows.

$$\begin{array}{cccccccc} x & y & z & u & v & w & P & b \\ \left[\begin{array}{ccccccc|c} 0 & 1 & 0 & 1 & -1 & 0 & 0 & 2 \\ \frac{1}{2} & 0 & 1 & -\frac{1}{2} & 1 & 0 & 0 & 19 \\ 0 & 0 & 0 & -1 & 0 & 1 & 0 & 8 \\ \hdashline 1 & 0 & 0 & 7 & 4 & 0 & 1 & 454 \end{array}\right] \end{array}$$

Since there are no negative entries in the last row, the process is complete. The maximum profit is \$454; it results from making no simple tables, 2 fancy tables, and 19 extra fancy tables. ■

Exercises 5.3

Solve the following problems by means of the simplex method.

1 Maximize $z = 2x + y$
   subject to $x + y \leq 1$, $x \geq 0$, $y \geq 0$.
2 Maximize $z = x + 3y$
   subject to $x + 2y \leq 4$, $x \geq 0$, $y \geq 0$.
3 Maximize $z = 3x + 2y$
   subject to $x + 2y \leq 8$, $2x + y \leq 7$, $x \geq 0$, $y \geq 0$.
4 Maximize $z = 4x + 5y$
   subject to $x + 3y \leq 15$, $3x + y \leq 13$, $x \geq 0$, $y \geq 0$.
5 Maximize $z = 4x + 4y$
   subject to $y \leq 3$, $x \leq 2$, $x \geq 0$, $y \geq 0$.
6 Maximize $z = 2x + 2y$
   subject to $y \leq 4$, $x \leq 5$, $x \geq 0$, $y \geq 0$.
7 Maximize $z = 5x + 4y$
   subject to $x + y \leq 4$, $2x + y \leq 6$, $2x - y \leq 4$, $x \geq 0$, $y \geq 0$.
8 Maximize $z = 2x + 3y$
   subject to $y \leq 3$, $x + y \leq 5$, $x - y \leq 1$, $x \geq 0$, $y \geq 0$.
9 Maximize $z = 3x + 2y$
   subject to $x + y \leq 2$, $x - y \leq 2$, $x \geq 0$, $y \geq 0$.
10 Maximize $z = 4x + 5y$
   subject to $y + 4x \leq 5$, $y - 3x \leq 5$, $x \geq 0$, $y \geq 0$.

11 Maximize $P = x + 2y + z$
subject to $x + y + z \leq 4$, $x - y - 2z \leq 3$, $x \geq 0$, $y \geq 0$, $z \geq 0$.

12 Maximize $P = 3x + y + z$
subject to $x + 3y + 2z \leq 5$, $-x + 2y + 3z \leq 6$, $x \geq 0$, $y \geq 0$, $z \geq 0$.

13 Maximize $P = 2x + y + 3z$
subject to $4x - y + z \leq 5$, $2x + 3y + z \leq 6$, $x \geq 0$, $y \geq 0$, $z \geq 0$.

14 Maximize $P = 2x + 4y + z$
subject to $3x + y - 2z \leq 8$, $x + y + 3z \leq 10$, $x \geq 0$, $y \geq 0$, $z \geq 0$.

15 Maximize $P = 4x + 5y + 3z$
subject to $x + y - 2z \leq 4$, $2x + y + 3z \leq 2$, $3x + y - 5z \leq 6$, $x \geq 0$, $y \geq 0$, $z \geq 0$.

16 Maximize $P = x + 2y + 3z$
subject to $2x + 3y + z \leq 6$, $x - 4y + z \leq 5$, $3x - y + z \leq 3$, $x \geq 0$, $y \geq 0$, $z \geq 0$.

17 Maximize $z = 2x_1 + x_2 + x_3 + 3x_4$
subject to $x_1 - x_2 + 2x_3 + x_4 \leq 5$, $2x_1 + 3x_2 - x_3 + x_4 \leq 3$, $x_1 - x_2 + 4x_3 + x_4 \leq 2$, $x_1 \geq 0$, $x_2 \geq 0$, $x_3 \geq 0$, $x_4 \geq 0$.

18 Maximize $z = x_1 + x_2 + 2x_3 + x_4$
subject to $x_1 + x_2 + x_3 + 3x_4 \leq 4$, $x_1 + 3x_2 + 2x_3 + x_4 \leq 6$, $x_1 + x_2 + x_3 + x_4 \leq 2$, $x_1 \geq 0$, $x_2 \geq 0$, $x_3 \geq 0$, $x_4 \geq 0$.

19 Suppose the Weir Company (in Example 3) increases the capacity of its manufacturing plant so that the cutting, assembly, and finishing lines are available for 120, 100, and 130 hours per week, respectively. Determine how many tables of each type should be made in order to maximize profit.

20 Suppose that the customer demand for the different kinds of tables forces Weir Company (in Example 3) to change its prices. Consequently, the profit on a simple table is $14 and the profit on a fancy table or extra fancy table is $12. Otherwise Weir Company's operation is the same as described in Example 3. Determine how many tables of each type should be made in order to maximize profit.

21 A grocery store has 100 pounds of unsold steak, 400 pounds of unsold pork roasts, and 500 pounds of unsold beef roasts that are getting old. The manager of the meat department decides to grind as much of this as possible to make hot dogs, hamburger patties, and sausage. The fractional composition of each is given in the following table.

	Steak	Beef Roasts	Pork Roasts	Cereal
Hot dogs	0	.40	.40	.20
Hamburgers	.40	.60	0	0
Sausage	0	0	.80	.20

For example, sausage is composed of 80% pork roasts and 20% cereal. The store has all the cereal it would ever need. How should the meat be used?

22 Suppose the grocery store in problem 21 has 200 pounds of unsold

steak, 300 pounds of unsold pork roasts, and 400 pounds of unsold beef roasts. How many pounds of hot dogs, hamburger patties, and sausage should be made in order to use as much of the unsold meat as possible?

23   A trucker can haul as much freight as he wants for a toy company, a tool company, or a crafts company. The table below shows volume and weight of the boxes used by each of the three companies; it also shows the revenue that the trucker earns for each box he takes.

	Volume	Weight	Revenue
Toy company	4	2	$1
Tool company	1	4	$.80
Crafts company	2	2	$.80

Volume is measured in cubic feet and weight is measured in pounds. The cargo capacity of the truck is 2200 cubic feet and 8000 pounds. How many boxes should the trucker take from each company to maximize his revenue? (Assume that the boxes can always be arranged in a way that completely fills the truck. Round off the numbers of boxes to the nearest integers.)

5.4   The Simplex Method for Finding a Minimum

Section 5.3 showed how the simplex method can be used to find a maximum for an objective function, provided the linear programming problem is in standard form. The examples and exercises for that section indicated some of the quantities that we seek to maximize: revenue, profit, use of production facilities, use of raw materials, and nutrition. On the other hand, we sometimes are interested in minimizing a quantity; the quantity to be minimized most often is the cost of something. Other quantities that we might want to minimize are investment risk, waste of raw materials, and lost production time.

A linear programming problem in three variables is said to have **standard form** provided there is an objective function

$$z = c_1x_1 + c_2x_2 + c_3x_3 \quad \text{with} \quad x_1, x_2, x_3, c_1, c_2, c_3 \geq 0$$

to be minimized and some constraints that can be written as linear inequalities like the ones that follow.

$$a_{11}x_1 + a_{12}x_2 + a_{13}x_3 \geq b_1$$
$$a_{21}x_1 + a_{22}x_2 + a_{23}x_3 \geq b_2$$
$$\cdots \cdots \cdots \cdots \cdots$$
$$a_{m1}x_1 + a_{m2}x_2 + a_{m3}x_3 \geq b_m.$$

The inequality signs in the constraints must be $\geq$ (as opposed to $\leq$ in a maximum problem). The simplex method works only for linear programming problems in standard form.

Let us examine the procedure for a problem involving only two independent variables. Suppose we want to minimize

$$w = 4x_1 + 3x_2$$

subject to the constraints $x_1 \geq 0$, $x_2 \geq 0$ and

$$4x_1 + \quad x_2 \geq 4,$$
$$4x_1 + 10x_2 \geq 20.$$

First, we note that this is a minimum linear programming problem in standard form. We list the coefficients from the inequalities and the equation for w just as they are, without introducing slack variables.

$$\begin{bmatrix} 4 & 1 & 4 \\ 4 & 10 & 20 \\ \cdot & \cdot & \cdot \\ 4 & 3 & 0 \end{bmatrix}$$

Then we interchange the rows and columns. We write the first row as the first column, the second row as the second column, etc. In the language of matrices, we take the **transpose** of the given matrix.

$$\begin{bmatrix} 4 & 4 & 4 \\ 1 & 10 & 3 \\ \cdot & \cdot & \cdot \\ 4 & 20 & 0 \end{bmatrix}$$

Now we examine the maximum linear programming problem corresponding to the preceding matrix. In the corresponding maximum linear programming problem, we shall maximize

$$z = 4y_1 + 20y_2$$

subject to the constraints $y_1 \geq 0$, $y_2 \geq 0$ and

$$4y_1 + 4y_2 \leq 4,$$
$$y_1 + 10y_2 \leq 3.$$

The original minimum problem is called the **primal problem,** and the maximum problem that we have introduced is called the **dual problem** corresponding to the primal problem. It is a consequence of the remarkable Duality Theorem stated next that the maximum value for z in the dual problem is equal to the minimum value for w in the primal problem.

Duality Theorem   The solution to a given minimum linear programming problem, provided it exists, equals the solution to the dual problem, which is a maximum linear programming problem.

Thus, in order to **minimize**

$$w = 4x_1 + 3x_2$$

subject to the constraints $x_1 \geq 0$, $x_2 \geq 0$ and

$$4x_1 + \quad x_2 \geq 4,$$
$$4x_1 + 10x_2 \geq 20,$$

we **maximize**

$$z = 4y_1 + 20y_2$$

subject to the constraints $y_1 \geq 0$, $y_2 \geq 0$ and

$$4y_1 + \quad 4y_2 \leq 4,$$
$$y_1 + 10y_2 \leq 3.$$

Note that the dual problem has standard form provided the original minimum problem has standard form. Now we solve the dual problem by the simplex method developed in Section 5.3. We introduce slack variables and rewrite the objective function.

$$4y_1 + \quad 4y_2 + u_1 = 4$$
$$y_1 + 10y_2 + u_2 = 3$$
$$-4y_1 - 20y_2 + \quad z = 0$$

The first simple tableau is

$$
\begin{bmatrix}
y_1 & y_2 & u_1 & u_2 & z & b \\
4 & 4 & 1 & 0 & 0 & 4 \\
1 & 10 & 0 & 1 & 0 & 3 \\
\hdots \\
-4 & -20 & 0 & 0 & 1 & 0
\end{bmatrix}.
$$

The pivot element is the 10, which is the second entry in the second column. The result of the pivot operation is

$$
\begin{bmatrix}
y_1 & y_2 & u_1 & u_2 & z & b \\
\frac{18}{5} & 0 & 1 & -\frac{2}{5} & 0 & \frac{14}{5} \\
\frac{1}{10} & 1 & 0 & \frac{1}{10} & 0 & \frac{3}{10} \\
\hdots \\
-2 & 0 & 0 & 2 & 1 & 6
\end{bmatrix}.
$$

The new pivot element is $\frac{18}{5}$, which is the first entry in the first column. The result of the pivot operation is

$$
\begin{bmatrix}
y_1 & y_2 & u_1 & u_2 & z & b \\
1 & 0 & \frac{5}{18} & -\frac{1}{9} & 0 & \frac{7}{9} \\
0 & 1 & -\frac{1}{36} & \frac{1}{9} & 0 & \frac{2}{9} \\
\hdots \\
0 & 0 & \frac{5}{9} & \frac{16}{9} & 1 & \frac{68}{9}
\end{bmatrix}.
$$

The simplex method is now completed, and the maximum value of z is $\frac{68}{9}$. By the Duality Theorem, the minimum value of w is $\frac{68}{9}$. Furthermore, the values of x_1 and x_2 that result in this minimum are reported as the bottom entries in the columns under u_1 and u_2, respectively. When $x_1 = \frac{5}{9}$ and $x_2 = \frac{16}{9}$, then w takes on its minimum value of $\frac{68}{9}$.

We now summarize the steps for applying the simplex method to a minimum linear programming problem in standard form.

Simplex Method for Minima

1. Be certain that the minimum problem has the standard form.
2. Without introducing slack variables, construct a matrix from the constraints and the formula for the objective function w.
3. Take the transpose of the matrix and write the dual problem indicated by the new matrix.
4. Find the maximum for z in the dual problem by the simplex method, as described in Section 5.3.
5. In the final tableau, the maximum value of z is equal to the minimum value for w. The values in the last row under the slack variables are the values that should be assigned to the independent variables, in order, to get the minimum w.

We illustrate the simplex method for minima in the following example.

EXAMPLE 1

Use the simplex method to minimize $w = x_1 + 2x_2$ subject to the constraints $x_1 \geq 0$, $x_2 \geq 0$ and

$$2x_1 + x_2 \geq 4,$$

$$x_1 + x_2 \geq 3,$$

$$x_1 + 2x_2 \geq 4.$$

Solution

Our problem has the standard form, so we construct the appropriate matrix.

$$\begin{bmatrix} 2 & 1 & 4 \\ 1 & 1 & 3 \\ 1 & 2 & 4 \\ \cdot & \cdot & \cdot \\ 1 & 2 & 0 \end{bmatrix}$$

The transpose matrix is

$$\begin{bmatrix} 2 & 1 & 1 & 1 \\ 1 & 1 & 2 & 2 \\ \cdot & \cdot & \cdot & \cdot \\ 4 & 3 & 4 & 0 \end{bmatrix}.$$

The dual problem corresponding to this matrix is given next. Maximize

$$z = 4y_1 + 3y_2 + 4y_3$$

subject to the constraints $y_1 \geq 0$, $y_2 \geq 0$, $y_3 \geq 0$ and

$$2y_1 + y_2 + y_3 \le 1$$

$$y_1 + y_2 + 2y_3 \le 2.$$

We construct the simple tableau for this maximum linear programming problem.

$$
\begin{array}{ccccccc}
y_1 & y_2 & y_3 & u_1 & u_2 & z & b \\
\left[\begin{array}{ccccccc}
2 & 1 & 1 & 1 & 0 & 0 & 1 \\
1 & 1 & 2 & 0 & 1 & 0 & 2 \\
\hdashline
-4 & -3 & -4 & 0 & 0 & 1 & 0
\end{array}\right]
\end{array}
$$

We choose as the pivot element the first entry in the third column. The result of the pivot operation is

$$
\begin{array}{ccccccc}
y_1 & y_2 & y_3 & u_1 & u_2 & z & b \\
\left[\begin{array}{ccccccc}
2 & 1 & 1 & 1 & 0 & 0 & 1 \\
-3 & -1 & 0 & -2 & 1 & 0 & 0 \\
\hdashline
4 & 1 & 0 & 4 & 0 & 1 & 4
\end{array}\right]
\end{array}.
$$

Since all entries in the final row are nonnegative, the process is complete. Take $x_1 = 4$ and $x_2 = 0$ to get the minimum value for w, which is 4. ∎

The next example gives a practical application of the simplex method for finding minima.

EXAMPLE 2

One ounce of food A provides 10 units of niacin and 5 units of riboflavin; it costs 10¢. One ounce of food B provides 5 units of niacin and 10 units of riboflavin; it costs 11¢. These two foods should be mixed so that the resulting diet contains at least 100 units of each nutrient. What mixture of these foods meets these nutrition requirements and costs the least?

Solution

Let x_1 and x_2 be the numbers of ounces of foods A and B, respectively, used in the mixture. The inequality that expresses the constraint that the mixture must contain at least 100 units of niacin is

$$10x_1 + 5x_2 \ge 100.$$

The constraint on the amount of riboflavin is written

$$5x_1 + 10x_2 \ge 100.$$

The cost function $C = .10x_1 + .11x_2$ is to be minimized. Thus, we want to minimize

$$C = .10x_1 + .11x_2 \quad \text{with} \quad x_1 \ge 0, x_2 \ge 0$$

subject to the constraints

$$10x_1 + 5x_2 \ge 100,$$

$$5x_1 + 10x_2 \ge 100.$$

Note that this is a minimum linear programming problem in standard form. To simplify the arithmetic, we rewrite the constraints after multiplying through each by $\frac{1}{5}$.

$$2x_1 + x_2 \geq 20,$$

$$x_1 + 2x_2 \geq 20$$

Now we construct the matrix corresponding to the given minimum problem.

$$\begin{bmatrix} 2 & 1 & 20 \\ 1 & 2 & 20 \\ \cdots & \cdots & \cdots \\ .10 & .11 & 0 \end{bmatrix}$$

We take the transpose of this matrix and construct the corresponding maximum linear programming problem.

$$\begin{bmatrix} 2 & 1 & .10 \\ 1 & 2 & .11 \\ \cdots & \cdots & \cdots \\ 20 & 20 & 0.00 \end{bmatrix} \qquad \begin{aligned} z &= 20y_1 + 20y_2 \\ 2y_1 + y_2 &\leq .10 \\ y_1 + 2y_2 &\leq .11 \end{aligned}$$

We write this problem in the usual form, using slack variables.

$$2y_1 + y_2 + u_1 = .10$$

$$y_1 + 2y_2 + u_2 = .11$$

$$-20y_1 - 20y_2 + z = 0$$

The corresponding simple tableau is

$$\begin{array}{cccccc} y_1 & y_2 & u_1 & u_2 & z & b \\ \begin{bmatrix} 2 & 1 & 1 & 0 & 0 & .10 \\ 1 & 2 & 0 & 1 & 0 & .11 \\ \cdots & \cdots & \cdots & \cdots & \cdots & \cdots \\ -20 & -20 & 0 & 0 & 1 & 0.00 \end{bmatrix}. \end{array}$$

We choose the first entry in the first column as the pivot element; the result of the pivot operation is

$$\begin{array}{cccccc} y_1 & y_2 & u_1 & u_2 & z & b \\ \begin{bmatrix} 1 & .5 & .5 & 0 & 0 & .05 \\ 0 & 1.5 & -.5 & 1 & 0 & .06 \\ \cdots & \cdots & \cdots & \cdots & \cdots & \cdots \\ 0 & -10 & 10 & 0 & 1 & 1.00 \end{bmatrix}. \end{array}$$

The pivot element in this tableau is 1.5, the second entry in the second column; the result of the pivot operation (approximating to three decimal places) is

$$\begin{bmatrix} y_1 & y_2 & u_1 & u_2 & z & b \\ 1 & 0 & .667 & -.334 & 0 & .030 \\ 0 & 1 & -.333 & .667 & 0 & .040 \\ \cdots & \cdots & \cdots & \cdots & \cdots & \cdots \\ 0 & 0 & 6.670 & 6.670 & 1 & 1.400 \end{bmatrix}.$$

This is the final tableau, and it shows that each of x_1 and x_2 should be chosen to be 6.67 for which the resulting minimum cost is $1.40. ■

We conclude this section with a practical example that involves four variables. This problem cannot be solved by the graphical method of Section 5.2, which can handle only two variables.

EXAMPLE 3

Two manufacturing plants supply tires to two retail stores. Each plant must produce at least 45 dozen tires per month in order to operate efficiently. Store no. 1 requires 40 dozen tires per month and Store no. 2 requires 50 dozen tires. The costs of shipping a dozen tires from Plant A to Store no. 1 and Store no. 2 are 40¢ and 50¢, respectively; from Plant B to Store no. 1 and Store no. 2 the costs are 45¢ and 45¢. Determine how many tires should be shipped from each plant to each store in order to minimize the shipping costs.

Solution
Let x_1 and x_2 denote the number of dozen tires shipped from Plant A to Store no. 1 and Store no. 2, respectively. Denote the shipments from Plant B to Store no. 1 and Store no. 2 by x_3 and x_4, respectively. The constraints resulting from the stores' requirements are

$$x_1 + x_3 \geq 40,$$
$$x_2 + x_4 \geq 50.$$

The constraints resulting from the plants' efficient operations are

$$x_1 + x_2 \geq 45,$$
$$x_3 + x_4 \geq 45.$$

The cost function is

$$C = .4x_1 + .5x_2 + .45x_3 + .45x_4.$$

This minimum problem has standard form, and it is represented by the following matrix.

$$\begin{bmatrix} 1 & 0 & 1 & 0 & 40 \\ 0 & 1 & 0 & 1 & 50 \\ 1 & 1 & 0 & 0 & 45 \\ 0 & 0 & 1 & 1 & 45 \\ \cdots & \cdots & \cdots & \cdots & \cdots \\ .4 & .5 & .45 & .45 & 0 \end{bmatrix}$$

The transpose of this matrix is

$$\begin{bmatrix} 1 & 0 & 1 & 0 & .4 \\ 0 & 1 & 1 & 0 & .5 \\ 1 & 0 & 0 & 1 & .45 \\ 0 & 1 & 0 & 1 & .45 \\ \cdots & \cdots & \cdots & \cdots & \cdots \\ 40 & 50 & 45 & 45 & 0.00 \end{bmatrix}.$$

The maximum problem corresponding to this matrix is

$$y_1 + y_3 \le .4$$
$$y_2 + y_3 \le .5$$
$$y_1 + y_4 \le .45$$
$$y_2 + y_4 \le .45$$

$$z = 40y_1 + 50y_2 + 45y_3 + 45y_4.$$

With slack variables, the problem takes the form

$$y_1 + y_3 + u_1 = .4$$
$$y_2 + y_3 + u_2 = .5$$
$$y_1 + y_4 + u_3 = .45$$
$$y_2 + y_4 + u_4 = .45$$

$$-40y_1 - 50y_2 - 45y_3 - 45y_4 + z = 0.$$

The simple tableau for this problem is

y_1	y_2	y_3	y_4	u_1	u_2	u_3	u_4	z	b
1	0	1	0	1	0	0	0	0	.4
0	1	1	0	0	1	0	0	0	.5
1	0	0	1	0	0	1	0	0	.45
0	1	0	1	0	0	0	1	0	.45
-40	-50	-45	-45	0	0	0	0	1	0.00

The pivot element is the fourth entry in the second column. The result of the pivot operation is

y_1	y_2	y_3	y_4	u_1	u_2	u_3	u_4	z	b
1	0	1	0	1	0	0	0	0	.4
0	0	1	-1	0	1	0	-1	0	.05
1	0	0	1	0	0	1	0	0	.45
0	1	0	1	0	0	0	1	0	.45
-40	0	-45	5	0	0	0	50	1	22.5

The new pivot element is the second entry in the third column. The result of the pivot operation is

$$\begin{bmatrix}
y_1 & y_2 & y_3 & y_4 & u_1 & u_2 & u_3 & u_4 & z & b \\
1 & 0 & 0 & 1 & 1 & -1 & 0 & 1 & 0 & .35 \\
0 & 0 & 1 & -1 & 0 & 1 & 0 & -1 & 0 & .05 \\
1 & 0 & 0 & 1 & 0 & 0 & 1 & 0 & 0 & .45 \\
0 & 1 & 0 & 1 & 0 & 0 & 0 & 1 & 0 & .45 \\
\cdots & & & & & & & & & \\
-40 & 0 & 0 & -40 & 0 & 45 & 0 & 5 & 1 & 24.75
\end{bmatrix}.$$

The pivot element in this tableau is the first entry in the first column. (We could use the first entry in the fourth column, but that would require more work.) The result of the pivot operation is

$$\begin{bmatrix}
y_1 & y_2 & y_3 & y_4 & u_1 & u_2 & u_3 & u_4 & z & b \\
1 & 0 & 0 & 1 & 1 & -1 & 0 & 1 & 0 & .35 \\
0 & 0 & 1 & -1 & 0 & 1 & 0 & -1 & 0 & .05 \\
0 & 0 & 0 & 0 & -1 & 1 & 1 & -1 & 0 & .10 \\
0 & 1 & 0 & 1 & 0 & 0 & 0 & 1 & 0 & .45 \\
\cdots & & & & & & & & & \\
0 & 0 & 0 & 0 & 40 & 5 & 0 & 45 & 1 & 38.75
\end{bmatrix}.$$

The last tableau indicates that the minimum is $w = \$38.75$, and it occurs when $x_1 = 40$, $x_2 = 5$, $x_3 = 0$, and $x_4 = 45$. ■

In this example, it is fairly easy to see why the solution takes the form that it does. The shipments with the highest cost per dozen tires are x_2, from Plant A to Store no. 2, so a minimum-cost solution will have the smallest possible value of x_2. Similarly, x_1 has the lowest cost per dozen, so we want that variable to be as large as possible. Looking at the constraints, we see that Store no. 2 requires 50 dozen tires per month, and it can get only 45 dozen from Plant B. Thus, the lowest-cost way to supply Store no. 2 is to make $x_4 = 45$ and $x_2 = 5$. Since the entire output of Plant B goes to Store no. 2, we have $x_3 = 0$; the entire supply for Store no. 1 comes from Plant A, so $x_1 = 40$.

Of course, manufacturing plants almost never supply only two outlets, so linear programming problems are rarely this simple. Fortunately, the simplex method is suitable for use on large computers, which can handle problems involving hundreds of variables.

Exercises   5.4

In each of the following problems, the given function is to be minimized subject to the given constraints. Construct the dual linear programming problem.

1   $w = 5x + 10y,$
$x + y \geq 2,$
$x \geq 0,\ y \geq 0$

2   $w = 20x + 8y,$
$x + y \geq 5,$
$x \geq 0,\ y \geq 0$

3 $w = 10x + 8y,$
   $4x + y \ge 4,$
   $2x + 4y \ge 9,$
   $x \ge 0, y \ge 0$

4 $w = 4x + 5y,$
   $3x + y \ge 6,$
   $x + 3y \ge 10,$
   $x \ge 0, y \ge 0$

5 $w = 6x_1 + 5x_2,$
   $5x_1 + x_2 \ge 10,$
   $x_1 + 2x_2 \ge 11,$
   $2x_1 + 10x_2 \ge 49,$
   $x_1 \ge 0, x_2 \ge 0$

6 $w = 9x_1 + 3x_2,$
   $10x_1 + x_2 \ge 10,$
   $2x_1 + 4x_2 \ge 21,$
   $x_1 + 5x_2 \ge 24,$
   $x_1 \ge 0, x_2 \ge 0$

In each of the following problems, minimize the given function subject to the given constraints.

7 $w = 3x + 8y,$
   $x + 10y \ge 10,$
   $x \ge 0, y \ge 0$

8 $w = 5x + 2y,$
   $8x + y \ge 8,$
   $x \ge 0, y \ge 0$

9 $w = x + y,$
   $x + y \ge 4$
   $x \ge 3, y \ge 0$

10 $w = 3x + 2y,$
    $2x + y \ge 5,$
    $x \ge 0, y \ge 2$

11 $w = x + y,$
    $4x + y \ge 5,$
    $x + 5y \ge 6,$
    $x \ge 0, y \ge 0$

12 $w = 5x + 4y,$
    $2x + y \ge 6,$
    $x + 2y \ge 6,$
    $x \ge 0, y \ge 0$

13 $w = 2x_1 + 3x_2,$
    $5x_1 + x_2 \ge 8,$
    $x_1 + x_2 \ge 4,$
    $x_1 + 3x_2 \ge 8,$
    $x_1 \ge 0, x_2 \ge 0$

14 $w = 4x_1 + 2x_2,$
    $8x_1 + x_2 \ge 13,$
    $x_1 + 2x_2 \ge 11,$
    $x_1 + 6x_2 \ge 27,$
    $x_1 \ge 0, x_2 \ge 0$

15 $w = x_1 + x_2 + x_3,$
    $x_1 \ge 0, x_2 \ge 0, x_3 \ge 0,$
    $2x_1 + x_2 + x_3 \ge 4,$
    $3x_2 + x_3 \ge 6$

16 $w = 2x_1 + 4x_2 + 3x_3,$
    $x_1 \ge 0, x_2 \ge 0, x_3 \ge 0,$
    $x_1 + x_2 \ge 2, x_2 + x_3 \ge 2$

17 $w = x_1 + 2x_2 + x_3,$
    $x_1 \ge 0, x_2 \ge 0, x_3 \ge 0,$
    $x_1 + x_3 \ge 1, x_1 + x_2 \ge 3,$
    $x_2 + x_3 \ge 2$

18 $w = 4x_1 + x_2 + 2x_3,$
    $x_1 \ge 0, x_2 \ge 0, x_3 \ge 0,$
    $x_2 + x_3 \ge 1,$
    $x_1 + x_3 \ge 2,$
    $x_1 + x_2 \ge 4$

Solve each of the following problems.

19 One ounce of food A provides 20 units of niacin and 16 units of thiamine; it costs 20¢. One ounce of food B provides 12 units of niacin and 16 units of thiamine; it costs 15¢. Determine the minimum cost of a diet of these two foods that supplies at least 240 units of each nutrient.

20 A grocery wholesaler uses a small truck with 20 cubic feet of refrigerated space and 10 cubic feet of uncooled space. Also, he uses a large truck with 20 cubic feet of cooled space and 30 cubic feet of uncooled space. The small truck costs $3.30 per mile to operate and the large truck costs $4.40 per mile. If the wholesaler needs another 120 cubic feet of each

type of cargo space, how many trucks of each type should he purchase in order to minimize costs?

21   A cafeteria offers fruit cocktail and peaches. The manager decides to buy at least 210 cans of fruit per month so that he will get a big discount from the supplier. The customer demand indicates that he should order at least twice as much fruit cocktail as peaches. A can of fruit cocktail costs $5 and a can of peaches costs $3. How many cans of each type should the manager order to minimize costs?

22   Solve the problem stated in Example 3 assuming that customer demand has changed in such a way that Store no. 1 requires only 35 dozen tires per month and Store no. 2 requires 55 dozen tires per month.

23   One ounce of food A provides 30 units of niacin, 20 units of thiamine, and 10 units of riboflavin; it costs 40¢. One ounce of food B provides 10 units of niacin, 30 units of thiamine, and 20 units of riboflavin; it costs 50¢. One ounce of food C provides 20 units of niacin, 10 units of thiamine, and 30 units of riboflavin; it costs 30¢. Determine the minimum cost of a diet of these three foods that supplies at least 200 units of each nutrient.

5.5   Problems Not in Standard Form (Optional)

The simplex method, as presented in Sections 5.3 and 5.4, applies only to problems in standard form. Since our intention was to present an introduction to linear programming, we considered very few complications. In this section, we shall briefly consider some linear programming problems that are not in standard form.

Suppose we want to maximize $z = x + 2y$, with $x \geq 0$, $y \geq 0$, subject to the constraints

$$5x + y \leq 5,$$

$$x + y \geq 1.$$

The second inequality is not a proper constraint for a maximum linear programming problem in standard form. We can make it look more like a proper constraint by multiplying each side by -1. Then we have

$$5x + y \leq 5,$$

$$-x - y \leq -1.$$

This is still not in standard form because the constant on the right side of the second inequality is negative. Nevertheless, we rewrite the problem in the usual way using slack variables.

$$5x + y + u_1 = 5$$

$$-x - y + u_2 = -1$$

$$-x - 2y + z = 0$$

The corresponding simple tableau is

$$\begin{array}{cccccc} x & y & u_1 & u_2 & z & b \\ \left[\begin{array}{cccccc} 5 & 1 & 1 & 0 & 0 & 5 \\ -1 & -1 & 0 & 1 & 0 & -1 \\ \cdots & \cdots & \cdots & \cdots & \cdots & \cdots \\ -1 & -2 & 0 & 0 & 1 & 0 \end{array}\right] \end{array}.$$

Because the original problem was not in standard form, we now have a negative entry in the b-column. To deal with this complication, we select as the pivot column some column that has a negative entry in the same row as the negative entry in the b-column. Then the pivot element is the entry in the pivot column, not in the last row, that produces the smallest nonnegative quotient with the corresponding element in the b-column. In our example, we could choose either the first entry in the first column or the second entry in the second column as the pivot element. We arbitrarily choose the second entry in the second column; the result of the pivot operation is

$$\begin{array}{cccccc} x & y & u_1 & u_2 & z & b \\ \left[\begin{array}{cccccc} 4 & 0 & 1 & 1 & 0 & 4 \\ 1 & 1 & 0 & -1 & 0 & 1 \\ \cdots & \cdots & \cdots & \cdots & \cdots & \cdots \\ 1 & 0 & 0 & -2 & 1 & 2 \end{array}\right] \end{array}.$$

Notice that all entries in the b-column are now nonnegative. The new pivot column is the u_2-column. The pivot element is the first entry in that column, since that is the only entry resulting in a nonnegative quotient. The result of the pivot operation is

$$\begin{array}{cccccc} x & y & u_1 & u_2 & z & b \\ \left[\begin{array}{cccccc} 4 & 0 & 1 & 1 & 0 & 4 \\ 5 & 1 & 1 & 0 & 0 & 5 \\ \cdots & \cdots & \cdots & \cdots & \cdots & \cdots \\ 9 & 0 & 2 & 0 & 1 & 10 \end{array}\right] \end{array}.$$

We get the answer from this final tableau in the usual way. The maximum value of z is 10, and it is obtained by taking $x = 0$ and $y = 5$.

The next example is a maximum problem with two improper constraints.

EXAMPLE 1

Find the maximum of the function

$$z = 3x + 2y$$

subject to the constraints $x \geq 0$, $y \geq 0$ and

$$x + \ y \leq 5,$$
$$3x + \ y \geq 4,$$
$$x + 4y \geq 5.$$

Solution

Multiply each side of each improper constraint by -1, and convert the constraints into equations.

$$x+\ y\le 5, \qquad\qquad x+y+u_1 = 5$$
$$-3x-\ y\le -4, \qquad -3x-y+u_2 = -4$$
$$-x-4y\le -5 \qquad -x-4y+u_3 = -5$$
$$\qquad\qquad\qquad\qquad -3x-2y+z = 0$$

We display the numbers in these equations in a simple tableau.

$$
\begin{array}{ccccccc}
x & y & u_1 & u_2 & u_3 & z & b \\
\end{array}
$$
$$
\left[
\begin{array}{ccccccc}
1 & 1 & 1 & 0 & 0 & 0 & 5 \\
-3 & -1 & 0 & 1 & 0 & 0 & -4 \\
-1 & -4 & 0 & 0 & 1 & 0 & -5 \\
\hdashline
-3 & -2 & 0 & 0 & 0 & 1 & 0
\end{array}
\right]
$$

We arbitrarily choose one of the negative entries in the b-column, say -4; we arbitrarily choose a column that contains a negative entry in the same row as -4, say -3 in the first column. We choose the pivot row according to which of the first three entries in the first column produces the smallest nonnegative quotient with the corresponding entry in the b-column. The pivot element is -3, and the result of the pivot operation is

$$
\begin{array}{ccccccc}
x & y & u_1 & u_2 & u_3 & z & b \\
\end{array}
$$
$$
\left[
\begin{array}{ccccccc}
0 & \frac{2}{3} & 1 & \frac{1}{3} & 0 & 0 & \frac{11}{3} \\
1 & \frac{1}{3} & 0 & -\frac{1}{3} & 0 & 0 & \frac{4}{3} \\
0 & -\frac{11}{3} & 0 & -\frac{1}{3} & 1 & 0 & -\frac{11}{3} \\
\hdashline
0 & -1 & 0 & -1 & 0 & 1 & 4
\end{array}
\right].
$$

We similarly deal with $-\frac{11}{3}$ appearing in the b-column. The pivot element is the third entry in the second column, and the result of the pivot operation is

$$
\begin{array}{ccccccc}
x & y & u_1 & u_2 & u_3 & z & b \\
\end{array}
$$
$$
\left[
\begin{array}{ccccccc}
0 & 0 & 1 & \frac{3}{11} & \frac{2}{11} & 0 & 3 \\
1 & 0 & 0 & -\frac{4}{11} & \frac{1}{11} & 0 & 1 \\
0 & 1 & 0 & \frac{1}{11} & -\frac{3}{11} & 0 & 1 \\
\hdashline
0 & 0 & 0 & -\frac{10}{11} & -\frac{3}{11} & 1 & 5
\end{array}
\right].
$$

Now all entries in the b-column are nonnegative, and we proceed by the usual scheme. We choose the pivot element to be the third entry in the fourth column. The result of the pivot operation is

$$
\begin{array}{ccccccc}
x & y & u_1 & u_2 & u_3 & z & b \\
\left[\begin{array}{ccccccc}
0 & -3 & 1 & 0 & 1 & 0 & 0 \\
1 & 4 & 0 & 0 & -1 & 0 & 5 \\
0 & 11 & 0 & 1 & -3 & 0 & 11 \\
\cdots & \cdots & \cdots & \cdots & \cdots & \cdots & \cdots \\
0 & 10 & 0 & 0 & -3 & 1 & 15
\end{array}\right].
\end{array}
$$

The new pivot element is the first entry in the fifth column, and the result of the pivot operation is

$$
\begin{array}{ccccccc}
x & y & u_1 & u_2 & u_3 & z & b \\
\left[\begin{array}{ccccccc}
0 & -3 & 1 & 0 & 1 & 0 & 0 \\
1 & 1 & 1 & 0 & 0 & 0 & 5 \\
0 & 2 & 3 & 1 & 0 & 0 & 11 \\
\cdots & \cdots & \cdots & \cdots & \cdots & \cdots & \cdots \\
0 & 1 & 3 & 0 & 0 & 1 & 15
\end{array}\right].
\end{array}
$$

The maximum value of z is 15, and it results when $x = 5$, $y = 0$. ■

We now summarize the modified simplex method for maximum linear programming problems that are not in standard form.

Modified Simplex Method

1. Multiply each side of each improper constraint by -1, so that it has the form of a proper constraint.
2. Convert each constraint into an equation by introducing a slack variable. Rewrite the equation defining the objective function.
3. Write the simple tableau for the given problem.
4. Arbitrarily select a negative entry in the b-column, and arbitrarily choose another negative entry in that same row. The column of this last chosen negative entry is the pivot column.
5. The pivot row is a row, other than the last row, such that the b-column entry of that row divided by that row's entry in the pivot column is the smallest such nonnegative quotient.
6. Perform the pivot operation using the indicated pivot element.
7. Repeat the process of selecting a pivot element until there are no negative entries in the b-column, except possibly the last entry.
8. When the b-column contains only nonnegative entries, use the simplex method summarized in Section 5.3.

The modified simplex method solves problems with more than two variables, as we illustrate in the next example.

EXAMPLE 2

Find the maximum of the function

$$z = x_1 + x_2 + x_3$$

subject to the constraints $x_1 \geq 0$, $x_2 \geq 0$, $x_3 \geq 0$ and

$$2x_1 + x_2 + 3x_3 \geq 1,$$
$$x_1 + 2x_2 + x_3 \leq 5.$$

Solution

Multiply each side of the first inequality, which is an improper constraint, by -1.

$$-2x_1 - x_2 - 3x_3 \leq -1,$$
$$x_1 + 2x_2 + x_3 \leq 5$$

Convert the constraints into equations.

$$-2x_1 - x_2 - 3x_3 + u_1 = -1$$
$$x_1 + 2x_2 + x_3 + u_2 = 5$$
$$-x_1 - x_2 - x_3 + z = 0$$

We display the numbers in the above equations in a simple tableau.

$$\begin{array}{ccccccc} x_1 & x_2 & x_3 & u_1 & u_2 & z & b \\ \left[\begin{array}{ccccccc} -2 & -1 & -3 & 1 & 0 & 0 & -1 \\ 1 & 2 & 1 & 0 & 1 & 0 & 5 \\ \cdots & \cdots & \cdots & \cdots & \cdots & \cdots & \cdots \\ -1 & -1 & -1 & 0 & 0 & 1 & 0 \end{array}\right] \end{array}$$

We choose the first entry in the second column as the pivot element. The result of the pivot operation is

$$\begin{array}{ccccccc} x_1 & x_2 & x_3 & u_1 & u_2 & z & b \\ \left[\begin{array}{ccccccc} 2 & 1 & 3 & -1 & 0 & 0 & 1 \\ -3 & 0 & -5 & 2 & 1 & 0 & 3 \\ \cdots & \cdots & \cdots & \cdots & \cdots & \cdots & \cdots \\ 1 & 0 & 2 & -1 & 0 & 1 & 1 \end{array}\right] \end{array}.$$

The b-column now contains only nonnegative entries, so we proceed with the simplex method. The new pivot element is the second entry in the fourth column, and the result of the pivot operation is

$$\begin{array}{ccccccc} x_1 & x_2 & x_3 & u_1 & u_2 & z & b \\ \left[\begin{array}{ccccccc} \frac{1}{2} & 1 & \frac{1}{2} & 0 & \frac{1}{2} & 0 & \frac{5}{2} \\ -\frac{3}{2} & 0 & -\frac{5}{2} & 1 & \frac{1}{2} & 0 & \frac{3}{2} \\ \cdots & \cdots & \cdots & \cdots & \cdots & \cdots & \cdots \\ -\frac{1}{2} & 0 & -\frac{1}{2} & 0 & \frac{1}{2} & 1 & \frac{5}{2} \end{array}\right] \end{array}.$$

The new pivot element is the first entry in the first column, and the pivot operation produces the final tableau below.

$$\begin{array}{ccccccc} x_1 & x_2 & x_3 & u_1 & u_2 & z & b \\ \left[\begin{array}{ccccccc} 1 & 2 & 1 & 0 & 1 & 0 & 5 \\ 0 & 3 & -1 & 1 & 2 & 0 & 9 \\ \cdots & \cdots & \cdots & \cdots & \cdots & \cdots & \cdots \\ 0 & 1 & 0 & 0 & 1 & 1 & 5 \end{array}\right] \end{array}$$

The maximum value of z is 5, and it results when $x_1 = 5$, $x_2 = 0$, $x_3 = 0$. ■

Linear programming is an extensive subject. We have attempted to give you a mastery of only the basic ideas and methods. A more exhaustive treatment might consider conditions that guarantee the success of the linear programming method, how to use a computer to execute the linear programming method, and some of the complicated problems that arise in modern applications.

Exercises 5.5

In each of the following problems, use the modified simplex method to maximize z subject to the indicated constraints.

1 $z = 2x + 3y$
   $4x + y \geq 4$,
   $x + y \leq 4$,
   $x \geq 0$, $y \geq 0$

2 $z = 5x + 2y$
   $8x + y \leq 8$,
   $4x + y \geq 4$,
   $x \geq 0$, $y \geq 0$

3 $z = x_1 + x_2$
   $x_1 + x_2 \leq 4$,
   $2x_1 + x_2 \geq 2$,
   $x \geq 0$, $y \geq 0$

4 $z = 3x_1 + 4x_2$
   $x_1 + x_2 \leq 5$,
   $x_1 + 3x_2 \geq 3$,
   $x_1 \geq 0$, $x_2 \geq 0$

5 $z = 3x + y$
   $5x + 2y \geq 10$,
   $2x + 5y \geq 10$,
   $x + y \leq 5$,
   $x \geq 0$, $y \geq 0$

6 $z = 5x + 3y$
   $2x + y \geq 6$,
   $2x + 3y \geq 10$,
   $6x + 5y \leq 30$,
   $x \geq 0$, $y \geq 0$

7 $z = 4x_1 + 2x_2 + 3x_3$
   $x_1 + x_2 + x_3 \geq 2$
   $x_1 + x_2 + x_3 \leq 5$
   $x_1 \geq 0$, $x_2 \geq 0$, $x_3 \geq 0$

8 $z = 5x_1 + 4x_2 + x_3$
   $20x_1 + 15x_2 + 12x_3 \leq 60$,
   $2x_1 + 3x_2 + 6x_3 \geq 6$,
   $x_1 \geq 0$, $x_2 \geq 0$, $x_3 \geq 0$

9 $z = 3x_1 + x_2 + 2x_3$
   $6x_1 + 2x_2 + 3x_3 \geq 6$,
   $6x_1 + 8x_2 + 12x_3 \leq 24$,
   $x_1 \geq 0$, $x_2 \geq 0$, $x_3 \geq 0$

10 $z = x_1 + x_2 + x_3$
   $2x_1 + 2x_2 + x_3 \leq 6$,
   $2x_1 + 2x_2 + x_3 \geq 4$,
   $x_1 \geq 0$, $x_2 \geq 0$, $x_3 \geq 0$

11 $z = x_1 + x_2 + x_3$
   $4x_1 + 5x_2 + 5x_3 \leq 20$,
   $3x_1 + x_2 + 3x_3 \geq 3$,
   $x_1 + 2x_2 + x_3 \geq 1$,
   $x_1 \geq 0$, $x_2 \geq 0$, $x_3 \geq 0$

12 $z = x_1 + 2x_2 + x_3$
   $3x_1 + x_2 + x_3 \geq 1$
   $x_1 + x_2 + 3x_3 \geq 1$,
   $x_1 + x_2 + x_3 \leq 5$,
   $x_1 \geq 0$, $x_2 \geq 0$, $x_3 \geq 0$

Review of Terms

Important Mathematical Terms
linear inequality, *p. 198*   simple tableau, *p. 221*

Review Problems

Show graphically in the plane the set of solutions for the given inequality or system of inequalities.

1 $y \geq 2x$

2 $x + y \leq 3$

3 $x - 4y + 4 > 0$

4 $2x + 5y - 10 < 0$

5 $x - y + 2 \geq 0$
  $x - y - 2 \leq 0$

6 $2x + y \geq 6$
  $x - y \leq 3$

7 $y - 3 \leq 0$
  $x - y \geq 0$

8 $x + 5y \leq 25$
  $x - y \leq 1$
  $x + y \geq 5$

Use the graphing method to solve each of the following linear programming problems.

9 Maximize $z = x + y$
  subject to $x \geq 0$, $y \geq 0$, $2x + 3y \leq 6$.

10 Minimize $z = 2x + 3y$
  subject to $x \geq 0$, $y \geq 0$, $2x + y \geq 3$, $x + 3y \geq 4$.

11 Maximize $z = 3x + 4y$
  subject to $x \geq 0$, $y \geq 0$, $-3x + 5y \leq 5$, $4x - y \leq 16$.

12 Minimize $z = x + y$
  subject to $x \geq 2$, $y \geq 2$, $x + y \leq 7$.

13 Maximize $z = 4x + 5y$
  subject to $x \geq 0$, $y \geq 0$, $4x - 3y \geq -2$, $3x + y \leq 18$, $x - 4y \leq -7$.

In each of the following problems, use the simplex method to find the maximum value for z subject to the given constraints.

14 $z = 2x + y$
  $5x + 2y \leq 10$,
  $x \geq 0$, $y \geq 0$

15 $z = 5x + 2y$
  $3x + 2y \leq 10$,
  $2x - y \leq 2$,
  $x \geq 0$, $y \geq 0$

16 $z = x + y$
$y \leq 4, x \leq 5,$
$x \geq 0, y \geq 0$

17 $z = x_1 + x_2 + x_3$
$5x_1 + 5x_2 + x_3 \leq 5,$
$5x_1 + x_2 + 5x_3 \leq 5,$
$x_1 \geq 0, x_2 \geq 0, x_3 \geq 0$

18 $z = x_1 + x_2 + x_3$
$x_1 + x_2 + x_3 \leq 3,$
$2x_1 + x_2 + x_3 \leq 4,$
$x_1 + x_2 + 2x_3 \leq 4$
$x_1 \geq 0, x_2 \geq 0, x_3 \geq 0$

In each of the following problems, use the simplex method to find the minimum value for w subject to the given constraints.

19 $w = x + y,$
$x + y \geq 3,$
$x \geq 0, y \geq 0$

20 $w = 3x + 2y,$
$4x + y \geq 5,$
$x + 4y \geq 5,$
$x \geq 0, y \geq 0$

21 $w = x + 2y,$
$3x + y \geq 5,$
$x + 3y \geq 5,$
$x + y \geq 3,$
$x \geq 0, y \geq 0$

22 $w = x_1 + x_2 + x_3,$
$x_1 + x_2 + 5x_3 \geq 5,$
$5x_1 + x_2 + x_3 \geq 5,$
$x_1 \geq 0, x_2 \geq 0, x_3 \geq 0$

23 $w = x_1 + 3x_2 + 2x_3,$
$x_1 + 3x_2 + x_3 \geq 6,$
$3x_1 + x_2 + x_3 \geq 6,$
$x_1 + x_2 + 3x_3 \geq 6,$
$x_1 \geq 0, x_2 \geq 0, x_3 \geq 0$

In each of the following problems use the modified simplex method to maximize z subject to the indicated constraints.

24 $z = x + 2y,$
$x + y \leq 5,$
$5x + 2y \geq 10,$
$x \geq 0, y \geq 0$

25 $z = x + y,$
$4x + 3y \leq 12,$
$4x + y \geq 5,$
$x + 4y \geq 5,$
$x \geq 0, y \geq 0$

26 $z = 3x_1 + 2x_2 + 4x_3,$
$x_1 + x_2 + x_3 \leq 5,$
$x_1 + x_2 + x_3 \geq 1,$
$x_1 \geq 0, x_2 \geq 0, x_3 \geq 0$

Solve each of the following problems using the methods of this chapter.

27 A manufacturer uses two production lines, assembly and finishing, to make two products, the economy model and the luxury model. The assembly line is available for 100 hours per week, while the finishing line is available for 80 hours. One economy model requires 10 hours of assembly and 4 hours of finishing; it earns $10 profit upon its sale. One luxury model requires 10 hours of assembly and 16 hours of finishing; it earns $18. Determine how many of each model should be manufactured for maximum profit.

28 A meat market has 400 pounds of beef and 200 pounds of pork that it has not sold. The owner decides to grind as much of the meat as possible to make hot dogs and hamburger patties. The hot dogs are 50% beef and 50% pork; the hamburger patties are 80% beef and 20% pork. How should the meat be used to minimize the amount left over?

29 One ounce of food A provides 20 units of protein and 5 units of vitamins; it costs 10¢. One ounce of food B provides 5 units of protein and 20 units of vitamins; it costs 8¢. These two foods should be mixed so that the resulting diet contains at least 100 units of protein and 100 units of vitamins. How should they be mixed to minimize the costs?

30 A stockbroker is dividing $20,000 between glamour stocks and growth stocks. The glamour stocks have an anticipated total return of 25% and a risk rating of 10. The growth stocks have an anticipated total return of 15% and a risk rating of 4. The broker decides that his investment must not have an average risk rating more than 7. How should the $20,000 be divided in order to maximize the anticipated total return?

31 A restaurant manager buys all of his beef stew and chili in cans from the same supplier. He decides to buy at least 100 cans per month of the two dishes so that he will get preferred customer treatment. The demand at the restaurant indicates that he should buy at least three times as much chili as beef stew. A can of chili costs $6 and a can of beef stew costs $4. How many cans of each should the manager order to minimize costs?

Social Science Applications

1 *Ecology*   A farmer has a tract of land that could support 50 cows as pastureland. In order to preserve the vegetation over the long term, he is considering substituting some sheep for cows. One cow can be replaced by five sheep. Because cows are more profitable, the farmer wants at least 20 of them. In order to protect the quality of the vegetation, he decides that he should replace at least ten cows with sheep. A cow is 10 times more profitable than a sheep, but no more than 25% of the cattle can be sent to market in 1 year. No more than 35% of the sheep can be sent to market in 1 year. Determine how many animals of each type the farmer should have for maximum profit.

2 *Sociology*   A department of social workers forms two teams to help its clients in different problem areas. Team A counsels clients on matters related to money, and Team B specializes in interpersonal relationships. Team A is available for 120 hours per week, and Team B is available for 100 hours per week. Clients are classified as type (1) and (2). Type (1) clients spend 1 hour per week with a member of Team A and 30 minutes per week with a member of Team B. Type (2) clients are advised for 1 hour per week by each team. How many clients of each type should the department schedule in order to maximize the total hours worked by the two teams.

3 *Psychology*   Two groups of mice are used in an experiment. Group A is given a special diet, and Group B is given a standard diet. Each mouse in

each group is trained in box (1) to press a lever to get food and in box (2) to go through a maze to get food. Each mouse in Group A spends 15 minutes in box (1) and 10 minutes in box (2) daily; each mouse in Group B spends 10 minutes in box (1) and 20 minutes in box (2) daily. If each box is available for 600 minutes daily, what is the maximum number of mice that can be trained daily?

4   *Public Administration*   A city water purification plant can use either of two processes. Process A reduces the total solids per liter to 500 milligrams and the chloride ions to 200 milligrams. Process B, which is twice as expensive as process A, reduces the total solids to 200 milligrams and the chloride ions to 150 milligrams. How many of each 100 liters of water should be treated by each process in order to minimize cost and produce water with not more than 500 milligrams of total solids or 250 milligrams of chloride ions per liter? Assume that the water purified by each process is thoroughly mixed before it leaves the plant.

CAREER PROFILE
Operations Research Analyst

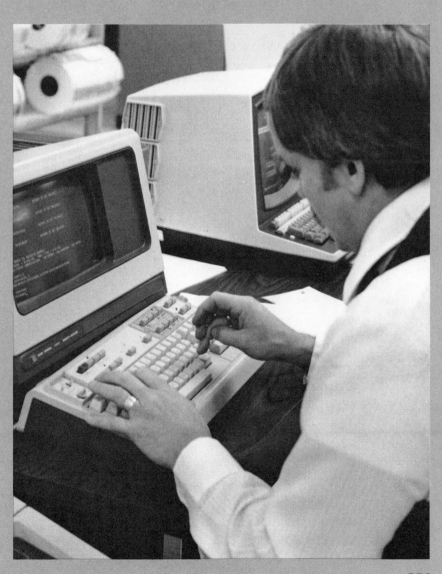

Operations Research (OR) is a relatively new field concerned with development and use of systematic and verifiable methods for deciding how best to design and operate systems, usually under the limiting conditions of scarce available resources.

During World War II, scientists and engineers were called upon to analyze **operational** problems including tactical employment of radar, convoy operations, antisubmarine operations, and strategic bombing and mining operations. After the war, "operations research" activity turned toward mathematical formalization of wartime procedures, and thus the scientific discipline of OR was born.

As OR techniques became well known, many analysts sought to apply these new methods to industrial situations. (At present, linear programming is the most widely known and used of OR procedures.) The number of applications and the number of facets of OR have grown rapidly and substantially. Now, large and small organizations of all types (business, industrial, educational, governmental) use OR techniques in applications such as inventory control, project planning and scheduling, long-range planning, optimal allocation of resources, and decision-making.

Some entry-level positions in OR are available to individuals with undergraduate degrees. A few universities offer bachelor's degrees in OR, usually within a college of engineering. Many OR analysts, however, come from undergraduate degree programs in mathematics or another major (such as economics, business administration, engineering, or computer science) with a strong mathematics orientation and a large number of required mathematics courses. The key mathematical tools needed for entering the field of OR are linear algebra, calculus, differential and difference equations, probability, statistics, and computer programming. Courses in simulation and in operations research (also known as "management science") show how these tools can be applied to a variety of decision problems.

Although a *few* entry-level positions are available in Operations Research for individuals with only a bachelor's degree, advanced training (usually a Master's Degree in Operation's Research) is a necessity for many jobs and an essential criterion for advancement.

An OR analyst with a Master's Degree in Operations Research can expect a starting salary in excess of $20,000. Enterprising and knowledgeable consultants in OR can earn annual salaries up to $100,000, depending on their educational backgrounds, problem-solving skills, initiative, and the client relationships that they are able to develop. The primary employers of OR specialists include government agencies, universities, the armed services, large corporations, and banking institutions.

Sources of Additional Information
Operations Research Society of America, 428 E. Preston Street, Baltimore, MD 21202. (Available from ORSA is a free booklet entitled "Careers in Operations Research," which includes a general description of the OR field together with specific job descriptions and profiles of typical OR analysts. This booklet also contains a list of educational programs in OR with the name of a contact person

for each program. Some of the information for this Career Profile was obtained from the ORSA booklet.)

The Mathematical Association of America, 1529 Eighteenth Street, N.W., Washington, DC 20036. (Career opportunities in Operations Research is a topic that is treated at length in the booklet, *Professional Opportunities in the Mathematical Sciences,* available for $1.50 from the MAA. This booklet emphasizes the role of mathematics in OR and provides a list of references for additional OR career information.)

6 Mathematics of Money

RATE OF EXCHANGE
KAUF
KØBERKURSER

PR · 100	SEDLER NOTEN – BILLS	TRAVELLERS CHECKS
STERLING	1760 KR.	1790,00 K
DOLLAR	735,00	744,00
DM.	186,00	
F.F.		
B. FRC.	14,50	
S. FRC.	171,50	
SV.	141,00	
NO.	102,00	
HFL	204,25	
LIRE	1,17	
ÖS. SCH.	27,75	
PESETAS	10,40	
CAN.DOLLAR	675,00	
MK.	175,50	

6.1 Interest

Few things in life are free. When you need money to start a business or simply to buy a new car, you can borrow it from a bank or some other lending agency. To compensate the lending agency for letting you have some of their money, you must agree to repay more than you borrow. The difference between what you repay and what you borrow is called the **interest.** For example, if your car suddenly develops a mechanical problem and you do not have the $100 necessary to pay the auto mechanic, you might borrow $100 from the bank by pledging to repay the money at the end of 30 days after you have received your next pay check. The bank would charge you a small fee for processing the paperwork involved in making the loan, and it would charge you some interest for the use of its money. For such a loan the bank might well indicate that it would charge you 18% per annum. Rates of interest are almost always stated as **per annum** or yearly rates. The phrase *per annum* means "by the year" in Latin. This means that the bank will charge you interest at a rate that would produce $18 at the end of 1 year if you kept the $100 for the entire year. Here is the computation.

interest = (interest rate) × (money borrowed)

or

$18 = (.18)($100)

Of course, you are not going to keep the $100 for an entire year, so you should not be required to pay the entire $18. Since you will keep the money for 1 month or one twelfth of a year, you should pay only one twelfth of the $18, or $$\frac{18}{12} = \$1.50$$. In general, if r is the per annum rate of interest, t is the length of time that you keep the borrowed money measured in fractions of a year, and p is the amount of money borrowed, then the interest i to be paid is given by the formula

$$i = prt.$$

The amount borrowed is usually called the **principal.**

EXAMPLE 1

Lucy Kilgo inherits $1000 from her deceased Aunt Agnes, and Lucy lends the money to her brother Tom for his college expenses. Tom borrows the money for 9 months and Lucy charges Tom interest at a rate of 12%. How much must Tom repay?

Solution

Assume that 12% is the per annum rate, since the problem does not say otherwise and rates are usually given that way. Convert 9 months to a fraction of a year by dividing by 12. Use the general equation to calculate the interest.

$$i = prt$$

$$i = (\$1000)(.12)\left(\frac{9}{12}\right)$$

$$i = \$90.$$

Remember, Tom must repay the $1000 plus $90 as compensation for using Lucy's money for 9 months. After all, Lucy could have had a good time spending that $1000. ■

The interest we have talked about so far is referred to as **simple interest** because all of the interest is paid when the principal of the loan is repaid, which makes for a simple computation. Anyone who borrows money should have a keen interest in how much he is going to have to pay at the end of the loan. It is easy to obtain a formula for the total amount repaid at the end of a loan requiring simple interest.

$$A \text{ (amount repaid)} = \text{principal} + \text{interest}$$

$$A = p + i$$

$$A = p + prt$$

$$\boxed{A = p(1 + rt).}$$

Some lenders charge for the money that they lend by withholding some of the loan money from the beginning. Suppose John discovers that he has dental problems requiring immediate attention and he goes to Easy Loan Company to borrow $200 for 1 year. The loan company says it will charge John a simple discount at the rate of 18%. This means that they will deduct ($200)(.18) = $36 from the $200. Consequently, at the time the loan is made, John will receive $200 − $36 = $164, although he will repay $200. Thus, a **simple discount** is the amount that will be repaid at the end of the loan minus the actual amount that is received at the beginning of the loan. As in the case of simple interest, the rate of discount is stated as a per annum percentage. If the duration of the loan is anything other than 1 year, then one must adjust for the time period. The formula for the amount of the discount is

$$d = prt$$

where d is the amount of the discount, p is the principal borrowed, r is the per annum rate of discount, and t is the length of time covered by the loan measured in fractions of a year. The amount actually received at the beginning of the loan, equal to $p - d$, is called the **proceeds** of the loan.

EXAMPLE 2

Tom Kilgo, who has been attending college and playing football without a scholarship, is given a generous scholarship because of his excellent play. Tom's sister Lucy, who previously lent him $1000 (see Example 1), now wants to borrow $1000 from Tom, who has already repaid her earlier loan. Tom lends Lucy $1000 for nine months with a simple discount at the rate of 12%. How much does Lucy receive and how much does she repay?

Solution
Of course, Lucy repays $1000. However, the amount that she receives is reduced by the discount, which is

$$d = prt$$

$$d = \$1000(.12)\left(\frac{9}{12}\right) = \$90.$$

Thus, Lucy actually receives $910. ■

In the two transactions between Lucy and Tom described in Examples 1 and 2, did Lucy and Tom come out even? Each charged $90 for the loan that was made. Nevertheless, they did not come out even because Tom had the use of $1000 for 9 months, whereas Lucy had the use of only $910. If Tom had lent Lucy $910 and charged simple interest for 9 months, and Lucy then repaid $1000 to Tom, what would have been the interest rate? We can find the answer by solving the simple interest formula for the unknown interest rate:

$$\$1000 = \$910\left(1 + r\frac{9}{12}\right) = 910 + 910\left(\frac{9}{12}\right)r$$

$$90 = 910\left(\frac{9}{12}\right)r.$$

Solving for r we get

$$r = \left(\frac{12}{9}\right)\frac{90}{910} \approx .132.$$

Thus, Tom's simple discount loan to Lucy had the same effect as if he had charged her 13.2% on a simple interest loan with a principal of $910. We say that 13.2% was the **effective rate of interest** on Tom's loan to Lucy. In general, **the effective interest rate equals the discount divided by the product of the proceeds and the length of the loan.** Many loans have a "loan origination fee" that makes the effective rate of interest much higher than the stated rate.

Exercises 6.1

In problems 1 through 10, find (a) the interest paid on the simple interest loan and (b) the total amount repaid to the lender.

1 $1000 for 5 months at 12%
2 $500 for 6 months at 18%
3 $1100 for $9\frac{1}{2}$ months at 10%
4 $240 for 11 months at 15%
5 $900 for 4 months at 20%
6 $3000 for 2 years at 12%
7 $5000 for 4 years at 10%
8 $10,000 for 5 years at 15%
9 $7500 for $3\frac{1}{2}$ years at 12%
10 $1000 for 10 years at 25%

In problems 11 through 20, find (a) the discount for the simple discount loan and (b) the proceeds actually received by the borrower.

11 $1000 for 6 months at 12%
12 $350 for 1 month at 24%
13 $9000 for 12 months at 18%
14 $100 for 2 months at 6%
15 $2000 for 10 months at 10%
16 $5000 for 2 years at 12%
17 $8000 for 1 year at 8%
18 $10,000 for 5 years at 15%
19 $2000 for $2\frac{1}{3}$ years at 12%
20 $100,000 for 5 years at 18%

In problems 21 through 26, find the effective interest rate for the indicated simple discount loan.

21 $1000 for 1 year at 18%
22 $100 for 6 months at 12%
23 $500 for 9 months at 15%
24 $100,000 for 5 years at 12%
25 $750 for 2 years at 6%
26 $400 for 3 months at 24%

Solve the following.

27 Determine the amount borrowed on a simple interest loan if the interest rate is 12%, the length of the loan is 6 months, and the interest is $100.

28 Determine the length of a loan of $1000 if the interest is $200 and the rate of interest is 10%.

29 Determine the interest rate on a loan of $100,000 for 18 months if the interest is $15,000.

30 Determine the amount borrowed on a simple interest loan if the interest rate is 18%, the length of the loan is 9 months, and the interest is $27.

31 Determine the length of a loan of $500 if the interest is $120 and the rate of interest is 12%.

32 Determine the interest rate on a loan of $600 for 30 months if the interest is $150.

6.2 Compound Interest, Effective Rate, and Present Value

In actual practice, a lender is usually quite eager to see some return on his loan, and interest is credited to the lender periodically before the end of the loan. If you lend $100 of your savings to Friendly Savings and Loan Association by opening a passbook account, then they will pay you interest at the rate of 5.5% and will credit the interest to your account quarterly. To compute the amount in your account at the end of 1 year requires four steps. At the end of the first quarter after your interest is paid to your account, its value is $100[1 + .055(\frac{1}{4})] = 100(1 + .01375)$. This amount becomes the principal for the second quarter and the value at the end of the second quarter is

$$100(1.01375)\left[1 + .055\left(\frac{1}{4}\right)\right] = 100(1.01375)[1 + .01375]$$

$$= 100(1.01375)^2.$$

This is the principal at the beginning of the third quarter, and at the end of the third quarter the value of the account is $100(1.01375)^3$. At the end of the year, the value of the account is $100(1.01375)^4 \approx 105.61$.

This is an example of compounding interest. In general, **compound interest** means that earned interest is credited to the lender periodically during the term of the loan, and the interest then becomes part of the principal. The total amount due the lender at the end of the loan period is called the **compound amount.** In view of the above computations, it is not surprising that the compound amount is given by the following formula.

$$A = p(1 + rt)^n$$

Here A stands for the compound amount, p is the principal, r is the per annum interest rate, t is the fraction of a year between payments of interest, and n is the number of times that interest is credited to the lender's account by the end of the loan. The compound amount A minus the principal p is the **amount of the compound interest.** The quantity $(1 + rt)^n$ can be calculated using a calculator or the tables in the appendix of this book.

EXAMPLE 1

Susy buys $1000 worth of stock in Reliable Mutual Fund that manages to increase the value of its stock by 16% per year for 10 years. At the end of that time, what is Susy's stock worth?

Solution

$$A = p(1 + rt)^n$$

$$A = \$1000\ [1 + .16(1)]^{10} \approx \$1000[4.41144]$$
$$\approx \$4,411.44 \quad \blacksquare$$

The length of time between payments of interest is called the **compound period,** and paying the interest is called **compounding.**

CALCULATOR EXAMPLE

Economy Savings and Loan Association pays interest on passbook accounts at the rate of 5.5%, and interest is credited to the account daily. If $100 is left in a passbook account for 1 year, what is the account worth? How much is the compound interest?

Solution

$$A = \$100\left[1 + .055\left(\frac{1}{365}\right)\right]^{365} \approx \$100[1.00015068]^{365}$$

$$\approx \$105.65$$

$$\text{compound interest} = A - p = \$105.65 - \$100.00 = \$5.65$$

The above power was computed on a calculator, since the exponent is larger than those in our tables. $\blacksquare$

When the number of compound periods is unknown, then the problem requires the use of logarithms, as in the next example.

EXAMPLE 2

How long does it take for a principal p to double if money is worth 18% compounded annually?

Solution
We must solve the following equation for n.

$$2p = p(1 + .18)^n$$
or
$$2 = (1.18)^n$$

$$\ln 2 = \ln[1.18^n] = n \ln 1.18$$

$$\frac{\ln 2}{\ln 1.18} = n$$

$$n \approx 4.19$$

Since interest is paid annually, it would be the end of the fifth year before the principal had doubled. $\blacksquare$

It is interesting to note that the daily compounding of Economy Savings and Loan Association in the Calculator Example made very little difference when compared to the quarterly compounding of Friendly Savings and Loan Association in the first paragraph of this section. For any loan, regardless of the actual compound period, we define the **effective interest rate** r_e to be the interest rate with a compound period of 1 year that will result in the same compound amount; thus, r_e is the solution to the equation

$$A = p(1 + r_e)^n$$

where A is the actual compound amount of the loan, p is the principal, and n is the number of years of the loan (so n may be a fraction or a whole number plus a fraction).

EXAMPLE 3

What is the effective interest rate in the Calculator Example?

Solution

$$A = p(1 + r_e)^n$$
$$\$105.65 = 100(1 + r_e)^1$$
$$.0565 = r_e \quad \blacksquare$$

For the sake of comparison, the effective interest rate of Friendly Savings and Loan Association is .0561. The moral is that **frequency of compounding,** (that is, the number of compound periods in 1 year) is not nearly as important as the interest rate. Throughout the 1970s, savings and loan associations were required by federal law to pay interest on passbook accounts at a rate not greater than 5.25%.

Another use for the compound amount formula is to calculate the value now of a financial obligation that is to be paid to you at some point in the future. Suppose Joe Radlow learns that he is going to inherit $10,000 from his deceased Uncle Bill, but it is going to take 5 years to settle the rather complicated estate. As it turns out, Joe needs some money now. Joe offers to make a contract with his brother Phillip to give to Phillip the $10,000 when it is available 5 years from now provided Phillip will pay Joe some cash now. How much should Phillip pay Joe? The answer depends on how valuable it is to have the cash for the next 5 years. Suppose it is agreed that 12% per annum with a compound period of 1 year is a fair interest rate. We must ask how much money Phillip would need to deposit in a bank at that interest rate so that the compound amount in 5 years would be $10,000. The answer is found by these computations.

$$p(1 + rt)^n = A$$
$$p = A(1 + rt)^{-n}$$
$$p = 10,000(1 + .12)^{-5}$$
$$p \approx \$5674.27$$

Thus, the present value of $10,000 that will be available 5 years from now is $5674.27. In general, the **present value** of the sum A at a future date is given by

$$p = A(1 + rt)^{-n}$$

where r is the interest rate, t is the length of a compound period measured in years, and n is the number of compound periods that will pass before the sum A is obtained. The quantity $(1 + rt)^{-n}$ can be calculated using a calculator or the tables in the appendix of this book.

EXAMPLE 4

Brown Office Supply Company has a debt of $80,000 due at Service Bank in 5 years, and Brown pays $10,000 now toward satisfying that debt. How much should Brown pay in 2 years to satisfy the obligation if money is worth 18% with annual compounding?

Solution

Since money is worth 18% with annual compounding, the $10,000 will result in the following compound amount for Service Bank 2 years from now.

$$A = p(1 + rt)^n$$

$$A = \$10{,}000(1 + .18)^2$$
$$= \$10{,}000(1.3924)$$
$$= \$13{,}924.00$$

Two years from now, the $80,000 debt will be due in another 3 years. Thus, 2 years from now the present value of the debt will be

$$p = A(1 + rt)^{-n}$$

$$p = \$80{,}000(1 + .18)^{-3}$$
$$\approx \$80{,}000(.608631)$$
$$\approx \$48{,}690.48$$

of which the first payment will account for $13,924.00. Thus, Brown's payment 2 years from now should be

$$\$48{,}690.48 - \$13{,}924.00 = \$34{,}766.48 \quad \blacksquare$$

Caution: It is very important to note that amounts of money that change hands at different times **cannot** be added or subtracted directly. Every transaction must be converted by the compound amount formula **to** a time in the future, or by the present value formula **from** a time in the future, so that all amounts are valued at the **same point in time.**

For the preceding example, it is easiest to convert both the first payment and the total debt to their values at the time 2 years from now, as shown in the time line diagram on the next page.

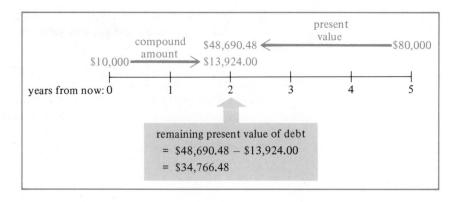

CALCULATOR EXAMPLE

The sum of $2500 is invested so that it earns 10% per annum with monthly compounding. How much is the investment worth after 20 years?

Solution

$$A = \$2500\left[1 + .10\left(\frac{1}{12}\right)\right]^{240} \approx \$2500[7.328074] \approx \$18,320.19 \quad \blacksquare$$

Exercises 6.2

In problems 1 through 8, find (a) the compound amount and (b) the interest on the loan described.

1   $1000 for 6 months at 12% with monthly compounding
2   $500 for 1 year at 18% with quarterly compounding
3   $800 for 50 weeks at 26% with weekly compounding
4   $100 for 2 years at 24% with monthly compounding
5   $5000 for 3 months at 10% with quarterly compounding
6   $10,000 for 10 years at 15% with yearly compounding
7   $1 for 20 years at 24% with semiannual compounding
8   $750 for 10 months at 12% with monthly compounding

In problems 9 through 16, find the present value of the indicated future payment assuming the interest rate and compounding that is given.

9    $1000 paid 2 years from now, 12% with yearly compounding
10   $500 paid 1 year from now, 15% with compounding every 4 months
11   $800 paid 6 months from now, 18% with monthly compounding
12   $100,000 paid 10 years from now, 10% with semiannual compounding
13   $100 paid 3 months from now, 12% with monthly compounding
14   $8000 paid 4 years from now, 24% with quarterly compounding

15 $750 paid one year from now, 18% with yearly compounding

16 $1,000,000 paid 25 years from now, 24% with yearly compounding

In problems 17 through 22, find the effective interest rate for the loan described.

17 $1000 for 1 year at 12% compounded monthly

18 $3000 for 1 year at 18% compounded quarterly

19 $500 for 1 year at 10% compounded semiannually

20 $1000 for 1 year at 6% compounded quarterly

21 $800 for 1 year at 24% compounded annually

22 $100 for 1 year at 15% compounded every 4 months

Solve the following.

23 How long does it take for a principal p to double if money is worth 12% compounded annually?

24 How long will it take for an investment of $100 to double if it grows at 20% per year?

25 Some credit card companies charge interest at the rate of $1\frac{1}{2}$% per month. What is the effective rate?

26 An investor can choose between one investment that pays 12% compounded monthly or another investment that pays 15% compounded yearly. Which investment is better?

27 A debt of $500 is due 5 years from now. If it is paid off 2 years from now, how much is paid? Money is worth 12% compounded annually.

28 A payment of $25,000 is made now on a debt of $100,000 due in 10 years. If money is worth 10% compounded annually, how much should be paid in 5 years to satisfy the debt?

29 An investment of $100 appreciates at an effective rate of 18% for 5 years and then appreciates at a rate of 10% for the next 5 years. What is the investment worth at the end of the 10 years?

30 On Mark's eighth birthday, his father put $5000 into a savings account paying 10% compounded quarterly. On Mark's eighteenth birthday, he wants to use the savings account to go to college. How much is in the account?

31 *Calculator Problem* Determine the compound amount provided the principal is $5000, the interest rate is 18% with quarterly compounding, and the loan is repaid after 30 years.

32 *Calculator Problem* The sum of $7800 is invested so that it earns 15% with monthly compounding. How much is the investment worth after 25 years?

33 *Calculator Problem* Determine the compound interest for a loan of $4000 for 5 years at 10% with monthly compounding.

34 *Calculator Problem* Determine the compound interest for a loan of $50,000 for 20 years at 15% with monthly compounding.

6.3   Inflation

Inflation has become such an important consideration that it was a major issue in the United States presidential election of 1980. There is much debate in economics as to the causes of inflation and the most accurate measure of inflation. Consideration of that debate and related discussions are left to economics textbooks. We want to consider here only the mathematics of inflation.

The most widely followed measure of inflation is the Consumer Price Index, which is compiled by the Department of Labor. The Department of Labor issues monthly reports on the rate of increase of the prices of the goods purchased by a "typical" American family. The mathematical understanding of these statistics is the same as understanding compound interest. For example, according to the figures of the Department of Labor, the rate of inflation for 1980 was 12%. Thus, a sports coat that cost $100 in January 1980 would typically cost $112 in January 1981. Of course, the price of a particular item might behave differently from the Consumer Price Index.

Many people plan and save to make big purchases like homes, so the effect of inflation on the price is important.

EXAMPLE 1

Bob and Alice were married recently and are living in a rented apartment. They picked out a subdivision that they really like with houses priced around $60,000. They plan to save for 5 years to get a down payment for one of those houses. (a) If inflation is 10% per annum for the 5 years that they are saving, what will be the price when they are ready to buy? (b) If inflation is only 8% per annum for the 5 years, what will be the price?

Solution
Let p be the present price of the house. Then the price after n years will be $A = p(1 + rt)^n$.

(a) $A = \$60,000[1 + .10(1)]^5$
    $= \$60,000[1.61051]$
    $\approx \$96,631$

(b) $A = \$60,000[1 + .08(1)]^5$
    $\approx \$60,000[1.46933]$
    $\approx \$88,160$ ■

When the unknown quantity in a problem is the number of compound periods, it is necessary to use logarithms to solve the equation, as shown in the next example.

EXAMPLE 2

How long does it take for prices to double if inflation persists at 12%?

Solution
Let p be a typical price; then the compound amount is $2p$ and the compound amount formula gives

$$2p = p[1 + .12]^n$$
$$2 = (1.12)^n$$
$$\ln 2 = \ln(1.12)^n = n \ln 1.12$$
$$n = \frac{\ln 2}{\ln 1.12} \approx 6.1.$$

A typical price doubles in 6.1 years when inflation persists at 12%. ■

The following table shows the rate of inflation for various years according to the Department of Labor.

1960	1.6%
1965	1.7%
1970	5.9%
1974	11.0%
1975	9.1%
1976	5.8%
1977	6.5%
1978	7.7%
1979	11.3%
1980	13.4%
1981	10.2%
1982	5.1%

Exercises 6.3

1 If the effective rate of inflation over the next 30 years is 10% and a loaf of bread now costs $1, how much will a loaf of bread cost 30 years from now?

2 Henry puts $10,000 in a savings account that pays 6% compounded quarterly for 10 years. If the effective rate of inflation for those 10 years is 10%, what will be the new cost of what was $10,000 worth of goods? What is the difference between the new cost of $10,000 worth of goods and the value of Henry's savings account?

3 If the Jones family now requires $10,000 per year in take-home pay to maintain a middle class life style and if the effective rate of inflation over the next 20 years is 8%, what take-home pay will the Jones family require in 20 years?

4 If gold now costs $500 per troy ounce and the effective rate of inflation over the next 10 years is 10%, what will gold cost in 10 years?

5 If silver now costs $13 per troy ounce and the effective rate of inflation over the next 10 years is 15%, what will silver cost in 10 years?

6 What is the price of a house, now costing $100,000, after 20 years of

inflation at an effective rate of 12%? How much less is the price of the house if inflation is only 8%?

7 What is the price of a car, now costing $8,000, after 20 years of inflation at an effective rate of 11%? How much less is the price of the car if inflation is only 6%?

8 What is the price of a suit, now costing $200, after 20 years of inflation at an effective rate of 10%? How much less is the price of the suit if inflation is only 7%?

9 What is the price of a college education, now costing $15,000, after 20 years of inflation at an effective rate of 12%? How much less is the price of a college education if inflation is only 9%?

10 How long does it take for prices to double if the effective rate of inflation is 11%?

11 How long does it take for prices to double if the effective rate of inflation is 9%?

12 In 1979 consumer prices in Italy went up about 21%. If this rate of inflation continues, how long will it take for their prices to double?

13 In 1979 consumer prices in Japan went up about 7%. If this rate of inflation continues, how long will it take for their prices to double?

14 In 1979 consumer prices in West Germany went up about 5%. If this rate of inflation continues, how long will it take for their prices to double?

15 In 1979 consumer prices in the United Kingdom went up about 19%. If this rate of inflation continues, how long will it take for their prices to double?

6.4   Ordinary Annuities

An **ordinary annuity** is any sequence of payments in which each payment is the same amount, the length of time between any two payments is always the same, and the interest is paid when each payment is made. If Jane puts $100 per month in her savings account at Commercial Bank, and she always makes the deposit on the last day of the month, then she has created an ordinary annuity. If Commercial Bank pays her interest on her account at the rate of 5% with monthly compounding, how much does she have in her account at the end of the year? To simplify the notation, let us write the compound amount formula from Section 6.2 as

$$A = p(1 + r)^n$$

where r is now the rate of interest per compound period (the per annum rate divided by the frequency of compounding) and n is the number of compound periods. The regular payment of the ordinary annuity is called the **rent.** Thus, the rent in Jane's ordinary annuity is $100 and the rate of interest per compound period is $\frac{.05}{12} \approx .004166$.

To compute the amount in Jane's account at the end of one year requires 12 steps. Her first payment of $100 has been in the account for 11 months at the end of the year and consequently the compound amount is $100(1 + .004166)^{11}$. Her second payment has been in the account for 10 months at the end of the year, so the compound amount is $100(1 + .004166)^{10}$. This pattern continues, and her last payment is made the last day of the year. Thus, it earns no interest. The value of her account at the end of the year is the sum of the compound amounts resulting from each of her 12 payments. Thus, the value of her account is the sum

$$\$100 + \$100(1.004166) + \$100(1.004166)^2 + \cdots + \$100(1.004166)^{11}$$

where the three dots indicate that some terms have been omitted but the omitted terms continue the pattern of the terms that are listed.

It is apparent that some better mathematical equipment is required to make this problem reasonable, since adding up the terms would be very tedious. Any sum of the form

$$S = a + aq + aq^2 + \cdots + aq^{n-1}$$

is said to be a **finite geometric series.** What distinguishes such sums is that each term after the first can be divided by the preceding term and the quotient is q. It is easy to find a simple formula for the value of S. Let's write out the terms of S, and below them write the result of multiplying through by $-q$:

$$S = a + aq + aq^2 + \cdots + aq^{n-1}$$
$$-Sq = -aq - aq^2 - aq^3 - \cdots - aq^n.$$

Now we find the sum of the two equations.

$$S - Sq = a + 0 + 0 + \cdots + 0 - aq^n$$
$$S(1 - q) = a - aq^n$$

Thus, if $1 - q \neq 0$, we have

$$S(1 - q) = a(1 - q^n)$$

or

$$S = \frac{a(1 - q^n)}{1 - q}.$$

This is the kind of simple formula that we wanted. For q greater than 1, we can avoid handling negative quantities by rewriting the formula as

$$S = \frac{a(q^n - 1)}{q - 1}.$$

If $q = 1$, this formula cannot be used because the denominator is 0. However, in that case each term equals a and the sum is na.

Now we have an easier way to get the value of Jane's account at Commercial Bank at the end of the year.

$$S = 100 + 100(1.004166) + \cdots + 100(1.004166)^{11}$$

is equivalent to

$$S = \frac{a(q^n - 1)}{q - 1} = \frac{100(1.004166^{12} - 1)}{1.004166 - 1}$$

$$\approx 100 \frac{.051154}{.004166} \approx 1227.89.$$

In general, if the rent of the annuity is R, the interest rate per compound period is r, and the number of payments in the annuity is n, then the value of the account at the end of the annuity, which is called the **sum of the annuity,** is given by the formula

$$S = R + R(1 + r) + \cdots + R(1 + r)^{n-1} = R\frac{(1 + r)^n - 1}{r}.$$

The quantity $[(1 + r)^n - 1]/r$ is abbreviated $s_{\overline{n}|r}$. Tables giving the value of $s_{\overline{n}|r}$ for various values of n and r can be found in the appendix. The symbol $s_{\overline{n}|r}$ is sometimes read "s angle n at r".

$$S = Rs_{\overline{n}|r}$$

EXAMPLE 1

ABC Company anticipates replacing their widget sorter, and they have signed a contract to buy a new sorter in 10 years for $100,000. They want to make equal annual payments into an account earning 12% interest so that the value of the account will be $100,000 at the end of 10 years. What should be the amount of their annual payment?

Solution
This is an example of an ordinary annuity where the rent is unknown but the sum of the annuity is known. Thus we must solve the equation

$$\$100,000 = R\frac{(1 + .12)^{10} - 1}{.12} \approx R(17.548735).$$

So $R = \$5698.42$ and the company should pay this amount into their account annually. An account such as this, which is designed to replace some expensive equipment, is called a **sinking fund.** ■

EXAMPLE 2

Sam rents an apartment for $200 per month and pays his rent on the last day of the month for the next month. If Sam has a 1-year lease and Sam's landlord is putting the rent into an investment with a return of 18% compounded monthly, what is the sum of the landlord's annuity at the end of the year?

Solution

$$S = R\frac{(1+r)^n - 1}{r} = \$200\frac{(1+.18/12)^{12} - 1}{.18/12}$$

$$S = \$200\frac{(1+.015)^{12} - 1}{.015} \approx \$2608.24 \quad \blacksquare$$

A common example of an annuity is found in what banks call install-ment loans. An **installment loan** is a loan agreement that has a borrower agree to pay a lender an annuity with a fixed rate of interest to repay a loan and to compensate the lender for making the loan. Suppose Luke Stone wants to buy a new car but he has only $500 to spend at the present. However, Luke has a good job and he figures he can pay $150 per month for 5 years to a bank to repay an auto loan. Neighborhood Bank agrees to give Luke an auto loan at 15% provided they can repossess the car, if necessary, to recover their money. Assuming Luke makes monthly payments of $150 to the bank, starting at the end of the first month, how much should the bank lend Luke?

To solve this problem in a straightforward way requires 60 steps. We must compute the present value of each of Luke's payments to Neighbor-hood Bank and then add up all of those present values to get the present value of such a loan contract. Here are the computations.

$$P = 150\left(1+\frac{.15}{12}\right)^{-1} + 150\left(1+\frac{.15}{12}\right)^{-2} + \cdots + 150\left(1+\frac{.15}{12}\right)^{-59} + 150\left(1+\frac{.15}{12}\right)^{-60}$$

Since this is a finite geometric series, we can use the sum formula previously developed for such series, provided we factor $150(1+\frac{.15}{12})^{-1}$ out of each term.

$$P = 150\left(1+\frac{.15}{12}\right)^{-1}\left[1+\left(1+\frac{.15}{12}\right)^{-1} + \cdots + \left(1+\frac{.15}{12}\right)^{-59}\right]$$

$$= 150\left(1+\frac{.15}{12}\right)^{-1}\frac{1-\left(1+\frac{.15}{12}\right)^{-60}}{1-\left(1+\frac{.15}{12}\right)^{-1}} \approx \$6305.19$$

So Luke should have a total of $500 + $6305.19 = $6805.19 to buy a new car.

In general, the **present value** of an ordinary annuity consisting of n payments of R dollars with interest per compound period at a rate of r is

$$P = R\frac{1-(1+r)^{-n}}{r} = Ra_{\overline{n}|r}.$$

The quantity $a_{\overline{n}|r}$ is read "a angle n at r," and it can be found in the appendix of this book.

EXAMPLE 3

Steve Pao finds a motorcycle that he wants to buy at Cycle Store. The motorcycle costs $2000 and Cycle Store will finance the purchase over 24 months at 12% with monthly compounding. If Steve is going to make payments at the end of each month for 2 years, how much will his payments be?

Solution

The present value of Steve's loan contract must be $2000 or Cycle Store would not give him the motorcycle. We know all of the numbers that go into the present-value formula above except the rent R. Hence, we solve the formula for R.

$$\$2000 = R\,\frac{1-(1+.01)^{-24}}{.01} \approx 21.243387R$$

$$R = \$94.15 \quad \blacksquare$$

One way that a corporation obtains money to meet its major expenses is by selling bonds. A **bond** is a written contract in which the corporation agrees to pay the purchaser the face value of the bond (normally $1000) at a specified future date called the **redemption date** (usually 10 to 25 years off). Also, the corporation pays interest to the purchaser at the end of each quarter until the redemption date; the rate of interest is printed on the bond along with the face value and the redemption date.

EXAMPLE 4

What is a fair price for a $1000 bond issued by Trusty Corporation that has 20 years until the redemption date? Assume that the bond pays interest at the rate of 8% and that a fair interest rate over those 20 years is 12%.

Solution

First, we compute the quarterly interest payments.

$$i = prt$$

$$i = \$1000(.08)\left(\frac{1}{4}\right) = \$20$$

Now, we compute the present value of those payments using $r = .12(\frac{1}{4}) = .03$ and $n = 20(4) = 80$.

$$P = R\,\frac{1-(1+r)^{-n}}{r}$$

$$P = \$20\,\frac{1-(1+.03)^{-80}}{.03} \approx \$20(30.200763)$$

$$\approx \$604.02$$

Finally, we compute the present value of the $1000 payment at the redemption date and add it to the present value just calculated.

$$P = A(1 + rt)^{-n}$$

$$P = \$1000 \left(1 + .12 \left(\frac{1}{4}\right)\right)^{-80} = \$1000(1.03)^{-80}$$

$$\approx \$1000(.093977) \approx \$93.98$$

current worth of bond = $604.02 + $93.98 = $698.00

A fair price for this bond is $698.00. ■

The moral of the preceding example is that the value of a bond drops when interest rates increase beyond the rate stated on the bond. Of course, the value of a bond increases when interest rates decline.

Exercises 6.4

In each of the problems 1 through 6, find the sum of the finite geometric series.

1 $1 + \dfrac{1}{2} + \left(\dfrac{1}{2}\right)^2 + \cdots + \left(\dfrac{1}{2}\right)^5$

2 $1 + 2 + (2)^2 + \cdots + (2)^6$

3 $\dfrac{1}{3} + \left(\dfrac{1}{3}\right)^2 + \left(\dfrac{1}{3}\right)^3 + \cdots + \left(\dfrac{1}{3}\right)^8$

4 $25 + 125 + \cdots + 3125$

5 $\dfrac{1}{4} + \dfrac{1}{16} + \cdots + \dfrac{1}{1024}$

6 $e^3 + e^4 + \cdots + e^9$

In each of the problems 7 through 12, find the sum of the given ordinary annuity.

7 $100 is paid each month for 2 years and the interest rate is 12%.
8 $200 is paid each month for 11 months and the interest rate is 18%.
9 $10,000 is paid yearly for 8 years and the interest rate is 10%.
10 R is $5000, r is 1%, and $n = 18$.
11 R is $150, r is .5%, and $n = 50$.
12 R is $100,000, r is 10%, and $n = 5$.

In each of the problems 13 through 18, find the present value of the given ordinary annuity.

13 $1000 is paid annually for 5 years and the interest rate is 9%.
14 $250 is paid annually for 10 years and the interest rate is 12%.
15 $100 is paid monthly for $1\frac{1}{2}$ years and the interest rate is 12%.
16 $10,000 is paid quarterly for 10 years and the interest rate is 16%.
17 R is $500, r is 2%, and $n = 20$.
18 R is $100,000, r is 20%, and $n = 3$.

Solve the following.

19 John can afford to pay $100 each month for 4 years. If the City Bank charges 12% with monthly compounding, how much can John afford to borrow?

20 Tom has a chance to buy a company that will pay him $100,000 each year for 20 years. If money is worth 15% with annual compounding, then what should Tom pay for the company?

21 Mary won a lottery and she gets to choose between receiving $50,000 now or receiving $7500 each year for 10 years. If money is worth 10% compounded annually, which should Mary choose?

22 Hustle Insurance Company sells a contract in which they agree to pay $100 at the end of each month for 4 years. If money is worth 12% compounded monthly, what should Hustle charge for its contract?

23 Sue has been paying $150 at the end of each month on her 5-year auto loan and she has made 24 payments. She has now inherited some money and she wants to pay off her loan. How much should she have to pay, if the interest rate is 15%?

24 Mr. Brown estimates it will cost him $16,000 to send his son to college 4 years from now. If he makes monthly deposits in an account paying 12% with monthly compounding, how much must his deposits be in order to result in $16,000 when his son is ready to start college?

25 Alice borrows $4000 from Ted and she pays him $1000 on the debt 1 year later. Also she pays him $1000 after 2 years and again after 3 years. At the end of the fourth year, she wants to settle the debt. If money is worth 15% compounded yearly, how much should she pay him?

26 It will cost Jones Company $250,000 to replace a certain machine 10 years from now and so they set up a sinking fund. Initially they make annual payments of $15,686.63 into a bank account receiving 10% interest with yearly compounding. After their fifth payment, the bank increases the interest to 15% with yearly compounding. How much should their remaining payments be?

27 Pete puts $200 each month in his account at the credit union which pays 6% with monthly compounding. After 4 years how much has accumulated in his account?

28 An **annuity due** is an annuity where the payments are made at the beginning of each compound period rather than at the end of the period. What is the sum of an annuity due consisting of 20 annual payments of $1000 if interest is paid at a rate of 18% with yearly compounding? (**Hint:** An annuity due is one initial payment plus an ordinary annuity.)

29 Find the present value for an annuity due consisting of 24 monthly payments of $100 with interest of 12% and monthly compounding.

30 What is the present value of an ordinary annuity consisting of 10 annual payments of $1000 with interest at 12% with annual compounding, if the first payment is 5 years from now? (**Hint:** Consider an ordinary annuity with 14 payments minus an ordinary annuity with 4 payments.)

31 What is a fair price for a $1000 bond that has 12 years until its redemption date if the bond's interest rate is 6% and a fair interest rate is 18%?

32 What is a fair price for a $1000 bond that has 12 years until its redemption date if the bond's interest rate is 8% and a fair interest rate is 16%?

33 If $300 is deposited at the end of each month for 4 years in an account earning 6% interest compounded monthly, what is the value of the account after 6 years?

34 What is the value now of an ordinary annuity that begins in 1 year and consists of monthly payments of $200 for 3 years if the interest rate is 12% compounded monthly?

35 Determine the amount that should be deposited at the end of each month for 4 years in an account earning 12% interest compounded monthly, so that the total in the account after 8 years is $100,000.

36 *Calculator Problem*   If $100 is deposited at the end of each month for 20 years in an account earning 12% compounded monthly, what is the value of the account at the time of the last payment?

37 *Calculator Problem*   What is the value of the account described in problem 36 10 years after the last deposit?

38 *Calculator Problem*   If you were offered an ordinary annuity that would pay you $150 at the end of each month for 15 years with interest computed at a rate of 18% with monthly compounding, what would be a fair price to pay for that annuity?

6.5   Mortgages

For most people the most important ordinary annuities in their lives are the mortgages that they pay on their homes. A **mortgage** is a loan secured by something like a house or building; the lender has a claim to the title or ownership of the house. Thus, if the borrower fails to make the payments, the lender can take possession of the house and sell it to obtain the money owed. The payments that the borrower makes constitute an ordinary annuity. **Equity** means the value of the house minus the amount owed on it.

EXAMPLE 1

Connie Fredrick is a successful store manager. She owns a house valued at $60,000 and she has made 36 monthly payments of $500 on her mortgage, which has an interest rate of 15%. Connie has 22 more years to pay on the mortgage. What is her equity in her home?

Solution

The easy way to make the computation is to find the present value of the annuity represented by her remaining payments and subtract that from $60,000. The interest rate per compound period is $\frac{.15}{12} = .0125$, and the number of periods (months) is $22(12) = 264$. Thus, we have

$$A = 500 \frac{1 - (1 + .0125)^{-264}}{.0125} \approx \$38,494.19$$

Thus, Connie's equity is $60,000 − $38,494.19 = $21,505.81. ■

Each payment made on a mortgage repays some of the principal and pays some interest. The rule is that, first, one deducts from the payment the interest on all loan funds not repaid for the period that the loan funds were held. Suppose that Connie's full mortgage was $39,037; for the month preceding her first $500 payment she had the benefits of that full amount. The interest that she owed for that month was

$$\$39,037 \left(\frac{.15}{12}\right) = \$487.96.$$

Thus, $500 − $487.96 = $12.04 went to repaying the principal. A loan is said to be **amortized** when a portion of each payment goes to repaying the principal. The total amount of principal repaid at any given time as a result of payments made is called the **amortization** of the loan.

EXAMPLE 2

If Connie made a 20% down payment when she bought her house, how much of her equity results from inflation increasing the value of her house and how much results from her repayment of the principal of the loan?

Solution
Since the full mortgage loan was $39,037 and the outstanding balance of the loan was calculated to be $38,494.19 in Example 1, the amortization is clearly

$39,037 − $38,494.19 = $542.81.

This is the amount of her equity resulting from her repayment of the principal of the loan. All of the rest of her equity must result from inflation and her down payment. Since her down payment was 20% of the price of the house, her mortgage loan must have been 80% of the price.

(.80)(price) = $39,037

$$\text{price} = \frac{\$39,037}{.80} = \$48,796.25$$

Connie's down payment must have been

(.20)($48,796.25) = $9,759.25.

Using the computation of Connie's equity from Example 1, we find inflation's contribution to Connie's equity to be

$21,505.81 − $542.81 − $9,759.25 = $11,203.75. ■

It is interesting to note that in the preceding example inflation made a greater contribution to Connie's equity than either her down payment or the amortization of her mortgage loan. And this happened during the first 3 years of ownership. Such computations as these have led many people to believe that owning a home is a good hedge against inflation.

EXAMPLE 3

Exactly 3 years ago, when Connie bought her house, Ralph bought the house next door for $49,000 and he paid for it in cash without a loan. His house, like Connie's house, is now worth $60,000. What is Ralph's rate of return on his investment? Is it higher than Connie's rate of return on her investment?

Solution

It is easy to compute Ralph's rate of return. He now has

$$\$60,000 - \$49,000 = \$11,000$$

more than he put in, so his rate of return is

$$\frac{11,000}{49,000} \approx .22 = 22\%.$$

In contrast, Connie put only $9,759.25 into her house initially and she put in $542.81 for the amortization of the loan. She now has

$$\$21,505.81 - \$9,759.25 - \$542.81 = \$11,203.75$$

more than she put in. So her rate of return is

$$\frac{\$11,203.75}{\$9759.25 + \$542.81} \approx 109\%.$$

Connie's rate of return is much higher. ■

These computations illustrate the idea of financial **leverage.** Leverage can be defined as making money with borrowed money. Connie is making money with the money that she borrowed on her home mortgage, while Ralph is making money with only his own money. Before anyone gets too excited about how clever Connie is, it should be pointed out that Connie pays $500 every month on her mortgage while Ralph pays nothing. Connie is paying a lot of interest on her borrowed money.

EXAMPLE 4

Maxi-Growth Corporation bought Local Company for $500,000. Maxi-Growth paid $200,000 at the time of purchase and signed a mortgage note agreeing to pay $124,904.69 on the first, second, and third anniversaries of the purchase date. Interest is charged at a rate of 12%. Make a table to show how each payment is divided into interest and repayment of principal.

Solution

Payment	Principal Before Payment	Interest	Repayment of Principal	Principal After Payment
1	300,000.00	36,000.00	88,904.69	211,095.31
2	211,095.31	25,331.44	99,573.25	111,522.06
3	111,522.06	13,382.65	111,522.04	.02

The .02 in the fifth column and the third row results from round-off error in making the computation. In actual practice, the last payment would be $124,904.71. ■

A table, like the preceding one, which shows how a loan is amortized on a payment-by-payment basis, is called an **amortization schedule.**

Historically a method that was used to help a home purchaser to buy a more expensive home was to extend the mortgage over a longer period of time. A typical home mortgage of the 1940s lasted about 10 years while a typical home mortgage of the 1970s lasted 25 years. When mortgage interest rates get sufficiently high, as in 1982, the home buyer's purchasing power is increased only slightly by extending the length of the mortgage. The next example demonstrates this.

EXAMPLE 5

A family wants to buy the best home that it can afford. The maximum monthly payment that the family can make is $600, and the lending institutions are all charging 16% on home mortgages. How much more can the family borrow with a 30-year mortgage than with a 25-year mortgage?

Solution

First we use the present value formula for an annuity with $R = \$600$, $r = .16(\frac{1}{12}) \approx .0133$, and $n = 25(12) = 300$.

$$A = R \frac{1 - (1 + r)^{-n}}{r} = \$600 \frac{1 - (1 + .0133)^{-300}}{.0133}$$

$$\approx \$600(73.589534) \approx \$44,153.72$$

Now we make the same computation with $n = 30(12) = 360$, and we compute the increase in A.

$$A = \$600 \frac{1 - (1 + .0133)^{-360}}{.0133} \approx \$600(74.362878)$$

$$\approx \$44,617.73$$

increase in mortgage $= \$44,617.73 - \$44,153.72 = \$464.01$

The increase in the present value of the mortgages, which is the amount that a lending institution would lend, is very small compared to the prices of homes. ■

Exercises 6.5

In each of the problems 1 through 4, make an amortization schedule for each indicated loan.

1 Five payments of $100 per month to retire a debt of $485.34 with interest at 12% compounded monthly.

2 Three payments of $10,000 per year to retire a debt of $22,832.25 with interest at 15% compounded yearly.

3 Four payments of $137.75 per quarter to retire a debt of $500 with interest at 16% compounded quarterly.

4 Three payments of $374.11 at the end of each 6 months to retire a debt of $1000 with interest at 12% compounded semiannually.

Solve the following.

5 The Smiths' house is worth $100,000 and they have 48 monthly payments of $400 remaining on their 9% mortgage. Find their equity.

6 The Taylors' house is worth $90,000 and they have 42 monthly payments of $360 remaining on their 8% mortgage. Find their equity.

7 The Johnsons' house cost them $50,000 ten years ago, and, as a result of inflation, it has appreciated at 10% each year. They have 50 monthly mortgage payments of $300 remaining on their 8% mortgage. Find their equity.

8 The Mitchells' house cost them $30,000 25 years ago and, as a result of inflation, it has appreciated at 6% each year. They have 50 monthly payments of $120 remaining on their 4% mortgage. Find their equity.

9 The Browns have 30 monthly payments of $900 left on their 15% mortgage, and they want to sell their house. How much will it cost the Browns to pay off their mortgage when their house is sold?

10 The Cobbs have 18 monthly payments of $240 left on their 9% mortgage, and they want to sell their home. How much will it cost to pay off the mortgage when their house is sold?

11 Sam pays $237 monthly on his auto loan, which has interest at a rate of 12% compounded monthly. Without constructing an amortization schedule, find how much of Sam's 30th payment is interest, assuming he must make 48 payments all together. (**Hint:** Find Sam's debt at the time the 29th payment is made by getting the present value of the remaining annuity.)

12 Joan pays $200 monthly on her auto loan, which has an interest rate of 12% compounded monthly. Find how much of Joan's 20th payment is interest, assuming she must make 36 payments all together. (**Hint:** See the above problem.)

13 The Smiths, from problem 5, made a down payment of $11,000 and got a mortgage for $47,664.65 to buy their house. How much of their equity results from inflation?

14 The Taylors, from problem 6, made a down payment of $9,000 and got a mortgage for $46,643.23 to buy their house. How much of their equity results from inflation?

15 *Calculator Problem* How much more money can a borrower obtain with a 30-year mortgage than a 25-year mortgage when interest rates are 17% and the borrower can pay only $700 per month?

16 *Calculator Problem* In the early 1970s, common mortgage interest rates were about 8%. How much more money could a borrower obtain

with a 30-year mortgage than a 25-year mortgage when interest rates were 8% and the borrower could pay only $350 per month?

Review of Terms

Important Mathematical Terms

interest, p. 256
per annum, p. 256
principal, p. 256
simple interest, p. 257
simple discount, p. 257
proceeds, p. 258
effective rate of interest, p. 258
compound interest, p. 260
compound amount, p. 260
compound period, p. 261

Important Terms from the Applications

inflation, p. 266
sinking fund, p. 270
installment loan, p. 271
bond, p. 272

compounding, p. 271
effective interest rate, p. 262
frequency of compounding, p. 260
present value, p. 263
ordinary annuity, p. 268
rent, p. 268
finite geometric series, p. 269
amortized, p. 276
amortization, p. 276
amortization schedule, p. 278

redemption date, p. 272
mortgage, p. 275
equity, p. 275
leverage, p. 277

Review Problems

1 What is the total amount repaid to the lender on a simple interest loan of $9000 for 3 years at 20%?

2 Find the proceeds of a simple discount loan of $5000 for 2 years at 15%.

3 Find the effective interest rate for a simple discount loan of $3000 for 1 year at 18%.

4 Determine the principal for a simple interest loan if the interest rate is 12%, the length of the loan is 10 months, and the interest is $400.

5 Determine the length of a simple discount loan of $8000 if the discount rate is 10% and the discount is $1000.

6 Determine the discount rate on a simple discount loan of $60,000 for 4 years if the discount is $36,000.

7 Find the compound amount for a loan of $10,000 for 1 year at 20% with quarterly compounding.

8 Find the compound interest on a loan of $1000 at 18% for 4 years with monthly compounding.

9 What is the present value of a payment of $1,000,000 25 years from now if an appropriate interest rate is 20% with annual compounding?

10 What is the effective interest rate for a loan of $6000 for 2 years at 12% with monthly compounding?

11 A debt of $10,000 is due 10 years from now. If it is paid off 5 years from now and the interest rate is 15% compounded annually, how much is the payment?

12 How long does it take for prices to double if the effective inflation rate is 6%?

13 How much difference does it make in the price of a diamond that now costs $10,000 whether the effective rate of inflation over the next 10 years is 10% or 20%?

14 Find the sum of the finite geometric series

$$1 + 4 + 16 + \cdots + 4096.$$

15 If $500 is deposited in a savings account quarterly earning interest at the rate of 12% compounded quarterly, what is the balance of the account after 10 years?

16 What is the present value of an ordinary annuity that pays $750 annually for 10 years and the interest rate is 20%?

17 What is a fair price for an investment that promises to pay $1000 at the end of each year for 10 years if an appropriate interest rate is 12% compounded annually?

18 How much should be required to pay off a 7-year loan after 36 of the monthly payments of $100 have been made? Assume the interest rate is 12%.

19 The Jones' house is worth $120,000, and the Jones have 38 monthly payments of $350 remaining on their 9% mortgage. What is the Jones' equity?

20 Make an amortization schedule for a loan of $90,000 retired by three annual payments of $38,765.83 with interest at 14% compounded yearly.

CAREER PROFILE
Bank Officer

Once banks were single-purpose institutions headed by a stately and well-dressed officer who moved with importance through the bank and community. Now, however, most banks are large, multipurpose organizations with numerous officers who specialize in many different areas.

Prospective bank officers need a bachelor's degree; they are then required to go through a management training program (which may last as long as 2 years) to learn the various areas of operation within the given banking organization. Many of these management trainees have graduated from undergraduate programs in business administration, often with a major in finance. Other suitable undergraduate majors include accounting, economics, political science, and statistics.

There are opportunities within banking for experts of almost all types. For example, personnel with backgrounds in agriculture, engineering, or nuclear physics may be important to a bank to direct its relations with specialized or high-technology industries.

A career as a bank officer is one in which the role of mathematics is usually indirect. A bank officer needs to be able to think logically and systematically and to be able to organize, remember, and evaluate large amounts of quantitative information. College courses in calculus, statistics, and finite mathematics provide background useful in understanding the financial formulas and patterns of analysis that are applied in banking. Some of these analyses are now performed automatically by calculators and computers.

After completing a management training program, a bank officer is likely to move into one of several banking specialties; these areas include positions as loan officers, trust officers, and operations officers.

All bank officers, and especially loan officers, need to know accounting. The loan officer must be able to analyze the financial statements of an individual or organization that requests financing. He also must be familiar with economics, with business operations and business cycles, and with commercial law.

Bank officers in trust management need knowledge of financial planning and investment strategies and of regulations and procedures for the administration of trusts and estates.

Experience with computers and electronic data processing is necessary for bank operations officers who plan, coordinate, and direct the work flow and communications patterns within the banking system. College courses in management information systems and operations research (also called "management science") are valuable for the prospective operations officer.

Once on the job, a banker is likely to continue his education with specialized courses, generally offered by professional organizations rather than by colleges and universities. For example, the American Bankers Association offers many home-study courses and seminars that enable bankers to learn while employed and to enhance their opportunities for promotions.

Banking is expected to be a rapidly growing industry throughout the

1980s. However, recent large numbers of college graduates in business administration have led to stiffening of competition for bank management positions.

Bank management trainees with bachelor's degrees may start at salaries around $1300 per month. Senior bank officers are likely to earn at least three times as much as starting salaries. Advancement in banking often is slow, however, and salaries are much lower in small or rural banks than in large or urban ones.

Sources of Additional Information
The following organizations provide general information about banking occupations and training opportunities:

American Banking Association, Bank Personnel Division, 1120 Connecticut Ave. N.W., Washington, DC 20036.
National Association of Bank Women, Inc., 500 N. Michigan Avenue, Chicago, IL 60611.
National Bankers Association, 499 S. Capitol Street, S.W., Washington, DC 20003

Occupational Outlook Handbook. Bureau of Labor Statistics, U.S. Department of Labor, Washington, DC 20212. (Revised every 2 years, this handbook provides information about job duties, working conditions, level and places of employment, education and training requirements, advancement possibilities, job outlook, and earnings for about 250 occupations. Some of the information for this Career Profile was obtained from this source.

7

Derivative

7.1 Slope of Straight Lines

One of the most useful ideas in elementary mathematics is the concept of slope. As pointed out in Section 2.3, a straight line can be characterized by the fact that the slope between any pair of points on it is always the same number. That number is the slope of the line. In the next section, we extend the idea of slope to graphs that are not straight lines. To facilitate that extension we now review "slope of a straight line" from an appropriate geometric point of view.

If (x_0, y_0) and (x_1, y_1) are two points on a given line, then the slope of the line is

$$\frac{y_1 - y_0}{x_1 - x_0}.$$

The difference $x_1 - x_0$ is called the **run;** it is how far we have to run our eyes to the left or to the right to get from (x_0, y_0) to (x_1, y_1). A positive difference means $x_1 > x_0$, and we must move our eyes from left to right, whereas a negative difference means that we must move our eyes from right to left. The difference $y_1 - y_0$ is called the **rise** because it is a measure of how far we must

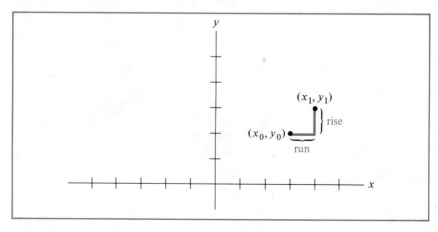

move our eyes up or down to get from (x_0, y_0) to (x_1, y_1). If the difference is positive then we must move our eyes up; whereas if it is negative we must move our eyes down. Rise and run are shown graphically in the picture on the preceding page. Thus, the **slope** is rise over run. As you read the graph from left to right, the line goes up or down according to the slope. A large positive slope means the line is going up fast and a small positive slope means the line is going up slowly. A negative slope means the line is going down. The graph shows six lines with the slope of each indicated next to the line.

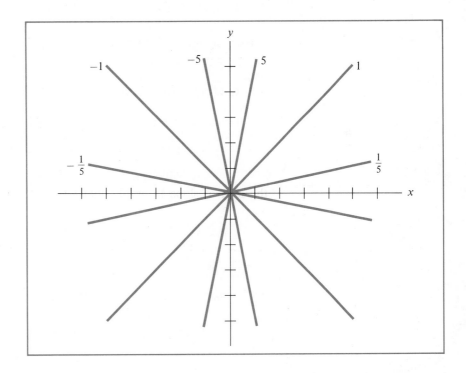

Once you know the slope of a line you know what the line looks like, and all that remains is to know where the line is placed relative to the coordinate system. When you also know one point on the line, then you know where the line is placed and the particular line is completely determined.

We illustrate the power of this geometric point of view by determining when two lines are parallel. By definition two lines are parallel provided they never cross. If one line is going up, as we read left to right, and the other line is going down, then clearly the lines cross. So either both lines are going up or else both are going down. If both lines are going up and one goes up faster than the other, that line will eventually overtake the other and the two will cross. So the two lines must be going up at the same rate; the two lines have the same slope. If the two lines are going down then the reasoning is similar. In general, **two nonvertical lines are parallel precisely when they have the same slope.**

EXAMPLE 1

Find an equation for the line parallel to the graph of $y = x$ and passing through (3, 2).

Solution
Since the equation $y = x$ is in slope-intercept form, we observe that the slope is 1. Since any line parallel to $y = x$ must have slope 1, we know that (x, y) lies on the desired line provided

$$\frac{y - 2}{x - 3} = 1.$$

Simplifying this we find that

$$y = x - 1$$

is an equation whose graph passes through (3, 2) and is parallel to $y = x$. ■

The next example justifies a fact given in Section 2.3.

EXAMPLE 2

What is the slope of every horizontal line?

Solution
The points (0, 0) and (1, 0) are on the x-axis, which is horizontal; so its slope is

$$\frac{0 - 0}{1 - 0} = 0.$$

Any horizontal line is parallel to the x-axis, so its slope is 0. ■

The next example indicates one practical application of the preceding facts about parallel lines.

EXAMPLE 3

A real estate developer has acquired the tract of land pictured. He wants to draw three lines parallel to the side boundaries so that four nice building lots result. The left and right sides of the figure are parallel to each other. Draw the lines for the developer.

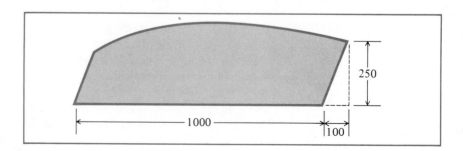

Solution

In order to use the ideas that we have developed, we draw a coordinate system with the x-axis along the bottom side of the tract and with the origin at the lower left corner, as pictured. The bottom side of the tract is easily divided into four equal segments by placing dots at 250, 500, and 750.

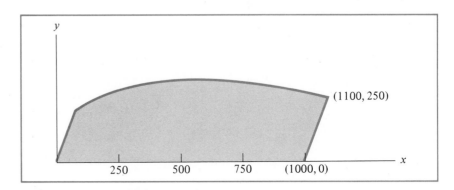

Using the corners adjacent to the right side of the tract we calculate the slope of the right side to be

$$\text{slope} = \frac{250 - 0}{1100 - 1000} = 2.5.$$

Thus, the first line to be drawn passes through (250, 0) and has slope 2.5; so a point-slope equation for that line is

$$y - 0 = 2.5(x - 250).$$

Similarly, we obtain the following point-slope equations for the second and third lines to be drawn.

$$y - 0 = 2.5(x - 500)$$

$$y - 0 = 2.5(x - 750).$$

For each of these three lines, we find a second point on the line and then sketch the parts of the lines inside the pictured tract.

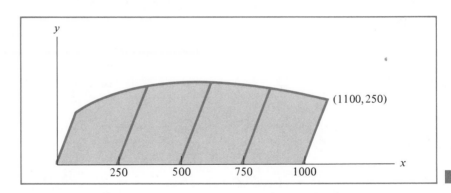

Suppose m is not 0 so the graph of $y = mx + b$ is not a horizontal line. It is known that a line **perpendicular** to that graph has slope $-1/m$ and all lines with slope $-1/m$ are perpendicular to that graph.

EXAMPLE 4

Find an equation for the line perpendicular to the graph of $y = 2x$ and passing through $(-1, 1)$.

Solution

Since the equation $y = 2x$ is in slope-intercept form, we know that its slope is 2. The slope of any perpendicular line is $-\frac{1}{2}$. The point (x, y) lies on the desired perpendicular line provided

$$y - 1 = -\frac{1}{2}(x + 1). \quad \blacksquare$$

Exercises 7.1

In problems 1 through 18, find a point-slope equation (if possible) for which the graph is parallel to the graph of the given equation and passes through the indicated point.

1   $y = 4x - 2$, $(0, 0)$
2   $y = x + 5$, $(2, 1)$
3   $y = -2x + 3$, $(1, 2)$
4   $y = .5x - 1$, $(-1, 1)$
5   $y = -.2x$, $(-1, 1)$
6   $y = -5x + 6$, $(0, 0)$
7   $0 = 2x + 4y + 8$, $(1, 1)$
8   $0 = 4x + 2y + 8$, $(-2, 1)$
9   $0 = 20x + 2y - 6$, $(-1, 1)$
10   $0 = 3x - 6y - 6$, $(3, 3)$
11   $0 = 3x - 27y + 3$, $(-2, -2)$
12   $0 = 5x - y + 2$, $(-2, -1)$
13   $y = 2$, $(4, 5)$
14   $y = -5$, $(-5, 2)$
15   $y = 1.5$, $(-3, -3)$
16   $x = 7$, $(0, 0)$
17   $x = -9$, $(2, 4)$
18   $x = -3.5$, $(-4, 2)$

Each of the questions 19 through 26 refer to the coordinate system with the lines labeled A, B, C, D, E, and F on the next page.

19   Which lines have positive slopes?
20   Which lines have negative slopes?
21   Which line has the largest positive slope?
22   Which line has the negative slope with the greatest magnitude?
23   Which line has the smallest positive slope?
24   Which line has the negative slope of smallest magnitude?
25   Which line has the slope closest to 1?
26   Which line has the slope closest to -1?

In problems 27 through 40, find the slope-intercept equation (if possible) for which the graph is perpendicular to the graph of the given equation and passes through the indicated point.

27   $y = x + 5$, $(7, 2)$
28   $y = 10x - 2$, $(3, -4)$

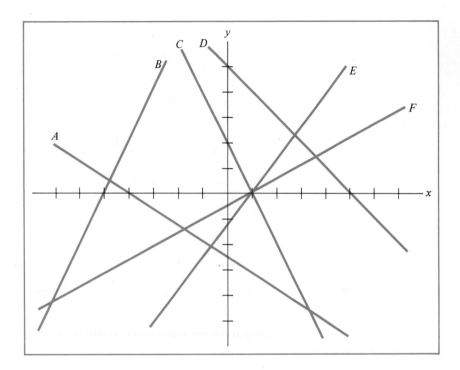

29 $y = -5x - 1,$ $(-2, 1)$	30 $y = -.1x + .5,$ $(-1, -3)$
31 $0 = x + 8y - 2,$ $(0, 0)$	32 $0 = 7x - y + 1,$ $(-2, -1)$
33 $0 = 9x + y - 3,$ $(-4, 5)$	34 $0 = x - 6y + 6,$ $(10, 10)$
35 $y = 3.2,$ $(1, 1)$	36 $y = -3,$ $(-2, 3)$
37 $y = 9,$ $(3, -5)$	38 $x = -2,$ $(1, 1)$
39 $x = 5,$ $(-1, -1)$	40 $x = 70,$ $(0, 0)$

Problems 41 through 46 refer to the real estate tract shown. The midpoint of the line segment connecting the points (a, b) and (c, d) has coordinates

$$\left(\frac{a + c}{2}, \frac{b + d}{2} \right).$$

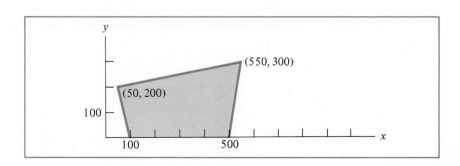

41   Find a point-slope equation for the line through the midpoint of the lower boundary that is parallel to the left side.

42   Find a point-slope equation for the line through the midpoint of the lower boundary that is parallel to the right side.

43   Find a point-slope equation for the line through the midpoint of the left side that is parallel to the lower boundary.

44   Find a point-slope equation for the line through the midpoint of the left side that is parallel to the upper boundary.

45   Find a point-slope equation for the line through the midpoint of the right side that is parallel to the lower boundary.

46   Find a point-slope equation for the line through the midpoint of the right side that is parallel to the upper boundary.

7.2   Slope of a Graph at a Point

The idea of slope is so helpful in understanding and handling straight lines that we are going to try to apply the idea to curves.* The obvious difficulty is that the slopes between different pairs of points on a given curve might well be different numbers. Indeed, if the slope between every pair of points on an unbroken curve were always the same number, then the curve would be a straight line.

After much thinking about this problem, we make the following simple observation. Any curve looks like a segment of a straight line if you take a small enough piece of the curve. The circle pictured next does not look like a line segment, but the tiny arc of the circle close around $(0, 1)$ looks a lot like a segment of the line $y = 1$.

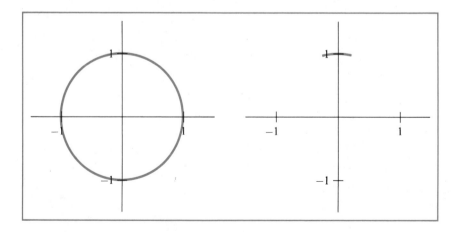

---

* For the sake of the student's intuition, we offer some observations leading to the use of tangent lines and derivatives.

The line that looks most like a given curve at a fixed point on the curve is called the **tangent line to the curve at the fixed point.** To be more precise, the tangent line has the property that it passes through the fixed point and most closely approximates a small segment of the curve close to the fixed point. We illustrate this idea by drawing the tangent lines to points A, B, C, and D on the curve.

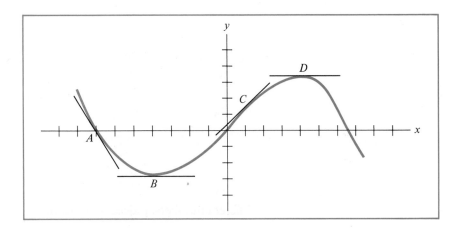

Since the tangent line looks like the curve close to the point where it is tangent, we can use the slope of the tangent line to describe what the curve is doing close to the point. For example, at point A the tangent line has negative slope and the curve is going down. At point B the curve is going neither up nor down; it is turning around, and the slope of the tangent line is 0. At point C the tangent line has positive slope and the graph is going up. At point D the curve is turning around again, and this is reflected in the fact that the slope of the tangent line is 0.

By using a lot of tangent lines we can even indicate a figure without ever sketching the figure itself. In the picture we indicate a circle and a graph similar to part of the graph of $y = x^2$ using many tangent lines.

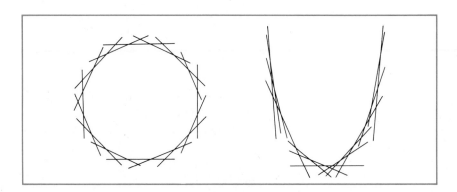

EXAMPLE 1

At the point $(1, 1)$ on the graph of $f(x) = x^2$, the slope of the tangent line is 2. Find the equation of the tangent line and compute the difference between $f(x)$ and the y-value of the tangent line for each of the following choices of x: .9, .99, .999, 1.1, 1.01, 1.001.

Solution
Since the tangent line passes through the point $(1, 1)$ and has slope 2, it must be the graph of the equation

$$y - 1 = 2(x - 1) \quad \text{or} \quad y = 2x - 1.$$

The table below shows the values of $f(x)$ and the difference.

x	Tangent Line	$f(x)$	$f(x) - y$-value
.9	.8	.81	.01
.99	.98	.9801	.0001
.999	.998	.998001	.000001
1.1	1.2	1.21	.01
1.01	1.02	1.0201	.0001
1.001	1.002	1.002001	.000001

The graph of $y = x^2$ and the tangent line at $(1, 1)$ are illustrated.

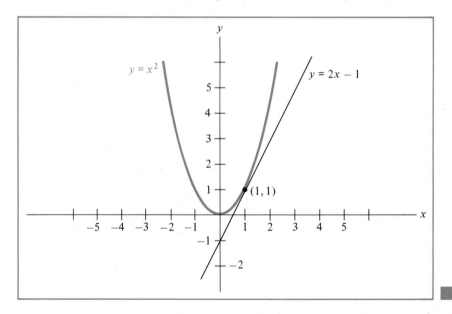

The fixed point on the graph that the tangent line must pass through is called the **point of tangency** and the slope of the tangent line is called the **slope of the graph** at the point of tangency.* Now we must find an easy way to

_____

* There are graphs that do not have a tangent line at every point on the graph. Such examples are presented in Section 7.6.

compute the slope of the tangent line. It is not satisfactory to draw the curve, draw the tangent lines and then estimate the slope of each tangent line. That is not analogous to our use of slope of a line. Given a linear function, we did not graph it in order to find its slope. Quite the contrary, we calculated the slope easily using only the linear function; then we used the slope to understand the graph.

After more thinking about this latest problem, we conclude that we can at least estimate the slope of the tangent line. Each picture in the illustration shows the graph of $f(x) = x^2$, the tangent line to the graph at the point $(1, 1)$, and a line passing through $(1, 1)$ and (x, y), some other point on the graph. In the top line of pictures (x, y) is chosen to the left of $(1, 1)$ and in the bottom line of pictures (x, y) is to the right of $(1, 1)$. As (x, y) is chosen closer and closer to $(1, 1)$, the line through (x, y) looks more and more like the tangent line to the graph at the point $(1, 1)$.

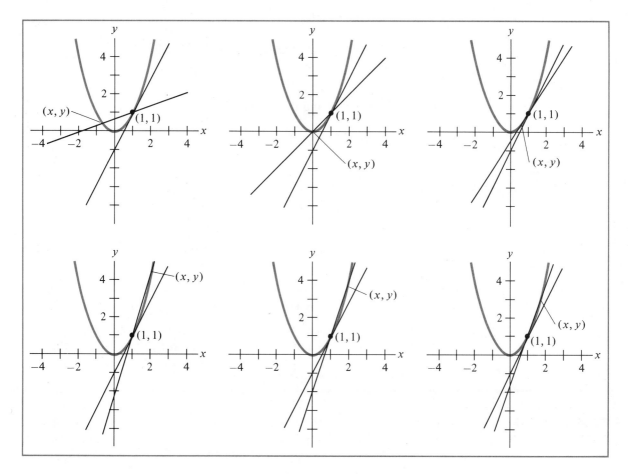

Each of these lines drawn through two points on the curve is called a **secant line.** Since the secant lines are getting closer and closer to the tangent line, the slopes of the secant lines are getting closer and closer to the slope of the tangent line. We can use the slope of a nearby secant line to estimate the slope of the tangent line.

EXAMPLE 2

Estimate the slope of the tangent line to the graph of $f(x) = x^2$ at $(1, 1)$ by using the following points from the graph: $(1.1, 1.21)$, $(1.01, 1.0201)$, and $(1.001, 1.002001)$.

Solution
The slope of the line passing through $(1, 1)$ and $(1.1, 1.21)$ is

$$\frac{1.21 - 1}{1.1 - 1} = 2.1.$$

The slope of the line through $(1, 1)$ and $(1.01, 1.0201)$ is

$$\frac{1.0201 - 1}{1.01 - 1} = 2.01.$$

The slope of the line through $(1, 1)$ and $(1.001, 1.002001)$ is

$$\frac{1.002001 - 1}{1.001 - 1} = 2.001.$$

Our three estimates are 2.1, 2.01, and 2.001, and each is closer to the actual slope of the tangent line than the preceding ones. ■

EXAMPLE 3

Without drawing a graph, estimate the slope of the tangent line to the graph of $f(x) = x^2$ at $(0, 0)$. Use the following points from the graph: $(0.1, 0.01)$, $(0.01, 0.0001)$, $(0.001, 0.000001)$, and $(0.0001, 0.00000001)$.

Solution
We compute the slope of the secant line through $(0, 0)$ and each of the indicated points.

$$\frac{0.01 - 0}{0.1 - 0} = 0.1, \qquad \frac{0.0001 - 0}{0.01 - 0} = 0.01, \qquad \frac{0.000001 - 0}{0.001 - 0} = 0.001, \qquad \frac{0.00000001 - 0}{0.0001 - 0} = 0.0001$$

In these computations we wrote "-0" only to show the steps involved, since subtracting 0 changes nothing. From these estimates, it appears that the slopes of the secant lines are getting closer and closer to 0. Thus, it appears that the slope of the tangent line at $(0, 0)$ is 0. ■

EXAMPLE 4

Speedy Courier Service is concerned about the gasoline consumption of their delivery trucks and hires an engineering consultant. The consultant advises them that the gallons of gas consumed on a typical 100-mile trip is

$$.01(x - 45)^2 + 5$$

where x is the average speed. If their current average speed is 70 miles per hour, at what rate do they change the number of gallons consumed by changing their speed by 1 mile per hour?

Solution

Let C be the gas consumed on a typical 100-mile trip; according to the consultant

$$C(x) = 0.01(x - 45)^2 + 5.$$

The slope of the tangent line to the graph of $C(x)$ at $(70, \ C(70)) = (70, 11.25)$ will indicate the rate at which the graph is going up. To approximate the slope of the tangent line we use a nearby point like $(70.1, C(70.1)) \approx (70.1, 11.30)$. The slope of the secant line through $(70, 11.25)$ and $(70.1, 11.30)$ is

$$\frac{11.30 - 11.25}{70.1 - 70} = .5.$$

Thus, increasing the average speed by 1 mile per hour increases the gas consumed by approximately .5 gallon and slowing down 1 mile per hour saves .5 gallon. ■

Soon we shall discover how to compute the slope of the tangent line exactly. First we must consider the process of passing from estimates of a number to obtaining the number precisely, which is the topic of the next section.

Exercises   7.2

In problems 1 through 18, find a point-slope equation for the tangent line to the graph of the given function at the given point. The slope of the tangent line is given. Compute the value of the function and the y-value of the tangent line at the x-value listed.

1 $f(x) = x^2$, $(-1, 1)$,   slope $= -2$, $x = -1.1$
2 $f(x) = x^2$, $(-1, 1)$,   slope $= -2$, $x = -.9$
3 $f(x) = x^4$, $(-1, 1)$,   slope $= -4$, $x = -1.1$
4 $f(x) = x^4$, $(-1, 1)$,   slope $= -4$, $x = -.9$
5 $f(x) = x^4$, $(0, 0)$,   slope $= 0$, $x = .1$
6 $f(x) = x^4$, $(0, 0)$,   slope $= 0$, $x = -.1$
7 $f(x) = x^4$, $(2, 16)$,   slope $= 32$, $x = 2.1$
8 $f(x) = x^4$, $(2, 16)$,   slope $= 32$, $x = 1.9$
9 $f(x) = x^3$, $(-1, -1)$,   slope $= 3$, $x = -1.1$
10 $f(x) = x^3$, $(-1, -1)$,   slope $= 3$, $x = -.9$
11 $f(x) = x^3$, $(0, 0)$,   slope $= 0$, $x = .1$
12 $f(x) = x^3$, $(0, 0)$,   slope $= 0$, $x = -.1$
13 $f(x) = x^3$, $(2, 8)$,   slope $= 12$, $x = 2.1$
14 $f(x) = x^3$, $(2, 8)$,   slope $= 12$, $x = 1.9$
15 $f(x) = 3x + 1$, $(0, 1)$,   slope $= 3$, $x = .1$
16 $f(x) = 3x + 1$, $(0, 1)$,   slope $= 3$, $x = -.1$
17 $f(x) = \dfrac{1}{x}$, $\left(2, \dfrac{1}{2}\right)$,   slope $= -\dfrac{1}{4}$, $x = 2.1$

18 $f(x) = \dfrac{1}{x}$, $\left(2, \dfrac{1}{2}\right)$,   slope $= -\dfrac{1}{4}$, $x = 1.9$

In problems 19 through 25, estimate the slope of the tangent line to the graph of the given function at the given point, using the indicated points from the graph of the function.

19 $f(x) = x^4$, at $(-1, 1)$,   use $(-1.1, 1.4641)$, $(-1.01, 1.0406)$
20 $f(x) = x^4$, at $(0, 0)$,   use $(.1, .0001)$, $(-.1, .0001)$
21 $f(x) = x^4$, at $(2, 16)$,   use $(2.1, 19.4481)$, $(2.01, 16.3224)$
22 $f(x) = x^3$, at $(-1, -1)$,   use $(-1.1, -1.331)$, $(-1.01, -1.0303)$
23 $f(x) = x^3$, at $(2, 8)$,   use $(2.1, 9.261)$, $(2.01, 8.1206)$

24 $f(x) = \dfrac{1}{x}$, at $\left(2, \dfrac{1}{2}\right)$, use $(2.1, .4762)$, $(2.01, .4975)$

25 $f(x) = \dfrac{1}{x}$, at $\left(\dfrac{1}{2}, 2\right)$, use $(.4, 2.5)$, $(.49, 2.0408)$

In problems 26 through 31, use some estimate for the slope of the tangent line to the graph of the given function at the given point to decide whether the slope of the graph is greater or less than 1.

26 $f(x) = \dfrac{1}{x}$,   at $(1, 1)$

27 $f(x) = \dfrac{x-1}{x+1}$,   at $(1, 0)$

28 $f(x) = x^2 - x + 3$,   at $(2, 5)$
29 $f(x) = x^2 - x + 3$,   at $(0, 3)$
30 $f(x) = x^3 - x^2 - x + 1$,   at $(1, 0)$
31 $f(x) = x^3 - x^2 - x + 1$,   at $(2, 3)$

7.3 Limit

We have considered numbers that get "closer and closer" to some value, but this language is not precise. We are going to say exactly what we mean by this phrase. The numbers

$$2.1, 2.01, 2.001, 2.0001, \ldots$$

are getting closer and closer to the number 2. This means that we can make the difference between the value 2 and a number in the list as close to 0 as we wish by going far enough along the list of numbers. We summarize this situation by saying that 2 is the **limit** of the given numbers.

Now we consider the limit of a function. As x gets closer and closer to 1, what is $g(x) = 2x$ getting close to? Consider the numbers

$$1.1, 1.01, 1.001, 1.0001, \ldots$$

and the numbers

.9, .99, .999, .9999, . . .

which are getting arbitrarily close to 1. These numbers are pictured on the following number line.

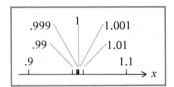

The corresponding values of $g(x)$ are tabulated next.

x	$g(x)$		x	$g(x)$
.9	1.8		1.1	2.2
.99	1.98		1.01	2.02
.999	1.998		1.001	2.002
.9999	1.9998		1.0001	2.0002

The values of $g(x)$ are pictured on the number line.

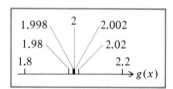

Clearly the corresponding values of $g(x)$ are getting arbitrarily close to 2. We say that 2 is the limit of $g(x)$ as x approaches 1. In general, we say that **L is the limit of $f(x)$ as x approaches a** provided $f(x)$ is arbitrarily close to L when x is sufficiently close to a, but not equal to a. As a sort of mathematical shorthand for this long phrase, we write

$$L = \lim_{x \to a} f(x).$$

Note that we must consider values of x that are both less than a and greater than a in defining the limit of a function. Since we never allow x to equal a, the value of $f(x)$ at $x = a$ is irrelevant to the limit $L = \lim_{x \to a} f(x)$. Such a limit sometimes does not exist, as indicated in Example 2.

EXAMPLE 1

Find $\lim_{x \to 0} (5x - 3)$, if it exists.

Solution

We choose numbers that are approaching 0 from both sides and we tabulate the corresponding values of $(5x - 3)$.

x	5x − 3
−.1	−3.5
−.01	−3.05
−.001	−3.005
−.0001	−3.0005

x	5x − 3
.1	−2.5
.01	−2.95
.001	−2.995
.0001	−2.9995

It is clear that -3 is the limit of $5x - 3$ as x approaches 0. This is illustrated by the graph.

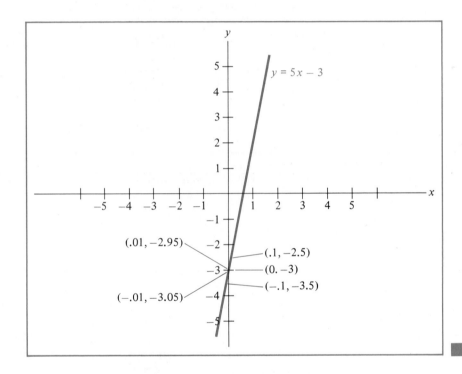

EXAMPLE 2

Find $\lim\limits_{x \to 0} \dfrac{1}{x}$, if it exists.

Solution

We choose numbers that are approaching 0 from both sides and we tabulate the corresponding values of $1/x$.

x	1/x
−.1	−10
−.01	−100
−.001	−1000
−.0001	−10,000

x	1/x
.1	10
.01	100
.001	1000
.0001	10,000

The corresponding values are not getting close to any number, so $\lim\limits_{x \to 0} 1/x$ does not exist. This is illustrated by the graph.

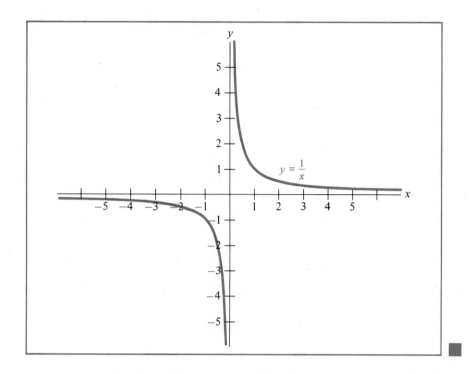

Sometimes it is difficult to guess a limit by just looking at the formula for the function, as in the next example.

CALCULATOR EXAMPLE

Guess the limit $\lim\limits_{x \to 2} f(x)$, where $f(x) = \dfrac{3x^2 - 12}{5x - 10}$, by calculating the values of $f(x)$ at numbers closer and closer to $x = 2$.

Solution
We choose numbers that are approaching 2 from both sides and we tabulate the corresponding values of $f(x)$.

x	$f(x)$	x	$f(x)$
1.9	2.34	2.1	2.46
1.99	2.394	2.01	2.406
1.999	2.3994	2.001	2.4006
1.9999	2.39994	2.0001	2.40006

The only reasonable guess is that $\lim\limits_{x \to 2} f(x)$ equals 2.4. ■

It is interesting to note that in the preceding example we were able to guess that $\lim_{x \to 2} f(x)$ equals 2.4 despite the fact $f(2)$ is not defined since

$$f(2) = \frac{3(2)^2 - 12}{5(2) - 10} = \frac{0}{0} \quad \text{is not defined.}$$

In Section 7.4 and later sections we shall discover that this happens often. Sometimes it is easy to find the limit, as in the next example.

EXAMPLE 3

Find $\lim_{x \to 2} f(x)$, where $f(x) = 5$ for all x.

Solution
Since the value of $f(x)$ is arbitrarily close to 5 (in fact, it is equal to 5) regardless of how x is chosen, the limit is 5. ∎

In general, if $f(x)$ is a constant function, that is, $f(x) = c$ for all x with c fixed, then $\lim_{x \to a} f(x) = c$ regardless of how a is chosen.

EXAMPLE 4

Determine $\lim_{x \to 0} f(x)$ and $\lim_{x \to 0} g(x)$ if they exist, where

$$f(x) = \begin{cases} 1/x & \text{for} \quad x \neq 0 \\ 1 & \text{for} \quad x = 0 \end{cases}, \quad g(x) = \begin{cases} x & \text{for} \quad x \neq 0 \\ 1 & \text{for} \quad x = 0 \end{cases}$$

Solution
From the computations made in Example 2, we see that $\lim_{x \to 0} f(x)$ does not exist. The value of $f(x)$ at $x = 0$ is not relevant, since we never let x equal 0 in the computation of the limit.

As x gets closer and closer to 0, then $g(x) = x$ gets closer and closer to 0. Thus, $\lim_{x \to 0} g(x)$ exists and equals 0, despite the fact that $g(0) = 1$. The value of $g(x)$ at $x = 0$ is not relevant in computing $\lim_{x \to 0} g(x)$. ∎

Note that $\lim_{x \to a} f(x)$ and the value of $f(x)$ at $x = a$ are completely independent of each other. We list the possibilities.

1. Both $\lim_{x \to a} f(x)$ and $f(a)$ exist, and they are equal. (See Examples 1 and 3.)

2. The limit $\lim_{x \to a} f(x)$ exists while $f(x)$ is not defined at $x = a$. (See the Calculator Example.)

3. The limit $\lim_{x \to a} f(x)$ fails to exist while $f(x)$ is not defined at $x = a$. (See Example 2.)

4. The limit $\lim_{x \to a} f(x)$ fails to exist while $f(x)$ is defined at $x = a$. (See $f(x)$ in Example 4.)

5. Both $\lim_{x \to a} f(x)$ and $f(a)$ exist, and they are not equal. (See $g(x)$ in Example 4.)

There are many ways that two given functions may be combined to make a new function. The limit of the new function as x approaches a can be calculated from the limits of the given functions. The following theorem shows how to do this for each of the basic algebraic operations.

Limit Theorem Provided $\lim_{x \to a} f(x)$ and $\lim_{x \to a} g(x)$ both exist, we have

1. $\lim_{x \to a}(f(x) + g(x)) = \lim_{x \to a} f(x) + \lim_{x \to a} g(x),$

2. $\lim_{x \to a}(f(x) - g(x)) = \lim_{x \to a} f(x) - \lim_{x \to a} g(x),$

3. $\lim_{x \to a}(f(x) \, g(x)) = \lim_{x \to a} f(x) \lim_{x \to a} g(x),$

4. $\lim_{x \to a}(f(x))^r = (\lim_{x \to a} f(x))^r$ for any positive number r,

5. $\lim_{x \to a} \dfrac{f(x)}{g(x)} = \dfrac{\lim_{x \to a} f(x)}{\lim_{x \to a} g(x)}$ provided $\lim_{x \to a} g(x) \neq 0.$

EXAMPLE 5

Find $\lim_{x \to 2}(3x + 5)$.

Solution
By part 1 of the limit theorem we have

$$\lim_{x \to 2}(3x + 5) = \lim_{x \to 2} 3x + \lim_{x \to 2} 5.$$

We determined in Example 3 that $\lim_{x \to 2} 5 = 5$. Thus, using part 3 of the Limit Theorem gives us

$$\lim_{x \to 2}(3x + 5) = (\lim_{x \to 2} 3)(\lim_{x \to 2} x) + 5.$$

Since 3 is a constant function, we have $\lim_{x \to 2} 3 = 3$. As x approaches 2 the limit of x is clearly 2, so $\lim_{x \to 2} x = 2$. Now we have

$$\lim_{x \to 2}(3x + 5) = (3)(2) + 5 = 11,$$

which is the answer. ∎

Example 5 suggests the following properties, which are special cases of parts 1 and 3 of the Limit Theorem.

6. $\lim_{x \to a}(f(x) + c) = c + \lim_{x \to a} f(x)$ for any constant c

7. $\lim_{x \to a}(c \, f(x)) = c(\lim_{x \to a} f(x))$ for any constant c

Although it is not necessary to write out each step in such detail as we have done in Example 5, it is wise to think through the steps when you are using part of the Limit Theorem. That way you make sure that you use the theorem correctly.

EXAMPLE 6

Find $\lim_{x\to 1}(2x-1)(x+3)$.

Solution
We find the limit with the following steps.

$$\lim_{x\to 1}(2x-1)(x+3) = \lim_{x\to 1}(2x-1)\,\lim_{x\to 1}(x+3) \qquad \text{by 3 of Limit Theorem}$$

$$= (\lim_{x\to 1}2x - \lim_{x\to 1}1)(\lim_{x\to 1}x + \lim_{x\to 1}3) \qquad \text{by 1 and 2 of Limit Theorem}$$

$$= [(\lim_{x\to 1}2)(\lim_{x\to 1}x) - 1](1+3)$$

$$= [(2)(1) - 1]4 = 4. \ \blacksquare$$

EXAMPLE 7

Find $\lim_{x\to 0}\dfrac{x+1}{x-1}$.

Solution

$$\lim_{x\to 0}\frac{x+1}{x-1} = \frac{\lim_{x\to 0}(x+1)}{\lim_{x\to 0}(x-1)} \qquad \text{by 5 of Limit Theorem}$$

$$= \frac{(\lim_{x\to 0}x)+1}{(\lim_{x\to 0}x)-1} \qquad \text{by 1 of Limit Theorem}$$

$$= \frac{0+1}{0-1} = -1. \ \blacksquare$$

There is a different kind of limit that is useful in understanding the graphs of certain functions. We say L is the limit of $f(x)$ as x increases without bound provided $f(x)$ is arbitrarily close to L when x is sufficiently large. The abbreviated version of this statement is

$$L = \lim_{x\to\infty}f(x).$$

Here the symbol ∞ stands for **infinity,** although "infinity" is not a number. The symbols $x\to\infty$ mean that x increases without bound.

CALCULATOR EXAMPLE

Find $\lim_{x\to\infty}\left(\dfrac{1}{2}\right)^x$, if it exists.

Solution
Examining the table of values below quickly convinces us that the limit is 0.

x	$(\frac{1}{2})^x$
3	.125
5	.03125
8	.00391
10	.00098
15	.00003

This table was constructed using a calculator, and the last three values of the function were rounded off to five decimal places.

Next we recall the graph of $f(x) = (\frac{1}{2})^x$.

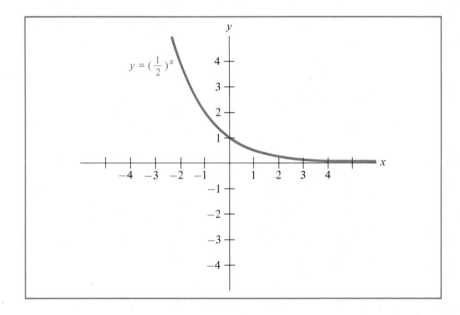

It is clear that the y-value of the graph gets arbitrarily close to 0 when you look far enough to the right. ∎

EXAMPLE 8

Find $\lim\limits_{x \to \infty} 2^x$, if it exists.

Solution
The table of values below convinces us that the limit does not exist.

x	2^x
3	8
5	32
8	256
10	1,024
15	32,768

Next we recall the graph of $f(x) = 2^x$.

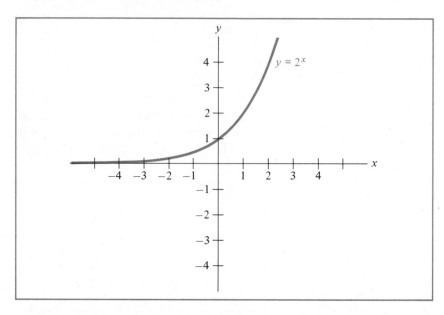

It is clear that the y-value of the graph does not get close to any number; indeed it increases without bound. ■

In general, if $a > 1$ then $\lim\limits_{x \to \infty} a^x$ does not exist, and if $0 < a < 1$ then $\lim\limits_{x \to \infty} a^x = 0$.

The following definition is analogous to the definition of the limit as x increases without bound. We say L is the limit of $f(x)$ as x decreases without bound provided $f(x)$ is arbitrarily close to L when x is a negative number with sufficiently large magnitude, and we write

$$L = \lim_{x \to -\infty} f(x).$$

In general, if $a > 1$ then $\lim\limits_{x \to -\infty} a^x = 0$, and if $0 < a < 1$ then $\lim\limits_{x \to -\infty} a^x$ does not exist.

EXAMPLE 9

A philanthropic foundation has decided to pay an annuity to help support a private school. On January 1 of each year the foundation will pay $20,000 to the school; we assume 12% per annum is an appropriate interest rate for the duration of the annuity. Initially the annuity was planned to continue for 50 years, but now the foundation is considering continuing it indefinitely. How much more will it cost for the annuity to continue indefinitely?

Solution

Let $A(n)$ be the present value of the annuity provided that it continues for n years. Using a formula developed in Section 6.4, we have

$$A(n) = 20,000 \frac{1 - (1.12)^{-n}}{0.12} = 20,000 a_{\overline{n}|.12}$$

For $n = 50$ we get the amount the foundation would have to invest to provide a 50-year annuity.

$$A(50) = 20,000 a_{\overline{50}|.12} = 166,089.97.$$

If the annuity continues indefinitely then its present value is

$$\lim_{n \to \infty} A(n) = \lim_{n \to \infty} \left(20,000 \frac{1 - (1.12)^{-n}}{0.12} \right)$$

$$= \lim_{n \to \infty} \left(166,666.67 \left(1 - \left(\frac{1}{1.12} \right)^n \right) \right)$$

$$= 166,666.67 \lim_{n \to \infty} \left(1 - \left(\frac{1}{1.12} \right)^n \right)$$

$$= 166,666.67(1 - 0) = 166,666.67.$$

Thus, the additional cost is

$$\$166,666.67 - \$166,089.97 = \$576.70. \quad \blacksquare$$

If the preceding example surprises you, then you should note that the present value of a $20,000 payment 51 years from now is

$$\$20,000(1.12)^{-51} = \$61.79$$

provided the interest rate is 12%.

The method of the previous example gives a formula for the present value of an annuity that continues indefinitely. If the rent of an annuity is R, the interest rate per compound period is r, and the duration of the annuity is t, then the present value is

$$A(t) = R \frac{1 - (1 + r)^{-t}}{r} = \frac{R}{r} \left(1 - \left(\frac{1}{1+r} \right)^t \right).$$

If the annuity continues indefinitely, then the present value A is

$$\lim_{t \to \infty} A(t) = \frac{R}{r} \lim_{t \to \infty} \left(1 - \left(\frac{1}{1+r} \right)^t \right) = \frac{R}{r}(1 - 0) = \frac{R}{r}.$$

An annuity that continues indefinitely is called a **perpetuity**.

The Federal Reserve, which controls the banking system, determines what percentage of a bank's assets can be lent to customers; in recent years that has been approximately 80%. This is called the **fractional banking system**. A consequence of this system is that most of the money supply consists of bookkeeping entries. The following example shows how a cash deposit is multiplied many times by loans.

EXAMPLE 10

Al takes $1000 from his home safe and deposits it at Reliable Bank. The bank then lends $(.80)(\$1000) = \$800$ to Bob with its customary requirement that

Bob put the money in a checking account at Reliable. The bank then lends (.80)($800) = $640 to Carl with the customary requirement. If this process continues indefinitely, what is the total amount of new money put into circulation as a result of Al's deposit?

Solution

After the transactions with Al, Bob, and Carl the new money in circulation is

$$1000 + (.8)(1000) + (.8)^2(1000) = 1000 + 800 + 640 = 2440.$$

After n such transactions, the total new money is

$$1000 + (.8)(1000) + \cdots + (.8)^{n-1}(1000).$$

Using the formula for the sum of a finite geometric series developed in Section 6.4, we get

$$\text{total new money} = 1000 \left(\frac{1 - .8^n}{1 - .8} \right).$$

If the transactions continue indefinitely, then the total new money is the limit of this function as n increases without bound.

$$\begin{aligned} \text{total new money} &= \lim_{n \to \infty} 1000 \left(\frac{1 - .8^n}{1 - .8} \right) \\ &= \frac{1000}{.2} \lim_{n \to \infty} (1 - .8^n) \\ &= 5000(1 - 0) = 5000 \end{aligned}$$

Thus, Al's "real money" is multiplied five times by the banking system. ■

Exercises 7.3

In problems 1 through 40, find the indicated limit.

1 $\lim_{x \to 3} 10$

2 $\lim_{x \to -2} e$

3 $\lim_{x \to -5} \sqrt{3}$

4 $\lim_{x \to 2} \dfrac{5}{2}$

5 $\lim_{x \to 7} \pi$

6 $\lim_{x \to 10} (x - 5)$

7 $\lim_{x \to -1} (3x + 2)$

8 $\lim_{x \to 3} (9 - 3x)$

9 $\lim_{x \to 1} (1.3x + 5.2)$

10 $\lim_{x \to -2} (6x^2 - 4x + 3)$

11 $\lim_{x \to 3} (x^2 - x - 3)$

12 $\lim_{x \to -3} (11x^2 + 12)$

13 $\lim_{x \to 10} (.01x^2 + 10x - 20)$

14 $\lim_{x \to 5} (7x - 1)(x + 2)$

15 $\lim_{x \to 4} (x + 3)(x - 4)$

16 $\lim_{x \to 1} 5x(6x - 4)$

17 $\lim\limits_{x \to 10} (.1x + .4)(.3x - .2)$

18 $\lim\limits_{x \to 6} \dfrac{3x + 2}{x - 5}$

19 $\lim\limits_{x \to 1} \dfrac{7x}{2x - 3}$

20 $\lim\limits_{x \to -1} \dfrac{11x + 9}{3x}$

21 $\lim\limits_{x \to 2} \dfrac{x(.1x + 2)}{.3x - .5}$

22 $\lim\limits_{x \to 1} (x^3 + x^2 + x + 1)$

23 $\lim\limits_{x \to -2} (x^3 - x)$

24 $\lim\limits_{x \to .1} (7x^3 + x^2)$

25 $\lim\limits_{x \to 0} \dfrac{x^2 + 2x + 1}{x - 5}$

26 $\lim\limits_{x \to 0} \dfrac{2x^2 - x + 4}{3x + 7}$

27 $\lim\limits_{x \to 0} \dfrac{3x^2 + 7}{x^2 + 1}$

28 $\lim\limits_{x \to 0} \dfrac{11x^2 - x + 2}{x^2 + 7}$

29 $\lim\limits_{x \to 0} \dfrac{3x + 4}{x^2 + x - 3}$

30 $\lim\limits_{x \to 0} \dfrac{-5x + 6}{3x^2 - 2x + 1}$

31 $\lim\limits_{x \to 0} \dfrac{x^3 + 5x}{x + 1}$

32 $\lim\limits_{x \to -1} \dfrac{-2x^3 + 3x^2 - 1}{x - 2}$

33 $\lim\limits_{x \to 4} \sqrt{x}$

34 $\lim\limits_{x \to 0} \sqrt{x^2 + 1}$

35 $\lim\limits_{x \to 0} \dfrac{\sqrt{x^4 + 4}}{\sqrt{x^4 + 1}}$

36 $\lim\limits_{x \to 1} \dfrac{\sqrt{9x^2 + 1}}{\sqrt{x}}$

37 $\lim\limits_{x \to 0} \dfrac{x^3 + x^2 - 2}{2x^3 + 1}$

38 $\lim\limits_{x \to 1} \dfrac{5x^3 + 4x^2 + 3}{x^3 + 10}$

39 $\lim\limits_{x \to 2} (x^4 + x^2 + 1)$

40 $\lim\limits_{x \to 1} (x^4 - x^3 + x - 1)$

In problems 41 through 57, find the indicated limit if it exists.

41 $\lim\limits_{x \to \infty} \dfrac{1}{x}$

42 $\lim\limits_{x \to -\infty} \dfrac{1}{x}$

43 $\lim\limits_{x \to \infty} \dfrac{1}{x^2}$

44 $\lim\limits_{x \to -\infty} \dfrac{1}{x^2}$

45 $\lim\limits_{x \to \infty} \left(\dfrac{10}{x^2} + \dfrac{5}{x} \right)$

46 $\lim\limits_{x \to \infty} \left(\dfrac{9}{x^3} + \dfrac{1}{x} \right)$

47 $\lim\limits_{x \to \infty} x^{-5}$

48 $\lim\limits_{x \to -\infty} x^{-5}$

49 $\lim\limits_{x \to \infty} x^6$

50 $\lim\limits_{x \to -\infty} x^6$

51 $\lim\limits_{x \to \infty} 11^x$

52 $\lim\limits_{x \to -\infty} 11^x$

53 $\lim\limits_{x \to \infty} .2^x$

54 $\lim\limits_{x \to -\infty} .2^x$

55 $\lim\limits_{x \to \infty} \sqrt{x}$

56 $\lim\limits_{x \to \infty} \dfrac{1}{\sqrt{x}}$

57 $\lim\limits_{x \to \infty} 5$

Solve the following.

58 Find the present value of a perpetuity with annual rent of $100 and interest rate 15%.

59 Find the present value of a perpetuity with annual rent of $1500 and interest rate 12%.

60 How much does it cost a millionaire to set up a perpetual annual scholarship award of $2500, if the millionaire's money would otherwise earn 10%?

61 The court orders Jim Stone to pay his former wife $5000 in alimony every year on the anniversary of the divorce decree. If this continued forever and the appropriate interest rate is 12%, what is the present value of Jim's obligation?

62 *Calculator Problem*   Guess the limit $\lim_{x \to -1} f(x)$, where $f(x) = \dfrac{2x^2 - 2}{3x + 3}$ by calculating the values of $f(x)$ at numbers closer and closer to $x = -1$.

63 *Calculator Problem*   Guess the limit $\lim_{x \to 3} f(x)$, where $f(x) = \dfrac{2x^2 - 18}{5x - 15}$ by calculating the values of $f(x)$ at numbers closer and closer to $x = 3$.

64 *Calculator Problem*   Guess the limit $\lim_{x \to \infty}(1/5)^x$, if it exists.

65 *Calculator Problem*   Guess the limit $\lim_{x \to \infty} .1^x$, if it exists.

66 *Calculator Problem*   Guess the limit $\lim_{x \to \infty} 5^x$, if it exists.

7.4 Definition of Derivative

In Section 7.2 we observed that secant lines can be used to approximate the tangent line to the graph of the function $f(x)$ at a fixed point $(a, f(a))$ on the graph of f. Indeed, the secant line approximated the tangent line more and more closely as the second point on the graph $(b, g(b))$ through which the secant line passed moved closer and closer to $(a, f(a))$. Thus, **the slope of the tangent line at $(a, f(a))$ is the limit of the slope of the secant line through $(a, f(a))$ and $(b, f(b))$ as b approaches a.** The tangent line to the graph of $f(x) = x^2$ at $(1, 1)$ is shown in the illustrations, along with a secant line passing through $(1, 1)$ and $(b, f(b))$ on the next page.

The slope of the secant line through $(a, f(a))$ and $(b, f(b))$ is the "rise over the run," or

$$\text{slope of secant line} = \frac{f(b) - f(a)}{b - a}.$$

We focus attention on how close b is to a by writing

$$h = b - a \quad \text{or} \quad b = a + h.$$

In this notation, the slope of the secant line is

$$\frac{f(a + h) - f(a)}{a + h - a} = \frac{f(a + h) - f(a)}{h}.$$

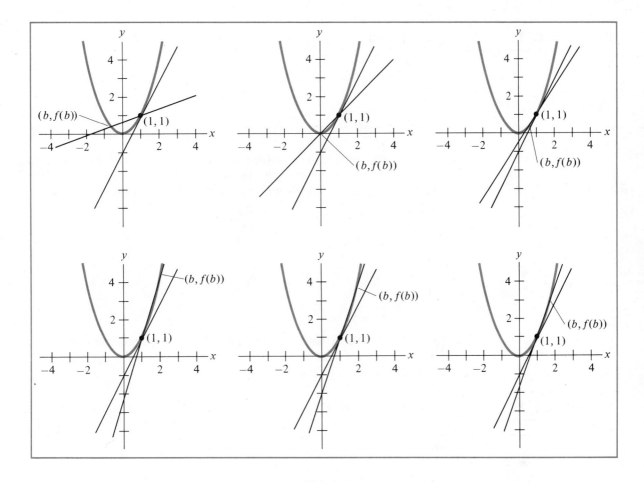

This last expression is called the **difference quotient** for $f(x)$ at $x = a$. The slope of the tangent line at $(a, f(a))$ is the limit of the slope of the secant line through $(a, f(a))$ and $(b, f(b)) = ((a + h), f(a + h))$ as b approaches a, or equivalently as $h \rightarrow 0$.

$$\text{slope of tangent line} = \lim_{h \to 0} \frac{f(a + h) - f(a)}{h}.$$

This number is so important that we give it a name.

Definition The **derivative of $f(x)$ at a** is defined to be the limit

$$\lim_{h \to 0} \frac{f(a + h) - f(a)}{h}$$

when this limit exists. The derivative is denoted by $f'(a)$, read as "f prime of a." For some choices of $f(x)$ and a, $f'(a)$ does not exist; such examples are given in Section 7.6.

The **derivative $f'(a)$** is the slope of the tangent line to the graph of $f(x)$ at $(a, f(a))$.

EXAMPLE 1

Find $f'(a)$ where $f(x) = x^2$ and $a = 1$.

Solution

In the definition we substitute $f(a + h) = (1 + h)^2$ and $f(a) = (1)^2$. Then we have

$$f'(1) = \lim_{h \to 0} \frac{(1 + h)^2 - 1^2}{h}$$

Our first inclination is to apply the Limit Theorem as we did many times in the previous section; however, the limit of the denominator is 0, so the Limit Theorem does not apply. We must first simplify the expression, and then use the Limit Theorem.

$$f'(1) = \lim_{h \to 0} \frac{(1 + 2h + h^2) - 1}{h}$$

$$= \lim_{h \to 0} \frac{h(2 + h)}{h}$$

$$= \lim_{h \to 0} (2 + h) = 2$$

In dividing out the h that appears in the numerator and the denominator we must be concerned that we are not dividing by 0. Because h approaches 0 but is never equal to 0, we can make the division. ■

EXAMPLE 2

Find $f'(a)$ where $f(x) = x^2$ and a is an arbitrary (fixed) number.

Solution

We use the same methods as in the previous example.

$$f'(a) = \lim_{h \to 0} \frac{(a + h)^2 - a^2}{h}$$

$$= \lim_{h \to 0} \frac{a^2 + 2ah + h^2 - a^2}{h}$$

$$= \lim_{h \to 0} \frac{h(2a + h)}{h}$$

$$= \lim_{h \to 0} (2a + h) = 2a$$ ■

Since there was nothing assumed about a in Example 2 except that it was fixed, we have obtained the general derivative formula below.

If $f(x) = x^2$ then $f'(x) = 2x$.

In obtaining this formula, we have defined a function $f'(x)$, called the **derivative function**; it gives the slope of the tangent line to the graph of $f(x)$ at any value of x. When $x = a$, then $f'(a)$ is the slope of the tangent line at $(a, f(a))$.

It would be convenient to describe this derivative function using only one equation. Provided x is the independent variable, the symbol **d/dx in front of a function means the derivative function**. That is,

$$\frac{d}{dx} f(x) = f'(x)$$

and $\dfrac{d}{dx} f(x) \Big|_{x=a}$ denotes $f'(a)$. Thus, the preceding derivative formula can be simply stated as

$$\boxed{\frac{d}{dx} x^2 = 2x}$$

EXAMPLE 3

Find $\dfrac{d}{dx} x^3$.

Solution

We choose an arbitrary fixed value for x, say a, and compute the derivative at a, which is denoted $\dfrac{d}{dx} x^3 \Big|_{x=a}$.

$$\frac{d}{dx} x^3 \Big|_{x=a} = \lim_{h \to 0} \frac{(a+h)^3 - a^3}{h} = \lim_{h \to 0} \frac{a^3 + 3a^2 h + 3ah^2 + h^3 - a^3}{h}$$

$$= \lim_{h \to 0} \frac{h(3a^2 + 3ah + h^2)}{h} = \lim_{h \to 0}(3a^2 + 3ah + h^2) = 3a^2$$

Since the value of a is arbitrary, the general rule is

$$\boxed{\frac{d}{dx} x^3 = 3x^2.}$$ ■

EXAMPLE 4

Find $\dfrac{d}{dx} c$.

Solution

Since the derivative at $x = a$ is the slope of the tangent line to the graph of $y = c$ at (a, c), this problem is easy. The graph of $y = c$ is a horizontal line and the tangent line at any point is again the horizontal line with slope 0. Thus,

$$\boxed{\frac{d}{dx} c = 0.}$$ ■

For our own peace of mind we also do the problem from the definition of derivative.

$$\frac{d}{dx} c \Big|_{x=a} = \lim_{h \to 0} \frac{c - c}{h} = \lim_{h \to 0} \frac{0}{h} = 0.$$

(Remember that h approaches 0 but never equals it; so $0/h = 0$ always.)

EXAMPLE 5

Find $\dfrac{d}{dx} \dfrac{1}{x}$.

Solution

We choose an arbitrary fixed value for x, say a, and compute the derivative at a.

$$\frac{d}{dx}\frac{1}{x}\bigg|_{x=a} = \lim_{h \to 0} \frac{\frac{1}{a+h} - \frac{1}{a}}{h} = \lim_{h \to 0} \frac{1}{h}\left(\frac{1}{a+h} - \frac{1}{a}\right)$$

$$= \lim_{h \to 0} \frac{1}{h}\left(\frac{a}{a(a+h)} - \frac{a+h}{a(a+h)}\right) = \lim_{h \to 0} \frac{1}{h}\left(\frac{-h}{a(a+h)}\right)$$

$$= \lim_{h \to 0} \frac{-1}{a(a+h)} = \frac{-1}{a(a)} = \frac{-1}{a^2}.$$

Thus, $\dfrac{d}{dx}\dfrac{1}{x} = \dfrac{-1}{x^2}$ or, more compactly,

$$\boxed{\frac{d}{dx}x^{-1} = -x^{-2}.}$$ ∎

In dealing with the definition of derivative, it is often helpful to factor the numerator. Some elementary factorizations are

$$x^2 - y^2 = (x - y)(x + y)$$

$$x^3 - y^3 = (x - y)(x^2 + xy + y^2)$$

$$x^4 - y^4 = (x - y)(x^3 + x^2y + xy^2 + y^3).$$

Each of these can be verified by computing the product on the right side and comparing it to the left side. Each is an example of the factorization

$$x^n - y^n = (x - y)(x^{n-1} + x^{n-2}y + \cdots + xy^{n-2} + y^{n-1})$$

where n is any positive integer. Here the three dots indicate that terms are omitted that continue the pattern of one lower power of x and one higher power of y. We verify the equation by computing and simplifying the product indicated on the right of the equation.

$$(x - y)(x^{n-1} + x^{n-2}y + \cdots + xy^{n-2} + y^{n-1})$$

$$= x^n + x^{n-1}y + \cdots + x^2y^{n-2} + xy^{n-1} - x^{n-1}y - \cdots - xy^{n-1} - y^n$$

$$= x^n + 0 + 0 + \cdots + 0 + 0 - y^n$$

$$= x^n - y^n$$

EXAMPLE 6

Find $\dfrac{d}{dx}x^n$, where n is a positive integer.

Solution

We choose an arbitrary fixed value for x, say a, and compute the derivative at a.

$$\frac{d}{dx} x^n \bigg|_{x=a} = \lim_{h\to 0} \frac{(a+h)^n - a^n}{h}$$

By factoring the numerator and simplifying, we have

$$\lim_{h\to 0} \frac{(a+h)^n - a^n}{h} = \lim_{h\to 0} \frac{((a+h) - a)((a+h)^{n-1} + (a+h)^{n-2}a + \cdots + (a+h)a^{n-2} + a^{n-1})}{h}$$

$$= \lim_{h\to 0} [(a+h)^{n-1} + (a+h)^{n-2}a + \cdots + (a+h)a^{n-2} + a^{n-1}]$$

$$= a^{n-1} + a^{n-2}a + \cdots + (a)a^{n-2} + a^{n-1}$$

$$= na^{n-1}$$

In the last step we concluded that there were n addends by noting that each addend beginning with the second one has a second factor that is a power of a starting with $a = a^1$ and going up to a^{n-1}. Consequently, there are $n-1$ addends after the first one. Since the value of a is arbitrary, we have discovered the general rule

$$\frac{d}{dx} x^n = nx^{n-1}.$$

∎

This rule, called the **Power Rule** (because it gives the derivative function for any power of x), can be applied to the functions of Examples 2, 3, and 5 to get the answers we have already found, but with less work. For example, if

$$f(x) = x^3, \quad \text{then} \quad n = 3 \quad \text{and} \quad \frac{d}{dx} f(x) = 3x^{3-1} = 3x^2$$

and if

$$f(x) = \frac{1}{x}, \quad \text{then} \quad n = -1 \quad \text{and} \quad \frac{d}{dx} f(x) = (-1)x^{-1-1} = -x^{-2}.$$

We shall use this rule often in our subsequent work. In the next chapter, we shall obtain other general rules for finding derivatives.

EXAMPLE 7

From Example 4 in Section 7.2, we recall that Speedy Courier Service is concerned about gasoline consumption. The gas consumed on a typical 100-mile trip is $.01(x - 45)^2 + 5$, where x is the average speed. What is the exact slope of the graph at $x = 70$?

Solution

If we define $C(x)$ by

$$C(x) = .01(x - 45)^2 + 5 = .01(x^2 - 90x + 2025) + 5$$
$$= .01x^2 - .9x + 25.25$$

then we must find $C'(70)$, the slope of the tangent line at $x = 70$.

$$C'(70) = \lim_{h \to 0} \frac{.01(70 + h)^2 - .9(70 + h) + 25.25 - .01(70)^2 + .9(70) - 25.25}{h}$$

$$= \lim_{h \to 0} \frac{.01(4900 + 140h + h^2 - 4900) - .9(70 + h - 70)}{h}$$

$$= \lim_{h \to 0} \frac{h(.01(140 + h) - .9)}{h}$$

$$= \lim_{h \to 0} (.5 + .01h) = .5$$

Thus, the slope of the tangent line is .5; if average speed x is decreased from 70 by 4 miles per hour, approximately 2 gallons of gas are saved on a typical 100-mile trip. ▪

CALCULATOR EXAMPLE

Guess the derivative of e^x at $x = 0$ by guessing the limit of the difference quotient

$$\frac{e^h - e^0}{h} = \frac{e^h - 1}{h}.$$

Compute it for $h = .1, .01, .001, .0001, -.1, -.01, -.001, -.0001$.

Solution

h	.1	.01	.001	.0001	$-.1$	$-.01$	$-.001$	$-.0001$
$\dfrac{e^h - 1}{h}$	1.05171	1.00502	1.00050	1.00005	.95163	.99502	.99950	.99995

The values in the second row of this table were rounded off to five decimal places. The only reasonable guess for the limit of the difference quotient is 1; thus, we guess that the derivative of e^x at $x = 0$ is 1. ▪

Exercises 7.4

Using the definition of derivative and the Limit Theorem, find a derivative formula for each of the functions in problems 1 through 30.

1 $f(x) = 5x$
2 $g(x) = 3x$
3 $f(x) = 7x + 9$
4 $g(x) = -3x - 6$
5 $h(x) = x^2 + 2x - 3$
6 $f(x) = -x^2 + x - 5$
7 $g(t) = t^2 - 9$
8 $h(t) = .1t^2 + .5t + .3$
9 $f(s) = \dfrac{2}{s}$
10 $g(s) = \dfrac{5}{s} - 3$
11 $h(p) = \dfrac{1}{p^2}$
12 $f(p) = \dfrac{1}{p^2} + \dfrac{1}{p}$

13 $g(x) = x^2 + \dfrac{1}{x}$

14 $h(x) = -5x + \dfrac{1}{x^2}$

15 $f(x) = \sqrt{11}$

16 $g(t) = 9^{1/3}$

17 $h(x) = \pi$

18 $f(x) = e$

19 $g(t) = \sqrt{t}$

20 $h(t) = 3\sqrt{t} + 1$

21 $f(x) = x + \sqrt{x}$

22 $g(x) = 3x^2 + 5\sqrt{x}$

23 $h(x) = x^3 + 6$

24 $f(x) = -13x^3$

25 $g(t) = t^3 - 5t^2$

26 $h(t) = 8t^3 + 2t$

27 $f(x) = (x+1)(x-1)$

28 $g(x) = x(x-2)$

29 $h(x) = \dfrac{2x^3 - 3x}{x}$

30 $f(x) = \dfrac{x^3 + x - 1}{2x}$

In problems 31 through 50, find an equation for the tangent line to the graph of the given function at the given point. You may use the derivative formulas that we have developed.

31 $f(x) = x^2$, at $(-2, 4)$

32 $f(x) = x^2$, at $(-1, 1)$

33 $f(x) = x^2$, at $(0, 0)$

34 $f(x) = x^2$, at $(1, 1)$

35 $f(x) = x^2$, at $(2, 4)$

36 $g(x) = x^3$, at $(-2, -8)$

37 $g(x) = x^3$, at $(-1, -1)$

38 $g(x) = x^3$, at $(0, 0)$

39 $g(x) = x^3$, at $(1, 1)$

40 $g(x) = x^3$, at $(2, 8)$

41 $h(x) = \dfrac{1}{x}$, at $\left(5, \dfrac{1}{5}\right)$

42 $h(x) = \dfrac{1}{x}$, at $\left(\dfrac{1}{5}, 5\right)$

43 $h(x) = \dfrac{1}{x}$, at $\left(-5, -\dfrac{1}{5}\right)$

44 $h(x) = \dfrac{1}{x}$, at $\left(-\dfrac{1}{5}, -5\right)$

45 $f(x) = x^4$, at $(2, 16)$

46 $f(x) = x^5$, at $(2, 32)$

47 $f(x) = x^6$, at $(2, 64)$

48 $f(x) = x^7$, at $(2, 128)$

49 $f(x) = x^8$, at $(2, 256)$

50 $f(x) = x^8$, at $(0, 0)$

51 ***Calculator Problem*** Guess the derivative of e^x at $x = 1$ by guessing the limit of the difference quotient

$$\frac{e^{1+h} - e}{h}.$$

Compute it for $h = .1, .01, .001, .0001, -.1, -.01, -.001, -.0001$.

52 ***Calculator Problem*** Guess the derivative of $\ln x$ at $x = 1$ by guessing the limit of the difference quotient

$$\frac{\ln(1+h) - \ln 1}{h} = \frac{\ln(1+h)}{h}.$$

Compute it for $h = .1, .01, .001, .0001, -.1, -.01, -.001, -.0001$.

53 ***Calculator Problem*** Guess the derivative of $\sqrt{x}$ at $x = 4$ by guessing the limit of the difference quotient

$$\frac{\sqrt{4+h} - 2}{h}.$$

Compute it for $h = .1, .01, .001, .0001, -.1, -.01, -.001, -.0001$.

7.5   Rate of Change Aspect of Derivative

The world is constantly changing. If we are going to be prepared for tomorrow, then we must understand the trends of today. Financial planning requires that we look at how things are changing and the rate at which they are changing. A useful elementary measure of trends is the idea of average change. If the cost of operating your car for the last six months was $38, $35, $36, $28, $31, and $30, respectively, then the average monthly cost was

$$\frac{38 + 35 + 36 + 28 + 31 + 30}{6} = \frac{198}{6} = 33$$

dollars. Thus, you might reasonably expect your auto costs to be $33 next month. However, note that the average of auto costs over the last three months is just

$$\frac{28 + 31 + 30}{3} = 29.67$$

dollars. Since each of these months reflects more recent driving experience, this average might be a better prediction of next month's costs.

Before calculus gained widespread use in business and economics, the **marginal change** in a function f at a was defined as

$$\frac{f(a + 1) - f(a)}{1} = f(a + 1) - f(a).$$

This is the average change in the value of f from a to $a + 1$. So this is approximately the rate at which the value of f is changing at a. However, much could happen to the value of f between a and $a + 1$, and so

$$\frac{f(a + h) - f(a)}{h}$$

with $0 < h < 1$ is a better description of what is happening at a. As h gets closer to 0, this average becomes an even better description of how the value of f is changing at a. The ultimate description of the rate of change in the value of f near a is

$$\lim_{h \to 0} \frac{f(a + h) - f(a)}{h} = f'(a).$$

Thus we define the **marginal change** in the function f at a to be $f'(a)$ and we say $f'(x)$ is the marginal change in f. **If the function f has a name like cost, revenue, or profit then we say $f'(x)$ is marginal cost, marginal revenue,** or **marginal profit**, respectively.

EXAMPLE 1

Let $C(x) = .01x^2 + 100$ be the cost of running a small yo-yo factory, expressed in hundreds of dollars, where x is the number of thousands of yo-yos produced. Find the marginal cost when 50 thousand units have been manu-factured and find $C(51) - C(50)$.

Solution
Marginal cost at $x = 50$ is

$$C'(50) = \lim_{h \to 0} \frac{.01(50 + h)^2 + 100 - .01(50)^2 - 100}{h}$$

$$= \lim_{h \to 0} .01 \frac{(50 + h)^2 - 50^2}{h} = .01 \lim_{h \to 0} \frac{2500 + 100h + h^2 - 2500}{h}$$

$$= .01 \lim_{h \to 0}(100 + h) = .01(100) = 1.$$

In contrast, we have

$$C(51) - C(50) = .01(51)^2 + 100 - .01(50)^2 - 100 = 26.01 - 25 = 1.01.$$

In this case, the average change in the value of $C(x)$ from 50 to 51 is a good description of how the value of $C(x)$ is changing at $x = 50$. ■

In the next chapter we shall encounter rules for calculating derivatives that will allow us to find the derivative of functions like $C(x) = .01x^2 + 100$ very quickly and very easily. In the meantime, to facilitate the handling of examples of marginal change, let us find a formula for the derivative of any function $f(x)$ with the form of a quadratic expression, that is, $f(x) = ax^2 + bx + c$ with a, b, and c fixed. When α is a fixed value, we have

$$f'(\alpha) = \lim_{h \to 0} \frac{f(\alpha + h) - f(\alpha)}{h}$$

$$= \lim_{h \to 0} \frac{a(\alpha + h)^2 + b(\alpha + h) + c - a\alpha^2 - b\alpha - c}{h}$$

$$= \lim_{h \to 0} \frac{a\alpha^2 + 2a\alpha h + ah^2 + bh - a\alpha^2}{h}$$

$$= \lim_{h \to 0}(2a\alpha + ah + b) = 2a\alpha + b.$$

Thus, $f'(x) = 2ax + b$.

EXAMPLE 2

The yo-yo factory in Example 1 sells its yo-yos to a wholesaler for 60¢ each. Find the revenue, marginal revenue, profit, and marginal profit functions.

Solution
If R is revenue in hundreds of dollars and x is how many thousands of yo-yos are sold, then

$$R(x) = \frac{.6(1000x)}{100} = 6x.$$

Since this has the form of a quadratic function with $a = 0$ and $b = 6$, the marginal revenue is

$$R'(x) = 6.$$

If P is profit then

$$P(x) = R(x) - C(x) = 6x - .01x^2 - 100.$$

Since this is a quadratic function with $a = -.01$ and $b = 6$, the marginal profit is

$$P'(x) = -.02x + 6. \ \blacksquare$$

We can now deduce a fundamental principle of business. Note that when it costs more to make a unit than can be obtained by selling the unit, then it is unprofitable to make it. Thus, manufacturing a product should be stopped before marginal cost exceeds marginal revenue. Provided all units made can be sold, **maximum profit** occurs when marginal revenue equals marginal cost. After we study curve sketching, the mathematical reason for this will be clear.

EXAMPLE 3

Using Examples 1 and 2, find how many thousands of yo-yos should be manufactured for maximum profit. Assume every yo-yo made is sold.

Solution

Since $C(x) = .01x^2 + 100$ is a quadratic function we see that its derivative is

$$C'(x) = .02x.$$

We set marginal revenue equal to marginal cost and then we solve

$$C'(x) = R'(x)$$
$$.02x = 6$$
$$x = \frac{6}{.02} = 300.$$

Maximum profit occurs when 300 thousand yo-yos are made.

$$P(300) = 6(300) - .01(300)^2 - 100 = 800$$

Using a calculator, you can check that values of x slightly larger and smaller than 300 give smaller profits. $\blacksquare$

Exercises   7.5

In problems 1 through 12, $R(x)$ and $P(x)$ are revenue and profit functions, respectively. Find marginal revenue or marginal profit as appropriate.

1   $R(x) = .5x$
3   $R(x) = .04x + 5$
5   $R(x) = \sqrt{x}$
7   $P(x) = -x^2 + x - 5$
9   $P(x) = -.01x^2 + x$
11 $P(x) = x^3$

2   $R(x) = 2x + 10$
4   $R(x) = .09x$
6   $R(x) = \sqrt[3]{x}$
8   $P(x) = -.5x^2 + .1x - 2$
10 $P(x) = -.08x^2 + 5x$
12 $P(x) = x^4$

In problems 13 through 20, $C(x)$ is a cost function. Find marginal cost at $x = 10$ and find $C(11) - C(10)$.

13 $C(x) = .008x^2 + 3$
15 $C(x) = .5x^2 + 100$
17 $C(x) = (x + 10)^2$
19 $C(x) = 15x + 20$

14 $C(x) = .009x^2 + 5$
16 $C(x) = .9x^2 + 12$
18 $C(x) = 2(x + 5)^2$
20 $C(x) = 27x + 36$

In problems 21 through 30, find the value of x that results in maximum profit for the given revenue function $R(x)$ and cost function $C(x)$.

21 $R(x) = x, \quad C(x) = .1x^2 + 5$
22 $R(x) = 5x, \quad C(x) = .3x^2 + 15$
23 $R(x) = 8x, \quad C(x) = x^2 + 1$
24 $R(x) = 12x, \quad C(x) = x^2 + 8$
25 $R(x) = .2x, \quad C(x) = .01x^2 + 2$
26 $R(x) = .1x, \quad C(x) = .03x^2 + 5$
27 $R(x) = .2x^2, \quad C(x) = .1x^2 + 10x + 20$
28 $R(x) = .1x^2, \quad C(x) = .01x^2 + x + 12$
29 $R(x) = 2x^2 + x, \quad C(x) = x^2 + 5x + 10$
30 $R(x) = 5x^2 + 2x, \quad C(x) = 2x^2 + 200x + 100$

7.6   Continuity and Differentiability

In understanding the behavior of a function, you must be especially careful about breaks in the graph. For example, in order to graph $f(x) = 1/x$ you must understand the behavior of $f(x)$ close to $x = 0$. Of course, f is not defined at $x = 0$ since division by 0 is not defined. Furthermore, as x approaches 0 and is positive (for example, x might be .1, .01, .001, .0001, etc.) then $f(x) = 1/x$ is positive and gets very large (specifically, 10, 100, 1000, 10,000, etc.). As x approaches 0 and is negative (for example, x might be $-.1$, $-.01$, $-.001$, $-.0001$, etc.) then $f(x) = 1/x$ is negative and gets very large in magnitude (specifically, -10, -100, -1000, $-10,000$, etc.). The graph of $f(x) = 1/x$ is shown on the next page.

In order to locate and understand breaks in the graph, it is necessary to locate and understand points on the graph that are not breaks. If the graph of $f(x)$ is not broken at a, then we might say that it is continuous at a. The following definition is more precise.

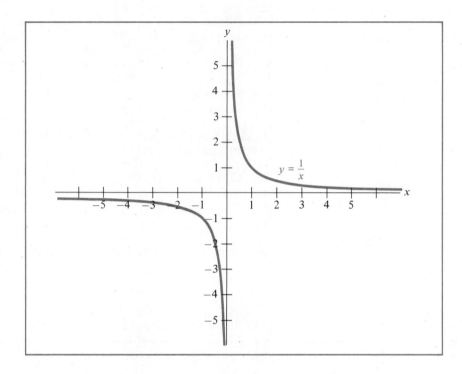

Definition   We say that $f(x)$ is **continuous** at $x = a$ provided

$$\lim_{x \to a} f(x) = f(a).$$

There are three ingredients to this rather terse definition. First, f must be defined at a, since we have written down $f(a)$. Second, the limit $\lim_{x \to a} f(x)$ must exist, since we have used it also. Third, the limit must equal $f(a)$. If g is not continuous at a, then we say that g is **discontinuous** at a or that a is a **discontinuity** for g. If a is a discontinuity for g and the magnitude of $g(x)$ gets arbitrarily large as x approaches a, then a is said to be an **infinite discontinuity** for g. So 0 is an infinite discontinuity for $f(x) = 1/x$.

EXAMPLE 1

United States demand for wheat as a function of price is approximated by

$$q = \frac{7}{p}$$

where p is measured in dollars per bushel and q is measured in billions of bushels. Graph this equation the way an economist usually graphs a demand curve (that is, plot p on the y-axis and q on the x-axis).

Solution
First we solve the equation for p so that it will be given as the dependent variable and q will be given as the independent variable.

$$qp = 7, \qquad p = \frac{7}{q}$$

Clearly p has an infinite discontinuity at $q = 0$. The graph is given. (Only positive values of p and q have meaning in this problem.)

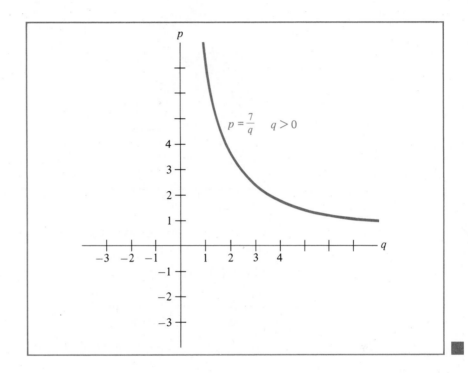

The graph of $p = 7/q$ does not have a tangent line at $q = 0$ since there is no point (p, q) on the graph with $q = 0$. Of course, dp/dq does not exist for $q = 0$.

A less dramatic type of discontinuity is present in the graph of some functions. Let $g(x)$ be the greatest integer not greater than x. This function is called the **greatest integer function** and is shown on the next graph. Here the hollow circle centered at the right-hand endpoint of each horizontal line segment indicates that the right-hand endpoint is omitted. For example, for x such that $1 \le x < 2$ the greatest integer not exceeding x is 1, but the greatest integer not exceeding 2 is 2. Thus, $(2, 1)$ is not on the graph and $(2, 2)$ is.

If $f(x)$ gets arbitrarily close to b as x approaches a from the right, then we say that b is the **limit from the right** and we write

$$\lim_{x \to a+} f(x) = b.$$

The symbols $x \to a+$ indicate that x assumes only values greater than a. If

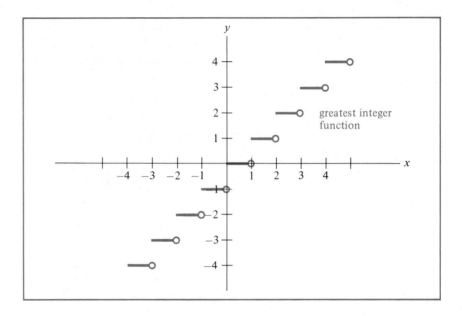

$f(x)$ gets arbitrarily close to c as x approaches a from the left, then c is the **limit from the left** and we write

$$\lim_{x \to a-} f(x) = c.$$

The symbols $x \to a-$ indicate that x assumes only values less than a. Note that for the greatest integer function g

$$\lim_{x \to k+} g(x) = k \quad \text{and} \quad \lim_{x \to k-} g(x) = k - 1$$

for every integer k. The graph of the greatest integer function does not have a tangent line for $x = 2$ or any other integer. (It does have a tangent line when x is any number that is not an integer.) The graph to the right of $(2, 2)$ looks like the line $y = 2$, while the graph to the left of $(2, 2)$ looks like the line $y = 1$. There is no single line that approximates the graph near $(2, 2)$.

If both $\lim\limits_{x \to a+} f(x)$ and $\lim\limits_{x \to a-} f(x)$ exist and are not equal, then we say that a is a **jump discontinuity** for f. Thus, every integer is a jump discontinuity for the greatest integer function. A practical example of a function with jump discontinuities is given next.

EXAMPLE 2

City Taxi Company charges its customers 50¢ for each ride plus 50¢ for each half mile or fraction of a half mile covered by the trip. Find the discontinuities for the fare as a function of the length of the ride and graph the fare function.

Solution
Let x be the length of the ride and note that the fare $f(x)$ jumps up 50¢ each

time x passes a new multiple of $\frac{1}{2}$. So the discontinuities are the positive multiples of $\frac{1}{2}$, and are shown on the graph.

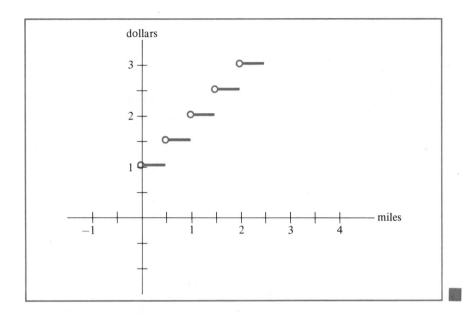

Other functions that behave this way include federal income tax as a function of taxable income and the price of postage as a function of the weight of the item mailed.

Another kind of discontinuity occurs in many examples. If $\lim\limits_{x \to a} f(x)$ exists but the limit is not equal to $f(a)$, then a is called a **removable discontinuity** for f. For example, define $f(x)$ by

$$f(x) = \frac{x^2 - 1}{x - 1}.$$

Then f is not continuous at $x = 1$, since $\frac{0}{0}$ is not defined. For $x \neq 1$ we have $x - 1 \neq 0$ and the following simplification is justified.

$$\frac{x^2 - 1}{x - 1} = \frac{(x - 1)(x + 1)}{x - 1} = x + 1 \qquad x \neq 1$$

Thus, $\lim\limits_{x \to 1} f(x)$ is exactly the same as $\lim\limits_{x \to 1} (x + 1)$, which exists and equals 2. The graph of $f(x)$ is shown on the next page. Again the hollow circle indicates that its center is omitted from the graph. The graph has neither a tangent line nor a derivative for $x = 1$ because $f(x)$ is not defined at $x = 1$.

Many functions are continuous at every number in their domain. For example, if $f(x)$ is any linear function, say $f(x) = mx + b$, then $f(x)$ is continuous at every number c.

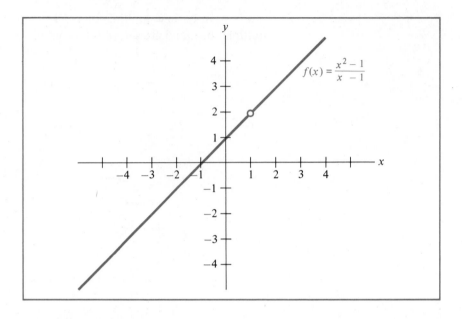

$$\lim_{x \to c} f(x) = \lim_{x \to c}(mx + b)$$

$$= \lim_{x \to c} mx + \lim_{x \to c} b$$

$$= m(\lim_{x \to c} x) + b = mc + b = f(c)$$

More generally, let $g(x)$ be a **polynomial** and recall that g can be written as

$$g(x) = a_n x^n + a_{n-1}x^{n-1} + \cdots + a_1 x + a_0 \qquad \text{with } a_n \neq 0$$

where $a_n, a_{n-1}, \ldots, a_1, a_0$ are constants and n is a nonnegative integer. Let c be some choice of x and note that

$$g(c) = a_n c^n + a_{n-1}c^{n-1} + \cdots + a_1 c + a_0.$$

Using the Limit Theorem, we compute $\lim_{x \to c} g(x)$.

$$\lim_{x \to c} g(x) = \lim_{x \to c}(a_n x^n + a_{n-1}x^{n-1} + \cdots + a_1 x + a_0)$$

$$= \lim_{x \to c} a_n x^n + \lim_{x \to c} a_{n-1}x^{n-1} + \cdots + \lim_{x \to c} a_1 x + \lim_{x \to c} a_0$$

$$= a_n \lim_{x \to c} x^n + a_{n-1} \lim_{x \to c} x^{n-1} + \cdots + a_1 \lim_{x \to c} x + a_0$$

$$= a_n c^n + a_{n-1}c^{n-1} + \cdots + a_1 c + a_0$$

$$= g(c)$$

Thus, $\lim_{x \to c} g(x)$ equals $g(c)$ and g is continuous at c. Since we used nothing about c except that it was a fixed choice of x, we have discovered the following theorem.

Theorem   Every polynomial is continuous at every number.

We say that $f(x)$ is a **rational function** provided it is the ratio of two polynomials, say $f(x) = p(x)/q(x)$ where $p(x)$ and $q(x)$ are polynomials. Thus, $f(c) = p(c)/q(c)$ makes sense for any choice of c such that $q(c) \neq 0$. Provided $q(c)$ is not 0, we can use the Limit Theorem and the above theorem to compute $\lim_{x \to c} f(x)$.

$$
\begin{aligned}
\lim_{x \to c} f(x) &= \lim_{x \to c} \frac{p(x)}{q(x)} \\
&= \frac{\lim_{x \to c} p(x)}{\lim_{x \to c} q(x)} \\
&= \frac{p(c)}{q(c)} = f(c) \qquad \text{provided } q(c) \neq 0
\end{aligned}
$$

This establishes the following theorem.

Theorem   Every rational function is continuous everywhere it is defined.

We conclude this section by considering the relation between "$f(x)$ is continuous at $x = a$" and "$f'(x)$ exists at $x = a$." The equation

$$
f(a + h) = h \left[\frac{f(a + h) - f(a)}{h} \right] + f(a) \qquad \text{provided } h \neq 0
$$

is easily verified by simplifying the right side. We use the above equation and the Limit Theorem to conclude that $\lim_{h \to 0} f(a + h) = f(a)$ whenever the limit

$$
\lim_{h \to 0} \frac{f(a + h) - f(a)}{h} \text{ exists.}
$$

$$
\begin{aligned}
\lim_{h \to 0} f(a + h) &= \lim_{h \to 0} \left(h \left[\frac{f(a + h) - f(a)}{h} \right] + f(a) \right) \\
&= \lim_{h \to 0} \left(h \left[\frac{f(a + h) - f(a)}{h} \right] \right) + \lim_{h \to 0} f(a) \\
&= (\lim_{h \to 0} h) \left(\lim_{h \to 0} \left[\frac{f(a + h) - f(a)}{h} \right] \right) + f(a) \\
&= (0)(f'(a)) + f(a) \\
&= f(a)
\end{aligned}
$$

In the next-to-last step, we used the fact that $f'(a)$ exists (that is, the limit of the difference quotient is a finite number). If h is chosen to be $x - a$, then the last step can be rewritten as

$$
\lim_{x \to a} f(x) = f(a),
$$

which is the definition of "$f(x)$ is continuous at $x = a$." We have proved that if $f'(a)$ exists, then $f(x)$ is continuous at $x = a$. It follows that if a is a discontinuity for $f(x)$, then $f'(a)$ does not exist.

It can happen that $f(x)$ is continuous at $x = a$ and, nevertheless, $f'(a)$ does not exist. If $f(x)$ is the absolute value function, that is $f(x) = |x|$, then $f(x)$ is continuous at $x = 0$. Nevertheless, $f'(0)$ does not exist, as we now show. If $h > 0$, then $f(0 + h) = |0 + h| = |h| = h$ and

$$\lim_{h \to 0+} \frac{f(0 + h) - f(0)}{h} = \lim_{h \to 0+} \frac{h - |0|}{h} = \lim_{h \to 0+} \frac{h}{h}$$
$$= \lim_{h \to 0+} 1 = 1.$$

If $h < 0$, then $f(0 + h) = |0 + h| = |h| = -h$ and

$$\lim_{h \to 0-} \frac{f(0 + h) - f(0)}{h} = \lim_{h \to 0-} \frac{-h - |0|}{h} = \lim_{h \to 0-} \frac{-h}{h}$$
$$= \lim_{h \to 0-} -1 = -1.$$

Since there is no single number that $\dfrac{f(0 + h) - f(0)}{h}$ is getting close to as h approaches 0, the limit $\lim_{h \to 0} \dfrac{f(0 + h) - f(0)}{h}$ does not exist. Thus, $f(x) = |x|$ does not have a derivative at $x = 0$.

It is interesting to note the peculiar corner that the graph of $f(x) = |x|$ forms at the point $(0, 0)$.

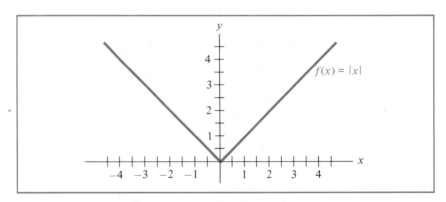

In general, a function $f(x)$ does not have a derivative at $x = a$ if the graph of $f(x)$ forms a corner at $(a, f(a))$.

In order to prove that $f(x) = |x|$ is continuous at $x = 0$, we must prove that

(1) $\lim_{x \to 0} |x| = |0| = 0.$

This is best accomplished by proving that

(2) $\lim_{x \to 0+} |x| = 0$ and $\lim_{x \to 0-} |x| = 0.$

It then follows that (1) must be true. In the first limit of (2), we need only consider $x > 0$, so $|x| = x$ and

$$\lim_{x \to 0+} |x| = \lim_{x \to 0+} x = 0.$$

In the second limit of (2), we need only consider $x < 0$, so $|x| = -x$ and

$$\lim_{x \to 0-} |x| = \lim_{x \to 0-} -x = -0 = 0.$$

We have now proved that (2) is true, and the truth of (1) follows. Thus, $f(x) = |x|$ is continuous at $x = 0$.

Summary

1   If $f'(a)$ exists, then $f(x)$ is continuous at $x = a$.
2   If $f(x)$ is not continuous at $x = a$, then $f'(a)$ does not exist.
3   If $f(x)$ is continuous at $x = a$, then $f'(a)$ may or may not exist.
4   If $f'(a)$ does not exist, then $f(x)$ may or may not be continuous at $x = a$.

Exercises 7.6

In problems 1 through 30, find each discontinuity (if any) for the given function and indicate whether it is an infinite discontinuity, a jump discontinuity, or a removable discontinuity.

1   $f(x) = 5$ 2 $g(x) = 9.3$
3   $h(x) = \pi$ 4 $f(x) = e$

5   $g(x) = \dfrac{1}{x} + 2$ 6 $h(x) = -37x + 99$

7   $f(x) = \dfrac{x + 1}{x - 1}$ 8 $g(x) = \dfrac{2x + 4}{2x - 4}$

9   $f(x) = 13x^2 + 21x - 6$ 10 $g(x) = 50(x^2 + 2)$

11   $h(x) = \dfrac{x^2 - 4}{x + 2}$ 12 $f(x) = \dfrac{x^2 - 9}{x - 3}$

13   $g(x) = (2x + 1)(3x - 2)$ 14 $h(x) = (x - 4)(5x + 2)$

15   $f(x) = \dfrac{x^3 - x}{x + 1}$ 16 $g(x) = \dfrac{2x^3 - 50x}{x + 25}$

17   $h(x) = \dfrac{(x + 1)(x + 3)}{(x - 2)(x - 3)}$ 18 $f(x) = \dfrac{2x^2 + 1}{x - 5}$

19   $g(x) = x^3 + 10x^2 + x - 7$ 20 $h(x) = 3x^3 + 11x + 5$

21   $f(x) = \dfrac{-2x^2 - 1}{x}$ 22 $g(x) = \dfrac{x^2 + 1}{x - 1}$

23   $h(x) = \dfrac{x^2 + x}{x}$

24   $f(x) = \dfrac{x^2 + 2x + 1}{x + 1}$

25   $g(x) = x^{126}$

26   $g(x) = x^{137} + 522x^{88}$

27   $f(x) = \begin{cases} -1 & \text{if } x < 0 \\ 1 & \text{if } x \geq 0 \end{cases}$

28   $g(x) = \begin{cases} x & \text{if } x < 0 \\ x + 1 & \text{if } x \geq 0 \end{cases}$

29   $h(x) = \begin{cases} x^2/x & \text{if } x \neq 0 \\ 0 & \text{if } x = 0 \end{cases}$

30   $f(x) = \begin{cases} x/x^2 & \text{if } x \neq 0 \\ 0 & \text{if } x = 0 \end{cases}$

In problems 31 through 40, show that the given function is continuous at the given point using the definition of a continuous function.

31   $f(x) = 3$ at $x = 0$

32   $g(x) = -39$ at $x = 5$

33   $h(x) = 3x - 3$ at $x = 1$

34   $f(x) = -5x + 10$ at $x = -1$

35   $g(x) = x^2 + x + 1$ at $x = 0$

36   $h(x) = 3x^2 + 4$ at $x = 2$

37   $f(x) = \dfrac{1}{x}$ at $x = 8$

38   $g(x) = \dfrac{5}{x - 4}$ at $x = 3$

39   $h(x) = \dfrac{2x + 3}{3x - 2}$ at $x = 1$

40   $f(x) = \dfrac{x^2 - 1}{x - 1}$ at $x = -1$

In problems 41 through 46, sketch the graph of the given function.

41   $f(x) = \dfrac{1}{x - 2}$

42   $g(x) = \dfrac{3}{3x + 9}$

43   $h(x) = \dfrac{5x^2 - 20}{x + 2}$

44   $f(x) = \dfrac{2x^2 - 72}{x - 6}$

45   $g(x) = \begin{cases} 1 & \text{if } x < 0 \\ 2 & \text{if } x \geq 0 \end{cases}$

46   $h(x) = \begin{cases} x & \text{if } x < 0 \\ -x & \text{if } x \geq 0 \end{cases}$

Review of Terms

Important Mathematical Terms

rise,   *p. 288*
run,   *p. 288*
slope of a line,   *p. 289*
parallel,   *p. 289*
perpendicular,   *p. 292*
tangent line,   *p. 295*
point of tangency,   *p. 296*
slope of a graph,   *p. 297*
secant line,   *p. 297*
limit,   *p. 300*
Limit Theorem,   *p. 305*
infinity,   *p. 306*
difference quotient,   *p. 313*
derivative,   *p. 313*

derivative function,   *p. 314*
Power Rule,   *p. 317*
continuous,   *p. 324*
discontinuous,   *p. 324*
discontinuity,   *p. 324*
infinite discontinuity,   *p. 324*
greatest integer function,   *p. 325*
limit from the right,   *p. 325*
limit from the left,   *p. 326*
jump discontinuity,   *p. 326*
removable discontinuity,   *p. 327*
polynomial,   *p. 328*
rational function,   *p. 329*

Important Terms from the Applications

perpetuity, *p. 309*	marginal revenue, *p. 320*
fractional banking system, *p. 309*	marginal profit, *p. 320*
marginal change, *p. 320*	maximum profit, *p. 322*
marginal cost, *p. 320*	

Review Problems

In problems 1 through 5, find a point-slope equation (if possible) for which the graph is parallel to the graph of the given equation and passes through the indicated point.

1 $y = .2x - 12$, $(1, 1)$
2 $y = -3x + 15$, $(0, 0)$
3 $y = 4$, $(2, 3)$
4 $x = -6$, $(-3, 3)$
5 $0 = 10x + 5y + 20$, $(0, 4)$

In problems 6 through 10, find the slope-intercept equation (if possible) for which the graph is perpendicular to the graph of the given equation and passes through the indicated point.

6 $y = .2x + 4$, $(5, 2)$
7 $y = -1.5x + 7$, $(-1, -1)$
8 $0 = 9x + 3y - 12$, $(3, -2)$
9 $y = -8$, $(1, 1)$
10 $x = 17$, $(0, 0)$

In problems 11 through 14, find a point-slope equation for the tangent line to the graph of the given function at the given point. The slope of the tangent line is given. Compute the value of the function and the y-value of the tangent line at the x-value listed.

11 $f(x) = 2x^4 - x^2$, $(0, 0)$,   slope $= 0$, $x = .1$
12 $f(x) = 2x^4 - x^2$, $(0, 0)$,   slope $= 0$, $x = -.1$
13 $f(x) = 2x^4 - x^2$, $(1, 1)$,   slope $= 6$, $x = 1.1$
14 $f(x) = 2x^4 - x^2$, $(1, 1)$,   slope $= 6$, $x = .9$

In problems 15 through 18, use some estimate for the slope of the tangent line to the graph of the given function at the given point to decide whether the slope of the graph is positive or negative.

15 $f(x) = x^3$,   at $\left(-\dfrac{1}{2}, -\dfrac{1}{8}\right)$

16 $g(x) = x^3 - 3x$,   at $(2, 2)$
17 $g(x) = x^3 - 3x$,   at $(0, 0)$
18 $h(x) = x^{-2}$,   at $(1, 1)$

In problems 19 through 30, find the indicated limit if it exists.

19 $\lim\limits_{x\to 0} -53$

20 $\lim\limits_{x\to 2}(-9x + 18)$

21 $\lim\limits_{x\to 0}(5x^2 + 3x - 6)$

22 $\lim\limits_{x\to 1}(x - 2)(x + 3)$

23 $\lim\limits_{x\to -1}\dfrac{3x + 2}{x^2 + 1}$

24 $\lim\limits_{x\to 2}\dfrac{8x + 7}{5x - 1}$

25 $\lim\limits_{x\to 8} \sqrt[3]{x}$

26 $\lim\limits_{x\to 0} \sqrt{x^4 + 16}$

27 $\lim\limits_{x\to \infty}\dfrac{1}{\sqrt[3]{x}}$

28 $\lim\limits_{x\to \infty} x^{1/4}$

29 $\lim\limits_{x\to \infty} .99^x$

30 $\lim\limits_{x\to \infty}\left(\dfrac{156}{x^4} + \dfrac{92}{x^2}\right)$

In problems 31 through 35, find a derivative formula for the indicated function using the definition of derivative and the Limit Theorem.

31 $f(x) = -2x - 9$

32 $g(s) = s^2 + 3s + 12$

33 $h(t) = \dfrac{4}{t^2}$

34 $f(x) = \sqrt{4x}$

35 $g(s) = \sqrt{13}$

In problems 36 through 40, find an equation for the tangent line to the graph of the given function at the given point. Use any derivative formulas you have learned.

36 $f(x) = x^2 + 2x + 1$,   at $(0, 1)$
37 $g(s) = s^2 + 4$,   at $(1, 5)$

38 $h(t) = t^{-2}$,   at $\left(\dfrac{1}{2}, 4\right)$

39 $f(x) = x^{100}$,   at $(1, 1)$
40 $f(x) = x^{100}$,   at $(0, 0)$

In problems 41 through 44, $P(x)$ is a profit function. Find marginal profit at $x = 5$ and find $P(6) - P(5)$.

41 $P(x) = -.1x^2 + x - 2$
43 $P(x) = 5x - 12$

42 $P(x) = x^2 - 10$
44 $P(x) = -.01x^3 + x^2 - 20$

In problems 45 through 46, find the value of x that results in maximum profit, given the revenue function $R(x)$ and the cost function $C(x)$

45 $R(x) = 5x$, $C(x) = .1x^2 + 18$
46 $R(x) = .01x^2$, $C(x) = .1x + 1$

In problems 47 through 52, find each discontinuity for the given function and indicate whether it is an infinite discontinuity, a jump discontinuity, or a removable discontinuity.

47 $f(x) = \dfrac{1}{\sqrt{11}}$

48 $g(x) = x(x - 1)(2x + 3)$

49 $h(x) = \dfrac{5x^3 + 4x^2 + 3x}{x}$

50 $f(x) = \dfrac{x^2 + 1}{x}$

51 $g(x) = \dfrac{6x}{x^2 + 1}$

52 $h(x) = \begin{cases} 1 & \text{if } x \neq 0 \\ 0 & \text{if } x = 0 \end{cases}$

Solve the following problems.

53 When a certain real estate tract is pictured on a coordinate system, the corners of the tract are $(1, 0)$, $(5, 0)$, $(7, 2)$, and $(3, 2)$. Find an equation for the line passing through the midpoint of the boundary along the x-axis, which is parallel to the side boundaries (which are parallel to each other).

54 Find the present value of a perpetuity with rent of $10,000 and an interest rate of 15%.

Social Science Applications

Newton's Method

We now explain a numerical procedure, called Newton's method, for determining the zeros of a given function $f(x)$. We seek the value $x = a$ that makes the equation $f(x) = 0$ true. Of course, $(a, f(a)) = (a, 0)$ is a point where the graph of $f(x)$ crosses the x-axis, as we illustrate.

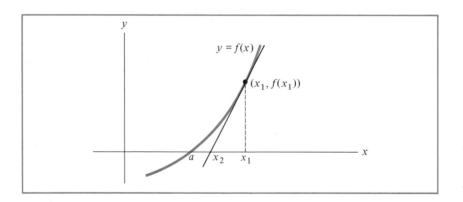

We begin by choosing $x = x_1$ to approximate the desired zero $x = a$. The x-coordinate of the point at which the tangent line to the graph of $y = f(x)$ at $(x_1, f(x_1))$ intersects the x-axis is our second approximation of the zero of $f(x) = 0$. As the graph indicates, x_2 is a better approximation of a than is x_1, provided x_1 is adequately chosen. Then we repeat the procedure using the tangent line to the graph of $y = f(x)$ at $(x_2, f(x_2))$. In this way, we construct the successive approximations

$$x_1, x_2, x_3, \ldots .$$

The equation for the tangent line to the graph of $y = f(x)$ at

$(x_n, f(x_n)) = (x_n, y_n)$ is

$$y - y_n = f'(x_n)(x - x_n).$$

When $y = 0$ in this equation, we have

$$x = x_n - \frac{y_n}{f'(x_n)}.$$

Thus, the next approximation of $x = a$ after $x = x_n$ is

$$\boxed{x_{n+1} = x_n - \frac{y_n}{f'(x_n)}.}$$

If we wish to approximate $x = a$ to two decimal places, then we construct approximations until the first two decimal places do not change.
For example, suppose we want to solve

$$x^3 + x - 9 = f(x) = 0.$$

Note that $f'(x) = 3x^2 + 1$. Taking $x_1 = 2$, we get $y_1 = f(x_1) = 1$ and $f'(x_1) = f'(2) = 13$. Using the boxed equation, we have

$$x_2 = x_1 - \frac{y_1}{f'(x_1)} = 2 - \frac{1}{13} \approx 1.923$$

and

$$x_3 = x_2 - \frac{y_2}{f'(x_2)} \approx 1.923 - \frac{.034}{12.094} \approx 1.920.$$

We see that a zero of the given function, to two decimal places, is $x = 1.92$.
Use Newton's method to solve each of the problems 1 through 6. Use a calculator, if possible.

1   Find a zero of $f(x) = x^3 - x - 1$ to one decimal place beginning with $x_1 = 1$.
2   Find a zero of $f(x) = x^{3/2} - x^{1/2} - 5$ to one decimal place beginning with $x_1 = 4$.
3   **Psychology**   Suppose a test subject can memorize x items from a given list in approximately $g(x) = x^{3/2} + x^{1/2}$ minutes. Approximated to one decimal place, how many items can this person memorize in 25 minutes? (**Hint:** Solve the equation $25 = x^{3/2} + x^{1/2}$, or equivalently, $x^{3/2} + x^{1/2} - 25 = h(x) = 0$.)
4   **Psychology**   Suppose a certain student can learn a string of x nonsense words in approximately $g(x) = .5x^3 - x + 1$ minutes. Approximated to one decimal place, how long a string can the student learn in 30 minutes? (**Hint:** Solve the equation $30 = .5x^3 - x + 1$, or equivalently, $x^3 - 2x - 58 = h(x) = 0$.)
5   **Urban Planning**   The population of a city t years after January 1, 1980, is given by

$$x(t) = 1,000,000e^{.1t}.$$

On January 1, 1980, the city has 333,000 housing units and the number of units is increasing at 3,000 per year. Determine (to one decimal place) when the population will equal 4 times the number of housing units. (**Hint:** Solve $4(333,000 + 3,000t) = 1,000,000e^{.1t}$ or equivalently, $f(t) = e^{.1t} - .012t - 1.332 = 0$, and use that the derivative of $e^{.1t}$ at $t = t_n$ is $.1e^{.1t_n}$.)

6 **Sociology**   A rumor that the Vice President of the United States has resigned begins in a city of 1,000,000 people. After t days the number of people spreading the rumor is given by

$$y(t) = \frac{1,000,000}{999e^{-2t} + 1}$$

The television and radio stations begin broadcasting special bulletins denying the rumor. After t days the number of people reached by the bulletins is given by

$$x(t) = 1,000,000(1 - e^{-.01t}).$$

How long will it take (to one decimal place) for the news media to reach as many people as there are spreading the rumor? (**Hint:** Solve the equation

$$1,000,000(1 - e^{-.01t}) = \frac{1,000,000}{999e^{-2t} + 1}.$$

After some simplification, we get the equivalent equation

$$f(t) = e^{1.99t} + 999e^{-.01t} - 999 = 0.$$

Use that the derivatives of $e^{1.99t}$ and $e^{-.01t}$ at $t = t_n$ are $1.99e^{1.99t_n}$ and $-.01e^{-.01t_n}$, respectively.)

Mathematics is one of the oldest and most basic of the sciences. Throughout recorded history, the activities of mathematicians have been divided into two classes: pure (or theoretical) mathematics and applied mathematics. Mathematicians involved with pure mathematics are concerned with the development and refinement of mathematical theories; those involved with applied mathematics deal with the development of mathematical strategies for solution of practical problems of science, business, industry, and government. In the past, these two categories of mathematical activity have overlapped, and they continue to do so today. In the development of calculus, for example, the methods used to study planetary motion and related matters were gradually unified into a theory; the theory, once developed, took on a life of its own, and eventual theoretical developments became useful in the solving of new and more difficult problems.

In present usage, the word "mathematician" is reserved primarily for individuals who hold a doctorate in a mathematical field. The major employers of pure mathematicians are colleges and universities; uncertainties about undergraduate enrollments in the 1980s suggest that there will be few, if any, new jobs available for academic mathematicians. Private industry and government agencies hire a limited number of applied mathematicians; they seek individuals with problem-solving skills in areas such as operations research, statistics, numerical analysis, mathematical physics, computer systems analysis, and market research.

As indicated, career opportunities for a "mathematician" generally require a Ph.D. and are limited in number. However, there are many significant career opportunities for an individual who obtains a bachelor's degree in mathematics. Although these career choices do not carry the title of "mathematician," the significant role of mathematics in educational preparation for these careers leads us to consider several of them here.

Career opportunities for mathematicians with a bachelor's degree include the following.

Secondary School Teaching   The need for preparing persons to live in ever-more-technological society and the recognition of mathematics as the foundation for such preparation have placed increased importance on the quality of training for mathematics teachers. The present shortage of qualified teachers is expected to continue through the 1980s. Teacher's salaries at present are modest ($12,000 to $25,000, with higher figures for large city systems), but salaries may improve in response to shortages.

A teacher of secondary school mathematics must know calculus, linear algebra, and geometry (Euclidean and transformation). Desirable additional studies include advanced courses in algebra and analysis and a course in the history of mathematics. A course that will enable the teacher to involve students in the process of applying mathematical knowledge to practical problem solving in areas such as business and the social sciences is extremely important.

Preprofessional Training   Professional schools in medicine, law, and government have joined schools of business administration in recognition of the value of an undergraduate major in mathematics in preprofessional preparation. These schools are looking for applicants with backgrounds that include knowledge of statistics, experience with computers, ability to develop and understand complex arguments, demonstration of problem-solving skills, and ability to work hard and to learn quickly. An undergraduate major in mathematics can develop all of these background prerequisites.

Interdisciplinary Research   In recent years, mathematical methods have moved into fields that formerly were nonquantitative disciplines, such as biology, management, psychology, sociology, and linguistics. At this time of transition, there are opportunities for an individual to use an undergraduate major in mathematics combined with a second major in one of these other fields as a strong preparation for graduate study and eventual research in the chosen nonmathematical field. Statistics has been the most important mathematical tool in these applied fields, but skills from probability, differential equations, optimization theory, computer science, abstract algebra, and topology have also been used.

Sources of Additional Information

Occupational Outlook Handbook. Bureau of Labor Statistics, U.S. Department of Labor, Washington, DC 20212. (Revised every 2 years, this handbook provides information about job duties, working conditions, level and places of employment, education and training requirements, advancement possibilities, job outlook, and earnings for about 250 occupations. Some of the information for this Career Profile was obtained from this source.)

Professional Opportunities in Mathematics. Mathematical Association of America, 1529 18th Street, N.W., Washington, DC 20036. (This booklet is available at a cost of $1.50 from the MAA. It describes career opportunities for individuals with mathematical training and includes a list of over 50 sources of additional information; special attention is given to career advice for women interested in entering mathematical professions. This MAA booklet was a source for some of the information in this Career Profile.)

8

Computing Derivatives

In order to make the derivative an effective tool for understanding graphs and for other applications, we must be able to readily compute the derivative formula for many functions. The strategy for accomplishing this is to view the formula for a function as being made from simpler components. For example, we regard $f(x) = 3x - 2$ as being the sum of $g(x) = 3x$ and $k(x) = -2$. We shall attempt to compute the derivative formula for f by getting the derivative formulas for g and k. From earlier work we know that the derivative of any constant function is 0 everywhere, and it is easy to compute a formula for $g'(x)$.

$$g'(x) = \lim_{h \to 0} \frac{g(x+h) - g(x)}{h} = \lim_{h \to 0} \frac{3(x+h) - 3x}{h}$$

$$= \lim_{h \to 0} 3 = 3$$

Now we can compute a formula for $f'(x)$.

$$f'(x) = \lim_{h \to 0} \frac{f(x+h) - f(x)}{h}$$

$$= \lim_{h \to 0} \frac{g(x+h) + k(x+h) - g(x) - k(x)}{h}$$

$$= \lim_{h \to 0} \left[\frac{g(x+h) - g(x)}{h} + \frac{k(x+h) - k(x)}{h} \right]$$

$$= \lim_{h \to 0} \frac{g(x+h) - g(x)}{h} + \lim_{h \to 0} \frac{k(x+h) - k(x)}{h} \quad \text{by the Limit Theorem}$$

$$= g'(x) + k'(x) \quad \text{by the definition of derivative}$$

$$= 3 + 0 = 3$$

The steps used above prove that, in general, the derivative of the sum of two functions is the sum of the derivatives. We summarize this statement in mathematical notation.

Sum Rule

$$\frac{d}{dx}\left(f(x) + g(x)\right) = \frac{d}{dx}f(x) + \frac{d}{dx}g(x).$$

Steps very similar to those used above show that the derivative of the difference of two functions is the difference of the derivatives.

Difference Rule

$$\frac{d}{dx}\left(f(x) - g(x)\right) = \frac{d}{dx}f(x) - \frac{d}{dx}g(x).$$

Recall that we have already established that $\frac{d}{dx}x^n = nx^{n-1}$ where n is a positive integer. In fact, this rule is true when n is any number at all. This rule, which is called the Power Rule, is one of the most useful derivative formulas. It will be proved in Section 8.5.

Power Rule

$$\frac{d}{dx}x^a = ax^{a-1} \qquad \text{for any fixed number } a.$$

Finally, we note that there is a very simple pattern to what happens to a derivative formula when the function is multiplied by a constant—the derivative is multiplied by the same constant.

Constant Multiple Rule

$$\frac{d}{dx}\left(cf(x)\right) = c\frac{d}{dx}f(x).$$

By applying these four rules, we can compute many derivatives very quickly.

EXAMPLE 1

Find a derivative formula for $f(x) = \sqrt{x}$.

Solution
Write $\sqrt{x}$ as $x^{1/2}$, and it becomes apparent that the Power Rule applies.

$$\frac{d}{dx}\sqrt{x} = \frac{d}{dx}x^{1/2} = \frac{1}{2}x^{-1/2} = \frac{1}{2}\left[\frac{1}{\sqrt{x}}\right] \quad \blacksquare$$

EXAMPLE 2

Find $\frac{d}{dx}x$.

Solution
Since the tangent line for any point on a straight line is the line itself and the

slope of $y = x$ is obviously 1, we know that $\dfrac{d}{dx} x = 1$. For our peace of mind, we observe that the Power Rule gives the same answer.

$$\frac{d}{dx} x = \frac{d}{dx} x^1 = 1x^0 = 1 \cdot 1 = 1$$

(Recall that $x^0 = 1$ for all x, by definition.) ■

EXAMPLE 3

Find a derivative formula for an arbitrary linear function written in the slope-intercept form, $g(x) = mx + b$.

Solution

$$\frac{d}{dx} g(x) = \frac{d}{dx} (mx) + \frac{d}{dx} b \qquad \text{by the Sum Rule}$$

$$= m \frac{d}{dx} x + \frac{d}{dx} b \qquad \text{by the Constant Multiple Rule}$$

$$= m \cdot 1 + 0 = m$$

This result should have been expected since the derivative of a function gives the slope of the graph of the function. ■

EXAMPLE 4

Find $\dfrac{d}{dx} \dfrac{1}{x}$.

Solution
We use the Power Rule.

$$\frac{d}{dx} \frac{1}{x} = \frac{d}{dx} x^{-1} = (-1)x^{-2} = -\frac{1}{x^2} \quad ■$$

EXAMPLE 5

Find $\dfrac{d}{dx} (5x^2 + 3x - 2)$.

Solution

$$\frac{d}{dx} (5x^2 + 3x - 2) = \frac{d}{dx} 5x^2 + \frac{d}{dx} 3x + \frac{d}{dx} (-2) \qquad \text{by the Sum Rule}$$

$$= 5 \frac{d}{dx} x^2 + 3 \frac{d}{dx} x + \frac{d}{dx} (-2) \qquad \text{by the Constant Multiple Rule}$$

$$= 5(2x) + 3(1) + 0$$
$$= 10x + 3 \quad ■$$

EXAMPLE 6

Find a derivative formula for an arbitrary quadratic function $g(x) = ax^2 + bx + c$.

Solution

$$\frac{d}{dx}(ax^2 + bx + c) = \frac{d}{dx}ax^2 + \frac{d}{dx}bx + \frac{d}{dx}c$$

$$= a\frac{d}{dx}x^2 + b\frac{d}{dx}x + 0$$

$$= 2ax + b \blacksquare$$

The approach of regarding the formula for a function as built from simpler components and using the elementary rules for derivatives is very powerful. It permits us to obtain a derivative formula for any polynomial. Recall that $f(x)$ is a polynomial provided it can be written as

$$f(x) = a_n x^n + a_{n-1}x^{n-1} + \cdots + a_1 x + a_0$$

where $a_n, a_{n-1}, \ldots, a_1, a_0$ are constants and n is a nonnegative integer. The derivative formula is obtained below.

$$\frac{d}{dx}f(x) = \frac{d}{dx}(a_n x^n + a_{n-1}x^{n-1} + \cdots + a_1 x + a_0)$$

$$= \frac{d}{dx}(a_n x^n) + \frac{d}{dx}(a_{n-1}x^{n-1}) + \cdots + \frac{d}{dx}(a_1 x) + \frac{d}{dx}a_0 \quad \text{by the Sum Rule}$$

$$= a_n\frac{d}{dx}x^n + a_{n-1}\frac{d}{dx}x^{n-1} + \cdots + a_1\frac{d}{dx}x + \frac{d}{dx}a_0 \quad \text{by the Constant Multiple Rule}$$

$$= a_n n x^{n-1} + a_{n-1}(n-1)x^{n-2} + \cdots + a_1 \quad \text{by the Power Rule}$$

(In the last step we used $\frac{d}{dx}x = 1$ and $\frac{d}{dx}a_0 = 0$.) Notice that $f'(x)$ is a polynomial with degree one less than the degree of $f(x)$. It would be very unwise to try to memorize the preceding formula; it is much easier to learn the method.

EXAMPLE 7

Find $\frac{d}{dx}(12x^{26} - 9x^{14} + 5x^3 + 7)\Big|_{x=1}$.

Solution
First, we compute the derivative function.

$$\frac{d}{dx}(12x^{26} - 9x^{14} + 5x^3 + 7) = \frac{d}{dx}(12x^{26}) + \frac{d}{dx}(-9x^{14}) + \frac{d}{dx}(5x^3) + \frac{d}{dx}7$$

$$= 12\frac{d}{dx}x^{26} - 9\frac{d}{dx}x^{14} + 5\frac{d}{dx}x^3 + 0$$

$$= 12(26x^{25}) - 9(14x^{13}) + 5(3x^2)$$

$$= 312x^{25} - 126x^{13} + 15x^2$$

Now we evaluate the derivative function.

$$\frac{d}{dx}(12x^{26} - 9x^{14} + 5x^3 + 7)\bigg|_{x=1} = 312(1)^{25} - 126(1)^{13} + 15(1)^2$$

$$= 201 \ \blacksquare$$

Anyone selling a product or service must be concerned with how demand is altered when the price is increased. Assume that the number of units sold q is a function of price p. A logical way to understand the response to price changes is to study the percentage change in demand divided by the percentage change in price. Let Δq and Δp denote the changes in q and p, respectively.

$$\frac{\text{percentage change in } q}{\text{percentage change in } p} = \frac{100\left(\dfrac{\Delta q}{q}\right)}{100\left(\dfrac{\Delta p}{p}\right)}$$

$$= \frac{\dfrac{\Delta q}{q}\,pq}{\dfrac{\Delta p}{p}\,pq} = \frac{p\,\Delta q}{q\,\Delta p} = \frac{p}{q}\frac{\Delta q}{\Delta p}$$

Any change in p can be written as a percentage and multiplied times the above ratio to get the corresponding percentage change in q. If we let $h = \Delta p$, then $\Delta q = q(p + h) - q(p)$, and the above ratio becomes

$$\frac{p}{q}\frac{q(p + h) - q(p)}{h}.$$

Take the limit of the above quantity as h approaches 0; the result is

$$s = \frac{p}{q}\frac{dq}{dp},$$

which is known as the **elasticity of demand.** For a given p and the resulting q, the elasticity of demand is the percentage rate at which q changes divided by the percentage change in p.

EXAMPLE 8

If $q = \dfrac{7}{p}$ is the relationship of price and demand for a certain commodity, what is a formula for the elasticity of demand?

Solution

$$s = \frac{p}{q}\frac{dq}{dp} = \frac{p}{q}\frac{d}{dp}(7p^{-1}) = \frac{p}{q}\,7\,\frac{d}{dp}\,p^{-1}$$

$$= \frac{p}{q}\,7(-p^{-2}) = -\frac{7p}{qp^2} = -\frac{7}{pq}$$

We can simplify this further by substituting $\dfrac{7}{p}$ for q.

$$s = -\frac{7}{p \cdot \dfrac{7}{p}} = -\frac{7}{7} = -1$$

The percentage change in demand is just the negative of the percentage change in price. So the demand goes down as much as the price goes up, in percentages. That property has made the above demand equation famous. ■

Empirical studies indicate that the elasticity of demand for an item such as food or medical service is about $-.2$, while the elasticity of demand for an item such an automobile or a home is about -1.2. The demand for the less essential items is more responsive to changes in price.

Exercises 8.1

In problems 1 through 40, find a derivative formula for the indicated function.

1 $f(x) = 11x - 7$ 2 $g(s) = 5s^2 + 35s + 2$

3 $h(t) = t^2 - 6t$ 4 $f(p) = 13p$

5 $g(x) = 3x^3 - x^2 + x + 15$ 6 $h(s) = \sqrt[3]{s}$

7 $f(s) = s\sqrt[3]{s}$ 8 $g(t) = t^2 + \sqrt{t}$

9 $h(x) = \dfrac{6}{x^5}$ 10 $f(x) = \dfrac{10}{x^2} - x^3$

11 $g(r) = \dfrac{1}{r}$ 12 $h(w) = \dfrac{-8}{\sqrt[3]{w}}$

13 $f(x) = \dfrac{5x^2}{\sqrt{x}}$ 14 $g(s) = \dfrac{s}{4\sqrt[3]{s}}$

15 $h(x) = 13$ 16 $f(q) = \dfrac{1}{9}$

17 $g(s) = s^{102} + s^{83}$ 18 $h(x) = 12x^{77} - 7x^{25} + x^{13}$

19 $f(t) = t^{42} + t^{-13} + \sqrt{t}$ 20 $g(x) = x^{1/5} - x^5 + x^{-9}$

21 $h(r) = (r + 1)(r - 1)$ 22 $f(p) = (p + 1)^2$

23 $g(x) = x(x - 2)(x + 1)$ 24 $h(s) = s^2(2s - 3)$

25 $f(x) = \sqrt{x}(x + 1)(x^{-1} + 2)$ 26 $g(x) = \dfrac{x(2x + 4)}{3}$

27 $h(t) = \dfrac{(t + 3)(t - 3)}{2}$ 28 $f(w) = \sqrt{7}$

29 $g(p) = \sqrt[3]{13}$ 30 $h(x) = \sqrt{x^3}$

31 $f(x) = \sqrt[3]{x^2}$ 32 $g(s) = \dfrac{s^4 - s^3 + s^2}{s}$

33 $h(x) = \dfrac{5x^3 + 2x^2 - x}{x}$ 34 $f(x) = \sqrt{5x} + \sqrt{3}$

35 $g(x) = ex^2 + e^3$ 36 $h(r) = r^{57/2}$

37 $f(x) = x^{13/3}$ 38 $g(s) = s^{5/11} + 13$

39 $h(x) = x^{6/11} x^{5/11}$ 40 $g(t) = t^{13/7} t^{-6/7}$

In problems 41 through 58, find the slope of the tangent line to the graph of the given function at the given point.

41 x^{-3}, at $\left(-2, -\dfrac{1}{8}\right)$ 42 x^{-3}, at $(-1, -1)$

43 x^{-3}, at $\left(-\dfrac{1}{2}, -8\right)$ 44 x^{-3}, at $\left(2, \dfrac{1}{8}\right)$

45 x^{-3}, at $(1, 1)$ 46 x^{-3}, at $\left(\dfrac{1}{2}, 8\right)$

47 $\sqrt{x}$, at $\left(\dfrac{1}{4}, \dfrac{1}{2}\right)$ 48 $\sqrt{x}$, at $(1,1)$

49 $\sqrt{x}$, at $(9, 3)$ 50 $2x^4 - x^2$, at $(-1, 1)$

51 $2x^4 - x^2$, at $\left(-\dfrac{1}{2}, -\dfrac{1}{8}\right)$ 52 $2x^4 - x^2$, at $(0, 0)$

53 $2x^4 - x^2$, at $\left(\dfrac{1}{2}, -\dfrac{1}{8}\right)$ 54 $2x^4 - x^2$, at $(1, 1)$

55 $.1x^2 - x$, at $(0, 0)$ 56 $.1x^2 - x$, at $(2, -1.6)$

57 $.1x^2 - x$, at $(5, -2.5)$ 58 $.1x^2 - x$, at $(8, -1.6)$

In problems 59 through 64, find an equation for the tangent line to the graph of the given function at the given point.

59 $f(x) = x^2 - x + 1$, at $(1, 1)$ 60 $g(x) = 3x + 6$, at $(1, 9)$

61 $h(x) = \sqrt{x}$, at $(4, 2)$ 62 $f(x) = \dfrac{5}{x}$, at $(1, 5)$

63 $g(x) = x^3$, at $(0, 0)$ 64 $h(x) = 2x^4 - x^2$, at $(0, 0)$

In problems 65 through 70, find the elasticity of demand for the given demand function.

65 $q = -p + 2$ 66 $q = -2p + 6$

67 $q = -.5p + 5$ 68 $q = p^{-1}$

69 $q = p^{-2}$ 70 $q = p^{-3}$

8.2 Generalized Power Rule

Our success in the last section encourages us to pursue the strategy of viewing the formula for a given function as made from simpler components. Now we consider functions $f(x)$ that arise from evaluating a function $y = u(x)$ and then evaluating a function $g(y)$ that uses the values of $u(x)$ as its independent variable.* For example, if $f(x) = \sqrt{x^2 + 1}$, we can view $f(x)$ as being $f(x) = g(u(x))$ where $g(y) = \sqrt{y}$ and $y = u(x) = x^2 + 1$. That is,

$$f(x) = \sqrt{x^2 + 1} = \sqrt{u(x)} = g(u(x)).$$

From formulas that we have already learned, we know that

$$g'(y) = \frac{d}{dy}\sqrt{y} = \frac{d}{dy}y^{1/2} = \frac{1}{2}y^{-1/2}$$

and

$$u'(x) = 2x.$$

Can $f'(x)$ be written in terms of $g'(y)$ and $u'(x)$? For comparison, we first compute $f'(x)$ from the definition of the derivative.

$$f'(x) = \lim_{h \to 0} \frac{f(x+h) - f(x)}{h} = \lim_{h \to 0} \frac{\sqrt{u(x+h)} - \sqrt{u(x)}}{h}$$

$$= \lim_{h \to 0} \left(\frac{\sqrt{u(x+h)} - \sqrt{u(x)}}{h} \cdot \frac{\sqrt{u(x+h)} + \sqrt{u(x)}}{\sqrt{u(x+h)} + \sqrt{u(x)}} \right)$$

$$= \lim_{h \to 0} \left(\frac{u(x+h) - u(x)}{h} \cdot \frac{1}{\sqrt{u(x+h)} + \sqrt{u(x)}} \right)$$

$$= \lim_{h \to 0} \frac{u(x+h) - u(x)}{h} \cdot \lim_{h \to 0} \frac{1}{\sqrt{u(x+h)} + \sqrt{u(x)}}$$

$$= u'(x) \frac{1}{\lim\limits_{h \to 0} \sqrt{u(x+h)} + \lim\limits_{h \to 0} \sqrt{u(x)}}$$

$$= u'(x) \frac{1}{\sqrt{\lim\limits_{h \to 0} u(x+h)} + \sqrt{u(x)}}$$

The last step follows from part 4 of the Limit Theorem.

To finish off the computation, we note that $u(x) = x^2 + 1$ is a polynomial, so it is continuous for every choice of x. Consequently, we know that

$$\lim_{h \to 0} u(x+h) = u(x).$$

And, we have

---

* Of course, g must be defined for all values in the range of u if f is to be defined for all x in the domain of u.

$$f'(x) = u'(x) \frac{1}{\sqrt{\lim_{h \to 0} u(x+h)} + \sqrt{u(x)}}$$

$$= u'(x) \frac{1}{\sqrt{u(x)} + \sqrt{u(x)}}$$

$$= u'(x) \frac{1}{2\sqrt{u(x)}}$$

$$= u'(x) \cdot \frac{1}{2} \cdot \frac{1}{\sqrt{u(x)}}$$

$$= \frac{1}{2} u'(x) u(x)^{-1/2}$$

$$= \frac{1}{2} u(x)^{-1/2} u'(x).$$

Again we consider whether $f'(x)$ can be written in terms of $g'(y)$ and $u'(x)$. Yes! We can recognize $g'(u(x))$ in the last expression.

$$f'(x) = \left[\frac{1}{2} u(x)^{-1/2} \right] u'(x)$$

$$f'(x) = g'(u(x)) \, u'(x).$$

Clearly this last formula is a quicker, easier way to calculate $f'(x)$ than using the definition of derivative as we did above.

In general, if $g(y)$ and $u(x)$ are any fixed choices of functions and the value of $f(x)$ is obtained by the formula

$$f(x) = g(u(x)),$$

then we say that $f(x)$ is a **composite function** or, specifically, that $f(x)$ is the **composite** of $g(y)$ and $u(x)$. The steps that we used in computing $f'(x)$ where $f(x)$ is $\sqrt{x^2 + 1}$ show that

$$\frac{d}{dx} (u(x))^{1/2} = \frac{1}{2} u(x)^{-1/2} \, u'(x).$$

The next theorem is considerably more difficult to prove.

Generalized Power Rule   For any real number a,

$$\frac{d}{dx} u(x)^a = a \, u(x)^{a-1} \, u'(x).$$

If $u(x)$ is the linear function $u(x) = x$, then $u'(x) = 1$, and the above formula becomes

$$\frac{d}{dx} x^a = a x^{a-1}.$$

This is the Power Rule given in the last section. Thus, for a particular choice of $u(x)$ the Generalized Power Rule becomes the Power Rule; clearly the new rule is "more general," that is, it applies in more situations than does the old Power Rule. This is why the term "generalized" is used.

EXAMPLE 1

Find $\dfrac{d}{dx} \sqrt[3]{x^3 + 3x^2 - 5x + 2}$.

Solution
We rewrite the function to facilitate the use of the Generalized Power Rule. It is easy to recognize that $u(x) = x^3 + 3x^2 - 5x + 2$ and $a = \frac{1}{3}$.

$$\frac{d}{dx} \sqrt[3]{x^3 + 3x^2 - 5x + 2} = \frac{d}{dx}(x^3 + 3x^2 - 5x + 2)^{1/3}$$

$$= \frac{1}{3}(x^3 + 3x^2 - 5x + 2)^{-2/3}\frac{d}{dx}(x^3 + 3x^2 - 5x + 2)$$

$$= \frac{1}{3}(x^3 + 3x^2 - 5x + 2)^{-2/3}(3x^2 + 6x - 5) \quad\blacksquare$$

EXAMPLE 2

Find $\dfrac{d}{dx}(x^2 - 1)^{102}$.

Solution
We use the Generalized Power Rule with $u(x) = x^2 - 1$ and $a = 102$.

$$\frac{d}{dx}(x^2 - 1)^{102} = 102(x^2 - 1)^{101}\frac{d}{dx}(x^2 - 1)$$

$$= 102(x^2 - 1)^{101}(2x)$$
$$= 204x(x^2 - 1)^{101} \quad\blacksquare$$

A very old law of economics is the **Law of Diminishing Returns,** which we now state. At some point in the expenditure of resources (for example, money, labor, time) we reach a point where each additional unit of resources spent produces less results than the units already spent. For example, a student might be able to learn 90% of the material in one chapter of a text in 10 hours; for him to learn 95% might require 20 hours and to learn 99% might require 40 hours. Each additional hour of study after the first 10 hours is not as productive as each of the first 10 hours. After 20 hours, each additional hour of study is even less productive. The Law of Diminishing Returns is essential to understanding such contemporary issues as air pollution, worker safety, and the taking of a census.

EXAMPLE 3

Suppose a chemical plant has installed a filter on each of its smokestacks that removes 10% of the pollutant passing out of the smokestack; suppose p, the percentage of pollutant removed, is expressed as a function of the additional hundreds of dollars spent on filtering, say x, by the formula

$$p(x) = \sqrt{x + 100}.$$

What is the marginal change in $p(x)$ when 20% of the pollutant has been removed? What is it when 50% has been removed? What is it when 90% has been removed?

Solution

First we find $p'(x)$, which is the marginal change in $p(x)$ when x hundreds of dollars have been spent.

$$\frac{d}{dx} p(x) = \frac{d}{dx} (x + 100)^{1/2} = \frac{1}{2} (x + 100)^{-1/2} \frac{d}{dx} (x + 100)$$

$$= \frac{1}{2} (x + 100)^{-1/2}$$

Now we find x such that $p(x) = 20$, and we substitute in the above derivative formula.

$$\sqrt{x + 100} = 20$$

$$x + 100 = 400$$

$$x = 300$$

$$p'(300) = \frac{1}{2} (300 + 100)^{-1/2} = \frac{1}{2} \cdot \frac{1}{20}$$

$$= \frac{1}{40} = .025$$

When 20% of the pollutant is being removed, the marginal change in $p(x)$ is .025, so an additional $100 spent on filtering removes .025% of pollutant. It would take $10,000 more to remove another 2.5% of pollutant.

We repeat the above computations for $p(x) = 50$%.

$$\sqrt{x + 100} = 50$$

$$x = 2400$$

$$p'(2400) = \frac{1}{2} (2400 + 100)^{-1/2} = \frac{1}{2} \cdot \frac{1}{50} = .01$$

When 50% of the pollutant is being removed, it requires the additional expenditure of $10,000 to remove another 1% of the pollutant.

We repeat the computations for $p(x) = 90$%.

$$\sqrt{x + 100} = 90$$

$$x = 8000$$

$$p'(8,000) = \frac{1}{2} (8,000 + 100)^{-1/2} = \frac{1}{2} \cdot \frac{1}{90} = .0056$$

When 90% of the pollutant is being removed, it requires the expenditure of $10,000 more to remove another .56% of the pollutant. ∎

Exercises 8.2

In problems 1 through 15, find explicit formulas for the composite functions $f(g(y))$ and $g(f(x))$ using the given formulas for $f(x)$ and $g(y)$.

1 $f(x) = 3x^2 + x - 1, \quad g(y) = \sqrt{y}$

2 $f(x) = 3x^2 + x - 1, \quad g(y) = \sqrt{y+1}$

3 $f(x) = 5x + 2, \quad g(y) = -3y$

4 $f(x) = 5x + 2, \quad g(y) = -3y + 6$

5 $f(x) = 2x^3 - x, \quad g(y) = \dfrac{1}{y}$

6 $f(x) = 2x^3 - x, \quad g(y) = \dfrac{1}{y+1}$

7 $f(x) = 3x + 6, \quad g(y) = \dfrac{1}{3}y - 2$

8 $f(x) = \dfrac{1}{2}x + 1, \quad g(y) = 2y - 2$

9 $f(x) = x^2 - 1, \quad g(y) = y^{110}$

10 $f(x) = x^2 + 2x + 1, \quad g(y) = y^{90}$

11 $f(x) = 2x^4 + x^2 - 1, \quad g(y) = e^y$

12 $f(x) = x^3 - 8, \quad g(y) = e^y$

13 $f(x) = \ln x, \quad g(y) = y^2 + 6$

14 $f(x) = \ln x, \quad g(y) = 2y^4 + 1$

15 $f(x) = \ln x, \quad g(y) = e^y$

In problems 16 through 23, define functions $g(y)$ and $u(x)$ so that the given function $f(x)$ is $g(u(x))$.

16 $f(x) = \sqrt{4x^2 + 9}$

17 $f(x) = \sqrt[3]{-3x + 5}$

18 $f(x) = (x^3 - 2x + 7)^{53}$

19 $f(x) = (x^2 - 13x + 15)^{95}$

20 $f(x) = e^{2x+3}$

21 $f(x) = e^5(e^x)^5$

22 $f(x) = \ln(x^2 + x + 1)$

23 $f(x) = \ln\left(\dfrac{x^4 + 2}{x^2 + 1}\right)$

Find the derivative formula for each of the following functions.

24 $g(t) = (5t - 3)^{23}$

25 $h(s) = (11s^2 - 13)^{35}$

26 $f(u) = \dfrac{(9 - u^2)^{19}}{5}$

27 $g(w) = \dfrac{12}{(w^2 - w - 2)^{43}}$

28 $h(r) = \sqrt[3]{(r^2 + 1)^5}$

29 $f(x) = \sqrt[5]{(x^3 - x)^3}$

30 $g(x) = \dfrac{1}{\sqrt{10 + x^2}}$

31 $h(p) = 5p + (p^2 - 3)^{29}$

32 $f(s) = (s + 2)^{15}(s - 2)^{15}$

33 $g(x) = \dfrac{(s^2 - 9)^{23}}{(s - 3)^{23}}$

34 $h(x) = ((x^4 + x^2)^{49} + 1)^{61}$

35 $f(s) = (\sqrt[3]{9s + 3} + 5)^{30}$

Solve the following.

36   Suppose the noise level in a factory is reduced p percent by an expenditure of x hundred dollars in addition to what has been spent, and $p = \sqrt{x + 400}$. What is the marginal change in $p(x)$ when the noise level is reduced 30%? What is it when the noise level is reduced 60%?

37   Suppose the dust level in a factory is reduced p percent by an expenditure of x hundred dollars in addition to what has been spent, and $p = \sqrt[3]{x + 1000}$. What is the marginal change in $p(x)$ when the dust level is reduced 20%? What is it when the dust level is reduced 50%?

8.3   Product and Quotient Rules

Another way that functions are constructed from simpler functions is by taking products and quotients of simpler functions. For example, if $h(x) = x(x + 1)^{25}$, we can choose $g(x) = (x + 1)^{25}$ and $f(x) = x$. Then we can see that

$$h(x) = x(x + 1)^{25} = f(x)\, g(x).$$

From the formulas that we have already learned, we know that

$$g'(x) = \frac{d}{dx}(x + 1)^{25} = 25(x + 1)^{24}$$

and

$$f'(x) = 1.$$

Can $h'(x)$ be written in terms of $f'(x)$ and $g'(x)$? It can, by using the following rule.

Product Rule

$$\frac{d}{dx}(f(x)\, g(x)) = f'(x)\, g(x) + f(x)\, g'(x).$$

Thus,

$$h'(x) = 1 \cdot (x + 1)^{25} + x \cdot 25(x + 1)^{24}$$
$$= (x + 1)^{25} + 25x(x + 1)^{24}.$$

In order to verify the Product Rule, we compute the derivative of the product $h(x) = f(x)\, g(x)$ from the definition of the derivative. We add and subtract $f(a)\, g(a + h)$ in the numerator, and then we simplify.

$$\frac{d}{dx}(f(x)\, g(x))\bigg|_{x=a} = \lim_{h \to 0} \frac{f(a + h)\, g(a + h) - f(a)\, g(a)}{h}$$

$$= \lim_{h \to 0} \frac{f(a + h)\, g(a + h) - f(a)\, g(a + h) + f(a)\, g(a + h) - f(a)\, g(a)}{h}$$

$$= \lim_{h \to 0} \left[\frac{f(a + h)\, g(a + h) - f(a)\, g(a + h)}{h} + f(a)\, \frac{g(a + h) - g(a)}{h} \right]$$

$$= \lim_{h \to 0} \left[\frac{f(a+h) - f(a)}{h} g(a+h) \right] + \lim_{h \to 0} \left[f(a) \frac{g(a+h) - g(a)}{h} \right]$$

$$= \lim_{h \to 0} \frac{f(a+h) - f(a)}{h} \lim_{h \to 0} g(a+h) + \lim_{h \to 0} f(a) \lim_{h \to 0} \frac{g(a+h) - g(a)}{h}$$

$$= f'(a) \lim_{h \to 0} g(a+h) + f(a) g'(a)$$

We have used the Limit Theorem and the definitions of $f'(a)$ and $g'(a)$.

We are close to verifying the Product Rule. The remaining step is to show that

$$\lim_{h \to 0} g(a+h) = g(a).$$

This is precisely the definition of "$g(x)$ is continuous at a." Of course, we are assuming that the limit

$$\lim_{h \to 0} \frac{g(a+h) - g(a)}{h}$$

exists, since we have written $g'(a)$ in the formula for $\frac{d}{dx}(f(x) g(x))\Big|_{x=a}$.

Thus, the remaining step is equivalent to proving that if a function is differentiable at $x = a$, then it is continuous at $x = a$. This was proved in Section 7.6. Thus, we have completed the verification of the Product Rule.

EXAMPLE 1

Find $\dfrac{d}{dx} x^3 \sqrt{x^2 + 1}.$

Solution

We define $f(x)$ and $g(x)$ by the formulas

$$f(x) = x^3, \qquad g(x) = \sqrt{x^2 + 1} = (x^2 + 1)^{1/2}$$

and we note that $f'(x)$ and $g'(x)$ are given by the formulas

$$f'(x) = 3x^2, \qquad g'(x) = \frac{1}{2}(x^2 + 1)^{-1/2}(2x) = \frac{x}{\sqrt{x^2 + 1}}.$$

We used the Generalized Power Rule to find $g'(x)$.

Now we find the indicated derivative by the Product Rule.

$$\frac{d}{dx} x^3 \sqrt{x^2 + 1} = \frac{d}{dx} f(x) g(x)$$

$$= f'(x) g(x) + f(x) g'(x)$$

$$= 3x^2 (x^2 + 1)^{1/2} + x^3 \frac{x}{\sqrt{x^2 + 1}}$$

$$= 3x^2 \sqrt{x^2 + 1} + \frac{x^4}{\sqrt{x^2 + 1}} \quad \blacksquare$$

It is not usually necessary to write all of the steps in such detail, as we illustrate next.

EXAMPLE 2

Find $\dfrac{d}{dx}(3x+2)^{15}(5x-1)^{18}$.

Solution

$$\frac{d}{dx}(3x+2)^{15}(5x-1)^{18} = \left[\frac{d}{dx}(3x+2)^{15}\right](5x-1)^{18} + (3x+2)^{15}\frac{d}{dx}(5x-1)^{18}$$

$$= [15(3x+2)^{14}(3)](5x-1)^{18} + (3x+2)^{15}18(5x-1)^{17}(5)$$
$$= 45(3x+2)^{14}(5x-1)^{18} + 90(3x+2)^{15}(5x-1)^{17}$$

It is interesting to note that the last derivative formula can be simplified by the following steps.

$$45(3x+2)^{14}(5x-1)^{18} + 90(3x+2)^{15}(5x-1)^{17} = 45(3x+2)^{14}(5x-1)^{17}[(5x-1)+2(3x+2)]$$
$$= 45(3x+2)^{14}(5x-1)^{17}[5x-1+6x+4]$$
$$= 45(3x+2)^{14}(5x-1)^{17}(11x+3)$$

This final formula is genuinely simpler than the first formula; given specific choices of x, values of the derivative can be computed more quickly and easily using the final formula. ■

Caution   Note that the derivative of the product of two functions is **not** the product of the derivatives of those functions. For example, the derivative of $x^2 = x \cdot x$ is $2x$, not $1 \cdot 1 = 1$.

We can use the Product Rule as many times as we need it. If there are n factors in the original product, we use the Product Rule $n - 1$ times, as the next example illustrates.

EXAMPLE 3

Find $\dfrac{d}{dx}((x+1)^{20}(x+3)^{21}(x-2)^{18})$.

Solution

$$\frac{d}{dx}((x+1)^{20}(x+3)^{21}(x-2)^{18})$$

$$= \frac{d}{dx}((x+1)^{20}[(x+3)^{21}(x-2)^{18}])$$

$$= \left[\frac{d}{dx}(x+1)^{20}\right](x+3)^{21}(x-2)^{18} + (x+1)^{20}\frac{d}{dx}[(x+3)^{21}(x-2)^{18}]$$

$$= \left[\frac{d}{dx}(x+1)^{20}\right](x+3)^{21}(x-2)^{18} + (x+1)^{20}\left[(x-2)^{18}\frac{d}{dx}(x+3)^{21} + (x+3)^{21}\frac{d}{dx}(x-2)^{18}\right]$$

$$= 20(x+1)^{19}(x+3)^{21}(x-2)^{18} + (x+1)^{20}[(x-2)^{18}21(x+3)^{20} + (x+3)^{21}18(x-2)^{17}]$$
$$= 20(x+1)^{19}(x+3)^{21}(x-2)^{18} + 21(x+1)^{20}(x+3)^{20}(x-2)^{18} + 18(x+1)^{20}(x+3)^{21}(x-2)^{17}$$

This answer can be simplified by the following steps.

$$20(x+1)^{19}(x+3)^{21}(x-2)^{18} + 21(x+1)^{20}(x+3)^{20}(x-2)^{18} + 18(x+1)^{20}(x+3)^{21}(x-2)^{17}$$
$$= (x+1)^{19}(x+3)^{20}(x-2)^{17}[20(x+3)(x-2) + 21(x+1)(x-2) + 18(x+1)(x+3)]$$
$$= (x+1)^{19}(x+3)^{20}(x-2)^{17}[20(x^2+x-6) + 21(x^2-x-2) + 18(x^2+4x+3)]$$
$$= (x+1)^{19}(x+3)^{20}(x-2)^{17}[59x^2 + 71x - 108] \ \blacksquare$$

Functions can also be constructed by taking the quotient of simpler functions, and we have a formula for computing the derivative of such a quotient.

Quotient Rule

$$\frac{d}{dx}\frac{f(x)}{g(x)} = \frac{f'(x)\,g(x) - g'(x)\,f(x)}{(g(x))^2}.$$

We can verify the Quotient Rule using rules that we already know.

$$\frac{d}{dx}\frac{f(x)}{g(x)} = \frac{d}{dx}f(x)\,(g(x))^{-1}$$

$$= f'(x)\,(g(x))^{-1} + f(x)\frac{d}{dx}(g(x))^{-1}$$

$$= f'(x)\,(g(x))^{-1} + f(x)\,(-1)\,(g(x))^{-2}\,g'(x)$$

$$= \frac{f'(x)\,g(x)}{g(x)^2} - \frac{f(x)\,g'(x)}{g(x)^2}$$

$$= \frac{f'(x)\,g(x) - f(x)\,g'(x)}{(g(x))^2}$$

In the preceding equation we used the Product Rule and the Generalized Power Rule.

EXAMPLE 4

Find $\dfrac{d}{dx}\dfrac{x+2}{3x-1}$.

Solution

$$\frac{d}{dx}\frac{x+2}{3x-1} = \frac{\left[\dfrac{d}{dx}(x+2)\right](3x-1) - (x+2)\dfrac{d}{dx}(3x-1)}{(3x-1)^2}$$

$$= \frac{1\cdot(3x-1) - (x+2)\cdot 3}{(3x-1)^2}$$

$$= \frac{3x-1-3x-6}{(3x-1)^2}$$

$$= \frac{-7}{(3x-1)^2} \ \blacksquare$$

EXAMPLE 5

Find $\dfrac{d}{dx}\dfrac{5x}{\sqrt{x^2+1}}$.

Solution

$$\frac{d}{dx}\frac{5x}{\sqrt{x^2+1}} = \frac{\sqrt{x^2+1}\frac{d}{dx}5x - 5x\frac{d}{dx}\sqrt{x^2+1}}{x^2+1}$$

$$= \frac{5\sqrt{x^2+1} - 5x\frac{d}{dx}(x^2+1)^{1/2}}{x^2+1}$$

$$= \frac{5\sqrt{x^2+1} - 5x\left[\frac{1}{2}(x^2+1)^{-1/2}(2x)\right]}{x^2+1}$$

$$= \frac{5\sqrt{x^2+1} - 5x^2(x^2+1)^{-1/2}}{x^2+1}$$

This answer is an improper fraction because the last factor in the numerator has a negative exponent. To simplify, we multiply the numerator and the denominator by $(x^2+1)^{1/2}$.

$$\frac{d}{dx}\frac{5x}{\sqrt{x^2+1}} = \frac{5\sqrt{x^2+1} - 5x^2(x^2+1)^{-1/2}}{x^2+1} \cdot \frac{(x^2+1)^{1/2}}{(x^2+1)^{1/2}}$$

$$= \frac{5(x^2+1) - 5x^2}{(x^2+1)(x^2+1)^{1/2}}$$

$$= \frac{5}{(x^2+1)^{3/2}}$$

$$= 5(x^2+1)^{-3/2} \ \blacksquare$$

Caution Note that the derivative of the quotient of two functions is **not** the quotient of the derivatives of the two functions. For example, the derivative of $x^2 = \frac{x^3}{x}$ is $2x$, not $\frac{3x^2}{1} = 3x^2$.

Sometimes we need the Product Rule and the Quotient Rule to do one problem.

EXAMPLE 6

Find $\dfrac{d}{dx}\dfrac{(x+3)^9(2x-5)^5}{(-x+2)^{12}}$.

Solution

$$\frac{d}{dx}\frac{(x+3)^9(2x-5)^5}{(-x+2)^{12}}$$

$$= \frac{(-x+2)^{12}\frac{d}{dx}[(x+3)^9(2x-5)^5] - (x+3)^9(2x-5)^5\frac{d}{dx}(-x+2)^{12}}{(-x+2)^{24}}$$

$$= \frac{(-x+2)^{12}[9(x+3)^8(2x-5)^5 + (x+3)^9 5(2x-5)^4(2)]}{(-x+2)^{24}} \frac{-(x+3)^9(2x-5)^5 12(-x+2)^{11}(-1)}{(-x+2)^{24}}$$

$$= \frac{9(x+3)^8(2x-5)^5(-x+2)^{12} + 10(x+3)^9(2x-5)^4(-x+2)^{12} + 12(x+3)^9(2x-5)^5(-x+2)^{11}}{(-x+2)^{24}}$$

This answer could be simplified by the same process that we used to simplify the answer in Example 3. Since we are not going to use the derivative formula any further, we do not bother to simplify it. ■

For any manufacturing company, the average cost of each unit made is important. Because the fixed costs are incurred before any units are produced, the company wants to make a lot of units and, thus, spread the fixed costs over many units. For many companies the variable cost begins to go up quickly at some point as the number of units produced increases: workers collect overtime pay, additional workers must be trained, and machines break down more often. Sometimes the increased demand for the raw materials involved forces prices up. Consequently, average cost per unit declines until the variable cost begins to increase so dramatically that the average cost per unit begins to increase. We can find the point where this occurs.

EXAMPLE 7

Suppose the cost of a manufacturing company is $C(x) = .1x^2 + 25$ in thousands of dollars where x is the hundreds of units made. Find formulas for the average cost $A(x)$ and the marginal average cost. Find the production at which the marginal average cost is zero.

Solution
Since $C(x)$ is the total cost of making x hundreds of units, the **average cost** for each hundred is

$$A(x) = \frac{C(x)}{x} = \frac{.1x^2 + 25}{x} = .1x + 25x^{-1}.$$

The **marginal average cost** is

$$A'(x) = .1 - 25x^{-2}.$$

We solve $A'(x) = 0$ for x.

$$.1 - 25x^{-2} = 0$$
$$.1 = 25x^{-2}$$
$$x^2 = 250$$
$$x = 15.8 \quad ■$$

A quick way to find where marginal average cost is zero is to find where marginal cost equals average cost, as the following computations show.

$$A'(x) = \frac{d}{dx}\frac{C(x)}{x} = \frac{C'(x)x - C(x)}{x^2}$$

$$0 = A'(x) = \frac{C'(x)x - C(x)}{x^2}$$

$$C(x) = C'(x)x$$

$$A(x) = \frac{C(x)}{x} = C'(x)$$

Exercises 8.3

Find the derivative formula for each of the following functions.

1 $f(x) = (4x + 2)(3x - 7)$
2 $g(x) = (x + 5)(-2x - 9)$
3 $h(x) = x(x^2 + x + 2)$
4 $f(s) = (s^2 + 1)(11s^2 + 5)$
5 $g(s) = (s + 9)(s^3 - 5s^2 + 2)$
6 $h(s) = (6 + s^2)(-12s^3 + 15s)$
7 $f(t) = (2t + 1)(t - 3)(5t + 2)$
8 $g(t) = (t - 1)(t + 2)(-7t + 4)$
9 $h(t) = (2t + 1)^{30}(-3t + 5)^{20}$
10 $f(r) = (r^2 + 13)^{18}(6r - 2)^{22}$
11 $g(r) = (3r - 1)^{15}(r + 1)^{23}(-4r + 2)^{30}$
12 $h(r) = (r + 3)^{26}(2r + 1)^{12}(r - 9)^{20}$
13 $f(p) = 5p(p^2 + 10) - (3p + 2)(12 - p)^8$
14 $g(p) = p^9 + (p + 1)^6(p - 2)^{10}$
15 $h(p) = 9p^3 - 6p^2 + 3p + \sqrt{7}$
16 $f(w) = 10w^2 - 2w + \pi$

17 $g(w) = \dfrac{3w - 5}{w + 7}$ 18 $h(w) = \dfrac{11}{5w + 6}$

19 $f(x) = \dfrac{-2 + 3x}{x^2 + x + 1}$ 20 $g(x) = \dfrac{5x + 2}{x^3 - 9x}$

21 $h(x) = \dfrac{8x^2 - 3}{x + 1}$ 22 $f(z) = \dfrac{4z^3 - z^2 + 11}{z}$

23 $g(z) = \dfrac{-8z^2 - 3z + 2}{5z^2 + 9}$ 24 $h(z) = \dfrac{12z^3 - z + 7}{z^3 + z^2 + 3}$

25 $f(x) = \dfrac{1}{3x^2 + 2x + 1}$ 26 $g(x) = \dfrac{2}{x^2 - 11x + 5}$

27 $h(x) = \dfrac{1}{x + 5} - (x + 1)^3(x - 2)^4$

28 $f(s) = \dfrac{3}{(s-2)^9} + (s^2 - 9)(s^2 + 7s - 3)$

29 $g(s) = \dfrac{4s + 5}{(s + 1)(s - 2)}$

30 $h(s) = \dfrac{s + 2}{(3s - 1)(2s + 4)}$

31 $f(x) = \dfrac{5x^2}{(x^2 + 1)(3x - 2)}$

32 $g(x) = \dfrac{3x - 1}{(-x + 5)(x^2 + 3)}$

33 $h(x) = \dfrac{(2x + 1)(x + 5)}{(x - 3)(4x + 1)}$

34 $f(r) = \dfrac{(r - 9)(3r + 5)}{(2r - 1)(2r + 1)}$

35 $g(r) = \dfrac{r}{r^2 + 1}$

36 $h(r) = \dfrac{3r + 2}{r}$

37 $f(x) = x\sqrt{x^2 + 1}$

38 $g(x) = (3x + 2)\sqrt{x + 5}$

39 $h(x) = \dfrac{\sqrt[3]{x + 8}}{2x + 3}$

40 $f(x) = \dfrac{\sqrt[3]{x}}{x^2 + 9x}$

41 $g(x) = \dfrac{1}{\sqrt{x}} + (x - 3)(5x^2 + 3)$

42 $h(x) = \dfrac{1}{\sqrt{x + 4}} - (x + 9)^6(2x + 1)$

43 $f(p) = p^4 - p^2 + 2$

44 $g(p) = 7p^3 + e$

45 $h(p) = \sqrt{\dfrac{p + 2}{3p + 4}}$

46 $f(x) = \sqrt{\dfrac{5x + 1}{x + 2}}$

47 $g(x) = \sqrt{(2x - 1)(3x + 4)}$

48 $h(x) = \sqrt{(x + 3)(x^2 + x + 1)}$

49 $f(v) = \sqrt{3v + \dfrac{5v}{v + 2}}$

50 $g(v) = \sqrt{v(v + 2)^{15} + 9}$

51 $h(v) = \sqrt{(v + 1)^{10}(2v - 1)^{12}}$

52 $f(x) = e^{7.3}$

53 $g(x) = \sqrt{\dfrac{(2x - 3)(x + 7)}{(-4x + 5)}}$

54 $h(x) = \sqrt{\dfrac{15x + 2}{(x - 2)(5x + 1)}}$

55 $f(x) = \dfrac{\sqrt{2x^2 + 4}}{\sqrt[3]{x^3 + 1}}$

56 $g(x) = \dfrac{\sqrt[3]{8x^3 + 27}}{\sqrt{x^2 + x + 1}}$

57 $h(x) = (x + 1)\sqrt[4]{x} + 13$

58 $f(x) = \sqrt{x}\sqrt[3]{x}\sqrt[4]{x} + 1017$

59 $g(x) = \dfrac{x^{2/3}x^{5/2}}{x^{3/7}} - 51$

60 $h(x) = 7^{16}(x - 345) + 12$

In problems 61 through 63, using the given cost function $C(x)$, find average cost $A(x)$, marginal average cost, and the production at which the marginal average cost is zero.

61 $C(x) = .1x^2 + 40$

62 $C(x) = .01x^2 + 16$

63 $C(x) = .01x^2 + 36$

8.4   Exponential Functions

Usually the function defined by the derivative formula looks quite different from the original function. We showed in Section 8.1 that the derivative function for any polynomial is again a polynomial with degree one less than the degree of the original polynomial. Thus, it is remarkable that there is a function with the property that its derivative function is the same as the original function. That function is the exponential function e^x introduced in Section 2.4. In order to obtain the derivative formula for e^x, we shall need to know the following limit, which is too difficult for us to verify.

$$\lim_{h \to 0} \frac{e^h - 1}{h} = 1$$

Now we compute the derivative of e^x at some fixed number $x = a$.

$$\frac{d}{dx} e^x \bigg|_{x=a} = \lim_{h \to 0} \frac{e^{a+h} - e^a}{h}$$

$$= \lim_{h \to 0} \frac{e^a e^h - e^a}{h} \qquad \text{by Rules of Exponents}$$

$$= \lim_{h \to 0} e^a \frac{e^h - 1}{h}$$

$$= \lim_{h \to 0} e^a \lim_{h \to 0} \frac{e^h - 1}{h} \qquad \text{by the Limit Theorem}$$

$$= e^a \cdot 1 = e^a$$

Since we assumed nothing about a except that it is a fixed number, we have proved the following rule.

$$\frac{d}{dx} e^x = e^x$$

EXAMPLE 1

Find $\dfrac{d}{dx} xe^x$.

Solution
We use the Product Rule and the formula for the derivative of e^x.

$$\frac{d}{dx} xe^x = \left(\frac{d}{dx} x\right) e^x + x \frac{d}{dx} e^x$$

$$= e^x + xe^x \quad \blacksquare$$

EXAMPLE 2

Find $\dfrac{d}{dx} e^{-2x}$.

Solution
We use the Rules of Exponents to rewrite e^{-2x}.

$$\frac{d}{dx}\,e^{-2x} = \frac{d}{dx}\,(e^x)^{-2}$$

$$= -2(e^x)^{-3}\,\frac{d}{dx}\,e^x \qquad \text{by Generalized Power Rule}$$

$$= -2e^{-3x}e^x$$
$$= -2e^{-3x+x} \qquad \text{by Rules of Exponents}$$
$$= -2e^{-2x} \quad \blacksquare$$

EXAMPLE 3

Find $\dfrac{d}{dx}\,e^{ax}$ where a is some fixed number.

Solution

$$\frac{d}{dx}\,e^{ax} = \frac{d}{dx}\,(e^x)^a$$

$$= a(e^x)^{a-1}\,\frac{d}{dx}\,e^x \qquad \text{by Generalized Power Rule}$$

$$= ae^{(a-1)x}e^x$$
$$= ae^{(a-1)x+x}$$
$$= ae^{ax} \quad \blacksquare$$

EXAMPLE 4

Find $\dfrac{d}{dx}\,e^{ax+5}$ where a is some fixed number.

Solution

$$\frac{d}{dx}\,e^{ax+5} = \frac{d}{dx}\,e^{ax}e^5 \qquad \text{by Rules of Exponents}$$

$$= e^5\,\frac{d}{dx}\,e^{ax}$$

$$= e^5\,ae^{ax} \qquad \text{by Example 3}$$

$$= ae^{ax+5} \quad \blacksquare$$

EXAMPLE 5

Find $\dfrac{d}{dx}\,e^{ax+b}$ where a and b are fixed numbers.

Solution

$$\frac{d}{dx}\,e^{ax+b} = \frac{d}{dx}\,e^{ax}e^b \qquad \text{by Rules of Exponents}$$

$$= e^b\,\frac{d}{dx}\,e^{ax}$$

$$= e^b\,ae^{ax} \qquad \text{by Example 3}$$
$$= ae^{ax+b} \quad \blacksquare$$

The formula

$$\frac{d}{dx} e^{ax+b} = ae^{ax+b},$$

obtained in Example 5, has an important application. Some banks and savings institutions advertise that they "compound your savings account continuously." If $A(t)$ is the amount of money in your account at time t, then we know that the rate of change of that amount at any particular instant t is $A'(t)$. If the account is compounded continuously at rate r, then the rate of change of $A(t)$ at any instant t must be $rA(t)$. Thus, it must be that

$$A'(t) = rA(t).$$

If we take $A(t) = e^{rt+b}$ then we have

$$A'(t) = re^{rt+b} = rA(t)$$

which is the **continuous compounding** property. Also note that

$$A(0) = e^{r \cdot 0 + b} = e^b$$

and

$$A(t) = e^{rt+b} = e^{rt}e^b = e^{rt}A(0).$$

Of course, $A(0)$ is the amount of the account when it is opened. Thus if a sum of money p is compounded continuously at a rate r then we can regard the **compound amount** at time t as given by the formula

$$A = pe^{rt}.$$

CALCULATOR EXAMPLE

If $100 is put in an account earning interest at a rate of 5.5% compounded continuously, how much is in the account at the end of 1 year?

Solution
In the formula above p is $100, r is .055, and t is 1. Using a calculator, we evaluate the relevant expression to two decimal place accuracy.

$$A = 100e^{.055}$$
$$= 100(1.0565) = 105.65 \ \blacksquare$$

It is interesting to note that if the account in the Calculator Example earned simple interest at the rate of 5.5%, then the compound amount would be $105.50 at the end of 1 year. Thus, continuous compounding only added $.15 to the account.

Exercises 8.4

Find the derivative formula for each of the following functions.

1   $f(x) = x^2 e^x$ 2 $g(x) = 4x^3 e^{-x}$

3   $h(x) = \dfrac{3e^x}{5x + 2}$

4   $f(x) = \dfrac{4x^2 + 2}{7e^x}$

5   $g(x) = \dfrac{1}{2} e^x + \dfrac{1}{2} e^{-x}$

6   $h(x) = \dfrac{1}{2} e^x - \dfrac{1}{2} e^{-x}$

7   $f(x) = e^{-2x+1}$

8   $g(x) = 9e^{7x-2}$

9   $h(x) = 14x^3 - xe^{2x} + 9e^{3x}$

10   $f(x) = 8e^{-x} + 11x - 5$

11   $g(x) = (e^x + 3)^{52}$

12   $h(x) = (5 - e^{2x})^{26}$

13   $f(x) = (e^x + e^{-x})^2$

14   $g(x) = (e^x - e^{-x})^2$

15   $h(x) = \sqrt{e^{2x} + 10}$

16   $f(x) = \dfrac{6}{\sqrt{e^x + 2}}$

17   $g(x) = \dfrac{x^2 - x + 1}{\sqrt{2e^x + 1}}$

18   $h(x) = \dfrac{\sqrt{e^x + x + 2}}{5x - 3}$

19   $f(x) = 110e^{6x+92}$

20   $g(x) = \dfrac{e^{3x-9}}{2e^{5x} + 1}$

Solve each of the following. Use the Table of Exponential Values in the back of this book or a calculator.

21   If an account with $500 earns interest at a rate of 10% compounded continuously, how much is in the account after 2 years?

22   If an account earns interest at a rate of 10% compounded continuously, how long does it take for the account to double?

23   An account containing $100 earns interest at the rate of 15% compounded annually for 10 years. How much is the account worth and how much more would it be worth if it earned interest at the rate of 15% with continuous compounding?

8.5   Chain Rule

In Section 8.2 we used the Generalized Power Rule to differentiate certain composite functions. Just as the Generalized Power Rule is applicable to more situations than the Power Rule, the Chain Rule is applicable to more situations than the Generalized Power Rule. So, the Chain Rule is more general than the Generalized Power Rule, which is more general than the Power Rule. The Chain Rule is a very powerful rule for computing the derivative of a composite function.

Chain Rule   If $h(x)$ is the composite of $f(y)$ and $g(x)$, so that $h(x) = f(g(x))$, then the derivative $h'(x)$ is given by the formula

$$h'(x) = f'(g(x)) \cdot g'(x)$$

Here $f(y)$ is differentiated and the derivative $f'(y)$ is evaluated at $g(x)$; that quantity is multiplied by $g'(x)$.

EXAMPLE 1

Find a formula for $h'(x)$ when $h(x) = f(g(x))$ and $f(y) = y^a$.

Solution
Note that

$$f'(y) = ay^{a-1}$$

by the Power Rule and, by simple substitution of $g(x)$ for y,

$$f'(g(x)) = a(g(x))^{a-1}.$$

Now we use the Chain Rule.

$$h'(x) = f'(g(x)) \cdot g'(x) = a(g(x))^{a-1}g'(x)$$

This is simply the Generalized Power Rule. ∎

Although the Generalized Power Rule is restricted to situations in which $f(y) = y^a$, the Chain Rule applies to $f(y)$ of any form.
For instance, one of the fundamental equations of Section 2.5 is

$$e^{\ln x} = x.$$

We can now compute the derivative of the composite function on the left side of the above equation. Using the same symbols as those used for the functions in the Chain Rule, we have $f(y) = e^y$ and $g(x) = \ln x$; thus, we get $f'(y) = e^y$ and $f'(g(x)) = e^{\ln x} = x$. We write out the Chain Rule, using $g'(x) = \dfrac{d}{dx} \ln x$ (for which we do not yet know a formula).

$$\frac{d}{dx} e^{\ln x} = x \frac{d}{dx} \ln x$$

Since the function $e^{\ln x}$ equals the function x, the derivative of $e^{\ln x}$ equals the derivative of x; that is,

$$\frac{d}{dx} e^{\ln x} = \frac{d}{dx} x = 1.$$

Substituting this result into the Chain Rule expression, we find that

$$x \frac{d}{dx} \ln x = 1.$$

Dividing both sides by x, we have

$$\boxed{\frac{d}{dx} \ln x = \frac{1}{x} = x^{-1}.}$$

We have obtained a formula for the derivative of $\ln x$.

EXAMPLE 2

Find $\dfrac{d}{dx} \ln(x^2 + 1)$.

Solution

We use the Chain Rule and the derivative of $\ln x$ evaluated at $x^2 + 1$.

$$\frac{d}{dx} \ln(x^2 + 1) = \frac{d}{dx} \ln(y)\Big|_{y=(x^2+1)} \frac{d}{dx}(x^2 + 1)$$

$$= \frac{1}{x^2 + 1} \frac{d}{dx}(x^2 + 1)$$

$$= \frac{1}{x^2 + 1} 2x = \frac{2x}{x^2 + 1} \quad\blacksquare$$

Using the Chain Rule and the derivative of $\ln x$, we next obtain a general formula for the derivative of the composite function $\ln u(x)$. Note that $\dfrac{d}{dy} \ln y = \dfrac{1}{y}$ evaluated at $y = u(x)$ is $\dfrac{1}{u(x)}$. Thus, we have

$$\frac{d}{dx} \ln u(x) = \frac{1}{u(x)} \frac{d}{dx} u(x) = \frac{u'(x)}{u(x)}.$$

It saves a lot of time to memorize the above formula rather than deriving it from the Chain Rule each time it is needed.

EXAMPLE 3

Find $\dfrac{d}{dx} \ln \left(\dfrac{3x + 2}{x - 5} \right)$.

Solution

We use the properties of the logarithm function to simplify the function before differentiating.

$$\frac{d}{dx} \ln \left(\frac{3x + 2}{x - 5} \right) = \frac{d}{dx} [\ln(3x + 2) - \ln(x - 5)]$$

$$= \frac{d}{dx} \ln(3x + 2) - \frac{d}{dx} \ln(x - 5)$$

$$= \frac{3}{3x + 2} - \frac{1}{x - 5} \quad\blacksquare$$

We can now verify the Power Rule for $x > 0$, without assuming that the exponent is a rational number. By the elementary rules of logarithms, we have

$$\ln x^a = a \ln x.$$

We compute the derivative of the left side using the last rule that we have obtained, and we set it equal to the derivative of the right side.

$$\frac{d}{dx} (\ln x^a) = \frac{d}{dx} (a \ln x)$$

$$\frac{\frac{d}{dx} x^a}{x^a} = a \frac{1}{x}$$

Now we solve the last equation for the unknown derivative $\dfrac{d}{dx}\, x^a$.

$$\frac{d}{dx}\, x^a = x^a\, a\, \frac{1}{x} = a\, x^{a-1}$$

A similar verification is possible for $x < 0$.

Using the Chain Rule and the derivative of e^x, we obtain a general formula for the derivative of the composite function $e^{u(x)}$. Note that $\dfrac{d}{dy}\, e^y = e^y$ evaluated at $u(x)$ is $e^{u(x)}$. Thus, we have

$$\frac{d}{dx}\, e^{u(x)} = e^{u(x)}\, \frac{d}{dx}\, u(x) = e^{u(x)} u'(x).$$

This formula should also be memorized rather than deriving it from the Chain Rule each time it is needed.

EXAMPLE 4

Find $\dfrac{d}{dx}\, e^{x^3 - x + 7}$.

Solution

$$\frac{d}{dx}\, e^{x^3 - x + 7} = e^{x^3 - x + 7}\, \frac{d}{dx}\, (x^3 - x + 7)$$
$$= e^{x^3 - x + 7}(3x^2 - 1)\ \blacksquare$$

EXAMPLE 5

Find $\dfrac{d}{dx}\, e^{ax + b}$ where a and b are fixed numbers.

Solution

This is the same problem as Example 5 of Section 8.4, but the Chain Rule makes it easier.

$$\frac{d}{dx}\, e^{ax + b} = e^{ax + b}\, \frac{d}{dx}\, (ax + b)$$
$$= e^{ax + b} a\ \blacksquare$$

In Section 2.5 we obtained the formula

$$a^x = (e^{\ln a})^x = e^{x \ln a}$$

where a is a fixed positive number. With this formula and the Chain Rule, we can compute the derivative of a^x.

$$\frac{d}{dx}\, a^x = \frac{d}{dx}\, e^{x \ln a} = e^{x \ln a}\, \frac{d}{dx}\, x \ln a$$
$$= e^{x \ln a}\ln a = (e^{\ln a})^x \ln a = a^x \ln a$$

Thus, we have the formula

$$\frac{d}{dx}\,a^x = a^x\ln a.$$

EXAMPLE 6

Find $\dfrac{d}{dx}\,2^x$.

Solution

$$\frac{d}{dx}\,2^x = 2^x\ln 2 \quad\blacksquare$$

One of the fundamental equations given in Section 2.5 is

$$a^{\log_a x} = x.$$

Taking the natural logarithm of both sides, we have

$$\ln(a^{\log_a x}) = \ln x$$

$$(\log_a x)\ln a = \ln x \qquad \text{since } \ln a^b = b \ln a$$

$$\log_a x = \frac{1}{\ln a}\,\ln x.$$

Now we can use this last formula to compute the derivative of $\log_a x$.

$$\frac{d}{dx}\,\log_a x = \frac{d}{dx}\left(\frac{1}{\ln a}\,\ln x\right) = \frac{1}{\ln a}\,\frac{d}{dx}\,\ln x$$

$$= \frac{1}{\ln a}\cdot\frac{1}{x}$$

Thus, we have the formula

$$\frac{d}{dx}\,\log_a x = \frac{1}{\ln a}\cdot\frac{1}{x}.$$

EXAMPLE 7

Find $\dfrac{d}{dx}\,\log_2 x$.

Solution

$$\frac{d}{dx}\,\log_2 x = \frac{1}{\ln 2}\cdot\frac{1}{x} \quad\blacksquare$$

There is another situation in which the Chain Rule is used, and the application looks a little different. Suppose that

$$z = f(y) \quad \text{and} \quad y = g(x).$$

When a choice for x is made, then y is determined by the second formula; this value of y then determines z according to the first formula. Thus, z is a function of x, and it is possible to get an explicit formula for z in terms of x by substituting $g(x)$ for y in the first formula. Sometimes this substitution process is tedious, however, and it is possible to find the derivative $(d/dx)(z)$ without doing the substitution. We usually write dz/dx instead of $(d/dx)(z)$, and we use the form of the Chain Rule given below.

$$\frac{dz}{dx} = \frac{dz}{dy}\frac{dy}{dx}$$

Using the notation $\dfrac{dz}{dx}\bigg|_{x=a}$ for the derivative $\dfrac{dz}{dx}$ at $x = a$, the above rule becomes

$$\frac{dz}{dx}\bigg|_{x=a} = \frac{dz}{dy}\bigg|_{y=g(a)} \frac{dy}{dx}\bigg|_{x=a}.$$

EXAMPLE 8

Find $\dfrac{dz}{dx}$ where $z = y^3 + 5y^2 + 2y - 3$ and $y = x^2 - x + 9$.

Solution

We use the form of the Chain Rule given just above.

$$\frac{dz}{dy} = 3y^2 + 10y + 2$$

$$\frac{dy}{dx} = 2x - 1$$

$$\frac{dz}{dx} = \frac{dz}{dy}\frac{dy}{dx} = (3y^2 + 10y + 2)(2x - 1) \quad \blacksquare$$

EXAMPLE 9

Find $\dfrac{dz}{dx}\bigg|_{x=0}$ where $z = y^3 + 5y^2 + 2y - 3$ and $y = x^2 - x + 9$.

Solution

We use the derivative formula obtained in Example 8, noting that when $x = 0$ we have $y = 9$.

$$\frac{dz}{dx}\bigg|_{x=0} = \frac{dz}{dy}\bigg|_{y=9} \frac{dy}{dx}\bigg|_{x=0}$$
$$= [3(9)^2 + 10(9) + 2][2(0) - 1]$$
$$= -335 \quad \blacksquare$$

EXAMPLE 10

The **savings rate** of Mr. Rogers, the sole owner of a small business, increases half as fast as his pretax business profit. If his monthly business profit is given by the function

$$P(x) = .6x - 400,$$

where x is the number of units sold, what is the rate of increase in his monthly savings as a function of the number of units sold?

Solution

If S represents Mr. Rogers' monthly savings, then S is some function of business profit P, so $S = S(P)$. Since his monthly savings increase at one half the rate of increase in profit, we know that

$$\frac{d}{dP} S(P) = .5.$$

It is easy to compute $P'(x)$.

$$P'(x) = .6$$

According to the Chain Rule, we have

$$\frac{dS}{dx} = \frac{dS}{dP}\frac{dP}{dx}$$
$$= (.5)(.6) = .3. \ \blacksquare$$

Be careful not to misinterpret the preceding example. It does not say that Mr. Rogers saves one half of his profit, but rather that he saves one half of any increase in his pretax profit. The *savings rate* of a person depends on the level of his income. People with high incomes who can easily handle the expenses of food, clothing, and shelter tend to save more money. To put it another way, as a person's income increases faster than inflation, the person tends to spend a smaller percentage of his income on necessities and save a larger percentage.

EXAMPLE 11

A manufacturing company has a limited market for its product. Supposing that its **revenue** is given by

$$R(x) = 8x - \frac{x^2}{500}$$

where x is the number of units produced in 1 week, what is the rate of increase in $R(x)$ when $x = 1000$ and production is increasing at the rate of 200 per week?

Solution

Since x is a function of time t, measured in weeks, R is a composite function of t. Since production is increasing at the rate of 200 per week, we know that $x'(t) = 200$ when $x = 1000$. We use the Chain Rule and then we substitute the preceding numbers.

$$\frac{dR}{dt} = \frac{dR}{dx}\frac{dx}{dt}$$

$$\frac{dR}{dt} = \left(8 - \frac{2x}{500}\right)\frac{dx}{dt}$$

Now we substitute.

$$\frac{dR}{dt} = 8(200) - \frac{1}{500} \cdot 2(1000)(200)$$

$$= 1600 - 800 = 800$$

Revenue is increasing at the rate of $800 per week. ■

Exercises 8.5

Find the derivative formula for each of the following functions.

1 $f(x) = \ln 5x$
2 $g(x) = \ln(2x + 1)$
3 $h(x) = \ln(x^2 + 5x + 2)$
4 $f(x) = \ln(7x^2 - 10)$
5 $g(x) = 9 \ln[(2x - 1)(3x + 2)]$
6 $h(x) = 4 \ln x(6x + 7)$
7 $f(x) = 2 \ln \sqrt{4x^2 + 9}$
8 $g(x) = 3 \ln \sqrt[3]{x^2 + x + 1}$
9 $h(x) = \ln(x^2 + 5x) - \ln x$
10 $f(x) = \ln(x^2 - 1) - \ln(x + 1)$

11 $g(x) = \ln \left(\dfrac{5x + 1}{2x + 3} \right)$
12 $h(x) = \ln \left(\dfrac{9}{3x + 6} \right)$

13 $f(x) = e^{3x}$
14 $g(x) = e^{-8x+3}$
15 $h(x) = e^{x^2 - 10x + 2}$
16 $f(x) = e^{x^3 + 4x}$
17 $g(x) = e^{x^2 + 1}\sqrt{x^2 + 1}$
18 $h(x) = e^{x\sqrt{x+1}}$
19 $f(x) = e^{(x+2)(3x-1)}$
20 $g(x) = e^{(x^2 - 1)(2x + 1)}$
21 $h(x) = e^{\ln x}$
22 $f(x) = e^{3 \ln x}$
23 $g(x) = 9xe^{2x}$
24 $h(x) = -7x^2 e^{5x+1}$
25 $f(x) = (e^{4x} + 3)^3$
26 $g(x) = (7e^{2x} + 1)^5$
27 $h(x) = (\ln(x^2 + 3x + 1))^5$
28 $f(x) = (\ln(7x + 2))^9$
29 $g(x) = e^{3x}\ln x$
30 $h(x) = 6e^x\ln(3x + 5)$
31 $f(x) = 5^x$
32 $g(x) = 4(2^x)$
33 $h(x) = 4^{2x+1}$
34 $f(x) = 3^{5x+3}$
35 $g(x) = 7^{x^2}$
36 $h(x) = 8^{3x^2+1}$
37 $f(x) = \log_3 x$
38 $g(x) = 7 \log_5 x$
39 $h(x) = \log_2(x^2 + 5x - 2)$
40 $f(x) = 5 \log_3(6x^2)$
41 $g(x) = 2 \ln(3x + 2)\log_2 8$
42 $h(x) = \ln(x^2 + 5)\log_3 27$

In problems 43 through 50, find dz/dx.

43 $z = 9y^2 + 3y - 1, \quad y = 7x + 9$
44 $z = y^3 + 11y^2 + 12, \quad y = 3x^2 + x + 1$
45 $z = e^{8y+2}, \quad y = \ln(x^4 + x^2 + 1)$
46 $z = 5e^{3y}, \quad y = \ln(x^2 + 25)$
47 $z = 2^y, \quad y = 5x^2 + 10x + 1$
48 $z = 3^y, \quad y = x^3 + 3x^2 - 5x + 6$
49 $z = \log_2 y, \quad y = \sqrt{x^2 + 4}$
50 $z = \log_3 y, \quad y = x^3 + x - 1$

Solve the following.

51   Mr. Paul has a struggling young business, and his monthly savings increase at .1 times the rate of increase in his pretax business profit. If his business profit is given by

$$P(x) = 5x - 1000$$

where x is the number of customers served, what is the rate of increase in his monthly savings as a function of the number of customers served?

52   Mr. Cook is the sole owner of a very successful business, and his savings increase at .7 times the rate of increase in his pretax profit. If his profit is given by

$$P(x) = 7.5x - .01x^2 - 100$$

where x is the number of units sold, what is the rate of increase in his savings as a function of the number of units sold?

53   The cost function for a business is

$$C(x) = 7x + 400$$

where x is the number of customers served in one week. What is the rate of increase in $C(x)$ when the number of customers is increasing at the rate of 50 per week?

8.6   Implicit Differentiation

In obtaining the derivative formulas up to this point we usually started with an **explicit** formula for the function to be differentiated. If dy/dx was sought, then we began with a formula for y in terms of x, for example

$$y = 5x^2 + 3x - 7.$$

If $g(x, y)$ is some expression in x and y, for example $g(x, y) = x^2y + 2xy^2 + y^2$, and (a, b) is an ordered pair that satisfies the equation $g(x, y) = 0$, that is $g(a, b) = 0$, then it is often possible to find dy/dx where $x = a$ and $y = b$ without solving for y in terms of x. We say that the equation $g(x, y) = 0$ **implicitly** gives y as a function of x, and finding dy/dx without solving for y is called **implicit differentiation.**

EXAMPLE 1

Find dy/dx when $x = 1$ and $y = -\frac{1}{3}$ provided y is a function of x implicitly defined by $x^2y + 2xy^2 + y^2 = 0$.

Solution

We differentiate each side of the equation while keeping in mind that y is the dependent variable and x is the independent variable.

$$\frac{d}{dx}x^2y + \frac{d}{dx}2xy^2 + \frac{d}{dx}y^2 = 0$$

$$\left[2xy + x^2\frac{dy}{dx}\right] + \left[2y^2 + 4xy\frac{dy}{dx}\right] + \left[2y\frac{dy}{dx}\right] = 0$$

We substitute 1 for x and $-\frac{1}{3}$ for y, and we solve the resulting equation.

$$-\frac{2}{3} + \frac{dy}{dx} + \frac{2}{9} - \frac{4}{3}\frac{dy}{dx} - \frac{2}{3}\frac{dy}{dx} = 0$$

$$\left(1 - \frac{4}{3} - \frac{2}{3}\right)\frac{dy}{dx} = \frac{4}{9}$$

$$-\frac{dy}{dx} = \frac{4}{9}$$

$$\frac{dy}{dx} = -\frac{4}{9} \quad \blacksquare$$

Let us note the geometric meaning of a derivative obtained by implicit differentiation. Suppose the equation $x^2 + y^2 = 1$ defines y as an implicit function of x for all x close to $(1/2, \sqrt{3}/2)$. The graph of $x^2 + y^2 = 1$ is the circle with its center at the origin and a radius of 1. The point $(1/2, \sqrt{3}/2)$ lies on this circle, and there is a tangent line to the circle at the point $(1/2, \sqrt{3}/2)$ as in the picture.

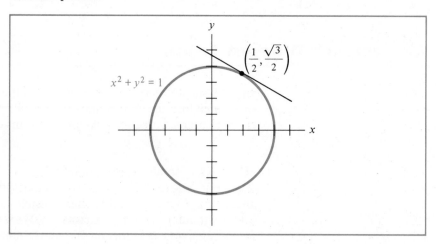

The derivative $y'(\frac{1}{2})$ is the slope of the tangent line shown. In particular, we have

$$2x + 2yy' = 0$$

$$y' = \frac{-x}{y}$$

$$y'\left(\frac{1}{2}\right) = -\left(\frac{2}{\sqrt{3}}\right)\left(\frac{1}{2}\right) = -\frac{1}{\sqrt{3}} \approx -.577.$$

EXAMPLE 2

Find a formula for dy/dx provided y is a function of x implicitly defined by

$$x^2y + 2xy^2 + y^2 = 5.$$

Solution

As in Example 1 we differentiate both sides of the equation, and we get

$$\left[2xy + x^2\frac{dy}{dx}\right] + \left[2y^2 + 4xy\frac{dy}{dx}\right] + \left[2y\frac{dy}{dx}\right] = 0.$$

Now we solve for dy/dx in terms of x and y.

$$(x^2 + 4xy + 2y)\frac{dy}{dx} = -2xy - 2y^2$$

$$\frac{dy}{dx} = \frac{-2xy - 2y^2}{x^2 + 4xy + 2y} \quad■$$

When solving a problem like Example 2, do not expect a formula for dy/dx in terms of x alone. Since we started with y implicitly given as a function of x, dy/dx will usually be obtained in terms of both x and y. Since y is a function of x, the formula for dy/dx makes it a function of x.

Always remember, in taking the derivative of each side of the equation that defines y implicitly as a function of x, that y is a function of x. Thus, to differentiate an expression such as y^4 requires the Generalized Power Rule, not the Power Rule.

EXAMPLE 3

Find a formula for dy/dx provided y is a function of x implicitly defined by

$$e^{xy} + xy^4 = 10.$$

Solution

We differentiate each side of the equation.

$$\frac{d}{dx}e^{xy} + \frac{d}{dx}xy^4 = 0$$

$$e^{xy}\frac{d}{dx}xy + y^4 + x\left(4y^3\frac{dy}{dx}\right) = 0$$

$$e^{xy}\left(y + x\frac{dy}{dx}\right) + y^4 + 4xy^3\frac{dy}{dx} = 0$$

Now we solve for dy/dx.

$$xe^{xy}\frac{dy}{dx} + 4xy^3\frac{dy}{dx} = -ye^{xy} - y^4$$

$$(xe^{xy} + 4xy^3)\frac{dy}{dx} = -ye^{xy} - y^4$$

$$\frac{dy}{dx} = \frac{-ye^{xy} - y^4}{xe^{xy} + 4xy^3} \quad■$$

EXAMPLE 4

Suppose a company has determined by careful analysis of its sales at different prices that at price p the demand for its product is

$$q = 10{,}000 - p^2.$$

Find the rate of *change in price p* resulting from a *change in demand.*

Solution

We want to find dp/dq, and we find it by implicit differentiation.

$$\frac{d}{dq} q = \frac{d}{dq} 10{,}000 - \frac{d}{dq} p^2$$

$$1 = -2p \frac{dp}{dq}$$

$$\frac{dp}{dq} = -\frac{1}{2p} \quad \blacksquare$$

Exercises 8.6

Find dy/dx assuming y is an implicit function of x.

1 $1 = x^2 + y^2$
2 $1 = 3x^2 + 9y^2$
3 $1 = 2x^2 - 4y^2$
4 $5 = x^2 - 4xy + y^3$
5 $y = x^3 + 5xy^2 + y^3$
6 $1 = x^3 y^3 + y$

7 $3 = 4x^4 y^4 + y^2 + x$
8 $x^3 = \dfrac{x - y}{x + y}$

9 $y^4 = \dfrac{x - 1}{x + y}$
10 $4 = \sqrt{x} + \sqrt{y}$

11 $1 = 9\sqrt{x} + \sqrt{y}$
12 $1 = \sqrt[3]{x} + 8\sqrt[3]{y}$
13 $1 = x \ln y + y \ln x$
14 $2 = \ln xy$
15 $\ln x = y^2 + 2y$
16 $3 = x \ln y + e^{xy}$
17 $1 = e^{x^2 + y^2}$
18 $2 = e^{x^3 y^3}$
19 $1 = e^x \ln y$
20 $5 = x \ln(x^2 + y^2)$
21 $1 = e^{xy} + (\ln y)^{10}$
22 $1 = x \ln(y + x^3) + y$

In problems 23 through 27, find the rate at which price p changes in response to a change in demand q provided that the given equation makes p an implicit function of q.

23 $q = 2500 - p^2$
24 $q = 64{,}000 - p^3$
25 $q = 200 - p$
26 $1000 = pq$
27 $40 = \sqrt{pq}$

Review of Terms

Review of Derivative Formulas

$$\frac{d}{dx} c = 0 \quad \text{where } c \text{ is any constant}$$

$$\frac{d}{dx} x^a = a x^{a-1} \quad \text{where } a \text{ is any real number}$$

$$\frac{d}{dx} [f(x) + g(x)] = f'(x) + g'(x)$$

$$\frac{d}{dx} [f(x) - g(x)] = f'(x) - g'(x)$$

$$\frac{d}{dx} c f(x) = c f'(x) \quad \text{where } c \text{ is any constant}$$

$$\frac{d}{dx} [f(x) g(x)] = f'(x) g(x) + f(x) g'(x)$$

$$\frac{d}{dx} \left[\frac{f(x)}{g(x)}\right] = \frac{f'(x) g(x) - g'(x) f(x)}{g(x)^2}$$

$$\frac{d}{dx} u^a = a u^{a-1} \frac{du}{dx} \quad \text{where } a \text{ is any real number}$$

$$\frac{dz}{dx} = \frac{dz}{du} \frac{du}{dx} \quad \text{where } z \text{ is a function of } u \text{ and } u \text{ is a function of } x$$

$$\frac{d}{dx} e^u = e^u \frac{du}{dx}$$

$$\frac{d}{dx} \ln u = \frac{1}{u} \frac{du}{dx}$$

$$\frac{d}{dx}\, a^u = a^u (\ln a)\,\frac{du}{dx}$$

$$\frac{d}{dx}\, \log_a u = \frac{1}{u}\left(\frac{1}{\ln a}\right)\frac{du}{dx}$$

Review Problems

In problems 1 through 35, find a derivative formula for the indicated function.

1   $f(s) = 20s^3 - s^2 + 5s + 12$

2   $g(t) = t^{3/2} + 5t^4 - 2t^{-1}$

3   $h(r) = \dfrac{11}{\sqrt{r}}$

4   $f(p) = 6p^{110} + \sqrt{p}$

5   $g(w) = (3w^2 + 1)^{30}$

6   $h(u) = \sqrt{u^3 + u^2 + u}$

7   $f(x) = (2x + 1)x$

8   $g(s) = (s - 3)(2s + 5)(3s - 7)$

9   $h(t) = (t + 6)^{20}(2t - 7)^{40}$

10   $f(r) = (r + 9)^{12}(2r - 4)^{24}(5r + 1)^8$

11   $g(p) = \dfrac{11p + 8}{5p - 6}$

12   $h(w) = \dfrac{5w + 2}{w^2 + 3w - 6}$

13   $f(z) = \dfrac{z^2 + 3z}{4z^2 + 8}$

14   $g(x) = \dfrac{1}{4x + 1} + (3x - 5)(7x + 2)^{15}$

15   $h(v) = \sqrt{\dfrac{8v + 1}{v + 3}}$

16   $f(r) = \dfrac{(r + 7)(3r - 2)}{(2r + 5)(r - 9)}$

17   $g(s) = \dfrac{\sqrt[3]{s^3 - 8}}{\sqrt{s^2 + 1}}$

18   $h(t) = e^{5t}$

19   $f(p) = p^2 e^{-2p}$

20   $g(u) = (10 - e^u)^{80}$

21   $h(x) = \dfrac{x^2 + 3x + 1}{e^x + e^{-x}}$

22 $f(u) = ue^u - u^3 + 9u^2$

23 $g(w) = \sqrt{e^{3w} + 1}$

24 $h(s) = 5e^{3s^2 - s + 11}$

25 $f(v) = \ln 7v$

26 $g(x) = \log_2 x$

27 $h(t) = \ln(9t + 5)$

28 $f(x) = 3 \ln[(2x + 1)(x - 6)]$

29 $g(r) = \ln \left(\dfrac{7r + 1}{r + 3} \right)$

30 $h(x) = 5^x$

31 $f(s) = s^3 e^{s+1} - e^{4s^2}$

32 $g(w) = e^w \ln w$

33 $h(r) = (\ln(5r^2 + 9))^{10}$

34 $f(x) = \log_5(x^2 + 3x + 1)$

35 $g(z) = 3^{2z+5}$

In problems 36 through 39, find an equation for the tangent line to the graph of the given function at the given point.

36 $f(x) = e^x$ at $(0, 1)$

37 $g(x) = \ln x$ at $(1, 0)$

38 $h(x) = 2x^2 + 3x - 5$ at $(0, -5)$

39 $f(x) = 5^x$ at $(1, 5)$

In problems 40 through 43, find dz/dx.

40 $z = 2e^{7y}, \quad y = -6x^2 + 2x + 3$

41 $z = \ln(y^2 + 1), \quad y = 13e^{2x}$

42 $z = y^4 + 3y^2 - 5, \quad y = 3x - 6$

43 $z = \sqrt{y + 2}, \quad y = 9x^2 + 7x + 5$

In problems 44 through 48, find dy/dx assuming that y is an implicit function of x.

44 $12 = 3x^2 + 4y^2$

45 $1 = x^5 y^5 - xy$

46 $1 = \ln(x^2 + y^2)$

47 $1 = e^{2x^2 + y^2}$

48 $y^2 = \dfrac{x + y}{x - y}$

Solve each of the following.

49 Find the elasticity of demand for the demand function $q = -1.5p + 3$.

50 Suppose the noise level in a plant is reduced p percent by spending an additional x hundreds of dollars and $p = \sqrt{x + 900}$. What is the marginal change in $p(x)$ when the noise level is reduced 40%? What is it when the noise level is reduced 70%.

51 Mr. Jung is sole owner of a business, and his savings increase at .4 times the rate of increase of his pretax profit. If his profit is given by

$$P(x) = 7x - 400$$

where x is the number of customers served, what is the rate of increase in his savings as a function of the number of customers served?

52 If an account with $1000 earns interest at a rate of 16% compounded continuously, how much is in the account after 5 years?

Social Science Applications

1 ***Psychology***   Suppose a person can memorize x items (for $x \geq 3$) from a certain list in approximately

$$f(x) = x\sqrt{x - 2}$$

minutes. At what rate is the person memorizing items, when 11 items are being memorized? What is the rate when 18 items are being memorized? (**Hint:** Compute $f'(11)$ and $f'(18)$.)

2 ***Urban Planning***   The population of a city t years after January 1, 1982, is given by

$$p(t) = 1,000,000 e^{.1t}.$$

Determine the rates of increase of the population on January 1, 1982, and on January 1, 1992, as percentages of the populations on those dates. (**Hint:** Compute $p'(0)/p(0)$ and $p'(10)/p(10)$.)

3 ***Sociology***   In a city of 1,000,000 people a certain rumor is spread rapidly. After t days the number of people spreading the rumor is given by

$$x(t) = \frac{1,000,000}{999 e^{-2t} + 1}.$$

Determine the rate of increase of the number of people spreading the rumor at the time that the rumor begins ($t = 0$) and after 5 days ($t = 5$). (**Hint:** Compute $x'(0)$ and $x'(5)$.)

4 ***Law Enforcement***   An automobile traveling at a speed of s (in miles per hour) will have a skid approximately b feet long, according to the formula

$$b = .041 s^2.$$

Obtain a formula for the rate of increase of the length of the skid, as a function of the speed.

5 ***Archeology***   The decay of carbon-14 is used to determine the age of ancient artifacts. The formula

$$x(t) = x_0 e^{-.00012t}$$

indicates the amount $x(t)$ of carbon-14 present after t years, assuming there was x_0 present when the organic materials in the artifact died. Obtain a formula for the rate of decay of carbon-14, as a percentage of the amount present. (**Hint:** Compute $x'(t)/x(t)$.)

6 ***Psychology***   Let x indicate the volume of a radio measured in decibels,

and let y indicate the smallest change in the volume that would be noticed by a certain listener. In a typical example we might have

$$y = .05x.$$

Show that the amount of a noticeable change in the volume (the quantity y) increases at a constant rate with respect to changes in x.

7 ***Ecology***   A plant emits sulfur dioxide into the surrounding air through its smokestack. Suppose the amount y of sulfur dioxide in 1 cubic foot of air (in appropriate units) is given by

$$y = \frac{.05}{x^2},$$

where x is the distance to the smokestack (in miles). Determine the rate of decrease of sulfur dioxide 1 mile from the smokestack. What is the rate 10 miles from the smokestack?

CAREER PROFILE
Urban or Regional Planner

Traffic problems, inadequate parking, and the high cost of providing systems of public transportation are among the most visible signs of the necessity for urban or regional planning.

A community, county, or city is likely to employ a planner when it recognizes the need to coordinate the many facets of its development to insure that it will be a place where people will want to live, work, and shop. The activities of planners are diverse: they examine the adequacy of school systems and health clinics to see whether these can meet expected future demands; they help to design transportation networks and parking facilities; they estimate needs for housing, business, and recreation sites, based on anticipation of population shifts and changes in economic conditions and people's preferences.

Some universities offer undergraduate programs in urban and regional planning, and a majority of planners graduate in these programs. Other suitable majors include economics, geography, and public administration. Courses in statistics and computer science are essential parts of career preparation. A planner must be able to design, conduct, and interpret surveys and to make wise use of data that are available from government and business sources. The ability to visualize the effects of plans and designs can be enhanced by taking courses in graphic arts and design. Computer graphics will become an increasingly important tool to help the planner see what plans may look like.

Advanced training, such as a master's degree in urban or regional planning is a key requirement for many of the planning positions available. In some cases, specialties are required. Major planning projects are likely to be the responsibility of a team whose members contribute expertise in areas such as landscape architecture, engineering, or statistics.

Most employers are looking for individuals with planning *experience*. Because of this, some universities advise students interested in a career in planning to supplement their academic courses with internships or work-study projects.

A planner's first position is likely to be a modest one, but a few years of experience in a small town or county planning project (possibly starting at an annual salary of no more than $14,000), combined with advanced study, can serve as a springboard to an exciting and profitable career (earning a salary of at least $30,000).

The penalties of failure to plan are evident all around us. Economic distress in downtown regions, scarcity of energy, and abundant pollution suggest the need for planning in our communities, cities, and regions. Thus urban and regional planning is an employment area that can be expected to grow. However, because planning is sometimes seen by communities as a luxury rather than a necessity, the growth of the profession will depend on the economic conditions of the communities and regions in which there will be potential employment. (In 1980 about 23,000 persons were employed as planners with most working for city, county, or regional planning agencies. Some others are employed in government agencies dealing with health care, housing, transportation, and environmental protection.)

Individuals who are flexible, patient, and effective in forging compromise when working with groups holding different views and in addition have strong skills in data analysis and can perceive spatial relationships with ease will find challenging opportunities in urban and regional planning.

Sources of Additional Information
American Planning Association, 1776 Massachusetts Avenue N.W., Washington, DC 20036. (This organization provides facts about careers in urban and regional planning and a list of schools offering training in these fields.)

Occupational Outlook Handbook. Bureau of Labor Statistics, U.S. Department of Labor, Washington, DC 20212. (Revised every two years, this handbook provides information about job duties, working conditions, level and places of employment, education and training requirements, advancement possibilities, job outlook, and earnings for about 250 occupations. Some of the information for this Career Profile was obtained from this source.)

9

Applications of the Derivative

9.1 First Derivative Test

Graphing functions is very important since it is the only way that we can look at all of the values of a given function at one time. Seeing all of the values, we can probably pick the largest value and the smallest value; we can see where the function is going up and where it is going down. Graphing is one of the most important tools for understanding such functions as cost, revenue, profit, demand, money supply, and prices.

No matter how many points we plot, we cannot be sure of the shape of the graph of a given function without the use of calculus. To illustrate the basic idea consider the plotted points on the first graph.

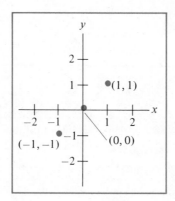

It appears that the graph that fits the points is a straight line through the origin with slope 1. In fact, however, the graph might be the graph of $f(x) = \frac{4}{3}x - \frac{1}{3}x^3$ given in the second illustration.

In order to know the shape of a graph, we must know the location of the points where the graph changes from going up to going down or vice versa. Such points, which can be found with the derivative, are called **turning points.** Let us illustrate the power of geometric reasoning using this idea. If an

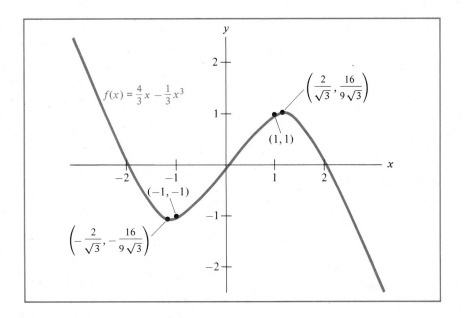

$$f(x) = \frac{4}{3}x - \frac{1}{3}x^3$$

$$\left(\frac{2}{\sqrt{3}}, \frac{16}{9\sqrt{3}} \right)$$

$$(1, 1)$$

$$(-1, -1)$$

$$\left(-\frac{2}{\sqrt{3}}, -\frac{16}{9\sqrt{3}} \right)$$

unbroken graph has no turning points and some part of the graph is going up, then, in fact, all of the graph must be going up. Between two consecutive turning points, an unbroken graph must be entirely going up or entirely going down since it cannot change from one to the other without passing through a turning point.

We say that the function **$f(x)$ is increasing from a to b** provided the entire graph of $f(x)$ from $(a, f(a))$ to $(b, f(b))$ is rising as we read from left to right; that is, if we choose any two numbers between a and b, say c and d, and if $c \leq d$, then $f(c) \leq f(d)$. We say that **$f(x)$ is decreasing from a to b** provided the entire graph of $f(x)$ from $(a, f(a))$ to $(b, f(b))$ is falling as we read left to right; that is, if we choose any two numbers between a and b, say c and d, and if $c \leq d$, then $f(c) \geq f(d)$. For example, the graph shown for $f(x) = x^2$ is decreasing between any negative number and 0; it is increasing between 0 and any positive number. See the graph on page 392.

We say that $(a, f(a))$ is a **relative minimum** provided none of the nearby points on the graph of $f(x)$ has a smaller y-value, that is, $f(x) < f(a)$ does not occur for x such that $c < x < d$ where c and d are some numbers satisfying the inequality $c < a < d$. On the graph of $f(x) = x^2$, $(0, 0)$ is a relative minimum, and $(-2/\sqrt{3}, -16/9\sqrt{3})$ is a relative minimum on the graph of $f(x) = \frac{4}{3}x - \frac{1}{3}x^3$ given earlier. We say that $(a, f(a))$ is an **absolute minimum** provided no point on the graph of $f(x)$ has a smaller y-value. Note that $(0, 0)$ is an absolute minimum for x^2, but $(-2/\sqrt{3}, -16/9\sqrt{3})$ is not an absolute minimum for $\frac{4}{3}x - \frac{1}{3}x^3$. We say that $(a, f(a))$ is a **relative maximum** provided none of the nearby points on the graph of $f(x)$ has a larger y-value. Furthermore, $(a, f(a))$ is an **absolute maximum** provided no point on the graph of $f(x)$ has a larger y-value. On the graph of $f(x) = \frac{4}{3}x - \frac{1}{3}x^3$, the point $(2/\sqrt{3}, 16/9\sqrt{3})$ is a relative maximum but not an absolute maximum, since $f(x)$

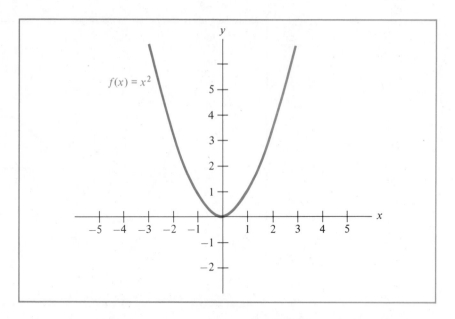

takes arbitrarily large values for x negative with sufficiently large magnitude. Note that any absolute minimum is also a relative minimum, and any absolute maximum is also a relative maximum.

If $(a, f(a))$ is a relative minimum (absolute minimum) then we say that $f(a)$ is a **relative minimum value (absolute minimum value)** for $f(x)$. If $(a, f(a))$ is a relative maximum (absolute maximum) then we say that $f(a)$ is a **relative maximum value (absolute maximum value)** for $f(x)$.

EXAMPLE 1

Describe where the graph of $f(x) = \frac{4}{3}x - \frac{1}{3}x^3$ is increasing and where it is decreasing.

Solution
We refer to the graph given earlier and we note that the graph is decreasing to the left of $-2/\sqrt{3}$ and to the right of $2/\sqrt{3}$. It is increasing between $-2/\sqrt{3}$ and $2/\sqrt{3}$. ∎

A relative minimum or an absolute minimum might occur at the endpoint of a graph, as is the case for $(0, 0)$ shown on the graph of $f(x) = \sqrt{x}$. A relative maximum or an absolute maximum might occur at the endpoint of a graph, also.

We say that a function $f(x)$ is **differentiable** provided $f'(x)$ exists for every x in the domain of $f(x)$. The fact that every polynomial is differentiable was shown in Section 8.1. The important First Derivative Test rests on the following observation. Any point that is a relative minimum or a relative maximum of the differentiable function $f(x)$, and does not occur at an endpoint of the graph, has a horizontal tangent line; consequently, $f'(x)$ is 0 at the x-coordinate of that point. We say that b is a **critical value** for $f(x)$

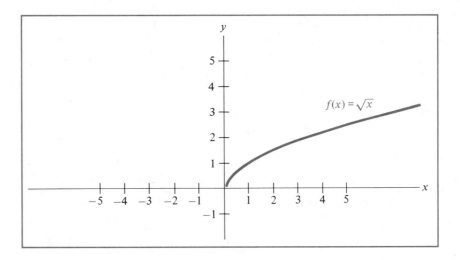

provided b is in the domain of $f(x)$ and either $f'(b) = 0$ or $f'(b)$ is not defined.*

First Derivative Test

1. Every relative minimum and relative maximum occurs at a critical value or at an endpoint of the graph.
2. If b is a critical value for $f(x)$ and $f'(a) < 0$ for a just to the left of b and $f'(c) > 0$ for c just to the right of b, then $(b, f(b))$ is a relative minimum.
3. If b is a critical value for $f(x)$ and $f'(a) > 0$ for a just to the left of b and $f'(c) < 0$ for c just to the right of b, then $(b, f(b))$ is a relative maximum.

At any critical value b for which $f'(b) = 0$ there are ordinarily four possibilities:

1. If the graph is decreasing to the left of b and increasing to the right of b, then it is similar to picture A; note $f'(x)$ is negative to the left of b and positive to the right of b. See page 394.
2. If the graph is decreasing both to the left of b and to the right of b, then it is similar to illustration B; note $f'(x)$ is negative to the left of b and to the right of b.
3. If the graph is increasing to the left of b and increasing to the right of b, then it is similar to picture C; note that $f'(x)$ is positive to the left and to the right of b.
4. If the graph is increasing to the left of b and decreasing to the right of b, then it is similar to picture D; note $f'(x)$ is positive to the left of b and negative to the right of b.

---

* There are many ways that the derivative of a function can fail to exist at a given point; most of these were considered in Section 7.6. Although most functions that arise in applications have a derivative at all points, it is a good practice to check that this is so when beginning a problem.

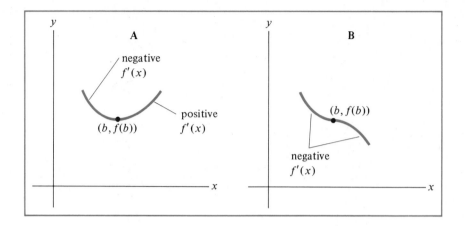

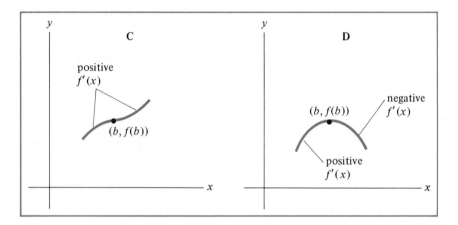

EXAMPLE 2

Find the critical values for $g(x) = -x^2 + 4x + 2$, and for each critical value, see whether it determines a relative minimum, a relative maximum, or neither.

Solution

Since both $g(x)$ and $g'(x)$ are defined for all x, every critical value of $g(x)$ must satisfy $g'(x) = 0$. So we differentiate $g(x)$, set the derivative equal to 0, and solve for x.

$$g'(x) = -2x + 4$$

$$0 = g'(x) = -2x + 4$$

$$2x = 4$$

$$x = 2$$

So $x = 2$ is the only critical value. If $a < 2$ (or $a - 2 < 0$), then

$$g'(a) = -2a + 4 = -2(a - 2)$$

must be positive since it is the product of two negative numbers. If $c > 2$ (or $c - 2 > 0$), then

$$g'(c) = -2(c - 2) < 0.$$

So the graph of $g(x)$ is increasing to the left of 2 and decreasing to the right of 2. Thus, $(2, g(2)) = (2, 6)$ is a relative maximum. Since there are no other critical values for $g(x)$ and, hence, no other turning points for the graph of $g(x)$, a quick look at picture D above convinces us that $(2, 6)$ is an absolute maximum for $g(x)$. ■

EXAMPLE 3

Find the critical values, relative minima, and relative maxima for $h(x) = x^3$.

Solution

$$h'(x) = 3x^2$$

$$0 = h'(x) = 3x^2$$

$$x = 0$$

So $x = 0$ is the only critical value. If $a < 0$ then $h'(a) = 3a^2$ is positive, and if $c > 0$ then $h'(c) = 3c^2$ is positive. So the graph of $h(x)$ is increasing both to the left and to the right of 0. Thus, $(0, 0)$ is not a turning point and the graph resembles picture C. ■

EXAMPLE 4

Find the critical values, relative minima, and relative maxima for

$$f(x) = -x^3 + 3x^2 - 3x + 1.$$

Solution
Note that $f(x)$ and $f'(x)$ exist for all x.

$$f'(x) = -3x^2 + 6x - 3$$

$$0 = f'(x) = -3(x^2 - 2x + 1) = -3(x - 1)^2$$

So $x = 1$ is the only critical value. If $a < 1$ (or $a - 1 < 0$), then $f'(a) = -3(a - 1)^2 < 0$, and if $c > 1$ (or $c - 1 > 0$), then $f'(c) = -3(c - 1)^2 < 0$. So the graph of $f(x)$ is decreasing to the left and to the right of 1. Thus, $(1, f(1)) = (1, 0)$ is not a turning point and the graph resembles picture B. ■

EXAMPLE 5

Find the critical values, relative minima, and relative maxima for $g(x) = \dfrac{3x + 5}{x - 2}$. Also, find where $g(x)$ is increasing and decreasing.

Solution

$$g'(x) = \frac{3(x - 2) - (3x + 5)}{(x - 2)^2} = \frac{-11}{(x - 2)^2}$$

This cannot equal 0 for any choice of x since a fraction equals 0 if and only if the numerator equals 0. Nevertheless, $x = 2$ might be a critical value because

$g'(x)$ does not exist for $x = 2$. Since $g(x)$ is not defined for $x = 2$, it is not a critical value and there is not a point $(2, g(2))$ on the graph of $g(x)$; thus, there is no relative minimum or relative maximum corresponding to $x = 2$. The function $g(x)$ has no relative minimum and no relative maximum. Since $g'(x)$ is negative for all x other than $x = 2$, the graph is decreasing to the left of 2 and to the right of 2. ■

Note Between any two consecutive critical values of $f(x)$, either $f(x)$ is increasing or it is decreasing, but it cannot change from one to the other. To the left of the smallest critical value and to the right of the largest critical value, either $f(x)$ is increasing or it is decreasing, but it cannot change from one to the other.

EXAMPLE 6

Find the critical values, relative minima, and relative maxima for $h(x) = (x + 1)^4(x - 4)^4$. Also find where $h(x)$ is increasing and decreasing.

Solution

$$h'(x) = 4(x + 1)^3(x - 4)^4 + (x + 1)^4 4(x - 4)^3$$
$$= 4(x + 1)^3(x - 4)^3[x - 4 + x + 1]$$
$$= 8(x + 1)^3(x - 4)^3\left(x - \frac{3}{2}\right)$$

The critical values are -1, 4, $\frac{3}{2}$. We evaluate $h(x)$ at each critical value, one convenient value to the left of -1, say $x = -2$, and one convenient value to the right of 4, say 5.

$$h(-2) = (-1)^4(-6)^4 = 36^2 = 1296$$
$$h(-1) = 0$$
$$h\left(\frac{3}{2}\right) = \left(\frac{5}{2}\right)^4\left(-\frac{5}{2}\right)^4 = \left(\frac{625}{16}\right)\left(\frac{625}{16}\right) = 1525.9 \quad \text{This number is rounded off.}$$
$$h(4) = 0$$
$$h(5) = 6^4 1^4 = 1296.$$

Since $h(-2)$ is larger than $h(-1)$, $h(x)$ is decreasing to the left of -1; since $h(-1)$ is less than $h(\frac{3}{2})$, $h(x)$ is increasing between -1 and $\frac{3}{2}$; since $h(\frac{3}{2})$ is larger than $h(4)$, $h(x)$ is decreasing between $\frac{3}{2}$ and 4; since $h(4)$ is less than $h(5)$, $h(x)$ is increasing to the right of 4. Clearly $(-1, 0)$ and $(4, 0)$ are relative minima and $(\frac{3}{2}, 1525.9)$ is a relative maximum. ■

For the curious reader, the graph of $h(x) = (x + 1)^4(x - 4)^4$ is sketched.
Practical applications of the material in this section are given in Sections 9.5 and 9.6.

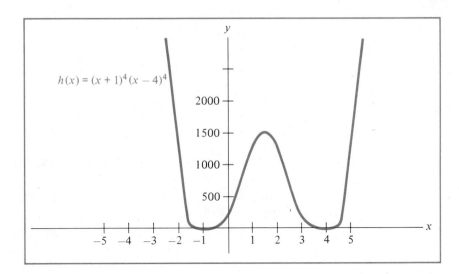

$$h(x) = (x + 1)^4(x - 4)^4$$

Exercises 9.1

In problems 1 through 26, find the critical values for the given function.

1 $f(x) = 3x - 2$

2 $g(x) = -5x + 10$

3 $h(x) = x^2 - 6$

4 $f(x) = 5x^2 + 13$

5 $g(x) = 3x^2 - 12x + 4$

6 $h(x) = 2x^2 + 2x - 11$

7 $f(x) = x^3 - 6x^2 + 12x - 8$

8 $g(x) = x^3 - 3x^2 + 3x + 1$

9 $h(x) = 2x^3 + 3x^2 - 12x + 1$

10 $f(x) = x^3 - 3x^2 - 9x + 2$

11 $g(x) = \dfrac{x + 2}{x - 1}$

12 $h(x) = \dfrac{x + 1}{3x + 1}$

13 $f(x) = 4x + 3 + \dfrac{1}{x}$

14 $g(x) = 3x + 10 + \dfrac{3}{x - 2}$

15 $h(x) = 7e^x$

16 $f(x) = 9e^{2x-1}$

17 $g(x) = e^x + e^{-x}$

18 $h(x) = e^x - e^{-x}$

19 $f(x) = -4 \ln x$

20 $g(x) = \ln 5(x - 1)$

21 $h(x) = 2x \ln x$

22 $f(x) = x^2 - 4 \ln x$

23 $g(x) = (x - 1)^8(x + 3)^{10}$

24 $h(x) = (2x + 1)^6(x - 5)^8$

25 $f(x) = 3\sqrt{x} - 13$

26 $g(x) = 1 + \dfrac{1}{\sqrt{x}}$

In problems 27 through 46, find the relative minima and relative maxima for the given function.

27 $f(x) = -2x + 1$

28 $g(x) = 9x + 12$

29 $h(x) = x^2 + 8x - 12$

30 $f(x) = -5x^2 + 10x - 35$

31 $g(x) = x^3 - 6x^2 + 12x - 8$

32 $h(x) = x^3 + 3x^2 + 3x + 1$

33 $f(x) = 2x^3 - 15x^2 + 10$

34 $g(x) = 2x^3 + 21x^2 + 72x - 8$

35 $h(x) = \dfrac{x}{2x + 3}$

36 $f(x) = \dfrac{2x - 4}{x + 5}$

37 $g(x) = 12x + \dfrac{4}{x - 1}$

38 $h(x) = 2x + \dfrac{1}{2x - 3}$

39 $f(x) = e^x + e^{-x}$

40 $g(x) = e^x - e^{-x}$

41 $h(x) = 5x \ln x$

42 $f(x) = 3x^2 - 6 \ln x$

43 $g(x) = (x - 2)^3(x + 1)^2$

44 $h(x) = (3x + 2)^2(x + 3)^3$

45 $f(x) = 2\sqrt{x} + 3$

46 $g(x) = \dfrac{5}{\sqrt{x}} + 8$

In problems 47 through 64, find where the function is increasing and where it is decreasing.

47 $f(x) = 9x - 12$

48 $g(x) = -3x + 10$

49 $h(x) = 1 + \dfrac{1}{x}$

50 $f(x) = \dfrac{3}{x^2} - 5$

51 $g(x) = 3x^2 - 12x + 4$

52 $h(x) = -5x^2 + 10x - 22$

53 $f(x) = x^3 - 6x^2 + 12x - 8$

54 $g(x) = x^3 + 3x^2 + 3x + 1$

55 $h(x) = \dfrac{x + 1}{x - 1}$

56 $f(x) = \dfrac{x - 1}{2x + 1}$

57 $g(x) = 2 \ln x$

58 $h(x) = -3 \ln(x + 2)$

59 $f(x) = (x - 1)^3(x + 2)^2$

60 $g(x) = (2x + 1)^2(x + 3)^3$

61 $h(x) = xe^x$

62 $f(x) = xe^{-x}$

63 $g(x) = e^{x^2}$

64 $h(x) = e^{x^3}$

9.2 Second Derivative Test

Since the derivative formula for a given function defines a function, which is called the **derivative function,** we could compute its derivative formula. For example, if $f(x) = .1x^2 - x + 25$, then $f'(x) = .2x - 1$ is the derivative function and

$$\frac{d}{dx} f'(x) = \frac{d}{dx}(.2x - 1) = .2$$

is the derivative formula for the derivative function. The function defined by this last formula is called the **second derivative** of the original function $f(x)$; it is denoted $\dfrac{d}{dx} f'(x)$ or $f''(x)$. We similarly define the **third derivative** $\dfrac{d}{dx} f''(x)$ or $f'''(x)$. If $f(x) = .1x^2 - x + 25$, then we have the following derivative functions.

$f'(x) = .2x - 1$

$f''(x) = .2$

$f'''(x) = 0.$

Another notation for higher derivatives is given below.

$$\frac{d^2}{dx^2} f(x) \qquad \text{for the second derivative}$$

and

$$\frac{d^3}{dx^3} f(x) \qquad \text{for the third derivative.}$$

EXAMPLE 1

Find the second derivative of each of the following functions: $f(x) = \dfrac{2x+1}{3x-1}$, $g(x) = e^{5x}$, $h(x) = \ln(x+3)$.

Solution

$$f'(x) = \frac{2(3x-1) - 3(2x+1)}{(3x-1)^2} = \frac{-5}{(3x-1)^2}$$

$$f''(x) = \frac{d}{dx} - 5(3x-1)^{-2} = 10(3x-1)^{-3} \cdot 3$$

$$f''(x) = 30(3x-1)^{-3}$$

$$g'(x) = 5e^{5x}$$

$$g''(x) = 25e^{5x}$$

$$\frac{d}{dx}\ln(x+3) = \frac{1}{x+3} = (x+3)^{-1}$$

$$\frac{d^2}{dx^2}\ln(x+3) = -(x+3)^{-2} \quad\blacksquare$$

The derivative of a function $g(x)$ tells us the rate of change of the value of $g(x)$ for each choice of x, as we discovered in Chapter 7. If $g(x)$ is $f'(x)$, then the derivative of $g(x)$, that is, $g'(x) = f''(x)$, tells us the rate of change of $f'(x)$ for each choice of x. Let us illustrate the significance of this by considering picture A of a relative minimum and picture B of a relative maximum. In illustration A, just to the left of b, the slope of the graph is negative and so

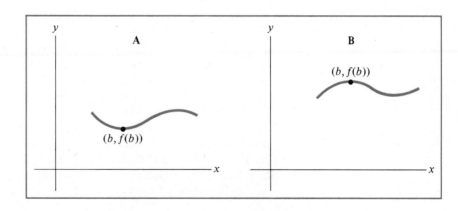

$f'(x)$ is negative; at $x = b$ we have $f'(b) = 0$; just to the right of b the slope of the graph is positive and so $f'(x)$ is positive. Thus, at a relative minimum, $f'(x)$ is changing from negative to zero to positive; $f'(x)$ is increasing at $x = b$ and so $f''(b)$ is positive. Looking at picture B it is easy to see that at a relative maximum $f'(x)$ is changing from positive to zero to negative; $f'(x)$ is decreasing at $x = b$ and so $f''(b)$ is negative. These observations are the basis of the Second Derivative Test.

Second Derivative Test

If b is a critical value for $f(x)$ and there are numbers a and c such that $a < b < c$ and $f'(x)$ exists for all x between a and c, then the following are true.

1. If $f''(b)$ is positive, then $(b, f(b))$ is a relative minimum.
2. If $f''(b)$ is negative, then $(b, f(b))$ is a relative maximum.
3. If $f''(b)$ is zero, then $(b, f(b))$ could be a relative minimum, a relative maximum, or neither of those.

EXAMPLE 2

Find the critical values, relative minima, and relative maxima for $f(x) = x^4 - 2x^2 + 1$.

Solution
Since $f(x)$ is a polynomial, it has a derivative for every x and the critical values are precisely where $f'(x)$ is 0.

$$f'(x) = 4x^3 - 4x$$

$$0 = f'(x) = 4x(x^2 - 1) = 4x(x - 1)(x + 1)$$

The critical values are the roots of this equation, or $x = 0, -1, +1$.

$$f''(x) = 12x^2 - 4$$

$$f''(-1) = 8, \qquad f''(0) = -4, \qquad f''(1) = 8$$

By the Second Derivative Test, $(-1, f(-1)) = (-1, 0)$ is a relative minimum, $(0, f(0)) = (0, 1)$ is a relative maximum and $(1, f(1)) = (1, 0)$ is a relative minimum. ■

We say that the graph of the function $f(x)$ is **concave up** between a and c, with $a < c$, provided any straight line segment connecting any two points on the graph of $f(x)$ between a and c lies entirely on or above the graph of $f(x)$. For example, in the first set of pictures the line segment connecting $(a, f(a))$ and $(c, f(c))$ lies entirely on or above the graph of $f(x)$ and any line segment connecting two points between a and c has the same property.

We say that the graph of the function $f(x)$ is **concave down** between a and c, with $a < c$, provided any straight line segment connecting two points on the graph of $f(x)$ between a and c lies entirely on or below the graph of $f(x)$. For example, in the second set of pictures the line segment connecting $(a, f(a))$ and $(c, f(c))$ lies entirely on or below the graph of $f(x)$ and any line segment connecting two points between a and c has the same property.

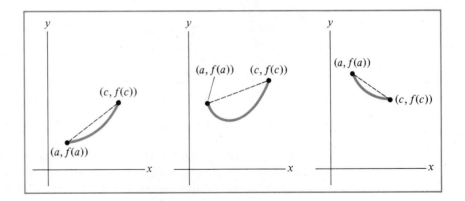

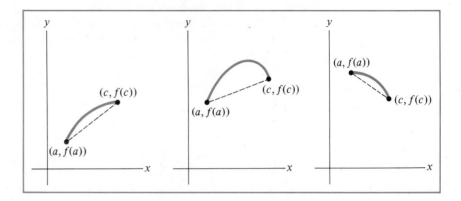

We say that the graph of $f(x)$ is **concave up at b** provided there are numbers a and c, with $a < b < c$, such that the graph of $f(x)$ is concave up between a and c. We say that the graph of $f(x)$ is **concave down at b** provided there are numbers a and c, with $a < b < c$, such that the graph of $f(x)$ is concave down between a and c. If $(b, f(b))$ is a relative minimum and it is not an endpoint, then the graph of $f(x)$ is concave up at b; if $(b, f(b))$ is a relative maximum and it is not an endpoint, then the graph of $f(x)$ is concave down at b. Thus, the Second Derivative Test is a special case of the Test for Concavity below.

Test for Concavity

If there are numbers a, b, and c such that $a < b < c$ and $f'(x)$ exists for all x between a and c, then the following are true.

1. If $f''(b)$ is positive, then the graph of $f(x)$ is concave up at b.
2. If $f''(b)$ is negative, then the graph of $f(x)$ is concave down at b.
3. If $f''(b)$ is zero, the graph of $f(x)$ could be concave up at b, concave down at b, or neither of those.

EXAMPLE 3

Determine whether the graph of $g(x) = x^3 - 3x^2 + 3x - 1$ is concave up or concave down at each of the numbers $-5, 0, 2, 5$.

Solution

We compute the second derivative of $g(x)$ and use the Test for Concavity.

$$g'(x) = 3x^2 - 6x + 3$$

$$g''(x) = 6x - 6 = 6(x - 1)$$

$$g''(-5) = -36, \qquad g''(0) = -6, \qquad g''(2) = 6, \qquad g''(5) = 24$$

The graph of $g(x)$ is concave down at -5 and 0; it is concave up at 2 and 5. ■

We say that $(b, f(b))$ is an **inflection point** provided the graph of $f(x)$ is concave up on one side of $(b, f(b))$ and concave down on the other side. Thus, there are numbers a and c, with $a < b < c$, such that the concavity of the graph between a and b is the opposite of the concavity between b and c.

It follows from the Test for Concavity that **the inflection points of a differentiable function are among the points $(b, f(b))$ such that $f''(b)$ is 0 or else $f''(b)$ does not exist.** In each of the four pictures, $(b, f(b))$ is an inflection point.

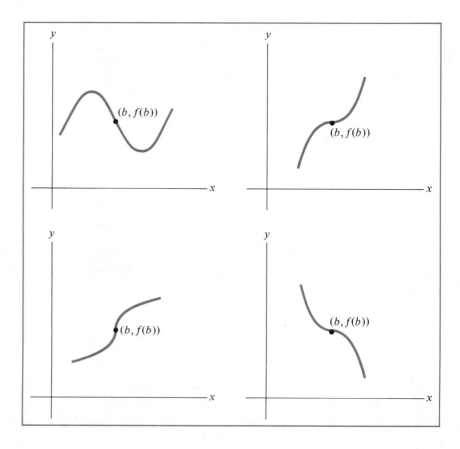

EXAMPLE 4

Determine whether the graph of $g(x) = x^3 - 3x^2 + 3x - 1$ has any inflection points.

Solution

Since $g(x)$ and $g'(x)$ are polynomials, $g''(x)$ exists for every choice of x and $g''(x)$ must be 0 at any inflection point. Thus, we compute $g''(x)$, set it equal to 0, and solve that equation.

$$g'(x) = 3x^2 - 6x + 3$$

$$g''(x) = 6x - 6 = 6(x - 1)$$

$$0 = 6(x - 1)$$

$$x = 1$$

The only possible inflection point is $x = 1$, and we examine the sign of $g''(x)$ to the left of 1 and to the right of 1 to see if the concavity changes. For $x < 1$, we have $x - 1 < 0$ and

$$g''(x) = 6(x - 1) < 0;$$

for $x > 1$, we have $x - 1 > 0$ and

$$g''(x) = 6(x - 1) > 0.$$

By the Test for Concavity, we know that the graph of $g(x)$ is concave down between a and 1 for any $a < 1$ and that the graph is concave up between 1 and c for any $c > 1$. Thus, 1 is an inflection point. ■

Caution   Some of the points $(b, f(b))$ such that $f''(b)$ exists and equals 0 might *not* be inflection points, as the next example shows.

EXAMPLE 5

Determine whether the graph of $f(x) = x^4$ has any inflection points.

Solution

Since $f'(x) = 4x^3$ and $f''(x) = 12x^2$ exist for every choice of x, it must be that $f''(x)$ is 0 at any inflection point. Clearly $f''(x) = 0$ holds only for $x = 0$, and so we need only test $x = 0$ to see if $(0, f(0)) = (0, 0)$ is an inflection point. For any $x \neq 0$, it is clear that $f''(x) = 12x^2$ is positive. For any $a < 0$, the graph is concave up between a and 0; for any $c > 0$, the graph is concave up between 0 and c. Thus, $(0, 0)$ is not an inflection point. ■

EXAMPLE 6

Find the critical values, relative minima, relative maxima, and inflection points for $g(x) = (x - 3)^2(x + 3)^2$.

Solution

First, we compute $g'(x)$ and determine where it is 0.

$$g'(x) = 2(x - 3)(x + 3)^2 + (x - 3)^2 2(x + 3)$$
$$= 2(x - 3)(x + 3)[(x + 3) + (x - 3)]$$
$$= 4(x - 3)(x + 3)x$$

Clearly the critical values are -3, 0, and 3. Now we compute $g''(x)$ and determine where it is 0, since those are the only values of x where we might find inflection points.

$$g'(x) = 4(x^3 - 9x)$$

$$g''(x) = 4(3x^2 - 9)$$
$$= 12(x^2 - 3)$$
$$= 12(x - \sqrt{3})(x + \sqrt{3})$$

Only $(-\sqrt{3}, g(-\sqrt{3}))$ and $(\sqrt{3}, g(\sqrt{3}))$ could be inflection points. We determine the concavity by examining the sign of $g''(x)$. For $x < -\sqrt{3}$ the factors of $g''(x)$ are

$$x + \sqrt{3} < 0, \qquad x - \sqrt{3} < 0;$$

thus, $g''(x)$ is positive and the graph of $g(x)$ is concave up to the left of $-\sqrt{3}$. For $-\sqrt{3} < x < \sqrt{3}$, we have the factors

$$x - \sqrt{3} < 0, \qquad 0 < x + \sqrt{3};$$

thus, $g''(x)$ is negative and the graph of $g(x)$ is concave down between $-\sqrt{3}$ and $\sqrt{3}$. For $x > \sqrt{3}$, the factors are

$$x - \sqrt{3} > 0, \qquad x + \sqrt{3} > 0;$$

thus, $g''(x)$ is positive and the graph of $g(x)$ is concave up to the right of $\sqrt{3}$. It follows that both $(-\sqrt{3}, g(-\sqrt{3})) = (-\sqrt{3}, 36)$ and $(\sqrt{3}, g(\sqrt{3})) = (\sqrt{3}, 36)$ are inflection points; $(-3, g(-3)) = (-3, 0)$ and $(3, g(3)) = (3, 0)$ are relative minima; and $(0, g(0)) = (0, 81)$ is a relative maximum. ■

For the curious reader, the graph of $g(x) = (x - 3)^2(x + 3)^2$ is sketched.

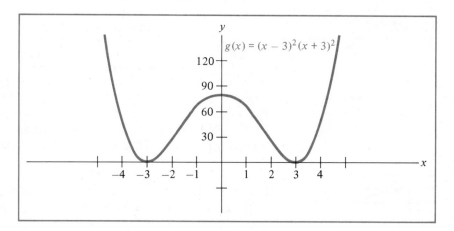

Practical applications of the material in this section are given in Sections 9.5 and 9.6.

Exercises 9.2

In problems 1 through 30, find the second derivative for the given function.

1 $f(s) = 20s^3 - s^2 + 5s + 12$

2 $g(t) = t^{3/2} + 5t^4 - 2t^{-1}$

3 $h(r) = \dfrac{11}{\sqrt[3]{r}}$

4 $f(p) = 6p^{110} + \sqrt{p}$

5 $g(w) = (3w^2 + 1)^{30}$

6 $h(x) = \sqrt{x^3 + x^2 + x}$

7 $f(x) = (2x + 1)x$

8 $g(s) = (s - 3)(2s + 5)(3s - 7)$

9 $h(t) = (t + 6)^{20}(2t - 7)^{40}$

10 $f(x) = (x + 9)^{12}(5x + 1)^8$

11 $g(x) = \dfrac{11x + 8}{5x - 6}$

12 $h(x) = \dfrac{x^2 + 3x}{4x^2 + 8}$

13 $g(x) = \dfrac{1}{\sqrt{4x + 1}} + (3x - 5)(7x + 9)^{15}$

14 $h(x) = \sqrt{\dfrac{8s + 1}{s + 3}}$

15 $h(t) = e^{5t}$

16 $f(p) = p^2 e^{-2p}$

17 $g(u) = (10 - e^u)^{80}$

18 $f(x) = xe^x + x^3 + 9x^2$

19 $g(w) = \sqrt{e^{3w} + 1}$

20 $h(x) = 5e^{3x^2 - x + 11}$

21 $f(x) = \ln 7x$

22 $g(x) = \log_2 x$

23 $h(t) = \ln(9t + 5)$

24 $f(x) = 3 \ln[(2x + 1)(x - 6)]$

25 $g(r) = \ln\left(\dfrac{7r + 1}{r + 3}\right)$

26 $h(x) = 5^x$

27 $g(w) = e^w \ln w$

28 $h(x) = (\ln(5x + 9))^{10}$

29 $f(x) = \log_5(x^2 + 3x + 1)$

30 $g(x) = 3^{2x+5}$

In problems 31 through 52, find the relative minima and relative maxima for the given function.

31 $f(x) = x^2 + 2x - 5$

32 $g(x) = 3x^2 + 1$

33 $h(x) = 10 - x^2$

34 $f(x) = -x^2 + 6x + 12$

35 $g(x) = 2x^3 + 3x^2 - 12x + 8$

36 $h(x) = 2x^3 - 21x^2 + 72x + 6$

37 $f(x) = 3x^4 - 8x^3 - 5$

38 $g(x) = -x^4 - 2x^3 + 9$

39 $h(x) = 4x^5 + x^4$

40 $f(x) = 5x^4 - 4x^5$

41 $g(x) = 3x + 6$

42 $h(x) = .8 - x$

43 $f(x) = \dfrac{2}{x - 5}$

44 $g(x) = \dfrac{4}{3x + 2}$

45 $h(x) = \dfrac{1}{x^2 - 2x + 1}$

46 $f(x) = \dfrac{1}{x^2 - 2x - 3}$

47 $g(x) = e^x + e^{-x}$

48 $h(x) = e^x - e^{-x}$

49 $f(x) = xe^x$

50 $g(x) = x \ln x$

51 $h(x) = (x - 1)^2(x + 2)^3$

52 $f(x) = (x + 5)^2(x - 3)^2$

In problems 53 through 68, find any inflection points that the given function might have.

53 $f(x) = 2x^5$

54 $g(x) = -3x^6$

55 $h(x) = x^2 + 3x + 4$

56 $f(x) = 5x^2 - 10x + 15$

57 $g(x) = 2x^3 + 3x^2 - 12x + 8$

58 $h(x) = 2x^3 - 21x^2 + 72x + 6$

59 $f(x) = 3x^4 - 8x^3 - 5$

60 $g(x) = -x^4 - 2x^3 + 9$

61 $h(x) = \dfrac{2}{x-5}$ 62 $f(x) = \dfrac{10}{5x+3}$

63 $g(x) = xe^x$ 64 $h(x) = xe^{-x}$
65 $f(x) = x \ln x$ 66 $g(x) = x \log x$
67 $h(x) = (x-1)^2(x+2)^3$ 68 $f(x) = (x+5)^2(x-3)^2$

9.3 Curve Sketching

We now have the important tools for sketching the graph of a given function. Usually in sketching a graph we try to accurately plot where the graph crosses the x-axis, which you recall from Chapter 2 is called the **x-intercept,** and where it crosses the y-axis, which is called the **y-intercept.**

EXAMPLE 1

Sketch the graph of $f(x) = x^2 + 8x + 12$.

Solution
We begin by setting the first derivative equal to 0 in order to locate all the critical values.

$$f'(x) = 2x + 8$$

$$0 = 2x + 8$$

$$-4 = x$$

The only critical value is $x = -4$. We apply the Second Derivative Test to this value.

$$f''(x) = 2, \qquad f''(-4) = 2$$

Since $f''(-4)$ is positive, $(-4, f(-4)) = (-4, -4)$ is a relative minimum. Since $f''(x)$ is positive for every x, the graph is concave up at every x, and there are no inflection points.
    To find the x-intercepts, we solve the equation $0 = y = f(x)$.

$$0 = x^2 + 8x + 12$$

$$0 = (x+6)(x+2)$$

$$x = -2, -6$$

To find the y-intercept, we set x equal to 0.

$$f(0) = 12$$

Thus, the x-intercepts are $(-2, 0)$ and $(-6, 0)$; the y-intercept is $(0, 12)$. Now we can sketch the graph.

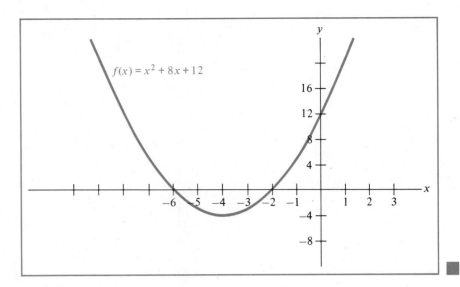

$f(x) = x^2 + 8x + 12$

EXAMPLE 2

Sketch the graph of $g(x) = -x^2 + 6x - 9$.

Solution

We set the first derivative equal to 0 and solve for x.

$$g'(x) = -2x + 6$$

$$0 = -2x + 6$$

$$3 = x$$

The only critical value is $x = 3$. The second derivative is

$$g''(x) = -2.$$

Since $g''(3)$ is negative, $(3, g(3)) = (3, 0)$ is a relative maximum. Since $g''(x)$ is negative for every x, the graph is concave down at every x and there are no inflection points.

To find the x-intercept, we solve $0 = y = f(x)$.

$$0 = -x^2 + 6x - 9$$

$$0 = x^2 - 6x + 9 = (x - 3)^2$$

$$x = 3$$

To find the y-intercept, we set x equal to 0.

$$g(0) = -9$$

The intercepts are $(3, 0)$ and $(0, -9)$. The graph is on page 408. ■

The graphs in Examples 1 and 2 are **parabolas;** it is not difficult to show that the graph of any **quadratic function** $f(x) = ax^2 + bx + c$ is a parabola that opens upward, as in Example 1, if a is positive and opens downward, as in Example 2, if a is negative.

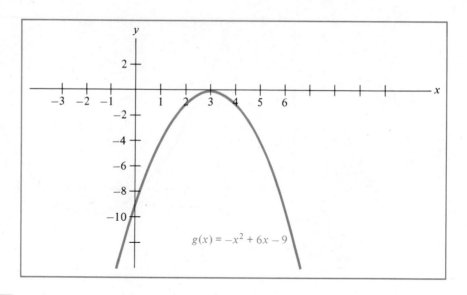

$g(x) = -x^2 + 6x - 9$

EXAMPLE 3

Sketch the graph of $f(x) = 2x^3 + 3x^2 - 12x + 8$.

Solution

The first derivative is

$$f'(x) = 6x^2 + 6x - 12$$
$$= 6(x^2 + x - 2)$$
$$= 6(x + 2)(x - 1).$$

The critical values are -2 and 1. Now we compute the second derivative.

$$f''(x) = 12x + 6$$
$$= 12\left(x + \frac{1}{2}\right)$$

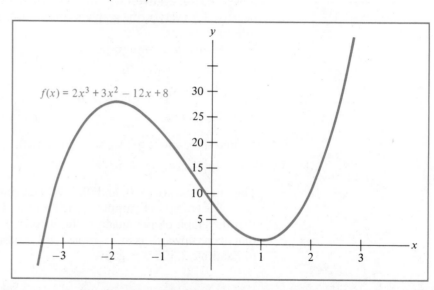

$f(x) = 2x^3 + 3x^2 - 12x + 8$

It is easy to see that $f''(-\frac{1}{2}) = 0$ and $f''(x) < 0$ for $x < -\frac{1}{2}$ (so $x + \frac{1}{2} < 0$) and $f''(x) > 0$ for $x > -\frac{1}{2}$ (so $x + \frac{1}{2} > 0$). Thus, $(-\frac{1}{2}, f(-\frac{1}{2})) = (-.5, 14.5)$ is an inflection point. Since $f''(-2)$ is negative, $(-2, f(-2)) = (-2, 28)$ is a relative maximum; since $f''(1)$ is positive, $(1, f(1)) = (1, 1)$ is a relative minimum.

The x-intercept is difficult to find, and we content ourselves with the y-intercept.

$$f(0) = 8 \ \blacksquare$$

EXAMPLE 4

Sketch the graph of $g(x) = x^3 + 3x^2 + 3x + 1$.

Solution

$$\begin{aligned} g'(x) &= 3x^2 + 6x + 3 \\ &= 3(x^2 + 2x + 1) \\ &= 3(x + 1)^2 \end{aligned}$$

The only critical value is $x = -1$.

$$\begin{aligned} g''(x) &= 6x + 6 \\ &= 6(x + 1) \end{aligned}$$

Clearly $g''(-1) = 0$; also $g''(x) < 0$ for $x < -1$ (so $x + 1 < 0$) and $g''(x) > 0$ for $x > -1$ (so $x + 1 > 0$). Thus, $(-1, g(-1)) = (-1, 0)$ is an inflection point and there are no relative minima or relative maxima. Note that $(-1, 0)$ is the x-intercept and $(0, g(0)) = (0, 1)$ is the y-intercept. It is helpful to plot the additional points, $(-3, g(-3)) = (-3, -8)$, $(-2, g(-2)) = (-2, -1)$, and $(1, g(1)) = (1, 8)$.

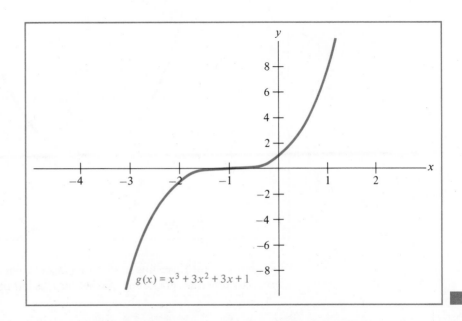

$g(x) = x^3 + 3x^2 + 3x + 1$

An idea that is useful occasionally is symmetry. We say that the graph of $f(x)$ is **symmetric about the y-axis** provided that whenever (a, b) is on the graph then $(-a, b)$ is on the graph also. **The graph of $f(x)$ is symmetric about the y-axis provided that $f(-x) = f(x)$ for every x in the domain of the function.**

EXAMPLE 5

Sketch the graph of $f(x) = e^x + e^{-x}$.

Solution

$$f'(x) = e^x - e^{-x}$$

$$0 = e^x - \frac{1}{e^x}$$

$$0 = (e^x)^2 - 1$$

$$1 = e^x$$

$$0 = x$$

The only critical value is $x = 0$.

$$f''(x) = e^x + e^{-x}$$

Since e^a is positive for any a, we see that $f''(x)$ is positive for all x. Thus, the graph is concave up at every x and $(0, f(0)) = (0, 2)$ is a relative minimum. There is no x-intercept.

Since $f(-x) = e^{-x} + e^{-(-x)} = e^x + e^{-x} = f(x)$, the graph is symmetric about the y-axis.

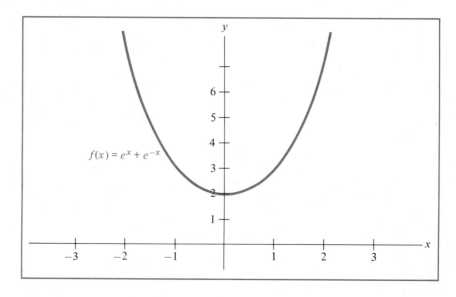

Note that a point travelling along the graph moving away from the y-axis goes up or down independent of whether it is left or right of the y-axis. ■

EXAMPLE 6

Sketch the graph of $g(x) = (x - 1)^2(x + 2)^2$.

Solution

$$g'(x) = 2(x - 1)(x + 2)^2 + 2(x - 1)^2(x + 2)$$
$$= 2(x - 1)(x + 2)[(x + 2) + (x - 1)]$$
$$= 2(x - 1)(x + 2)(2x + 1)$$

The critical values are $-2, -\frac{1}{2}$, and 1. Because examining the second deriva-tive is clearly going to be tedious, we use the last method given in Section 9.1. We evaluate $g(x)$ at each of the critical values, a convenient value less than -2, say -3, and a convenient value greater than 1, say 2.

$$g(-3) = (-4)^2(-1)^2 = 16$$

$$g(-2) = 0$$

$$g\left(-\frac{1}{2}\right) = \left(-\frac{3}{2}\right)^2 \left(\frac{3}{2}\right)^2 = \frac{81}{16}$$

$$g(1) = 0$$

$$g(2) = 1^2 \cdot 4^2 = 16$$

The graph is decreasing between -3 and -2, and it is increasing between -2 and $-\frac{1}{2}$; thus, $(-2, 0)$ is a relative minimum. The graph is decreasing be-tween $-\frac{1}{2}$ and 1; thus $(-\frac{1}{2}, \frac{81}{16}) \approx (-.5, 5.06)$ is a relative maximum. The graph is increasing between 1 and 2; thus, $(1, 0)$ is a relative minimum. The x-intercepts are $(-2, 0)$ and $(1, 0)$ and the y-intercept is $(0, g(0)) = (0, 4)$.

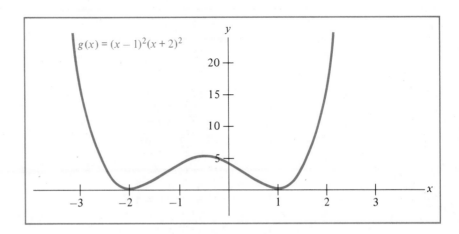

$g(x) = (x - 1)^2(x + 2)^2$

There must be an inflection point between -2 and $-\frac{1}{2}$ and another one between $-\frac{1}{2}$ and 1. If we need to locate these, we can compute the second derivative, set it equal to 0, and solve the equation. ∎

Procedure for Sketching the Graph of f(x)

1. Compute $f'(x)$; compute $f''(x)$ if that is not too tedious.
2. Find the critical values by solving the equation $f'(x) = 0$ and by examining the formula for $f'(x)$.
3. Find the relative minima and relative maxima by examining each critical value a. Use one of the following tests:
   a. If $f''(a) > 0$, then $(a, f(a))$ is a relative minimum; if $f''(a) < 0$, then $(a, f(a))$ is a relative maximum; if $f''(a) = 0$, then no information is gained.
   b. If, for x near to $a, f'(x) < 0$ when $x < a$ and $f'(x) > 0$ when $x > a$, then $(a, f(a))$ is a relative minimum; if, for x near to $a, f'(x) > 0$ when $x < a$ and $f'(x) < 0$ when $x > a$, then $(a, f(a))$ is a relative maximum.
   c. Examine values of $f(x)$ to the left and to the right of a to determine whether $(a, f(a))$ is a relative minimum or a relative maximum.
4. Find each point a that might be an inflection point by solving the equation $f''(x) = 0$ and examining the formula for $f''(x)$. Then examine the sign of $f''(x)$ for x on either side of a to determine whether the concavity changes at $x = a$.
5. Consider whether the graph is symmetric about the y-axis.
6. Find the x-intercept and the y-intercept. Make a table of values, if necessary.

It should be noted that often it is difficult to determine the x-intercepts of a given function, as step 6 calls for. In such a circumstance, we either forego finding the x-intercepts or else we use a calculator to implement some procedure like Newton's method presented at the end of Chapter 7.

Exercises 9.3

Sketch the graph of each of the following functions.

1 $f(x) = 2x + 1$
2 $g(x) = -3x - 2$
3 $f(x) = x^2 - 3x - 4$
4 $g(x) = x^2 + x - 6$
5 $f(x) = -x^2 - 5x - 6$
6 $g(x) = -x^2 + 3x - 2$
7 $f(x) = x^2 - 4x + 4$
8 $g(x) = x^2 + 2x + 1$
9 $f(x) = x^3 + 6x^2 + 12x + 8$
10 $g(x) = x^3 - 3x^2 + 3x - 1$
11 $f(x) = -x^3 - 9x^2 - 27x - 27$
12 $g(x) = -x^3 + 6x^2 - 12x + 8$
13 $f(x) = 2x^3 - 3x^2 - 12x + 1$
14 $g(x) = x^3 - 12x + 2$

15   $f(x) = -x^3 + 27x - 10$
16   $g(x) = -2x^3 + 3x^2 + 36x - 12$
17   $f(x) = 3x^4 - 4x^3 - 72x^2 + 30$
18   $g(x) = 3x^4 - 8x^3 - 18x^2 + 4$
19   $f(x) = 3x^5 - 5x^3 + 1$
20   $g(x) = 3x^5 - 20x^3 + 12$
21   $f(x) = x^4 - 2x^2 + 1$
22   $g(x) = x^4 - 8x^2 + 16$
23   $f(x) = x^4 + 8x^2 + 16$
24   $g(x) = x^4 + 2x^2 + 1$
25   $f(x) = e^x - e^{-x}$
26   $g(x) = 2e^x + 8e^{-x}$
27   $f(x) = (x + 3)^2(x - 2)^2$
28   $g(x) = (x - 3)^2(x + 2)^2$
29   $f(x) = (x + 1)^3(x - 2)^2$
30   $g(x) = (x - 1)^3(x + 2)^2$
31   $h(x) = xe^x$
32   $f(x) = xe^{-x}$
33   $g(x) = e^{x^2}$
34   $h(x) = e^{x^3}$

9.4   Curve Sketching with Asymptotes (Optional)

Now we shall develop our last tool for sketching graphs. Recall that in Section 7.3, we defined what it means for L to be the limit of $f(x)$ as x increases without bound or as x decreases without bound, which we denoted $\lim_{x \to \infty} f(x) = L$ or $\lim_{x \to -\infty} f(x) = L$. The line $y = a$ is a **horizontal asymptote** for the graph of the function $f(x)$ provided at least one of the equations below holds.

$$\lim_{x \to \infty} f(x) = a, \qquad \lim_{x \to -\infty} f(x) = a$$

The next example illustrates the graphical significance of a horizontal asymptote.

EXAMPLE 1

Sketch the graph of $f(x) = \dfrac{x^2 - 1}{x^2 + 1}$.

Solution

$$f'(x) = \frac{2x(x^2 + 1) - 2x(x^2 - 1)}{(x^2 + 1)^2} = \frac{4x}{(x^2 + 1)^2}$$

Note that $f'(x)$ is 0 precisely when x is 0; since $x^2 + 1$ is never 0 (indeed, $x^2 + 1 \geq 1$), the only critical value is $x = 0$. It is easy to apply the First

Derivative Test since $f'(x) < 0$ for $x < 0$ and $f'(x) > 0$ for $x > 0$. Thus, $(0, f(0)) = (0, -1)$ is a relative minimum. The x-intercepts are 1 and -1.

At this point, we might erroneously conclude that the graph of $f(x) = (x^2 - 1)/(x^2 + 1)$ looks like the graph of $g(x) = x^2 - 1$ shown in the illustration.

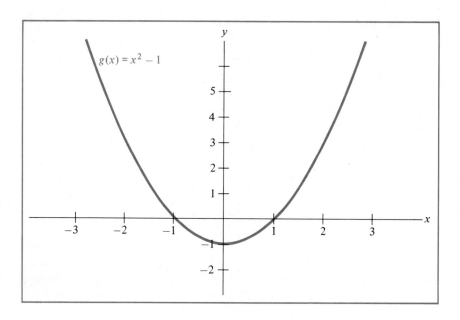

However, since $1 > -1$, we have $x^2 + 1 > x^2 - 1$ and so $1 > (x^2 - 1)/(x^2 + 1)$; the y-values on the graph of $f(x)$ can never equal or exceed 1. Thus, the graph of $f(x)$ is dramatically different from the graph of $g(x)$. We note that the value of $f(x)$ gets very close to 1 for any x with large magnitude. In fact, we have

$$\lim_{x \to \infty} f(x) = \lim_{x \to \infty} \frac{x^2 - 1}{x^2 + 1} = \lim_{x \to \infty} \frac{1 - \dfrac{1}{x^2}}{1 + \dfrac{1}{x^2}}$$

$$= \frac{1 - 0}{1 + 0} = 1$$

and

$$\lim_{x \to -\infty} f(x) = \lim_{x \to -\infty} \frac{x^2 - 1}{x^2 + 1} = \lim_{x \to -\infty} \frac{1 - \dfrac{1}{x^2}}{1 + \dfrac{1}{x^2}}$$

$$= \frac{1 - 0}{1 + 0} = 1.$$

The graph of $f(x)$ is given in the second illustration. In general, it is wise to sketch the horizontal asymptote as a dotted line. Note that the graph of $f(x)$ gets closer and closer to the horizontal asymptote as $x \to \infty$ and as $x \to -\infty$.

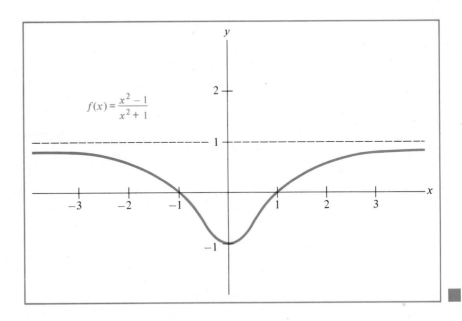

CALCULATOR EXAMPLE

Sketch the graph of $f(x) = e^{-x}$.

Solution

$$f'(x) = -e^{-x}$$

Since e^a is positive for all choices of a, $f'(x)$ is negative for all x and there are no critical values. We note that the value of $f(x)$ gets close to 0 when large positive numbers are chosen for x.

x	1	2	3	4	5	8	10	15
$\dfrac{1}{e^x}$	.36788	.13534	.04979	.01832	.00674	.00034	.00005	.00000

The second row of the table was computed to five decimal place accuracy. The table strongly suggests that

$$\lim_{x \to \infty} e^{-x} = \lim_{x \to \infty} \frac{1}{e^x} = 0.$$

Thus, $y = 0$ or the x-axis is a horizontal asymptote. The graph is given.

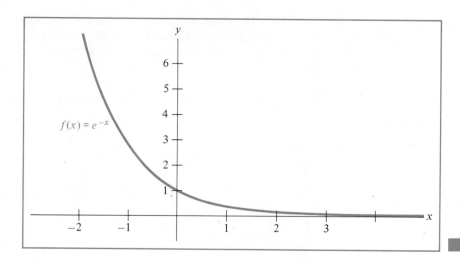

Recall that in Section 7.6 we defined what it means for L to be the limit of $f(x)$ as x approaches a from the right or as x approaches a from the left, which we denoted $\lim\limits_{x\to a+} f(x) = L$ or $\lim\limits_{x\to a-} f(x) = L$. In that definition, L is a fixed real number; we now define a very similar notation where L is replaced by the symbol ∞ for infinity. **If $f(x)$ gets arbitrarily large as x approaches a from the right (or from the left), then we write**

$$\lim_{x\to a+} f(x) = \infty \qquad (\text{or } \lim_{x\to a-} f(x) = \infty).$$

If $f(x)$ gets arbitrarily large in magnitude with a negative sign as x approaches a from the right (or from the left) then we write

$$\lim_{x\to a+} f(x) = -\infty \qquad (\text{or } \lim_{x\to a-} f(x) = -\infty).$$

It should not be surprising that the idea of vertical asymptote is also useful. The line $x = a$ is a **vertical asymptote** for the graph of the function $f(x)$ provided at least one of the statements below is true.

$$\lim_{x\to a+} f(x) = \infty, \qquad \lim_{x\to a-} f(x) = \infty, \qquad \lim_{x\to a+} f(x) = -\infty, \qquad \lim_{x\to a-} f(x) = -\infty$$

The next example illustrates the graphical significance of a vertical asymptote.

EXAMPLE 2

Sketch the graph of $f(x) = \dfrac{2x - 1}{x - 1}$.

Solution

$$f'(x) = \frac{2(x - 1) - (2x - 1)}{(x - 1)^2} = \frac{-1}{(x - 1)^2}$$

So $f'(x)$ is negative for all $x \neq 1$ and it does not exist at $x = 1$. Of course, $f(x)$ does not exist at $x = 1$; thus, the graph is broken into two pieces and each piece is decreasing.

Now we consider the graph for x close to 1. For $x > 1$ we have $x > \frac{1}{2}$ and

$$x - 1 > 0, \qquad x - \frac{1}{2} > 0, \qquad 2x - 1 > 0;$$

thus, $f(x)$ is positive. In view of the limits

$$\lim_{x \to 1} 2x - 1 = 2, \qquad \lim_{x \to 1} x - 1 = 0$$

we see that $f(x)$ gets arbitrarily large with a positive sign as x approaches 1 from the right, that is, $\lim_{x \to 1+} f(x) = \infty$.

For $\frac{1}{2} < x < 1$ we have

$$0 < x - \frac{1}{2} \quad \text{or} \quad 0 < 2x - 1, \quad \text{and} \quad x - 1 < 0;$$

thus, $f(x)$ is negative. We have

$$\lim_{x \to 1-} f(x) = -\infty.$$

Thus, $x = 1$ is a vertical asymptote.

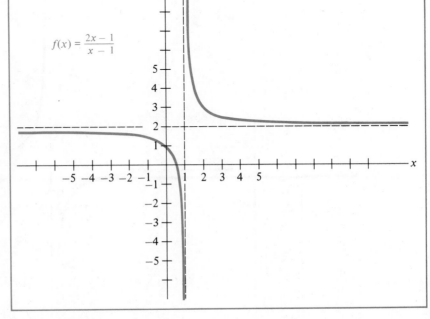

x	$f(x)$
-2	$\frac{5}{3}$
-1	$\frac{3}{2}$
0	1
$\frac{1}{2}$	0
$\frac{3}{4}$	-2
$\frac{5}{4}$	6
$\frac{3}{2}$	4
2	3
3	$\frac{5}{2}$

We note that the graph has a horizontal asymptote, also.

$$\lim_{x \to \infty} f(x) = \lim_{x \to \infty} \frac{2x-1}{x-1} = \lim_{x \to \infty} \frac{2 - \dfrac{1}{x}}{1 - \dfrac{1}{x}}$$

$$= \frac{2-0}{1-0} = 2$$

and

$$\lim_{x \to -\infty} f(x) = \lim_{x \to -\infty} \frac{2x-1}{x-1} = \lim_{x \to -\infty} \frac{2 - \dfrac{1}{x}}{1 - \dfrac{1}{x}}$$

$$= \frac{2-0}{1-0} = 2$$

We sketched the asymptotes and the graph on page 417. ■

EXAMPLE 3

Sketch the graph of $f(x) = 1/x$.

Solution

$$f'(x) = \frac{d}{dx} x^{-1} = -x^{-2} = \frac{-1}{x^2}$$

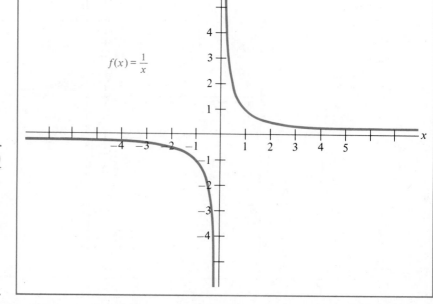

$f(x) = \dfrac{1}{x}$

x	$f(x)$
-2	$-\frac{1}{2}$
-1	-1
$-\frac{1}{2}$	-2
$-\frac{1}{4}$	-4
$\frac{1}{4}$	4
$\frac{1}{2}$	2
1	1
2	$\frac{1}{2}$

The derivative $f'(x)$ is negative for all $x \neq 0$ and it does not exist for $x = 0$. The function is not defined at $x = 0$; thus, the graph is broken into two pieces and each piece is decreasing. We note that the following limits hold.

$$\lim_{x \to -\infty} f(x) = \lim_{x \to -\infty} \frac{1}{x} = 0$$

$$\lim_{x \to \infty} f(x) = \lim_{x \to \infty} \frac{1}{x} = 0$$

$$\lim_{x \to 0+} f(x) = \lim_{x \to 0+} \frac{1}{x} = \infty$$

$$\lim_{x \to 0-} f(x) = \lim_{x \to 0-} \frac{1}{x} = -\infty$$

Thus, $y = 0$ or the x-axis is a horizontal asymptote, and $x = 0$ or the y-axis is a vertical asymptote. We sketched the asymptotes and graph on p. 418. ■

Note that **the graph of $f(x)$ has a vertical asymptote $x = a$ for each a that is an infinite discontinuity** as discussed in Section 7.6.

EXAMPLE 4

Sketch the graph of $f(x) = \dfrac{x^2 + 1}{x^2 - 1}$.

Solution

$$f'(x) = \frac{2x(x^2 - 1) - 2x(x^2 + 1)}{(x^2 - 1)^2} = \frac{-4x}{(x^2 - 1)^2}$$

The derivative is 0 precisely when $x = 0$, and it is not defined at $x = 1$ and $x = -1$. The function is also not defined at $x = 1$ and $x = -1$; thus, the graph is broken into three pieces. It is increasing for $x < 0$ and it is decreasing for $x > 0$ (except, of course, at the discontinuities). The First Derivative Test shows that $(0, f(0)) = (0, -1)$ is a relative maximum.

The fact that $y = 1$ is a horizontal asymptote follows from the limits below.

$$\lim_{x \to \infty} f(x) = \lim_{x \to \infty} \frac{x^2 + 1}{x^2 - 1} = \lim_{x \to \infty} \frac{1 + \dfrac{1}{x^2}}{1 - \dfrac{1}{x^2}}$$

$$= \frac{1 + 0}{1 - 0} = 1$$

and

$$\lim_{x \to -\infty} f(x) = \lim_{x \to -\infty} \frac{x^2 + 1}{x^2 - 1} = \lim_{x \to -\infty} \frac{1 + \dfrac{1}{x^2}}{1 - \dfrac{1}{x^2}}$$

$$= \frac{1 + 0}{1 - 0} = 1.$$

Since $f(x)$ has infinite discontinuities at $x = 1$ and $x = -1$ where the denominator is 0, we suspect the lines $x = 1$ and $x = -1$ are vertical asymptotes. First, we factor the denominator as $x^2 - 1 = (x - 1)(x + 1)$. For $x < -1$, we have $x < 1$ and

$$x + 1 < 0, \qquad x - 1 < 0;$$

since $x^2 - 1$ is positive and $x^2 + 1$ is always positive, $f(x)$ is positive and we have

$$\lim_{x \to -1-} f(x) = \infty.$$

For $-1 < x < 1$, we have $0 < x + 1$ and $x - 1 < 0$; thus, $f(x)$ is negative and we have

$$\lim_{x \to -1+} f(x) = -\infty \quad \text{and} \quad \lim_{x \to 1-} f(x) = -\infty.$$

For $1 < x$, we have $-1 < x$ and $0 < x - 1$, $0 < x + 1$; thus, $f(x)$ is positive and we have

$$\lim_{x \to 1+} f(x) = \infty.$$

We sketch the asymptotes and graph, noting that the graph is symmetric about the y-axis since $f(-x) = f(x)$.

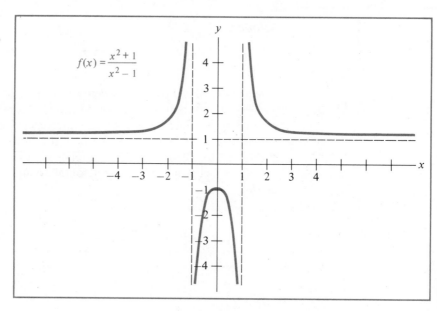

x	f(x)
-3	$\frac{9}{8}$
-2	$\frac{5}{3}$
$-\frac{3}{2}$	$\frac{13}{5}$
$-\frac{1}{2}$	$-\frac{5}{3}$

$$f(x) = \frac{x^2 + 1}{x^2 - 1}$$

Exercises 9.4

Sketch the graph of each of the following functions.

1 $f(x) = \dfrac{x^2}{x^2 + 1}$

2 $g(x) = \dfrac{x^2 - 4}{x^2 + 1}$

3 $f(x) = \dfrac{x^2 + 2}{x^2 + 5}$ 4 $g(x) = \dfrac{x^2 + 9}{x^2 + 4}$

5 $f(x) = \dfrac{x - 3}{x + 2}$ 6 $g(x) = \dfrac{x + 1}{x + 5}$

7 $f(x) = \dfrac{8x}{x - 3}$ 8 $g(x) = \dfrac{6x}{2x + 3}$

9 $f(x) = \dfrac{x^2}{x^2 - 1}$ 10 $g(x) = \dfrac{x^2}{x^2 - 4}$

11 $f(x) = \dfrac{x^2 + 3}{x^2 - 4}$ 12 $g(x) = \dfrac{x^2 + 5}{x^2 - 1}$

13 $f(x) = \dfrac{\sqrt{x} - 1}{\sqrt{x} + 1}$ 14 $g(x) = \dfrac{\sqrt{x}}{\sqrt{x} + 5}$

15 $f(x) = \dfrac{1}{x^2}$ 16 $g(x) = \dfrac{2}{x^2} + 3$

17 $f(x) = \dfrac{1}{x^3}$ 18 $g(x) = \dfrac{5}{x^3} + 1$

19 $f(x) = \dfrac{6}{x - 2}$ 20 $g(x) = \dfrac{3}{x + 5}$

21 $f(x) = \dfrac{4}{(x - 1)^2}$ 22 $g(x) = \dfrac{7}{(x + 3)^2}$

23 $f(x) = \dfrac{e^x}{e^x + 1}$ 24 $g(x) = \dfrac{e^{-x}}{e^{-x} + 2}$

25 $f(x) = 2x + \dfrac{1}{x}$ 26 $g(x) = 3x + \dfrac{2}{x}$

9.5 Algebraic Optimization Problems

The techniques for sketching graphs are now applied to practical problems. In each problem, we shall begin by following the procedure for sketching the graph of $f(x)$ that was given at the end of Section 9.3. Then we go one step further in order to find an absolute minimum or an absolute maximum rather than a relative minimum or a relative maximum. An **optimum** is a point that is either an absolute minimum or an absolute maximum.

EXAMPLE 1

Suppose that the profit of a certain service company is given by the function

$$P(x) = 9x - 200 - .2x^{3/2}$$

where x is the number of customers served. How many customers should be served for maximum profit?

Solution

We must find the point on the graph of $P(x)$ that has the largest y-value; thus, we must find an absolute maximum. This requires that we determine the appearance of the graph of $P(x)$.

$$P'(x) = 9 - .3x^{1/2}$$

$$0 = 9 - .3x^{1/2}$$

$$.3x^{1/2} = 9$$

$$x^{1/2} = 30$$

$$x = 900$$

$$P''(x) = -.15x^{-1/2}$$

$$P''(900) = -.005 < 0$$

According to the Second Derivative Test, $(900, P(900)) = (900, 2500)$ is a relative maximum. Since there are no critical values to the left of 900 and the graph of $P(x)$ is increasing just to the left of 900, the graph must be increasing at every nonnegative $x < 900$. Similarly the graph must be decreasing at every $x > 900$. Thus, $(900, 2500)$ is the absolute maximum and the maximum profit of \$2500 occurs when 900 customers are served. ■

Now we shall carry out the procedure with less explanation.

EXAMPLE 2

A movie theatre's management is considering reducing the price of tickets from \$5.50 in order to get more customers. After checking with other theatres and mailing out some questionnaires, they decide that the average number of customers per day q is given by the function

$$q(x) = 100 + 100x$$

where x is the amount the ticket price is reduced. Find the ticket price that results in maximum revenue.

Solution

After a reduction of x, the ticket price is $(5.50 - x)$ and the revenue per day is

$$R(x) = (5.50 - x)q(x) = (5.50 - x)(100 + 100x)$$

$$R(x) = 550 + 450x - 100x^2$$

$$R'(x) = 450 - 200x$$

$$0 = 450 - 200x$$

$$200x = 450$$

$$x = 2.25$$

$$R''(x) = -200$$

$$R''(2.25) < 0$$

According to the Second Derivative Test, $(2.25, R(2.25)) = (2.25, 1056.25)$ is a relative maximum. Since there are no critical values other than 2.25, $(2.25, 1056.25)$ is the absolute maximum, and the maximum revenue results when the ticket price is $\$5.50 - \$2.25 = \$3.25$. ∎

EXAMPLE 3

Suppose $C(x) = .01x^2 + 100$ is the cost of running a yo-yo factory in hundreds of dollars where x is how many thousands of yo-yos are produced. Determine the value of x that results in minimum average cost.

Solution

Recall from Section 8.3 that average cost is $A(x) = \dfrac{C(x)}{x}$.

$$A(x) = .01x + 100x^{-1}$$

$$A'(x) = .01 - 100x^{-2}$$

$$0 = .01 - 100x^{-2}$$

$$100x^{-2} = .01$$

$$x^{-2} = .0001$$

$$x^2 = 10,000$$

$$x = 100$$

$$A''(x) = 200x^{-3}$$

$$A''(100) > 0$$

According to the Second Derivative Test, $(100, A(100)) = (100, 2)$ is a relative minimum. Since there are no critical values other than 100, $(100, 2)$ is the absolute minimum, and the minimum average cost results from producing 100 thousand yo-yos. ∎

EXAMPLE 4

A company receives some profit from franchises that it sold for a manufacturing process. The company decides to also operate a manufacturing plant of its own. The company's profit in thousands of dollars as a function of x, the number of units produced (in hundreds), is

$$P(x) = x^2 - 6x - 19 \qquad 0 \le x \le 10.$$

The reason for the inequality $x \le 10$ is that the company promised its franchisees not to ever manufacture more than 1000 units. Determine the value of x that results in maximum profit.

Solution

We must determine what point on the graph of $P(x)$ has the greatest y-value. To determine the general shape of the graph, we find the critical values.

$$P'(x) = 2x - 6$$
$$0 = 2x - 6$$
$$x = 3$$

$$P''(x) = 2$$
$$P''(3) = 2 > 0$$

There is a relative minimum at the only critical value, $x = 3$. The graph is decreasing to the left of $x = 3$ and increasing to the right of $x = 3$; thus, the maximum value occurs at one of the endpoints $x = 0$ or $x = 10$. Since

$$P(0) = 19 \quad \text{and} \quad P(10) = 59,$$

the maximum profit occurs when $x = 10$. ■

We can now better explain why marginal cost equals marginal revenue when profit is a maximum. A common sense derivation was given for this in Section 7.5. Let x be the number of units manufactured and assume that all units made are sold. Recall that profit $P(x)$ is revenue $R(x)$ minus cost $C(x)$. If the business is well conceived, then $P(0)$ will not be the maximum profit, and $(0, P(0))$ is the only endpoint for the graph of $P(x)$. Thus, any absolute maximum occurs at a critical value a; provided $R(x)$ and $C(x)$ are differentiable, this requires that

$$0 = P'(a) = R'(a) - C'(a)$$
or
$$C'(a) = R'(a).$$

EXAMPLE 5

A company with revenue of \$100,000 for its last completed month spent \$5000 on advertising during that month. Its research indicates that whenever it doubles its advertising budget its revenue increases 20%. How much should it spend on advertising in order to maximize revenue minus advertising expenditure?

Solution

If x denotes the number of times the advertising budget has doubled then the current budget is $2^x(5000)$. According to the company's research, their revenue should be $1.2^x(100,000)$ and we want to maximize

$$R(x) = 1.2^x(100,000) - 2^x(5,000).$$
$$R'(x) = 100,000(1.2^x)\ln 1.2 - 5,000(2^x)\ln 2$$
$$0 = 100,000(1.2^x)\ln 1.2 - 5,000(2^x)\ln 2$$
$$(2^x)\ln 2 = 20(1.2^x)\ln 1.2$$
$$\left(\frac{2}{1.2}\right)^x = \frac{2^x}{1.2^x} = \frac{20 \ln 1.2}{\ln 2}$$
$$1.67^x = 5.26 \qquad \text{(approximated to two decimal places)}$$

$$x \ln 1.67 = \ln(1.67^x) = \ln 5.26$$

$$x = \frac{\ln 5.26}{\ln 1.67} = 3.24$$

This is the only critical value; to determine whether it is a relative maximum, we compute $R(3)$, $R(3.24)$, and $R(4)$.

$$R(3) = 1.73(100,000) - 8(5,000) = 133,000$$

$$R(3.24) = 1.81(100,000) - 9.45(5,000) = 133,750$$

$$R(4) = 2.08(100,000) - 16(5,000) = 127,000$$

This indicates that $(3.24, R(3.24))$ is a relative maximum. Since there are no critical values other than 3.24, $(3.24, R(3.24))$ is the absolute maximum, and to maximize revenue minus advertising expenditure the company should spend $2^{3.24}$ ($5,000) = $47,250 for advertising. ∎

EXAMPLE 6

Each time a certain retailer reorders merchandise there are *fixed costs* related to bookkeeping, readying a warehouse, and handling. These costs, which are $200 per reorder, are increased as the number of reorders is increased. The annual cost of storing the merchandise (called the **carrying cost**) is $10 times the average number of units stored in inventory. Find the number of units that should be ordered with each reorder to minimize the sum of inventory costs and reorder costs. Assume that units from inventory are used at a constant rate and that 100,000 units are required each year.

Solution
Let x be the number of units obtained by each reorder. Since inventory is x just after reordering and 0 just before reordering, the average number of units in inventory is $x/2$ and the inventory costs are $10(x/2)$. The number of reorders is $100,000/x$ and the reordering costs are

$$200\left(\frac{100,000}{x}\right).$$

Thus, the sum of the costs is

$$C(x) = 5x + 20,000,000x^{-1}.$$
$$0 = C'(x) = 5 - 20,000,000x^{-2}$$
$$5 = 20,000,000x^{-2}$$
$$x^2 = 4,000,000$$
$$x = 2,000$$

$$C''(x) = 40,000,000x^{-3}$$
$$C''(2,000) = .005 > 0$$

According to the Second Derivative Test, $(2,000, C(2,000)) = (2,000,$

20,000) is a relative minimum. Since there are no critical values other than 2,000, (2,000, 20,000) is the absolute minimum, and the minimum costs result from obtaining 20,000 units by each of 50 orders. ■

Exercises 9.5

1 Suppose that the profit of a certain retailer is given by the function

$$P(x) = 10x - 500 - .1x^2$$

where x is the number of units sold. How many units should be sold for maximum profit?

2 Suppose that the profit of a certain manufacturer is given by the function

$$P(x) = 5x - .01e^x$$

where x is the number of thousand units made. For maximum profit how many units should be made?

3 A cable television company charges $15 per month. Its research indicates that the number of its subscribers is given by the function

$$q(x) = 10,000 + 800x$$

where x is the amount that it reduces its monthly subscription fee. Determine what the fee should be for the company to receive maximum revenue.

4 A commuter airlines charges $40 per passenger and its research suggests that the number of its passengers per week is

$$q(x) = 450 + 50x$$

where x is the amount it reduces the ticket price. Determine what the price should be for maximum revenue.

5 Suppose the cost in dollars of running a certain factory is

$$C(x) = .005x^3 + 80$$

where x is the number of units produced. Determine the value of x that results in minimum average cost.

6 Suppose the cost of running a certain business in dollars is

$$C(x) = x^2 - 6x + 100$$

where x is the number of customers served. Determine the value of x that results in minimum average cost.

7 The factory in problem 5 sells each unit for $216. Determine how many units should be produced for maximum profit to result assuming that all units made are sold.

8 The business in problem 6 receives $20 in revenue from its typical customer. Determine how many customers it should serve in order to maximize its profit.

9   Suppose a firm with current revenue of $25,000 and an advertising budget of $1000 discovers that doubling its advertising expenditure increases its revenue by 10%. How much should it spend on advertising in order to maximize revenue minus advertising expenditure?

10   Suppose a business with current revenue of $50,000 and an advertising budget of $1000 discovers that increasing its advertising budget by 50% increases its revenue by 5%. How much should it spend on advertising in order to maximize revenue minus advertising expenditure?

11   For a certain business, the reorder costs of placing an order for more merchandise are $150 and the cost of keeping one unit stored for 1 year is $18. If 2400 units are required for each year and units are used from inventory at a constant rate, how many units should be ordered with each reorder to minimize the sum of inventory costs and reorder costs?

12   A manufacturer has fixed costs of $500 to ready its plant for one production period, and the cost of storing one unit in inventory for one year is $20. Each year a total of 5000 units is required. How many production periods should be scheduled if the sum of inventory costs and production fixed costs is to be minimized?

13   What is the largest product xy that is possible if x and y are nonnegative numbers and $x + y = 10$?

14   What is the smallest product xy that is possible if x and y are nonnegative numbers and $x + y = 10$?

15   Find an absolute minimum for the quadratic function $f(x) = ax^2 + bx + c$ where a, b, and c are constants and a is positive.

16   Find an absolute maximum for $f(x) = ax^2 + bx + c$ where a, b and c are constants and a is negative.

9.6   Geometric Optimization Problems

The problems treated in Section 9.5 were called algebraic optimization problems because each problem used a term like "profit," "average cost," or "revenue" and that term had an associated algebraic formula. The function to be optimized was indicated by the terms mentioned in the problem. Each problem in this section involves additional steps because we must discover a formula for the function to be optimized. In each problem it is necessary to draw a picture and consider the geometric relations of the quantities involved in the problem.

EXAMPLE 1

A store that sells outdoor furniture wants to set up a display patio in front of the store. To accommodate the furniture to be displayed, the patio must cover 600 square feet. The side of the patio along the road will have a nice redwood fence that costs $12 per linear foot, and the other three sides will have a fence costing $6 per linear foot. Determine the dimensions of the rectangular patio which minimize the total cost of the fence.

Solution

First we draw a picture of the patio and we label each side with a variable, since many configurations are possible.

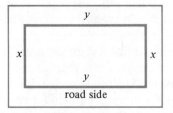

Now we decide what is to be optimized, and we find a formula for it. Cost of the fence C is to be minimized, and a formula for it is

$$C = 12y + 6(2x + y).$$

Note that C is written in terms of the two variables x and y; we want a function of one variable.

Now we find a *constraint* on what rectangular patios are acceptable, and we use that constraint to get rid of one of the variables.

$$600 = \text{area of patio} = xy$$

$$600x^{-1} = y$$

Substitute for y in the cost function and then find the absolute minimum, as we are accustomed to doing.

$$C = 12(600x^{-1}) + 6(2x + 600x^{-1})$$

$$C(x) = 10{,}800x^{-1} + 12x$$

$$C'(x) = -10{,}800x^{-2} + 12$$

$$0 = -10{,}800x^{-2} + 12$$

$$x^2 = 900 \quad \text{and} \quad x = 30$$

Although $x = -30$ is also a critical value, we neglect it because distances cannot be negative.

$$C''(x) = 21{,}600x^{-3}$$

$$C''(30) > 0$$

$(30, C(30))$ is a relative minimum.

Since 30 is the only critical value, $(30, C(30)) = (30, 720)$ is the absolute minimum, and the patio should be 30 by 20 feet. ∎

The steps used to solve Example 1 give a general procedure for solving any similar problem.

Procedure for Solving Geometric Optimization Problems

1. Draw a picture.
2. Label the quantities that vary.
3. Write down formulas that relate those quantities.
4. Decide what quantity is to be optimized and write down a formula for it.
5. Write down the *constraints* on the problem.
6. Use the constraints to get a formula with one independent variable.
7. Find the required optimum for the given function.

We use the given procedure in each of the remaining examples.

EXAMPLE 2

A new store wants to mail advertising announcements to nearby residences. The announcement must be printed on a stiff rectangular piece of paper; there must be 72 square inches of printed material with 1-inch margins on the sides and 2-inch margins at the top and bottom. What is the smallest piece of paper that can be used?

Solution

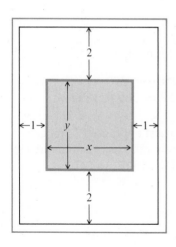

$$A = \text{area of announcement} = (x + 2)(y + 4)$$

$$xy = \text{area of printed matter}$$

$$72 = xy$$

$$72x^{-1} = y$$

$$A(x) = (x + 2)(72x^{-1} + 4)$$

$$A(x) = 80 + 4x + 144x^{-1}$$

$$A'(x) = 4 - 144x^{-2}$$

$$0 = 4 - 144x^{-2}$$

$$4 = 144x^{-2}$$

$$x^2 = 36$$

$$x = 6$$

$$A''(x) = 288x^{-3}$$

$$A''(6) > 0$$

$(6, A(6))$ is a relative minimum

Since 6 is the only critical value for $x > 0$, $(6, A(6)) = (6, 128)$ is the absolute minimum and the piece of paper should be 8 by 16 inches. ■

EXAMPLE 3

The owner of a miniature golf course wants to enclose the largest rectangular park that he can using 1000 feet of fence on three sides and a straight drainage ditch on the fourth side. What should be the dimensions of the rectangle?

Solution

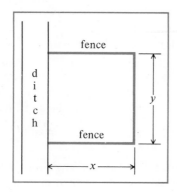

$A = $ length of fence $= 2x + y$

$xy = $ area of park

$1000 = 2x + y$

$1000 - 2x = y$

$A(x) = x(1000 - 2x)$

$A(x) = 1000x - 2x^2$

$A'(x) = 1000 - 4x$

$0 = 1000 - 4x$

$4x = 1000$

$x = 250$

$A''(x) = -4$

$A''(250) < 0$

$(250, A(250))$ is a relative maximum

Since 250 is the only critical value, $(250, A(250)) = (250, 125{,}000)$ is the absolute maximum, and the park should be 250 by 500 feet. ∎

EXAMPLE 4

The length of a package plus its circumference must not exceed 72 inches or else U.S. Parcel Post will not accept it. What are the dimensions of the largest rectangular package with square bottom that Parcel Post will accept?

Solution

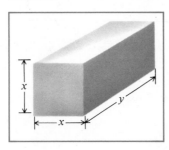

$V = $ volume $= x^2 y$

$y + 4x = $ length $+$ circumference.

$72 = y + 4x$

$72 - 4x = y$

$V(x) = x^2(72 - 4x)$

$V(x) = 72x^2 - 4x^3$

$V'(x) = 144x - 12x^2$

$0 = 144x - 12x^2 = 12x(12 - x)$

critical values 0, 12

$V''(x) = 144 - 24x$

$V''(12) < 0, \qquad V''(0) > 0$

$(12, V(12))$ is a relative maximum

Clearly $(12, V(12)) = (12, 3456)$ is the absolute maximum, and the package should be 12 by 12 by 24 inches. ■

EXAMPLE 5

A manufacturing plant at A (see illustration) delivers its finished product to a warehouse at B by carrying it across the river by boat and along the other side of the river by railroad. If the boat cost \$10 per mile and the railroad \$5 per mile, what is the least costly route for transporting the product from A to B?

Solution

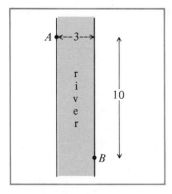

The shortest route from A to B would be along the straight line through A and B. That route is probably not the cheapest because the boat is so much more expensive than the railroad. Going directly across the river would get the most use out of the railroad that is cheap but that would require traveling a maximum distance. Let x be the distance from the boat's docking point D to B. The triangle ACD is a right triangle and the Pythagorean Theorem says that the square of the distance from A to D (the hypotenuse) equals $9 + (10 - x)^2$ (the sum of the squares of the two legs).

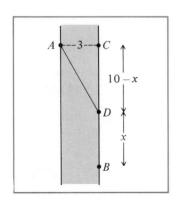

$$\text{cost of trip} = 10[9 + (10 - x)^2]^{1/2} + 5x$$

$$C(x) = 10[9 + (10 - x)^2]^{1/2} + 5x$$

$$C'(x) = 5[9 + (10 - x)^2]^{-1/2}2(10 - x)(-1) + 5$$

$$0 = -10(10 - x)[9 + (10 - x)^2]^{-1/2} + 5$$

$$10(10 - x)[9 + (10 - x)^2]^{-1/2} = 5$$

$$10(10 - x) = 5[9 + (10 - x)^2]^{1/2}$$

$$2(10 - x) = [9 + (10 - x)^2]^{1/2}$$

$$4(10 - x)^2 = 9 + (10 - x)^2$$

$$3(10 - x)^2 = 9$$

$$(10 - x)^2 = 3$$

$$10 - x = \sqrt{3} \text{ or } -\sqrt{3}$$

$$x = 10 \pm \sqrt{3}$$

$$\approx 11.7 \text{ or } 8.3$$

The route with $x = 11.7$ is unnecessarily long and cannot possibly be the least costly route. A correct answer must have $0 \le x \le 10$. We choose convenient values less than and greater than $x = 8.3$.

$$C'(6) = -3, \qquad C'(10) = 5$$

By the First Derivative Test $(8.3, C(8.3)) = (8.3, 76)$ is a relative minimum and, since $x = 8.3$ is the only critical value with $0 \le x \le 10$, it must be that $(8.3, 76)$ is the absolute minimum. ■

EXAMPLE 6

A chemical manufacturer needs a container that is a right circular cylinder with a volume of 16π cubic feet to ship a certain chemical to its customers.

Determine the dimensions of the container that uses the least amount of material.

Solution

This problem requires that we know that the volume of a right circular cylinder is the area of the base πr^2 times the height h (see the first illustration). The problem also requires that we compute the surface area of the cylinder. The area of each of the circles that form the top and the base is πr^2, but we must also compute the lateral area. In order to compute the lateral area, we imagine cutting the lateral surface vertically and opening it out as indicated in the next pictures.

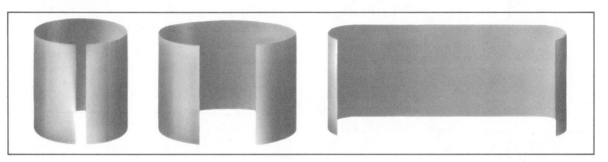

After the cylinder is cut, opened, and spread out, it is a rectangle h tall and $2\pi r$ wide, since $2\pi r$ is the circumference of the circle. Now we can solve the problem.

$$16\pi = \pi r^2 h = \text{volume}$$

$$S = \text{surface area} = 2\pi r^2 + 2\pi rh$$

$$h = \frac{16\pi}{\pi r^2} = 16r^{-2}$$

$$S(r) = 2\pi r^2 + 2\pi r(16r^{-2})$$

$$S(r) = 2\pi r^2 + 32\pi r^{-1}$$

$$S'(r) = 4\pi r - 32\pi r^{-2}$$

$$0 = 4\pi r - 32\pi r^{-2}$$

$$32\pi r^{-2} = 4\pi r$$

$$8 = \frac{32\pi}{4\pi} = r^3$$

$$r = 2$$

$$S''(r) = 4\pi + 64\pi r^{-3}$$

$$S''(2) > 0$$

$(2, S(2))$ is a relative minimum.

Since 2 is the only critical value, $(2, S(2)) = (2, 24\pi)$ is the absolute minimum; the container should have $r = 2$ feet and $h = 4$ feet. ■

Exercises 9.6

1   A rectangular patio is to be constructed with a brick wall costing $20 per linear foot on three sides and a concrete block wall costing $15 per linear foot on the fourth side. If 56 square feet must be enclosed, what are the dimensions of the least expensive patio?

2   Do the preceding problem with a brick wall on only one side, concrete block wall on the remaining three sides, and 168 square feet enclosed.

3   An ad is to be printed on a rectangular poster that is 200 square inches in area. If the side margins must be 1 inch each and the margins at the top and bottom must be 2 inches each, what is the largest area that the printed matter can cover?

4   Do the preceding problem with margins of 2 inches on the sides and a 3-inch margin at the top and at the bottom; also, the rectangular poster has an area of 216 square inches.

5   Using an outside wall of his warehouse, which is 100 feet long, as one side and 196 feet of fence on the other three sides, the warehouse owner wants to enclose the largest rectangular storage yard possible. What should be the dimensions of the yard?

6   Do the preceding problem with only 88 feet of fence available.

7   The length of a package plus its circumference must not exceed 108 inches or else the United Parcel Service will not accept it. What are the dimensions of the largest rectangular package with square bottom that United Parcel Service will accept?

8   If United Parcel Service decided not to accept a package with length plus circumference exceeding 88 inches, what would be the largest rectangular package with square bottom that it would accept? (See problem 7.)

9   Determine the dimensions of the largest package in the shape of a right circular cylinder that U.S. Parcel Post will accept (see Examples 4 and 6).

10   Determine the dimensions of the largest package in the shape of a right circular cylinder that United Parcel Service will accept (see problem 7 and Example 6).

11   A container is to be made in the shape of a right circular cylinder with total surface area of 24π. Determine the dimensions of the container if the volume is to be as large as possible (see Example 6).

12   Do the preceding problem assuming that the surface area is to be 150π.

13   A 150-foot section of fence is to be shaped into a circle to enclose a play

yard or it is to be cut into two pieces with one piece shaped into a circle and the other piece shaped into a square. How should the fence be used to enclose a maximum area?

14 In the preceding problem, assume that a minimum area is to be enclosed because the fence is part of an exhibit on a crowded display floor. How should the fence be used?

15 A rectangular box with a square base and no top is to be constructed to have a volume of 32 cubic inches. Determine the dimensions of the box that has minimum total surface area.

16 Do the preceding problem assuming that the box has a top and the volume should be 1000 cubic inches.

17 A rectangular box with square base and no top is to be constructed with a total surface area of 192 square inches. Determine the dimensions of the box with maximum volume.

18 Do the preceding problem assuming that the box has a top and the surface area is 216 square inches.

19 A fence is to be constructed along the perimeter of a rectangular area of 600 square feet and then the rectangle is to be divided into two equal areas by another fence parallel to two sides of the rectangle. Determine the dimensions requiring the minimum length of fence.

20 Do the preceding problem assuming that the rectangular area is 1800 square feet and it should be divided into three equal areas by two additional fences parallel to two sides of the rectangle.

Review of Terms

Important Mathematical Terms

turning points, *p. 390*
increasing from *a* to *b*, *p. 391*
decreasing from *a* to *b*, *p. 391*
relative minimum, *p. 391*
absolute minimum, *p. 391*
relative maximum, *p. 391*
absolute maximum, *p. 391*
relative minimum value, *p. 392*
relative maximum value, *p. 392*
absolute minimum value, *p. 392*
absolute maximum value, *p. 392*
First Derivative Test, *p. 393*
critical value, *p. 392*
derivative function, *p. 398*
second derivative, *p. 398*
third derivative, *p. 398*
Second Derivative Test, *p. 400*

concave up, *p. 400*
concave down, *p. 400*
Test for Concavity, *p. 401*
inflection point, *p. 402*
x-intercept, *p. 406*
y-intercept, *p. 406*
parabola, *p. 407*
quadratic function, *p. 407*
symmetric about the *y*-axis, *p. 410*
horizontal asymptote, *p. 413*
vertical asymptote, *p. 418*
infinite discontinuity, *p. 421*
optimum, *p. 421*
constraint, *p. 429*

Important Terms from the Applications

revenue, *p. 422*   marginal cost, *p. 424*

Review Problems

In problems 1 through 8, find the relative minima and the relative maxima for the given function.

1 $f(x) = 5 - 2x$

2 $g(x) = 3 + 6x - x^2$

3 $h(x) = x^3 + 3x^2 + 3x - 6$

4 $f(x) = 2x^3 - 3x^2 - 36x + 40$

5 $g(x) = \dfrac{x^2 + 3}{x - 1}$

6 $h(x) = x - e^x$

7 $f(x) = (x - 1)^3(x + 1)^2$

8 $g(x) = \sqrt[3]{x}$

In problems 9 through 12, determine where the function is increasing and where it is decreasing.

9 $f(x) = \dfrac{4}{x} + x$

10 $g(x) = 2x^3 - 3x^2 - 36x + 40$

11 $h(x) = x - \ln x$

12 $f(x) = (x - 1)^3(x + 1)^2$

Find the second derivative of each of the following functions.

13 $f(x) = 3x^{5/3} + 3x^2 + 2x^{-1}$

14 $g(x) = \sqrt{x^2 + 4}$

15 $h(x) = (2x + 1)^8(x - 3)^{12}$

16 $f(x) = e^{x^2+3}$

17 $g(x) = \dfrac{5x - 3}{x + 4}$

18 $h(x) = 4 \ln(x + 7)$

19 $f(x) = 2 \log_3 x$

20 $g(x) = 4^x$

In problems 21 through 24, find any inflection points that the given function might have.

21 $f(x) = x^3 + 3x^2 + x$

22 $g(x) = 2x^3 - 15x^2 + 36x$

23 $h(x) = x^2 - 2e^x$

24 $f(x) = (x + 1)^3(x + 2)^4$

Sketch the graph of each of the following functions.

25 $f(x) = x^2 - 4x + 3$

26 $g(x) = 2x^3 - 15x^2 - 36x$

27 $h(x) = x^3 + 3x^2 + x$

28 $f(x) = (x - 2)^2(x + 1)^2$

29 $g(x) = \dfrac{1}{5} x^5 + \dfrac{2}{3} x^3 + x$

30 $h(x) = x^4 - 18x^2 + 10$

31 $f(x) = \dfrac{x + 1}{x - 1}$

32 $g(x) = 3e^{2x}$

33 $h(x) = \dfrac{5}{x} - 3$

34 $f(x) = \dfrac{1}{x^2 - 4}$

Solve each of the following problems.

35 Suppose that the profit of a certain business is given by the function

$$P(x) = 30x - .1x^3 - 100$$

where x is the number of customers served. Determine the value of x that results in maximum profit.

36 In a certain orchard it is observed that if 30 peach trees are planted per acre then each tree will yield 25 pounds. How many trees should be planted per acre if each additional tree reduces the yield per tree by one half of a pound?

37 An apartment complex consisting of 400 two bedroom apartments has 300 of them rented and the rent is $360 per month. Management's research indicates that if the rent is reduced by x dollars then the number of apartments rented q will be $q = (5/4)x + 300$ for $0 \le x \le 80$. Determine the rent that results in maximum revenue.

38 Suppose the cost for running a certain service business is $C(x) = 50x + e^{.01x}$ where x is the number of customers served. Determine the value of x that results in minimum average cost.

39 An ad is to be printed on a rectangular page that is 100 square inches in area. If every margin must be 1 inch, what is the largest area that the printed matter can cover?

40 A fence is to be constructed along the perimeter of a rectangular area of 264 square feet and then the rectangle is to be divided into two equal areas by another fence parallel to two sides of the rectangle. If the fence that divides the rectangle costs $10 per linear foot and the fence on the sides costs $6 per linear foot, what is the minimum cost of the fences?

41 A rectangular yard for small children is to be made using 40 feet of fence for two adjacent sides and two long straight walls of a school building that come together at a right angle. Determine the dimensions of the yard with maximum area.

42 A rectangular box is to be made with a volume of 576 cubic inches; the base is to be twice as wide as it is tall. Determine the dimensions of the box that uses the least amount of material.

43 A trash can is to be made in the shape of a right circular cylinder with bottom and without top. If the volume must be 8π cubic feet, what are the dimensions of the can that uses the least amount of material?

44 A window is to be made in the shape of a semicircle with diameter d on top of a rectangle with width d. Determine the dimensions of the window with maximum surface area if the perimeter of the window must be 12 feet.

Social Science Applications

1 *Psychology*   A worker can learn to perform x different tasks in approximately $f(x) = x^{3/2} + 6x^{1/2}$ minutes. Determine the number of tasks that result in the maximum rate of learning by the worker. (**Hint:** Find x such that $f'(x)$ is a minimum.)

2 *Psychology*   There is evidence to indicate that the strength of a habit is a function of the number of times x that the habit is repeated. Let y be the likelihood that the habitual activity is repeated according to habit on a scale of 0 to 100. (Here 0 indicates the habit will be broken, and 100 indicates that the activity will always be completed according to habit.) Assume y is given by

$$y = f(x) = 100(1 - e^{-.05x}).$$

For what x is the function $f(x)$ increasing? Does $f(x)$ assume a maximum value?

3 *Public Administration*   A commuter train that makes a 100 mile trip can maintain any average speed s in the range $30 \leq s \leq 60$. Suppose the fuel costs per hour for the trip are approximately equal to $s^{3/2}$, and the other costs are $180 per hour. What should the average speed be to minimize the total costs of the trip?

4 *Public Administration*   Between 1974 and 1984, the number of people in a small city who used public transportation is approximately equal (in thousands) to

$$p(t) = 2t^3 - 3t^2 - 36t + 300$$

where $t = 0$ corresponds to January 1, 1980. Determine the maximum number of people who used public transportation between 1974 and 1984.

5 *Sociology*   During the 20 years from 1964 to 1984, the population of a certain city grew, stagnated, and declined. If the population $p(t)$ (in hundreds of thousands) is approximated by

$$p(t) = 5t - .25t^2 + 10, \qquad 0 \leq t \leq 20$$

where t is the number of years past January 1, 1964. Determine when the population was a maximum.

6 *Sociology*   An important rumor begins in a city of 1,000,000 people, and after t days the number of people spreading the rumor is given by

$$y(t) = \frac{1,000,000}{999e^{-2t} + 1}.$$

For what t is the function $y(t)$ increasing? Does $y(t)$ assume a maximum value?

7 *Ecology*   A factory emits sulfur dioxide into the surrounding air through its smokestack. Suppose the amount y of sulfur dioxide in one cubic foot of air (in appropriate units) is given by

$$y = f(x) = \frac{.05}{x^2}.$$

where x is the distance to the smokestack (in miles). For what x is the function $f(x)$ decreasing? Does $f(x)$ assume a minimum value?

CAREER PROFILE
Market Research
Analyst

"What products do consumers want?"
"If we produce Product X, will consumers buy it?"
"What price will consumers pay for Product Y?"
"What advertising strategies will 'sell' Product Z?"

Questions like these call for the talents of a market research analyst. As
you can guess, market research is an interesting field; one of its drawbacks,
however, is that employment fluctuates with the economy. When the
economic outlook is good and business activity and personal incomes are
on the rise, marketing experts are in demand. Optimistic manufacturers
seek to expand their product lines, and market research analysts are
employed to help them attract buyers. On the other hand, in periods of
economic recession, consumers restrict their buying to necessities and
clever marketing strategies are less profitable, leading to the hiring of
fewer market research workers. A good market research analyst has a
variety of skills. A typical background includes a bachelor's degree in
marketing, but graduates with majors in economics, psychology, and sociology (all with strong backgrounds in quantitative research methods) also
qualify for many positions.

Because a primary goal of marketing is to satisfy the consumer,
research analysts spend a lot of time learning about consumer preferences
and buying habits. These analysts plan, design, implement, and analyze
the results of surveys. Basic preparation for this includes college courses in
applied statistics.

To understand the various psychological factors that influence
peoples' preferences, course work in psychology is helpful. The study of
psychology, together with on-the-job experience, aids the analyst in
predicting probable differences between what consumers **say** they prefer
(in response to a survey question about a potential product) and what
they actually **do** prefer (as indicated by their purchases once products are
on the market).

Certain marketing campaigns are successful when aimed at one
social or economic group and fail when directed at another group.
Anticipation of these differences is aided by study of sociology and
economics.

Along with strong quantitative research skills and a background in
psychology, sociology, and economics, the market research analyst must
have excellent communication skills. Development of surveys requires
careful use of language. Slight differences in wording, such as "**Would you
like** to purchase Product X?" instead of "**Will you** purchase Product X?",
can lead to significantly different survey results. The ability to deliver
clear and convincing reports of research findings, both orally and in
writing, is another essential quality of the successful analyst. English composition and speech courses thus should be part of a college program.

College graduates who wish to enter the market research field may
find their first jobs in a variety of places: advertising agencies, financial
institutions, insurance companies, government planning agencies, or

(most likely) in the marketing research departments of large firms. The graduate who has had an opportunity to participate in a student research internship will have gained practical experience that is extremely valuable in landing a job.

It is typical for a newly hired analyst to begin as a trainee or junior analyst, assisting a senior analyst in one or more aspects of a market research project. Graduate study usually is required for advancement; a frequently held degree is the MBA, although advanced degrees in psychology or sociology also are useful.

In 1980, salaries of beginning market researchers ranged from $12,000 to $17,000. In that same year, market research directors with over 15 years' experience had a salary average of over $50,000. Although the demand for market researchers is uncertain because it rises and falls with the economy, the field is an interesting one, drawing on communication, quantitative, and social science skills; thus, market research continues to attract a lot of new people and the competition for jobs is keen.

Sources of Additional Information
American Marketing Association, 250 Wacker Street, Chicago, IL 60606. (A pamphlet, "Careers in Marketing," is available from the AMA.)

Occupational Outlook Handbook. Bureau of Labor Statistics, U.S. Department of Labor, Washington, DC 20212. (Revised every 2 years, this handbook provides information about job duties, working conditions, level and places of employment, education and training requirements, advancement possibilities, job outlook, and earnings for about 250 occupations. Some of the information for this Career Profile was obtained from this source.)

10 Antiderivatives and the Definite Integral

10.1 Getting Antiderivatives from Differentiation Formulas

There are many situations where we need to find an unknown function. Sometimes it is easier to find a formula involving the function and its derivative than it is to find a formula for the function itself. For example, if a savings account with $A(t)$ dollars at time t receives continuous compounding at the rate of 10%, then the rate of growth of $A(t)$, that is $A'(t)$, equals $.10\, A(t)$. Thus, we have a formula for the derivative $A'(t) = .10\ A(t)$ and more work is required to find a formula for $A(t)$, as was shown in Section 8.4. Sometimes it is easier to find a formula that is consistent with marginal revenue figures or marginal cost figures than it is to find a formula that fits revenue figures or cost figures. Consequently, we are going to develop a proficiency at getting a formula for a function from a given formula for the derivative.

The function $F(x)$ is an **antiderivative** for the function $f(x)$ provided $F'(x) = f(x)$ for all x in the domain of $f(x)$. It means exactly the same thing to say that "the derivative of $F(x)$ is $f(x)$" or to say that "$F(x)$ is an antiderivative of $f(x)$;" the statements say the same thing from different points of view. A standard notation for the most general antiderivative of $f(x)$ is

$$\int f(x)\ dx.$$

Here the elongated s symbol $\int$ in front of $f(x)$, which is called the **integral sign,** and the "dx" must be used together, since the integral sign has no meaning by itself. The most general antiderivative of $f(x)$ is sometimes called the **indefinite integral of $f(x)$.**

Many differentiation formulas can be used to produce facts about antiderivatives. For example, the following formula is a slight variation on the Power Rule

$$\frac{d}{dx}\frac{1}{a}x^a = x^{a-1}, \qquad a \neq 0.$$

This rule can easily be rewritten as a fact about antiderivatives. An antiderivative for x^{a-1} is

$$\frac{1}{a} x^a, \qquad a \neq 0.$$

To make this fact easier to use, we let the symbol b represent $a - 1$; since $b = a - 1$, it follows that $b + 1 = a$. Thus, an antiderivative for x^b is

$$\frac{1}{b+1} x^{b+1}, \qquad b + 1 \neq 0 \quad (\text{or } b \neq -1).$$

This fact tells us that $\frac{1}{2}x^2$ is an antiderivative for x^1 or x. We should observe that there are other antiderivatives for x; for example, $\frac{1}{2}x^2 + 5$ is an antiderivative for x, and $\frac{1}{2}x^2 + 9$ is another one. Indeed, if C is any constant then $d/dx \left(\frac{1}{2}x^2 + C\right) = x$, so $\frac{1}{2}x^2 + C$ is an antiderivative for x. The general fact behind this observation is given in the next theorem.

Theorem   If $F(x)$ is some antiderivative for $f(x)$, then all antiderivatives are obtained from the formula $F(x) + C$ by all possible choices of the constant C, which is called the **constant of integration**. Thus, the most general antiderivative of $f(x)$ is given by the formula

$$\int f(x) \, dx = F(x) + C.$$

Using the above theorem and the facts obtained from the Power Rule earlier, we can now describe all antiderivatives of x^b.

Power Rule for Antiderivatives

$$\int x^b \, dx = \frac{1}{b+1} x^{b+1} + C, \qquad b \neq -1$$

EXAMPLE 1

Find all antiderivatives of $f(x) = \sqrt{x}$.

Solution
First we rewrite the function in exponential form and then we apply the Power Rule.

$$\int \sqrt{x} \, dx = \int x^{1/2} \, dx = \frac{2}{3} x^{3/2} + C \quad \blacksquare$$

In the same way that we changed the Power Rule into a rule for antiderivatives, we can change other differentiation rules into antidifferentiation rules. The Sum Rule

$$\frac{d}{dx} (F(x) + G(x)) = F'(x) + G'(x)$$

can be written as

$$(F(x) + G(x)) + C = \int (F'(x) + G'(x))\, dx$$

Since $F(x)$ is an antiderivative for $F'(x)$ and $G(x)$ is an antiderivative for $G'(x)$, we can rewrite the above formula and reverse the sides to get the following basic formula.

Sum Rule for Antiderivatives

$$\int (f(x) + g(x))\, dx = \int f(x)\, dx + \int g(x)\, dx$$

Since the formulas for $\int f(x)\, dx$ and $\int g(x)\, dx$ involve constants of integration, we have not bothered to add another constant of integration.

Since the derivative of a constant c times a function $f(x)$ is given by the formula

$$\frac{d}{dx}\, cf(x) = cf'(x),$$

it is not difficult to discover the next formula.

Constant Multiple Rule for Antiderivatives

$$\int cf(x)\, dx = c \int f(x)\, dx$$

EXAMPLE 2

Find all antiderivatives of $g(x) = 5x^3 - 2x^2 + 3x + 7$.

Solution

$$\int (5x^3 - 2x^2 + 3x + 7)\, dx = \int 5x^3\, dx + \int -2x^2\, dx + \int 3x\, dx + \int 7\, dx \qquad \text{by the Sum Rule}$$

$$= 5 \int x^3\, dx - 2 \int x^2\, dx + 3 \int x\, dx + 7 \int 1\, dx \qquad \begin{array}{l}\text{by the Constant} \\ \text{Multiple Rule}\end{array}$$

$$= 5 \left(\frac{1}{4} x^4\right) - 2 \left(\frac{1}{3} x^3\right) + 3 \left(\frac{1}{2} x^2\right) + 7x + C \qquad \begin{array}{l}\text{by the Power} \\ \text{Rule}\end{array}$$

$$= \frac{5}{4} x^4 - \frac{2}{3} x^3 + \frac{3}{2} x^2 + 7x + C$$

Here we have used the fact that x is an antiderivative for 1 or, equivalently, that the derivative of x is 1. Note that we do not write four constants (one for each antiderivative calculated); rather, we consider all four collected into one single constant C. ■

EXAMPLE 3

Find all antiderivatives of $h(x) = \dfrac{9}{x^2} + 3$.

Solution

$$\int \left(\frac{9}{x^2} + 3 \right) dx = \int \frac{9}{x^2}\, dx + \int 3\, dx$$

$$= 9 \int x^{-2}\, dx + 3 \int 1\, dx$$

$$= 9 \left(\frac{1}{-1} x^{-1} \right) + 3x + C$$

$$= -9x^{-1} + 3x + C \quad\blacksquare$$

There is only one power of x for which we do not know an antiderivative; we do not know an antiderivative for x^{-1}. If we try to apply the Power Rule, we must write

$$\frac{1}{-1+1} x^{-1+1}$$

which is not defined since it requires division by zero. To find an antiderivative for x^{-1}, we must recall each of our differentiation formulas until we find one that gives x^{-1} as the derivative. Since the proper formula is $\dfrac{d}{dx} \ln x = \dfrac{1}{x}$, we have the following rule.

Logarithm Rule for Antiderivatives

$$\int x^{-1}\, dx = \int \frac{1}{x}\, dx = \ln x + C, \qquad x > 0$$

Recall that $\ln x$ is defined only for positive x, so this formula only holds for such x.*

Recall that the derivative of e^x is e^x; it follows that the derivative of $(1/a)e^{ax}$ is e^{ax}, where a is any nonzero constant. The next formula follows from that observation.

Exponential Rule for Antiderivatives

$$\int e^{ax}\, dx = \frac{1}{a} e^{ax} + C, \qquad a \neq 0$$

EXAMPLE 4

Find all antiderivatives of $f(x) = \dfrac{2}{x^3} - \dfrac{3}{x} + 9$.

---

* Since $\ln(-x)$ is defined for $x < 0$ and $d/dx\, \ln(-x) = 1/x$, we see that $d/dx\, \ln|x| = 1/x$ for $x \neq 0$. For the more complicated domain indicated by $x \neq 0$, we have $\int x^{-1}\, dx = \ln|x| + C$.

Solution

$$\int \left(\frac{2}{x^3} - \frac{3}{x} + 9 \right) dx = \int \frac{2}{x^3} \, dx + \int \frac{-3}{x} \, dx + \int 9 \, dx$$

$$= 2 \int x^{-3} \, dx - 3 \int x^{-1} \, dx + 9 \int 1 \, dx$$

$$= 2 \left(\frac{1}{-2} x^{-2} \right) - 3 \ln x + 9x + C$$

$$= -x^{-2} - 3 \ln x + 9x + C \ \blacksquare$$

Sometimes for brevity's sake we shall write "evaluate $\int f(x) \, dx$" when we mean "find a formula for all antiderivatives of $f(x)$."

EXAMPLE 5

Evaluate $\int \frac{1}{2} (e^x + e^{-x}) \, dx.$

Solution

$$\int \frac{1}{2} (e^x + e^{-x}) \, dx = \int \frac{1}{2} e^x \, dx + \int \frac{1}{2} e^{-x} \, dx$$

$$= \frac{1}{2} \int e^x \, dx + \frac{1}{2} \int e^{-x} \, dx$$

$$= \frac{1}{2} e^x - \frac{1}{2} e^{-x} + C$$

$$= \frac{1}{2} (e^x - e^{-x}) + C \ \blacksquare$$

These rules for finding antiderivatives are very useful. The first three rules permit us to obtain a formula for the antiderivatives of any polynomial. Recall that $f(x)$ is a polynomial provided it can be written as

$$f(x) = a_n x^n + a_{n+1} x^{n-1} + \cdots + a_1 x + a_0, \qquad a_n \neq 0$$

where $a_n, a_{n-1}, \ldots, a_1, a_0$ are constants and n is a nonnegative integer. The formula for the antiderivatives of $f(x)$ is

$$\int f(x) \, dx = \int (a_n x^n + a_{n-1} x^{n-1} + \cdots + a_1 x + a_0) \, dx$$

$$= \int a_n x^n \, dx + \int a_{n-1} x^{n-1} \, dx + \cdots + \int a_1 x \, dx + \int a_0 \, dx$$

$$= a_n \int x^n \, dx + a_{n-1} \int x^{n-1} \, dx + \cdots + a_1 \int x \, dx + a_0 \int 1 \, dx$$

$$= a_n \frac{x^{n+1}}{n+1} + a_{n-1} \frac{x^n}{n} + \cdots + a_1 \frac{x^2}{2} + a_0 x + C.$$

Notice that $\int f(x) \, dx$ is a polynomial with degree one more than the degree of $f(x)$. It would be very unwise to try to memorize this formula; it is much easier to learn the method.

EXAMPLE 6

For a certain business, marginal costs are observed to be $C'(x) = .4x - .1$, where x is the number of customers served, and the fixed costs are 100. Determine a formula for costs $C(x)$.

Solution

Since $C(x)$ is an antiderivative of $C'(x)$, we can find it by finding a formula for all antiderivatives of $C'(x)$ and then choosing the arbitrary constant appropriately.

$$\int C'(x)\, dx = \int (.4x - .1)\, dx$$

$$= \int .4x\, dx + \int -.1\, dx$$

$$= .4 \int x\, dx - .1 \int 1\, dx$$

$$= .4 \frac{x^2}{2} - .1x + c$$

$$= .2x^2 - .1x + c$$

Thus, we know that $C(x) = .2x^2 - .1x + c$ for some choice of the constant c. When no customers are served, the only costs are the fixed costs, so we have

$$100 = C(0) = c.$$

The formula for costs must be

$$C(x) = .2x^2 - .1x + 100. \quad\blacksquare$$

EXAMPLE 7

Suppose marginal revenue for a plant is observed to be $R'(x) = \frac{1}{2}(3 + x^{-1/2})$, where x is the number of units made. Determine a formula for $R(x)$.

Solution

$$\int R'(x)\, dx = \int \frac{1}{2}(3 + x^{-1/2})\, dx$$

$$= \int \frac{3}{2}\, dx + \int \frac{1}{2} x^{-1/2}\, dx$$

$$= \frac{3}{2} \int 1\, dx + \frac{1}{2} \int x^{-1/2}\, dx$$

$$= \frac{3}{2} x + \frac{1}{2}(2x^{1/2}) + c$$

Thus, $R(x) = \frac{3}{2}x + x^{1/2} + c$ for some choice of c. Since there is no revenue when no units are made, we have

$$0 = R(0) = \frac{3}{2}(0) + (0)^{1/2} + c$$

$$0 = c$$

and it must be that

$$R(x) = \frac{3}{2}x + x^{1/2}. \ \blacksquare$$

Summary

1   $\displaystyle \int (f(x) + g(x))\, dx = \int f(x)\, dx + \int g(x)\, dx$

2   $\displaystyle \int cf(x)\, dx = c \int f(x)\, dx$

3   $\displaystyle \int x^b\, dx = \frac{1}{b+1} x^{b+1} + C, \qquad b \neq -1$

4   $\displaystyle \int x^{-1}\, dx = \ln x + C, \qquad x > 0$

5   $\displaystyle \int e^{ax}\, dx = \frac{1}{a} e^{ax} + C, \qquad a \neq 0$

Exercises 10.1

Find a formula for the antiderivatives of each of the following functions.

1   $f(x) = 5$

3   $h(x) = e$

5   $g(x) = 3x - 5$

7   $f(x) = x^2 - x + 1$

9   $h(x) = x^3 + 5x^2 - 6x + 3$

11   $g(x) = 5x^4 + 3x^2 - 8$

13   $f(x) = \sqrt[3]{x}$

15   $h(x) = x^{5/2}$

17   $g(x) = \dfrac{10}{x^3}$

19   $f(x) = 9x^{-3}$

21   $h(x) = \dfrac{-2}{x^{3/2}} + x$

23   $g(x) = \dfrac{3}{\sqrt{x}}$

2   $g(x) = -11$

4   $f(x) = e^3$

6   $h(x) = 5x + 2$

8   $g(x) = 3x^2 - 2x + 1$

10   $f(x) = x^3 + 10x$

12   $h(x) = x^4 + 2x^3 - x$

14   $g(x) = 4\sqrt{x}$

16   $f(x) = 3x^{7/2}$

18   $h(x) = \dfrac{4}{x^2}$

20   $g(x) = 7x^{-2} + 3$

22   $f(x) = 4 - \dfrac{1}{x^{5/2}}$

24   $h(x) = 5 + \dfrac{1}{\sqrt[3]{x}}$

25 $f(x) = (x-1)(x+1)$

26 $g(x) = (x-2)(x+2)$

27 $h(x) = x(2x+3)$

28 $f(x) = x(x^2+1)$

29 $g(x) = \dfrac{x^2+1}{x}$

30 $h(x) = \dfrac{x^3-x^2+3}{x}$

31 $f(x) = \dfrac{(x+1)(x-2)}{x}$

32 $g(x) = \dfrac{(3x+2)(x-1)}{x}$

33 $h(x) = x^\pi$

34 $f(x) = 6x^e$

35 $g(x) = x^{4.2} - \dfrac{3}{x^{1.6}}$

36 $h(x) = 4x^{2.1} + \dfrac{1}{x^{5.2}}$

37 $f(x) = 7e^x$

38 $g(x) = 9e^{-x}$

39 $h(x) = e^{2x} + e^{-x} + 10$

40 $f(x) = .4e^{10x} - e^{5x}$

41 $g(x) = \dfrac{6}{e^{3x}} + 8$

42 $h(x) = 12 - \dfrac{4}{e^{2x}}$

43 $f(x) = e^5e^x - 1$

44 $g(x) = e^2 + e^6e^{-x}$

45 $h(x) = (e^x)^2 - e^x$

46 $f(x) = (e^x)^2 + 3e^{-x}$

47 $g(x) = x^2 - e^{3x} + 11$

48 $h(x) = x^3 - 5x + e^{5x}$

49 $f(x) = e^4e^{2x}e^{-x}$

50 $g(x) = 9e^{3x}e^{-6}$

51 $h(x) = \sqrt{x} - x^{7/3} + 4e^{2x}$

52 $f(x) = x^{5/2} - e^{5x} - x^{-3}$

53 $g(x) = \dfrac{e^{9x} + e^{6x}}{e^{3x}}$

54 $h(x) = \dfrac{5e^{4x} - 10}{5e^x}$

55 $f(x) = (e^x - 1)(e^x + 1)$

56 $g(x) = (e^{3x} - 2)(e^{3x} + 2)$

57 $h(x) = e^{-2x}(e^{5x} - 7)$

58 $f(x) = (12e^x - 3)e^{3x}$

59 $g(x) = (e^{6x} - 1)^2 + \sqrt{x}$

60 $h(x) = (2 + e^{2x})^2 - 11$

In problems 61 through 68, evaluate the given indefinite integral and find the particular antiderivative that satisfies the given equation.

61 $\displaystyle\int (.6x - .3)\,dx = C(x)$ and $C(0) = 220$

62 $\displaystyle\int (.2x - .01)\,dx = C(x)$ and $C(0) = 150$

63 $\displaystyle\int (3x^2 - .4x - .6)\,dx = C(x)$ and $C(0) = 400$

64 $\displaystyle\int (2.1x^2 - x - .1)\,dx = C(x)$ and $C(0) = 510$

65 $\displaystyle\int 5\,dx = R(x)$ and $R(0) = 0$

66 $\displaystyle\int 12\,dx = R(x)$ and $R(0) = 0$

67 $\displaystyle\int (2x - 1)\,dx = R(x)$ and $R(0) = 0$

68 $\displaystyle\int x^{1/2}\, dx = R(x)$ and $R(0) = 0$

69 For a certain business, marginal costs are observed to be $C'(x) = .6x + 2$, where x is the number of customers served, and the fixed costs are 50. Determine a formula for costs $C(x)$.

70 For a manufacturer, marginal costs are observed to be $C'(x) = 10 + .1x$, where x is the number of units manufactured, and the fixed costs are 230. Determine a formula for costs $C(x)$.

71 Suppose the marginal revenue for a particular plant is observed to be $R'(x) = 2 + x^{1/2}$, where x is the number of units made. Determine a formula for $R(x)$.

72 For a certain business, marginal revenue is observed to be $R'(x) = .9x$, where x is the number of customers served. Determine a formula for $R(x)$.

10.2 Substitution with the Generalized Power Rule

Just as we used the Power Rule for differentiation to get a rule for finding antiderivatives, we might try to get a rule for antiderivatives from the Generalized Power Rule. The formula

$$\frac{d}{dx}\frac{1}{a}u(x)^a = u(x)^{a-1}u'(x)$$

can be rewritten as

$$\int u(x)^{a-1}u'(x)\, dx = \frac{1}{a}u(x)^a + C.$$

If we again use b as a symbol for $a - 1$ and note that $b + 1 = a$, then we obtain the following.

Generalized Power Rule for Antiderivatives

$$\int u(x)^b u'(x)\, dx = \frac{1}{b+1}u(x)^{b+1} + C$$

EXAMPLE 1

Evaluate $\displaystyle\int \sqrt{x+2}\, dx$.

Solution
We rewrite the problem in exponential notation and apply the above rule, with $u(x) = x + 2$ and $u'(x) = 1$.

$$\int \sqrt{x+2}\ dx = \int (x+2)^{1/2}\ dx$$

$$= \int (x+2)^{1/2}1\ dx$$

$$= \frac{2}{3}(x+2)^{3/2} + C \ \blacksquare$$

The function between the integral sign and the "*dx*" is called the **integrand.** It is possible to change the integrand by means of the Constant Multiple Rule for Antiderivatives. The formula

$$c \int f(x)\ dx = \int cf(x)\ dx$$

can be rewritten as

$$\boxed{\int f(x) = \frac{1}{c} \int cf(x)\ dx, \qquad c \ne 0.}$$

Thus, we can multiply the integrand by any nonzero constant provided we multiply in front of the integral sign by the reciprocal of that constant. This step is often necessary to obtain the $u'(x)$ for use with the Generalized Power Rule.

EXAMPLE 2

Evaluate $\displaystyle\int \sqrt{x^2+1}\ x\ dx$.

Solution
When we try to use the Generalized Power Rule for Antiderivatives with $u(x) = x^2 + 1$, we find that $u'(x) = 2x$. Since the factor in the integrand is only x, we multiply the integrand by 2 and multiply in front of the integral sign by $\frac{1}{2}$. Then we can evaluate the antiderivative.

$$\int \sqrt{x^2+1}\ x\ dx = \int (x^2+1)^{1/2}x\ dx$$

$$= \frac{1}{2}\int (x^2+1)^{1/2}2x\ dx$$

$$= \frac{1}{2}\left(\frac{2}{3}\right)(x^2+1)^{3/2} + C$$

$$= \frac{1}{3}(x^2+1)^{3/2} + C \ \blacksquare$$

Caution   The formula

$$\int f(x)\ dx = \frac{1}{c}\int cf(x)\ dx$$

is true only if c is a nonzero constant. The equation is not true if c is a nonconstant function of x.

EXAMPLE 3

Evaluate $\displaystyle\int \frac{5x}{\sqrt{3x^2 + 4}}\, dx$

Solution

Anticipating the use of the Generalized Power Rule for Antiderivatives, we compute the derivative of the more complicated expression under the square root sign to see how to adjust the constant factor in the numerator. Let $u(x) = 3x^2 + 4$. We find that

$$u'(x) = \frac{d}{dx}(3x^2 + 4) = 6x.$$

Thus, we must remove the factor 5 and replace it with the factor 6.

$$\int \frac{5x}{\sqrt{3x^2 + 4}}\, dx = 5 \int \frac{x}{\sqrt{3x^2 + 4}}\, dx$$

$$= \frac{5}{6} \int \frac{6x}{\sqrt{3x^2 + 4}}\, dx$$

$$= \frac{5}{6} \int (3x^2 + 4)^{-1/2} 6x\, dx$$

$$= \frac{5}{6} (2)(3x^2 + 4)^{1/2} + C$$

$$= \frac{5}{3} \sqrt{3x^2 + 4} + C \quad \blacksquare$$

EXAMPLE 4

Evaluate $\displaystyle\int (5x^2 + 1)^{10} 3x\, dx.$

Solution

Let $u(x) = 5x^2 + 1$; then $u'(x) = 10x$.

$$\int (5x^2 + 1)^{10} 3x\, dx = 3 \int (5x^2 + 1)^{10} x\, dx$$

$$= \frac{3}{10} \int (5x^2 + 1)^{10} 10x\, dx$$

$$= \frac{3}{10} \left(\frac{1}{11}\right) (5x^2 + 1)^{11} + C$$

$$= \frac{3}{110} (5x^2 + 1)^{11} + C \quad \blacksquare$$

Sometimes it can be difficult to pick out what is $u(x)^b$ and what is $u'(x)$ in the application of the Generalized Power Rule for Antiderivatives. It is usually best to let $u(x)$ represent the most complicated expression in the integrand that is raised to a power. (Of course, the "power" may be $b = 1$, or b may be negative.)

EXAMPLE 5

Evaluate $\displaystyle\int \frac{(\sqrt{x}+3)^8}{\sqrt{x}}\, dx$.

Solution

We compute the derivative of the expression $u(x) = \sqrt{x} + 3$.

$$\frac{d}{dx}(\sqrt{x}+3) = \frac{d}{dx}(x^{1/2}+3) = \frac{1}{2}x^{-1/2} = \frac{1}{2}\cdot\frac{1}{\sqrt{x}}.$$

Now it is clear how to use the factor in the denominator of the integrand.

$$\int \frac{(\sqrt{x}+3)^8}{\sqrt{x}}\, dx = \int (x^{1/2}+3)^8 x^{-1/2}\, dx$$

$$= 2\int (x^{1/2}+3)^8 \frac{1}{2}x^{-1/2}\, dx$$

$$= 2\left(\frac{1}{9}\right)(x^{1/2}+3)^9 + C$$

$$= \frac{2}{9}(\sqrt{x}+3)^9 + C \quad\blacksquare$$

Sometimes $u'(x)$ can be more than just a power of x, as we shall illustrate.

EXAMPLE 6

Evaluate $\displaystyle\int \frac{x+1}{(x^2+2x)^5}\, dx$.

Solution

Note that if $u(x) = x^2 + 2x$, then

$$u'(x) = \frac{d}{dx}(x^2+2x) = 2x + 2 = 2(x+1)$$

Now it is clear how to adjust the constants.

$$\int \frac{x+1}{(x^2+2x)^5}\, dx = \int (x^2+2x)^{-5}(x+1)\, dx$$

$$= \frac{1}{2}\int (x^2+2x)^{-5}(2x+2)\, dx$$

$$= \frac{1}{2}\left(-\frac{1}{4}\right)(x^2+2x)^{-4} + C$$

$$= -\frac{1}{8}(x^2+2x)^{-4} + C \quad\blacksquare$$

There are some problems that require only algebraic manipulations but strongly resemble problems that we solved with the Generalized Power Rule for Antiderivatives.

EXAMPLE 7

Evaluate $\displaystyle\int (4x^3 + 1)^2 x\,dx$.

Solution

As always, we compute the derivative of the expression raised to a power, $u(x) = 4x^3 + 1$.

$$u'(x) = \frac{d}{dx}(4x^3 + 1) = 12x^2$$

We could introduce the constant 12, but we could never change the factor x into x^2; thus, we cannot use the Generalized Power Rule for Antiderivatives. However, we can change the form of the integrand.

$$\int (4x^3 + 1)^2 x\,dx = \int (16x^6 + 8x^3 + 1)x\,dx$$

$$= \int (16x^7 + 8x^4 + x)\,dx$$

$$= 16\frac{x^8}{8} + 8\frac{x^5}{5} + \frac{x^2}{2} + C$$

$$= 2x^8 + \frac{8}{5}x^5 + \frac{1}{2}x^2 + C \quad\blacksquare$$

EXAMPLE 8

Evaluate $\displaystyle\int \frac{\sqrt{x}+2}{\sqrt{x}}\,dx$.

Solution

The easiest solution is just algebraic.

$$\int \frac{\sqrt{x}+2}{\sqrt{x}}\,dx = \int \left(\frac{\sqrt{x}}{\sqrt{x}} + \frac{2}{\sqrt{x}}\right)dx$$

$$= \int (1 + 2x^{-1/2})\,dx$$

$$= x + 2(2)x^{1/2} + C$$

$$= x + 4\sqrt{x} + C \quad\blacksquare$$

The Generalized Power Rule for Antiderivatives has extended our ability to work with marginal costs and marginal revenues, as shown next.

EXAMPLE 9

Suppose the marginal costs for a certain business are $C'(x) = \dfrac{3x^2}{2\sqrt{x^3 + 400}}$ and the fixed costs are 20. Find a formula for costs.

Solution

Let $u(x) = x^3 + 400$, so $u'(x) = 3x^2$. Then, we have

$$C(x) = \int \frac{3x^2}{2\sqrt{x^3 + 400}}\,dx = \frac{1}{2}\int (x^3 + 400)^{-1/2}3x^2\,dx$$

$$= \frac{1}{2}(2)(x^3 + 400)^{1/2} + c$$

$$= \sqrt{x^3 + 400} + c.$$

Since

$$20 = C(0) = \sqrt{400} + c = 20 + c$$

we know that $c = 0$ and

$$C(x) = \sqrt{x^3 + 400} \quad\blacksquare$$

Steps for Applying the Generalized Power Rule for Antiderivatives

1. Identify the factor in the integrand that is $u(x)^b$.
2. Compute $u'(x)$.
3. Examine the integrand divided by the factor $u(x)^b$ to see if it differs from $u'(x)$ by only a constant factor. If it does not, then find another method; if it does, then proceed.
4. Adjust the constants so the integrand has the form $u(x)^b u'(x)$.
5. Integrate according to

$$\int u(x)^b u'(x)\,dx = \frac{1}{b+1}\,u(x)^{b+1} + C.$$

Exercises 10.2

Evaluate each of the following.

1 $\displaystyle\int \sqrt{x+5}\,dx$

2 $\displaystyle\int \sqrt[3]{x+8}\,dx$

3 $\displaystyle\int (2x+1)^{10}\,dx$

4 $\displaystyle\int (3x+7)^8\,dx$

5 $\displaystyle\int \frac{1}{\sqrt{5x+1}}\,dx$

6 $\displaystyle\int \frac{1}{\sqrt[3]{2x+3}}\,dx$

7 $\displaystyle\int \frac{9}{(4x+3)^{12}}\,dx$

8 $\displaystyle\int \frac{1}{(6x+1)^9}\,dx$

9 $\displaystyle\int \sqrt{4x^2+9}\,x\,dx$

10 $\displaystyle\int \sqrt[3]{x^2+8}\,x\,dx$

11 $\displaystyle\int (x^2 + 7)^{11}x\ dx$

12 $\displaystyle\int (1 - 3x^2)^9 x\ dx$

13 $\displaystyle\int \frac{x}{\sqrt{4x^2 + 3}}\ dx$

14 $\displaystyle\int \frac{x}{\sqrt[3]{x^2 + 5}}\ dx$

15 $\displaystyle\int \frac{3x}{(x^2 + 4)^{14}}\ dx$

16 $\displaystyle\int \frac{8x}{(3x^2 + 1)^6}\ dx$

17 $\displaystyle\int (x^2 + 1)^2\ dx$

18 $\displaystyle\int (2x^2 - 3)^2\ dx$

19 $\displaystyle\int (x^3 - 2)^2 x\ dx$

20 $\displaystyle\int (5 - x^3)^2 4x\ dx$

21 $\displaystyle\int (3x + 2)x\ dx$

22 $\displaystyle\int (5 - 3x)x^2\ dx$

23 $\displaystyle\int \frac{x + 2}{\sqrt{x^2 + 4x}}\ dx$

24 $\displaystyle\int \frac{4x + 6}{\sqrt[3]{x^2 + 3x}}\ dx$

25 $\displaystyle\int (3x^2 + 6x - 8)^5(x + 1)\ dx$

26 $\displaystyle\int (x^2 - 2x + 4)^8(x - 1)\ dx$

27 $\displaystyle\int \frac{2x + 1}{\sqrt{x^2 + x + 1}}\ dx$

28 $\displaystyle\int \frac{x + 2}{\sqrt[5]{x^2 + 4x + 4}}\ dx$

29 $\displaystyle\int \frac{5x + 1}{(5x^2 + 2x + 1)^4}\ dx$

30 $\displaystyle\int \frac{6x - 3}{(x^2 - x + 6)^5}\ dx$

31 $\displaystyle\int \frac{(2\sqrt{x} + 1)^5}{\sqrt{x}}\ dx$

32 $\displaystyle\int \frac{(\sqrt{x} - 1)^9}{\sqrt{x}}\ dx$

33 $\displaystyle\int \frac{2\sqrt{x} + 1}{\sqrt{x}}\ dx$

34 $\displaystyle\int \frac{\sqrt{x} - 1}{\sqrt{x}}\ dx$

35 $\displaystyle\int (\sqrt{x} + 3)\sqrt{x}\ dx$

36 $\displaystyle\int 3\sqrt{x}(2\sqrt{x} + 5)\ dx$

37 $\displaystyle\int \frac{\sqrt[3]{x} + 9}{\sqrt[3]{x^2}}\ dx$

38 $\displaystyle\int \frac{5 - \sqrt[3]{x}}{\sqrt[3]{x^2}}\ dx$

39 $\displaystyle\int \frac{(\sqrt[3]{x} + 2)^5}{\sqrt[3]{x^2}}\ dx$

40 $\displaystyle\int \frac{(5 - \sqrt[3]{x})^8}{\sqrt[3]{x^2}}\ dx$

41 $\displaystyle\int \frac{4x^2 + 3x + 7}{6x}\ dx$

42 $\displaystyle\int \frac{x^2 - 8x + 15}{2x}\ dx$

43 $\displaystyle\int \frac{x^3 - 6}{x^2}\ dx$

44 $\displaystyle\int \frac{7 - 2x^3}{4x^2}\ dx$

45 $\displaystyle\int \frac{3x^2 - 3}{x + 1}\ dx$

46 $\displaystyle\int \frac{4x^2 - 9}{2x + 3}\ dx$

47 $\displaystyle\int \frac{2x - 8}{2 + \sqrt{x}}\ dx$

48 $\displaystyle\int \frac{1 - x}{\sqrt{x} - 1}\ dx$

$$49 \quad \int (x^{16} + 53)^{10} x^{15} \, dx \qquad\qquad 50 \quad \int (113 - x^{33})^5 x^{32} \, dx$$

In problems 51 through 54, evaluate the given indefinite integral and find the particular antiderivative that satisfies the given equation.

$$51 \quad \int (4x - 3)^{10} \, dx = R(x) \text{ and } R(1) = 0$$

$$52 \quad \int \sqrt{x^2 + 5} \, x \, dx = R(x) \text{ and } R(0) = 0$$

$$53 \quad \int \frac{x}{(x^2 + 1)^9} \, dx = P(x) \text{ and } P(0) = -\frac{1}{16}$$

$$54 \quad \int \frac{1}{(7x + 2)^{20}} \, dx = P(x) \text{ and } P(0) = 0$$

10.3 Substitution with the Exponential and Logarithmic Rules and Solving Differential Equations

Two more differentiation formulas lead to basic tools for computing antiderivatives. As indicated in Chapter 8, the Chain Rule and the derivative of e^x are all that is needed to get

$$\frac{d}{dx} e^{u(x)} = e^{u(x)} u'(x).$$

Reversing the sides, we get the antiderivative rule below.

Generalized Exponential Rule

$$\int e^{u(x)} u'(x) \, dx = e^{u(x)} + C$$

In Chapter 8, we used the Chain Rule and the derivative of $\ln x$ to get

$$\frac{d}{dx} \ln u(x) = \frac{u'(x)}{u(x)}.$$

Rewriting this formula with the sides reversed, we obtain the rule below.

Generalized Logarithm Rule

$$\int \frac{u'(x)}{u(x)} \, dx = \ln u(x) + C$$

Since $\ln y$ is defined only for y positive, $u(x)$ must be positive for each x in order for $\ln u(x)$ to be defined.

EXAMPLE 1

Evaluate $\int e^{-x}\,dx$.

Solution

We compute the derivative of the exponent $u(x) = -x$.

$$u'(x) = \frac{d}{dx}(-x) = -1$$

We must have this as a factor in order to use the Generalized Exponential Rule, so we adjust the constant factor in the integrand.

$$\int e^{-x}\,dx = (-1)\int e^{-x}(-1)\,dx$$

$$= (-1)e^{-x} + C$$
$$= -e^{-x} + C \quad \blacksquare$$

EXAMPLE 2

Evaluate $\int \frac{x}{x^2 + 9}\,dx$.

Solution

We compute the derivative of the denominator,

$$u'(x) = \frac{d}{dx}(x^2 + 9) = 2x$$

because this is what we must have in the numerator in order to use the Generalized Logarithm Rule. Then we adjust the constant factor in the integrand.

$$\int \frac{x}{x^2 + 9}\,dx = \frac{1}{2}\int \frac{2x}{x^2 + 9}\,dx$$

$$= \frac{1}{2}\ln(x^2 + 9) + C \quad \blacksquare$$

Now we do some examples without so much explanation.

EXAMPLE 3

Evaluate $\int e^{x^2 + 5}x\,dx$.

Solution

$$\frac{d}{dx}(x^2 + 5) = 2x$$

$$\int e^{x^2 + 5}x\,dx = \frac{1}{2}\int e^{x^2 + 5}2x\,dx$$

$$= \frac{1}{2}e^{x^2 + 5} + C \quad \blacksquare$$

EXAMPLE 4

Evaluate $\int \dfrac{9x+3}{3x^2+2x+8}\,dx$.

Solution

$$\frac{d}{dx}(3x^2+2x+8) = 6x+2 = 2(3x+1)$$

$$\int \frac{9x+3}{3x^2+2x+8}\,dx = \int \frac{3(3x+1)}{3x^2+2x+8}\,dx$$

$$= \frac{3}{2}\int \frac{2(3x+1)}{3x^2+2x+8}\,dx$$

$$= \frac{3}{2}\ln(3x^2+2x+8) + C \quad\blacksquare$$

We can now solve many practical problems. Whenever a positive quantity Q is a function of time t and its rate of change $Q'(t)$ is a nonzero constant k times $Q(t)$, then we can obtain a formula for $Q(t)$ using the Generalized Logarithm Rule.

$$Q'(t) = kQ(t)$$

$$\frac{Q'(t)}{Q(t)} = k$$

$$\int \frac{Q'(t)}{Q(t)}\,dt = \int k\,dt = kt + C_1$$

Since

$$\int \frac{Q'(t)}{Q(t)}\,dt = \ln Q(t) + C_2$$

we have the equation

$$\ln Q(t) = kt + C.$$

(We have collected both C_1 and C_2 into the single constant C.)

We use one of the basic equations in Section 2.5 to get

$$Q(t) = e^{\ln Q(t)} = e^{kt+C}.$$

Finally, we simplify the preceding equation using the rules for exponents. Since C is an arbitrary constant, e^C is an arbitrary positive constant, which we denote by A.

$$Q(t) = e^{kt}e^C = Ae^{kt}$$

Whenever $Q'(t) = kQ(t)$, then $Q(t)$ is given by the above formula for some choice of the constant A.

A **differential equation** is an equation involving an unknown function and some of its derivatives. An example of a differential equation is

$$Q'(t) = kQ(t).$$

We say that we have **solved** the differential equation when we find a function that makes the equation true. The method that we used to solve

$$Q'(t) = kQ(t)$$

is called **separation of variables.**

EXAMPLE 5

A savings account that began with $500 receives continuous compounding at a rate of 10%. Find a formula for the amount in the account at any later time.

Solution

At time t the rate of change of the account $S'(t)$ is 10% of the amount in the account $S(t)$. Thus, we have the differential equation $S'(t) = .1S(t)$, which we solve by separation of variables.

$$\frac{S'(t)}{S(t)} = .1$$

$$\int \frac{S'(t)}{S(t)} \, dt = \int .1 \, dt$$

$$\ln S(t) = .1t + C$$

$$S(t) = e^{\ln S(t)} = e^{.1t + C} = Ae^{.1t}$$

Since $S(0) = 500$, we have $500 = Ae^0$; thus, $A = 500$, and the answer is

$$S(t) = 500e^{.1t}. \quad \blacksquare$$

There are many unstable radioactive substances that are constantly disintegrating at a rate equal to a constant times the amount of the substance present. Uranium-235 is such a substance.

EXAMPLE 6

A plant generating electricity from nuclear power uses uranium-235, which has a half-life of 710 million years. That means it takes 710 million years for half of any quantity of uranium-235 to disintegrate. If the plant has 100 pounds now, find a formula for how much it will have at any later time.

Solution

If $Q(t)$ is the amount of uranium-235 at time t, then $Q'(t) = kQ(t)$ since this is true for any radioactive substance. The derivation preceding Example 5 shows that

$$Q(t) = Ae^{kt}.$$

Clearly $Q(0)$ equals A, and we have

$$.5Q(0) = Q(710) = Q(0)e^{710k}$$

$$.5 = e^{710k}$$

$$\ln .5 = 710k$$

$$k = \frac{\ln .5}{710} \approx -.000976$$

$$Q(t) = 100e^{-.000976t}$$

Note that a positive value of k indicates an increase in $Q(t)$, as t increases, and a negative value of k indicates a decrease. ■

Evidence indicates that the number of repeat customers for a heavily advertised new product increases initially at a rate equal to a constant times the number of potential customers who are not repeat customers.

EXAMPLE 7

An advertising firm believes that a new drink that it is promoting has 25 million potential repeat customers. Suppose the advertising campaign is begun when the product is first offered to the public, and after 10 weeks it has 5 million repeat customers. Find a formula for the number of repeat customers at any time during the first year.

Solution
Let $N(t)$ be the number of millions of repeat customers t weeks after the campaign begins. For some constant k the following is true.

$$N'(t) = k(25 - N(t))$$

We use the method of separation of variables.

$$\frac{N'(t)}{25 - N(t)} = k$$

$$\int \frac{N'(t)}{25 - N(t)} \, dt = \int k \, dt$$

$$-\ln(25 - N(t)) = kt + C$$

$$\ln(25 - N(t)) = -kt - C$$

$$e^{\ln(25-N(t))} = e^{-kt-C}$$

$$25 - N(t) = e^{-kt}e^{-C}$$

$$N(t) = 25 - e^{-C}e^{-kt} = 25 - Ae^{-kt}$$

Since $N(0) = 0$, we have $A = 25$ and

$$N(t) = 25 - 25e^{-kt}.$$

Now we use the fact that $N(10) = 5$ to determine the constant k.

$$5 = N(10) = 25 - 25e^{-10k}$$

$$-20 = -25e^{-10k}$$

$$.8 = e^{-10k}$$

$$\ln .8 = \ln(e^{-10k}) = -10k$$

$$k = \frac{\ln .8}{-10} \approx .022$$

The answer is

$$N(t) \approx 25 - 25e^{-.022t}. \quad \blacksquare$$

We conclude this section with a more difficult application of the Generalized Logarithm Rule.

EXAMPLE 8

Evaluate $\displaystyle\int \frac{1}{x \ln x}\,dx$.

Solution
By rewriting the integral we recognize that $1/x$ is the derivative of $\ln x$.

$$\int \frac{1}{x \ln x}\,dx = \int \frac{1}{\ln x}\frac{1}{x}\,dx = \int \frac{\dfrac{1}{x}}{\ln x}\,dx$$
$$= \ln(\ln x) + C \quad \blacksquare$$

Exercises   10.3

Evaluate each of the following.

1 $\displaystyle\int e^{2x}\,dx$

2 $\displaystyle\int e^{-3x}\,dx$

3 $\displaystyle\int 4e^{-5x}\,dx$

4 $\displaystyle\int 8e^{-2x}\,dx$

5 $\displaystyle\int 3e^{x^2}x\,dx$

6 $\displaystyle\int 7e^{-x^2}x\,dx$

7 $\displaystyle\int e^{4x^2}x\,dx$

8 $\displaystyle\int e^{-3x^2}x\,dx$

9 $\displaystyle\int e^{2x^3}x^2\,dx$

10 $\displaystyle\int e^{5x^3}x^2\,dx$

11 $\displaystyle\int 6e^{x^3}x^2\,dx$

12 $\displaystyle\int -4e^{x^3}x^2\,dx$

13 $\displaystyle\int e^{\ln x}\,dx$

14 $\displaystyle\int 3e^{2\ln x}\,dx$

15 $\displaystyle\int (e^{7x} + xe^{3x^2} + 5)\,dx$

16 $\displaystyle\int (xe^{x^2} - e^{3x} + 2)\,dx$

17 $\displaystyle\int x^e\,dx$

18 $\displaystyle\int x^{3e}\,dx$

19 $\displaystyle\int \frac{e^{7x} + 3}{e^x}\,dx$

20 $\displaystyle\int \frac{4 - e^{2x}}{e^{-x}}\,dx$

21 $\displaystyle\int x\sqrt{e^{x^2+1}}\,dx$

22 $\displaystyle\int x\sqrt[3]{e^{x^2+4}}\,dx$

23 $\displaystyle\int xe^{\ln(x+2)}\,dx$

24 $\displaystyle\int x^2 e^{\ln(x+5)}\,dx$

25 $\displaystyle\int \frac{x}{5x^2 + 2}\,dx$

26 $\displaystyle\int \frac{x}{3x^2 - 7}\,dx$

27 $\displaystyle\int \frac{5x}{9 - 4x^2}\,dx$

28 $\displaystyle\int \frac{-4x}{10 + 3x^2}\,dx$

29 $\displaystyle\int \frac{1}{9x - 1}\,dx$

30 $\displaystyle\int \frac{1}{7x + 5}\,dx$

31 $\displaystyle\int \frac{2}{6 - 5x}\,dx$

32 $\displaystyle\int \frac{5}{11 + 9x}\,dx$

33 $\displaystyle\int \frac{x + 2}{x^2 + 4x + 3}\,dx$

34 $\displaystyle\int \frac{x - 1}{x^2 - 2x + 5}\,dx$

35 $\displaystyle\int \frac{x^2 - 1}{x(x^2 - 3)}\,dx$

36 $\displaystyle\int \frac{x^2 + 4x}{x^2(x + 6)}\,dx$

37 $\displaystyle\int \frac{2x + 5}{x^2}\,dx$

38 $\displaystyle\int \frac{5x^3 - 3}{2x}\,dx$

39 $\displaystyle\int \frac{(3\sqrt{x} + 2)^2}{\sqrt{x}}\,dx$

40 $\displaystyle\int \frac{(5 - \sqrt{x})^2}{\sqrt{x}}\,dx$

41 $\displaystyle\int \frac{\ln x}{x}\,dx$

42 $\displaystyle\int \frac{\ln(x + 2)}{3x + 6}\,dx$

43 $\displaystyle\int \frac{(\ln x)^2}{x}\,dx$

44 $\displaystyle\int \frac{(\ln x)^3}{x}\,dx$

45 $\displaystyle\int \frac{\ln(2x)}{x}\,dx$

46 $\displaystyle\int \frac{3\ln(5x)}{x}\,dx$

Solve the following.

47 If a savings account that receives continuous compounding at 10% begins the year with $100, how much is in the account at the end of the year?

48 If a savings account that receives continuous compounding at 12%

begins the year with $750, how much is in the account at the end of the year?

49 How long does it take for a savings account to double if it receives continuous compounding at a rate of 10%?

50 How long does it take for a savings account to double if it receives continuous compounding at a rate of 12%?

51 Uranium-234 has a half-life of 248 thousand years. If a plant begins with 10 pounds of uranium-234, how much is left at any later time?

52 Uranium-238 has a half-life of 4.5 billion years. If a plant begins with 5 pounds of uranium-238, how much is left at any later time?

53 A new bread is believed to have 100 million potential repeat customers, and it is introduced with a big advertising campaign. If there are 10 million repeat customers after 10 weeks, find a formula for the number of repeat customers at any time during the first year.

54 A new headache remedy is believed to have 50 million potential repeat customers and it is introduced with a big advertising campaign. If there are 2 million repeat customers after 20 weeks, find a formula for the number of repeat customers at any time during the first year.

10.4 The Definite Integral

Areas are considered in elementary plane geometry. The area of a rectangle is defined as the product of the lengths of two adjacent sides, and the area of a triangle is proved to be one half of the product of the base and the height.

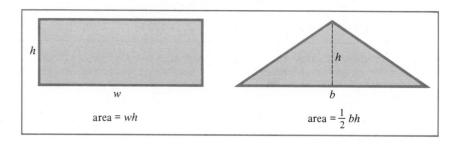

area = wh   area = $\frac{1}{2}bh$

When we draw the graph of marginal revenue for a business, the area of the region between the graph and the x-axis can be interpreted as the total revenue. For instance, suppose a factory has a constant marginal revenue, say $R'(x) = 1.5$, where x is the number of units made. Then the revenue per unit is simply 1.5, and the total revenue for n units is $1.5n$. You can see from the picture that the shaded region, bounded by the graph of $R'(x)$, the x-axis, and the vertical lines $x = 0$ and $x = n$ has an area of $1.5n$. Thus, **the area of the enclosed region is numerically equal to the total revenue.**

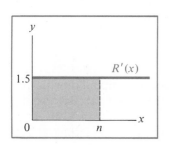

You can verify for yourself that the same result is true when the marginal revenue function's graph is an inclined straight line and the enclosed region is a triangle; the total revenue is still numerically equal to the area of the region. The interpretation is equally valid when the graph is that of marginal cost, marginal profit, marginal savings, or a number of similar quantities.

Things are not so simple, however, when the graph that forms a boundary of the region is not a straight line. With few exceptions, we do not have a formula from geometry to tell us the area of the region, so we would not be able to determine the value of the total revenue (or cost, etc.) by this method (although the area, whatever it is, is still numerically equal to the total revenue, cost, or other quantity).

The theory of integral calculus can give us a mechanism for calculating areas of regions of almost any shape. In order to begin developing this process, we define the concept of the definite integral.

Definition 1 Provided $f(x)$ is continuous and nonnegative at every x such that $a \le x \le b$, we define the **definite integral of $f(x)$ from a to b** to be the area of the region bounded above by the graph of $f(x)$, below by the x-axis, and on the left and right by the vertical lines $x = a$ and $x = b$. We denote the definite integral by

$$\int_a^b f(x) \, dx.$$

The difference between the **notation** for the definite integral and that for the antiderivative (or indefinite integral) of $f(x)$ is just the addition of a and b to the integral sign. For a particular definite integral, the letters a and b will be replaced by numbers. The lower number is called the **lower limit of integration** and the upper number is called the **upper limit of integration.** As in the antiderivative notation, $f(x)$ called the **integrand,** and the integral sign and the dx must be considered together.

Despite the similarities in notation, the definite integral is quite a different concept from the antiderivative. Whereas $\int f(x) \, dx$ represents a *family of functions,* the definite integral $\int_a^b f(x) \, dx$ is a *number.* We shall see in the next section that these two very different concepts are related to each other by the Fundamental Theorem of Integral Calculus, so the similarity of notation is justified; but you must keep in mind the difference between them.

EXAMPLE 1

Graph the function $f(x) = x^2$ and shade the region whose area is $\displaystyle\int_1^3 x^2 \, dx$.

Solution
Since $f'(x) = 2x$ and $f''(x) = 2$, the only critical value is $x = 0$, and $(0, 0)$ is an absolute minimum. This implies that $f(x)$ is nonnegative for all x between $a = 1$ and $b = 3$. The region is shown in the figure on the next page.

Sometimes we have a fully bounded region without using one of the side boundaries. Looking at the graph in Example 1, it is clear that $\int_0^3 x^2 \, dx$ is

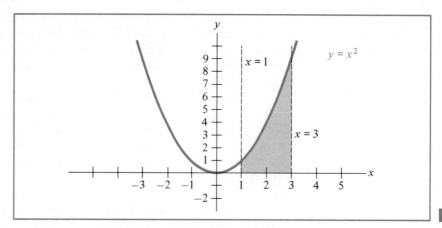

x	$f(x)$
0	0
-1	1
-2	4
1	1
2	4

the area of a region bounded by the x-axis, the graph of x^2, and the line $x = 3$. The line $x = 0$ is not required as part of the boundary.

Sometimes neither side boundary is required, as indicated in the next example.

EXAMPLE 2

Graph the function $f(x) = 4 - x^2$ and shade the region whose area is $\int_{-2}^{2} (4 - x^2)\, dx.$

Solution
Since $f'(x) = -2x$ and $f''(x) = -2$, the only critical value is $x = 0$, and $(0, 4)$ is an absolute maximum. The limits of integration are $x = -2$ and $x = 2$, where the graph of $f(x)$ crosses the x-axis.

x	$f(x)$
0	4
-1	3
-2	0
1	3
2	0

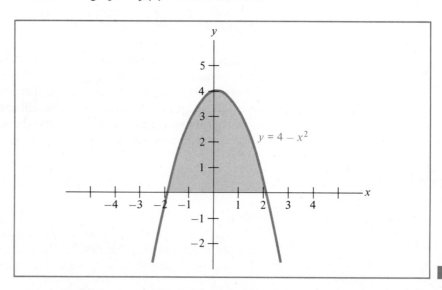

Definition 1 provided us with a definition of the definite integral only when $f(x)$ is nonnegative for all x such that $a \le x \le b$. We can make a similar definition for areas that lie entirely below the x-axis as follows.

Definition 2   Provided $f(x)$ is continuous and nonpositive at every x such that $a \leq x \leq b$, we define $\int_a^b f(x)\, dx$ to be (-1) times the area of the region bounded below by the graph of $f(x)$, above by the x-axis, and on the left and right by the vertical lines $x = a$ and $x = b$.

EXAMPLE 3

Graph $y = x - 4$ and shade the region whose area is $-\displaystyle\int_0^4 (x - 4)\, dx$.

Solution
Since y is a linear function of x, the graph is a straight line and we need only plot two points. Notice that the vertical line $x = 4$ is not part of the boundary, since the graph of the equation crosses the x-axis there.

x	y
0	-4
4	0

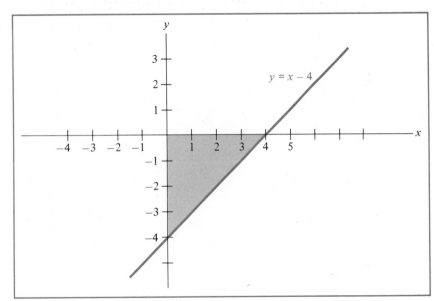

Since the shaded region in Example 3 is a triangle and the area of a triangle is one half the base times the height, we can compute $\int_0^4 (x - 4)\, dx$. The area of the triangle is 8, so the value of the definite integral is -8.

It may happen, for certain functions $f(x)$ and certain choices of the limits of integration, that $f(x)$ takes on both positive and negative values. In that case, neither Definition 1 nor Definition 2 applies. However, each of those definitions may apply to entire sections of the graph of $f(x)$. The following definition extends the notion of the definite integral to many of these cases.

Definition 3   Let $c_1, c_2, \ldots, c_n$ be the only values of x between a and b with the property that $f(x) = 0$ and assume we have listed these values from left to right (that is, $a < c_1 < c_2 < \cdots < c_n < b$). If $f(x)$ is continuous at every x such that $a \leq x \leq b$, then we define $\int_a^b f(x)\, dx$ by

$$\int_a^b f(x)\, dx = \int_a^{c_1} f(x)\, dx + \cdots + \int_{c_n}^b f(x)\, dx$$

Each of the integrals in the preceding sum is defined by either Definition 1 or Definition 2. Note that $f(x)$ could not assume some positive values and some negative values unless it crossed the x-axis,* and it does not cross the x-axis between a and b except at $c_1, c_2, \ldots, c_n$.

EXAMPLE 4

For the definite integral $\displaystyle\int_{-4}^{3} x \, dx$, write the equation in Definition 3; on the graph of $y = x$, shade each region whose area is involved in the definition.

Solution
Since the graph of $y = x$ is a straight line, we plot only two points.

x	y
0	0
3	3

Since $f(x) = x$ crosses the x-axis only at $x = 0$, we have $n = 1$ and $c_1 = 0$ in Definition 3. The defining equation is

$$\int_{-4}^{3} x \, dx = \int_{-4}^{0} x \, dx + \int_{0}^{3} x \, dx.$$

Because each of the two shaded regions is a triangle, we can easily calculate the area of each.

$$\int_{-4}^{0} x \, dx = -8, \qquad \int_{0}^{3} x \, dx = 4.5$$

Thus, the sum is

$$\int_{-4}^{3} x \, dx = -3.5. \quad \blacksquare$$

---

* This fact, which is a consequence of the continuity of $f(x)$, is a special case of the Intermediate Value Theorem: if $f(x)$ is continuous for all x such that $a \le x \le b$ and if $f(a) \ne f(b)$, then $f(x)$ takes on every value between $f(a)$ and $f(b)$.

Definitions 1, 2, and 3 together give a precise definition of the definite integral in terms of the intuitive concept of area; unfortunately, those definitions are a little long and cumbersome. Consequently, we now give a less precise, quick definition that includes all of Definitions 1, 2, and 3.

Quick Definition   The definite integral $\int_a^b f(x)\, dx$ is the sum formed from the areas of the regions bounded by the graph of $f(x)$, the x-axis, and the lines $x = a$ and $x = b$, where the area of any region above the x-axis is preceded by a plus sign and the area of any region below the x-axis is preceded by a minus sign.

EXAMPLE 5

For the definite integral $\displaystyle\int_{-2}^{\sqrt{3}} (x^3 - 3x)\, dx$, write the equation in Definition 3; on the graph of $y = x^3 - 3x$, shade each region whose area is involved in the definition.

Solution

In order to sketch the graph of $y = f(x) = x^3 - 3x$, we use the full procedure developed in Section 9.3.

$$f'(x) = 3x^2 - 3 = 3(x^2 - 1) = 3(x - 1)(x + 1)$$

The only critical values are $x = -1$ and $x = 1$.

$$f''(x) = 6x$$

$$f''(-1) < 0 \quad \text{and} \quad f''(1) > 0$$

It follows that $(-1, f(-1)) = (-1, 2)$ is a relative maximum and $(1, f(1)) = (1, -2)$ is a relative minimum. Setting $f(x) = 0$, we find

$$0 = x^3 - 3x = x(x^2 - 3) = x(x - \sqrt{3})(x + \sqrt{3}).$$

The x-intercepts are $-\sqrt{3}$, 0, and $\sqrt{3}$.

x	$f(x)$
-1	2
1	-2
$-\sqrt{3}$	0
$\sqrt{3}$	0
0	0
-2	-2

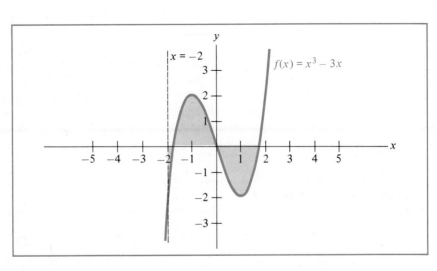

The defining equation is

$$\int_{-2}^{\sqrt{3}} (x^3 - 3x)\, dx = \int_{-2}^{-\sqrt{3}} (x^3 - 3x)\, dx + \int_{-\sqrt{3}}^{0} (x^3 - 3x)\, dx + \int_{0}^{\sqrt{3}} (x^3 - 3x)\, dx. \ \blacksquare$$

In Section 7.6, we showed that every polynomial $f(x)$ is continuous at every x. Consequently, $\int_a^b f(x)\, dx$ is defined for any limits of integration when $f(x)$ is a polynomial. If the integrand is not a polynomial, then we must be alert for discontinuities.

EXAMPLE 6

Graph the integrand of the definite integral $\int_2^5 \dfrac{1}{x}\, dx$ for $x > 0$ and shade the region whose area is indicated by the definite integral.

Solution
Since $f'(x) = -x^{-2}$ is negative for all $x \neq 0$, we conclude that $f(x) = 1/x$ is decreasing for $x > 0$. Here it is important that 0, the discontinuity of $1/x$, is not between 2 and 5; the definite integral is defined because $1/x$ is continuous at every x such that $2 \le x \le 5$.

x	$f(x)$
1	1
2	.5
4	.25
5	.20

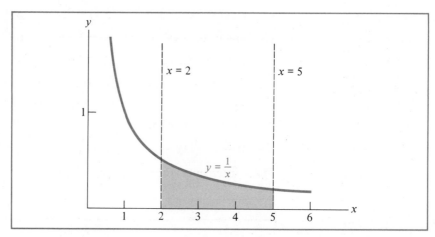

In this section we have exploited the fact that almost everyone has an intuitive concept of area; almost everyone is comfortable talking about the area of a region even when a formula for that area is not clear. **All of the observations in this section are required if the definite integral is to be used as a device for computing area.** In Section 10.7 we justify the use of the term "area."

Exercises 10.4

In problems 1 through 24, graph the integrand and shade the region whose area is indicated by the definite integral.

1 $\displaystyle\int_0^2 x^2\,dx$

2 $\displaystyle\int_0^1 x^2\,dx$

3 $\displaystyle\int_{-1}^1 x^2\,dx$

4 $\displaystyle\int_{-2}^2 x^2\,dx$

5 $\displaystyle\int_0^2 x^3\,dx$

6 $\displaystyle\int_0^1 x^3\,dx$

7 $\displaystyle\int_{-1}^1 x^3\,dx$

8 $\displaystyle\int_{-2}^2 x^3\,dx$

9 $\displaystyle\int_0^2 x^4\,dx$

10 $\displaystyle\int_0^1 x^4\,dx$

11 $\displaystyle\int_{-3}^3 (9 - x^2)\,dx$

12 $\displaystyle\int_{-1}^1 (1 - x^2)\,dx$

13 $\displaystyle\int_{-1}^3 (9 - x^2)\,dx$

14 $\displaystyle\int_{-2}^1 (1 - x^2)\,dx$

15 $\displaystyle\int_{-2}^2 (x^2 - 4)\,dx$

16 $\displaystyle\int_{-5}^5 (x^2 - 25)\,dx$

17 $\displaystyle\int_{-1}^2 (x^2 - 4)\,dx$

18 $\displaystyle\int_{-2}^5 (x^2 - 25)\,dx$

19 $\displaystyle\int_1^2 \frac{1}{x}\,dx$

20 $\displaystyle\int_1^5 \frac{1}{x}\,dx$

21 $\displaystyle\int_{-1}^{3/2} (2x^3 - 3x^2)\,dx$

22 $\displaystyle\int_0^{3/2} (2x^3 - 3x^2)\,dx$

23 $\displaystyle\int_{-1}^1 (2x^3 - 3x^2)\,dx$

24 $\displaystyle\int_{-1}^2 (2x^3 - 3x^2)\,dx$

In problems 25 through 36, graph the integrand and determine the value of the definite integral by determining the areas of the regions involved in the definition of the definite integral.

25 $\displaystyle\int_0^4 (3x - 2)\,dx$

26 $\displaystyle\int_{-1}^1 (3x - 2)\,dx$

27 $\displaystyle\int_2^8 (.5x + 4)\,dx$

28 $\displaystyle\int_{-4}^4 (.5x + 4)\,dx$

29 $\displaystyle\int_0^3 (4 - 2x)\,dx$

30 $\displaystyle\int_2^4 (4 - 2x)\,dx$

31 $\displaystyle\int_1^3 8\,dx$

32 $\displaystyle\int_2^4 e\,dx$

33 $\displaystyle\int_0^1 3\pi\,dx$

34 $\displaystyle\int_{-2}^2 -11\,dx$

35 $\displaystyle\int_0^3 |x|\,dx$

36 $\displaystyle\int_{-2}^2 |x|\,dx$

37 The marginal revenue of a certain company is $R'(x) = .5x$, where x is the number of units sold. Graph the marginal revenue function and calculate the definite integral $\int_0^{100} R'(x)\,dx$. Is this equal to the total revenue for $x = 100$?

38 The marginal cost of a certain company is $C'(x) = .2x + 2$, where x is the number of units made. Graph the marginal cost function and calculate the definite integral $\int_0^{100} C'(x)\,dx$. Is this equal to the total cost for $x = 100$?

39 The marginal profit of a certain company is $P'(x) = .3x - 2$, where x is the number of units made and sold. Graph the marginal profit function and calculate the definite integral $\int_0^{100} P'(x)\,dx$. Is this equal to the total profit for $x = 100$?

10.5   The Fundamental Theorem of Integral Calculus

The Fundamental Theorem of Integral Calculus allows us to find the value of any definite integral provided that we can find an antiderivative for the integrand.

Fundamental Theorem of Integral Calculus   If $f(x)$ is continuous at every x such that $a \le x \le b$ and if $F(x)$ is any antiderivative for $f(x)$, then

$$\int_a^b f(x)\,dx = F(b) - F(a).$$

Later we shall give some indication of why this remarkable theorem is true, but first we want to demonstrate its usefulness.

EXAMPLE 1

Evaluate $\displaystyle\int_{-1}^{2} x^3\,dx$.

Solution

Antiderivatives for x^3 are given by the Power Rule for Antiderivatives.

$$\int x^3\,dx = \frac{1}{4} x^4 + C$$

Since we need only one antiderivative, we take $\dfrac{1}{4} x^4$. According to the Fundamental Theorem of Integral Calculus, we have

$$\int_{-1}^{2} x^3\,dx = \frac{1}{4}(2)^4 - \frac{1}{4}(-1)^4$$

$$= 4 - \frac{1}{4} = 3.75. \ \blacksquare$$

It is comforting to note that we get the same answer in the preceding example with any antiderivative that we use, since

$$\left[\frac{1}{4}(2)^4 + C\right] - \left[\frac{1}{4}(-1)^4 + C\right] = 4 - \frac{1}{4} = 3.75.$$

The above steps show that the constant C "cancels itself."

Notation   For convenience we use the symbols $F(x)|_a^b$ to mean $F(b) - F(a)$. This notation will allow us to write down the relevant antiderivative before we evaluate it.

EXAMPLE 2

Evaluate $\displaystyle\int_0^1 e^{2x}\, dx.$

Solution
First we find the antiderivatives for the integrand.

$$\int e^{2x}\, dx = \frac{1}{2}\int e^{2x} 2\, dx = \frac{1}{2}e^{2x} + C$$

Then we use one of the antiderivatives to evaluate the definite integral.

$$\int_0^1 e^{2x}\, dx = \frac{1}{2}e^{2x}\Big|_0^1 = \frac{1}{2}e^2 - \frac{1}{2}e^0$$

$$= \frac{1}{2}e^2 - \frac{1}{2} \ \blacksquare$$

Now that we can evaluate the definite integral, we can find the area of each region obtained in the examples in Section 10.4.

EXAMPLE 3

Find the area of the region bounded above by the graph of $f(x) = x^2$ and below by the x-axis and between $x = 1$ and $x = 3$, as shown in Example 1 of Section 10.4.

Solution
Since $f(x)$ is positive and continuous for every x such that $1 \le x \le 3$, the indicated area equals $\int_1^3 x^2\, dx$. Thus, we evaluate the definite integral.

$$\int_1^3 x^2\, dx = \frac{1}{3}x^3\Big|_1^3 = 9 - \frac{1}{3} = 8\frac{2}{3}. \ \blacksquare$$

EXAMPLE 4

Find the area of the region bounded above by the graph of $f(x) = 4 - x^2$ and below by the x-axis, as shown in Example 2 in Section 10.4.

Solution
Since the region is entirely above the x-axis, it equals the definite integral $\int_{-2}^2 (4 - x^2)\, dx$, which we evaluate as follows:

$$\int_{-2}^{2} (4 - x^2) \, dx = \left(4x - \frac{1}{3} x^3 \right) \Big|_{-2}^{2}$$

$$= \left(8 - \frac{8}{3} \right) - \left(-8 + \frac{8}{3} \right) = \frac{32}{3} \quad \blacksquare$$

EXAMPLE 5

Find the sum of the areas of the two regions bounded by the x-axis, the graph of $y = x$, the graph of $x = -4$, and the graph of $x = 3$, as shown in Example 4 in Section 10.4.

Solution

If we simply calculated the definite integral $\int_{-4}^{3} x \, dx$, then we would *not* get the sum of the areas of the two regions because $\int_{-4}^{0} x \, dx$ is (-1) times the area below the x-axis, as the graph for Example 4 in 10.4 shows. We must add (-1) times $\int_{-4}^{0} x \, dx$ to $\int_{0}^{3} x \, dx$ in order to get the answer.

$$\int_{0}^{3} x \, dx - \int_{-4}^{0} x \, dx = \left(\frac{1}{2} x^2 \right) \Big|_{0}^{3} - \left(\frac{1}{2} x^2 \right) \Big|_{-4}^{0}$$

$$= \frac{9}{2} - 0 - [0 - 8] = \frac{9}{2} + 8$$

$$= 12.5 \quad \blacksquare$$

EXAMPLE 6

Find the area of the region bounded above by the x-axis and below by the graph of $y = x^3 - 3x$ between $x = 0$ and $x = \sqrt{3}$, as shown in Example 5 in Section 10.4.

Solution

Since the region is entirely below the x-axis, it equals (-1) times the integral $\int_{0}^{\sqrt{3}} (x^3 - 3x) \, dx$; we calculate the answer next.

$$-\int_{0}^{\sqrt{3}} (x^3 - 3x) \, dx = -\left(\frac{1}{4} x^4 - \frac{3}{2} x^2 \right) \Big|_{0}^{\sqrt{3}}$$

$$= -\left[\left(\frac{9}{4} - \frac{9}{2} \right) - (0) \right] = -\left[-\frac{9}{4} \right]$$

$$= \frac{9}{4} \quad \blacksquare$$

Each of the following properties of the definite integral is easily established using the Fundamental Theorem of Integral Calculus.

Properties of the Definite Integral

1. $\int_{a}^{b} cf(x) \, dx = c \int_{a}^{b} f(x) \, dx$ for any constant c

2. $\int_{a}^{b} (f(x) + g(x)) \, dx = \int_{a}^{b} f(x) \, dx + \int_{a}^{b} g(x) \, dx$

3. $\int_a^b (f(x) - g(x)) \, dx = \int_a^b f(x) \, dx - \int_a^b g(x) \, dx$

4. $\int_a^b f(x) \, dx = \int_a^c f(x) \, dx + \int_c^b f(x) \, dx$ provided $a \le c \le b$

To verify property 1, we let $F(x)$ be an antiderivative for $f(x)$; since $cF(x)$ is an antiderivative for $cf(x)$, we have

$$\int_a^b cf(x) \, dx = cF(b) - cF(a) = c(F(b) - F(a)) = c \int_a^b f(x) \, dx.$$

The other properties have similar verifications.

EXAMPLE 7

Evaluate $\int_0^1 (x^3 + 4x^2 - 2x + 3) \, dx$.

Solution
We use the Properties of the Definite Integral to simplify the given problem.

$$\int_0^1 (x^3 + 4x^2 - 2x + 3) \, dx = \int_0^1 x^3 \, dx + \int_0^1 4x^2 \, dx + \int_0^1 -2x \, dx + \int_0^1 3 \, dx$$

$$= \int_0^1 x^3 \, dx + 4 \int_0^1 x^2 \, dx - 2 \int_0^1 x \, dx + 3 \int_0^1 1 \, dx$$

$$= \frac{1}{4} x^4 \Big|_0^1 + 4 \left(\frac{1}{3} x^3 \right) \Big|_0^1 - 2 \left(\frac{1}{2} x^2 \right) \Big|_0^1 + 3x \Big|_0^1$$

$$= \frac{1}{4} - 0 + \frac{4}{3} - 0 - 1 + 0 + 3 - 0$$

$$= \frac{43}{12} = 3 \frac{7}{12} \quad \blacksquare$$

Why the Fundamental Theorem of Integral Calculus Is True

We now want to give some explanation as to why the Fundamental Theorem of Integral Calculus is true. Our remarks are not intended as a verification, but rather as a hint to the relationship between antiderivative and the area of a region in the plane.

Let $A(s)$ be the area of the region bounded above by the graph of $f(x) = x^2$ and below by the x-axis between $x = 1$ and $x = s$, as shown in the illustration. We intend to vary the right boundary of the region and to consider the area of each resulting region. (See page 478.)

In order to calculate the derivative of the area function $A(s)$, we choose another value $x = s + h$, and we consider the difference quotient

$$\frac{1}{h} [A(s + h) - A(s)].$$

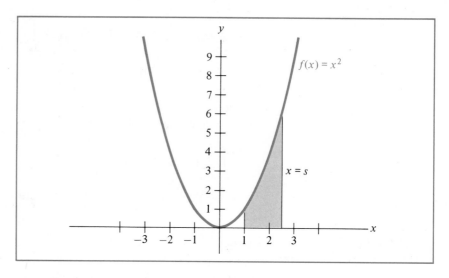

According to property 4 for definite integrals, we have

$$A(s + h) = \int_1^{s+h} x^2 \, dx = \int_1^s x^2 \, dx + \int_s^{s+h} x^2 \, dx.$$

We use this to simplify the difference quotient.

$$\frac{1}{h}[A(s + h) - A(s)] = \frac{1}{h}\left[\int_1^s x^2 \, dx + \int_s^{s+h} x^2 \, dx - \int_1^s x^2 \, dx\right]$$

$$= \frac{1}{h}\int_s^{s+h} x^2 \, dx$$

Using the next picture, we obtain estimates that are sufficient to calculate the limit of the above difference quotient in this special case.

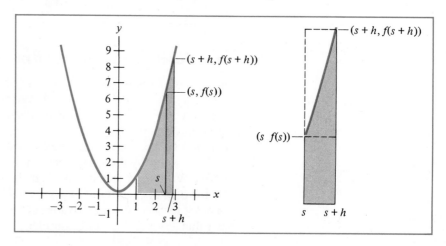

The area under the graph of $f(x)$ and above the x-axis between $x = s$ and $x = s + h$ is greater than the area of the rectangle formed by drawing a horizontal line through $(s, f(s))$; it is less than the area of the rectangle formed by the x-axis and the graphs of $x = s$, $x = s + h$, $y = f(s + h)$. The smaller rectangle has dimensions of h and $f(s)$ while the larger rectangle has dimensions of h and $f(s + h)$. Consequently, we have the following inequalities.

$$\frac{1}{h}(hf(s)) \le \frac{1}{h}\int_s^{s+h} x^2\, dx \le \frac{1}{h}(hf(s+h))$$

$$s^2 = f(s) \le \frac{1}{h}\int_s^{s+h} x^2\, dx \le f(s+h) = (s+h)^2$$

By the Limit Theorem of Section 7.3, we know that

$$\lim_{h\to 0}(s+h)^2 = (\lim_{h\to 0}(s+h))^2 = s^2$$

Since $\dfrac{1}{h}\displaystyle\int_s^{s+h} x^2\, dx$ is squeezed between s^2 and $(s+h)^2$ while $(s+h)^2$ converges to s^2, it must be that

$$\lim_{h\to 0}\frac{1}{h}\int_s^{s+h} x^2\, dx = s^2 = f(s)$$

so

$$\lim_{h\to 0}\frac{1}{h}[A(s+h) - A(s)] = f(s)$$

and, by the definition of the derivative, we have

$$A'(s) = f(s).$$

The area function $A(s)$ is an antiderivative for the integrand $f(s)$! This is the remarkable property that accounts for the Fundamental Theorem of Integral Calculus. Referring again to the figure on page 478, you can see that $A(1)$ must be zero (since it is the area of a region of zero width). Thus, $A(s) = A(s) - A(1)$ for any choice of s. If $F(s)$ is some convenient antiderivative for $f(s)$, then $A(s) = F(s) + C$ for some choice of the constant C. By the definition of the definite integral, we have $\int_1^s f(x)\, dx = A(s)$; thus,

$$\int_1^s f(x)\, dx = A(s) = A(s) - A(1)$$
$$= F(s) + C - F(1) - C$$
$$= F(s) - F(1).$$

Although we have justified the Fundamental Theorem of Integral Calculus only for the special case of $f(x) = x^2$ and for the fixed left boundary $x = 1$, it is possible to prove it in general for any continuous function $f(x)$ and any pair of boundaries.

We conclude this section with two practical applications of the definite integral. The first problem (Example 8) is oversimplified due to the limitations of time and space. After the problem, we discuss refinements that are applicable to more difficult problems.

EXAMPLE 8

A prospective buyer would like to determine the area of the tract of land shown in the illustration. What is the area?

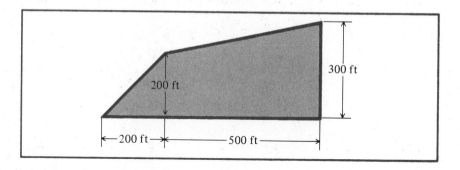

Solution

In order to find the area of the tract, we introduce a coordinate system and we label each upper vertex with its coordinates.

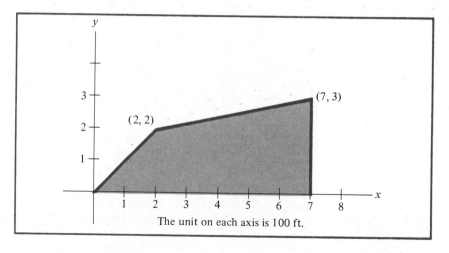

The unit on each axis is 100 ft.

The area of the triangle with vertices at $(0, 0)$, $(2, 0)$, and $(2, 2)$ is $(\frac{1}{2})(2)(2) = 2$; it suffices to find the area of the remaining trapezoid and add that to 2. The top boundary of the trapezoid is the line passing through $(2, 2)$ and $(7, 3)$ and we find the slope-intercept equation for that line below.

$$\frac{y-2}{x-2} = \frac{3-2}{7-2}$$

$$y - 2 = (x - 2)\left(\frac{1}{5}\right) = \frac{1}{5}x - \frac{2}{5}$$

$$y = \frac{1}{5}x + \frac{8}{5}$$

The area of the trapezoid is

$$\int_2^7 \left(\frac{1}{5}x + \frac{8}{5}\right) dx = \left(\frac{1}{5} \cdot \frac{x^2}{2} + \frac{8}{5}x\right)\bigg|_2^7 = 12.5$$

Thus, the area of the tract is $2 + 12.5 = 14.5$. ■

With some refinement, the method given in Example 8 for calculating the area of a tract of land is very useful. In order to work with a curved boundary, the boundary is usually divided into a number of shorter curves. Then for each short curve a linear function or a quadratic function is chosen such that part of its graph approximates the short curve; such a function is said to be a **model** for the short curve. The sum of the appropriate definite integrals of the model functions approximates the area under the curved boundary. The process of finding such model functions is beyond the scope of this text; however, the following example should suggest the spirit of the process.

EXAMPLE 9

A prospective buyer would like to determine the area of the tract of land shown. What is the area?

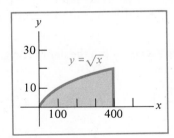

Solution
Since someone has already been kind enough to find a function whose graph coincides with the upper boundary, our job is easy. The area of the indicated region is the definite integral of $f(x) = \sqrt{x}$ from $x = 0$ to $x = 400$.

$$\int_0^{400} \sqrt{x}\, dx = \int_0^{400} x^{1/2}\, dx$$

$$= \frac{2}{3} x^{3/2}\bigg|_0^{400}$$

$$= \frac{2}{3}(8000)$$

$$= \frac{16,000}{3}$$ ■

Our second application concerns finding an average value for a function that assumes an infinite number of values. Recall that the average of four numbers, say x_1, x_2, x_3, x_4, is the one number $(x_1 + x_2 + x_3 + x_4)/4$; the average of n numbers, say $x_1, x_2, \ldots, x_n$, is the one number $(x_1 + x_2 + \cdots + x_n)/n$. The average is important; it is a single number that represents the several numbers that were averaged. The **average value of the function** $f(x)$ from $x = a$ to $x = b$, which we denote by $M(f)$, is defined by the equation

$$M(f) = \frac{\int_a^b f(x)\, dx}{b - a}$$

This single number represents all the values that $f(x)$ assumes from $x = a$ to $x = b$.

EXAMPLE 10

The consumption of oil by the free world from 1976 through 1980 is approximated quite accurately by the function

$$f(t) = .1t^2 + 1.45t + 48.4$$

where t is the number of years since the beginning of 1976. Here the value of $f(t)$ is the number of millions of barrels used per day. Assuming that this function continues to be an accurate approximation, find the average daily use of oil for the period from 1976 to 2000.

Solution

We compute the average from the definition given earlier. Note that t varies from 0 (in 1976) to 24 (in 2000).

$$M(f) = \frac{1}{24} \int_0^{24} (.1t^2 + 1.45t + 48.4)\, dt$$

$$= \frac{1}{24} \left[(.1)\left(\frac{1}{3} t^3\right) + (1.45)\left(\frac{1}{2} t^2\right) + 48.4t \right]\Big|_0^{24}$$

$$= \frac{1}{24} [460.8 + 417.6 + 1161.6]$$

$$= \frac{1}{24} [2040] = 85$$

The average daily consumption of oil by the free world during the period 1976 through 2000 should be 85 million barrels. ■

Exercises 10.5

Using the properties of the definite integral, evaluate each of the following.

1 $\displaystyle\int_0^1 (x^2 - x + 1)\, dx$

2 $\displaystyle\int_0^1 (3x^2 - 2x + 1)\, dx$

3 $\displaystyle\int_{-1}^{0} (x^3 + 5x^2 - 6x + 3)\, dx$

4 $\displaystyle\int_{-1}^{0} (x^3 + 10x)\, dx$

5 $\displaystyle\int_{-2}^{2} (5x^4 + 3x^2 - 8)\, dx$

6 $\displaystyle\int_{-2}^{2} (x^4 + 2x^3 - x)\, dx$

7 $\displaystyle\int_{0}^{8} \sqrt[3]{x}\, dx$

8 $\displaystyle\int_{0}^{9} x^{5/2}\, dx$

9 $\displaystyle\int_{1}^{2} \frac{10}{x^3}\, dx$

10 $\displaystyle\int_{1}^{2} \left(\frac{4}{x^2} + 5\right) dx$

11 $\displaystyle\int_{0}^{3} (x-1)(x+1)\, dx$

12 $\displaystyle\int_{-1}^{2} x(2x+3)\, dx$

13 $\displaystyle\int_{1}^{4} \frac{(x+1)(x-2)}{x}\, dx$

14 $\displaystyle\int_{1}^{2} \frac{(3x+2)(x-1)}{x}\, dx$

15 $\displaystyle\int_{0}^{1} (e^{2x} + e^{-x} + 3)\, dx$

16 $\displaystyle\int_{0}^{1} \left(\frac{6}{e^{3x}} + 4\right) dx$

17 $\displaystyle\int_{0}^{2} \frac{e^{9x} + e^{6x}}{e^{3x}}\, dx$

18 $\displaystyle\int_{0}^{2} \frac{e^{4x} - 2}{e^{x}}\, dx$

19 $\displaystyle\int_{-2}^{1} (e^x - 1)(e^x + 1)\, dx$

20 $\displaystyle\int_{-1}^{1} e^{-2x}(e^{3x} - 5)\, dx$

21 $\displaystyle\int_{-1}^{4} \sqrt{x+5}\, dx$

22 $\displaystyle\int_{0}^{19} \sqrt[3]{x+8}\, dx$

23 $\displaystyle\int_{0}^{2} \sqrt{4x^2+9}\, x\, dx$

24 $\displaystyle\int_{0}^{7} \sqrt[3]{x^2+1}\, x\, dx$

25 $\displaystyle\int_{0}^{1} \frac{3x}{(x^2+4)^{14}}\, dx$

26 $\displaystyle\int_{0}^{1} \frac{8x}{(3x^2+1)^6}\, dx$

27 $\displaystyle\int_{-1}^{1} (x^3-2)^2 x\, dx$

28 $\displaystyle\int_{-1}^{1} (5-x^3)^2\, 4x\, dx$

29 $\displaystyle\int_{1}^{4} \frac{(2\sqrt{x}+1)^5}{\sqrt{x}}\, dx$

30 $\displaystyle\int_{1}^{4} \frac{(\sqrt{x}-1)^9}{\sqrt{x}}\, dx$

31 $\displaystyle\int_{0}^{9} (\sqrt{x}+3)\sqrt{x}\, dx$

32 $\displaystyle\int_{0}^{9} 3\sqrt{x}(2\sqrt{x}+5)\, dx$

33 $\displaystyle\int_{0}^{4} \frac{x}{x^2+9}\, dx$

34 $\displaystyle\int_{0}^{2} \frac{x}{3x^2+1}\, dx$

35 $\displaystyle\int_{0}^{3} \frac{1}{3x+2}\, dx$

36 $\displaystyle\int_{3}^{5} \frac{1}{2x-5}\, dx$

37 $\displaystyle\int_{1}^{4} \frac{2x+5}{x^2}\, dx$

38 $\displaystyle\int_{2}^{3} \frac{5x^3-3}{2x}\, dx$

39 $\displaystyle\int_{-1}^{1} e^{x^2}x\, dx$

40 $\displaystyle\int_{0}^{1} 2e^{x^3}x^2\, dx$

In each of the following problems, determine the area of the region bounded by the x-axis and the graphs of the indicated equations.

41 $y = x^2$, $x = 4$
42 $y = x^2$, $x = 9$
43 $y = x^2$, $x = 1$, $x = 4$
44 $y = x^2$, $x = 1$, $x = 9$
45 $y = x^3$, $x = 8$
46 $y = x^3$, $x = 1$, $x = 8$
47 $y = 9 - x^2$
48 $y = 16 - x^2$
49 $y = x^2 - 4$
50 $y = x^2 - 25$

51 $y = \dfrac{1}{x}$, $x = 1$, $x = 5$
52 $y = \dfrac{1}{x}$, $x = 2$, $x = 4$

53 $y = e^x$, $x = 0$, $x = 1$
54 $y = e^{-x}$, $x = 0$, $x = 1$

55 $y = \dfrac{1}{x}$, $x = 1$, $x = e$
56 $y = \dfrac{1}{x}$, $x = \dfrac{1}{e}$, $x = 1$

In each of the following problems, determine the sum of the areas of the two regions bounded by the x-axis and the graphs of the indicated equations.

57 $y = 9 - x^2$, $x = 4$
58 $y = 16 - x^2$, $x = 6$
59 $y = x^2 - 4$, $x = 4$
60 $y = x^2 - 25$, $x = 8$
61 $y = x^3$, $x = -2$, $x = 1$
62 $y = x^3$, $x = -1$, $x = 1$
63 $y = x$, $x = -1$, $x = 1$
64 $y = -x$, $x = -2$, $x = 3$
65 $y = e^x - 1$, $x = -1$, $x = 2$
66 $y = e^x - 1$, $x = -2$, $x = 1$

Determine the area of each tract of land pictured in problems 67 through 70. All lengths are in feet.

67

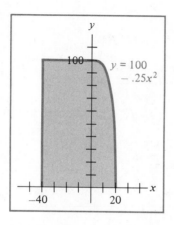

68

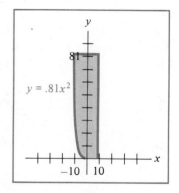

69

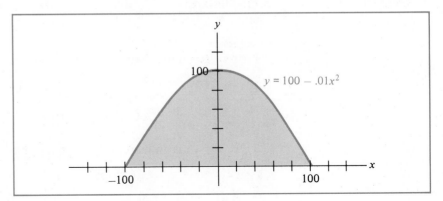

70

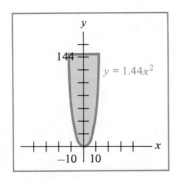

Find the average value for each of the following functions between the indicated values for the independent variable.

71 $f(t) = t^2 + .1t + 5$ from $t = 0$ to $t = 10$
72 $g(t) = .5t^2 + .2t + 20$ from $t = 1$ to $t = 5$
73 $f(x) = 5x - 3$ from $x = 0$ to $x = 10$
74 $g(x) = 8 - 2x$ from $x = -2$ to $x = 12$
75 $f(t) = 105$ from $t = 0$ to $t = 1000$
76 $g(t) = -13$ from $t = -11$ to $t = 57$
77 $p(x) = 100x - .1x^2 - 200$ from $x = 50$ to $x = 60$
Here $p(x)$ is the profit in dollars of manufacturing x units.
78 $C(x) = .5x^2 + 110$ from $x = 10$ to $x = 25$
Here $C(x)$ is the cost in dollars of manufacturing x units.

10.6   More on Areas of Regions

In this section, we consider regions with boundaries that are more complicated than those of regions considered in Sections 10.4 and 10.5.

EXAMPLE 1

Find the area of the region bounded by the x-axis and the graphs of $y = \sqrt{x}$ and $y = \sqrt{4 - x}$.

Solution

First we graph the given functions.

$$y = \sqrt{x} = x^{1/2}$$

$$\frac{dy}{dx} = \frac{1}{2} x^{-1/2} > 0$$

Thus, $y = \sqrt{x}$ is an increasing function for $x > 0$.

$$y = \sqrt{4 - x} = (4 - x)^{1/2}$$

$$\frac{dy}{dx} = \frac{1}{2} (4 - x)^{-1/2}(-1) < 0$$

Thus, $y = \sqrt{4 - x}$ is a decreasing function for $4 - x > 0$ or $4 > x$, and it is defined only for $4 \geq x$.

x	$\sqrt{x}$	x	$\sqrt{4-x}$
0	0	4	0
1	1	3	1
4	2	0	2
9	3	−5	3

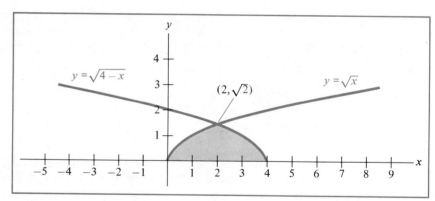

Next, we find the coordinates of each point where two boundaries meet.

$$\sqrt{x} = 0 \qquad \sqrt{4 - x} = 0$$

$$x = 0 \qquad 4 - x = 0$$

$$4 = x$$

The graphs of $\sqrt{x}$ and $\sqrt{4 - x}$ cross the x-axis at $(0, 0)$ and $(4, 0)$, respectively.

$$\sqrt{x} = \sqrt{4 - x}$$

$$x = 4 - x$$

$$2x = 4$$

$$x = 2$$

The graphs of $\sqrt{x}$ and $\sqrt{4 - x}$ intersect at $(2, \sqrt{2})$. We must compute the area of the given region in two steps, since the upper boundary changes at $x = 2$;

we compute the area under the graph of $y = \sqrt{x}$ from $x = 0$ to $x = 2$ and the area under the graph of $y = \sqrt{4-x}$ from $x = 2$ to $x = 4$.

$$\text{region's area} = \int_0^2 \sqrt{x}\, dx + \int_2^4 \sqrt{4-x}\, dx$$

$$= \int_0^2 x^{1/2}\, dx + \int_2^4 (4-x)^{1/2}\, dx$$

$$= \frac{2}{3} x^{3/2} \Big|_0^2 - \int_2^4 (4-x)^{1/2}(-1)\, dx$$

$$= \frac{2}{3} (\sqrt{2})^3 - \frac{2}{3} (4-x)^{3/2} \Big|_2^4$$

$$= \frac{4\sqrt{2}}{3} + \frac{2}{3} (2)^{3/2}$$

$$= \frac{4\sqrt{2}}{3} + \frac{4\sqrt{2}}{3} = \frac{8\sqrt{2}}{3} \quad \blacksquare$$

Some regions are bounded both above and below by the graphs of functions. In such cases, it is necessary to determine the limits of integration from the points of intersection of the two graphs.

EXAMPLE 2

Find the area of the region bounded by the graphs of $y = x^2$ and $y = 8 - x^2$.

Solution
First, we sketch the graphs involved; second, we find the coordinates of each point where two boundaries intersect.

$$y = x^2$$

$$\frac{dy}{dx} = 2x$$

$$0 = 2x$$

$$0 = x$$

$$\frac{d^2y}{dx^2} = 2$$

There is a relative minimum at $(0, 0)$, and $x = 0$ is the only critical value.

$$y = 8 - x^2$$

$$\frac{dy}{dx} = -2x$$

$$0 = -2x$$

$$0 = x$$

$$\frac{d^2y}{dx^2} = -2$$

There is a relative maximum at $(0, 8)$, and $x = 0$ is the only critical value.

x	x^2	x	$8 - x^2$
0	0	0	8
1	1	1	7
2	4	2	4
-1	1	-1	7
-2	4	-2	4

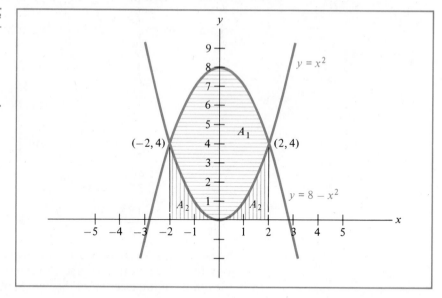

$$x^2 = 8 - x^2$$

$$2x^2 = 8$$

$$x^2 = 4$$

$$x = 2, -2$$

The graphs intersect at $(-2, 4)$ and $(2, 4)$.

It is not difficult to see that the area under the graph of $y = 8 - x^2$ from $x = -2$ to $x = 2$ is the sum of the area of the region A_1 with horizontal shading and the area of the region A_2 with vertical shading. Since the area of the region A_2 with vertical shading is $\int_{-2}^{2} x^2 \, dx$, the area of the region A_1 with horizontal shading must be

$$\int_{-2}^{2} (8 - x^2) \, dx - \int_{-2}^{2} x^2 \, dx.$$

We determine this number by using the properties of definite integrals and finding appropriate antiderivatives.

$$\int_{-2}^{2} (8 - x^2) \, dx - \int_{-2}^{2} x^2 \, dx = \int_{-2}^{2} (8 - x^2) \, dx + \int_{-2}^{2} -x^2 \, dx$$

$$= \int_{-2}^{2} (8 - x^2 - x^2) \, dx = \int_{-2}^{2} (8 - 2x^2) \, dx$$

$$= \left(8x - \frac{2}{3} x^3 \right) \Big|_{-2}^{2} = 16 - \frac{16}{3} - \left(-16 + \frac{16}{3} \right)$$

$$\text{area of } A_1 = \frac{64}{3} \ \blacksquare$$

The preceding example illustrates a general fact that is summarized in the next theorem. Be sure to note that the theorem does not require any part of either graph to be above the x-axis. The theorem is verified by adding an appropriately large constant to each function so that an identical region is obtained entirely above the x-axis.

Theorem   If the graph of $f(x)$ lies above the graph of $g(x)$ from $x = a$ to $x = b$, then the area of the region between the graphs of $f(x)$ and $g(x)$ from $x = a$ to $x = b$ is

$$\int_a^b (f(x) - g(x))\, dx$$

provided the definite integral is defined.

Next we give a procedure for finding the area of a given region; this procedure is appropriate for all regions that have been discussed in this section and in Sections 10.4 and 10.5.

Procedure for Finding the Area of a Region

1. Sketch the graphs that bound the region.
2. Find the coordinates of each point where two boundaries intersect.
3. If the x-axis is the lower boundary, and the graph of $f(x)$ from $(a, f(a))$ to $(b, f(b))$ is the upper boundary, then the area is $\int_a^b f(x)\, dx$. (If the roles of the x-axis and $f(x)$ are reversed, then the area is $-\int_a^b f(x)\, dx$.)
4. If the x-axis is the lower boundary, and the upper boundary consists of the graph of $f(x)$ from $(a, f(a))$ to $(c, f(c))$ and the graph of $g(x)$ from $(c, g(c)) = (c, f(c))$ to $(b, g(b))$, then the area is

$$\int_a^c f(x)\, dx + \int_c^b g(x)\, dx.$$

(If, instead, the x-axis is the upper boundary, then the area is the negative of the above sum.)

5. If the graph of $f(x)$ from $(a, f(a))$ to $(b, f(b))$ is the upper boundary and the graph of $g(x)$ from $(a, g(a))$ to $(b, g(b))$ is the lower boundary, then the area is

$$\int_a^b (f(x) - g(x))\, dx.$$

EXAMPLE 3

Find the area of the region bounded by the graphs of $y = x^2 - 2$ and $y = x$.

Solution

We follow the procedure given above, using steps 1, 2, and 5. The graph of $y = x$ is a straight line; the two points $(-1, -1)$ and $(1, 1)$ determine the graph. To sketch the graph of $y = x^2 - 2$, we note the following.

$$y = x^2 - 2$$

$$\frac{dy}{dx} = 2x$$

$$0 = 2x$$

$$0 = x$$

$$\frac{d^2y}{dx^2} = 2 > 0$$

There is a relative minimum at $(0, -2)$, and $x = 0$ is the only critical value.

x	$y = x^2 - 2$
-2	2
-1	-1
0	-2
1	-1
2	2

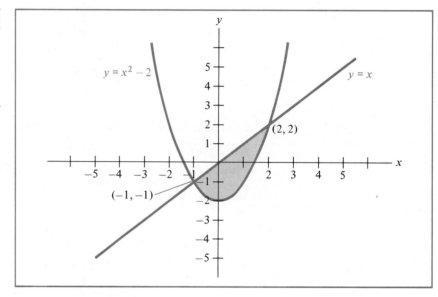

$$x = x^2 - 2$$

$$0 = x^2 - x - 2 = (x - 2)(x + 1)$$

The graphs cross at $(-1, -1)$ and $(2, 2)$, and the line $y = x$ is the upper boundary. We use step 5 of the procedure for finding an area of a region.

$$\text{area of region} = \int_{-1}^{2} [x - (x^2 - 2)] \, dx$$

$$= \int_{-1}^{2} (x - x^2 + 2) \, dx$$

$$= \left(\frac{1}{2} x^2 - \frac{1}{3} x^3 + 2x \right) \Bigg|_{-1}^{2}$$

$$= \left(2 - \frac{8}{3} + 4 \right) - \left(\frac{1}{2} + \frac{1}{3} - 2 \right)$$

$$= 4.5 \ \blacksquare$$

EXAMPLE 4

Determine the total area of the regions bounded by the graphs of $y = x^3 - 3x$ and $y = -x$.

Solution

Following the procedure given earlier, we sketch the graphs of the two given functions. The graph of $y = -x$ is the straight line passing through $(-1, 1)$ and $(1, -1)$.

$$y = x^3 - 3x$$

$$\frac{dy}{dx} = 3x^2 - 3 = 3(x^2 - 1) = 3(x - 1)(x + 1)$$

$$\frac{d^2y}{dx^2} = 6x$$

$$\left.\frac{d^2y}{dx^2}\right|_{x=-1} = -6 < 0 \qquad \left.\frac{d^2y}{dx^2}\right|_{x=1} = 6 > 0$$

There is a relative maximum at $(-1, 2)$ and a relative minimum at $(1, -2)$. The only critical values are $x = -1$ and $x = 1$.

$$x^3 - 3x = -x$$

$$x^3 - 2x = 0$$

$$(x^2 - 2)x = 0$$

$$x = 0, \pm\sqrt{2}$$

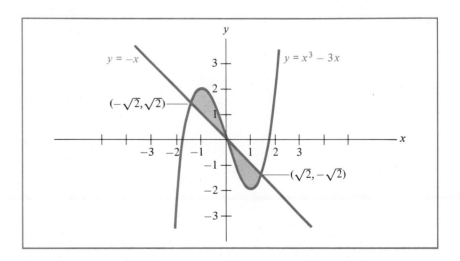

The curve and the line intersect at $(-\sqrt{2}, \sqrt{2})$, $(0, 0)$, and $(\sqrt{2}, -\sqrt{2})$.

The two regions must be handled separately because the graph of $y = -x$ is the lower boundary for one and the upper boundary for the other.

$$\text{sum of areas} = \int_{-\sqrt{2}}^{0} (x^3 - 3x + x)\, dx + \int_{0}^{\sqrt{2}} (-x - x^3 + 3x)\, dx$$

$$= \int_{-\sqrt{2}}^{0} (x^3 - 2x)\, dx + \int_{0}^{\sqrt{2}} (2x - x^3)\, dx$$

$$= \left(\frac{1}{4} x^4 - x^2\right)\Bigg|_{-\sqrt{2}}^{0} + \left(x^2 - \frac{1}{4} x^4\right)\Bigg|_{0}^{\sqrt{2}}$$

$$= -(1 - 2) + (2 - 1)$$

$$= 2 \quad \blacksquare$$

Sometimes vertical lines are part of the boundaries.

EXAMPLE 5

Find the area of the region between the graphs of $y = 2 - x$ and $y = 1/x$ from $x = 1$ to $x = 3$.

Solution
The graph of $y = 2 - x$ is the straight line passing through $(0, 2)$ and $(2, 0)$; the graph of $y = 1/x$ is familiar. We do not sketch graphs in the third quadrant since the indicated region is in the first and fourth quadrants.

x	$1/x$
.25	4
.5	2
1	1
2	.5
4	.25

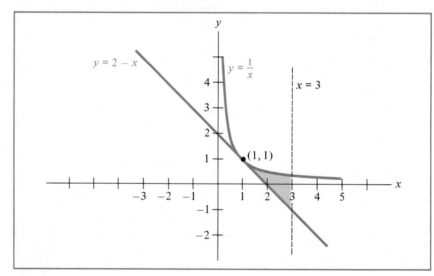

$$2 - x = \frac{1}{x}$$

$$2x - x^2 = 1$$

$$0 = x^2 - 2x + 1 = (x - 1)^2$$

$$x = 1$$

The graphs intersect only at $x = 1$. We now apply step 5.

$$\text{area} = \int_1^3 \left(\frac{1}{x} - 2 + x\right) dx = \int_1^3 (x^{-1} - 2 + x)\, dx$$

$$= \left(\ln x - 2x + \frac{1}{2} x^2\right)\Big|_1^3$$

$$= \left(\ln 3 - 6 + \frac{9}{2}\right) - \left(\ln 1 - 2 + \frac{1}{2}\right)$$

$$= \ln 3 \ \blacksquare$$

The procedure given earlier in this section does not apply to all regions; rather, it applies to three of the most common kinds of regions. The example below shows how we can combine parts 4 and 5 from that procedure.

EXAMPLE 6

Find the area of the region bounded by the graphs of $y = 6 - 2x$, $y = .5x + 3$, and $y = x^2 - 4x + 3$.

Solution
We sketch the graph of each equation and determine the points of intersection. The graph of $y = 6 - 2x$ is the straight line passing through $(0, 6)$ and $(3, 0)$; the graph of $y = .5x + 3$ is the straight line passing through $(0, 3)$ and $(4, 5)$. Now we consider the remaining function.

$$y = x^2 - 4x + 3$$

$$\frac{dy}{dx} = 2x - 4 = 2(x - 2)$$

$$\frac{d^2y}{dx^2} = 2$$

There is a relative minimum at $(2, -1)$, and $x = 2$ is the only critical value.

$$6 - 2x = .5x + 3$$

$$3 = 2.5x$$

$$x = 1.2$$

The straight lines intersect at $(1.2, 3.6)$.

$$6 - 2x = x^2 - 4x + 3$$

$$0 = x^2 - 2x - 3$$

$$0 = (x - 3)(x + 1)$$

$$x = 3, -1$$

The line with negative slope intersects the curve at $(-1, 8)$ and $(3, 0)$.

x	y
0	3
1	0
2	−1
3	0
4	3

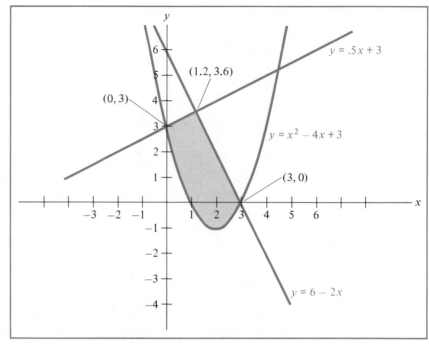

$$.5x + 3 = x^2 - 4x + 3$$
$$0 = x^2 - 4.5x$$
$$0 = x(x - 4.5)$$
$$x = 0, 4.5$$

The line with positive slope intersects the curve at $(0, 3)$ and $(4.5, 5.25)$. Looking at the graph, we see that we can discard the points $(-1, 8)$ and $(4.5, 5.25)$, which are outside the region of interest.

The area must be found in two steps because the upper boundary changes at $x = 1.2$.

$$\text{area} = \int_0^{1.2} (.5x + 3 - x^2 + 4x - 3) \, dx + \int_{1.2}^3 (6 - 2x - x^2 + 4x - 3) \, dx$$
$$= \int_0^{1.2} (4.5x - x^2) \, dx + \int_{1.2}^3 (3 - x^2 + 2x) \, dx$$
$$= \left(2.25x^2 - \frac{1}{3} x^3 \right) \Big|_0^{1.2} + \left(3x - \frac{1}{3} x^3 + x^2 \right) \Big|_{1.2}^3$$
$$= (3.24 - .576) + (9 - 9 + 9) - (3.6 - .576 + 1.44)$$
$$= 7.2 \; \blacksquare$$

The procedure for finding the area of a region can be refined and extended in many ways; the extended procedure applies to more complicated regions than those shown in the preceding six examples.

In Section 3.1 we considered how supply and demand determine the market price p_0 and the market demand x_0 for a particular commodity. The demand function for the commodity, denoted $p = D(x)$, gives the price p that consumers will pay when total consumer demand is x. The supply function for the commodity, denoted $p = S(x)$, gives the price that a producer will charge for the commodity when total consumer demand is x. The point of intersection of the graphs of these two functions is the equilibrium point; the coordinates of that point are the demand and price that will occur in a free market. The demand at the equilibrium point we called the "market demand" and the price at the equilibrium point we called the "market price."

Some consumers of this commodity would be willing to pay more than the market price p_0. Consequently, each of these consumers saves money as a result of the free market system. The total amount saved by all consumers is called the **consumers' surplus (CS).** On the combined graphs of the demand and supply functions, the area of the shaded region is the CS. The shaded region is bounded above by the graph of $p = D(x)$ and below by the graph of $p = p_0$; the area of the shaded region is

$$CS = \int_0^{x_0} (D(x) - p_0)\, dx.$$

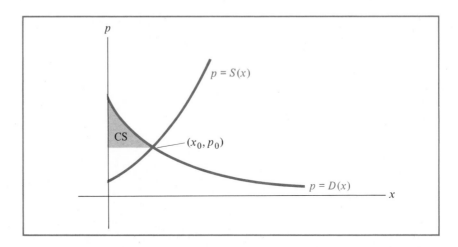

Thus, we have a formula for computing the value of the consumers' surplus.

Some producers would offer this commodity for sale even if the price were lower than p_0. Consequently, each of these producers makes more money as a result of the free market system. The total amount of the additional revenue is called the **producers' surplus (PS).** On the following combined graphs of the demand and supply functions, the area of the shaded region is the PS. The shaded region is bounded above by the graph of $p = p_0$ and below by the graph of $p = S(x)$; the area of the shaded region is

$$PS = \int_0^{x_0} (p_0 - S(x))\, dx.$$

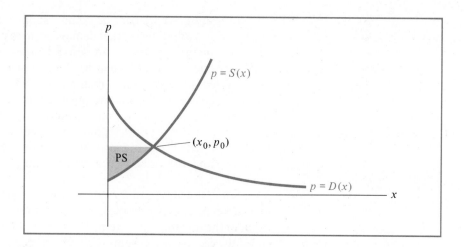

Thus, we have a formula for computing the value of the producers' surplus.

EXAMPLE 7

Determine the consumers' surplus and the producers' surplus for the demand function $p = D(x) = 7 - x$ and the supply function $p = S(x) = 2x + 1$.

Solution

We graph the two functions and determine the point of intersection for the two graphs.

x	$D(x)$	x	$S(x)$
0	7	0	1
7	0	3	7

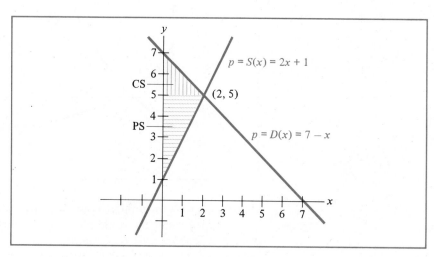

$$7 - x = 2x + 1$$
$$6 = 3x$$
$$2 = x_0$$
$$p_0 = D(2) = 7 - 2 = 5$$

The consumers' surplus is

$$CS = \int_0^2 (7 - x - 5)\, dx = \int_0^2 (2 - x)\, dx$$
$$= \left(2x - \frac{1}{2} x^2\right)\Big|_0^2 = (4 - 2)$$
$$= 2.$$

The producers' surplus is

$$PS = \int_0^2 (5 - 2x - 1)\, dx = \int_0^2 (4 - 2x)\, dx$$
$$= (4x - x^2)\Big|_0^2 = (8 - 4)$$
$$= 4. \ \blacksquare$$

Exercises 10.6

Find the area of the region bounded by the graphs of the given equations.

1 $y = 0, \quad y = x, \quad y = \sqrt{6 - x}$
2 $y = 0, \quad y = 4 - x, \quad y = \sqrt{x + 8}$
3 $y = 0, \quad y = 6 - 2x, \quad y = x$
4 $y = 0, \quad y = 2 - x, \quad y = x$
5 $y = 0, \quad y = x + 2, \quad y = \dfrac{4}{x} - 1, \quad x \geq -2$
6 $y = 0, \quad y = 2x + 1, \quad y = \dfrac{4}{x} - 1, \quad x \geq -.5$
7 $y = 0, \quad y = 6 - x, \quad y = x^2$
8 $y = 0, \quad y = x + 12, \quad y = x^2$
9 $y = 6 - x, \quad y = x^2$
10 $y = x + 12, \quad y = x^2$
11 $y = x - 3, \quad y = 3 - x^2$
12 $y = -3x, \quad y = 4 - x^2$
13 $y = x^2 - 4x + 4, \quad y = -x^2 + 4x + 4$
14 $y = x^2 + 2x + 1, \quad y = -x^2 - 2x + 1$
15 $y = 5 - 4x, \quad y = \dfrac{1}{x}$
16 $y = 4 - 3x, \quad y = \dfrac{1}{x}$
17 $y = x + 1, \quad y = 1 + 3x - x^3$
18 $y = 1, \quad y = 1 + 3x - x^3$
19 $y = x, \quad y = x^2 + 2, \quad x = -1, \quad x = 1$
20 $y = -x, \quad y = x^2 + 2, \quad x = -2, \quad x = 2$

21   $y = 1 - x, \quad y = e^x, \quad x = 2$

22   $y = 1, \quad y = e^x, \quad x = 3$

23   $y = -x, \quad y = e^{-x}, \quad x = -2, \quad x = 0$

24   $y = x, \quad y = e^{-x}, \quad x = -1, \quad x = 0$

25   $y = x - 2, \quad y = \dfrac{-1}{x}, \quad x = 1, \quad x = e$

26   $y = 2 - 2x, \quad y = \dfrac{-1}{x}, \quad x = 2$

27   $y = 4 - x^2, \quad y = -x, \quad y = x$

28   $y = 4 - x^2, \quad y = x - 2, \quad y = -x - 4, \quad y \geq -3$

29   $y = x^3 - 3x, \quad y = 3x - x^3$

30   $y = \dfrac{1}{2}(e^x + e^{-x}), \quad y = \dfrac{1}{2}(e^x - e^{-x}), \quad x = 0, \quad x = 1$

In each problem below, determine the consumers' surplus and the producers' surplus for the given demand function and supply function.

31   $p = S(x) = .5x + 1, \quad p = D(x) = 4 - x$

32   $p = S(x) = .5x, \quad p = D(x) = 5 - .5x$

33   $p = S(x) = x + 1, \quad p = D(x) = 5 - x$

34   $p = S(x) = x + 1, \quad p = D(x) = 7 - .5x$

35   $p = S(x) = x, \quad p = D(x) = \dfrac{1}{x + .5}$

36   $p = S(x) = .5x, \quad p = D(x) = \dfrac{1}{x + .5}$

10.7   Justifying the Term "Area" (Optional)

In Section 10.4 we used the intuitive concept of area to define the definite integral. In Section 10.5 we used the Fundamental Theorem of Integral Calculus to evaluate the definite integral, and we gave an illustrative demonstration as to why the Fundamental Theorem of Integral Calculus is true. There is another more fundamental way to justify defining the definite integral in terms of the areas of regions. This approach is based on the area of a rectangle. There is no ambiguity as to the area of a rectangle; it is the product of the lengths of two adjacent sides. Again we shall *only give illustrative arguments for some special cases* since the general proof is beyond the scope of this text.

We are going to approximate the area of the region under the graph of the function $y = f(x)$ from a to b by constructing rectangles contained in the region with one side along the x-axis. The sum of the areas of these rectangles is our approximation for the area of the region. For example, the first picture shows the region with two approximating rectangles of equal width.

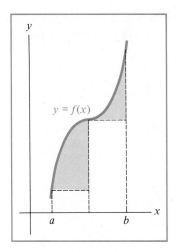

Here we have shaded the part of the region that is not contained in the rectangles to indicate how much difference there is between the area of the region and the sum of the areas of the rectangles. We shall refer to the area of the region minus the sum of the areas of the rectangles as the **error.** Now we use four rectangles of equal width inscribed in the given region, and we note how the approximation improves and the error decreases.

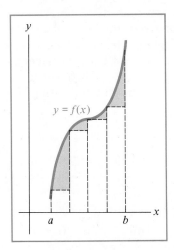

The sum of the areas of the approximating rectangles is called a **lower sum** or a **lower Riemann sum.*** Each rectangle is called a **lower rectangle.** Since the rectangles are contained in the given region, the area of the region is at least as large as any lower sum. The two pictures indicate that the lower sums get larger as the number of rectangles increases and their widths de-

---

* G.F.B. Riemann was a famous German mathematician who gave a careful definition of the definite integral in the nineteenth century.

crease; consequently, the error is getting smaller. When we use eight rectangles of equal width the approximation is quite good.

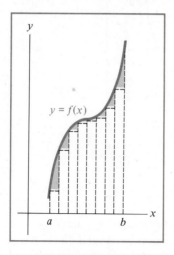

The limit of the lower sums as the number of rectangles increases without bound, and the width of each rectangle approaches zero, is the area of the region. Since the definite integral $\int_a^b f(x)\,dx$ is defined to be the area of the region, the definite integral is the limit of the lower sums. This shows that there is no ambiguity in the statement "the definite integral $\int_a^b f(x)\,dx$ is the area of the region under the graph of $y = f(x)$ from a to b."

When we obtain a formula for the coordinates of the vertices of the approximating rectangles, then we can get a formula for the lower sums *in this illustrative special case.* When we use n approximating rectangles with the same width, then the width of each rectangle must be

$$w = \frac{b-a}{n}.$$

Since the first vertex of an approximating rectangle is at $x = a$, the second vertex must be at $x = a + w$ and the third must be at $x = (a + w) + w = a + 2w$. In fact, the x-coordinates of the vertices of the approximating rectangles are

$$a,\ a + w,\ a + 2w,\ \ldots,\ a + (n-1)w,\ a + nw = b.$$

Each approximating rectangle extends up to the graph, and the left side of the rectangle is the side that meets the graph. The height of the first rectangle is $f(a)$; the height of the second rectangle is $f(a + w)$; the height of the third one is $f(a + 2w)$, and so forth. Thus, the areas of the approximating rectangles from left to right are

$$wf(a),\ wf(a + w),\ \ldots,\ wf(a + [n-1]w).$$

Thus, the lower sum for n approximating rectangles is

$$wf(a) + wf(a + w) + \cdots + wf(a + [n-1]w).$$

There is a standard method for abbreviating sums like the preceding one. We write a capital **sigma, Σ,** from the Greek alphabet, to indicate that a sum is to be computed; after the sigma we give the form of the terms to be summed. For the preceding lower sum we write

$$\sum_{k=0}^{n-1} wf(a + kw).$$

The equation below the sigma indicates that k will equal 0 in the first term and will increase by one from any term to the next term. The "$n - 1$" above the sigma indicates that k equals $n - 1$ in the last term. Our conclusion that the limit of the lower sums, as the number of rectangles increases without bound, is the definite integral can be written as

$$\int_a^b f(x)\, dx = \lim_{n \to \infty} \sum_{k=0}^{n-1} wf(a + kw).$$

EXAMPLE 1

Use the sigma notation to write the sum of the first 30 positive integers, $1 + 2 + \cdots + 29 + 30$.

Solution
Each term has the form k, where k is a positive integer; k equals 1 in the first term, and it equals 30 in the last term. In sigma notation this sum is

$$\sum_{k=1}^{30} k. \ \blacksquare$$

There is a formula that allows us to easily calculate such sums as the one in Example 1. The following formula gives the sum of the first n positive integers in terms of n.

$$\sum_{k=1}^{n} k = \frac{n(n + 1)}{2}$$

It follows that

$$\sum_{k=1}^{30} k = \frac{30(31)}{2} = 465.$$

EXAMPLE 2

Use the sigma notation to write the sum of the squares of the first 20 positive integers $1^2 + 2^2 + \cdots + 19^2 + 20^2$.

Solution
Each term has the form k^2, where k is a positive integer; k equals 1 in the first term, and it equals 20 in the last term. In sigma notation this sum is

$$\sum_{k=1}^{20} k^2. \ \blacksquare$$

There is a formula that allows us to easily calculate such sums as the one in Example 2. The following formula gives the sum of the squares of the first n positive integers in terms of n.

$$\sum_{k=1}^{n} k^2 = \frac{n(n+1)(2n+1)}{6}$$

It follows that

$$\sum_{k=1}^{20} k^2 = \frac{20(21)(41)}{6} = 2870.$$

EXAMPLE 3

Use the sigma notation to write the lower sum for the function $f(x) = x^2$ from $x = 0$ to $x = 2$ using four approximating rectangles of equal width.

Solution

We sketch the familiar graph of this function. Since the width of each rectangle should be $w = \frac{2}{4} = \frac{1}{2}$, and the x-coordinate of the first vertex of the first rectangle is 0, the vertices are at $x = 0, \frac{1}{2}, 1, \frac{3}{2}, 2$. We construct the approximating rectangles. As usual, we have shaded the error.

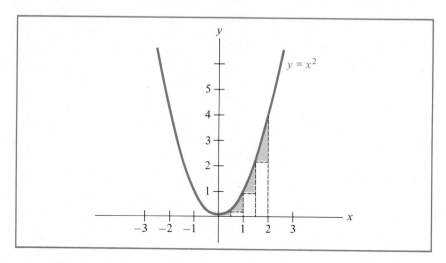

The first rectangle has height 0, so it is just a line segment. The lower sum is

$$\sum_{k=0}^{n-1} wf(a+kw) = \sum_{k=0}^{3} \frac{1}{2} f\left(0 + \frac{1}{2}k\right)$$

$$= \sum_{k=0}^{3} \frac{1}{2}\left(\frac{k}{2}\right)^2$$

$$= 0 + \frac{1}{2}\left(\frac{1}{2}\right)^2 + \frac{1}{2}(1)^2 + \frac{1}{2}\left(\frac{3}{2}\right)^2$$

$$= \frac{7}{4} = 1.75. \ \blacksquare$$

CALCULATOR EXAMPLE

Compute the lower sum for the function $f(x) = x^2$ from $x = 0$ to $x = 2$ using 10 approximating rectangles of equal width.

Solution
Since the width of each rectangle is $w = \frac{2}{10} = .2$, and the x-coordinate of the first vertex of the first rectangle is 0, the vertices are at $x = 0, .2, .4, .6, .8, 1.0,$ $1.2, 1.4, 1.6, 1.8, 2.0$. The height of each rectangle is the value of $f(x) = x^2$ at the x-coordinate of the left vertex of each rectangle; the width of each rectangle is .2.

$$\text{lower sum} = .2[0 + (.2)^2 + (.4)^2 + (.6)^2 + (.8)^2 + 1 + (1.2)^2 + (1.4)^2 + (1.6)^2 + (1.8)^2]$$
$$= .2(11.4) = 2.28 \quad \blacksquare$$

It should not be surprising that we can also approximate the area of the region under the graph of the function $y = f(x)$ from a to b by using rectangles of equal width that extend above the graph. The sum of the areas of such rectangles is said to be the **upper sum** for $f(x)$ from $x = a$ to $x = b$ corresponding to the given approximating rectangles. Each rectangle is called an **upper rectangle.** *For the same graph used earlier,* we picture the approximations using two rectangles, four rectangles, and eight rectangles as shown. We shade the region that is contained in the rectangles but is outside the region under the graph of $f(x)$ from $x = a$ to $x = b$. The difference between the upper sum and the area of the region under the graph is the area of the shaded region in each picture.

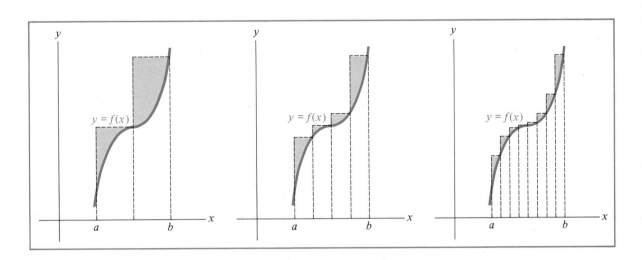

The height of each approximating rectangle is the value of $f(x)$ at the x-coordinate of the right vertex of that rectangle. When we obtained a formula for the lower sums, we concluded that the x-coordinates of the vertices are

$$a, a + w, a + 2w, \ldots, a + (n - 1)w, a + nw = b$$

where $w = (b - a)/n$. Thus, the heights of the approximating rectangles in the upper sum, from left to right, are

$$f(a + w), f(a + 2w), \ldots, f(b).$$

It follows that the upper sum is

$$wf(a + w) + wf(a + 2w) + \cdots + wf(b),$$

or in sigma notation

$$\text{upper sum} = \sum_{k=1}^{n} wf(a + kw).$$

The sequence of pictures suggests that the limit of the upper sums, as n increases without bound and w approaches zero, is the area of the region below the graph of $f(x)$ from $x = a$ to $x = b$. In the sigma notation we have

$$\int_{a}^{b} f(x) \, dx = \lim_{n \to \infty} \sum_{k=1}^{n} wf(a + kw).$$

We conclude this discussion by considering a sum that is neither a lower sum nor an upper sum. Choose numbers $x_1, \ldots, x_n$ such that

$$a \le x_1 \le a + w, \ldots, a + (n - 1)w \le x_n \le a + nw = b$$

where $w = (b - a)/n$, as in the preceding computations. The sum

$$wf(x_1) + wf(x_2) + \cdots + wf(x_n) = \sum_{j=1}^{n} wf(x_j)$$

is simply called a **Riemann sum.** It is the sum of the areas of n rectangles of equal width, as illustrated.

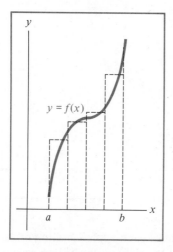

When this picture is compared to the earlier pictures of the lower sum with four rectangles and the upper sum with four rectangles, it is clear that each of these rectangles is between a lower rectangle and an upper rectangle.

Consequently, this Riemann sum is not less than the lower sum with four rectangles and not greater than the upper sum with four rectangles. Each Riemann sum is trapped between a lower sum and an upper sum. It follows that

$$\int_a^b f(x)\, dx = \lim_{n\to\infty} \sum_{j=1}^n wf(x_j).$$

In the preceding illustrative example, we have suggested much of the proof for the following theorem.

Theorem   Let $f(x)$ be a function that is continuous at every x such that $a \le x \le b$. As the number of rectangles in each Riemann sum increases without bound, and the width of each rectangle approaches zero, the limit of the Riemann sums exists and

$$\int_a^b f(x)\, dx = \lim_{n\to\infty} \sum_{j=1}^n wf(x_j).$$

Now we present an example to show how the preceding theorem can be used to evaluate a given definite integral.

EXAMPLE 4

Evaluate the definite integral $\int_0^2 x^2\, dx$ by computing the limit of appropriate Riemann sums.

Solution
The x-coordinate of the first vertex of the first rectangle is 0, and we use rectangles with a width of $\frac{1}{m}$. Thus, the x-coordinates of the vertices of our rectangles are $0, \frac{1}{m}, \frac{2}{m}, \ldots, \frac{2m-1}{m}, \frac{2m}{m} = 2$. These have the form

$$0 + \frac{k}{m} \quad \text{for} \quad k = 0, 1, \ldots, 2m-1, 2m.$$

According to the preceding theorem, we can use any Riemann sums to compute the definite integral. For simplicity we evaluate $f(x)$ at the x-coordinate of the right bottom vertex of each rectangle. We use these Riemann sums in the following computation.

$$\int_0^2 x^2\, dx = \lim_{m\to\infty} \sum_{k=1}^{2m} \frac{1}{m} f\left(0 + \frac{k}{m}\right)$$
$$= \lim_{m\to\infty} \sum_{k=1}^{2m} \frac{1}{m}\left(\frac{k}{m}\right)^2$$
$$= \lim_{m\to\infty} \sum_{k=1}^{2m} \frac{k^2}{m^3}$$
$$= \lim_{m\to\infty} \frac{1}{m^3} \sum_{k=1}^{2m} k^2$$

This last step results from factoring $\dfrac{1}{m^3}$ out of each of the $2m$ addends indicated by the notation $\displaystyle\sum_{k=1}^{2m} \dfrac{k^2}{m^3}$. Now we use the formula given after Example 2 for computing the sum of the squares of the first n positive integers. We replace n in that formula with $2m$, that is,

$$\sum_{k=1}^{2m} k^2 = \frac{2m(2m+1)(4m+1)}{6}.$$

Thus, we have

$$
\begin{aligned}
\int_0^2 x^2 \, dx &= \lim_{m \to \infty} \frac{1}{m^3} \cdot \frac{2m(2m+1)(4m+1)}{6} \\
&= \lim_{m \to \infty} \frac{1}{6} \cdot \frac{2m(2m+1)(4m+1)}{m^3} \\
&= \lim_{m \to \infty} \frac{1}{6} \left(\frac{2m}{m}\right)\left(\frac{2m+1}{m}\right)\left(\frac{4m+1}{m}\right) \\
&= \lim_{m \to \infty} \frac{1}{6} (2) \left(2 + \frac{1}{m}\right)\left(4 + \frac{1}{m}\right) \\
&= \frac{1}{6}(2)(2)(4) = \frac{8}{3}. \quad \blacksquare
\end{aligned}
$$

Exercises   10.7

Write each of the following sums in sigma notation.

1   $1^2 + 2^2 + \cdots + 9^2$

2   $1 + \dfrac{1}{2} + \cdots + \dfrac{1}{20}$

3   $(1)(2) + (2)(3) + \cdots + (12)(13)$

4   $\dfrac{1}{2} + \dfrac{2}{3} + \cdots + \dfrac{30}{31}$

5   $(1)(3) + (2)(3) + \cdots + (40)(3)$

6   $(1)(2) + (2)(2) + \cdots + (50)(2)$

7   $[2(1) + 3] + [2(2) + 3] + \cdots + [2(18) + 3]$

8   $[5 + 7(1)] + [5 + 7(2)] + \cdots + [5 + 7(25)]$

9   $5 + 6 + 7 + \cdots + 49$

10   $11 + 12 + 13 + \cdots + 28$

Approximate each of the following definite integrals by computing the lower sum corresponding to four lower rectangles of equal width.

11   $\displaystyle\int_0^2 x^3 \, dx$

12   $\displaystyle\int_0^2 x^4 \, dx$

13 $\displaystyle\int_0^4 x\,dx$

14 $\displaystyle\int_0^4 2x\,dx$

15 $\displaystyle\int_{-2}^2 e^x\,dx$

16 $\displaystyle\int_0^4 (e^x + 1)\,dx$

17 $\displaystyle\int_{-4}^0 (4 - x^2)\,dx$

18 $\displaystyle\int_{-2}^0 (2 - x^2)\,dx$

19 $\displaystyle\int_{-4}^0 e^{-x}\,dx$

20 $\displaystyle\int_0^4 (4 - x)\,dx$

Approximate each of the following definite integrals by computing the upper sum corresponding to four upper rectangles of equal width.

21 $\displaystyle\int_2^4 2x\,dx$

22 $\displaystyle\int_0^4 3x\,dx$

23 $\displaystyle\int_{-4}^0 x^4\,dx$

24 $\displaystyle\int_0^4 x^3\,dx$

25 $\displaystyle\int_0^2 (e^x + 2)\,dx$

26 $\displaystyle\int_{-1}^1 (4 - e^{-x})\,dx$

27 $\displaystyle\int_0^2 (4 - x)\,dx$

28 $\displaystyle\int_0^4 (8 - 2x)\,dx$

29 $\displaystyle\int_{-2}^2 (4 - x^2)\,dx$

30 $\displaystyle\int_{-2}^2 (2 - x^2)\,dx$

Review of Terms

Important Mathematical Terms
antiderivative, *p. 444*
integral sign, *p. 444*
indefinite integral, *p. 444*
constant of integration, *p. 445*
Power Rule for Antiderivatives, *p. 445*
Sum Rule for Antiderivatives, *p. 446*
Constant Multiple Rule for Antiderivatives, *p. 446*
Logarithm Rule for Antiderivatives, *p. 447*
Exponential Rule for Antiderivatives, *p. 447*
Generalized Power Rule for Antiderivatives, *p. 452*
integrand, *p. 453*
Generalized Exponential Rule, *p. 459*
Generalized Logarithm Rule, *p. 459*
differential equation, *p. 462*
solved, *p. 462*
separation of variables, *p. 462*
definite integral, *p. 467*
lower limit of integration, *p. 467*
upper limit of integration, *p. 467*

Fundamental Theorem of Integral Calculus, *p. 474*
average value of a function, *p. 482*
error, *p. 499*
lower Riemann sum, *p. 499*
lower rectangle, *p. 499*
sigma Σ, *p. 501*
upper sum, *p. 503*
upper rectangle, *p. 503*
Riemann sum, *p. 504*

Important Terms from the Applications

marginal costs, *p. 449*   consumers' surplus, *p. 495*
marginal revenue, *p. 449*   producers' surplus, *p. 495*
continuous compounding, *p. 462* market demand, *p. 495*
half-life, *p. 462*   market price, *p. 495*
model, *p. 481*   equilibrium point, *p. 495*

Review Problems

Evaluate each of the following.

1 $\int 3e^2 \, dx$

2 $\int (e + 9) \, dx$

3 $\int (3x - 5) \, dx$

4 $\int (.1x^2 - .3x + .6) \, dx$

5 $\int (x^3 - \sqrt[3]{x}) \, dx$

6 $\int \left(\frac{8}{x} + x^{9/4}\right) dx$

7 $\int x^{(e+2)} \, dx$

8 $\int 3x(5 + 2x) \, dx$

9 $\int \frac{(x - 1)(x + 2)}{x} \, dx$

10 $\int (9e^x + 3e) \, dx$

11 $\int [5(e^x)^3 - e^{-x}] \, dx$

12 $\int \sqrt{2x + 3} \, dx$

13 $\int (5x + 1)^{11} \, dx$

14 $\int \sqrt{4x^2 + 25} \, x \, dx$

15 $\int \frac{x}{\sqrt[3]{x^2 - 9}} \, dx$

16 $\int (x^2 + 2)^2 \, dx$

17 $\int \frac{4x + 18}{\sqrt{x^2 + 9x}} \, dx$

18 $\int (x^2 - x + 6)^5(10x - 5) \, dx$

19 $\int \frac{\sqrt[3]{x} + 5}{\sqrt[3]{x}} \, dx$

20 $\int \frac{(\sqrt{x} + 5)^9}{\sqrt{x}} \, dx$

21 $\int \frac{\sqrt{x} + 5}{\sqrt{x}} \, dx$

22 $\int \frac{x^3 - 6}{x^2} \, dx$

23 $\displaystyle\int (x^{19} + 20)x^{18}\, dx$

24 $\displaystyle\int 10e^{2x}\, dx$

25 $\displaystyle\int e^{4x^2}x\, dx$

26 $\displaystyle\int e^{3\ln x}\, dx$

27 $\displaystyle\int x(e^{x^2+6})^{5/2}\, dx$

28 $\displaystyle\int \frac{x}{4x^2 + 9}\, dx$

29 $\displaystyle\int \frac{3x}{x^2 - 5}\, dx$

30 $\displaystyle\int \frac{\ln x}{5x}\, dx$

31 $\displaystyle\int \frac{(2\sqrt{x} - 1)^2}{\sqrt{x}}\, dx$

32 $\displaystyle\int \frac{2x^2 - 8}{x(x^2 - 4)}\, dx$

33 $\displaystyle\int \frac{11}{3x + 5}\, dx$

34 $\displaystyle\int \frac{(\ln 5x)^2}{x}\, dx$

In problems 35 and 36, evaluate the indefinite integral and find the particular antiderivative that satisfies the given equation.

35 $\displaystyle\int (x^3 - x^2 + x - 3)\, dx = f(x)$ and $f(0) = 10$

36 $\displaystyle\int (\sqrt{x} + 3)\, dx = g(x)$ and $g(0) = 6$

Use the properties of the definite integral to evaluate each of the following.

37 $\displaystyle\int_0^1 (x^3 - 5x)\, dx$

38 $\displaystyle\int_0^4 \sqrt{x}\, dx$

39 $\displaystyle\int_0^1 \left(3e^{2x} - \frac{2}{e^x}\right) dx$

40 $\displaystyle\int_1^3 \frac{1}{2x + 3}\, dx$

Find the area of the region bounded by the graphs of the given equations.

41 $y = x^2$, $y = 0$, $x = 2$

42 $y = 25 - x^2$, $y = 0$

43 $y = \dfrac{5}{x}$, $y = 0$, $x = 1$, $x = e$

44 $y = 3x$, $y = 0$, $x = 3$

45 $y = x^2$, $y = 6 - x$

46 $y = 5 - x^2$, $y = x + 3$

47 $y = \sqrt{x + 4}$, $y = 2 - x$, $y = 0$

48 $y = 3 - 2x$, $y = \dfrac{1}{x}$

49 $y = \sqrt{x + 2}$, $y = x$, $y = -x$

Find the average value for each of the following functions and the indicated values for the independent variable.

50 $f(x) = x^2 + 3x + 5$ from $x = 0$ to $x = 4$

51 $g(x) = 3x + 2$ from $x = 1$ to $x = 10$

Determine the consumers' surplus and the producers' surplus for the given demand function and supply function.

52   $p = S(x) = 2x, \quad p = D(x) = 6 - x$

53   $p = S(x) = \dfrac{1}{6}x, \quad p = D(x) = \dfrac{1}{x+1}$

Approximate each of the following definite integrals by computing the lower sum corresponding to six lower rectangles of equal width.

54   $\displaystyle\int_0^3 (9 - x^2)\, dx$ 55 $\displaystyle\int_0^3 x^3\, dx$

56   $\displaystyle\int_{-3}^3 (x + 3)\, dx$

Solve the following.

57   How long does it take for a savings account to double if it receives continuous compounding at a rate of 18%?

58   Uranium-234 has a half-life of 248 thousand years. If a government plant begins the year with 100 pounds, how much does the plant have left after 10 years?

59   ***Calculator Problem***   Approximate the following definite integral by computing the lower sum corresponding to 12 lower rectangles of equal width.

$$\int_0^3 (9 - x^2)\, dx$$

60   ***Calculator Problem***   Approximate the following definite integral by computing the lower sum corresponding to 12 lower rectangles of equal width.

$$\int_0^3 x^3\, dx$$

61   ***Calculator Problem***   Approximate the following definite integral by computing the lower sum corresponding to 12 lower rectangles of equal width.

$$\int_{-3}^3 (x + 3)\, dx$$

Social Science Applications

Lorenz Curves

A standard device for showing the income distribution of a society, such as the United States, North America, or the world, is a Lorenz curve. Along the x-axis, we indicate percentiles of families by income. For example, 40 indicates all of the families in the bottom 40%, according to their family income. Along the y-axis, we show the percentage of the sum of all family incomes that is earned by that percentile group. On the Lorenz curve for the United

States, we see that the bottom 60 percentile of families earn only 35% of the total income.

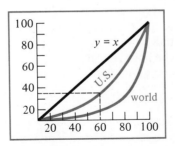

The curve for the world indicates that the bottom 60 percentile in the world earns only 10% of the world's income. The straight dashed line $y = x$ shows what the Lorenz curve would be if there was perfect equality in family incomes. The area between the line $y = x$ and the Lorenz curve for a given society is an accepted standard measure of the income inequality that exists in that society. Political scientists sometimes relate the income inequality of a country to the prospects of a stable government for the country.

1   Determine the area between the line $y = x$ and the graph of $g(x) = .0208x^2 - 1.08x$. The graph of $g(x)$ is a rough approximation of the Lorenz curve for the world. Thus, the computed area is a measure of the inequality of income distribution throughout the world.

2   Determine the area between the line $y = x$ and the graph of

$$h(x) = .00005x^3 + .0035x^2 + .15x.$$

The graph of $h(x)$ is a rough approximation of the Lorenz curve for the United States. Thus, the computed area is a measure of the inequality of income distribution in the U.S.

3   ***Archeology (Carbon Dating)***   The rate of decay of radioactive carbon-14 at time t is $(-.00012)$ times the amount present at time t, denoted $P(t)$. Thus, we have the differential equation

$$P'(t) = -.00012\, P(t).$$

Solve this equation for $P(t)$ by means of separation of variables using P_0 to represent the amount of carbon-14 present at time $t = 0$. (Compare your answer to problem 1 of the Social Science Applications for Chapter 2.)

4   ***Psychology (Learning Curves)***   A computer programmer learns to type instructions on the display screen of a computer terminal. At the very beginning, the programmer can type no words per minute but she rapidly increases the number of words per minute that she can type. As she gains a mastery of the process her progress slows. If $n(t)$ is the number of words per minute that she can input, then

$$n'(t) = .340 - .004n(t).$$

Use separation of variables to solve this differential equation for $n(t)$. (Compare your answer to problem 2 of the Social Science Applications for Chapter 2.)

5 *Sociology* Important news events, such as the winner of the U.S. presidential election, are communicated primarily by the mass media, such as television and newspapers. The number $n(t)$ of people in a city of 1,000,000 who have been informed of the event, t hours after it occurred, typically increases at a rate indicated by the differential equation

$$n'(t) = .03(1,000,000 - n(t)).$$

Solve this equation by separation of variables using the fact that $n(0) = 0$. (Compare your answer to problem 5 of the Social Science Applications for Chapter 2.)

6 *Public Administration* In a town with a population of 150,000, a contagious disease is observed to be infecting 1% of the healthy inhabitants per week. Thus, if $p(t)$ represents the number of people infected then we have

$$p'(t) = .01(150,000 - p(t)).$$

Assume that no public action is taken so the disease continues to spread according to the differential equation. Solve for $p(t)$ by separation of variables using that $p(0) = 0$.

7 *Urban Studies* During the period from 1970 to 1980, the population of Houston grew at an annual rate of 3.69%. Let $H(t)$ denote the population of Houston t years after June 30, 1970, and assume that

$$H'(t) = .0369H(t).$$

Determine $H(t)$ by separation of variables using that $H(0) = 1,999,316$.

CAREER PROFILE
Cryptographer

Cryptography (the art of writing or deciphering messages in code) is an important but little known application of mathematics. Cryptography methods have not been publicized because, until recent times, its most frequent application has been in the codebreaking activities of secret intelligence organizations.

Recent technological advances, however, have brought a need for coding and decoding methods in science, engineering, and business, and the field of cryptography has grown rapidly and has entered the public domain.

Large-scale development of computers has placed importance on error-detecting and error-correcting codes for maintaining integrity of electronic information. The long distance transmission of signals through space, for example, from a satellite to a land station, has led to the need for decoding strategies that eliminate errors in data transmission that result from radiation and other distorting influences. Mathematical techniques have been used to accomplish these tasks.

The need for security measures to prevent unauthorized access to computer files has been another stimulus to the development of mathematical methods for cryptography.

To obtain employment with government agencies, such as the Central Intelligence Agency (CIA) and the National Security Agency (NSA), for basic intelligence work, a strong undergraduate major in mathematics provides good preparation. Courses in probability and statistics, calculus, and linear and abstract algebra (including theory of finite fields) are important. Good English language skills are essential, and facility with foreign languages can be very helpful. Musical training, which emphasizes the translation of thoughts into music and vice versa, can also be valuable preparation for coding and decoding tasks.

To prepare for work in computer security that is at the forefront of cryptography today, considerable educational background is required. Extensive knowledge of number theory and computer science is essential. (A Ph.D. in one of these fields with a great deal of study in the other would be good preparation.) Courses in "information theory," often offered by departments of engineering, also are valuable educational tools for this exciting, highly demanding field of mathematical application.

Cryptography is a field of the future. Employment opportunities are almost certain to grow, but their nature and number are hard to predict. An undergraduate mathematics student with an interest in cryptography should read widely in electrical engineering, computer, and mathematics periodicals to keep abreast of the employment opportunities that become available.

Sources of Additional Information
Central Intelligence Agency, Washington, DC 20212

National Security Agency, Office of Employment (M32R), Fort George G. Meade, MD 20755.

11 Techniques of Integration

There are some differentiation formulas that we did not rewrite as antiderivative formulas in Chapter 10. One such formula is the product formula

$$\frac{d}{dx}\left(f(x)g(x)\right) = f'(x)g(x) + f(x)g'(x).$$

Formally taking antiderivatives of both sides in the above formula gives us

$$f(x)g(x) = \int \frac{d}{dx}\left(f(x)g(x)\right)\,dx = \int f'(x)g(x)\,dx + \int f(x)g'(x)\,dx$$

or, by subtraction,

$$\int f(x)g'(x)\,dx = f(x)g(x) - \int f'(x)g(x)\,dx.$$

This last equation is known as the **Integration by Parts Formula for Antiderivatives.** Because of the Fundamental Theorem of Calculus, we know that

$$\int_a^b f(x)g'(x)\,dx = f(x)g(x)\Big|_a^b - \int_a^b f'(x)g(x)\,dx.$$

This is the **Integration by Parts Formula for Definite Integrals.**

The use of integration by parts depends on being able to identify a given integrand as the product of two functions, in the form $f(x)g'(x)$, where it is easy to find $g(x)$ from $g'(x)$. Once this is done, we can use the Integration by Parts Formula. The remaining task is to find an antiderivative for $f'(x)g(x)$. The rest of this section consists of examples illustrating the technique.

EXAMPLE 1

Evaluate $\displaystyle\int xe^x\,dx$.

Solution

If we can arrange to have $g(x) = e^x$ and $f'(x) = 1$, the antiderivative on the right-hand side of the Integration by Parts Formula will be just $\int e^x \, dx = e^x + C$. Thus, on the left-hand side of the formula for antiderivatives, we take $f(x)$ to be x and $g'(x)$ to be e^x. Below we use that formula.

$$f(x) = x \qquad g'(x) = e^x$$
$$f'(x) = 1 \qquad g(x) = e^x$$
$$\int xe^x \, dx = xe^x - \int 1e^x \, dx$$
$$= xe^x - e^x + C \quad \blacksquare$$

Let u and v be understood to be functions of x. The formulas below are terse and convenient.

Integration by Parts Formula

$$\int uv' \, dx = uv - \int u'v \, dx$$
$$\int_a^b uv' \, dx = uv \Big|_a^b - \int_a^b u'v \, dx$$

The next example indicates another situation where it is helpful to differentiate one factor of the integrand.

EXAMPLE 2

Evaluate $\int x \ln x \, dx$.

Solution
Let $u = \ln x$ so that we can work with $u' = 1/x$ instead of the more difficult function $\ln x$. Then $v' = x$ and $v = x^2/2$. According to the Integration by Parts Formula, we have

$$\int x \ln x \, dx = \frac{1}{2} x^2 \ln x - \int \left(\frac{x^2}{2} \right) \left(\frac{1}{x} \right) dx$$
$$= \frac{1}{2} x^2 \ln x - \frac{1}{2} \int x \, dx$$
$$= \frac{1}{2} x^2 \ln x - \frac{1}{4} x^2 + C. \quad \blacksquare$$

The next example shows that integration by parts is useful in some problems despite the fact that the integrand is not a product.

EXAMPLE 3

Evaluate $\int \ln x \, dx$.

Solution
As in Example 2, we are determined to differentiate $\ln x$, so we let $u = \ln x$.

The only possibility for v' is $v' = 1$. Then $u' = 1/x$ and $v = x$. We have

$$\int \ln x \, dx = x \ln x - \int x \left(\frac{1}{x}\right) dx$$
$$= x \ln x - x + C. \ \blacksquare$$

The next example is representative of a type of practical application. It is oversimplified for the sake of brevity and clarity.

EXAMPLE 4

An advertising agency promises a retail sales chain that the graph of its marginal revenue will look like the graph of $(t + 1)^2 \ln(t + 1)$, where t is the duration of a new ad campaign. Assuming that revenue is 1 million dollars when the campaign begins, determine the revenue function.

Solution
Since the revenue function is an antiderivative of $(t + 1)^2 \ln(t + 1)$, we obtain a formula for all antiderivatives. Let $u = \ln(t + 1)$ and $v' = (t + 1)^2$; note that $u' = 1/(t + 1)$ and $v = (t + 1)^3/3$. Using integration by parts, we have

$$\int (t + 1)^2 \ln(t + 1) \, dt = \frac{1}{3}(t + 1)^3 \ln(t + 1) - \int \frac{1}{3}(t + 1)^2 \, dt$$

$$= \frac{1}{3}(t + 1)^3 \ln(t + 1) - \frac{1}{9}(t + 1)^3 + C.$$

For some choice of the constant C the revenue after t months of the new ad campaign, $R(t)$, is given by

$$R(t) = \frac{1}{3}(t + 1)^3 \ln(t + 1) - \frac{1}{9}(t + 1)^3 + C.$$

Since the revenue was 1 million dollars when the campaign was begun, we know that

$$1{,}000{,}000 = R(0) = 0 - \frac{1}{9} + C$$

or

$$C = 1{,}000{,}000 + \frac{1}{9}.$$

Thus, the revenue function is given by

$$R(t) = \frac{1}{3}(t + 1)^3 \ln(t + 1) - \frac{1}{9}(t + 1)^3 + \frac{1}{9} + 1{,}000{,}000. \ \blacksquare$$

The next example is very similar to the first example, although it looks different.

EXAMPLE 5

Evaluate $\int x\sqrt{x + 3} \, dx$.

Solution

Let $u = x$, $v' = \sqrt{x+3}$, and note that $u' = 1$, $v = \frac{2}{3}(x+3)^{3/2}$. Thus, we have

$$\int x\sqrt{x+3}\ dx = \frac{2}{3}x(x+3)^{3/2} - \int \frac{2}{3}(x+3)^{3/2}\ dx$$

$$= \frac{2}{3}x(x+3)^{3/2} - \frac{4}{15}(x+3)^{5/2} + C.\ ■$$

Caution   A poor choice of u and v' in the original integrand can lead to a new integral that is no easier to evaluate than the first one. For instance, in Example 5, we could have chosen $u = \sqrt{x+3}$ and $v' = x$, so that $u' = 1/2\sqrt{x+3}$ and $v = \frac{1}{2}x^2$. The Integration by Parts Formula then gives

$$\int x\sqrt{x+3}\ dx = \frac{1}{2}x^2\sqrt{x+3} - \int \frac{x^2}{4\sqrt{x+3}}\ dx.$$

The integral on the far right is just as difficult to evaluate as was the original integral.

As we develop more techniques for finding antiderivatives in this section and the next, it is important to keep in mind the techniques already developed. One of the fundamental skills in finding antiderivatives is recognizing which technique applies to a given problem.

Exercises   11.1

Determine each of the indicated antiderivatives.

1   $\displaystyle\int xe^{3x}\ dx$

2   $\displaystyle\int 4xe^{-x}\ dx$

3   $\displaystyle\int 5te^{6t-4}\ dt$

4   $\displaystyle\int 2te^{3t+9}\ dt$

5   $\displaystyle\int 3x^2e^x\ dx$

6   $\displaystyle\int x^2e^{5x}\ dx$

7   $\displaystyle\int (t-1)^2e^{t+1}\ dt$

8   $\displaystyle\int (t+1)^2e^{t-1}\ dt$

9   $\displaystyle\int xe^{x^2}\ dx$

10   $\displaystyle\int x^2e^{x^3}\ dx$

11   $\displaystyle\int 3x\ln(5x)\ dx$

12   $\displaystyle\int 7x\ln(2x)\ dx$

13   $\displaystyle\int \sqrt{t}\ \ln(3t)\ dt$

14   $\displaystyle\int \frac{(\ln t)}{\sqrt{t}}\ dt$

15   $\displaystyle\int (x^2+4)\ln x\ dx$

16   $\displaystyle\int (x^2+x+1)\ln x\ dx$

17 $\displaystyle\int \ln(9t)\, dt$ **18** $\displaystyle\int 8 \ln\left(\frac{t}{3}\right) dt$

19 $\displaystyle\int 3x\sqrt{x+5}\, dx$ **20** $\displaystyle\int x\sqrt{4x+3}\, dx$

21 $\displaystyle\int x^2(x+2)^{2/3}\, dx$ **22** $\displaystyle\int 5x^2(4x+1)^{1/4}\, dx$

23 $\displaystyle\int 7x^3 e^x\, dx$ **24** $\displaystyle\int x^3 e^{8x}\, dx$

Use the Fundamental Theorem of Calculus to find the value of each definite integral below.

25 $\displaystyle\int_0^1 9xe^{2x}\, dx$ **26** $\displaystyle\int_0^3 e^{7x}\, dx$

27 $\displaystyle\int_1^5 t \ln(4t)\, dt$ **28** $\displaystyle\int_1^4 3 \ln(5t)\, dt$

29 $\displaystyle\int_0^5 x\sqrt{x+4}\, dx$ **30** $\displaystyle\int_{-1}^1 x\sqrt{x^2+1}\, dx$

11.2 Dividing Polynomials and Elementary Partial Fractions

For certain types of problems it is useful to reverse the process of combining several fractions into one fraction with a common denominator. We go from the combined fraction with the common denominator to the sum of simpler fractions. For example,

$$\frac{3}{x^2-4} = \frac{3}{(x-2)(x+2)}$$

can be regarded as the sum of two simpler fractions

$$\frac{a}{x-2} + \frac{b}{x+2}$$

where a and b are constants to be determined. Combining these simple fractions, we have

$$\frac{a(x+2) + b(x-2)}{(x-2)(x+2)} = \frac{(a+b)x + 2(a-b)}{x^2-4}.$$

Now, we equate the sum of the two simple fractions and the original combined fraction; we solve for a and b.

$$\frac{3}{x^2-4} = \frac{(a+b)x + 2(a-b)}{x^2-4}$$

or

$$3 = (a+b)x + 2(a-b)$$

Since the left side has no power or x and the constant is 3, it must be that

$$0 = a + b$$

$$3 = 2(a - b).$$

Simplifying slightly, we get

$$0 = a + b$$

$$\frac{3}{2} = a - b.$$

Adding the two equations, we find that $a = \frac{3}{4}$, $b = -\frac{3}{4}$. We have discovered that

$$\frac{3}{x^2 - 4} = \frac{3}{4}\left(\frac{1}{x - 2}\right) - \frac{3}{4}\left(\frac{1}{x + 2}\right).$$

The process of going from the combined fraction to the sum of simpler fractions is called the technique of **partial fractions.** This process is useful in determining some antiderivatives, as we indicate in the next example.

EXAMPLE 1

Evaluate $\displaystyle\int \frac{3}{x^2 - 4}\,dx$.

Solution
From the work in the first paragraph, we know that

$$\frac{3}{x^2 - 4} = \frac{3}{4}\left(\frac{1}{x - 2}\right) - \frac{3}{4}\left(\frac{1}{x + 2}\right).$$

Thus, we have

$$\int \frac{3}{x^2 - 4}\,dx = \int \left[\frac{3}{4}\left(\frac{1}{x - 2}\right) - \frac{3}{4}\left(\frac{1}{x + 2}\right)\right]dx$$

$$= \frac{3}{4}\int \frac{1}{x - 2}\,dx - \frac{3}{4}\int \frac{1}{x + 2}\,dx$$

$$= \frac{3}{4}\ln(x - 2) - \frac{3}{4}\ln(x + 2) + C. \ \blacksquare$$

The next example involves the same ideas but is slightly more complicated.

EXAMPLE 2

Evaluate $\displaystyle\int \frac{5}{x^3 + x^2 - 6x}\,dx$.

Solution
First, we factor the denominator.

$$x^3 + x^2 - 6x = x(x^2 + x - 6) = x(x + 3)(x - 2)$$

Each factor becomes the denominator for a simple fraction.

$$\frac{a}{x} + \frac{b}{x+3} + \frac{c}{x-2} = \frac{a(x+3)(x-2) + bx(x-2) + cx(x+3)}{x(x+3)(x-2)}$$

$$= \frac{a(x^2 + x - 6) + b(x^2 - 2x) + c(x^2 + 3x)}{x(x+3)(x-2)}$$

$$= \frac{(a+b+c)x^2 + (a-2b+3c)x - 6a}{x(x+3)(x-2)}$$

We set the sum of the simple fractions equal to the original fraction, and we solve for a, b, and c.

$$\frac{5}{x^3 + x^2 - 6x} = \frac{(a+b+c)x^2 + (a-2b+3c)x - 6a}{x(x+3)(x-2)}$$

or

$$5 = (a+b+c)x^2 + (a-2b+3c)x - 6a$$

We must solve the system

$$0 = a + b + c$$

$$0 = a - 2b + 3c$$

$$5 = -6a.$$

The solution is $a = -\frac{5}{6}$, $b = \frac{1}{3}$, and $c = \frac{1}{2}$. Now, we can rewrite the original antiderivative.

$$\int \frac{5}{x^3 + x^2 - 6x}\,dx = \int \left[-\frac{5}{6}\left(\frac{1}{x}\right) + \frac{1}{3}\left(\frac{1}{x+3}\right) + \frac{1}{2}\left(\frac{1}{x-2}\right) \right]dx$$

$$= -\frac{5}{6}\int \frac{1}{x}\,dx + \frac{1}{3}\int \frac{1}{x+3}\,dx + \frac{1}{2}\int \frac{1}{x-2}\,dx$$

$$= -\frac{5}{6}\ln x + \frac{1}{3}\ln(x+3) + \frac{1}{2}\ln(x-2) + C \ \blacksquare$$

Note   The technique of partial fractions applies only to problems where the integrand is the ratio of polynomials; these functions are called **rational functions.** Furthermore, the technique of partial fractions requires that the degree of the denominator exceeds the degree of the numerator.

When the degree of the denominator does not exceed the degree of the numerator, the correct first step is to *divide the denominator polynomial into the numerator polynomial.* We illustrate this in the next example.

EXAMPLE 3

Evaluate $\displaystyle\int \frac{2x^2}{x^2 - 4}\,dx.$

Solution

Since the degree of the numerator equals the degree of the denominator, we begin by dividing the denominator into the numerator.

$$\begin{array}{r} 2 \\ x^2 - 4\overline{\smash{\big)}\,2x^2} \\ \underline{2x^2 - 8} \\ 8 \end{array}$$

Thus, we see that

$$\frac{2x^2}{x^2 - 4} = 2 + \frac{8}{x^2 - 4}.$$

The skeptic can verify this equation by taking the common denominator for the terms on the right side and computing the sum.

We have converted the original problem into two simpler problems.

$$\begin{aligned} \int \frac{2x^2}{x^2 - 4}\, dx &= \int \left(2 + \frac{8}{x^2 - 4} \right) dx \\ &= \int 2\, dx + \int \frac{8}{x^2 - 4}\, dx \\ &= 2x + 8 \int \frac{1}{x^2 - 4}\, dx \end{aligned}$$

We use partial fractions to determine the last antiderivative.

$$\frac{1}{x^2 - 4} = \frac{1}{(x - 2)(x + 2)} = \frac{a}{x - 2} + \frac{b}{x + 2} = \frac{a(x + 2) + b(x - 2)}{(x - 2)(x + 2)}$$

$$\frac{1}{x^2 - 4} = \frac{(a + b)x + 2(a - b)}{(x - 2)(x + 2)}$$

$$1 = (a + b)x + 2(a - b)$$

$$0 = a + b$$

$$\frac{1}{2} = a - b$$

$$a = \frac{1}{4}, \qquad b = -\frac{1}{4}$$

$$\begin{aligned} \int \frac{1}{x^2 - 4}\, dx &= \int \left[\frac{1}{4}\left(\frac{1}{x - 2} \right) - \frac{1}{4}\left(\frac{1}{x + 2} \right) \right] dx \\ &= \frac{1}{4} \int \frac{1}{x - 2}\, dx - \frac{1}{4} \int \frac{1}{x + 2}\, dx \\ &= \frac{1}{4} \ln(x - 2) - \frac{1}{4} \ln(x + 2) + C \end{aligned}$$

By substituting, we obtain the solution to the original problem.

$$\int \frac{2x^2}{x^2 - 4}\, dx = 2x + 2 \ln(x - 2) - 2 \ln(x + 2) + C \quad \blacksquare$$

Sometimes the process of dividing the denominator into the numerator suffices to make the problem easy. We illustrate this.

EXAMPLE 4

Evaluate $\int \dfrac{x^2 + 3}{x - 2}\, dx$.

Solution

Since the degree of the numerator exceeds the degree of the denominator, we begin by dividing the denominator into the numerator.

$$
\begin{array}{r}
x + 2 \\
x - 2 \overline{\big)\, x^2 + 3} \\
\underline{x^2 - 2x } \\
2x + 3 \\
\underline{2x - 4} \\
7
\end{array}
$$

Thus, we have

$$\frac{x^2 + 3}{x - 2} = x + 2 + \frac{7}{x - 2}.$$

Now, we can rewrite the original problem as two easy problems.

$$
\begin{aligned}
\int \frac{x^2 + 3}{x - 2}\, dx &= \int \left(x + 2 + \frac{7}{x - 2} \right) dx \\
&= \int (x + 2)\, dx + 7 \int \frac{1}{x - 2}\, dx \\
&= \frac{x^2}{2} + 2x + 7 \ln(x - 2) + C \quad \blacksquare
\end{aligned}
$$

At this point, we have used the partial fractions technique only on rational functions where the denominator is the product of linear factors. Since some polynomials of degree two are not the product of linear factors, there is another kind of partial fractions representation different from anything that we have done. We illustrate the basic ideas in the next example.

EXAMPLE 5

Evaluate $\int \dfrac{5x^2 + 15}{x^3 + 5x}\, dx$.

Solution

Since the degree of the denominator exceeds the degree of the numerator, we do not divide the denominator into the numerator. The first step is to factor the denominator.

$$x^3 + 5x = x(x^2 + 5)$$

We see that the second factor cannot be written as the product of linear factors, because there is no real value of x that makes $x^2 + 5$ equal 0. Indeed, the square of any real number is nonnegative, so $x^2 \geq 0$; consequently, $x^2 + 5 \geq 5$.

For the given integrand, we seek a partial fractions representation like the following.

$$\frac{5x^2 + 15}{(x^2 + 5)x} = \frac{ax + b}{x^2 + 5} + \frac{c}{x}$$

We multiply both sides of the above equation by the common denominator, and we solve for the unknown constants.

$$5x^2 + 15 = (ax + b)x + c(x^2 + 5)$$
$$= (a + c)x^2 + bx + 5c$$

We equate the constants and the coefficients of x^2, x on both sides of this equation.

$$5 = a + c$$
$$0 = b$$
$$15 = 5c$$

The solution to this system of equations is $a = 2$, $b = 0$, $c = 3$.

Now, we can find the desired antiderivative.

$$\int \frac{5x^2 + 15}{x^3 + 5x} \, dx = \int \left(\frac{2x}{x^2 + 5} + \frac{3}{x} \right) dx$$
$$= \int \frac{2x}{x^2 + 5} \, dx + \int \frac{3}{x} \, dx$$
$$= \ln(x^2 + 5) + 3 \ln x + C \ \blacksquare$$

There is one more type of problem that we want to solve in this introduction to the partial fractions technique. The next example shows what to do when the denominator has a repeated linear factor.

EXAMPLE 6

Evaluate $\int \dfrac{x^2 + 2x + 9}{x^3 - 3x + 2} \, dx$.

Solution

Since the degree of the denominator exceeds the degree of the numerator, we do not divide the denominator into the numerator. Since no factorization of the denominator is apparent, we try to find a root of the equation

$$x^3 - 3x + 2 = 0$$

by trial and error. When we substitute 1 for x into the equation, we discover that the equation is true. Thus, $x - 1$ is a factor of the denominator; we find the remaining factor by division.

$$
\begin{array}{r}
x^2 + x - 2 \\
x - 1 \overline{\smash{\big)}\ x^3 \qquad\ - 3x + 2} \\
\underline{x^3 - x^2} \\
x^2 - 3x \\
\underline{x^2 -\ \ x} \\
-2x + 2 \\
\underline{-2x + 2}
\end{array}
$$

We now know that the factorization of the denominator is

$$
\begin{aligned}
x^3 - 3x + 2 &= (x - 1)(x^2 + x - 2) \\
&= (x - 1)(x + 2)(x - 1) \\
&= (x - 1)^2(x + 2).
\end{aligned}
$$

Because the linear factor $x - 1$ is repeated in the factorization of $x^3 - 3x + 2$, we must use a different representation than the one for distinct factors. The form of this representation is

$$
\frac{x^2 + 2x + 9}{x^3 - 3x + 2} = \frac{a}{(x - 1)^2} + \frac{b}{x - 1} + \frac{c}{x + 2}
$$

We multiply each side by the common denominator and solve for the unknown constants.

$$
\begin{aligned}
x^2 + 2x + 9 &= a(x + 2) + b(x - 1)(x + 2) + c(x - 1)^2 \\
&= a(x + 2) + b(x^2 + x - 2) + c(x^2 - 2x + 1) \\
&= (b + c)x^2 + (a + b - 2c)x + (2a - 2b + c)
\end{aligned}
$$

$$
1 = b + c
$$

$$
2 = a + b - 2c
$$

$$
9 = 2a - 2b + c
$$

The solution to this system of linear equations is $a = 4$, $b = 0$, $c = 1$. Now, we can find the indicated antiderivative.

$$
\begin{aligned}
\int \frac{x^2 + 2x + 9}{x^3 - 3x + 2}\, dx &= \int \left(\frac{4}{(x - 1)^2} + \frac{1}{x + 2} \right) dx \\
&= \int \frac{4}{(x - 1)^2}\, dx + \int \frac{1}{x + 2}\, dx \\
&= -4(x - 1)^{-1} + \ln(x + 2) + C \ \blacksquare
\end{aligned}
$$

Below, we summarize the method developed in the previous examples.

The Partial Fractions Technique

1. If the degree of the denominator does not exceed the degree of the numerator, then divide the denominator into the numerator.

2. Factor the denominator. Each distinct factor requires additional terms in the partial fractions representation of the integrand.
3. If $(x - a)^m$ is the highest power of $(x - a)$ that is a factor of the denominator, then add the terms

$$\frac{a_m}{(x-a)^m} + \cdots + \frac{a_2}{(x-a)^2} + \frac{a_1}{x-a}$$

to the partial fractions representation of the integrand.
4. If $ax^2 + bx + c$ is a factor of the denominator, and $ax^2 + bx + c$ is not the product of linear factors, then add

$$\frac{a_1 x + a_0}{ax^2 + bx + c}$$

to the partial fractions representation for the integrand.
5. Set the original fraction equal to the sum of the terms in the partial fractions representation. Multiply through by the common denominator, and collect terms in like powers of x.
6. Equate the coefficients of like powers of x on each side of the equation resulting from step 5, and solve the system of linear equations for the unknown constants.
7. Substitute the constants in the partial fractions representation to arrive at a sum of simple fractions, and integrate each fraction separately. Remember to include a constant of integration, C.

Exercises 11.2

Evaluate each of the following antiderivatives.

1. $\int \frac{4}{1-x^2} \, dx$

2. $\int \frac{30}{x^2-9} \, dx$

3. $\int \frac{7x+6}{x^2+x-6} \, dx$

4. $\int \frac{4x-19}{x^2+x-12} \, dx$

5. $\int \frac{9x-8}{3x^2-10x-8} \, dx$

6. $\int \frac{8x-19}{2x^2-7x+3} \, dx$

7. $\int \frac{x-2}{x^2-4x+13} \, dx$

8. $\int \frac{x+1}{x^2+2x+5} \, dx$

9. $\int \frac{16x-10}{x^3-3x^2-10x} \, dx$

10. $\int \frac{2x^2-16x-12}{x^3+x^2-12x} \, dx$

11. $\int \frac{7x^2+11x-6}{x^3+6x^2} \, dx$

12. $\int \frac{5x^2-5x-12}{x^3-3x^2} \, dx$

13. $\int \frac{3x^2-12x-11}{x^3-6x^2-11x-40} \, dx$

14. $\int \frac{3x^2+18x+16}{x^3+9x^2+16x+14} \, dx$

15 $\displaystyle\int \frac{x^2 + 9}{x^3 + 6x} dx$ 16 $\displaystyle\int \frac{6x^2 + 15}{x^3 + 3x} dx$

17 $\displaystyle\int \frac{4x^2 - 14x + 15}{x^3 - 4x^2 + 5} dx$ 18 $\displaystyle\int \frac{3x^2 + 21x + 40}{x^3 + 6x^2 + 10x} dx$

19 $\displaystyle\int \frac{4x^2 - 14x + 14}{x^3 - 5x^2 + 9x - 5} dx$ 20 $\displaystyle\int \frac{5x^2 + 26x + 36}{x^3 + 7x^2 + 16x + 10} dx$

11.3 More on Substitution

The Chain Rule for differentiation says that if u is a function of x and F is a function of u, then

$$\frac{d}{dx} F(u) = F'(u)u'(x).$$

As we have transformed other differentiation formulas into antiderivative formulas, we can transform this one. From the above equation, we get

$$\int F'(u)u'(x) \, dx = F(u) + C.$$

Letting f be a symbol for F' and using the first theorem in Section 10.1, the last equation becomes

$$\int f(u)u'(x) \, dx = \int f(u) \, du.$$

This is called the **substitution formula.**

We have already used the substitution formula for the three choices of $f(u)$ shown below in Sections 10.2 and 10.3.

$$\int u^n u' \, dx = \int u^n \, du = \frac{u^{n+1}}{n + 1} + C$$

$$\int \frac{1}{u} u' \, dx = \int \frac{1}{u} \, du = \ln u + C$$

$$\int e^u u' \, dx = \int e^u \, du = e^u + C$$

In these three equations $f(u)$ is u^n, $1/u$, and e^u, respectively; the substitution formula verifies the answer that we already know.

In these previous uses, we could readily identify the function $f(u)$. There is another type of problem in which $f(u)$ is not readily identified. We rewrite the substitution formula in an alternate form appropriate to this new type of problem. We exchange the sides of the substitution formula; we use x and t in place of u and x, respectively.

$$\int f(x)\,dx = \int f(x)\,\frac{dx}{dt}\,dt \qquad *$$

The next example indicates how this alternate form of the substitution formula is used.

EXAMPLE 1

Evaluate $\displaystyle\int \frac{1}{1 + \sqrt{x}}\,dx$.

Solution

Let $t = \sqrt{x}$ and note that $t^2 = x$. Replace dx with

$$\frac{dx}{dt}\,dt = 2t\,dt.$$

Thus, we get

$$\int \frac{1}{1 + \sqrt{x}}\,dx = \int \frac{2t}{1 + t}\,dt.$$

According to the first rule in the partial fractions technique, we divide the denominator into the numerator.

$$\int \frac{1}{1 + \sqrt{x}}\,dx = \int \left(2 - \frac{2}{t + 1}\right) dt$$
$$= 2t - 2\ln(t + 1) + C$$
$$= 2\sqrt{x} - 2\ln(\sqrt{x} + 1) + C \;\blacksquare$$

In the previous example, the procedure begins with the identification of t, the new independent variable. Then, we write the old variable x as a function of the new variable, and we replace dx according to the substitution formula. Once an antiderivative is found, all occurrences of t are replaced by their equivalent expressions in x.

The next example will further demonstrate the procedure for using the alternate form of the substitution formula. Often we choose t so that an expression involving roots is replaced by an expression involving powers.

EXAMPLE 2

Evaluate $\displaystyle\int \frac{\sqrt[6]{x}}{\sqrt[6]{x} + \sqrt[3]{x}}\,dx$.

Solution

Let $t = \sqrt[6]{x}$ and note that $\sqrt[3]{x} = x^{1/3} = x^{2/6} = t^2$. Replace dx with

$$\frac{dx}{dt}\,dt = \frac{d}{dt}\,t^6\,dt = 6t^5\,dt.$$

---

* To facilitate the use of this formula, dx is sometimes called the *differential* and defined to be $\dfrac{dx}{dt}\,dt$.

Thus, we get

$$\int \frac{\sqrt[6]{x}}{\sqrt[6]{x} + \sqrt[3]{x}} \, dx = \int \frac{t}{t + t^2} \, 6t^5 \, dt$$

$$= \int \frac{6t^6}{(t + 1)t} \, dt = \int \frac{6t^5}{t + 1} \, dt.$$

From Section 11.2, we know that the appropriate step is to divide the denominator into the numerator.

$$
\begin{array}{r}
6t^4 - 6t^3 + 6t^2 - 6t + 6 \\
t + 1\overline{\smash{\big)}\, 6t^5 } \\
\underline{6t^5 + 6t^4} \\
- 6t^4 \\
\underline{- 6t^4 - 6t^3} \\
6t^3 \\
\underline{6t^3 + 6t^2} \\
- 6t^2 \\
\underline{- 6t^2 - 6t} \\
6t \\
\underline{6t + 6} \\
- 6
\end{array}
$$

It follows that

$$\int \frac{6t^5}{t + 1} \, dt = \int \left(6t^4 - 6t^3 + 6t^2 - 6t + 6 - \frac{6}{t + 1} \right) dt$$

$$= \frac{6}{5} t^5 - \frac{3}{2} t^4 + 2t^3 - 3t^2 + 6t - 6 \ln(t + 1) + C.$$

The answer to the original problem is

$$\int \frac{\sqrt[6]{x}}{\sqrt[6]{x} + \sqrt[3]{x}} \, dx = \frac{6}{5} x^{5/6} - \frac{3}{2} x^{2/3} + 2x^{1/2} - 3x^{1/3} + 6x^{1/6} - 6 \ln(x^{1/6} + 1) + C. \quad \blacksquare$$

The kind of problem that requires the alternate form of the substitution formula frequently involves the techniques developed in Section 11.2. Below is a slightly different problem.

EXAMPLE 3

Evaluate $\int x^3 \sqrt{x^2 + 4} \, dx$.

Solution

Let $t = \sqrt{x^2 + 4}$ and note that $x = \sqrt{t^2 - 4}$ and $x^3 = \sqrt{t^2 - 4} \, (t^2 - 4)$. Replace dx with

$$\frac{dx}{dt} \, dt = \frac{1}{2} (t^2 - 4)^{-1/2} 2t \, dt = \frac{t}{\sqrt{t^2 - 4}} \, dt$$

Thus, we get

$$\int x^3 \sqrt{x^2 + 4} \, dx = \int \sqrt{t^2 - 4} \, (t^2 - 4)t \, \frac{t}{\sqrt{t^2 - 4}} \, dt$$

$$= \int (t^2 - 4)t^2 \, dt$$

$$= \int (t^4 - 4t^2) \, dt$$

$$= \frac{1}{5} t^5 - \frac{4}{3} t^3 + C$$

$$= \frac{1}{5} (x^2 + 4)^{5/2} - \frac{4}{3} (x^2 + 4)^{3/2} + C. \quad \blacksquare$$

The following example is a remarkable instance of converting square roots into powers.

EXAMPLE 4

Evaluate $\displaystyle\int \left(\frac{1 + x}{2 + x}\right)^{1/2} dx$.

Solution

Let $t = \left(\dfrac{1 + x}{2 + x}\right)^{1/2}$ and solve for x as follows:

$$t^2 = \frac{1 + x}{2 + x}$$

$$(2 + x)t^2 = 1 + x$$

$$t^2 x - x = 1 - 2t^2$$

$$x = \frac{1 - 2t^2}{t^2 - 1}.$$

Replace dx with

$$\frac{dx}{dt} \, dt = \frac{-4t(t^2 - 1) - 2t(1 - 2t^2)}{(t^2 - 1)^2} \, dt = \frac{2t}{(t^2 - 1)^2} \, dt.$$

Thus, we get

$$\int \left(\frac{1 + x}{2 + x}\right)^{1/2} dx = \int \frac{2t^2}{(t^2 - 1)^2} \, dt,$$

and this requires partial fractions.

$$\frac{2t^2}{(t^2 - 1)^2} = \frac{2t^2}{(t - 1)^2(t + 1)^2} = \frac{a}{t - 1} + \frac{b}{(t - 1)^2} + \frac{c}{t + 1} + \frac{d}{(t + 1)^2}$$

$$2t^2 = a(t - 1)(t + 1)^2 + b(t + 1)^2 + c(t + 1)(t - 1)^2 + d(t - 1)^2.$$

It is routine to discover that

$$a = \frac{1}{2}, \quad b = \frac{1}{2}, \quad c = -\frac{1}{2}, \quad d = \frac{1}{2}.$$

It follows that

$$\int \frac{2t^2}{(t^2-1)^2} \, dt = \int \left[\frac{1}{2}\left(\frac{1}{t-1}\right) + \frac{1}{2}\left(\frac{1}{(t-1)^2}\right) - \frac{1}{2}\left(\frac{1}{t+1}\right) + \frac{1}{2}\left(\frac{1}{(t+1)^2}\right) \right] dt$$

$$= \frac{1}{2}\ln(t-1) - \frac{1}{2}\left(\frac{1}{t-1}\right) - \frac{1}{2}\ln(t+1) - \frac{1}{2}\left(\frac{1}{t+1}\right) + C$$

$$= \frac{1}{2}\ln(t-1) - \frac{1}{2}\ln(t+1) - \frac{1}{2}\left(\frac{1}{t-1} + \frac{1}{t+1}\right) + C$$

$$= \frac{1}{2}\ln(t-1) - \frac{1}{2}\ln(t+1) - \frac{t}{t^2-1} + C$$

After substituting for t in terms of x and simplifying the last term, the answer to the original problem is

$$\int \left(\frac{1+x}{2+x}\right)^{1/2} dx = \frac{1}{2}\ln\left(\left[\frac{1+x}{2+x}\right]^{1/2} - 1\right) - \frac{1}{2}\ln\left(\left[\frac{1+x}{2+x}\right]^{1/2} + 1\right) + \sqrt{(1+x)(2+x)} + C. \blacksquare$$

Substitution in Definite Integrals

Recall from Section 10.5 that the average value of the function $f(x)$ from $x = a$ to $x = b$ is defined to be

$$A(f) = \frac{1}{b-a} \int_a^b f(x) \, dx.$$

As we develop more techniques of integration, there are more and more functions for which we can compute the average value. This is illustrated in the next example.

EXAMPLE 5

Over a 6-month period the price of a certain commodity is

$$p(x) = x\sqrt{x+1} + 10$$

where x varies from $x = 0$ to $x = 6$. Find the average price of the commodity over the period.

Solution
We must calculate

$$A(p) = \frac{1}{6} \int_0^6 (x\sqrt{x+1} + 10) \, dx$$

$$= \frac{1}{6} \int_0^6 x\sqrt{x+1} \, dx + \frac{1}{6} \int_0^6 10 \, dx.$$

In order to compute the antiderivative

$$\int x\sqrt{x+1}\ dx,$$

we let $t = \sqrt{x+1}$ and we note that $x = t^2 - 1$. Replacing dx with $2t\ dt$, we have

$$\int x\sqrt{x+1}\ dx = \int (t^2 - 1)2t^2\ dt = \int (2t^4 - 2t^2)\ dt$$

$$= \frac{2}{5}t^5 - \frac{2}{3}t^3 + C$$

$$= \frac{2}{5}(x+1)^{5/2} - \frac{2}{3}(x+1)^{3/2} + C$$

Using this antiderivative, we compute $A(p)$.

$$A(p) = \frac{1}{6}\int_0^6 x\sqrt{x+1}\ dx + \frac{1}{6}\int_0^6 10\ dx$$

$$= \frac{1}{6}\left[\frac{2}{5}(x+1)^{5/2} - \frac{2}{3}(x+1)^{3/2}\right]\Big|_0^6 + 10$$

$$= \frac{1}{6}\left[\frac{2}{5}(7)^{5/2} - \frac{2}{3}(7)^{3/2} - \frac{2}{5} + \frac{2}{3}\right] + 10$$

$$\approx 16.63 \ \blacksquare$$

In evaluating definite integrals by the method of substitution, there is a variation of the procedure that is sometimes easier to carry out. It is based on the observation that, just before substituting for t in terms of x, we have found an antiderivative for $f(t)$ in which $t = g(x)$ is the independent variable. If we find the values of t corresponding to $x = a$ and $x = b$, which are $t_1 = g(a)$ and $t_2 = g(b)$, we can insert these values into the antiderivative directly and avoid the work of substituting for t in terms of x. That is, if $F(t)$ is an antiderivative of $f(t)$, then

$$\boxed{\int_a^b f(x)\ dx = \int_{g(a)}^{g(b)} f(x(t))\frac{dx}{dt}\ dt = F(g(b)) - F(g(a)).}$$

The final example shows how this works in practice.

EXAMPLE 6

Evaluate $\displaystyle\int_0^2 x^3\sqrt{x^2+4}\ dx$.

Solution
In Example 3 we used $t = g(x) = \sqrt{x^2+4}$ to find an antiderivative for this integrand. The limits of integration, $a = 0$ and $b = 2$, correspond to

$$g(0) = 2 \quad \text{and} \quad g(2) = \sqrt{8} = 2\sqrt{2}.$$

The antiderivative in terms of t was found to be

$$\int x^3 \sqrt{x^2 + 4}\, dx = \frac{1}{5} t^5 - \frac{4}{3} t^3 + C,$$

so the definite integral is

$$\int_0^2 x^3 \sqrt{x^2 + 4}\, dx = \left(\frac{1}{5} t^5 - \frac{4}{3} t^3 \right)\Big|_2^{2\sqrt{2}}$$

$$= \frac{128}{5} \sqrt{2} - \frac{64}{3} \sqrt{2} - \left(\frac{32}{5} - \frac{32}{3} \right)$$

$$= \frac{64}{15} (\sqrt{2} + 1). \ \blacksquare$$

Exercises   11.3

Use the original substitution formula, if necessary, to determine each anti-derivative below.

1 $\int \sqrt{x^2 + 5}\, x\, dx$

2 $\int \sqrt{3x^2 + 1}\, 4x\, dx$

3 $\int \sqrt[3]{5x + 2}\, dx$

4 $\int \sqrt[3]{7x + 9}\, dx$

5 $\int (2x + 6)^{50}\, dx$

6 $\int (3x - 2)^{61}\, dx$

7 $\int (x^2 + 2)^{20}\, x\, dx$

8 $\int (5x^2 - 4)^{40}\, x\, dx$

9 $\int \frac{x}{x^2 + 10}\, dx$

10 $\int \frac{2x}{1 + 3x^2}\, dx$

11 $\int \frac{1}{5x - 1}\, dx$

12 $\int \frac{3}{9x + 7}\, dx$

13 $\int \frac{x}{(x^2 + 6)^{12}}\, dx$

14 $\int \frac{4}{(2 + 3x)^{18}}\, dx$

15 $\int \frac{x^2 + 2x + 2}{x^3 + 3x^2 + 6x + 1}\, dx$

16 $\int \frac{7x^2}{x^3 + 5}\, dx$

17 $\int e^{x^2} x\, dx$

18 $\int e^{3x}\, dx$

19 $\int e^{-5x}\, dx$

20 $\int e^{2x^3} x^2\, dx$

21 $\int \frac{e^{\sqrt{x}}}{\sqrt{x}}\, dx$

22 $\int e^{\ln x}\, dx$

23 $\displaystyle\int \frac{\ln x}{x}\, dx$

24 $\displaystyle\int \frac{3(\ln x)^2}{x}\, dx$

25 $\displaystyle\int \frac{1}{x\ln x}\, dx$

26 $\displaystyle\int \frac{1}{x(\ln x)^5}\, dx$

Use the alternate form of the substitution formula, if necessary, to find each antiderivative below.

27 $\displaystyle\int \frac{x^3}{(x^2+1)^{3/2}}\, dx$

28 $\displaystyle\int \frac{1+x}{\sqrt{1-x}}\, dx$

29 $\displaystyle\int x^2\sqrt{x+3}\, dx$

30 $\displaystyle\int 5x^2\sqrt{x+1}\, dx$

31 $\displaystyle\int \frac{1}{x-x^{2/3}}\, dx$

32 $\displaystyle\int \frac{x^5}{\sqrt{1+x^3}}\, dx$

33 $\displaystyle\int \frac{4-x}{2-\sqrt{x}}\, dx$

34 $\displaystyle\int \frac{x-1}{x+\sqrt{x}}\, dx$

35 $\displaystyle\int x^5\sqrt{x^2+1}\, dx$

36 $\displaystyle\int \frac{3x}{x+\sqrt{x}}\, dx$

37 $\displaystyle\int \frac{x+5}{x\sqrt{x+1}}\, dx$

38 $\displaystyle\int \frac{\sqrt{x+1}-4}{\sqrt{x+1}+2}\, dx$

39 $\displaystyle\int \frac{1}{2+\sqrt{x}}\, dx$

40 $\displaystyle\int \frac{5}{3+2\sqrt{x}}\, dx$

41 $\displaystyle\int \frac{3-\sqrt{x}}{3+\sqrt{x}}\, dx$

42 $\displaystyle\int \frac{2}{x^{1/2}-x^{1/3}}\, dx$

43 $\displaystyle\int \frac{1}{x(1-\sqrt{x})}\, dx$

44 $\displaystyle\int \frac{1}{x(2-\sqrt[3]{x})}\, dx$

45 $\displaystyle\int \frac{x+2\sqrt[3]{x^2}-2\sqrt[3]{x}}{3x(x+1)}\, dx$

46 $\displaystyle\int \frac{4}{3+e^x}\, dx$

47 $\displaystyle\int \left(\frac{x+3}{x-1}\right)^{1/2} dx$

48 $\displaystyle\int \frac{1}{x(x^8+2)}\, dx$

11.4 Integral Tables

Most people who use integrals on a regular basis have occasion to use a table of integrals. A table of integrals is simply a list of antiderivatives that have been computed and tabulated for reference. To make the list as useful as possible, some of the integrands involve arbitrary constants. Below are eight formulas like those in the 34 pages of formulas in *Standard Mathematical Tables* published by the Chemical Rubber Publishing Company. The list uses the widely accepted convention of writing $\int \dfrac{dx}{g(x)}$, instead of

$\int \dfrac{1}{g(x)} \, dx.$

To save space, the usual constant of integration has not been added to each antiderivative.

1. $\displaystyle\int \frac{dx}{\sqrt{x^2 \pm a^2}} = \ln(x + \sqrt{x^2 \pm a^2})$

2. $\displaystyle\int \frac{dx}{x\sqrt{x^2 + a^2}} = -\frac{1}{a} \ln\left(\frac{a + \sqrt{x^2 + a^2}}{x}\right)$

3. $\displaystyle\int \frac{dx}{x\sqrt{a^2 - x^2}} = -\frac{1}{a} \ln\left(\frac{a + \sqrt{a^2 - x^2}}{x}\right)$

4. $\displaystyle\int \frac{dx}{x^2\sqrt{a^2 - x^2}} = -\frac{\sqrt{a^2 - x^2}}{a^2 x}$

5. $\displaystyle\int \frac{dx}{\sqrt{a + bx}} = \frac{2\sqrt{a + bx}}{b}$

6. $\displaystyle\int \frac{x \, dx}{\sqrt{a + bx}} = -\frac{2(2a - bx)}{3b^2} \sqrt{a + bx}$

7. $\displaystyle\int \sqrt{a + bx} \, dx = \frac{2}{3b} \sqrt{(a + bx)^3}$

8. $\displaystyle\int x\sqrt{a + bx} \, dx = -\frac{2(2a - 3bx)\sqrt{(a + bx)^3}}{15b^2}$

It requires some practice to recognize the formula in a table of integrals that solves a given problem. The ability to recognize that two expressions have the same form is a useful basic mathematical skill. We illustrate the process.

EXAMPLE 1

Use the short table of integrals to evaluate

$$\int \frac{5}{2x\sqrt{3 - x^2}} \, dx.$$

Solution
First, we factor out some constants, and we write the antiderivative in the same form as those in our short table.

$$\int \frac{5}{2x\sqrt{3 - x^2}} \, dx = \frac{5}{2} \int \frac{dx}{x\sqrt{3 - x^2}}$$

Since x^2 appears under the radical, we must use one of the first four formulas. Since the radical is multiplied by x, we must use either the second formula or third formula. Since our problem has "$-x^2$," the only formula that we might use is formula 3. To use formula 3, we must regard $3 - x^2$ as $a^2 - x^2$; thus, it must be that $3 = a^2$ or $a = \sqrt{3}$. According to formula 3, the answer is

$$\frac{5}{2}\left(-\frac{1}{\sqrt{3}}\right)\ln\left(\frac{\sqrt{3}+\sqrt{3-x^2}}{x}\right) + C.$$

(Although the table omits C to save space, we must remember to write it in the answer to each problem.) ■

In using a table of integrals, it is sometimes necessary to employ the elementary process of **completing the square.** Given an expression

$$x^2 + bx + c,$$

we add and subtract $(b/2)^2$; this introduces the square of a linear term.

$$x^2 + bx + \frac{b^2}{4} + \left(c - \frac{b^2}{4}\right) = \left(x + \frac{b}{2}\right)^2 + \left(c - \frac{b^2}{4}\right)$$

We illustrate the usefulness of this process in the next example.

EXAMPLE 2

Use the short table of integrals to evaluate

$$\int \frac{1}{\sqrt{x^2 - 4x + 5}}\, dx.$$

Solution
Since x^2 appears under the radical and there is no power of x except under the radical, our antiderivative most resembles the one in formula 1. However, the radical in formula 1 does not involve a first power of x, and our radical does. To remove the "x-term" in our problem, we complete the square.

$$x^2 - 4x + 5 = x^2 - 4x + 4 + 1 = (x - 2)^2 + 1$$

Setting $u = x - 2$, we can write our antiderivative as

$$\int \frac{1}{\sqrt{(x-2)^2 + 1}}\, dx = \int \frac{1}{\sqrt{u^2 + 1}}\, du.$$

Now it is clear that the answer is given in formula 1 by using the plus sign and letting $a = 1$. The answer is

$$\ln(u + \sqrt{u^2 + 1}) + C = \ln(x - 2 + \sqrt{x^2 - 4x + 5}) + C. \quad ■$$

Sometimes the constants are distracting, as indicated in the next example.

EXAMPLE 3

Use the short table of integrals to evaluate

$$\int \frac{\sqrt{11}}{\sqrt{x + 8}}\, dx.$$

Solution

Despite the radical sign, the numerator is a constant and we factor it out.

$$\int \frac{\sqrt{11}}{\sqrt{x+8}}\,dx = \sqrt{11}\int \frac{dx}{\sqrt{x+8}}$$

Since x appears in our problem only under the radical, we must use either formula 5 or formula 7. Since our radical is in the denominator, formula 5 is the appropriate one. We must have

$$x+8 = a+bx$$

and so $a=8$, $b=1$. According to formula 5, the answer is

$$\sqrt{11}\,2\sqrt{8+x}+C. \quad \blacksquare$$

Even when we are using a table of integrals, we must not forget the elementary rules for taking antiderivatives.

EXAMPLE 4

Evaluate $\displaystyle\int \frac{4x+1}{\sqrt{3+x}}\,dx.$

Use the short table of integrals, if it is helpful.

Solution

We use an elementary rule for antiderivatives, and we rewrite the problem.

$$\int \frac{4x+1}{\sqrt{3+x}}\,dx = \int \frac{4x}{\sqrt{3+x}}\,dx + \int \frac{1}{\sqrt{3+x}}\,dx$$

$$= 4\int \frac{x\,dx}{\sqrt{3+x}} + \int \frac{dx}{\sqrt{3+x}}$$

The next to last antiderivative above requires formula 6; the last one requires formula 5.

$$4\int \frac{x\,dx}{\sqrt{3+x}} + \int \frac{dx}{\sqrt{3+x}} = -\frac{8(6-x)}{3}\sqrt{3+x}+2\sqrt{3+x}+C$$

The right side of the last equation is the answer. $\quad\blacksquare$

Exercises 11.4

Use the short table of integrals to determine each antiderivative below.

1 $\displaystyle\int \frac{3}{2x\sqrt{4-x^2}}\,dx$

2 $\displaystyle\int \frac{5}{x\sqrt{9+x^2}}\,dx$

3 $\displaystyle\int 3\sqrt{5x+7}\,dx$

4 $\displaystyle\int \frac{9}{\sqrt{x+4}}\,dx$

5 $\displaystyle\int \frac{\sqrt{18}}{x\sqrt{5-3x^2}}\,dx$

6 $\displaystyle\int \frac{e}{x\sqrt{7+2x^2}}\,dx$

7 $\displaystyle\int \frac{\sqrt{19}}{\sqrt{5+6x}}\,dx$

8 $\displaystyle\int \frac{e^2x}{\sqrt{x+11}}\,dx$

9 $\displaystyle\int \frac{x+1}{x\sqrt{16+x^2}}\,dx$

10 $\displaystyle\int \frac{x+1}{x^2\sqrt{16-x^2}}\,dx$

11 $\displaystyle\int \frac{2x+3}{\sqrt{9+x}}\,dx$

12 $\displaystyle\int (5x+3)\sqrt{x+4}\,dx$

13 $\displaystyle\int \frac{1}{\sqrt{x^2+6x+5}}\,dx$

14 $\displaystyle\int \frac{1}{\sqrt{x^2+4x+6}}\,dx$

15 $\displaystyle\int \frac{1}{\sqrt{4x^2+4x}}\,dx$

11.5 Approximating the Definite Integral

Sometimes we cannot find an antiderivative to use in evaluating a given definite integral. For example, we cannot construct an antiderivative for e^{-x^2} from the functions that we know. In such a situation, we can use some elementary schemes for approximating the value of the definite integral. For the sake of simplicity in explaining the schemes, we assume that $f(x) \geq 0$ for $a \leq x \leq b$; thus, the definite integral

$$\int_a^b f(x)\,dx$$

is the area below the graph of $f(x)$, above the x-axis, and between the lines $x = a$, $x = b$.

One scheme for approximating such a definite integral is based on the area of a trapezoid. A trapezoid is a four-sided figure with at least one pair of sides parallel. We picture a trapezoid with the lengths of three sides indicated by a, b, and c.

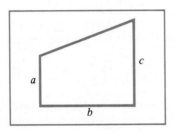

To compute the area of the trapezoid, we draw a line parallel to the base so that the figure is divided into a rectangle and a right triangle as shown in the second figure. Thus, the area of the trapezoid is

$$ab + \frac{1}{2}(c-a)b = \frac{1}{2}ab + \frac{1}{2}cb = \frac{1}{2}(a+c)b.$$

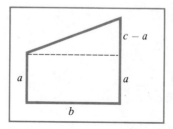

In order to approximate the definite integral, we divide the interval $a \le x \le b$ into n subintervals, each having length $h = (b-a)/n$. Let x_k be $a + kh$, with $k = 0, 1, \ldots, n$, and note that $x_0 = a$, $x_n = b$. The third illustration gives a representative picture showing how we construct a trapezoid over each subinterval by drawing a line from $(x_k, f(x_k))$ to $(x_{k+1}, f(x_{k+1}))$.

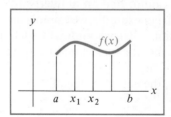

By the additive property of the definite integral, we get

$$\int_a^b f(x)\, dx = \int_a^{x_1} f(x)\, dx + \int_{x_1}^{x_2} f(x)\, dx + \cdots + \int_{x_{n-1}}^b f(x)\, dx.$$

To approximate the definite integral $\int_a^b f(x)\, dx$ according to the trapezoidal rule, we replace each definite integral $\int_{x_k}^{x_{k+1}} f(x)\, dx$ with the area of the trapezoid formed by joining $(x_k, f(x_k))$ to $(x_{k+1}, f(x_{k+1}))$. That area is

$$A_k = \frac{1}{2}(f(x_k) + f(x_{k+1}))h.$$

Thus, the approximation for the definite integral is

$$\frac{1}{2}(f(a) + f(x_1))h + \frac{1}{2}(f(x_1) + f(x_2))h + \cdots + \frac{1}{2}(f(x_{n-1}) + f(b))h.$$

We note that $f(x_1)$ appears twice in the above sum; in fact, each value of $f(x)$ except $f(a)$ and $f(b)$ appears twice. This observation permits us to give the following formula for approximating the definite integral.

The Trapezoidal Rule

$$\int_a^b f(x)\, dx \approx h\left[\frac{1}{2}f(a) + f(x_1) + \cdots + f(x_{n-1}) + \frac{1}{2}f(b)\right]$$

EXAMPLE 1

Use the Trapezoidal Rule with $n = 4$ to approximate the definite integral

$$\int_1^3 \frac{1}{1+x^2}\, dx.$$

Solution

The length of the subintervals is $h = \dfrac{(3-1)}{4} = \dfrac{1}{2}$, so

$$x_0 = 1, \qquad x_1 = \frac{3}{2}, \qquad x_2 = 2, \qquad x_3 = \frac{5}{2}, \qquad x_4 = 3.$$

Rounding off each calculation to two decimal places, we get the following approximation.

$$\int_1^3 \frac{1}{1+x^2}\, dx \approx \frac{1}{2}\left(\frac{1}{4} + \frac{4}{13} + \frac{1}{5} + \frac{4}{29} + \frac{1}{20}\right)$$
$$\approx .5(.25 + .31 + .20 + .4 + .05)$$
$$\approx .48 \quad \blacksquare$$

Using methods beyond the scope of this book, we determine that the exact value of the definite integral in Example 1, to two decimal places, is .46. So the Trapezoidal Rule gave a pretty good approximation with $n = 4$. In general, the larger the value of n then the better the resulting approximation.

CALCULATOR EXAMPLE

Use the Trapezoidal Rule with $n = 8$ to approximate the definite integral

$$\int_1^3 \frac{1}{1+x^2}\, dx.$$

Solution

We round off each calculation to three decimal places. The length of the subintervals is $\dfrac{(3-1)}{8} = \dfrac{1}{4}$, so

$$x_0 = 1, \qquad x_1 = \frac{5}{4}, \qquad x_2 = \frac{3}{2}, \qquad x_3 = \frac{7}{4}, \qquad x_4 = 2,$$

$$x_5 = \frac{9}{4}, \qquad x_6 = \frac{5}{2}, \qquad x_7 = \frac{11}{4}, \qquad x_8 = 3.$$

Thus, we have

$$\int_1^3 \frac{1}{1+x^2}\, dx \approx \frac{1}{4}(.250 + .390 + .308 + .246 + .200 + .165 + .138 + .117 + .050)$$
$$= .466 \quad \blacksquare$$

The exact value of the definite integral in the preceding example, to three decimal places, is .464. Notice that increasing n from 4 to 8 improved the accuracy of the approximation.

There is another scheme for approximating a given definite integral by use of segments of the graphs of parabolas. This method, which is called Simpson's Rule, generally results in more accurate approximations than those obtained by the Trapezoidal Rule with the same number of subintervals. In order to use Simpson's Rule the number of subintervals must be an even number.

Simpson's Rule

$$\int_a^b f(x)\,dx \approx \frac{h}{3}\,[f(a) + 4f(x_1) + 2f(x_2) + \cdots + 2f(x_{n-2}) + 4f(x_{n-1}) + f(b)]$$

where $x_k = a + kh$, with $k = 0, 1, \ldots, n$ and $h = (b - a)/n$ with n an even integer.

Note that the coefficients in Simpson's Rule begin with 1, 4 and end with 4, 1; in between the coefficients alternate between 2 and 4 except that they begin and end with 2. For the sake of comparison, we redo the problem in Example 1 using Simpson's Rule.

EXAMPLE 2

Use Simpson's Rule with $n = 4$ to approximate the definite integral

$$\int_1^3 \frac{1}{1 + x^2}\,dx.$$

Solution

The length of the subintervals is $\dfrac{3 - 1}{4} = \dfrac{1}{2}$, so

$$x_0 = 1, \qquad x_1 = \frac{3}{2}, \qquad x_2 = 2, \qquad x_3 = \frac{5}{2} \qquad x_4 = 3.$$

Rounding off each calculation to three decimal places, we get the following approximation.

$$\int_1^3 \frac{1}{1 + x^2}\,dx \approx \frac{1}{6}\left[\frac{1}{2} + 4\left(\frac{4}{13}\right) + 2\left(\frac{1}{5}\right) + 4\left(\frac{4}{29}\right) + .1\right]$$

$$\approx \frac{1}{6}\,[.500 + 1.231 + .400 + .552 + .100]$$

$$\approx .464 \; \blacksquare$$

The approximation in the preceding example is exactly the value of the definite integral to three decimal places.

EXAMPLE 3

Compute the approximate value of ln 4 by approximating the value of

$$\int_1^4 \frac{1}{x}\, dx = \ln x\Big|_1^4 = \ln 4$$

with Simpson's Rule, and $n = 6$.

Solution

The length of the subintervals is $\dfrac{4-1}{6} = \dfrac{1}{2}$, so

$$x_0 = 1, \quad x_1 = \frac{3}{2}, \quad x_2 = 2, \quad x_3 = \frac{5}{2}, \quad x_4 = 3, \quad x_5 = \frac{7}{2}, \quad x_6 = 4.$$

Rounding off each calculation to three decimal places, we get the following approximation.

$$\int_1^4 \frac{1}{x}\, dx \approx \frac{1}{6}\left[1 + 4\left(\frac{2}{3}\right) + 2\left(\frac{1}{2}\right) + 4\left(\frac{2}{5}\right) + 2\left(\frac{1}{3}\right) + 4\left(\frac{2}{7}\right) + \frac{1}{4}\right]$$

$$\approx \frac{1}{6}[1 + 2.667 + 1 + 1.6 + .667 + 1.143 + .25]$$

$$\approx 1.388 \ \blacksquare$$

The exact value of ln 4 to three decimal places is 1.386.

Exercises 11.5

Use the Trapezoidal Rule and the indicated value of n to approximate each definite integral below.

1 $\displaystyle\int_0^2 \frac{1}{1+x^2}\, dx, \quad n = 4$

2 $\displaystyle\int_0^1 \frac{1}{1+x^2}\, dx, \quad n = 4$

3 $\displaystyle\int_1^3 \frac{1}{x}\, dx, \quad n = 4$

4 $\displaystyle\int_1^3 \frac{1}{x}\, dx, \quad n = 8$

5 $\displaystyle\int_2^4 \frac{1}{x^2-1}\, dx, \quad n = 2$

6 $\displaystyle\int_2^4 \frac{1}{x^2-1}\, dx, \quad n = 4$

Use Simpson's Rule and the indicated value of n to approximate each definite integral below.

7 $\displaystyle\int_0^2 x^2\, dx, \quad n = 4$

8 $\displaystyle\int_0^2 x^2\, dx, \quad n = 6$

9 $\displaystyle\int_0^3 \frac{1}{x+2}\, dx, \quad n = 6$

10 $\displaystyle\int_{-1}^1 \frac{1}{x+2}\, dx, \quad n = 4$

11 $\displaystyle\int_3^5 \frac{1}{x^2-4}\, dx, \quad n = 4$

12 $\displaystyle\int_0^2 \frac{1}{x^2+4}\, dx, \quad n = 4$

13 ***Calculator Problem*** Approximate $\displaystyle\int_{-1}^{1} \frac{1}{\sqrt{4-x^2}}\,dx$ according to Simpson's Rule with $n = 4$.

14 ***Calculator Problem*** Approximate $\displaystyle\int_{0}^{2} e^{-x^2}\,dx$ according to Simpson's Rule with $n = 4$.

15 ***Calculator Problem*** Compute the approximate value of ln 5 by approximating the value of

$$\int_{1}^{5} \frac{1}{x}\,dx = \ln 5$$

with Simpson's Rule, and $n = 8$.

11.6   Improper Integrals

If $f(x)$ is nonnegative for $x \geq a$, then $\int_a^b f(x)\,dx$ is the area under the graph of $f(x)$, above the x-axis, and between the lines $x = a$ and $x = b$. It is possible that the limit $\lim_{b\to\infty} \int_a^b f(x)\,dx$ exists. If it does exist, then we define $\int_a^\infty f(x)\,dx$ by

$$\int_a^\infty f(x)\,dx = \lim_{b\to\infty} \int_a^b f(x)\,dx.$$

We say that $\int_a^\infty f(x)\,dx$ is the area under the graph of $f(x)$, above the x-axis, and to the right of the line $x = a$, despite the fact that this region is unbounded to the right.

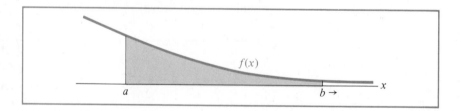

Note   In general, regardless of the sign of $f(x)$, $\int_a^\infty f(x)\,dx$ **is defined to be**

$$\lim_{b\to\infty} \int_a^b f(x)\,dx.$$

The integral $\int_a^\infty f(x)\,dx$ is said to exist when the above limit exists; the integral does not exist when the limit does not exist.

Analogously, $\int_{-\infty}^b f(x)\,dx$ **is defined to be**

$$\lim_{a\to-\infty} \int_a^b f(x)\,dx.$$

The integral is said to exist provided the limit exists; the integral does not exist when the limit does not exist. The integrals

$$\int_a^\infty f(x)\,dx \quad \text{and} \quad \int_{-\infty}^b f(x)\,dx$$

are called **improper integrals**.

EXAMPLE 1

Evaluate the improper integral

$$\int_1^\infty x^{-2}\,dx,$$

if it exists.

Solution

First, we evaluate the integral $\int_1^b x^{-2}\,dx$.

$$\int_1^b x^{-2}\,dx = -x^{-1}\Big|_1^b = 1 - \frac{1}{b}$$

Then, we compute the appropriate limit.

$$\int_1^\infty x^{-2}\,dx = \lim_{b\to\infty}\int_1^b x^{-2}\,dx = \lim_{b\to\infty}\left(1 - \frac{1}{b}\right) = 1$$

The improper integral exists, and its value is 1. ∎

Sometimes a given improper integral does not exist.

EXAMPLE 2

Evaluate the improper integral

$$\int_{-\infty}^{-2} \frac{1}{\sqrt{-x}}\,dx,$$

if it exists.

Solution

First, we evaluate the integral $\int_a^{-2} \frac{1}{\sqrt{-x}}\,dx$.

$$\int_a^{-2} \frac{1}{\sqrt{-x}}\,dx = -\int_a^{-2} -(-x)^{-1/2}\,dx = -2(-x)^{1/2}\Big|_a^{-2}$$
$$= 2\sqrt{-a} - 2\sqrt{2}$$

Then, we compute the appropriate limit.

$$\int_{-\infty}^{-2} \frac{1}{\sqrt{-x}}\,dx = \lim_{a\to-\infty}\int_a^{-2} \frac{1}{\sqrt{-x}}\,dx = \lim_{a\to-\infty}(2\sqrt{-a} - 2\sqrt{2})$$

Since the value of $\sqrt{-a}$ increases without bound as the magnitude of a

increases without bound, we see that the above limit does not exist. Thus, the given improper integral does not exist. ■

The two preceding notions of improper integral can be combined into one integral. We define $\int_{-\infty}^{+\infty} f(x)\,dx$ to be

$$\lim_{\substack{b\to+\infty \\ a\to-\infty}} \int_a^b f(x)\,dx.$$

The integral is said to exist provided that the limit exists; the integral does not exist when the limit does not exist.

For the improper integral $\int_a^\infty f(x)\,dx$ to be defined requires that $f(x)$ is continuous at all $x \ge a$. The integral $\int_{-\infty}^b f(x)\,dx$ is defined only for $f(x)$ that is continuous at all $x \le b$. In order for the integral $\int_{-\infty}^{+\infty} f(x)\,dx$ to be defined, $f(x)$ must be continuous at x equal to any real number.

There is another kind of improper integral. If $x = a$ is a discontinuity for $f(x)$, and $f(x)$ is continuous at every other value of x such that $a \le x \le b$, then **we define $\int_a^b f(x)\,dx$ by**

$$\int_a^b f(x)\,dx = \lim_{r\to a+} \int_r^b f(x)\,dx.$$

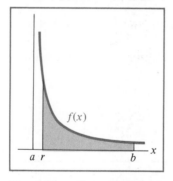

If $x = b$ is a discontinuity for $f(x)$, and $f(x)$ is continuous at every other value of x such that $a \le x \le b$, then **we define $\int_a^b f(x)\,dx$ by**

$$\int_a^b f(x)\,dx = \lim_{r\to b-} \int_a^r f(x)\,dx.$$

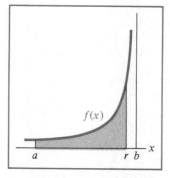

As before, **if the limit exists then we say that the improper integral exists;** otherwise, we say that the improper integral does not exist.
We illustrate how limits of this type are computed.

EXAMPLE 3

Evaluate the improper integral

$$\int_0^3 \frac{1}{\sqrt{3-x}}\, dx,$$

if it exists.

Solution
First, we note that the integral is improper because $1/\sqrt{3-x}$ is discontinuous at $x = 3$, and it is continuous at all other values of x such that $3 - x \geq 0$, or $3 \geq x$.

$$\int_0^3 \frac{1}{\sqrt{3-x}}\, dx = \lim_{r \to 3-} \int_0^r \frac{1}{\sqrt{3-x}}\, dx$$

$$= \lim_{r \to 3-} \left. -2(3-x)^{1/2} \right|_0^r$$

$$= \lim_{r \to 3-} (2\sqrt{3} - 2\sqrt{3-r})$$

$$= 2\sqrt{3} \quad \blacksquare$$

Next, we give an example of an improper integral of this type that does not exist.

EXAMPLE 4

Evaluate the improper integral

$$\int_5^{10} \frac{1}{x-5}\, dx,$$

if it exists.

Solution
The only discontinuity of $1/(x-5)$ is $x = 5$. Consequently, the given improper integral is defined by

$$\int_5^{10} \frac{1}{x-5}\, dx = \lim_{r \to 5+} \int_r^{10} \frac{1}{x-5}\, dx$$

$$= \lim_{r \to 5+} \left. \ln(x-5) \right|_r^{10}$$

$$= \lim_{r \to 5+} [\ln 5 - \ln(r-5)].$$

Clearly $\lim\limits_{r \to 5+} (r-5) = 0$; so the two limits below are equal if they both exist, and if one does not exist then neither does the other.

$$\lim_{r \to 5+} \ln(r-5), \qquad \lim_{x \to 0+} \ln x$$

By considering the graph of $\ln x$ for x close to $x = 0$, we decide that the above limits do not exist. Thus, the improper integral does not exist. ■

The last topic discussed in Section 10.6 was consumers' surplus and producers' surplus. We are now able to consider a demand equation that was too difficult for us earlier.

EXAMPLE 5

Determine the consumers' surplus for the demand function $p = D(x) = x^{-1/3}$ and the supply function $p = S(x) = x^{5/3}$.

Solution

We graph the two functions and determine the point of intersection for the two graphs.

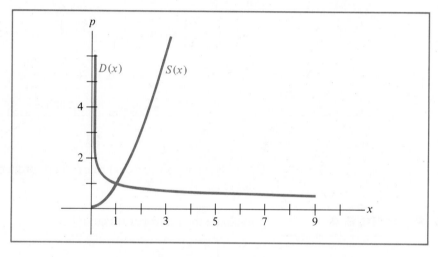

$$x^{-1/3} = x^{5/3}$$

$$1 = x^{6/3} = x^2$$

$$x_0 = 1$$

$$p_0 = D(1) = 1$$

$$\text{consumers' surplus} = \int_0^{x_0} (D(x) - p_0)\, dx = \int_0^{x_0} (x^{-1/3} - 1)\, dx$$

$$= \lim_{r \to 0+} \int_r^1 (x^{-1/3} - 1)\, dx$$

$$= \lim_{r \to 0+} \left[\frac{3}{2} x^{2/3} - x \right]\Big|_r^1$$

$$= \lim_{r \to 0+} \left[\frac{3}{2} - 1 - \frac{3}{2} r^{2/3} + r \right]$$

$$= \frac{1}{2} \quad ■$$

Exercises   11.6

Evaluate each improper integral below, if it exists.

1. $\displaystyle\int_4^\infty x^{-3/2}\,dx$ 2. $\displaystyle\int_8^\infty x^{-4/3}\,dx$

3. $\displaystyle\int_1^\infty x^{-2/3}\,dx$ 4. $\displaystyle\int_2^\infty x^{-1/2}\,dx$

5. $\displaystyle\int_{-\infty}^0 e^x\,dx$ 6. $\displaystyle\int_{-\infty}^1 e^{x^3}x^2\,dx$

7. $\displaystyle\int_1^\infty \frac1x\,dx$ 8. $\displaystyle\int_3^\infty \frac{1}{x-2}\,dx$

9. $\displaystyle\int_{-\infty}^{-2} \frac{1}{(3x+4)^2}\,dx$ 10. $\displaystyle\int_{-\infty}^{-3} \frac{1}{(2x+5)^2}\,dx$

11. $\displaystyle\int_2^\infty \frac{1}{x^2+x-2}\,dx$ 12. $\displaystyle\int_5^\infty \frac{1}{x^2-x-6}\,dx$

Evaluate each integral below, if it exists.

13. $\displaystyle\int_4^8 \frac{1}{\sqrt{x-4}}\,dx$ 14. $\displaystyle\int_2^{11} \frac{1}{\sqrt{x-2}}\,dx$

15. $\displaystyle\int_0^3 \frac{1}{\sqrt{x+1}}\,dx$ 16. $\displaystyle\int_0^5 \frac{1}{\sqrt{x+4}}\,dx$

17. $\displaystyle\int_1^2 \frac{1}{x-1}\,dx$ 18. $\displaystyle\int_6^8 \frac{1}{x-3}\,dx$

19. $\displaystyle\int_0^{27} x^{-2/3}\,dx$ 20. $\displaystyle\int_0^1 x^{-3/4}\,dx$

21. $\displaystyle\int_0^8 x^{-4/3}\,dx$ 22. $\displaystyle\int_0^9 x^{-3/2}\,dx$

23. $\displaystyle\int_1^2 \frac{1}{x^2+x-2}\,dx$ 24. $\displaystyle\int_3^6 \frac{1}{x^2-x-6}\,dx$

In problems 25 and 26, determine the consumers' surplus for the given demand function and supply function.

25. $p = D(x) = x^{-1/2}, \quad p = S(x) = x^{3/2}$
26. $p = D(x) = x^{-2/3}, \quad p = S(x) = x^{4/3}$

Review of Terms

Important Mathematical Terms
Integration by Parts Formula,   *p. 517*
partial fractions,   *p. 521*

Review Problems

Determine each antiderivative below.

1 $\displaystyle\int 2e^{9x}\, dx$

2 $\displaystyle\int xe^{2x}\, dx$

3 $\displaystyle\int \ln(4x)\, dx$

4 $\displaystyle\int \frac{(\ln x)^3}{x}\, dx$

5 $\displaystyle\int 3x \ln x\, dx$

6 $\displaystyle\int \frac{1}{(2+x)^2}\, dx$

7 $\displaystyle\int x^2 e^{-5x}\, dx$

8 $\displaystyle\int \frac{1+\sqrt{x}}{1-\sqrt{x}}\, dx$

9 $\displaystyle\int xe^{7x^2}\, dx$

10 $\displaystyle\int \frac{x+1}{x+5}\, dx$

11 $\displaystyle\int x\sqrt{x+2}\, dx$

12 $\displaystyle\int \sqrt{2x+5}\, dx$

13 $\displaystyle\int \frac{5x}{x^2+3}\, dx$

14 $\displaystyle\int x^3 \sqrt{x^2+1}\, dx$

15 $\displaystyle\int \frac{1}{e^x+1}\, dx$

16 $\displaystyle\int \frac{x^2+x+2}{x^3-x^2+x-1}\, dx$

17 $\displaystyle\int \frac{3x^2-2x+1}{x^3-x^2+x-1}\, dx$

18 $\displaystyle\int \frac{1}{x+x^{1/2}}\, dx$

19 $\displaystyle\int \frac{1}{x^2+x-20}\, dx$

20 $\displaystyle\int \frac{x+2}{x^2+3x}\, dx$

21 $\displaystyle\int \frac{3x^2+5x+2}{x^3+2x^2}\, dx$

Evaluate each improper integral below, if it exists.

22 $\displaystyle\int_{1}^{\infty} x^{-5/4}\, dx$

23 $\displaystyle\int_{-\infty}^{-1} x^{-4/5}\, dx$

24 $\displaystyle\int_{0}^{\infty} e^{-x}\, dx$

25 $\displaystyle\int_{-\infty}^{0} \frac{1}{(7x-2)^3}\, dx$

26 $\displaystyle\int_{5}^{6} \frac{1}{\sqrt{x-5}}\, dx$

27 $\displaystyle\int_{-2}^{0} (x+2)^{-3/2}\, dx$

28 $\displaystyle\int_{2}^{3} \frac{x+1}{x^2+2x-8}\, dx$

29 Use the Trapezoidal Rule with $n=4$ to approximate the value of the definite integral.

$$\int_{1}^{3} \frac{1}{x^2+4}\, dx$$

30 Use Simpson's Rule with $n=4$ to approximate the value of the definite integral.

$$\int_{0}^{2} \frac{1}{x+3}\, dx$$

31 ***Calculator Problem***   Use Simpson's Rule with $n=10$ to approximate the value of ln 6.

Social Science Applications

Normal Curves

Many times data such as test scores and heights and weights of people are reported by giving a graph called a normal curve. For example, the normal curve given next shows the relative frequency with which different intelligence quotients (or IQ's) occur as measured by the Stanford-Binet scale.

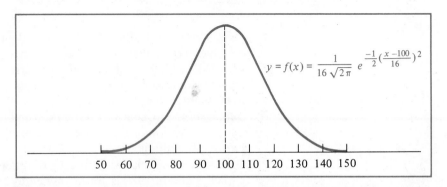

$$y = f(x) = \frac{1}{16\sqrt{2\pi}}\, e^{\frac{-1}{2}\left(\frac{x-100}{16}\right)^2}$$

On the x-axis, we have listed the possible IQ's from 50 to 150, and the height of the graph of $y=f(x)$ above the score indicates the fraction of the population with that IQ. For example, about 2.5% of the population should have an IQ of exactly 100, which is regarded as perfect normalcy.

We can determine the percentage of the population with IQ's between a and b by computing the definite integral $\int_a^b f(x)\,dx$. For example, the fraction of the population with IQ's between 90 and 110 (the range that is considered "normal") is $\int_{90}^{110} f(x)\,dx$. Because we cannot find an antiderivative for the integrand $f(x)$, we must use some procedure like Simpson's Rule to evaluate the definite integral.

Below, we apply Simpson's Rule. We divide the interval from 90 to 110 into four subintervals, each with width 5.

$$\int_{90}^{110} \frac{1}{16\sqrt{2\pi}}\, e^{-\frac{1}{2}\left(\frac{x-100}{16}\right)^2}\, dx \approx \frac{5}{3}[.0205 + 4(.0237) + 2(.0249) + 4(.0237) + .0205]$$

$$\approx .4673$$

Thus, we conclude that approximately 46.73% of the population falls in the range considered "normal."

Use Simpson's Rule with four intervals and the integrand

$$y = f(x) = \frac{1}{16\sqrt{2\pi}}\, e^{-\frac{1}{2}\left(\frac{x-100}{16}\right)^2}$$

to solve problems 1 and 2. Use a calculator or the appropriate table to evaluate the exponential function.

1 **Psychology** Determine the fraction of the population with IQ's in the interval 120 to 140.

2 **Psychology** Determine the fraction of the population with IQ's in the interval 50 to 70.

3 **Anthropology** The relative frequency of different cranial widths (in millimeters) for the modern adult of western civilization is given by the graph of

$$y = f(x) = \frac{1}{5\sqrt{2\pi}}\, e^{-\frac{1}{2}\left(\frac{142-x}{5}\right)^2}.$$

Here x indicates widths from 120 to 160 and y indicates the fraction of adult western civilization with that particular cranial width. Use Simpson's Rule with four subintervals to determine the fraction of adults with cranial widths from 132 to 152. Use a calculator or the appropriate table to evaluate the exponential function.

4 **Anthropology** Using the information and methods of problem 3, determine the fraction of adult western civilization with cranial widths from 137 to 147 millimeters.

5 **Sociology** The relative frequency of different family incomes for the United States in 1980 is approximated by the graph of

$$y = f(x) = \frac{1}{12\sqrt{2\pi}}\, e^{-\frac{1}{2}\left(\frac{24-x}{12}\right)^2}.$$

Here x indicates thousands of dollars of family income from 0 to 48 and y indicates the fraction of U.S. families with that particular income. Use Simpson's Rule with four subintervals to determine the fraction of families with incomes from 12 to 36 thousands of dollars. Use a calculator or the appropriate table to evaluate the exponential function.

6 *Sociology* Using the information and methods of problem 5, determine the fraction of U.S. families with incomes from 24 to 48 thousands of dollars.

CAREER PROFILE
Actuary

Q: What is an actuary?

A: An actuary is a professional who is expert at the design, financing, and operation of insurance plans of all kinds, and of annuity and welfare plans.

The answer stated above was given by a past president of the Society of Actuaries, John Bragg (1974). The formality of Bragg's definition conveys some of the no-nonsense formality that is characteristic of the actuarial profession. Career actuaries are dedicated, hard-working professionals who combine intelligence (including a flair for mathematics) with intensive educational training and close adherence to a strong code of ethics. The work of actuaries affects the financial security of many people; because of this, actuaries must bring both integrity and intelligence to their work.

An actuary needs good mathematical aptitude. His or her undergraduate education should include a thorough background in calculus, linear algebra, probability, statistics, computer science, and data processing. Beyond these mathematical studies, courses in accounting, economics, English composition, finance, marketing, and liberal arts are important in providing a broad foundation for the many judgments an actuary must make.

For a college student who is considering a career as an actuary, an Actuarial Aptitude Test (available from the Society of Actuaries, whose address is given below) is available. Although this test is designed to be taken at about the sophomore level of college, it may be taken at any level.

To become a fully qualified actuary means becoming a Fellow of the Society of Actuaries. The status of Fellow is obtained by passing a series of nine examinations given by the Society. These exams begin with one that covers mathematics up through calculus; subsequent exams move upward through mathematical fields (probability and statistics, mathematics of finance, and numerical analysis) and then into specialized insurance areas. The Society provides study materials for the post-college exams.

Tasks that an actuary performs are illustrated by the following list:

1. design and price a new type of life insurance policy with benefits varying with the cost of living;
2. recommend changes to modernize benefits in health insurance policies and put a price tag on these changes;
3. study the effect of various health habits (such as regular exercise) on mortality, and recommend any needed changes in the company's selection of risks;
4. develop pension plans (an enormous task that requires extensive data gathering and data analysis, calculation, and planning to insure that employer and employee contribution rates will provide adequate and equitable funding to allow these funds to remain viable over the long term).

Although employment of actuaries is influenced by the health of the insurance industry, there has been a perennial shortage of actuaries and this is likely to continue in the near future. About 8000 persons worked as actuaries in 1980. Many were located in insurance company headquarters in New York, Hartford, Chicago, Philadelphia, and Boston. New college graduates entering the insurance field without having passed any actuarial exams earn around $14,000. Because prospective actuaries may take several exams while still in college, potential starting salaries are much higher. Actuarial Fellows earn salaries averaging over $50,000 per year.

Sources of Additional Information
Some information for this Career Profile was obtained from each of the following publications:

Professional Opportunities in the Mathematical Sciences. Mathematical Association of America, 1529 18th Street, N.W., Washington, DC 20036. (This booklet is available at a cost of $1.50 from the MAA. It describes career opportunities for individuals with mathematical training and includes a list of over 50 sources of additional information.)

"So You're Good at Math—Then Consider a Career as an Actuary." Society of Actuaries, 208 South LaSalle Street, Chicago, IL 60604. (This pamphlet is available free from the Society as are a booklet describing the actuarial exams and a list of colleges and universities offering specific actuarial courses.)

Occupational Outlook Handbook. Bureau of Labor Statistics, U.S. Department of Labor, Washington, DC 20212. (Revised every 2 years, this handbook provides information about job duties, working conditions, level and places of employment, education and training requirements, advancement possibilities, job outlook, and earnings for about 250 occupations.)

12 | Multivariable Calculus

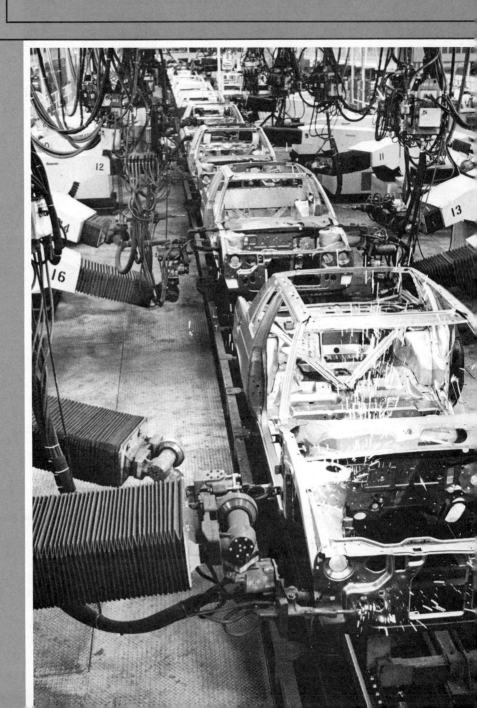

12.1 Functions of Several Variables

Many useful functions involve more than one variable. The area of a rectangle, A, is the product of its width x and its height y. So we have $A = xy$, and both x and y can vary. If a chair manufacturer sells x chairs in its simple style for \$120 each and y chairs in its fancy style for \$200 each, then the revenue function for this manufacturer is

$$R(x, y) = 120x + 200y.$$

Typical functions of two variables are given by the formulas

$$f(x, y) = x^2 + 2xy + y^2, \qquad g(x, y) = e^{xy} + 3x$$
$$V = \pi r^2 h, \qquad S = 2\pi rh.$$

A function may depend on more than two variables. The volume of a rectangular solid, V, is the product of the three dimensions; we have $V = xyz$. Other examples of functions of several variables are

$$f(x, y, z) = 3x + 2y - 5z, \qquad g(x, y, z) = \frac{x}{y} + 4z$$

$$A = P(1 + rt), \qquad V = \frac{1}{2}\,bhw.$$

It is possible to graph a function $f(x, y)$ with two independent variables. We let z represent the value of the function; then, we construct three mutually perpendicular number lines as shown next. These number lines, or **axes,** define three mutually perpendicular planes, called the **coordinate planes.**

Any point P in xyz-space can be described by an ordered triple (x_0, y_0, z_0) that reports its x-coordinate, y-coordinate, and z-coordinate, respectively. The z-coordinate measures how far P is above, or below, the xy-plane. The x-coordinate measures the distance of the point from the yz-plane, and the y-coordinate measures its distance from the xz-plane. A line perpendicu-

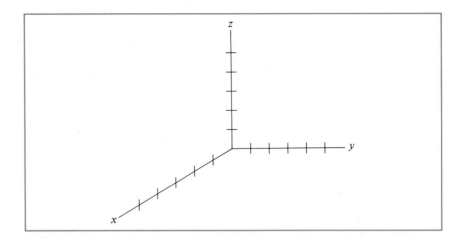

lar to the xy-plane and passing through P must also pass through the point $(x_o, y_0, 0)$. This is illustrated in the second picture and the subsequent example.

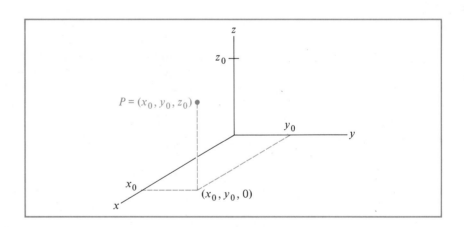

EXAMPLE 1

Determine the value of the function

$$z = f(x, y) = x^2 + 2xy + y^2$$

at $(x, y) = (1, 2)$ and plot the corresponding point in xyz-space.

Solution

We substitute 1 for x and 2 for y to get the desired value.

$$z = f(1, 2) = 1 + 4 + 4 = 9$$

We plot this point.

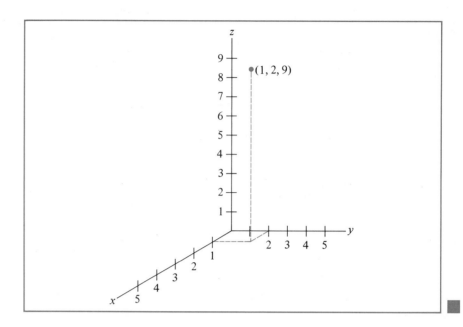

The process of describing a point by its coordinates is reversible. Corresponding to any ordered triple (x_0, y_0, z_0) is a unique point P in xyz-space. The graph of the function $f(x, y)$ is the picture in xyz-space of the triples $(x, y, f(x, y))$ for all choices of x and y in the domain of the function. The next example shows how such a graph can be obtained.

EXAMPLE 2

Sketch the graph of

$$z = f(x, y) = 2 - 2x - \frac{2y}{3}.$$

Solution

The points in the xy-plane are precisely the points (x, y, z) with $z = 0$; thus, the graph of $z = 0$ is the xy-plane. Similarly, the xz-plane is the graph of $y = 0$, and the yz-plane is the graph of $x = 0$. To show a portion of the graph of the function $f(x, y)$, we locate the intersection of the graph with each coordinate plane.

By taking $z = 0$ in the equation defining $f(x, y)$, we get

$$0 = 2 - 2x - \frac{2y}{3} \quad \text{or} \quad 6x + 2y = 6,$$

which is the intersection of the graph with the xy-plane. By setting $x = 0$ in the defining equation, we get

$$z = 2 - \frac{2y}{3} \quad \text{or} \quad 2y + 3z = 6,$$

which is the intersection of the graph with the yz-plane. By setting $y = 0$ in the defining equation, we obtain the intersection with the xz-plane.

$$z = 2 - 2x \quad \text{or} \quad 2x + z = 2$$

Each of these three equations describes a straight line. Next, we show the intersections of the graph with the three coordinate planes. Clearly, the graph is a plane. Since the x, y, and z variables are not restricted, the graph extends indefinitely in all directions; the part shown in the picture is only a representative portion.

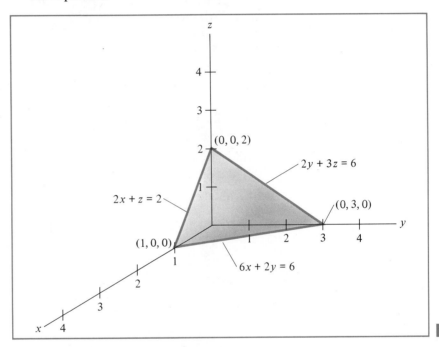

The intersections of a given graph with the coordinate planes are called the **traces** of the graph. By taking $z = 0$ in the equation that we are graphing, we get the trace in the xy-plane. To get the traces in the yz-plane and the xz-plane, we take $x = 0$ and $y = 0$, respectively. We ordinarily draw the traces only for *nonnegative x, y, and z*.

EXAMPLE 3

Sketch the graph of

$$z = 5 \sqrt{1 - \frac{x^2}{4} - \frac{y^2}{9}}.$$

Solution

We set $z = 0$, square both sides, and simplify; the graph of the resulting equation is the trace in the xy-plane.

$$\frac{x^2}{4} + \frac{y^2}{9} = 1$$

By setting $x = 0$, we get the trace in the yz-plane.

$$\frac{y^2}{9} + \frac{z^2}{25} = 1$$

We get the trace in the xz-plane by taking $y = 0$.

$$\frac{x^2}{4} + \frac{z^2}{25} = 1$$

Next we sketch the traces. In the given equation, z must be nonnegative (since it is a multiple of a square root), and we must have $-2 \le x \le 2$, $-3 \le y \le 3$, $0 \le z \le 5$. The picture on the right attempts to show all such points as a surface in xyz-space.

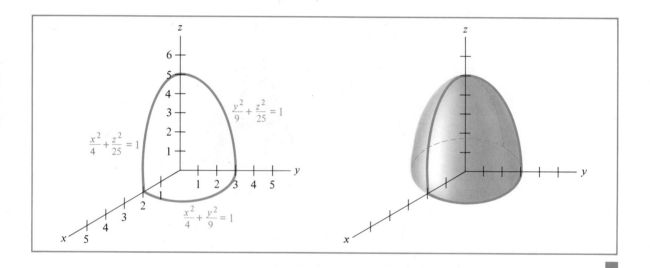

The graph in Example 3 looks like half of a watermelon. A figure like a watermelon is called an **ellipsoid.***

EXAMPLE 4

Sketch the graph of

$$z = x^2 + y^2.$$

Solution

We set $z = 0$ to get the trace in the xy-plane. Since the square of any real number is nonnegative, the only solution to the equation

$$0 = x^2 + y^2$$

is $x = 0$, $y = 0$. The trace in the xy-plane is the one point $(0, 0, 0)$. To get the

---

* In general, the graph of $x^2/a^2 + y^2/b^2 = 1$ in the plane, with a and b constants, is called an ellipse. Thus, each trace of an ellipsoid is an ellipse.

traces in the yz-plane and the xz-plane we take $x = 0$ and $y = 0$, respectively. The graph of each of the equations

$$z = y^2, \qquad z = x^2$$

is a parabola. We sketch a portion of the traces next; they extend indefinitely upward. At the right of the illustration we show a portion of the surface that is the graph of the given function.

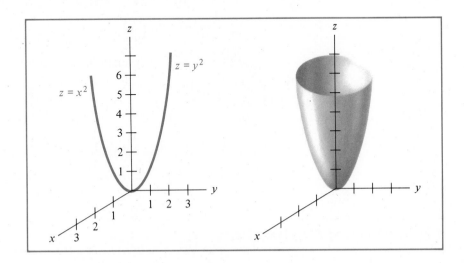

In order to get more insight into the shape of the graph, note that we can choose any fixed positive value of z, say $z = 4$, and find the intersection of the graph of the given function with that horizontal plane. In this case the intersection has the equation

$$4 = x^2 + y^2$$

which describes a circle of radius 2, centered on the z-axis and lying in the plane $z = 4$. The intersection with the plane $z = 9$ is a circle of radius 3, and so forth. ∎

Exercises   12.1

For each function below, determine the indicated value and plot the corresponding point in xyz-space.

1   $f(x, y) = 3x + 4y, \quad f(2, 1)$

2   $g(x, y) = \dfrac{x}{y} + 2, \quad g(6, 2)$

3   $f(x, y) = 5x^2y - xy^2, \quad f(1, -2)$

4   $g(x, y) = 2x^3y - x^2y^2 + xy^3, \quad g(-1, -1)$

5 $f(x, y) = \dfrac{3x}{\sqrt{y^2 + 1}}, \quad f(-2, 0)$

6 $g(x, y) = \dfrac{\sqrt{x^2 + 4}}{2y + 1}, \quad g(\sqrt{5}, 2)$

7 $f(x, y) = 9e^{xy}, \quad f(0, 4)$

8 $g(x, y) = 7e^{x^2 - y^2}, \quad f(3, 3)$

9 $f(x, y) = 2 \ln x - \ln y, \quad f(e, e)$

10 $g(x, y) = \ln(xy)\ln\left(\dfrac{x}{y}\right), \quad g(5, 5)$

Sketch the graph of each function given below.

11 $f(x, y) = 3$ 12 $g(x, y) = -2$

13 $f(x, y) = 5x$ 14 $g(x, y) = y$

15 $f(x, y) = x + y$ 16 $g(x, y) = 8x + 4y$

17 $f(x, y) = \sqrt{1 - 4x^2 - 4y^2}$ 18 $g(x, y) = \dfrac{\sqrt{144 - 16x^2 - 9y^2}}{12}$

19 $f(x, y) = 4x^2 + y^2$ 20 $g(x, y) = x^2 + 9y^2$

12.2 Partial Derivatives

If we fix the value of y equal to the constant y_0, then the function $f(x, y)$ becomes a function of x alone. The derivative of that function at $x = x_0$ is said to be the partial derivative of $f(x, y)$ with respect to x at (x_0, y_0).

Analogously, if we fix the value of x equal to the constant x_0 then the function $f(x, y)$ becomes a function of y alone. The derivative of that function at $y = y_0$ is said to be the partial derivative of $f(x, y)$ with respect to y at (x_0, y_0). The formal definition is below.

Formal Definition   The **partial derivative** of $z = f(x, y)$ with respect to x at (x_0, y_0), denoted $f_x(x_0, y_0)$ or $\dfrac{\partial z}{\partial x}\Big|_{(x_0, y_0)}$, is

$$\lim_{h \to 0} \frac{f(x_0 + h, y_0) - f(x_0, y_0)}{h},$$

provided the limit exists.

The partial derivative of $z = f(x, y)$ with respect to y at (x_0, y_0), denoted $f_y(x_0, y_0)$ or $\dfrac{\partial z}{\partial y}\Big|_{(x_0, y_0)}$, is

$$\lim_{h \to 0} \frac{f(x_0, y_0 + h) - f(x_0, y_0)}{h},$$

provided the limit exists.

We next illustrate how to compute these partial derivatives.

EXAMPLE 1

Compute the partial derivatives of
$$R(x, y) = 120x + 200y.$$

Solution

In order to compute $R_x(x_0, y_0)$, we fix the value of y equal to y_0. By the usual rules of differentiation, we get
$$R_x(x, y_0) = 120.$$

In order to compute $R_y(x_0, y_0)$, we fix the value of x equal to x_0. By the usual rules of differentiation, we get
$$R_y(x_0, y) = 200.$$

Thus, we have
$$R_x(x_0, y_0) = 120 \quad \text{and} \quad R_y(x_0, y_0) = 200. \quad \blacksquare$$

In Example 1, notice that it is not necessary to formally set $y = y_0$. We can simply treat y as a constant when we are computing the derivative with respect to x. Similarly, we can treat x as a constant when we compute the derivative with respect to y. Since the resulting formula holds for any choice of (x_0, y_0), we simply write x, instead of x_0, and y, instead of y_0. We demonstrate this simplified process.

EXAMPLE 2

Compute the partial derivatives of
$$f(x, y) = x^2 + 3xy + y^2.$$

Solutions

Thinking of y as a constant, we take the derivative with respect to x by the usual rules.

$$f_x(x, y) = \frac{\partial}{\partial x} x^2 + \frac{\partial}{\partial x} 3xy + \frac{\partial}{\partial x} y^2$$
$$= 2x + 3y + 0$$
$$= 2x + 3y$$

Thinking of x as a constant, we take the derivative with respect to y by the usual rules.

$$f_y(x, y) = \frac{\partial}{\partial y} x^2 + \frac{\partial}{\partial y} 3xy + \frac{\partial}{\partial y} y^2$$
$$= 0 + 3x + 2y$$
$$= 3x + 2y \quad \blacksquare$$

We now demonstrate how to compute the partial derivatives for some less elementary functions.

EXAMPLE 3

Compute the partial derivatives of each of the following functions.
$$f(x, y) = x \ln y - 11y, \qquad g(x, y) = e^{xy} + 5x, \qquad S = 2\pi rh.$$

Solution

$$f_x(x, y) = \frac{\partial}{\partial x} x(\ln y) - \frac{\partial}{\partial x} 11y \qquad f_y(x, y) = \frac{\partial}{\partial y} x(\ln y) - \frac{\partial}{\partial y} 11y$$

$$= 1 \cdot \ln y - 11 \cdot 0 \qquad\qquad\qquad = x \cdot \frac{1}{y} - 11 \cdot 1$$

$$= \ln y \qquad\qquad\qquad\qquad\qquad = \frac{x}{y} - 11$$

$$g_x(x, y) = \frac{\partial}{\partial x} e^{xy} + \frac{\partial}{\partial x} 5x \qquad g_y(x, y) = \frac{\partial}{\partial y} e^{xy} + \frac{\partial}{\partial y} 5x$$

$$= ye^{xy} + 5 \cdot 1 \qquad\qquad\qquad = xe^{xy} + 5 \cdot 0$$

$$= ye^{xy} + 5 \qquad\qquad\qquad\qquad = xe^{xy}$$

$$S_r(r, h) = \frac{\partial}{\partial r} 2\pi rh \qquad\qquad S_h(r, h) = \frac{\partial}{\partial h} 2\pi rh$$

$$= 2\pi h \qquad\qquad\qquad\qquad = 2\pi r \quad\blacksquare$$

Each partial derivative has a natural geometric interpretation. The plane that is perpendicular to the y-axis and intersects it at $y = y_0$ is the graph of the equation $y = y_0$; consequently, that plane is the collection of points in xyz-space with y-coordinate equal to y_0. When we restrict the value of y, for the function $f(x, y)$, to the constant y_0, we are considering the values of the function at points in the plane $y = y_0$. The surface that is the graph of $z = f(x, y)$ intersects the plane $y = y_0$ in a curve, which is the graph of $z = f(x, y_0)$. The partial derivative $f_x(x_0, y_0)$ is the derivative of the single-

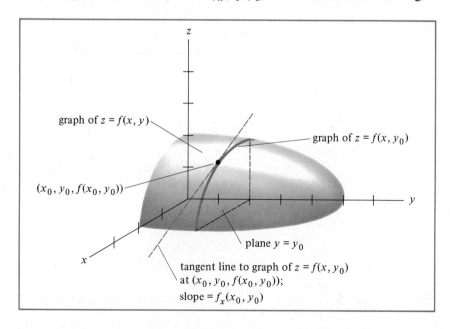

variable function $f(x, y_0)$ with respect to x; thus, $f_x(x_0, y_0)$ is the slope of the tangent line to the graph of $z = f(x, y_0)$ at the point $(x_0, y_0, f(x_0, y_0))$. This is illustrated in the picture at the bottom of page 568.

Similarly, we see that $f_y(x_0, y_0)$ is the slope of the tangent line to the graph of $z = f(x_0, y)$ at the point $(x_0, y_0, f(x_0, y_0))$. This is illustrated next.

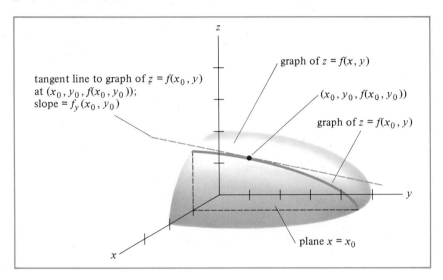

If $R(x, y)$ is the revenue function for some business, then $R_x(x, y)$ and $R_y(x, y)$ are the **marginal revenue functions.** Similarly, the partial derivatives of the cost and profit functions are the **marginal cost** and **marginal profit** functions, respectively.

EXAMPLE 4

The revenue function for a certain business is $R(x, y) = 3x^2 + 7y$ where x is the number of service contracts sold and y is the number of products sold. Determine the values of the marginal revenue functions for $x = 10$, $y = 5$.

Solution
We compute the partial derivatives by the methods indicated earlier.

$$R_x(x, y) = 6x, \qquad R_y(x, y) = 7$$

Taking $x = 10$ and $y = 5$ in these formulas, we get the desired values.

$$R_x(10, 5) = 60, \qquad R_y(10, 5) = 7 \quad \blacksquare$$

We have defined two partial derivatives for a function with two independent variables. It is easy to see how the idea is extended for a function with more than two independent variables. Suppose f is a function of n independent variables, where n is some integer larger than 1; denote the independent variables by $x_1, x_2, \ldots, x_n$. The **partial derivative of** $f(x_1, x_2, \ldots, x_n)$ **with respect to** x_1 **at** $(a_1, a_2, \ldots, a_n)$ is the derivative of the function $f(x_1, a_2, \ldots, a_n)$ at $x_1 = a_1$. If every x_j, except x_r, is set equal to the corresponding a_j and the resulting function of the single independent variable x_r is differentiated, then we obtain the **partial derivative of $f(x_1$, $x_2, \ldots, x_n)$ with respect to x_r at $(a_1, a_2, \ldots, a_n)$.** We demonstrate how to take partial derivatives of functions with three or more independent variables.

EXAMPLE 5

Compute the partial derivatives of each of the following functions.

$$f(x_1, x_2, x_3) = x_1^2 x_2 + x_2^3 x_3 + x_1 x_3, \qquad g(x_1, x_2, x_3) = \left(\frac{x_1}{x_2}\right)^{14}$$

$$h(w, x, y, z) = we^{xy} + wz, \qquad A = P(1 + rt)$$

Solution

$$f_{x_1}(x_1, x_2, x_3) = 2x_1 x_2 + x_3, \qquad f_{x_2}(x_1, x_2, x_3) = x_1^2 + 3x_2^2 x_3, \qquad f_{x_3}(x_1, x_2, x_3) = x_2^3 + x_1$$

$$g_{x_1}(x_1, x_2, x_3) = 14\left(\frac{x_1}{x_2}\right)^{13}\left(\frac{1}{x_2}\right), \; g_{x_2}(x_1, x_2, x_3) = 14\left(\frac{x_1}{x_2}\right)^{13}(x_1)(-x_2)^{-2}, \; g_{x_3}(x_1, x_2, x_3) = 0$$

$$h_w(w, x, y, z) = e^{xy} + z, \qquad h_x(w, x, y, z) = we^{xy}y,$$

$$h_y(w, x, y, z) = we^{xy}x, \qquad h_z(w, x, y, z) = w$$

$$\frac{\partial A}{\partial P} = 1 + rt, \qquad \frac{\partial A}{\partial r} = Pt, \qquad \frac{\partial A}{\partial t} = Pr \quad \blacksquare$$

Since each partial derivative of a function of several independent variables is again a function of several independent variables, we can compute the partial derivatives of the partial derivatives. When we take the partial derivative of a function with respect to one of its independent variables, the result is called a **first order partial derivative.** When we take a partial derivative of a first order partial derivative, the result is called a **second order partial derivative** or **second partial derivative.**

Note   The notation for the second partial derivatives of the function $z = f(x, y)$ may take several forms. The partial derivative with respect to y of the partial derivative with respect to x is denoted.

$$\frac{\partial}{\partial y}\frac{\partial z}{\partial x} \quad \text{or} \quad \frac{\partial^2 z}{\partial y \, \partial x} \quad \text{or} \quad f_{xy}(x, y).$$

The partial derivative with respect to x of the partial derivative with respect to y is denoted

$$\frac{\partial}{\partial x}\frac{\partial z}{\partial y} \quad \text{or} \quad \frac{\partial^2 z}{\partial x \, \partial y} \quad \text{or} \quad f_{yx}(x, y).$$

The other two second partial derivatives are denoted

$$\frac{\partial}{\partial x}\frac{\partial z}{\partial x} \quad \text{or} \quad \frac{\partial^2 z}{\partial x^2} \quad \text{or} \quad f_{xx}(x, y)$$

and

$$\frac{\partial}{\partial y}\frac{\partial z}{\partial y} \quad \text{or} \quad \frac{\partial^2 z}{\partial y^2} \quad \text{or} \quad f_{yy}(x, y).$$

EXAMPLE 6

Compute all of the second partial derivatives of

$$f(x, y) = x^2 + 5xy + y^2.$$

Solution

First, we compute the first partial derivatives.

$$f_x(x, y) = 2x + 5y, \qquad f_y(x, y) = 5x + 2y$$

Then, we compute the partial derivatives of each of these functions.

$$f_{xx}(x, y) = 2, \qquad f_{xy}(x, y) = 5$$
$$f_{yx}(x, y) = 5, \qquad f_{yy}(x, y) = 2 \ \blacksquare$$

For most functions $f(x, y)$ and most points (x_0, y_0) encountered in elementary mathematics, we find that

$$f_{xy}(x_0, y_0) = f_{yx}(x_0, y_0).$$

In advanced calculus it is proved that **these two second partial derivatives are equal provided they exist and are continuous at (x_0, y_0).** These conditions are satisfied by almost all of the functions that we shall consider.

Exercises 12.2

Compute all of the first order partial derivatives for each function below.

1 $f(x, y) = 10x - 22y$
2 $g(x, y) = 3.5y + 1.7x$
3 $f(x, y) = x^2$
4 $g(x, y) = 5y^3$
5 $f(x, y) = x^2 y^3 + 4xy^2 - x^3$
6 $g(x, y) = x^4 y^5 - x^3 y + xy^3$
7 $f(x, y) = \sqrt{x^2 + 4y^2}$
8 $g(x, y) = \sqrt{xy}$

9 $f(x, y) = \dfrac{3 + x^2}{1 + y^2}$
10 $g(x, y) = \dfrac{1 - y + y^2}{x^2 + 5}$

11 $f(x, y) = \dfrac{x^3 y^2 + 2x^2}{y + x^3}$
12 $g(x, y) = \dfrac{x^2 - 2xy + y^2}{x + y}$

13 $f(x, y) = \sqrt{\dfrac{x}{y + 1}}$
14 $g(x, y) = \left(\dfrac{x^3 - 1}{y^2 + 2}\right)^{1/3}$

15 $f(x, y) = e^{x^2 + y^2}$
16 $g(x, y) = xe^{xy} + 11y$
17 $f(x, y) = 5 \ln(x^2 + y^4)$
18 $g(x, y) = y \ln(x^2 + 1) + 3x$

19 $f(x, y) = 9e^x \ln y$

20 $g(x, y) = ye^{x - \ln y}$

21 $V = xyz$

22 $S = 2xz + 2yz + 2xy$

23 $w = ze^{x+y}$

24 $w = e^z \ln(3x^2 + y^2)$

25 $u = e^{x/y}\sqrt{z^2 + 1} - wxy$

26 $u = \sqrt{w^2 + x^2} \ln(y^2 + z^2)$

Compute all of the second order partial derivatives for each function below.

27 $f(x, y) = e$

28 $g(x, y) = 17$

29 $f(x, y) = 23x - 6y$

30 $g(x, y) = \dfrac{x}{5} + 2y$

31 $f(x, y) = x^3y^5 + x^6y^4$

32 $g(x, y) = xy - x^3y^3$

33 $f(x, y) = \dfrac{y}{x^2 + 1}$

34 $g(x, y) = \dfrac{4x}{5 + y^2}$

35 $f(x, y) = e^{xy}$

36 $g(x, y) = \ln(x^2 + y^4)$

Solve each of the following.

37 Obtain the marginal revenue functions, assuming that revenue is $R(x, y) = 10x + y^2 + 3y$.

38 Obtain the marginal cost functions, assuming that cost is $C(x, y) = x + .5y^2 + 100$.

39 Obtain the marginal profit functions, assuming that profit is $P(x, y) = 9x + .5y^2 + 3y - 100$.

40 If C is the curve formed by the intersection of the graph of $z = \sqrt{25 - x^2 - y^2}$ with the plane $y = 3$, what is the slope of the tangent line to C at $(0, 3, 4)$?

41 What is the slope of the tangent line to the curve C in problem 40 at $(3, 3, \sqrt{7})$?

42 If C is the curve formed by the intersection of the graph of $z = x^2 + y^2$ with the plane $x = 1$, what is the slope of the tangent line to C at $(1, 0, 1)$?

43 What is the slope of the tangent line to the curve C in problem 42 at $(1, 3, 10)$?

12.3 Maxima and Minima

In Chapter 9 we showed the usefulness of the concepts of relative maximum and relative minimum for functions of one variable. These concepts are also helpful when working with functions of two variables. We say that $(x_0, y_0, f(x_0, y_0))$ is a **relative minimum** provided none of the nearby points on the graph of $z = f(x, y)$ has a smaller z-value. The point $(x_0, y_0, f(x_0, y_0))$ is an **absolute minimum** provided no point on the graph of $z = f(x, y)$ has a smaller z-value. We say that $(x_0, y_0, g(x_0, y_0))$ is a **relative maximum** provided none of the nearby points on the graph of $z = g(x, y)$ has a larger z-value. The point $(x_0, y_0, g(x_0, y_0))$ is an **absolute maximum** provided no point on the graph of $z = g(x, y)$ has a larger z-value. Picture A shows a relative minimum at $(x_0, y_0, f(x_0, y_0))$ and picture B shows a relative maximum at $(x_0, y_0, g(x_0, y_0))$.

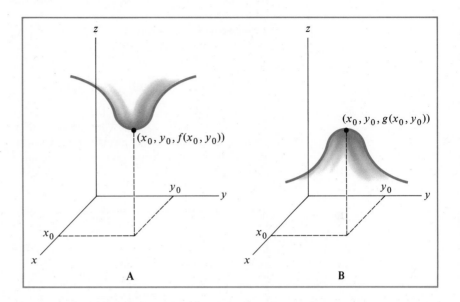

A   B

If the graphs pictured in A and B are all of the graphs of $z = f(x, y)$ and $z = g(x, y)$ then the relative minimum is an absolute minimum, and the relative maximum is an absolute maximum.

If $g(x_0, y_0) \geq g(x, y)$ for all (x, y) near (x_0, y_0) then $g(x_0, y_0) \geq g(x_0, y)$ for all y near y_0. Since the graph of $g(x_0, y)$ is the curve formed by intersecting the graph of $z = g(x, y)$ with the plane $x = x_0$, we see that $(x_0, y_0, g(x_0, y_0))$ is a relative maximum for $g(x_0, y)$. Provided the first partial derivatives of $g(x, y)$ exist, this means that $g_y(x_0, y_0) = 0$. This is illustrated in the second picture.

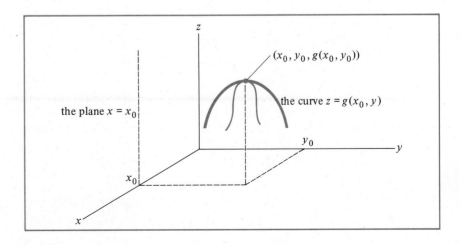

Similarly, we can conclude that $g_x(x_0, y_0) = 0$ when $(x_0, y_0, g(x_0, y_0))$ is a relative maximum. Analogous arguments show that $f_x(x_0, y_0) = 0$ and $f_y(x_0, y_0) = 0$ when $(x_0, y_0, f(x_0, y_0))$ is a relative minimum for $f(x, y)$. We summarize these observations in the next theorem.

Theorem 1   Assume that $f_x(x_0, y_0)$ and $f_y(x_0, y_0)$ exist. If $(x_0, y_0, f(x_0, y_0))$ is a relative extremum, that is, it is either a relative maximum or a relative minimum, then

$$f_x(x_0, y_0) = 0 \quad \text{and} \quad f_y(x_0, y_0) = 0.$$

A point (x_0, y_0) where $f_x(x_0, y_0) = 0$ and $f_y(x_0, y_0) = 0$ is called a **critical value.** We must be careful when we use the preceding theorem because not every critical value gives a relative extremum. A critical value that does not give a relative extremum gives a **saddle point.** The picture shows a representative saddle point.

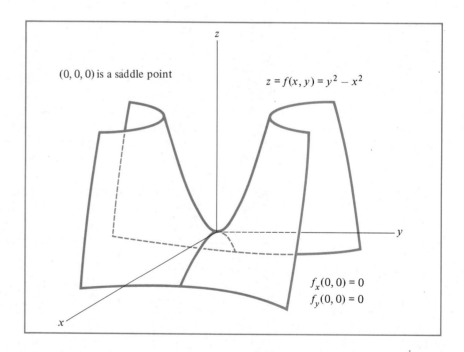

(0, 0, 0) is a saddle point

$z = f(x, y) = y^2 - x^2$

$f_x(0, 0) = 0$
$f_y(0, 0) = 0$

Geometrically, the traces of $f(x, y)$ is the xz- and yz- planes are parabolas, and the tangent line to each parabola is horizontal at the origin. However, among the points near $(0, 0, 0)$, some have positive z-values and some have negative z-values. Thus, the origin is neither a relative maximum nor a relative minimum.

EXAMPLE 1

Find the critical values for the function

$$z = f(x, y) = 3\sqrt{1 - x^2 - y^2}.$$

Solution

We compute the first partial derivatives

$$\frac{\partial z}{\partial x} = 3\left(\frac{1}{2}\right)(1 - x^2 - y^2)^{-1/2}(-2x) = \frac{-3x}{\sqrt{1 - x^2 - y^2}}$$

$$\frac{\partial z}{\partial y} = \frac{-3y}{\sqrt{1 - x^2 - y^2}}$$

We set the first partial derivatives equal to zero, and we solve those equations.

$$\frac{-3x}{\sqrt{1 - x^2 - y^2}} = 0 \quad \text{or} \quad 3x = 0 \quad \text{or} \quad x = 0$$

$$\frac{-3y}{\sqrt{1 - x^2 - y^2}} = 0 \quad \text{or} \quad 3y = 0 \quad \text{or} \quad y = 0$$

The only critical value is $(0, 0)$. (Since the largest possible value of $1 - x^2 - y^2$ is 1, the largest possible value of z is $3\sqrt{1} = 3$. Thus, $(0, 0, 3)$ is an absolute maximum.) ∎

In Chapter 9, we found that we can frequently use the Second Derivative Test to determine whether a critical value is a relative extremum. For a function of two variables there is an analogous test, but it is somewhat more complicated.

Second Derivative Test for a Function of Two Variables Let (x_0, y_0) be a critical value for the function $z = f(x, y)$. Assume that all of the second order partial derivatives of $f(x, y)$ exist and are continuous at (x_0, y_0) and all nearby points; let

$$A = f_{xx}(x_0, y_0), \qquad B = f_{xy}(x_0, y_0), \qquad C = f_{yy}(x_0, y_0)$$

Then, the following statements are true.

1. If $AC - B^2 > 0$ and $A < 0$, then $(x_0, y_0, f(x_0, y_0))$ is a relative maximum.
2. If $AC - B^2 > 0$ and $A > 0$, then $(x_0, y_0, f(x_0, y_0))$ is a relative minimum.
3. If $AC - B^2 < 0$, then $(x_0, y_0, f(x_0, y_0))$ is not a relative extremum. (It is a saddle point.)
4. If $AC - B^2 = 0$ then $(x_0, y_0, f(x_0, y_0))$ might be a relative extremum, or it might not be a relative extremum.

We illustrate the use of the Second Derivative Test in the next two examples.

EXAMPLE 2

Find all of the critical values of

$$z = f(x, y) = x^2 + y^2 - 2x - 6y + 10,$$

and classify them to the extent possible using the Second Derivative Test.

Solution

To find the critical values, we compute the first partial derivatives and set them equal to zero.

$$0 = f_x(x, y) = 2x - 2, \qquad 0 = f_y(x, y) = 2y - 6$$
$$1 = x \qquad\qquad\qquad 3 = y$$

The only critical value is $(1, 3)$.

We compute the second order partial derivatives and evaluate them at $(1, 3)$.

$$f_{xx}(x, y) = 2, \qquad f_{yy}(x, y) = 2, \qquad f_{xy}(x, y) = 0$$
$$A = f_{xx}(1, 3) = 2, \qquad B = f_{xy}(1, 3) = 0, \qquad C = f_{yy}(1, 3) = 2$$

Thus, we have

$$AC - B^2 = 4 > 0$$
and
$$A = 2 > 0.$$

It follows from the Second Derivative Test that $(1, 3, 0)$ is a relative minimum. ■

Of course, the Second Derivative Test does not always work, as the next example shows.

EXAMPLE 3

Find all of the critical values of

$$z = f(x, y) = x^3 + y^3 + 3x^2 + 3x + 1,$$

and classify them to the extent possible using the Second Derivative Test.

Solution

We follow the procedure used in Example 2.

$$0 = f_x(x, y) = 3x^2 + 6x + 3, \qquad 0 = f_y(x, y) = 3y^2$$
$$0 = 3(x + 1)^2 \qquad\qquad 0 = y$$
$$-1 = x$$

The only critical value is $(-1, 0)$.

$$f_{xx}(x, y) = 6x + 6, \qquad f_{yy}(x, y) = 6y \qquad f_{xy}(x, y) = 0$$
$$A = f_{xx}(-1, 0) = 0 \qquad B = f_{xy}(-1, 0) = 0, \qquad C = f_{yy}(-1, 0) = 0$$

Thus, we have

$$AC - B^2 = 0,$$

and we can draw no conclusion about the nature of the critical value. ■

The next example gives one practical application of Theorem 1.

EXAMPLE 4

A business needs a rectangular box with a volume of 125 cubic inches in order to ship its product. Determine the dimensions of the box that has the least surface area and, consequently, costs the least. (Assume that there is an absolute minimum for surface area.)

Solution

If the dimensions of the box are x, y, and z, then the volume is $V = xyz$ and the surface area is $A = 2xy + 2xz + 2yz$. We set $V = 125$, and we solve for z in terms of x and y. Substituting that into the formula for A, we get the function to be minimized.

$$125 = xyz, \qquad z = 125x^{-1}y^{-1}$$

$$A = 2xy + 2x(125x^{-1}y^{-1}) + 2y(125x^{-1}y^{-1})$$

$$A = 2xy + 250y^{-1} + 250x^{-1}$$

The absolute minimum for the surface area must be a relative minimum; thus, the absolute minimum occurs at a critical value. We determine the critical values.

$$\frac{\partial A}{\partial x} = 2y - 250x^{-2}$$

$$\frac{\partial A}{\partial y} = 2x - 250y^{-2}$$

$$0 = 2y - 250x^{-2} \qquad\qquad 0 = 2x - 250y^{-2}$$

$$125x^{-2} = y \qquad\qquad 125y^{-2} = x$$

Substituting one of these last equations into the other one, we find the only critical value.

$$125(125x^{-2})^{-2} = x$$

$$125^{-1}x^4 = x$$

$$x^3 = 125$$

$$x = 5$$

$$y = 125(5)^{-2} = 5$$

$$z = 125 \, 5^{-1}5^{-1} = 5$$

The absolute minimum must occur at the only critical value, $(5, 5)$. ∎

We are now able to consider more difficult demand equations, as we demonstrate in the next example.

EXAMPLE 5

A company sells two similar breakfast cereals, and sales of one tend to diminish sales of the other. The number of boxes of cereal A (in thousands) that can be sold is given by

$$f(x, y) = 44 - 3x + y,$$

where x is the price of cereal A and y is the price of cereal B. The number of boxes of cereal B that can be sold is given by

$$g(x, y) = 60 + x - 2y.$$

Determine the prices of the two cereals that result in maximum revenue. (Assume there is an absolute maximum for revenue.)

Solution
We note that revenue is given by

$$R(x, y) = xf(x, y) + yg(x, y)$$
$$= -3x^2 + 44x + 2xy - 2y^2 + 60y.$$

The maximum must be a relative maximum, so it must occur at a critical value. We compute the first partial derivatives of $R(x, y)$, and we find the critical values.

$$\frac{\partial R}{\partial x} = -6x + 44 + 2y$$

$$\frac{\partial R}{\partial y} = 2x - 4y + 60$$

$$0 = -3x + y + 22$$
$$0 = x - 2y + 30$$

Solving these equations simultaneously, we get $x = 14.8$, $y = 22.4$. The maximum revenue is

$$R(14.8, 22.4) = 997.6. \quad \blacksquare$$

Exercises   12.3

Determine all of the critical values of each function below.

1 $f(x, y) = 3x^2 - 2x + y^2 + 6y + 12$
2 $g(x, y) = x^2 + 4x + 25 - 5y^2 + 5y$
3 $f(x, y) = x^2 + 10xy + y^2$
4 $g(x, y) = 3y^2 - 6xy - x^2$
5 $f(x, y) = x^2 + 2x + 6xy + 4y + y^2 + 19$
6 $g(x, y) = 5x^2 - 12x + 2xy - 3y^2 + 112$

7 $f(x, y) = xy - \dfrac{2}{x} + \dfrac{3}{y}$

8 $g(x, y) = 3x - 6xy - \dfrac{1}{y}$

9 $f(x, y) = x^3 - 9x + y^2 + 6y + 52$

10 $g(x, y) = x^2 - 6xy + y^3 + 27$

Find all of the critical values of each function below, and classify them to the extent possible using the Second Derivative Test.

11 $f(x, y) = 4x^2 + 2xy - 5y^2$

12 $g(x, y) = 3x^2 - 9xy - y^2$

13 $f(x, y) = 5y^2 + 4y - x^2 - 2x + 12$

14 $g(x, y) = y^2 - y + 56 + x^2 + 4x$

15 $f(x, y) = \dfrac{6}{x} - xy + \dfrac{2}{y}$

16 $g(x, y) = 4xy - \dfrac{1}{x} - \dfrac{1}{y}$

17 $f(x, y) = 6x - 2x^3 + 28 + y^2 - 2xy$

18 $g(x, y) = y^3 - 3y - x^2 + 4xy + 96$

19 $f(x, y) = 2x^3 + 3x^2 - 12x + 10 + 2y^3 + 3y^2 - 72y$

20 $g(x, y) = x^3 - 3x + 27 + y^3 - 12y$

Solve each of the following. In each problem, assume that the desired minimum or maximum exists.

21 A rectangular box with no top must have a volume of 500 cubic inches. What are the dimensions of the box with the least surface area?

22 A rectangular box with a top and a bottom must have a volume of 250 cubic inches. The material for the sides costs 1¢ per square inch; the material for the top and bottom costs 2¢ per square inch. Determine the dimensions of the least expensive box.

23 A rectangular box with top and bottom must have a surface area equal to 6 square feet. What are the dimensions of the box with the maximum volume?

24 A rectangular box with no top must have a surface area equal to 12 square feet. What are the dimensions of the box with the maximum volume?

25 A business sells two dissimilar products. Product no. 1 sells for x dollars, and the business sells $10{,}000 - 5x$ units of it. The business can sell $8{,}000 - 4y$ units of product no. 2 at a price of y dollars. What should be the prices of the two products to maximize revenue?

26 Suppose a retailer sells $20{,}000 - 5x$ units of the deluxe model of its featured product at a price of x dollars; it sells $15{,}000 - 30y$ units of its inexpensive model of the same product at a price of y dollars. Determine the prices x and y for maximum revenue.

12.4 Lagrange Multipliers

In Section 12.3, we found an extremum of a given function $h(x, y)$ by determining the critical values of $h(x, y)$, and applying the Second Derivative Test. In Example 4 and similar problems, we are given an equation $g(x, y, z) = 0$ in addition to the function $f(x, y, z)$; the desired extremum of $f(x, y, z)$ must satisfy the equation. In those problems, we use the equation $g(x, y, z) = 0$ to solve for one variable, say z, in terms of the other variables, say $z = u(x, y)$. Then, we determine the critical values of the function $h(x, y) = f(x, y, u(x, y))$.

There is a very useful alternative procedure called the **method of Lagrange Multipliers.** Lagrange's method is often simpler than the previous method, because it does not require that we solve the equation $g(x, y, z) = 0$ for one variable in terms of the others. In Lagrange's method, we take a multiplier, usually represented by the Greek letter λ (lambda), and we form the new function

$$h(x, y, z) = f(x, y, z) - \lambda g(x, y, z).$$

Then, we solve the equations

$$\frac{\partial h}{\partial x}(x, y, z) = 0, \qquad \frac{\partial h}{\partial y}(x, y, z) = 0, \qquad \frac{\partial h}{\partial z}(x, y, z) = 0$$

$$g(x, y, z) = 0$$

simultaneously, to find the desired extremum of $f(x, y, z)$. The method is demonstrated in the next example. For the sake of simplicity, we use functions of two variables rather than three variables.

EXAMPLE 1

Determine a point (x, y) such that the value of $f(x, y) = xy$ is an absolute maximum for points satisfying the equation $x^2 + xy + y^2 = 12$.

Solution
In this case, $g(x, y) = x^2 + xy + y^2 - 12$ and it would be unpleasant to solve $g(x, y) = 0$ for one variable in terms of the other. We use Lagrange's method. Thus, we must compute the partial derivatives of

$$h(x, y) = xy - \lambda(x^2 + xy + y^2 - 12).$$

$$\frac{\partial h}{\partial x} = y - \lambda(2x + y)$$

$$\frac{\partial h}{\partial y} = x - \lambda(x + 2y)$$

We solve each of the equations

$$\frac{\partial h}{\partial x} = 0, \qquad \frac{\partial h}{\partial y} = 0$$

for λ in terms of x and y. Setting these two expressions equal, we determine y in terms of x.

$$y - \lambda(2x + y) = 0, \qquad x - \lambda(x + 2y) = 0$$

$$\frac{y}{2x + y} = \lambda, \qquad \frac{x}{x + 2y} = \lambda$$

$$\frac{y}{2x + y} = \frac{x}{x + 2y}$$

$$y(x + 2y) = x(2x + y)$$

$$xy + 2y^2 = 2x^2 + xy$$

$$y^2 = x^2$$

$$y = \pm x$$

We note that when $y = -x$, we have

$$f(x, y) = -x^2;$$

this cannot possibly lead to a maximum value, since $x^2 \geq 0$ and $-x^2 \leq 0$. We discard the possibility that $y = -x$. Using $y = x$, we substitute in the one remaining equation.

$$x^2 + xy + y^2 = 12$$

$$x^2 + x(x) + x^2 = 12$$

$$x^2 = 4$$

$$x = \pm 2$$

Both $(x, y) = (2, 2)$ and $(x, y) = (-2, -2)$ satisfy the equation

$$x^2 + xy + y^2 = 12.$$

Since

$$f(2, 2) = 4 = f(-2, -2),$$

$(2, 2, 4)$ and $(-2, -2, 4)$ are both absolute maxima. ■

The condition that must be satisfied by the extremum of the given function $f(x, y)$ is called the **constraint**. It is possible to solve problems with more than one constraint. Below we outline the procedure for applying the method of Lagrange Multipliers to a function of two variables with one constraint.

Lagrange Multipliers for a Function of Two Variables

1. If we seek an extremum for $f(x, y)$ subject to the constraint $g(x, y) = 0$, then we form the function

$$h(x, y) = f(x, y) - \lambda g(x, y).$$

2. We solve each of the equations

$$\frac{\partial h}{\partial x}(x, y) = 0, \qquad \frac{\partial h}{\partial y}(x, y) = 0$$

for λ, and we equate the two expressions for λ.
3. From the equation obtained in part 2, we get one of the variables x, y in terms of the other, and we substitute accordingly in the constraint.
4. The equation obtained in part 3 is used to get values for each of the variables x, y.

Any manufacturing operation requires some capital for such things as machinery, buildings, and vehicles; it also requires some labor for assembly lines, repairs, supervision, etc. In general, some labor can be replaced by additional capital; one worker with a robot might be able to accomplish more than two workers. Sometimes some capital can be replaced by additional labor; a welding machine might be replaced by two welders. Consequently, the number of units produced in one year is a function of both capital and labor. The next example illustrates the application of Lagrange Multipliers to such a production function.

EXAMPLE 2

The total annual production of a factory as a function of capital and labor is given by

$$p(x, y) = 20x^{5/2}y^{1/2}$$

where x is capital measured in units of $20,000 and y is labor measured in man-years. Assume that one man-year (that is, one man working for one year) costs $8000. The total amount to be invested in capital and labor is $400,000. Determine how those funds should be divided between capital and labor in order to achieve absolute maximum production.

Solution
The constraint can be written as

$$20,000x + 8,000y = 400,000$$

or

$$5x + 2y - 100 = 0.$$

Now we follow the steps of Lagrange's method

$$h(x, y) = 20x^{5/2}y^{1/2} - \lambda(5x + 2y - 100)$$

$$\frac{\partial h}{\partial x} = 50x^{3/2}y^{1/2} - 5\lambda = 0$$

$$\frac{\partial h}{\partial y} = 10x^{5/2}y^{-1/2} - 2\lambda = 0$$

so

$$\lambda = 10x^{3/2}y^{1/2}$$
$$\lambda = 5x^{5/2}y^{-1/2}$$

It follows that

$$10x^{3/2}y^{1/2} = 5x^{5/2}y^{-1/2}$$

or

$$2y = x.$$

Substituting this into the equation $5x + 2y - 100 = 0$, we get

$$5(2y) + 2y - 100 = 0$$

$$y = \frac{25}{3}$$

$$x = \frac{50}{3}.$$

We note that $(\frac{50}{3}, \frac{25}{3}, p(\frac{50}{3}, \frac{25}{3}))$ cannot be an absolute minimum, since $p(0, 0) = 0$. It is the absolute maximum. Thus, $\$20,000(\frac{50}{3}) = \$333,333.33$ should be invested in capital, and $\$8,000(\frac{25}{3}) = \$66,666.67$ in labor. ■

In the preceding problem, it is interesting to note that

$$\frac{\frac{\partial p}{\partial x}\left(\frac{50}{3}, \frac{25}{3}\right)}{20,000} = \frac{25\sqrt{2}}{72} = \frac{\frac{\partial p}{\partial y}\left(\frac{50}{3}, \frac{25}{3}\right)}{8,000}.$$

Assume that production, p, is written as a function of the number of units of capital, x, and the number of units of labor, y. The **Principle of Substitution,** in economics, asserts that capital and labor will be substituted one for another until production is as cost efficient as possible, and

$$\frac{\frac{\partial p}{\partial x}}{\text{one unit of capital}} = \frac{\frac{\partial p}{\partial y}}{\text{the price of one unit of labor}}$$

holds when production is as cost efficient as possible. We call $\partial p/\partial x$ the **marginal productivity of capital;** $\partial p/\partial y$ is the **marginal productivity of labor.**

Lagrange Multipliers can be used to obtain an extremum for a function of three or more variables. As the number of variables increases, the difficulty of solving for the extremum increases dramatically. We illustrate this with a very elementary example.

EXAMPLE 3

Determine the dimensions of a rectangular box with a top that has an absolute maximum volume and a surface area of 96 square inches.

Solution
Let x, y, and z denote the width, depth, and height, respectively, of the rectangular box. We are seeking an absolute maximum for the volume

$$V = xyz \quad \text{with} \quad x > 0, \quad y > 0, \quad z > 0.$$

The constraint is the equation below setting the surface area equal to 96 square inches.

$$0 = 2xy + 2xz + 2yz - 96$$

We form the function

$$h(x, y, z) = xyz - \lambda(2xy + 2xz + 2yz - 96).$$

We compute each first partial derivative of this function, and then we set each equal to zero.

$$\frac{\partial h}{\partial x} = yz - \lambda(2y + 2z) = 0$$

$$\frac{\partial h}{\partial y} = xz - \lambda(2x + 2z) = 0$$

$$\frac{\partial h}{\partial z} = xy - \lambda(2x + 2y) = 0$$

(1) $\lambda = \dfrac{yz}{2y + 2z}$

(2) $\lambda = \dfrac{xz}{2x + 2z}$

(3) $\lambda = \dfrac{xy}{2x + 2y}$

From equations (1) and (2), we get

$$\frac{yz}{2y + 2z} = \frac{xz}{2x + 2z}$$

or

$$yz(2x + 2z) = xz(2y + 2z)$$

or

$$y = x.$$

Similarly, from equations (2) and (3), we get

$$z = y.$$

Thus, $y = x$ and $z = x$; we substitute in the constraint according to these equations.

$$0 = 2x^2 + 2x^2 + 2x^2 - 96$$

$$96 = 6x^2$$

$$x^2 = 16$$

$$x = 4$$

The absolute maximum is

$$(4, 4, 4, V(4, 4, 4)) = (4, 4, 4, 64). \ \blacksquare$$

Exercises 12.4

In each problem below, use Lagrange Multipliers to determine the indicated extremum for f subject to the given constraint.

1 absolute maximum, $f(x, y) = xy$, $2x + 3y = 6$
2 absolute minimum, $f(x, y) = 3xy$, $4x - y = 1$
3 absolute maximum, $f(x, y) = x + y$, $x^2 + y^2 = 1$
4 absolute minimum, $f(x, y) = 2x + y$, $x^2 + y^2 = 4$
5 absolute minimum, $f(x, y) = x^2 + y^2$, $4x^2 + y^2 - 2y = 3$
6 absolute maximum, $f(x, y) = x^2 + y^2$, $x^2 + 4x + 4y^2 = 0$
7 absolute minimum, $f(x, y) = xy$, $x^2 + 9y^2 - 18y = 0$
8 absolute maximum, $f(x, y) = 2xy$, $x^2 - 2x + y^2 = 0$
9 absolute minimum, $f(x, y, z) = 5xyz$, $x + y - z = 3$
10 absolute maximum, $f(x, y, z) = xyz$, $3x + y + 2z = 6$
11 absolute maximum, $f(x, y, z) = x + 2y + 2z$, $x^2 + y^2 + z^2 = 1$
12 absolute minimum, $f(x, y, z) = 3x + 3y + z$, $x^2 + y^2 + z^2 = 9$
13 absolute maximum, $f(x, y, z) = x^2 + y^2 + z^2$, $9x^2 + 4y^2 + 36z^2 = 36$
14 absolute minimum, $f(x, y, z) = x^2 + y^2 + z^2$, $4x^2 - 8x + 4y^2 + 8y + z^2 + 4 = 0$

Solve each of the following using Lagrange Multipliers.

15 Suppose annual production, p, is given as a function of capital, x, and labor, y, by

$$p = 18x^{5/3}y^{1/2}.$$

A unit of capital is \$50,000, and a unit of labor is a man-year, which costs \$10,000. If \$200,000 is to be invested in the production process, what is the best way to divide it between capital and labor?

16 Rework problem 15 with

$$p = 4x^2y.$$

Let a unit of capital be \$30,000 and let a unit of labor cost \$20,000.

17 Determine the dimensions of a rectangular box with no top that has maximum volume and a surface area of 75 square inches.

18 Find the point (x, y) on the circle $x^2 + y^2 = 1$ such that the square of the distance from (x, y) to $(2, 3)$, that is, $(x - 2)^2 + (y - 3)^2$, is an absolute minimum. In other words, find the point on the circle that is closest to $(2, 3)$.

19 Find the point on the parabola $y^2 = x$ that is closest to $(\frac{1}{2}, 2)$. (**Hint:** Read problem 18.)

20 Find the point on the line $y = 2x$ that is closest to $(3, 1)$. (**Hint:** Read problem 18.)

Review of Terms

Review Problems

Sketch the graph of each function given below.

1 $f(x, y) = 3.2$

2 $g(x, y) = 2x + 1$

3 $f(x, y) = 4y^2$

4 $g(x, y) = \dfrac{x}{3} + 3y$

5 $h(x, y) = x^2 + y^2$

Compute all of the first order partial derivatives for each function below.

6 $f(x, y) = .4x^2 + .1x - y^2 + .5y + .8$
7 $g(x, y) = x^2 - 3xy + 5y^2 - \sqrt{11}$
8 $f(x, y) = (3x^2 + y^2)^{1/3}$

9 $g(x, y) = \dfrac{y^2 + 5y - 6}{x^2 + 4}$

10 $f(x, y) = \sqrt{\dfrac{x^2 + 1}{y^2 + 9}}$

11 $g(x, y) = xe^{x^2 + y}$
12 $f(x, y) = \ln(5x^2 + y^{-2})$
13 $g(x, y) = e^{y - \ln 2} \ln(x^2 + 2x + 1)$

Compute all of the second order partial derivatives for each function below.

14   $f(x, y) = \sqrt{13}\, e^3$
16   $h(x, y) = e^{x^2 + y^2}$

15   $g(x, y) = x^2 y + xy^2 - xy$

Find all of the critical values of each function below, and classify them to the extent possible using the Second Derivative Test.

17   $f(x, y) = x^2 + xy - 4y^2$
18   $g(x, y) = 4x^2 - xy + 3y^2$

19   $f(x, y) = \dfrac{3}{x} + y^2 - 6y + 18$

20   $g(x, y) = 8x - xy - \dfrac{4}{y}$

21   $h(x, y) = 2x^2 - 4xy + 2y^3 - y^2 - 12y + 7$

In each problem below, use Lagrange Multipliers to determine the indicated extremum for f subject to the given constraint.

22   absolute minimum, $f(x, y) = -5xy,\quad x + y = 1$
23   absolute maximum, $f(x, y) = xy,\quad x^2 + 4y^2 = 4$
24   absolute minimum, $f(x, y, z) = 2x + y + 2z,\quad x^2 + y^2 + z^2 = 4$
25   absolute maximum, $f(x, y, z) = x^2 + y^2 + z^2,\quad 4x^2 + y^2 + z^2 - 2z = 0$

Solve each of the following. Whenever an extremum is sought, assume that the desired extremum exists.

26   A rectangular box with a top and a bottom must have a volume of 45 cubic inches. The material for the sides costs 3¢ per square inch; the material for the top and bottom costs 5¢ per square inch. Determine the dimensions of the least expensive box.

27   Suppose a retailer sells $2000 - 10x$ units of the deluxe model of his featured product at a price of x dollars; he sells $10,000 - 25y$ units of his inexpensive model of the same product at a price of y dollars. Determine what x and y should be for maximum revenue.

28   Suppose annual production, p, is given as a function of capital, x, and labor, y, by

$$p = 60x^{6/5}y^{4/5}.$$

A unit of capital is \$15,000, and a unit of labor is a man-year, which costs \$12,000. If \$150,000 is to be invested in the production process, how should the funds be divided between capital and labor in order for p to be an absolute maximum?

29   Find the point on the circle $x^2 - 4x + y^2 = 0$ that is the closest to $(4, 2)$.

30   Determine the largest possible product of three positive numbers, provided the sum of the numbers is 3.

Social Science Applications

1 ***Law Enforcement*** The length d of the skid of a car is a function of the speed s and the skid resistance r. The formula is

$$d = .033 \frac{s^2}{r}.$$

The skid resistance is determined by the weight of the car, the steepness of the grade, and the condition of the tires. (The skid resistance is calculated by making some test skids at 30 miles per hour.) Determine d for $s = 60$ and $r = .5$.

2 ***Law Enforcement*** Using the facts and notation in problem 1, determine d for $s = 60$ and $r = 1$.

3 ***Law Enforcement*** Use the facts and notation in problem 1. Determine the rate of change of d as s changes and r is fixed for $s = 60$ and $r = 1$.

4 ***Law Enforcement*** Use the facts and notation in problem 1. Determine the rate of change of d as s changes and r is fixed for $s = 100$ and $r = 1$.

5 ***Psychology*** A behavioral psychologist tests white mice with a maze to determine how many inches along the correct path a mouse can reach in a fixed period of time. After many trials, she concludes that the distance y covered along the correct path is a function of the time allowed for the test t and the total amount of time x that the mouse has previously spent in the maze. Specifically, we have

$$y = ct(\sqrt{x} + 1)$$

where c is a constant that varies from one mouse to another. For a mouse with $c = 2$, determine y for $t = 5$ and $x = 0$.

6 ***Psychology*** Using the facts and notation in problem 5, determine y for $c = 1$, $t = 5$, and $x = 16$.

7 ***Psychology*** Use the facts and notation in problem 5. Determine the rate of change of y as t changes and x is fixed for $c = 1$, $t = 10$, and $x = 0$.

8 ***Psychology*** Use the facts and notation in problem 5. Determine the rate of change of y as x changes and t is fixed for $c = 1$, $t = 5$, and $x = 9$.

9 ***Sociology*** A sociologist studies the growth of a certain sprawling sunbelt city. He observes that there are two independent factors in the growth. The city is continually adding to its total area by adding suburbs that were previously excluded, and the number of people living on each acre of city ground is steadily increasing. The 1980 population was one million, and the sociologist concludes that the subsequent population p (in thousands) is given by

$$p = 1,000 + 10t^2 d,$$

where t is the amount of time (in years) past January 1, 1980, and d is the average number of people permitted to live on one acre of city ground according to current zoning regulations. Determine p for $t = 1$ and $d = .4$.

10 ***Sociology***   Using the facts and notation in problem 9, determine p for $t = 5$ and $d = .8$.

11 ***Sociology***   Use the facts and notation in problem 9. Determine the rate of change of p as t changes and d is fixed for $t = 4$ and $d = .5$.

12 ***Sociology***   Use the facts and notation in problem 9. Determine the rate of change of p as d changes and t is fixed for $t = 4$ and $d = .5$.

CAREER PROFILE
Engineer (Industrial)

Engineers take scientific and mathematical theories and procedures and use them to solve practical problems. They work to design machinery, products, systems, and processes for efficient and economical use.

Altogether there are over 25 engineering specialties; those that contain the largest numbers are electrical, mechanical, civil, chemical, and aerospace.

A bachelor's degree in engineering is the usual prerequisite for beginning engineering jobs. However, a college graduate with a degree in a natural science or mathematics also qualifies for many entry-level positions. As with most fields, an advanced degree is desirable for gaining promotion. Graduate work often is necessary to learn new technology; in fact, some specialties such as nuclear, environmental, and biomedical engineering are taught mainly at the graduate level.

All engineering students need to take courses in the physical sciences, and several mathematics courses are essential as well: linear algebra, differential and integral calculus, and differential equations. Many practicing engineers express the wish that they had taken more mathematics courses as part of their undergraduate preparation. Mathematical topics such as probability and statistics, Fourier analysis, and tensor analysis provide valuable engineering tools. The training of electrical engineers requires an especially heavy concentration of quantitative courses. Discrete mathematics, graph theory, and information theory are subjects that offer important background for the design and development of computing and communications equipment.

Computer science courses have become staples of an engineering curriculum; in particular, the engineer needs to know how to develop and evaluate computer algorithms that accomplish the mathematical problem-solving methods that are the core of engineering education.

Industrial engineers are the group of engineers with the widest variety of employment opportunities. They have the responsibility to determine the most effective ways for their organization to use its people, machines, and materials. Unlike other engineers, who deal with equipment and structures and products and processes, industrial engineers are concerned with people and with methods of business. College courses in business management and operations research are important ingredients of an industrial engineer's career preparation.

Among the variety of tasks that an industrial engineer may be expected to perform are the following:

1. design data processing systems;
2. develop management control systems to aid in financial planning and cost analysis;
3. design quality control procedures;
4. design systems for the distribution of goods and services;
5. select optimal locations for plants and warehouses, looking for the best combination of sources of raw materials, labor, transportation, and taxes;

6. develop wage and salary administration systems and job evaluation programs.

As the list indicates, industrial engineers need to take college courses in data processing and management. Often these engineers are ideally situated to advance to management positions, because their engineering duties are closely related to management functions. Industrial engineers are employed by more different industries than individuals in other branches of engineering. For example, they work for insurance companies, banks, construction and mining firms, public utilities, hospitals, retail organizations, and government agencies.

Engineering is second only to teaching as the largest profession in the United States. In 1980 about 1.2 million persons were employed as engineers (with over 300,000 in electrical and over 100,000 in industrial engineering). In that year, average starting salaries for engineers exceeded $20,000. Experienced engineers can earn over $35,000, and those who leave engineering to advance to management earn much more.

Sources of Additional Information
Information on engineering careers, including engineering school requirements, is available from these organizations:

Engineering Manpower Commission of America, Association of Engineering Societies, 345 E. 47th Street, New York, NY 10017. (At this same address, one may contact the Society of Women Engineers for special information about engineering opportunities for women.)
National Society of Professional Engineers, 2029 K Street, N.W., Washington, DC 20006.

Occupational Outlook Handbook. Bureau of Labor Statistics, U.S. Department of Labor, Washington, DC 20212. (Revised every 2 years, this handbook provides information about job duties, working conditions, level and places of employment, education and training requirements, advancement possibilities, job outlook, and earnings for about 250 occupations. Some of the information for this Career Profile was obtained from this source.)

13 | Sets and Counting

13.1   Sets and Related Operations

The terminology of sets to be introduced in this section is part of the basic language of probability and some other areas of mathematics. A **set** is a collection of objects. Although the objects may be anything (cities, people, movies), we shall usually be interested in sets that are collections of numbers. The set that consists of the numbers 0, 2, and 4 can be written

$$\{0, 2, 4\};$$

here, the numbers belonging to the set are listed between the braces { }. Each object that is in a given set is called a **member** or **element** of the set. The numbers 0, 2, and 4 are the elements of the set exhibited above. Capital letters, such as A, B, and C, are often used to designate particular sets. We write $a \in A$ to indicate that a is an element of the set A; we say that *a* **belongs to** *A*. For example, it is clear that

$$0 \in \{0, 2, 4\}.$$

We write $b \not\in A$ to indicate that b is not an element of the set A. For example, it is clear that

$$1 \not\in \{0, 2, 4\}.$$

If a set has no elements, then it is called the **null set** or **empty set,** and it is denoted by $\varnothing$. (This is not the same as the number zero or the set containing only zero; that set has one member, 0, so it is not empty.)

Sometimes a set is indicated by enclosing within the braces a property that indicates the desired elements. For example, the expression

$$\{x \mid x \text{ is an even nonnegative integer less than 5}\}$$

is read "the set of all x such that x is an even nonnegative integer less than 5"; note that the vertical bar means "such that." We can verify by listing all of the members that this is the same set given earlier.

EXAMPLE 1

Define a set such that 1 belongs to the set. Define another set such that every nonnegative integer belongs to the set.

Solution

There are many sets that have 1 as an element. A very simple set that has 1 as an element is

$\{1\}.$

The above set has no elements except 1!

A set that has every nonnegative integer as an element is

$\{x|x$ is a nonnegative integer$\}.$

This same set is sometimes given with the notation

$\{0, 1, 2, \ldots\},$

where the three dots indicate that the remaining elements continue the pattern of the elements already given. ■

If every element of the set A is an element of B, then we say that A is a **subset** of B, and we write

$A \subset B.$

For any set B, the statements below are true.

$B \subset B, \quad \phi \subset B.$

If the sets A and B have exactly the same elements, then we say that they are **equal,** and we write

$A = B.$

It does not matter whether the elements of A are listed in the same order as the elements of B. For example, we have

$\{0, 2, 4\} = \{4, 2, 0\}.$

In fact, it does not matter whether an element is listed more than once. For example, we have

$\{0, 2, 4\} = \{0, 2, 0, 4, 0\}.$

All that matters is that A and B have exactly the same elements.

If A is a subset of C and A is not equal to C, then we say A is a **proper subset** of C. It is easy to see that if A is a proper subset of C, then C must have at least one element that does not belong to A. In fact, **two sets A and B are equal precisely when the statement**

$A \subset B \quad \text{and} \quad B \subset A$

is true.

EXAMPLE 2

List all possible subsets of the sets $\{a, b\}$ and $\{0, 2, 4\}$.

Solution
The subsets of $\{a, b\}$ are

$$\varnothing, \quad \{a\}, \quad \{b\}, \quad \{a, b\}.$$

Note that the given set has two elements, and there are four subsets.
    The subsets of $\{0, 2, 4\}$ are

$$\varnothing, \quad \{0\}, \quad \{2\}, \quad \{4\}, \quad \{2, 4\}, \quad \{0, 4\}, \quad \{0, 2\}, \quad \{0, 2, 4\}.$$

Note that the given set has three elements, and there are eight subsets. ■

        In any discussion or problem, the set of all elements under considera-
tion is called the **universal set**; it is denoted by U. If A is a subset of U, then the
complement of A, denoted A', is the set of all elements that are in U and
not in A.
        The set of all elements belonging to either A or B is called the **union** of A
and B; it is denoted $A \cup B$. For example, the union of $A = \{0, 2, 4\}$ and
$B = \{1, 2, 3\}$ is

$$A \cup B = \{0, 1, 2, 3, 4\}.$$

The **intersection** of A and B, denoted $A \cap B$, is the set of all elements belonging to both A and B. For example, the intersection of the sets A and B above is
$A \cap B = \{2\}$. In the picture on the left, we have represented A and B with two
circles and we have shaded $A \cup B$. In the picture on the right, we have shaded
$A \cap B$. In the pictures, the universal set U consists of all the elements being
considered in whatever problem we are trying to solve.

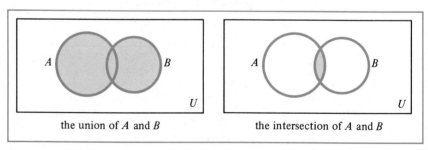

the union of A and B the intersection of A and B

        Pictures, like the preceding ones, used to understand the relation of
some sets are called **Venn diagrams**. If A and B have no elements in common,
then their intersection is the null set. For example, when $A = \{0, 2, 4\}$ and
$B = \{1, 3\}$, we have $A \cap B = \varnothing$.
        The next example illustrates the ideas of complement, union, and
intersection.

EXAMPLE 3

Let $A = \{1, 2, 3, 4, 5\}$, $B = \{2, 3, 4\}$, and $C = \{a, d, f\}$. Determine each of
the following sets:

$$A \cup B, \quad B \cup C, \quad A \cap B, \quad B \cap C, \quad B', \quad A'.$$

Solution

Since the universal set U consists of all elements under consideration, we have

$$U = \{1, 2, 3, 4, 5, a, d, f\}.$$

Below, we indicate the desired sets.

$$A \cup B = \{1, 2, 3, 4, 5\}$$

$$B \cup C = \{2, 3, 4, a, d, f\}$$

$$A \cap B = \{2, 3, 4\}$$

$$B \cap C = \phi$$

$$B' = \{1, 5, a, d, f\}$$

$$A' = \{a, d, f\} \ \blacksquare$$

The next example shows how more complicated sets can be represented by Venn diagrams.

EXAMPLE 4

Indicate the set $(A \cap B') \cup C$ in a Venn diagram.

Solution

In the left picture of the next illustration, we shade the set $A \cap B'$; in the right picture, we shade the set $(A \cap B') \cup C$, where A, B, and C are represented by the indicated circles.

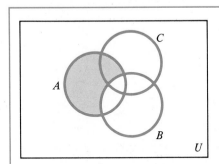 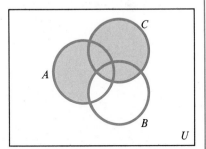 ∎

It is easy to see from the definitions or from Venn diagrams that

$$(A \cup B) \cup C = A \cup (B \cup C)$$

and

$$(A \cap B) \cap C = A \cap (B \cap C).$$

Consequently, we write

$$A \cup B \cup C \quad \text{and} \quad A \cap B \cap C;$$

it does not matter how these two sets are computed since the result is the same.

The following data will be used in Example 5. The rate of inflation is measured by the Consumer Price Index, and the rate of increase in the money supply is measured by M1.

Year	Inflation Rate (%)	Money Supply Growth Rate (%)
1970	5.9	6.7
1971	4.3	7.1
1972	3.3	7.3
1973	6.2	4.9
1974	11.0	4.6
1975	9.1	5.5
1976	5.8	7.5
1977	6.5	8.2
1978	7.6	7.8
1979	11.5	6.4
1980	13.4	6.8

EXAMPLE 5

Let the sets A, B, C, and D be defined by

$A = \{x|x$ is a year with inflation rate exceeding 6%$\}$

$B = \{x|x$ is a year with inflation rate exceeding 9%$\}$

$C = \{x|x$ is a year with money supply growth exceeding 6%$\}$

$D = \{x|x$ is a year with money supply growth exceeding 7%$\}$.

Describe each of the sets A, B, C, D, $A \cap C$, and $B \cup D$ by listing the elements between braces.

Solution
From the preceding data we determine A, B, C, and D.

$A = \{1973, 1974, 1975, 1977, 1978, 1979, 1980\}$

$B = \{1974, 1975, 1979, 1980\}$

$C = \{1970, 1971, 1972, 1976, 1977, 1978, 1979, 1980\}$

$D = \{1971, 1972, 1976, 1977, 1978\}$

Using these descriptions of A, B, C, and D, we determine $A \cap C$ and $B \cup D$.

$A \cap C = \{1977, 1978, 1979, 1980\}$

$B \cup D = \{1971, 1972, 1974, 1975, 1976, 1977, 1978, 1979, 1980\}$ ■

Exercises   13.1

In problems 1 through 6, define a set with the indicated property.

1 The number 0 belongs to the set, and the set has only one element.

2 The numbers 5 and 6 belong to the set, and the set has only two elements.

3 Every nonnegative integer divisible by 3 belongs to the set, and every element of the set is a nonnegative integer divisible by 3.

4 Every nonnegative integer divisible by 5 belongs to the set, and every element of the set is a nonnegative integer divisible by 5.

5 The set contains the first five letters of the alphabet and no other elements.

6 The set contains the last three letters of the alphabet and no other elements.

Determine all possible subsets of each set below.

7 $\varnothing$ 8 $\{z\}$ 9 $\{0, 5\}$ 10 $\{a, b, 3\}$ 11 $\{4, 5, 6, e\}$

Let $A = \{1, 2, 3, 4, 5, 6, 7, 8\}$, $B = \{1, 3, 5, 7\}$, $C = \{2, 4, 6, 8\}$, and $D = \{1, 2, 5, 6\}$. Determine each set indicated below.

12 $A \cup B$	13 $A \cup C$
14 $B \cup C$	15 $B \cup D$
16 $A \cap D$	17 $B \cap D$
18 $B \cap C$	19 A'
20 B'	21 C'
22 D'	23 $B' \cup C$
24 $B \cup C'$	25 $A' \cap D$
26 $A' \cap C$	27 $D' \cup B$
28 $D' \cup C$	29 $(B \cap A) \cup C$
30 $(C \cap A) \cup B$	31 $(B \cup C) \cup D$
32 $(B \cup C) \cap D$	

Let $U = \{$all people$\}$, $C = \{$all people good at their jobs$\}$, $E = \{$all lazy people$\}$, $I = \{$all college instructors$\}$, and $S = \{$all college students$\}$. Describe each set below by identifying precisely those properties possessed by the elements of the set.

33 $I \cap C$	34 $I \cap C'$
35 $S \cap C$	36 $S \cap C'$
37 $S \cup E$	38 $I \cap E$
39 $E \cap S \cap C$	40 $(E' \cap I) \cup S$

Use a Venn diagram, as in Example 4, to indicate each of the sets below.

41 $A \cap B \cap C$	42 $A \cap B \cap C'$
43 $A' \cap B' \cap C$	44 $(A \cup B)' \cap C$
45 $A' \cap B' \cap C'$	46 $(A \cup B \cup C)'$
47 $(A' \cup B) \cap C$	48 $(A' \cup B') \cap C$
49 $A' \cup B' \cup C'$	50 $(A \cap B \cap C)'$

51 Define A, B, C, and D as in Example 5, and define E by

$$E = \{x + 2 | x \text{ is a year with money supply growth exceeding } 7\%\}.$$

Describe the sets $B \cap E$ and $B \cap D$ by indicating the elements of each.

13.2 Counting

Each of us learned to count as a young child. First, we learned the positive integers, 1, 2, 3, . . . , and then we learned to use those numbers to determine how many objects there were in some given set. We determined how many coins were on a given table or how many people were in a given picture. We count the coins on the table by picking out one coin and saying the number 1; then, we pick out another coin and we say the number 2; this process continues until we have run out of coins on the table. The last positive integer that we speak is the number of coins on the table. In order to handle practical problems like those given in Examples 3 and 4 of this section, we must achieve greater skill and precision at counting.

Definition We say that the set **S has m elements,** written $n(S) = m$, provided we can match the distinct elements of S with the elements of $\{1, 2, \ldots, m\}$ so that we exhaust the elements of S and do not use any element twice. If the set S has m elements, for some positive integer m, then S is said to be a **finite set.** Any set that is not finite is said to be **infinite.** The empty set is a finite set with 0 elements.

Now we illustrate how the concepts in the preceding definition are used.

EXAMPLE 1

Determine which of the following sets are finite and which are infinite. Identify the number of elements in each finite set.

$$A = \{a, b, c, \ldots, x, y, z\}, \quad B = \{2, 4, 6, \ldots\}, \quad D = \left\{ .5, \frac{1}{2}, \frac{.3}{.6} \right\},$$

$$C = \{x \,|\, x \text{ is a real number} \quad \text{and} \quad 2 < x < 4\}.$$

Solution
We recognize that set A consists of the letters of the alphabet. Thus, A is finite and $n(A) = 26$. Set B consists of all of the positive even integers. Thus, B is an infinite set. Set C consists of all the real numbers greater than 2 and less than 4; this is an infinite set. Set D has only one distinct element, since $.5 = \frac{1}{2} = \frac{.3}{.6}$. Thus, we have $n(D) = 1$, despite the fact that three symbols appear between the braces in the definition of D. ∎

The next theorem states one of the many quantitative formulas that relate the numbers of elements in newly constructed sets to the numbers of elements in the original given sets.

Theorem If A and B are finite sets, then

$$n(A \cup B) = n(A) + n(B) - n(A \cap B).$$

We do not prove the theorem. Instead, we examine Venn diagrams that suggest the truth of the theorem. To count the number of distinct elements in $A \cup B$, we first count the elements in A and the elements in B. Then, we add $n(A)$ to $n(B)$, and we subtract the number of elements that were counted twice, namely $n(A \cap B)$. Each element of $A \cap B$ was counted once when we counted A, and again when we counted B.

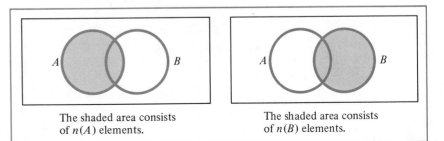

The shaded area consists of $n(A)$ elements.

The shaded area consists of $n(B)$ elements.

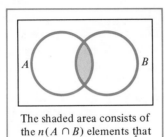

The shaded area consists of the $n(A \cap B)$ elements that were counted twice.

The next example shows how to use a Venn diagram to determine the numbers of elements in complicated sets.

EXAMPLE 2

Suppose set A has 8 elements, B has 12 elements, and C has 10 elements. Furthermore, suppose that the numbers of elements in $A \cap B$, $A \cap C$, $B \cap C$, and $A \cap B \cap C$ are 4, 2, 3, and 1, respectively. Determine the numbers of elements in $A \cap B' \cap C'$, $B \cap A' \cap C'$, and $C \cap A' \cap B'$.

Solution
We begin by drawing a Venn diagram that shows the three sets A, B, and C.

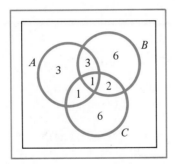

We have written the number 1 in the region that represents $A \cap B \cap C$, since $n(A \cap B \cap C) = 1$. Since $n(A \cap B) = 4$, we put 3 in the part of $A \cap B$ outside of $A \cap B \cap C$. Similarly, we put 1 in the part of $A \cap C$ outside of $A \cap B \cap C$; we put 2 in the part of $B \cap C$ outside of $A \cap B \cap C$. Since A has 8 elements and the regions inside A that are already labeled with numbers have 5 elements all together, it must be that the remaining region of A has 3 elements. Similarly, we fill in the last numbers for B and C.

The set $A \cap B' \cap C'$ is represented by the region inside A and outside of both B and C. Thus, we know that $n(A \cap B' \cap C')$, is 3. Similarly, we observe that $n(B \cap A' \cap C')$ and $n(C \cap A' \cap B')$ are each equal to 6. ■

The technique used in the previous example has many applications, as we illustrate in the next example.

EXAMPLE 3

A certain publisher is interested in the reading habits of the people in a certain small city. A survey indicates that the number of subscribers (in thousands) to *Time* is 25, the number of subscribers to *Newsweek* is 18, and the number of subscribers to *U.S. News and World Report* is 12. The number of subscribers to both *Time* and *Newsweek* is 10, the number subscribing to both *Time* and *U.S. News and World Report* is 1, and the number subscribing to both *Newsweek* and *U.S. News and World Report* is 2. The number of people with subscriptions to all three magazines is 1. Determine the number of people subscribing to exactly one of these magazines.

Solution

This problem is similar to Example 2 with A, B, and C denoting the sets of people subscribing to *Time, Newsweek,* and *U.S. News and World Report,* respectively. Again, we make a Venn diagram, and we indicate the numbers of elements in each region of the Venn diagram. We begin with $A \cap B \cap C$; then, we go to $A \cap B \cap C'$, $A \cap C \cap B'$, and $B \cap C \cap A'$. Finally, we indicate the numbers of elements in $A \cap B' \cap C'$, $B \cap A' \cap C'$, and $C \cap A' \cap B'$.

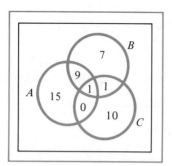

The total number of people subscribing to exactly one of these periodicals is $15 + 7 + 10 = 32$ (thousand). ■

The next example presents another applied problem that is similar in method to Example 3.

EXAMPLE 4

A trade association wants to determine the number of families who own more than one television set and do not own either a video game or a video recorder. These families are regarded as potential purchasers of video devices. There is a brief questionnaire attached to the warranty notification card enclosed with each video game and recorder sold. Using the questionnaires from a certain urban area, the trade association determines the numbers of families (in thousands) who own certain combinations of equipment. Below we list each combination along with the number of owners (in thousands):

(a) a video recorder, 12;
(b) a video recorder, no video game, and only one television set, 2;
(c) a video game, 60;
(d) a video game, no video recorder, and only one television set, 15;
(e) a video recorder and a video game, 9;
(f) a video recorder, a video game, and more than one television set, 8;
(g) more than one television set, 50.

Determine the number sought by the trade association.

Solution

Let the set A be all families who own a video recorder, B those who own more than one television set, and C those who own a video game. We make a Venn diagram and indicate the number of families represented by each region, one by one.

The set suggested by (f) has the characteristics of all three sets, so we know that $n(A \cap B \cap C) = 8$; this goes in the central region of the diagram. Next, we note that the set suggested by (e) has the properties of $A \cap C$, and it has 9 members; 8 of those are in the central region, so $n(A \cap C \cap B') = 9 - 8 = 1$.

Now, the set suggested by (a) is exactly A, so $n(A) = 12$. Also, the set suggested by (b) is $A \cap B' \cap C'$, with 2 members. Thus, the region in A that we have not yet filled in must have $n(A \cap B \cap C') = 12 - 8 - 1 - 2 = 1$ element. Similarly, the set suggested by (c) is exactly C, with $n(C) = 60$, so the remaining region in C has $n(C \cap B \cap A') = 60 - 15 - 8 - 1 = 36$ elements.

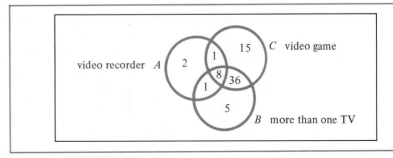

Finally, the set suggested by (g) is exactly B, with $n(B) = 50$, so we compute $n(B \cap A' \cap C') = 50 - 36 - 8 - 1 = 5$.

We have discovered that 5000 families own more than one television set and do not own either a video recorder or a video game. ■

The counting skills developed in this section and the next two sections will be used extensively in the chapter on probability.

Exercises   13.2

Determine whether each set below is finite or infinite. If the set is finite, give the number of elements.

1 $\{1, a, b, 6\}$
2 $\{d, x, w, 4\}$

3 $\left\{2, c, y, \dfrac{10}{5}\right\}$

4 $\left\{7, \dfrac{1}{4}, w, .125\right\}$

5 $\{1, a, 2, a, 3, \ldots\}$
6 $\{-4, -5, -6, \ldots\}$
7 $\{-5, -10, -15, \ldots\}$
8 $\{4, 8, 12, \ldots\}$
9 $\{x \mid x^2 - 4 = 0\}$
10 $\{x \mid 9x^2 - 1 = 0\}$
11 $\{x \mid x^2 + 5 = 0 \text{ and } x \text{ is a real number}\}$
12 $\{x \mid 3x^2 + 2 = 0 \text{ and } x \text{ is a real number}\}$
13 $\{x \mid \sqrt{x} \text{ is an integer}\}$
14 $\{x \mid \sqrt{x} \text{ is an even integer}\}$

Suppose A has 20 elements, B has 15 elements, and C has 18 elements. Furthermore, suppose that the numbers of elements in $A \cap B, A \cap C, B \cap C$, and $A \cap B \cap C$ are 8, 10, 6, and 3, respectively. Determine the number of elements in each set below.

15 $A \cap B' \cap C'$ 16 $B \cap A' \cap C'$
17 $C \cap A' \cap B'$ 18 $A \cap B \cap C'$
19 $A \cap C \cap B'$ 20 $B \cap C \cap A'$

Suppose $A \cap B' \cap C'$ has 10 elements, $B \cap A' \cap C'$ has 6 elements, and $C \cap A' \cap B'$ has 8 elements. Furthermore, suppose that the numbers of elements in $A \cap B \cap C', A \cap C \cap B', B \cap C \cap A'$, and $A \cap B \cap C$ are 8, 3, 6, and 1, respectively. Determine the number of elements in each set below.

21 A 22 B
23 C 24 $A \cap B$
25 $A \cap C$ 26 $B \cap C$

Suppose A has 45 elements, B has 60 elements, and C has 50 elements. Furthermore, suppose that the numbers of elements in $A \cap B \cap C'$, $A \cap C \cap B'$, $B \cap C \cap A'$, and $A \cap B \cap C$ are 10, 15, 16, and 8, respectively. Determine the number of elements in each set below.

27 $A \cap B' \cap C'$ 28 $B \cap A' \cap C'$

29 $C \cap A' \cap B'$ 30 $A \cap B$

31 $A \cap C$ 32 $B \cap C$

33 The manager of a small fast food restaurant kept a record of his customers' breakfast orders for one week. He discovered the following:

200 ordered eggs,
180 ordered sausage,
150 ordered hotcakes,
100 ordered eggs and sausage,
 80 ordered eggs and hotcakes,
 60 ordered sausage and hotcakes,
 30 ordered sausage, eggs, and hotcakes.

Determine the number of customers

a. who ordered eggs only,
b. who ordered sausage only,
c. who ordered hotcakes only,
d. who ordered eggs and sausage only,
e. who ordered eggs and hotcakes only,
f. who ordered sausage and hotcakes only.

34 Some enrollment totals at Rockford College for the fall quarter are the following:

150 enrolled in statistics,
640 enrolled in mathematics,
310 enrolled in computer science,
 90 enrolled in both statistics and mathematics,
 60 enrolled in both statistics and computer science,
200 enrolled in both mathematics and computer science,
 50 enrolled in all three areas.

Determine the numbers of students enrolled in

a. statistics only,
b. mathematics only,
c. computer science only,
d. statistics and mathematics, and not computer science
e. computer science and mathematics, and not statistics
f. computer science and statistics, and not mathematics.

35 A research organization is studying the impact of increasing gasoline prices on automobile purchases. A survey of some homes indicates that the respondents own the following cars:

70 own large cars exclusively,
560 own medium sized cars exclusively,
570 own small cars exclusively,
150 own both large and medium sized cars,
200 own both large and small cars,
180 own both medium sized and small cars,
70 own at least one car of each size.

Determine how many of the respondents

a. own a large car,
b. own a medium sized car,
c. own a small car,
d. own large and medium sized cars, but no small car,
e. own large and small cars, but no medium sized car,
f. own medium sized and small cars, but no large car.

13.3 Permutations and the Multiplication Principle

We begin with an example of a common type of counting problem.

EXAMPLE 1

A busy executive must travel from Atlanta to Boston, and then to Chicago. The travel agent finds four acceptable flights from Atlanta to Boston and three acceptable flights from Boston to Chicago. Determine the total number of possible routes for the full trip.

Solution
We use a picture to organize the possible routes. Let A, B, and C represent Atlanta, Boston, and Chicago, respectively. We show four curves, representing the possible flights, from A to B and three curves from B to C.

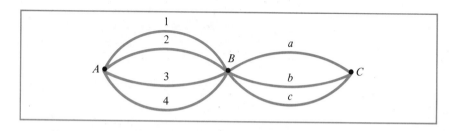

We label each flight from A to B with a number, and we designate each flight from B to C with a letter. From A to C, there are twelve possible routes symbolized by

$$1a, \quad 2a, \quad 3a, \quad 4a, \qquad 1b, \quad 2b, \quad 3b, \quad 4b, \qquad 1c, \quad 2c, \quad 3c, \quad 4c.$$

Here, $4b$ means use flight 4 from A to B and then flight b from B to C. ∎

The preceding example illustrates the Multiplication Principle for a two-step task, which we now state.

Multiplication Principle for Two-Step Tasks   Suppose the completion of a task requires two separate consecutive steps. If the first step can be completed in m ways and, for each of these, the second step can be completed in n ways, then the full task can be completed in $m \cdot n$ ways.

In Example 1, we discovered the Multiplication Principle by drawing a picture appropriate to our problem. Now, we solve another example by direct application of the Multiplication Principle.

EXAMPLE 2

The board of directors of a corporation must choose a new chairman and a new president from three candidates named Adams, Brooks, and Cole. How many different ways can the board fill these two positions?

Solution
The task confronting the board of directors consists of two steps. The first step is to select a new chairman from the three candidates, and it can be completed in three ways. The second step is to choose a president from the remaining two candidates, and it can be completed in two ways. According to the Multiplication Principle, the board can complete its full task in $3 \cdot 2 = 6$ ways. ■

Even in problems like Example 2, it is possible to construct a helpful picture. The next picture is called a **tree diagram,** and it illustrates how the board of directors, in Example 2, can choose a chairman and a president from the candidates, Adams, Brooks, and Cole.

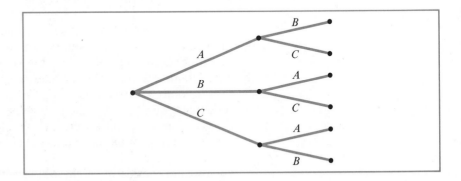

A, B, and C represent Adams, Brooks, and Cole, respectively. Reading left to right, the first three lines represent the three possible choices for chairman. The top line indicates that A is chosen; the middle line indicates that B is chosen; the bottom line indicates that C is chosen. After choosing A for chairman, either B or C could be chosen president. In the picture, that is symbolized by the top line branching out into lines labeled B and C. Similarly, the middle and bottom lines branch out to indicate the remaining possible choices. A tree diagram can be used to depict the possible ways to complete any task that consists of a finite sequence of steps.

It should not be surprising that the Multiplication Principle applies to tasks that have more than two consecutive steps. Below we state the general principle.

General Multiplication Principle   Suppose the completion of a task requires the completion of p separate consecutive steps. Suppose the first step can be completed in m_1 ways; for each of these, the second step can be completed in m_2 ways; for each of these, the third step can be completed in m_3 ways; etc. Then the full task can be completed in

$$m_1 \cdot m_2 \cdot m_3 \cdot \cdots \cdot m_p$$

ways.

In the next example, we determine the number of ways to complete a three-step task, by applying the General Multiplication Principle.

EXAMPLE 3

Determine how many ordered triples (a, b, c) can be formed with a, b, and c chosen from the set $\{1, 2, 3, 4, 5, 6, 7, 8, 9\}$.

Solution
The three steps of this task are the choices of a, b, and c. There are 9 ways to choose a; after a is chosen, there are 8 possible choices for b; after a and b have been chosen, there remain 7 digits that might be selected for c. Thus, the full task can be completed in

$$9 \cdot 8 \cdot 7 = 504$$

ways, so 504 different ordered triples can be formed. ■

The ordered triples formed in Example 3 are called the permutations of the 9 symbols, 1, 2, . . . , 8, 9, taken 3 at a time. A **permutation of n objects taken r at a time** is an arrangement of r of the n objects in a specific order. The problem in Example 2 amounted to determining the number of permutations of the 3 "objects," Adams, Brooks, Cole, taken 2 at a time. Here it is understood that the first object in the arrangement is the choice for chairman and the second object in the arrangement is the choice for president. The tree diagram after Example 2 shows graphically all of the permutations for the problem in Example 2.

When we are considering permutations either explicitly or implicitly, as in Examples 2 and 3, we frequently encounter products like $3 \cdot 2 \cdot 1$. For the sake of brevity, we define **n factorial,** denoted **$n!$**, to be the product

$$n(n - 1)(n - 2) \cdot \cdot \cdot (2)(1).$$

Thus, n factorial is the product of the first n positive integers. It is easy to see that $n(n - 1)! = n!$ for any positive integer $n \geq 2$; in order to extend this relation to $n = 1$, **we define 0! to be 1.** Now we can state a general theorem that identifies the number of permutations that can be formed from a given set.

Permutation Theorem If $P(n, r)$, with $r \le n$, denotes the number of permutations of n elements taken r at a time, then

$$P(n, r) = \frac{n!}{(n - r)!} = n(n - 1)(n - 2) \cdots (n - r + 2)(n - r + 1).$$

To see that the second part of this equality is true, notice the cancellation of factors in the numerator and denominator:

$$\frac{n!}{(n - r)!} = \frac{n(n - 1) \cdots (n - r + 1)(n - r)(n - r - 1) \cdots (2)(1)}{(n - r)(n - r - 1) \cdots (2)(1)}$$

For example,

$$P(5, 2) = \frac{5!}{(5 - 2)!} = \frac{5!}{3!} = \frac{5 \cdot 4 \cdot 3 \cdot 2 \cdot 1}{3 \cdot 2 \cdot 1} = 5 \cdot 4 = 20.$$

The truth of the Permutation Theorem can be established by using the General Multiplication Theorem, although we will not do it here.

EXAMPLE 4

Determine the number of permutations of 10 objects taken 5 at a time; determine the number of permutations of 5 objects taken 2 at a time.

Solution
In view of the Permutation Theorem, we need only compute the numbers

$$P(10, 5) \quad \text{and} \quad P(5, 2).$$

We note that

$$P(10, 5) = \frac{10!}{(10 - 5)!} = \frac{10!}{5!} = 10 \cdot 9 \cdot 8 \cdot 7 \cdot 6 = 30{,}240$$

and

$$P(5, 2) = \frac{5!}{(5 - 2)!} = \frac{5!}{3!} = 5 \cdot 4 = 20. \ \blacksquare$$

Now we demonstrate how the Permutation Theorem facilitates the handling of practical problems.

EXAMPLE 5

A retailer has a display case with a highly visible top shelf, a moderately visible middle shelf, and a hard-to-see bottom shelf. The retailer must choose 3 perfumes from 8 that are offered for display. How many different ways can the choice be made?

Solution
In this problem, the retailer chooses 3 objects in a specific order from a set of 8 objects. Thus, the number of ways to make the choice is the number of permutations of 8 objects taken 3 at a time. The answer is

$$P(8, 3) = \frac{8!}{5!} = 8 \cdot 7 \cdot 6 = 336. \ \blacksquare$$

We conclude this section by noting that some problems that cannot be solved by means of the Permutation Theorem can be solved using the General Multiplication Principle. Example 1, shown before, and Example 6, below, illustrate this fact.

EXAMPLE 6

A radio manufacturer uses a 5-digit serial number to identify each radio that it produces. How many radios can be manufactured before this scheme must be modified?

Solution

Constructing a serial number is a five-step task. Each step consists of choosing one of the digits, 0, 1, . . . , 8, 9. Since each step can be completed 10 ways, the construction of a serial number can be completed in

$$10 \cdot 10 \cdot 10 \cdot 10 \cdot 10 = 10^5 = 100,000$$

ways.

Because each of the digits 0, 1, . . . , 8, 9 can be used repeatedly, we are constructing more than just the permutations of 10 objects taken 5 at a time. If we were restricted, for some reason, to serial numbers in which all of the digits were different, the answer would be the number of permutations of 10 digits taken 5 at a time, or

$$P(10, 5) = 10 \cdot 9 \cdot 8 \cdot 7 \cdot 6 = 30,240.$$

This is much smaller than the 100,000 possibilities with repeated digits. ■

Exercises   13.3

Determine the value of each number indicated below.

1	0!	2	4!
3	7!	4	8!
5	$P(8, 1)$	6	$P(7, 1)$
7	$P(5, 2)$	8	$P(6, 2)$
9	$P(6, 4)$	10	$P(5, 3)$
11	$P(10, 7)$	12	$P(9, 6)$

Determine how many ordered triples can be formed using the elements of each set below.

13	{2, 4, 6}	14	{1, 3, 5}
15	{5, 6, 7, 8, 9}	16	{1, 2, 3, 4, 5}
17	{1, a, 2, b, 3, c, 4, d}	18	{7, 8, 9, w, x, y, z}

Use the General Multiplication Principle to solve each problem below.

19   If there are 3 paths from A to B and 5 paths from B to C, then how many ways can you go from A to C?

20 If there are 7 paths from A to B and 5 paths from B to C, then how many ways can you go from A to C?

21 How many serial numbers can be constructed, if each consists of a letter of the alphabet followed by 4 single-digit numbers?

22 How many serial numbers can be constructed, if each consists of 2 letters of the alphabet followed by 3 single-digit numbers?

23 How many 7-digit telephone numbers are there that begin with 542?

24 How many 7-digit telephone numbers are there that end with 11?

Solve each of the following problems.

25 The board of directors of a corporation must choose a new president, executive vice president, and comptroller. If there are 7 candidates for these 3 positions, how many ways can the positions be filled?

26 A political party must choose its candidates for president and vice president. If there are 10 people being considered for these two ballot positions, how many ways can the choices be made?

27 A display window has 4 distinct sections. If there are 12 possible exhibits that might be used in any section of the display window, how many different ways could the window be filled?

28 If a display window has 3 distinct sections and there are 5 possible exhibits to use in the window, how many different ways can the window be filled?

29 An accounting test has 10 true-false questions followed by 10 multiple choice questions, each with 3 possible answers. How many ways could the test be answered?

30 An English test has 12 true-false questions followed by 8 multiple choice questions, each with 3 possible answers. How many ways could the test be answered?

31 A man has 5 sports coats, and for each sports coat he has 3 compatible pairs of slacks. How many different ways can he combine his sports coats and slacks?

32 A woman has 6 blouses, and for each blouse she has 4 compatible skirts. How many different ways can she combine her skirts and blouses?

33 An automobile license plate contains 6 characters, either letters or numbers or both. If the characters may be repeated, how many different license plates can be issued?

34 Considering the license plates in problem 33, how many plates could be issued if all of the characters on any one plate must be different?

13.4 Combinations and the Binomial Theorem

In the previous section we considered two types of arrangements: arrangements in which objects could be repeated (such as the serial numbers in Example 6) and arrangements in which all of the objects are distinct (such as the choice of perfumes in Example 5). An important point to note about permutations of distinct objects, as described by the Permutation Theorem, is that the *order* of the objects matters. For example, we regard the arrangement (3, 2, 1) as different from (1, 2, 3) or (2, 1, 3), or any of the other permutations.

In some applications, however, we are interested only in *which* objects are in the grouping, and *not* in how they are ordered. If we consider the number of ways to make change for a dollar, for example, we distinguish 3 quarters, 2 dimes, and 1 nickel from 4 quarters; but it makes no sense to distinguish 2 dimes, 1 nickel, and 3 quarters from 3 quarters, 2 dimes, and 1 nickel. Thus, counting groupings without regard for order is a different problem from those we have discussed so far.

Our first example is a counting problem that involves the construction of sets without regard for the order in which the elements are chosen.

EXAMPLE 1

A corporation has 10 members of the board of directors. How many ways can the board of directors choose a 5-person finance committee from its members?

Solution

Until the committee elects its officers, there is nothing to distinguish one committee position from another. Thus, we are concerned with the number of ways to choose a set with 5 elements from a set with 10 elements.

If each committee position were unique then the order of selection of the 5 members would be important; there would be

$$P(10, 5) = \frac{10!}{5!} = 10 \cdot 9 \cdot 8 \cdot 7 \cdot 6 = 30{,}240$$

such committees. One way to form a typical committee, with unique committee positions, is to choose 5 members and then arrange the 5 in an order that indicates specific committee positions. Since the number of ways to arrange the 5 chosen members is the number of permutations of 5 objects taken 5 at a time, the General Multiplication Principle shows that

$$P(10, 5) = xP(5, 5),$$

where x is the number of ways to choose the 5 persons. Solving this equation for x, we find that

$$x = \frac{P(10, 5)}{P(5, 5)} = \frac{30{,}240}{120} = 252. \ \blacksquare$$

In Example 1, the 5-member finance committee chosen from the 10-member board of directors is an example of a combination of 10 objects taken 5 at a time.

Definition   A **combination of n objects taken r at a time** is a set with r distinct elements, each chosen from the n objects; the order of selection is neglected.

Using the Permutation Theorem and the General Multiplication Principle, we can obtain a general formula for the number of combinations that can be formed from a given set.

Combination Theorem The number of combinations of n objects taken r at a time, denoted $C(n, r)$, is given by

$$C(n, r) = \frac{P(n, r)}{r!}.$$

To show that this formula is true, we consider how to construct all permutations of n objects taken r at a time. The construction of all such permutations can be regarded as a two-step task. The first step is to select a combination of r objects, and the second step is to arrange the r objects in a specific order. The first step can be completed in $C(n, r)$ ways, and the second step can be completed in

$$P(r, r) = \frac{r!}{0!} = r!$$

ways. By the General Multiplication Principle, we see that

$$P(n, r) = C(n, r)P(r, r) = C(n, r)r!$$

or, solving for $C(n, r)$,

$$C(n, r) = \frac{P(n, r)}{r!}.$$

In the next example, we illustrate how to make the computation indicated in the Combination Theorem.

EXAMPLE 2

Determine the number of combinations of 10 objects taken 5 at a time; determine the number of combinations of 5 objects taken 2 at a time.

Solution
According to the Combination Theorem, the number of combinations of 10 objects taken 5 at a time is

$$
\begin{aligned}
C(10, 5) &= \frac{P(10, 5)}{5!} = \frac{1}{5!} \cdot \frac{10!}{5!} \\
&= \frac{1}{5!} \cdot \frac{10 \cdot 9 \cdot 8 \cdot 7 \cdot 6 \cdot 5 \cdot 4 \cdot 3 \cdot 2 \cdot 1}{5 \cdot 4 \cdot 3 \cdot 2 \cdot 1} \\
&= \frac{10 \cdot 9 \cdot 8 \cdot 7 \cdot 6}{5 \cdot 4 \cdot 3 \cdot 2 \cdot 1} \\
&= \frac{30{,}240}{120} = 252.
\end{aligned}
$$

According to the Combination Theorem, the number of combinations of 5 objects taken 2 at a time is

$$C(5, 2) = \frac{P(5, 2)}{2!} = \frac{1}{2!} \cdot \frac{5!}{3!}$$

$$= \frac{1}{2!} \cdot \frac{5 \cdot 4 \cdot \cancel{3} \cdot \cancel{2} \cdot \cancel{1}}{\cancel{3} \cdot \cancel{2} \cdot \cancel{1}}$$

$$= \frac{5 \cdot 4}{2 \cdot 1} = 10. \quad \blacksquare$$

Next, we solve a practical problem by direct application of the Combination Theorem.

EXAMPLE 3

A real estate developer promises to award 2 prizes, from a list of 5 prizes, to each person who tours a new resort property. How many different pairs of prizes might a prospective buyer receive?

Solution
In this problem, we are concerned with a combination of 5 objects taken 2 at a time, since the order in which the prizes are awarded is insignificant. From Example 2, we know that the number of combinations of 5 objects taken 2 at a time is

$$C(5, 2) = \frac{P(5, 2)}{2!} = 10,$$

so there are 10 different pairs of prizes. ◼

Counting techniques are very useful in analyzing games of chance. This is illustrated in the next example.

EXAMPLE 4

If the dealer gives each poker player 5 cards from a deck of 52, how many hands might a player be given?

Solution
In this problem, we are concerned with a combination of 52 objects taken 5 at a time, since the order in which the cards are dealt is insignificant. We compute the number of combinations of 52 objects taken 5 at a time.

$$C(52, 5) = \frac{P(52, 5)}{5!} = \frac{1}{5!} \cdot \frac{52!}{47!}$$

$$= \frac{1}{5!} \cdot 52 \cdot 51 \cdot 50 \cdot 49 \cdot 48$$

$$= \frac{311{,}875{,}200}{120}$$

$$= 2{,}598{,}960$$

There are 2,598,960 different hands that a poker player might be dealt! ◼

Each of the numbers $C(n, r)$ is called a **binomial coefficient;** frequently, the number $C(n, r)$ **is denoted** $\binom{n}{r}$. These numbers occur in the Binomial

Theorem, which is a rule indicating how to raise the sum of two terms to a power. For small exponents, the Binomial Theorem is not required; by direct computation, we determine that

$$(x + y)^2 = x^2 + 2xy + y^2$$

$$(x + y)^3 = x^3 + 3x^2y + 3xy^2 + y^3$$

$$(x + y)^4 = x^4 + 4x^3y + 6x^2y^2 + 4xy^3 + y^4.$$

In each of these equations, the right side is said to be the **binomial expansion** of the power indicated on the left side. For higher powers the Binomial Theorem is very useful.

Binomial Theorem   For n a positive integer, we have

$$(x + y)^n = \binom{n}{0} x^n + \binom{n}{1} x^{n-1}y + \binom{n}{2} x^{n-2}y^2 + \cdots + \binom{n}{n-1} xy^{n-1} + \binom{n}{n} y^n.$$

The right side of this formula is said to be the binomial expansion of the power on the left. Reading the terms in the binomial expansion from left to right, we notice that the power of x decreases by 1 from each term to the next, and the power of y increases by 1 from each term to the next. The sum of the exponents of x and y is n for each term in the expansion. It is routine to verify that the Binomial Theorem gives the same binomial expansions for $(x + y)^2$, $(x + y)^3$, and $(x + y)^4$ that were obtained earlier by direct computation.

EXAMPLE 5

Obtain the binomial expansion for $(x + y)^5$.

Solution
Before we can give the binomial expansion, we must compute the binomial coefficients $\binom{5}{0}, \binom{5}{1}, \binom{5}{2}, \binom{5}{3}, \binom{5}{4}, \binom{5}{5}$.

$$\binom{5}{0} = \frac{P(5, 0)}{0!} = \frac{5!}{5!} = 1$$

$$\binom{5}{1} = \frac{P(5, 1)}{1!} = \frac{5!}{4!} = \frac{5 \cdot 4 \cdot 3 \cdot 2 \cdot 1}{4 \cdot 3 \cdot 2 \cdot 1} = 5$$

$$\binom{5}{2} = \frac{P(5, 2)}{2!} = \frac{1}{2!} \cdot \frac{5!}{3!} = \frac{1}{2} \cdot \frac{5 \cdot 4 \cdot 3 \cdot 2 \cdot 1}{3 \cdot 2 \cdot 1} = 10$$

$$\binom{5}{3} = \frac{P(5, 3)}{3!} = \frac{1}{3!} \cdot \frac{5!}{2!} = \frac{1}{2!} \cdot \frac{5 \cdot 4 \cdot 3 \cdot 2 \cdot 1}{3 \cdot 2 \cdot 1} = 10$$

$$\binom{5}{4} = \frac{P(5, 4)}{4!} = \frac{1}{4!} \cdot \frac{5!}{1!} = \frac{5 \cdot 4 \cdot 3 \cdot 2 \cdot 1}{4 \cdot 3 \cdot 2 \cdot 1} = 5$$

$$\binom{5}{5} = \frac{P(5, 5)}{5!} = \frac{1}{5!} \cdot \frac{5!}{0!} = \frac{5!}{5!} = 1$$

Now it is clear that

$$(x + y)^5 = x^5 + 5x^4y + 10x^3y^2 + 10x^2y^3 + 5xy^4 + y^5. \blacksquare$$

With the Binomial Theorem, we can determine the number of subsets that can be formed from a finite set with n elements. We demonstrate the process for a set S with 3 elements. Any subset of S has 0, 1, 2, or 3 elements. The numbers of subsets with 0, 1, 2, and 3 elements are $\binom{3}{0}, \binom{3}{1}, \binom{3}{2}$, and $\binom{3}{3}$, respectively. Thus, the total number of subsets of S is

$$\binom{3}{0} + \binom{3}{1} + \binom{3}{2} + \binom{3}{3}.$$

We could compute each of these binomial coefficients and then sum them; however, there is an easier way to get the answer. Notice that the binomial expansion of $(1 + 1)^3$ is given by

$$(1 + 1)^3 = \binom{3}{0} 1^3 + \binom{3}{1} 1^2 \cdot 1 + \binom{3}{2} 1 \cdot 1^2 + \binom{3}{3} 1^3$$

$$= \binom{3}{0} + \binom{3}{1} + \binom{3}{2} + \binom{3}{3}.$$

In this case, the power is easier to compute than the binomial expansion. The power is

$$(1 + 1)^3 = 2^3 = 8.$$

Thus, the sum of the binomial coefficients is 8; we have determined that a set with 3 elements has 8 subsets.

The method that we used in the preceding can be used to prove the Subset Theorem given next.

Subset Theorem   A finite set with n elements has 2^n subsets.

The previous theorem is easy to use, as we illustrate.

EXAMPLE 6

Determine the number of subsets of the set $\{1, a, 2, b, 3, c, 4, d\}$.

Solution

The given set has 8 elements, and so it has $2^8 = 256$ subsets. ◼

The Combination Theorem can be used together with the General Multiplication Principle to solve problems like the one in our next example.

EXAMPLE 7

A production committee is formed by choosing 2 people from each of 3 production lines. The numbers of people on each line are 9, 12, and 10, respectively. How many ways can the committee be chosen?

Solution

The number of ways to choose 2 people from the 9 on the first production line is $C(9, 2)$. Similarly, the numbers of ways to choose the other members are $C(12, 2)$ and $C(10, 2)$. By the General Multiplication Principle the total number of ways to staff the committee is

$$C(9, 2)C(12, 2)C(10, 2) = 106,920. \blacksquare$$

Exercises 13.4

Determine the value of each number below.

1 $C(2, 2)$ 2 $C(3, 3)$
3 $C(3, 2)$ 4 $C(5, 4)$
5 $C(8, 6)$ 6 $C(4, 2)$
7 $C(8, 2)$ 8 $C(9, 3)$

Determine the number of subsets of each set below.

9 $\emptyset$ 10 $\{1\}$
11 $\{a, 2\}$ 12 $\{3, 4, c\}$
13 $\{x, y, z, 5\}$ 14 $\{2, 4, 6, 8, v\}$
15 $\{1, 3, d, 7, w, 9\}$ 16 $\{1, 2, 3, 4, 5, 6\}$

Obtain the first four terms of the binomial expansion for each power below.

17 $(x + y)^6$ 18 $(x + y)^7$
19 $(x + y)^8$ 20 $(x + y)^9$

Solve each of the following.

21 How many ways can a club with 9 members choose a 4-member committee?
22 How many ways can a team with 11 members choose 2 co-captains?
23 How many ways can you choose 3 prizes from a list of 10 prizes?
24 How many ways can you choose 2 prizes from a list of 7 prizes?
25 If 2 bills are withdrawn from an envelope containing one bill of each of the denominations $1, $5, $10, $20, $100, how many different amounts of money might be taken from the envelope?
26 If 3 bills are withdrawn from an envelope containing one bill of each of the denominations $1, $5, $10, $20, $50, how many different amounts of money might be taken from the envelope?
27 If a card game requires that each player be dealt 7 cards, how many different hands might a player receive?
28 In the game of bridge each player is dealt 13 cards. How many different hands might a player receive?
29 How many ways can the starting 5 basketball players be chosen from a squad of 12 players?
30 How many ways can the starting 11 football players be chosen from a squad of 35 players?
31 If a regular pizza has 2 toppings and there are 6 to choose from, how many different regular pizzas can be made?

32 A club has 8 male members and 9 female members. The club elects a committee of 2 men and 2 women. How many different committees might be elected?

33 A menu allows you to choose 1 meat from 3 offered, 3 vegetables from 6 offered, and 1 dessert from 4 offered. How many different meals can you choose?

34 A poker player is dealt 2 hearts, 2 spades, and 1 diamond. How many different hands might the player be holding?

Review of Terms

Important Mathematical Terms

set, *p. 596*	finite set, *p. 602*
member, *p. 596*	infinite set, *p. 602*
element, *p. 596*	counting problem, *p. 603*
belongs to, *p. 596*	tree diagram, *p. 609*
null set, *p. 596*	Multiplication Principle, *p. 609*
empty set, *p. 596*	permutation, *p. 610*
subset, *p. 597*	*n* factorial (*n*!), *p. 610*
equal sets, *p. 597*	Permutation Theorem, *p. 611*
proper subset, *p. 597*	combinations, *p. 614*
universal set, *p. 598*	Combination Theorem, *p. 615*
complement, *p. 598*	binomial coefficient, *p. 616*
union, *p. 598*	binomial expansion, *p. 617*
intersection, *p. 598*	Binomial Theorem, *p. 617*
Venn diagrams, *p. 598*	Subset Theorem, *618*

Important Terms from the Applications

inflation rate, *p. 599*	money supply growth rate, *p. 600*

Review Problems

1 Define a set with the property that every element is the square of an integer and does not exceed 100.

2 List all possible subsets of the set {3, *a*, *x*, 4, *y*}.

Let *A*, *B*, and *C* be defined by

$$A = \{1, 2, 3, 4, 5, 6\}, \qquad B = \{4, 5, 6, 7, 8\}, \qquad C = \{3, 4, 5, 6\}.$$

Determine each set below.

3 $A \cap B \cap C$

4 $A \cap B \cap C'$

5 $(A \cup B) \cap C'$

6 $(A \cap C') \cup (B \cap C')$

7 $(A \cap B') \cup C$

8 Use a Venn diagram to depict the set $A \cap B \cap C'$.

Determine the value of each number below.

9	0!	10	5!
11	$P(6, 6)$	12	$P(5, 2)$
13	$P(10, 7)$	14	$C(5, 5)$
15	$C(5, 1)$	16	$C(8, 5)$

17 Determine how many ordered triples can be formed using the elements of the set $\{3, a, x, 4, y\}$.

18 Determine the number of subsets of the set $\{3, a, x, 4, y\}$.

19 Obtain the last four terms of the binomial expansion for $(x + y)^6$.

Suppose A has 30 elements, B has 20 elements, and C has 10 elements. Furthermore, suppose that the numbers of elements in $A \cap B$, $A \cap C$, $B \cap C$, and $A \cap B \cap C$ are 12, 8, 4, and 2, respectively. Determine the number of elements in each set below.

20 $A \cap B' \cap C'$

21 $C \cap A' \cap B'$

22 $A \cap C \cap B'$

23 The manager of a small fast food restaurant kept a record of his customers' lunch orders for one week. He discovered the following:

210 ordered French fries,
170 ordered hamburgers,
85 ordered hot dogs,
130 ordered hamburgers and French fries,
50 ordered hot dogs and French fries,
40 ordered hamburgers and hot dogs,
20 ordered hamburgers, hot dogs, and French fries.

Determine the number of customers

a. who ordered French fries only,
b. who ordered hamburgers only,
c. who ordered hot dogs only.

Solve each of the following with the counting methods of this chapter.

24 If there are 5 paths from A to B and 4 paths from B to C, then how many ways can you go from A to C?

25 How many serial numbers can be constructed provided each consists of 3 letters of the alphabet followed by 3 single-digit numbers?

26 How many ways can a class of 25 choose a 6-member committee?

27 How many different 7-card hands can be dealt from a 52-card deck?

28 Ten people lined up in front of the ticket office at a theatre. How many different ways might they have lined up?

29 Of the 50 companies belonging to a certain trade association, 40 made a profit and 10 did not. How many different ways could that happen?

30 A poker player is dealt 3 clubs and 2 diamonds. How many different hands might the player be holding?

Social Science Applications

1 *Law Enforcement* In Georgia an auto license consists of 3 letters of the alphabet followed by 3 single-digit numbers. If the license of a getaway car begins with EUN8, what is the largest number of cars that might be the getaway car?

2 *Medicine* The classification of a person's blood begins with A provided the A antigen is present; the classification includes B provided the B antigen is present. If the blood contains the Rh antigen, then the classification ends with +; otherwise, it ends with −. List all possible blood types using these letters.

3 *Political Science* A committee is formed to draft a bill that compromises the differences between two versions passed by the House of Representatives and the Senate. The committee should consist of two persons from each of the two congressional units. How many ways can the committee be formed? (Assume there are 100 senators and 435 representatives.)

4 *Political Science* Voting records in a small city show that in the last election the numbers of voters (in thousands) voting for governor, registering as Democrats, and registering as Republicans were 50, 20, and 15, respectively. For each election, each voter can register as exactly one of the following: Democrat, Republican, or Independent. The number of registered Democrats voting in the governor's race and the number of registered Republicans (in thousands) voting in that race were 18 and 12, respectively. How many registered Independents voted in the governor's race?

5 *Sociology* A sociologist wants to choose 10 families, living in a certain building, to complete a questionnaire. If there are 25 families living in that building, how many different ways might the 10 families be chosen?

CAREER PROFILE
Statistician

"There are three kind of lies: lies, damned lies and statistics."

—Benjamin Disraeli (1804–1881)

Disraeli's statement from long ago points to a challenge for statisticians that still exists today, the challenge to analyze large volumes of data in an objective fashion and to obtain reliable and useful results.

Medical science calls on statisticians to evaluate the results of different treatments. Government funding decisions, such as the decision of whether to continue Federal support for remedial reading programs, are often tied to statistical evaluation of program effectiveness. Charges of discrimination in employment are often supported (and rebutted) by statistical arguments.

Because different ways of organizing data can lead to different results, debates that involve statistics are not necessarily easy to resolve. (For example, the following can happen in college admissions: in the college as a whole, the admissions rate for women is higher than it is for men, but in each department the admissions rate for men is higher than it is for women. Which, if either, sex has been discriminated against?) A statistician must not only know mathematical methods but also must be able to make sound judgments about which methods are best in a variety of decision situations.

Most professional statisticians have received the major part of their statistical training at the graduate level. Although many universities and colleges offer undergraduate programs in statistics, it is typical for students to move into employment or graduate studies from an undergraduate major in mathematics or from an applied field such as economics or one of the sciences. Mathematics courses in differential and integral calculus, probability theory, and statistical methods are a necessity. Experience in application of quantitative methods to the analysis and solution of complex problems is vital also.

In many undergraduate programs, such as engineering, biology, economics, psychology, and sociology, students are qualified for broader employment opportunities by carrying statistics as a minor. A concentration in statistics is useful in diverse areas such as weather forecasting, quality control for manufactured products, analysis of the evidence for biological evolution, study of social attitudes, and forecasting needs for natural resources and labor.

Individuals who obtain a job in statistics with only a bachelor's degree can expect to spend most of their time on routine work or under the supervision of an experienced statistician. Opportunities for promotion to positions of responsibility are open primarily to those with advanced degrees.

Widespread use of computers has had important effects on the practice of statistics. Calculations that once took weeks of labor now can be done in just a few seconds. Learning to use computers is an important part of a statistician's job preparation, even for the individual who takes only a minor in statistics at the undergraduate level.

Study of statistics is valuable not only for career training but also as preparation for being an intelligent consumer. Every purchaser of goods and service can profit from the ability to assess advertising claims; every individual who manages his own finances can benefit from the ability to organize and analyze data; wise decisions about diet and health can be made on the basis of an increasing volume of statistical evidence.

A strong background in statistics together with a thorough knowledge of a field to which it can be applied is expected to be a good preparation for employment throughout the 1980s. Such experts are not always called statisticians, however; for example, a statistician working with anthropological data may be called an anthropologist. Although titles may vary, the expertise is valued: an individual who has earned a Ph.D. in statistics can expect a starting position with a salary of $25,000 or higher.

About half of the 26,500 individuals employed as statisticians in 1980 worked in private industry (primarily for manufacturing, finance, and insurance companies). About one third worked for government agencies. Most of the employment for statisticians is available in metropolitan areas.

Statistics is a demanding profession, requiring strong quantitative skills together with the exercise of patience and care in their application. Individuals who can meet these demands can find an application of statistics that will combine with almost any other interest that they may have, for example, law, medicine, government service, meteorology, space exploration, business, or education.

Sources of Additional Information
American Statistical Association. 806 15th Street, N.W., Washington, DC 20005. (Available from the ASA is a free booklet entitled "Careers in Statistics;" it describes a variety of jobs available to statisticians, names schools that offer programs in statistics, and lists organizations that may be contacted for further career information. Some of the information for this Career Profile was obtained from the ASA booklet.)

Occupational Outlook Handbook. Bureau of Labor Statistics, U.S. Department of Labor, Washington, DC 20212. (Revised every 2 years, this handbook provides information about job duties, working conditions, level and places of employment, education and training requirements, advancement possibilities, job outlook, and earnings for about 250 occupations. Some of the information for this Career Profile was obtained from this source.)

14 Probability

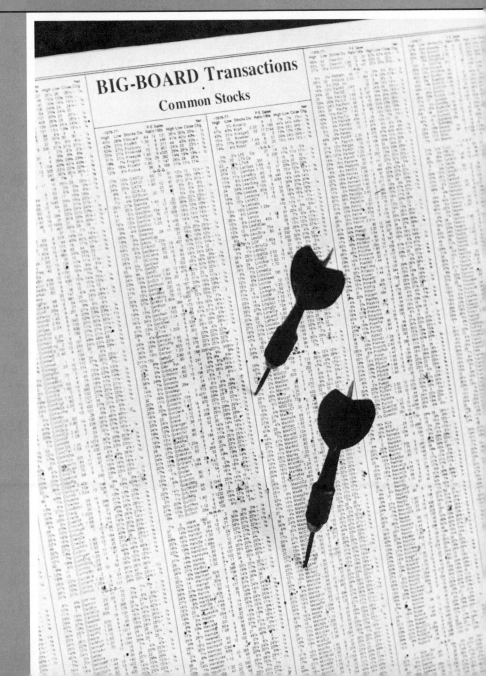

14.1   Basic Properties

Few things are certain in the modern-day world. Not only are we uncertain whether it will rain tomorrow, we are also uncertain whether interest rates will increase or decrease during the next year. We cannot be sure whether the economy will flourish or stagnate during the next year. The theory of probability was developed so that we could deal with uncertainty in a rational systematic manner.

An activity that does not have one predetermined result is called an **experiment,** and each possible result is called an **outcome.** Each repetition of the experiment is called a **trial.** Often we are concerned with experiments for which each outcome is equally likely. For example, the experiment of interest might be the toss of a coin, which has two possible outcomes. The coin might land so that the head is showing, which we describe as "landing on heads," or the coin might land on tails. Provided the coin has not been altered so that it favors landing on one particular side, we say that it is a **fair coin.** For a fair coin, we regard each outcome as equally likely.

The set of all outcomes for a given experiment is called the **sample space S.** Any subset of S is called an **event E;** we say that E occurs provided one of the outcomes belonging to E occurs. In the experiment of tossing a coin once, the sample space is $\{H, T\}$, where H indicates that the coin lands on heads and T indicates that the coin lands on tails. The event $E = \{H\}$ occurs provided the coin lands on heads. **When all outcomes are equally likely, then the probability of the event E occurring, denoted $P(E)$, is defined by**

$$P(E) = \frac{\text{number of outcomes in } E}{\text{total number of outcomes in } S} = \frac{n(E)}{n(S)}.$$

This is called the **objective assignment of probabilities.**

The significance of this number, $P(E)$, is the following: If the experiment is repeated N times, where N is some "large" number, then we expect the event E to occur $NP(E)$ times. In any given series of repetitions, E may occur slightly more or less often than this prediction states. However, as N becomes very large (thousands of millions of repetitions), the agreement will become better and better.

We illustrate these concepts in the next example.

EXAMPLE 1

Determine the probability that when we toss a fair coin 3 successive times, it lands on heads each time.

Solution
This experiment is a task with 3 steps, and each step can be completed either of 2 ways. The General Multiplication Principle indicates that there are

$$2 \cdot 2 \cdot 2 = 8$$

possible outcomes. We use a tree diagram to determine them.

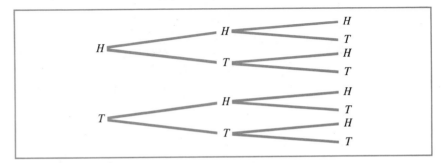

The sample space is

$$\{HHH, \quad HHT, \quad HTH, \quad HTT, \quad THH, \quad THT, \quad TTH, \quad TTT\}.$$

(Note that HHT is a different outcome from THH.)
We are interested in the occurrence of the event

$$E = \{HHH\}.$$

The probability of E occurring is

$$P(E) = \frac{n(E)}{n(S)} = \frac{1}{8} = .125.$$

Thus, we expect that about $\frac{1}{8}$ of a large number of repetitions of the experiment would result in the outcome HHH. ∎

Note that for any experiment and any event E, the probability of E occurring satisfies the inequalities

$$0 \le P(E) \le 1.$$

The next example shows how $P(E)$ might equal 0 or 1.

EXAMPLE 2

Suppose a fair coin is tossed once. List all possible events and determine the probability of each.

Solution

The sample space is $S = \{H, T\}$, and every subset of S is an event. Since S has two elements, the Subset Theorem indicates that S has $2^2 = 4$ subsets. Thus, there are 4 possible events. The events are

$$\emptyset, \quad \{H\}, \quad \{T\}, \quad \{H, T\}.$$

These events correspond, respectively, to (a) neither heads nor tails (we rule out the possibility of the coin landing on edge or a bird snatching the coin from midair); (b) landing on heads; (c) landing on tails; and (d) landing on one of the sides, either heads or tails. We calculate the probability of each event.

$$P(\emptyset) = \frac{0}{2} = 0, \qquad P(\{H, T\}) = \frac{2}{2} = 1.0, \qquad P(\{H\}) = \frac{1}{2} = .5, \qquad P(\{T\}) = \frac{1}{2} = .5 \quad \blacksquare$$

Since events are sets, we can form new events by means of the set operations. If E and F are events for an experiment with sample space S, then $E \cup F$ is also an event for that experiment. Since $E \cup F$ consists of the outcomes belonging to E and the outcomes belonging to F, if either E occurs or F occurs (or both occur) then $E \cup F$ occurs. Since $E \cap F$ consists of those outcomes that belong to both E and F, if $E \cap F$ occurs then both E and F occur. The next theorem gives an equation relating the probabilities of the events E, F, $E \cup F$, and $E \cap F$.

Theorem   Let E and F be events for a certain experiment with sample space S. Then the probabilities of E, F, $E \cup F$, and $E \cap F$ satisfy the following equation

$$P(E \cup F) = P(E) + P(F) - P(E \cap F).$$

We shall verify this formula only for the case that all outcomes are equally likely. According to the theorem in Section 13.2, we have the following relation among the four sets

$$n(E \cup F) = n(E) + n(F) - n(E \cap F).$$

If we divide each side of this equation by $n(S)$, then we get the equation asserted in the theorem.

$$\frac{n(E \cup F)}{n(S)} = \frac{n(E)}{n(S)} + \frac{n(F)}{n(S)} - \frac{n(E \cap F)}{n(S)}$$

or

$$P(E \cup F) = P(E) + P(F) - P(E \cap F)$$

We illustrate the use of the preceding theorem in our next example.

EXAMPLE 3

A group testing a new product consists of 9 men and 11 women; exactly 10 of the people are younger than 30, and exactly 4 men are younger than 30. Determine the probability that a randomly chosen member of the group is either a man or a person younger than 30.

Solution

Let S consist of the 20 people in the test group. Let E consist of the men in the group, and let F consist of the people younger than 30. Then $E \cap F$ consists of the men younger than 30, and $E \cup F$ consists of the persons who are either men or younger than 30. We use the previous theorem to get $P(E \cup F)$.

$$P(E \cup F) = P(E) + P(F) - P(E \cap F)$$
$$= \frac{9}{20} + \frac{10}{20} - \frac{4}{20} = \frac{15}{20} = .75 \ \blacksquare$$

In the next example, we shall use the counting methods as they are used in experiments with a large number of outcomes. Frequently, we can determine the probability of an indicated event without explicitly constructing the sample space.

EXAMPLE 4

A fair coin is tossed 4 consecutive times. Determine the probability that exactly 2 heads result.

Solution

We regard this experiment as a four-step task. Since each step can be completed in exactly 2 ways, the General Multiplication Principle indicates that there are $2 \cdot 2 \cdot 2 \cdot 2 = 16$ outcomes. We shall solve the problem without explicitly listing the 16 outcomes, although that could be accomplished with a tree diagram similar to the one in Example 1.

We need to count the outcomes that indicate exactly 2 heads. Thus, the outcomes that we must count are all distinct permutations of $HHTT$. Note that some permutations of $HHTT$ do not result in a different outcome. For example, if we interchange the first H with the second H, we get the same outcome.

To form a typical outcome of the type that we are counting, we choose 2 of the 4 coin tosses and enter H for each of those tosses. We enter T for each of the other 2 tosses. Thus, the number of such outcomes is the number of ways to choose a subset with 2 elements (the positions for the H's) from a set with 4 elements (the 4 positions corresponding to the 4 coin tosses). This number is $C(4, 2) = 6$, and the answer is $\frac{6}{16} = \frac{3}{8} = .375.$ $\blacksquare$

In some experiments the probability of a given event is not determined by simply counting the outcomes belonging to that event and dividing by the number of outcomes in the sample space. Certain outcomes may seem more likely than others. On occasion, different probabilities are assigned to the different outcomes according to a subjective appraisal of the factors involved. This is called the **subjective assignment of probabilities;** it is permitted provided the following properties hold.

1. The probability assigned to every outcome is nonnegative.
2. The sum of the probabilities of all of the outcomes is 1.
3. The probability assigned to any event is the sum of the probabilities of the outcomes belonging to that event.

We illustrate the subjective assignment of probabilities in the next example.

EXAMPLE 5

An economist, who analyzes the New York Stock Exchange for a stock brokerage firm, assigns the following probabilities to each possible movement of the Dow Jones Industrial Average (D.J.I.A.) over the next 12 months.

Movement of the D.J.I.A.	Probability of the Movement
1. Up 100 points or more	.20
2. Up between 50 and 100 points	.30
3. Up 50 points or less	.20
4. Down 50 points or less	.15
5. Down between 50 and 100 points	.10
6. Down 100 points or more	.05
Total	1.00

Assuming that the economist is correct, determine the probability that the D.J.I.A. will go up more than 50 points. What is the probability that the D.J.I.A. will be higher in a year?

Solution

The event E that the D.J.I.A. goes up more than 50 points consists of the outcomes 1 and 2. Thus, the probability of E is the sum of the probabilities of those 2 outcomes.

$$P(E) = P(1) + P(2)$$
$$= .20 + .30$$
$$= .50$$

The event F that the D.J.I.A. will be higher in a year consists of the outcomes 1, 2, and 3. Thus, the probability of F is the sum of the probabilities of those 3 outcomes.

$$P(F) = P(1) + P(2) + P(3)$$
$$= .20 + .30 + .20$$
$$= .70 \quad \blacksquare$$

It should be noted that the theorem given earlier in this section is true when the probabilities are assigned subjectively, although the verification given applies only when all outcomes are equally likely. A more general verification is possible.

For many experiments we can compute the expected value of some unknown that interests us. Suppose the experiment is to toss a fair coin 3 successive times, and we are interested in the exact number of times that the coin lands on heads. The possibilities are 0, 1, 2, and 3. From the sample space constructed in Example 1, we know that the probability of 0 heads is $\frac{1}{8}$, since TTT is the only outcome for which that occurs. Similarly, we find the probabilities of 1, 2, and 3 heads to be $\frac{3}{8}$, $\frac{3}{8}$, and $\frac{1}{8}$, respectively. The expected value for the number of heads is

$$0 \cdot \frac{1}{8} + 1 \cdot \frac{3}{8} + 2 \cdot \frac{3}{8} + 3 \cdot \frac{1}{8} = \frac{12}{8} = \frac{3}{2} = 1.5.$$

This means that if this experiment were repeated many times, the average number of heads obtained per experiment (the total number of heads obtained divided by the number of times the experiment was repeated) should be approximately 1.5.

Although the expected value gives an idea of what value to expect, it might be impossible for the expected value to occur when the experiment is performed once. We cannot possibly get 1.5 heads on 3 successive coin tosses. The expected value gives an insight into the likelihood of different outcomes rather than picking out the one most likely outcome.

Suppose the only outcomes for some experiment are $x_1, x_2, \ldots, x_n$ and the probabilities of each of these outcomes are $p_1, p_2, \ldots, p_n$, respectively. Then the **expected value** is defined to be

$$E = x_1 p_1 + x_2 p_2 + \cdots + x_n p_n.$$

We illustrate this idea in the next example.

EXAMPLE 6

On your 30th birthday you receive a letter from Mutual Insurance Company offering you a life insurance policy. The policy pays your beneficiary $10,000 if you die before your 40th birthday, and it has a one-time premium of $750. Assume that the probability that you will be alive on your 40th birthday is .95 (so the probability that you will die first is .05). Use the expected value of this policy to decide if it is a good proposition.

Solution

If you die before you are 40 then the policy is worth $10,000 to your beneficiary, but you have already paid $750; the probability of that is .05. If you reach your 40th birthday then the policy is worth $-$750 (you lose your premium and your beneficiary gets nothing); the probability of that is .95. Thus, the expected value of the policy is

$$E = (10,000 - 750)(.05) - 750(.95)$$
$$= 500 - 750 = -250$$

On the basis of expected value only, the policy is a bad proposition. ◼

Exercises 14.1

Construct an appropriate sample space for each of the following experiments.

1 Two consecutive coin tosses
2 Four consecutive coin tosses

  3 Rolling one die
  4 Rolling two dice
  5 Rolling one die two consecutive times
  6 Randomly choosing a day of the week
  7 Randomly choosing a month of the year
  8 Randomly choosing a letter of the alphabet

Suppose that a fair coin is tossed 3 consecutive times. Determine the probability of each of the following events.

  9 The coin lands on heads exactly once.
 10 The coin lands on tails exactly once.
 11 The coin lands on tails at least once.
 12 The coin lands on heads at least once.
 13 The coin lands on heads exactly twice.
 14 The coin lands on tails exactly twice.

Suppose two fair dice are rolled. Determine the probability of each of the following events.

 15 Each die shows one dot.
 16 Each die shows 6 dots.
 17 The sum of dice is 4.
 18 The sum of dice is 5.
 19 Both dice show the same number.
 20 The two dice show different numbers.

Suppose that a fair coin is tossed 4 consecutive times. Determine the probability of each of the following events.

 21 It lands on heads exactly once.
 22 It lands on tails exactly once.
 23 It lands on tails at least twice.
 24 It lands on heads at least twice.
 25 It lands on heads more than it lands on tails.
 26 It lands on tails more than it lands on heads.

Suppose that 2 cards are randomly chosen from a standard deck of 52 cards. Determine the probability of each of the following events.

 27 Both cards are diamonds.
 28 Both cards are clubs.
 29 Each card is a three.
 30 Each card is a five.
 31 The two cards belong to the same suit.
 32 The two cards are the same kind (both fives or queens or aces, etc).

After years of running a used car dealership, the manager makes the following subjective assignment of probabilities. The first column shows the number of cars sold during the week, and the second column shows the probability of that number of sales.

Number of Sales for the Week	Probability
6 or more	.05
5	.10
4	.20
3	.30
2	.20
1	.10
0	.05

Determine the probabilities of each of the following events.

33 One sale or more
34 Three sales or more
35 Five sales or less
36 Two sales or less
37 Either 5 sales or 1 sale
38 Either 0 sales or 6 or more sales

We say that the **odds** for event E are "a to b" provided $a/b = P(E)/P(E')$ where E' is the complement of E, using the sample space as the universe. Usually, a and b are taken from the simplest possible fraction. For example, the odds of a fair coin landing on heads are 1 to 1, since $\frac{1}{1} = \frac{.5}{.5}$. Give the odds for each of the following events.

39 A coin toss lands on tails.
40 A die lands on 2.
41 A randomly drawn card from a 52-card deck is the 3 of clubs.
42 Two successive coin tosses land on heads.
43 Three successive coin tosses land on heads.
44 Four successive coin tosses land on heads.
45 What is the expected value of the number of heads obtained with 4 successive tosses of a fair coin?
46 What is the expected value of the number of tails with 5 successive tosses of a fair coin?

14.2   Pairs of Events

One of the simplest relationships that a pair of events can have is that one is the complement of the other. Let S be the sample space for some experiment, and consider the events E and E'. Since $E \cap E' = \varnothing$ and $P(\varnothing) = 0$, the theorem in Section 14.1 gives us

$$P(S) = P(E \cup E') = P(E) + P(E') - P(\varnothing)$$
or
$$P(S) = P(E) + P(E').$$

Since every outcome belongs to S, $P(S)$ must be 1. Thus, we have

$$1 = P(E) + P(E')$$

or

$$P(E') = 1 - P(E).$$

We illustrate the use of the preceding formula in the next example.

EXAMPLE 1

What is the probability that a single card drawn randomly from a deck of 52 cards is not the ace of spades?

Solution

If E consists of all outcomes such that the ace of spades is drawn, then

$$P(E) = \frac{1}{52},$$

since there is only one ace of spades. Note that E' consists of the outcomes such that the card drawn is not the ace of spades, and

$$P(E') = 1 - P(E)$$

$$= 1 - \frac{1}{52}$$

$$= \frac{51}{52}. \quad \blacksquare$$

We say that E and F are **mutually exclusive events** provided that $E \cap F = \varnothing$. Thus, E and E' are mutually exclusive events. The next theorem follows from the theorem given in Section 14.1.

Theorem 1   If E and F are mutually exclusive events then

$$P(E \cup F) = P(E) + P(F).$$

Of course, E and F can be mutually exclusive events without F being E'. We illustrate this in the next example.

EXAMPLE 2

What is the probability that a single card drawn randomly from a deck of 52 cards is either a king or a queen?

Solution

Let E consist of the outcomes such that a king is drawn, and let F consist of the outcomes such that a queen is drawn. We are asked to determine the probability of $E \cup F$. Clearly, $E \cap F = \varnothing$ and

$$P(E \cup F) = P(E) + P(F)$$

$$= \frac{4}{52} + \frac{4}{52}$$

$$= \frac{2}{13} \approx .154. \quad \blacksquare$$

Two different events might be linked to one another by facts that we are given about the experiment. We indicate one way this might happen in the next example.

EXAMPLE 3

The quality control department for a car-truck manufacturer randomly selects a vehicle from a warehouse containing 50 cars and 100 trucks. Suppose that 1 of every 5 cars has faulty brakes and 1 of every 20 trucks has faulty brakes. What is the probability that the selected vehicle has faulty brakes? Suppose the selected vehicle is a car, what is the probability that it has faulty brakes?

Solution

Since there are 10 cars and 5 trucks with faulty brakes, the probability that a randomly selected vehicle has faulty brakes is

$$\frac{15}{150} = .1.$$

If we know that the vehicle selected is a car, then the probability that it has faulty brakes is

$$\frac{10}{50} = \frac{1}{5} = .2. \blacksquare$$

The preceding example shows that the effective sample space is sometimes altered by the occurrence of a certain event. The probability of event E, given that event F has occurred, is called the **conditional probability** of E given that F has occurred; it is denoted $P(E|F)$. Since we know that F has already occurred, the effective sample space is just the set of outcomes in F, and there are $n(F)$ of them. We are now interested in how many of those outcomes are contained in E also, that is $n(E \cap F)$. If all outcomes are equally likely, then

$$P(E|F) = \frac{n(E \cap F)}{n(F)}.$$

When we divide both the numerator and the denominator of the fraction on the right by $n(S)$, we get a formula for $P(E \cap F)$ in terms of the probabilities of $E \cap F$ and F.

$$P(E|F) = \frac{n(E \cap F)}{n(F)} = \frac{n(E \cap F)/n(S)}{n(F)/n(S)}$$

$$P(E|F) = \frac{P(E \cap F)}{P(F)} \qquad \text{provided } P(F) \neq 0$$

This last formula is true for probabilities assigned subjectively, also.
We give a further example of conditional probability.

EXAMPLE 4

Two cards are drawn randomly, one at a time, from a standard deck of 52 cards. What is the probability that both of the cards are diamonds? Given that the first card is a diamond, what is the probability that both cards are diamonds?

Solution

The number of outcomes is the number of ways that a set with 2 elements can be chosen from a set with 52 elements. That number is

$$C(52, 2) = \frac{52 \cdot 51}{2} = 1326.$$

The number of outcomes consisting only of diamonds is the number of ways that a set with 2 elements can be chosen from a set with 13 elements (the set of diamonds). That number is

$$C(13, 2) = \frac{13 \cdot 12}{2} = 78.$$

The probability that both cards are diamonds is

$$\frac{C(13, 2)}{C(52, 2)} = \frac{78}{1326} \approx .059.$$

If we know that the first card drawn is a diamond then the probability that both cards are diamonds changes. Let F consist of all of the outcomes such that the first card drawn is a diamond, and let E consist of all of the outcomes such that both cards drawn are diamonds. We must compute $P(E|F)$. We note that $E \cap F = E$ and $P(F) = \frac{13}{52} = \frac{1}{4} = .25$. Thus, we have

$$P(E|F) = \frac{P(E \cap F)}{P(F)} = \frac{P(E)}{.25} = 4P(E).$$

In the first paragraph, we computed $P(E)$; thus, we get

$$P(E|F) = 4P(E) = 4\left(\frac{78}{1326}\right) = \frac{312}{1326} \approx .235. \quad \blacksquare$$

Let E and F be events for an experiment with sample space S. We say that E and F are **independent** provided that

$$P(E|F) = P(E) \quad \text{and} \quad P(F|E) = P(F).$$

The next theorem gives a useful formula for independent events.

Theorem 2   The events E and F are independent if and only if

$$P(E \cap F) = P(E)P(F).$$

First, we show that the formula holds. According to the definition of independence, we have

$$P(E) = P(E|F).$$

The formula for conditional probability shows that

$$P(E) = P(E|F) = \frac{P(E \cap F)}{P(F)}$$

or

$$P(F)P(E) = P(E \cap F).$$

An analogous argument starting with $P(F|E)$ shows that

$$P(E)P(F) = P(E \cap F).$$

This proves that the formula in the theorem holds whenever E and F are independent.

Assume the formula in the theorem holds and divide both sides by $P(F)$. We shall show that E and F are independent.

$$\frac{P(E \cap F)}{P(F)} = P(E)$$

By the conditional probability formula, we have

$$P(E|F) = \frac{P(E \cap F)}{P(F)} = P(E).$$

An analogous argument shows that

$$P(F|E) = P(F),$$

and so E and F are independent. This verifies the theorem.

The preceding theorem is very useful, as we demonstrate in the next example.

EXAMPLE 5

Suppose a fair coin is tossed twice. What is the probability of getting 2 heads?

Solution

The outcome of the second coin toss is clearly independent of the outcome of the first coin toss. Let E correspond to "the first coin lands on heads," and let F correspond to "the second coin lands on heads." Then $E \cap F$ corresponds to "both times the coin lands on heads," and we have

$$P(E \cap F) = P(E)P(F) = \frac{1}{2} \cdot \frac{1}{2} = \frac{1}{4} = .25. \quad \blacksquare$$

The definition of independent events and Theorem 2 can be extended in a natural way to any number of events. If E_1, E_2, and E_3 are 3 independent events, then

$$P(E_1 \cap E_2 \cap E_3) = P(E_1)P(E_2)P(E_3).$$

In general, if $E_1, E_2, \ldots, E_n$ are n independent events, then

$$\boxed{P(E_1 \cap E_2 \cap \cdots \cap E_n) = P(E_1)P(E_2) \cdots P(E_n).}$$

We shall refer to the preceding formula as the **Independent Event Formula,** and we demonstrate its usefulness in the next example.

EXAMPLE 6

Suppose a fair coin is tossed 5 times. What is the probability of getting 5 heads?

Solution

Clearly the outcome of any coin toss is independent of the outcomes of the other coin tosses. Let E_1 correspond to "the first coin toss results in heads;" let E_2 correspond to "the second coin toss results in heads;" etc. Then $E_1 \cap E_2 \cap \cdots \cap E_5$ corresponds to "every toss of the coin results in heads," and we have

$$P(E_1 \cap E_2 \cap \cdots \cap E_5) = P(E_1)P(E_2) \cdots P(E_5)$$
$$= \frac{1}{2} \cdot \frac{1}{2} \cdot \frac{1}{2} \cdot \frac{1}{2} \cdot \frac{1}{2}$$
$$= \frac{1}{32} \approx .031. \blacksquare$$

The previous example shows how the concept of independent events provides an alternative to some tedious counting methods. Next, we give a practical application of the independent event formula.

EXAMPLE 7

The quality control department for an auto manufacturer has determined the probability of a malfunction in each of four systems. In the chart, each system is listed on the left, and the probability of a malfunction in the system is listed on the right.

System	Probability of a Malfunction
Brakes	.01
Electrical	.10
Mechanical	.08
Engine	.05

What is the probability of getting a car with no malfunction in any of the four systems? (Assume that the quality of each system is independent of the others.)

Solution

Let $E_1, E_2, E_3,$ and E_4 correspond to "there is a malfunction in the brakes," "there is a malfunction in the electrical system," "there is a malfunction in the mechanical system," and "there is a malfunction in the engine," respectively. According to the first formula in this section, the probabilities of the complements of these events are

$$P(E_1') = 1 - .01 = .99,$$
$$P(E_2') = 1 - .10 = .90,$$
$$P(E_3') = 1 - .08 = .92,$$
$$P(E_4') = 1 - .05 = .95.$$

We assume that E_1', E_2', E_3', and E_4' are independent events. By the independent event formula, we get

$$
\begin{aligned}
P(E_1' \cap E_2' \cap E_3' \cap E_4') &= P(E_1')P(E_2')P(E_3')P(E_4') \\
&= (.99)(.90)(.92)(.95) \\
&\approx .779.
\end{aligned}
$$

Thus, the probability of getting a car with no malfunction in any of the four systems is .779. ■

When it is not intuitively clear whether two events are independent, then Theorem 2 can be used to test whether they are, as we illustrate.

EXAMPLE 8

Each of three assembly lines produce 1000 units of a company's product. Thirty of the 3000 units are defective, and 12 of the units produced by Assembly Line A are defective. A unit is chosen randomly from the 3000 units. Are the following two events independent?

A. The unit selected is defective.
B. The unit selected was made by Assembly Line A.

Solution
According to Theorem 2, we have

$$P(A \cap B) = P(A)P(B),$$

precisely when events A and B are independent. Thus, we calculate each of the probabilities $P(A)$, $P(B)$, and $P(A \cap B)$; then we test whether the above equation is true.

$$P(A) = \frac{30}{3000} = \frac{1}{100}$$

$$P(B) = \frac{1000}{3000} = \frac{1}{3}$$

$$P(A \cap B) = \frac{12}{3000} = \frac{1}{250}$$

Note that $P(A \cap B) = \frac{1}{250}$ is not the product of $P(A) = \frac{1}{100}$ and $P(B) = \frac{1}{3}$. Thus, the events A and B are not independent. ■

Exercises 14.2

Use the formula in Theorem 1 to compute the probability of each of the following events.

1 Two successive coin tosses result either in 2 heads or in 2 tails.
2 In two successive coin tosses, either heads followed by tails results or tails followed by heads results.
3 When a die is rolled 2 successive times, either 1 comes up twice or 6 comes up twice.
4 When a die is rolled 2 successive times, either 2 followed by 3 results or 3 followed by 2 results.
5 A letter selected randomly from the alphabet is either A or Z.
6 A letter selected randomly from the alphabet is either J or S.
7 When 5 cards are dealt from a deck of 52 cards, all cards are spades or all cards are clubs.
8 When 3 cards are dealt from a deck of 52 cards, all cards are diamonds or all cards are hearts.

A fair coin is tossed 3 successive times. Determine the conditional probability of each of the following events.

9 Given that the first 2 tosses result in heads, what is the probability of 3 heads?
10 Given that the first 2 tosses had different results, what is the probability of getting 2 or more heads?
11 Given that the first toss results in tails, what is the probability of 3 tails?
12 Given that the first toss results in heads, what is the probability that the last 2 tosses result in heads?

Five cards are dealt from a standard deck of 52 cards. Determine the probability of each of the following events.

13 All 5 cards are spades.
14 All 5 cards are black.
15 All 5 cards are spades, given that the first 3 are spades.
16 All 5 cards are black, given that the first 4 are black.
17 All 5 cards belong to the same suit.
18 All 5 cards belong to the same suit, given that the first 2 cards belong to the same suit.
19 Two cards are hearts and 3 cards are clubs.
20 Two cards are hearts and 3 cards are clubs, given that the first 3 cards are clubs.
21 Each card is either a 4 or a 5.
22 Each card is either a 4 or 5, given that the first four cards are 5's.

A pair of fair dice are rolled. Determine the probability of each of the following events.

23 The sum of the dice is 4 or less.
24 The sum of the dice is 4 or less, given that one die is 3.
25 The sum of the dice is 5 or more.
26 The sum of the dice is 5 or more, given that one die is 4.
27 The sum of the dice is 6.
28 The sum of the dice is 6, given that the first die is 6.

A fair coin is tossed 5 successive times. Use the concept of independent events to determine the probability of each of the following events.

29 The first 3 tosses result in heads and the last 2 tosses result in tails.
30 The first 3 tosses result in tails and the last 2 tosses result in heads.
31 The first 4 tosses result in tails and the last toss results in heads.
32 The first 4 tosses result in heads and the last toss results in tails.
33 The first toss lands on heads.
34 The last toss lands on tails.
35 The last toss lands on heads, given that each of the first 4 tosses results in heads.
36 The last toss lands on heads, given that each of the first 4 tosses results in tails.

Suppose E and F are events for some experiment with sample space S, and suppose that

$$P(E) = .5, \qquad P(F) = .4, \qquad P(E \cap F) = .3.$$

Determine the probability of each of the following events.

37 $E \cup F$
38 E'
39 F'
40 $(E \cap F)'$
41 $E' \cap F'$ (**Hint:** $(A \cup B)' = A' \cap B'$.)

Solve each of the following.

42 A shipment of new slacks consists of 100 pairs in each of 3 styles, A, B, and C. One out of every 20 pairs in style A is flawed; one out of every 10 in style B is flawed; one out of every 25 in style C is flawed. If 10 pairs are selected randomly to be examined, what is the probability that one or more of those pairs is flawed? Suppose 5 of the pairs selected are style A and the other 5 are style C; what is the probability that a pair is flawed?
43 How do the answers in problem 42 change if only one out of every 50 pairs in style A is flawed?
44 A box contains 40 red transistors and 60 blue transistors. One out of every 8 red transistors is bad, and one out of every 10 blue transistors is bad. If 4 transistors are randomly chosen from the box and tested, what is the probability that a bad one was chosen? Suppose that 3 of the transistors chosen are blue and 1 is red; what is the probability that a bad one was chosen?
45 Suppose that 2 blue transistors and 3 red transistors are chosen from the

box in problem 44, what is the probability that a bad transistor was chosen?

46   A business student must pass each of 3 courses in which he is enrolled in order to graduate. He decides the probabilities of passing mathematics, economics, and accounting are .8, .9, and .6, respectively. What is the probability that the student will graduate on time?

47   Suppose the correct probabilities of the student in problem 46 passing mathematics, economics, and accounting are .5, .4, and .3, respectively. What is the correct probability that the student will graduate on time?

48   An airplane manufacturer has 3 independent assembly lines working on parts of its new airplane. The company president believes that the probabilities of the 3 lines finishing on schedule are .90, .85, and .95. What is the probability that the new airplane will be assembled on schedule?

49   What is the answer to problem 48, if the correct probabilities for the assembly lines finishing on schedule are .80, .80, and .90?

14.3   Bayes' Formula

Sometimes we can use probability to investigate the possible cause of an event that has already taken place. An important tool in such an investigation is Bayes' Formula.

Bayes' Formula for Two Mutually Exclusive Events

$$P(F|E) = \frac{P(F)P(E|F)}{P(F)P(E|F) + P(F')P(E|F')}$$

In this formula F and F' are mutually exclusive events, where F' denotes the complement of F.

The usefulness of this formula is best explained by example.

EXAMPLE 1

The state accounting board that administers the Certified Public Accountant Exam says the probability of a competent accountant passing the exam is .87. The board says the probability of an incompetent accountant passing the exam is .005. The results of the exam in recent years convince the board that 40% of those taking the exam are competent accountants. What is the probability that someone who passes the exam is a competent accountant?

Solution
In this problem the experiment is a typical person registering for the CPA. Exam and completing the exam. The person who registers may be competent or incompetent at accounting. If F is the event that the person is competent, then the probability $P(F)$ is .40 and $P(F') = .6$. If E is the event that the person passes the exam then $P(E|F) = .87$, since $P(E|F)$ is the probability of a competent accountant passing the exam. Similarly, we have

$P(E|F') = .005$, since $P(E|F')$ is the probability of an incompetent accountant passing the exam.

We are seeking $P(F|E)$, the probability that a person who passes the exam is a competent accountant. According to Bayes' Formula, we have

$$P(F|E) = \frac{P(F)P(E|F)}{P(F)P(E|F) + P(F')P(E|F')}$$

$$= \frac{(.4)(.87)}{(.4)(.87) + (.6)(.005)} \approx .99$$

Thus, if a person passes the exam, the probability that the person is competent is .99. ◼

Before we can verify Bayes' Formula for Two Mutually Exclusive Events, we must demonstrate the elementary formula

$$E = (E \cap F) \cup (E \cap F').$$

The explanation is easier to follow, if you consider a Venn diagram.

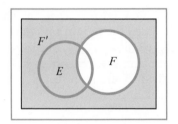

The events E and F are indicated by the circles, and F' is indicated by the shaded area. Note that $E \cap F$, $E \cap F'$, and $(E \cap F) \cup (E \cap F')$ are parts of E; what must be determined is that $(E \cap F) \cup (E \cap F')$ is all of E. Any outcome x belonging to E either belongs to F (and so $x \in E \cap F$), or else x does not belong to F, in which case we have $x \in F'$ and $x \in E \cap F'$. Thus, every outcome in E belongs to either $E \cap F$ or $E \cap F'$, and $(E \cap F) \cup (E \cap F')$ is all of E. This shows that the preceding equation is true.

Now we can verify Bayes' Formula for Two Mutually Exclusive Events. It is easy to see that $(E \cap F)$ and $(E \cap F')$ are mutually exclusive events. According to Theorem 1 in Section 14.2, we have

$$P(E) = P([E \cap F] \cup [E \cap F']) \qquad (1)$$
$$= P(E \cap F) + P(E \cap F').$$

We shall substitute in the last equation for $P(E \cap F)$ and $P(E \cap F')$. The conditional probability formula

$$P(E|F) = \frac{P(E \cap F)}{P(F)}$$

gives us

$$P(F)P(E|F) = P(E \cap F),\tag{2}$$

and similarly

$$P(F')P(E|F') = P(E \cap F').\tag{3}$$

Substituting in (1) according to (2) and (3), we get

$$P(E) = P(F)P(E|F) + P(F')P(E|F').$$

Finally, we use the last equation and (2) to rewrite the conditional probability formula

$$P(F|E) = \frac{P(F \cap E)}{P(E)}$$

$$= \frac{P(F)P(E|F)}{P(F)P(E|F) + P(F')P(E|F')}.$$

This verifies Bayes' Formula for Two Mutually Exclusive Events.

The next example is a practical problem that is quite different from the first example.

EXAMPLE 2

A prototype for a new lawn mower has chronic starting problems. The engineer who designed the mower says that the probability of an improper mixture of gasoline and air is .04. She says that the probability of starting problems when the mixture is incorrect is .9, and the probability of starting problems when the mixture is correct is .1. What is the probability that the gas-air mixture is incorrect?

Solution

In this problem the experiment consists of making attempts to start the lawn mower and then checking the correctness of the gas-air mixture. Let E be the event corresponding to "the mower has chronic starting problems," and let F be the event corresponding to "the gas-air mixture is incorrect." The probability of starting problems when the gas-air mixture is incorrect is $P(E|F) = .9$. The probability of starting problems when the gas-air mixture is correct is $P(E|F') = .1$. Since we know that $P(F) = .04$ and $P(F') = 1 - .04 = .96$, we get the desired answer from Bayes' Formula.

$$P(F|E) = \frac{P(F)P(E|F)}{P(F)P(E|F) + P(F')P(E|F')}$$

$$= \frac{(.04)(.9)}{(.04)(.9) + (.96)(.1)}$$

$$\approx .27 \quad \blacksquare$$

We shall now explain another form of Bayes' Formula. Suppose we have three mutually exclusive events F_1, F_2, F_3 in the sample space S, and

$$S = F_1 \cup F_2 \cup F_3.$$

It can be shown that for any event E we have

$$E = (E \cap F_1) \cup (E \cap F_2) \cup (E \cap F_3)$$

and
$$P(E) = P(E \cap F_1) + P(E \cap F_2) + P(E \cap F_3).$$

If $F_1 = F$, $F_2 = F'$, and $F_3 = \phi$, then the above equation is the same as our starting point in the derivation of Bayes' Formula prior to Example 1. Using the conditional probability formula as in that derivation, we see that

$$P(E) = P(F_1)P(E|F_1) + P(F_2)P(E|F_2) + P(F_3)P(E|F_3)$$

and
$$P(E \cap F_1) = P(F_1)P(E|F_1).$$

Substituting in the formula

$$P(F_1|E) = \frac{P(F_1 \cap E)}{P(E)},$$

we get

$$P(F_1|E) = \frac{P(F_1)P(E|F_1)}{P(F_1)P(E|F_1) + P(F_2)P(E|F_2) + P(F_3)P(E|F_3)}.$$

This is **Bayes' Formula for Three Mutually Exclusive Events.** Similarly, we obtain formulas for $P(F_2|E)$ and $P(F_3|E)$.

Now we solve a practical problem with the previous formula.

EXAMPLE 3

A stock analyst believes that only three factors could explain an increase in the Dow Jones Industrial Average (D.J.I.A.) over the past year. He believes that whenever interest rates are perceived as declining, the probability of an increase in the D.J.I.A. is .9; whenever consumer spending is perceived as increasing, the probability of an increase in the D.J.I.A. is .7; and whenever Congress is perceived as pro-business, the probability of an increase is .6. Furthermore, the analyst believes that the probabilities that each of these events occurred during the past year are .6, .6, and .2, respectively. What was the most likely cause of the increase in the D.J.I.A.?

Solution
Let F_1, F_2, and F_3 be the events corresponding to "interest rates are perceived as declining," "consumer spending is perceived as increasing," and "Congress is perceived as pro-business," respectively. Let E be the event corresponding to "the D.J.I.A. increases." We must use Bayes' Formula to compute

$$P(F_1|E), \qquad P(F_2|E), \qquad P(F_3|E).$$

We are given the conditional probabilities

$$P(E|F_1) = .9, \qquad P(E|F_2) = .7, \qquad P(E|F_3) = .6.$$

Also, we are given the probabilities

$$P(F_1) = .6, \qquad P(F_2) = .6, \qquad P(F_3) = .2.$$

Now we make the appropriate substitutions in Bayes' Formula for Three Mutually Exclusive Events.

$$P(F_1|E) = \frac{P(F_1)P(E|F_1)}{P(F_1)P(E|F_1) + P(F_2)P(E|F_2) + P(F_3)P(E|F_3)}$$

$$= \frac{(.6)(.9)}{(.6)(.9) + (.6)(.7) + (.2)(.6)}$$

$$= .5$$

$$P(F_2|E) = \frac{P(F_2)P(E|F_2)}{P(F_1)P(E|F_1) + P(F_2)P(E|F_2) + P(F_3)P(E|F_3)}$$

$$= \frac{(.6)(.7)}{(.6)(.9) + (.6)(.7) + (.2)(.6)}$$

$$\approx .39$$

$$P(F_3|E) = \frac{P(F_3)P(E|F_3)}{P(F_1)P(E|F_1) + P(F_2)P(E|F_2) + P(F_3)P(E|F_3)}$$

$$= \frac{(.2)(.6)}{(.6)(.9) + (.6)(.7) + (.2)(.6)}$$

$$\approx .11$$

The most likely cause of the increase in the D.J.I.A. was F_1, the perception that interest rates were declining. ∎

Sometimes the formula for $P(E)$, preceding Bayes' Formula for Three Mutually Exclusive Events, is useful in solving problems. We illustrate this.

EXAMPLE 4

Three plants, which we denote by A, B, and C, produce 60%, 20%, and 20%, respectively, of a certain product. The percentages of the units produced by A, B, and C that are faulty are 1%, 2%, and 3%, respectively. What is the probability of a randomly chosen unit of this product being faulty?

Solution
Let F_1, F_2, and F_3, respectively, be the events that the chosen unit was produced at A, B, and C, respectively. From the preceding information we know that

$$P(F_1) = .6, \qquad P(F_2) = .2, \qquad P(F_3) = .2.$$

Let E be the event "the chosen unit is faulty." From the information in the problem, we know that

$$P(E|F_1) = .01, \qquad P(E|F_2) = .02, \qquad P(E|F_3) = .03.$$

Now we use the formula for $P(E)$, preceding Bayes' Formula for Three Mutually Exclusive Events.

$$P(E) = P(F_1)P(E|F_1) + P(F_2)P(E|F_2) + P(F_3)P(E|F_3)$$
$$= (.6)(.01) + (.2)(.02) + (.2)(.03)$$
$$= .016$$

The probability that the chosen unit is faulty is .016. ∎

As you might guess, there is a version of Bayes' Formula for Four Mutually Exclusive Events. In fact, if n is a positive integer, then the version of Bayes' Formula for n Mutually Exclusive Events is

$$P(F_i|E) = \frac{P(F_i)P(E|F_i)}{P(F_1)P(E|F_1) + P(F_2)P(E|F_2) + \cdots + P(F_n)P(E|F_n)}.$$

Exercises 14.3

Let E and F be events for an experiment with sample space S. In each of the following problems, find $P(F|E)$ given the other probabilities.

1 $P(F) = .5$, $P(E|F) = .8$, $P(E|F') = .1$
2 $P(F) = .1$, $P(E|F) = .5$, $P(E|F') = .5$
3 $P(F') = .3$, $P(E|F) = .1$, $P(E|F') = .6$
4 $P(F') = .6$, $P(E|F) = .9$, $P(E|F') = .9$
5 $P(E) = .2$, $P(F \cap E) = .1$
6 $P(E) = .5$, $P(F \cap E) = .3$
7 $P(E') = .7$, $P(E \cap F) = .1$
8 $P(E') = .6$, $P(E \cap F) = .2$
9 $P(E \cap F) = .2$, $P(E|F) = .4$, $P(E|F') = .6$
10 $P(E \cap F) = .6$, $P(E|F) = .8$, $P(E|F') = .4$

Let F_1, F_2, F_3, and E be the events for an experiment with sample space S. In each of the following problems, determine each indicated probability given that F_1, F_2, and F_3 are mutually exclusive events such that $S = F_1 \cup F_2 \cup F_3$ and

$$P(F_1) = .2, \qquad P(F_2) = .3, \qquad P(F_3) = .5,$$
$$P(E|F_1) = .4, \qquad P(E|F_2) = .6, \qquad P(E|F_3) = .4.$$

11 $P(F_1|E)$
12 $P(F_2|E)$
13 $P(F_3|E)$
14 $P(E)$
15 $P(E')$
16 $P(F_1'|E)$ (**Hint:** $P(E \cap F_1') = P(E \cap F_2) + P(E \cap F_3)$.)
17 $P(F_2'|E)$
18 $P(F_3'|E)$

Solve each of the following.

19   The state bar association, which administers the law exam that determines if a person is allowed to practice law, says the probability of a competent attorney passing the exam is .84. The probability of an incompetent attorney passing the exam is .01, and 30% of those taking the exam are competent attorneys. What is the probability that someone who passes the exam is a competent attorney?

20   In problem 19, what is the probability that someone who passes the exam is an incompetent attorney?

21   In Example 1, what is the probability that someone who passes the CPA Exam is an incompetent accountant?

22   In Example 2, what is the probability that the gas-air mixture is correct?

23   A motor on an assembly line suddenly quits. The probability that less than normal voltage would reach the motor is .3, and the probability that the motor would quit with less than normal voltage is .9. The probability that the motor would suddenly quit with normal voltage is .2. What is the probability that the cause of the motor quitting is less than normal voltage?

24   In problem 23, what is the probability that the cause of the motor quitting is something other than less than normal voltage?

25   Using the information supplied in Example 4, determine the probability that a randomly chosen unit that is faulty was produced at Plant A. Determine the probability that it was produced at Plant B; find the probability that it was produced at Plant C.

26   Suppose the probability of a stagnant economy over the next year is .6, the probability of a booming economy is .2, and the probability of a recession is .2. Furthermore, suppose that the probabilities of a good return on an investment strategy based on commodities are .2, .9, and .001, respectively, in each of the preceding economic circumstances. If the commodity strategy does result in a good return, what is the probability that the economy was booming?

27   In problem 26, what is the probability that the economy was in a recession when the commodity strategy produced a good return?

14.4   Bernoulli Processes

Many multistep experiments are merely repetitions of one simple experiment. For example, tossing a coin 20 times is an experiment with 2^{20}, or 1,048,576, outcomes; however, this is merely 20 repetitions of the simple experiment of tossing a coin once. Any such experiment is called a **Bernoulli process** if it has the following three properties.

1. The multistep experiment consists of repeating one simple experiment with only two outcomes.
2. The probability of each outcome of the simple experiment is the same in each repetition.
3. The repetitions are independent events.

These processes are named after Jacob Bernoulli (1654–1705), a Swiss mathematician who made extensive contributions to probability. The simple experiment is said to be the **trial,** and the Bernoulli process consists of **repeated trials.**

We shall use Example 1 to show how this point of view permits a nice solution to a problem that would otherwise be very tedious.

EXAMPLE 1

Suppose a coin is tossed 10 times. What is the probability of getting 5 heads and 5 tails?

Solution

Since the number of outcomes is $2^{10} = 1024$, we want to avoid constructing the sample space. Clearly this experiment consists of repeated trials where one trial is a single coin toss, so it is a Bernoulli process.

One of the outcomes that would result in 5 heads and 5 tails is

$$H\ H\ H\ H\ H\ T\ T\ T\ T\ T,$$

where H stands for heads and T stands for tails. The probability of this particular outcome can be calculated by using the independent event formula given in Section 14.2. The probability is

$$\left(\frac{1}{2}\right)^{10} = \frac{1}{1,204} \approx .00098.$$

However, the problem expressed in Example 1 does not ask for the probability of the particular outcome

$$H\ H\ H\ H\ H\ T\ T\ T\ T\ T.$$

It asks for the probability of the event E that consists of *all* the outcomes that enumerate 5 heads and 5 tails. Each of those outcomes uses the symbol H five times and the symbol T five times. What varies from outcome to outcome is the location of the H's (and, hence, the location of the T's). Since we have 10 repeated trials, a typical outcome in E is completely determined when we select which 5 trials result in H. The number of ways to choose a set with 5 elements from a set with 10 elements is

$$\binom{10}{5} = \frac{10!}{5!5!} = \frac{10 \cdot 9 \cdot 8 \cdot 7 \cdot 6}{5 \cdot 4 \cdot 3 \cdot 2} = 252.$$

Thus, E contains 252 outcomes and the probability of any one of them is $\frac{1}{1,204}$.

For each outcome, construct a set that consists of just that outcome. These are mutually exclusive events, and the union of them is E. It follows that the probability of E is the sum of the probabilities of each of these outcomes. Thus, we have

$$P(E) = 252\left(\frac{1}{1,024}\right) = \frac{63}{256} \approx .246.$$

The probability of getting 5 heads and 5 tails is .246. ■

The methods used in Example 1 are precisely the methods needed to prove Theorem 1, which we give without proof. In order to deal with the great diversity of experiments possible, we agree to designate one of the outcomes of the trial as **success,** and the other outcome as **failure.**

Theorem 1   If p is the probability of success in a single trial, the probability of r successes and $n - r$ failures in n repeated trials is

$$P(r \text{ successes}) = \binom{n}{r} p^r (1 - p)^{n-r}.$$

You may recognize that $P(r \text{ successes})$ is one of the terms in the binomial expansion for $(p + q)^n$ where $q = 1 - p$. In fact, Bernoulli processes are sometimes called "binomial processes." If we add together the probabilities of each possible number of successes, we get

$$\binom{n}{0} (1 - p)^n + \binom{n}{1} p(1 - p)^{n-1} + \cdots + \binom{n}{n-1} p^{n-1}(1 - p) + \binom{n}{n} p^n.$$

This is exactly the binomial expansion of $(p + [1 - p])^n$. Clearly $(p + [1 - p])^n = 1^n = 1$; thus, we have observed that the sum of the preceding probabilities is 1. This result is as it must be, since for every outcome of the experiment there is some number of successes!

In the remaining examples we exploit Theorem 1 without any further comment as to why it is true.

EXAMPLE 2

Suppose a die is rolled 6 successive times. What is the probability of getting a 4 exactly 3 times?

Solution

A single roll of the die is a trial, and we let "success" represent getting a 4. We want to find the probability of 3 successes and 3 failures in 6 repeated trials. The probability of getting a 4 on a single roll of the die is $\frac{1}{6}$, and the probability of getting something else is $\frac{5}{6}$. According to Theorem 1, the probability of 3 successes in 6 repeated trials is

$$\binom{6}{3}\left(\frac{1}{6}\right)^3\left(\frac{5}{6}\right)^3 = 20\left(\frac{5^3}{6^6}\right) = \frac{2500}{46,656} \approx .05. \ \blacksquare$$

Now we give a practical example.

EXAMPLE 3

The probability that any unit produced by a certain plant is defective is .02. If 12 units are randomly selected, what is the probability that exactly 5 of them are defective?

Solution

Selecting a single unit is the trial, and we let success indicate that the chosen unit is defective. The probability of success on a single trial is .02, and the probability of failure is .98. According to Theorem 1, the probability of 5 successes and 7 failures in 12 repeated trials is

$$\binom{12}{5}(.02)^5(.98)^7 = 792(.02)^5(.98)^7$$

$$\approx .0000022. \ \blacksquare$$

We conclude this section with a more difficult practical example.

EXAMPLE 4

Hustle Life Insurance Company determines that the probability of one of its representatives selling a policy to a prospect at a face-to-face meeting is .05. If a sales representative has face-to-face meetings with 20 prospects in a given week, what is the probability that he sells policies to 3 or more prospects?

Solution
The trial in this experiment is the face-to-face meeting with a prospect in order to sell a policy. Success means that a sale was made, and failure means that it was not. The probability of success on any trial is .05, and there are 20 repeated trials.

After a moment of thought, we decide it would be easiest to compute the probability that the representative sells policies to 2 or fewer prospects and subtract that from 1; a direct computation of the probabilities of 3, 4, 5, . . . , 19, 20 sales would be more tedious. The probabilities of no successes, exactly 1 success and exactly 2 successes are

$$\binom{20}{0}(.95)^{20}, \qquad \binom{20}{1}(.05)(.95)^{19}, \qquad \text{and} \qquad \binom{20}{2}(.05)^2(.95)^{18},$$

respectively. To get the probability of 2 or fewer successes, we must calculate each of these numbers and then add them together.

$$\binom{20}{0}(.95)^{20} = (.95)^{20} \approx .358$$

$$\binom{20}{1}(.05)(.95)^{19} \approx 20(.05)(.377) = .377$$

$$\binom{20}{2}(.05)^2(.95)^{18} \approx 190(.0025)(.397) = .189$$

Thus, the answer is

$$P(3 \text{ or more successes}) = 1 - (.358 + .377 + .189)$$
$$= 1 - .924 = .076. \ \blacksquare$$

Exercises 14.4

A coin is tossed 8 successive times. Determine the probability of each of the following events.

1 Four heads and 4 tails

2 Three heads and 5 tails

3 Two heads and 6 tails
4 One heads and 7 tails
5 No heads and 8 tails
6 No tails and 8 heads
7 One tails and 7 heads
8 Two tails and 6 heads

A die is rolled 5 successive times. Determine the probability of each of the following events.

9 Three 2's
10 Two 3's
11 One 6
12 Four 5's
13 Five 1's
14 Six 4's

An experiment consists of 6 repeated trials and the probability of success on each trial is .8. Determine the probability of each of the following events.

15 One success
16 Six successes
17 No successes
18 Five successes
19 Three successes
20 Four successes

Solve each of the following.

21 The probability of a transistor made by Reliable Electronics being faulty is .001. If 6 of its transistors are randomly selected, what is the probability that exactly 1 of them is faulty?

22 In problem 21, what is the probability of getting exactly 2 faulty transistors?

23 In problem 21, what is the probability of getting no faulty transistors?

24 In problem 21, what is the probability of 2 or fewer faulty transistors?

25 The manager of an automobile dealership observes that his sales staff sells a car to 1 out of every 25 prospects who enter the showroom. If 5 prospective customers enter the showroom on a given day, what is the probability of exactly 1 sale?

26 In problem 25, what is the probability of exactly 2 sales?

27 In problem 25, what is the probability of no sales?

28 In problem 25, what is the probability of 3 or more sales?

29 A store owner observes that 1 of every 10 people walking by his store enters. Thus, he decides that the probability of a passerby entering the store is .1. What is the probability that exactly 2 of the next 8 passersby enter the store?

30 In problem 29, what is the probability that 1 passerby enters the store?

31 In problem 29, what is the probability that no passerby enters the store?

32 In problem 29, what is the probability that 3 or more passersby enter the store?

14.5   Markov Processes (Optional)

A Markov process is a sharp contrast to a Bernoulli process. A Bernoulli process consists of repeated trials, each of which is independent of the others. In a **Markov process** the result of each trial is determined by the result of the previous trial, according to the intrinsic properties of the experiment.

In order to describe a Markov process, we construct a row vector with nonnegative entries summing to 1, which is called a **state vector.** For example, (.01 .99) is a state vector. The state vector for an experiment gives the probabilities of the possible outcomes of a single trial. Each Markov process requires a square matrix with nonnegative entries having the property that the entries in any row sum to 1. This matrix is called the **transition matrix.** For example, the matrix

$$\begin{bmatrix} .8 & .2 \\ .4 & .6 \end{bmatrix}$$

is a transition matrix. The transition matrix must have the appropriate order to allow us to compute the product of the state vector and the transition matrix, like

$$(.01 \; .99) \begin{bmatrix} .8 & .2 \\ .4 & .6 \end{bmatrix} = (.404 \; .596).$$

In a Markov process, the result of any trial is expressed as a state vector and the result of the next trial is obtained by multiplying the state vector by the transition matrix.

Example 1 will clarify the nature of Markov processes and show how they arise.

EXAMPLE 1

A fast food restaurant introduces a new pork sandwich and begins to advertise it a few weeks later. The restaurant owner hopes to switch customers from the current menu, consisting of beef sandwiches like hamburgers, to the new pork sandwich. The price of beef has risen while the price at which hamburgers can be sold has been held down by intense competition; consequently, the restaurant can make more profit on the pork sandwich. Initially, the probability of a customer ordering the pork sandwich is only .01. The initial state vector is (.01 .99), which indicates the probabilities of an order for the pork sandwich and an order for a beef sandwich.

An advertising campaign designed to switch customers of the restaurant to the pork sandwich is begun. The following weekly pattern emerges. About 80% of those ordering the pork sandwich continue to order it, and 40% of those who have not tried it will order it. These facts are summarized in the chart.

		This Week	
		Pork	Beef
Previous Week	Pork	.8	.2
	Beef	.4	.6

Determine the probability of a customer ordering the pork sandwich after 3 weeks.

Solution

The initial state vector is (.01 .99). We obtain the state vector at the end of the first week by multiplying the state vector by the transition matrix.

$$(.01\ .99)\begin{bmatrix} .8 & .2 \\ .4 & .6 \end{bmatrix} = (.404\ .596)$$

For simplicity, we shall round off each number to two decimal places. Thus, the state vector at the beginning of the second week is (.40 .60). The state vector at the end of the second week is obtained

$$(.40\ .60)\begin{bmatrix} .8 & .2 \\ .4 & .6 \end{bmatrix} = (.56\ .44).$$

The state vector at the end of the third week is

$$(.56\ .44)\begin{bmatrix} .8 & .2 \\ .4 & .6 \end{bmatrix} = (.624\ .376) \approx (.62\ .38).$$

After 3 weeks, the probability of a customer ordering the pork sandwich is .62. ■

If we continue to follow the results of the advertising campaign described in Example 1, an interesting pattern emerges. We make a table to show the state vector at the end of each week for 8 weeks. We round off the numbers to four decimal places.

At the End of Week	State Vector
1	(.4040 .5960)
2	(.5616 .4384)
3	(.6246 .3754)
4	(.6499 .3501)
5	(.6599 .3401)
6	(.6640 .3360)
7	(.6656 .3344)
8	(.6662 .3338)

It appears that the impact of the ad campaign is greatly diminished by the end of the fifth week. With the numbers rounded off to two decimal places, the last two state vectors are the same vector (.67 .33). Applying the transition matrix to this state vector and rounding off to two decimal places, we have

$$(.67\ .33)\begin{bmatrix} .8 & .2 \\ .4 & .6 \end{bmatrix} = (.67\ .33).$$

A state vector with the property that it is unaltered when multiplied by the transition matrix is called the **steady state vector** for the Markov process.

Now that we suspect that a Markov process has a steady state vector, we can think of an easier way to find it. If the transition matrix for the Markov process is

$$\begin{bmatrix} .8 & .2 \\ .4 & .6 \end{bmatrix},$$

then the steady state vector $(x \quad 1-x)$ has the property that

$$(x \quad 1-x)\begin{bmatrix} .8 & .2 \\ .4 & .6 \end{bmatrix} = (x \quad 1-x).$$

This matrix equation gives us two equations, which are easily solved.

$$.8x + .4(1 - x) = x$$

$$.2x + .6(1 - x) = 1 - x$$

or

$$-.6x + .4 = 0$$

$$.6x - .4 = 0$$

Clearly the solution is

$$.6x = .4$$

$$x = \frac{2}{3} \approx .67.$$

We conclude that the steady state vector is

$$(x \quad 1 - x) \approx (.67 \ .33).$$

The steady state vector gives us the ultimate result of a given Markov process. The next example demonstrates further the method for finding the steady state vector. Each multiplication of the state vector times the transition matrix is called a **transition.**

EXAMPLE 2

The initial state vector for a Markov process is (.3 .7) and the transition matrix is

$$\begin{bmatrix} .6 & .4 \\ .5 & .5 \end{bmatrix}.$$

Find the state vector after 2 transitions, and determine the steady state vector.

Solution

We compute two transitions.

$$(.3 \ .7)\begin{bmatrix} .6 & .4 \\ .5 & .5 \end{bmatrix} = (.53 \ .47)$$

$$(.53 \ .47)\begin{bmatrix} .6 & .4 \\ .5 & .5 \end{bmatrix} = (.553 \ .447)$$

After 2 transitions, the state vector is (.553 .447).

Now we compute the steady state vector by the method given in the two paragraphs preceding this example.

$$(x \quad 1-x)\begin{bmatrix} .6 & .4 \\ .5 & .5 \end{bmatrix} = (x \quad 1-x)$$

$$.6x + .5(1-x) = x$$

$$.4x + .5(1-x) = 1 - x$$

or

$$x = \frac{5}{9}, \quad 1 - x = \frac{4}{9}$$

The steady state vector is $(\frac{5}{9} \frac{4}{9}) \approx (.56 \ .44)$. ■

We conclude our examples with a practical application of steady state vectors with three possible outcomes.

EXAMPLE 3

A survey of car buyers shows that 60% buy a General Motors car, 20% buy a Ford car, and 20% buy some other make of car. The chart shows the loyalties of car buyers. For example, if a person bought a G.M. car last time then the probability that the next purchase will be a G.M. car is .85; the probability that the next purchase will be a Ford car is .05; and the probability that the next purchase will be some other make of car is .10.

		Next Car Purchased		
		G.M.	Ford	Other
Last Car Purchased	G.M.	.85	.05	.10
	Ford	.20	.60	.20
	Other	.20	.10	.70

If none of these factors is altered, what will be the division of the automobile market after the next purchases, and what will be the market division ultimately?

Solution

The current state vector is (.60 .20 .20) and the transition matrix is

$$\begin{bmatrix} .85 & .05 & .10 \\ .20 & .60 & .20 \\ .20 & .10 & .70 \end{bmatrix}.$$

We multiply the state vector by the transition matrix to get the market division after the next round of purchases.

$$(.6 \ .2 \ .2)\begin{bmatrix} .85 & .05 & .1 \\ .2 & .6 & .2 \\ .2 & .1 & .7 \end{bmatrix} = (.59 \ .17 \ .24)$$

These indicate that 59% of the buyers purchase G.M. cars, 17% purchase Ford cars, and 24% purchase other makes.

In order to determine the steady state vector, we let x, y, and z be nonnegative numbers such that $x + y + z = 1$. Now, we must solve the equation

$$(x\ y\ z)\begin{bmatrix} .85 & .05 & .1 \\ .2 & .6 & .2 \\ .2 & .1 & .7 \end{bmatrix} = (x\ y\ z).$$

Equating entries on both sides and solving the resulting systems, we get the following.

$$.85x + .2y + .2z = x$$
$$.05x + .6y + .1z = y$$
$$.1x + .2y + .7z = z$$

or

$$-15x + 20y + 20z = 0$$
$$5x - 40y + 10z = 0$$
$$10x + 20y - 30z = 0$$

or

$$-3x + 4y + 4z = 0$$
$$x - 8y + 2z = 0$$
$$x + 2y - 3z = 0$$

or

$$z = 2y \quad \text{and} \quad x = 4y$$

Since

$$1 = x + y + z = 4y + y + 2y = 7y,$$

we get

$$x = \frac{4}{7} \approx .57, \qquad y = \frac{1}{7} \approx .14, \qquad z = \frac{2}{7} \approx .29.$$

Ultimately, 57% of car buyers purchase G.M. cars, 14% purchase Ford cars, and 29% purchase other makes. ∎

Exercises 14.5

In each problem below, use the given transition matrix and the initial state vector to compute the state vector after two transitions. Also, determine the steady state vector.

1 $\begin{bmatrix} .2 & .8 \\ .5 & .5 \end{bmatrix}$, $(.4\ .6)$

2 $\begin{bmatrix} .6 & .4 \\ .4 & .6 \end{bmatrix}$, $(.5\ .5)$

3 $\begin{bmatrix} 1 & 0 \\ .4 & .6 \end{bmatrix}$, (.2 .8)   4 $\begin{bmatrix} .5 & .5 \\ 0 & 1 \end{bmatrix}$, (.6 .4)

5 $\begin{bmatrix} 1 & 0 \\ 0 & 1 \end{bmatrix}$, (.1 .9)   6 $\begin{bmatrix} 1 & 0 & 0 \\ 0 & 1 & 0 \\ 0 & 0 & 1 \end{bmatrix}$, (.2 .2 .6)

7 $\begin{bmatrix} .1 & .1 & .8 \\ .2 & .6 & .2 \\ .4 & .2 & .4 \end{bmatrix}$, (.3 .3 .4)   8 $\begin{bmatrix} .3 & .4 & .3 \\ .4 & .2 & .4 \\ .8 & .1 & .1 \end{bmatrix}$, (.1 .8 .1)

9 $\begin{bmatrix} .2 & .2 & .6 \\ 0 & 1 & 0 \\ .5 & .3 & .2 \end{bmatrix}$, (0 .4 .6)   10 $\begin{bmatrix} .1 & .8 & .1 \\ .6 & .2 & .2 \\ 0 & 0 & 1 \end{bmatrix}$, (.5 .2 .3)

Identify which of the following matrices might be transition matrices.

11 $\begin{bmatrix} .3 & .7 \\ .9 & .1 \end{bmatrix}$ 12 $\begin{bmatrix} .5 & .5 \\ .4 & .6 \end{bmatrix}$

13 $\begin{bmatrix} .5 & .4 \\ .5 & .6 \end{bmatrix}$ 14 $\begin{bmatrix} .3 & .9 \\ .7 & .1 \end{bmatrix}$

15 $\begin{bmatrix} 0 & 1 \\ 1 & 0 \end{bmatrix}$ 16 $\begin{bmatrix} 0 & 0 & 1 \\ 0 & 1 & 0 \\ 1 & 0 & 0 \end{bmatrix}$

17 $\begin{bmatrix} .1 & .2 & .3 \\ .2 & .3 & .4 \\ .3 & .4 & .5 \end{bmatrix}$ 18 $\begin{bmatrix} .4 & .3 & .4 \\ .2 & .5 & .4 \\ .4 & .2 & .2 \end{bmatrix}$

Identify which of the following might be a state vector.

19 (.6 .4)   20 (0 1)
21 (−.2 1.2)   22 (1.5 −.5)
23 (.7 .4)   24 (.6 .6)
25 (0 1 0)   26 (0 0 1)
27 (.4 .3 .5)   28 (.6 .6 −.2)

Solve each of the following.

29 An advertising campaign is designed to switch the customers of a beer company from the moderately priced beer to the higher priced beer. Each month 25% of those who drink the cheaper beer try the expensive beer, and 85% of those who try it continue to drink it. Initially, only 10% of the customers drink the expensive beer. How many of the customers drink it after 1 month?

30 In problem 29, what percentage of the customers drink the expensive beer after 2 months?

31 In problem 29, what percentage of the customers drink the expensive beer after 3 months?

32 In problem 29, what percentage of the customers will ultimately drink the expensive beer?

33 A survey of soft drink consumers indicates that 70% of the soft drinks consumed are colas. Each year 25% of the noncola drinkers try a cola, and 90% of those who try a cola continue to drink it. Determine the percentage of cola drinkers 1 year later.

34 In problem 33, determine the percentage of cola drinkers after 2 years.

35 In problem 33, determine the percentage of cola drinkers after 3 years.

36 In problem 33, what percentage of the soft drink consumers will ultimately drink a cola?

Review of Terms

Review Problems

A fair coin is tossed 5 consecutive times. Determine each of the following.

1 The number of outcomes in the sample space
2 The probability of 5 tails
3 The probability of 4 heads followed by 1 tails
4 The probability of 2 heads and 3 tails
5 The probability of 4 heads and 1 tails
6 The probability of more heads than tails
7 The probability of 1 or more heads
8 The probability that the fifth toss is heads, given that the first 4 tosses were heads

A fair die is rolled 4 consecutive times. Determine the probability of each of the following.

9   One followed by 2 followed by 3 followed by 4
10 Six followed by 5 followed by 4 followed by 3
11 Exactly two 5's
12 Exactly three 6's
13 The sum of the resulting numbers is 5 or less.
14 The sum of the resulting numbers is 7 or less.
15 The fourth roll lands on 6, after each of the first 3 rolls landed on 6
16 How many outcomes belong to the sample space for this experiment?

Five cards are dealt randomly from a standard deck of 52 cards. Determine the probability of each of the following.

17 Three kings and 2 queens
18 Four aces and 1 king
19 Five spades
20 Four spades and 1 diamond
21 The fifth card is a club, given that the first 4 cards are clubs.
22 All the cards are hearts, given that all the cards are red.

Let F_1, F_2, F_3, and E be events for an experiment with sample space S. Determine each indicated probability, given that F_1, F_2, and F_3 are mutually exclusive events such that $S = F_1 \cup F_2 \cup F_3$ and

$$P(F_1) = .4 \qquad P(F_2) = .2 \qquad P(F_3) = .4$$

$$P(E|F_1) = .5 \qquad P(E|F_2) = .4 \qquad P(E|F_3) = .6.$$

23 $P(F_2')$ 24 $P(E)$
25 $P(F_2|E)$ 26 $P(E|F_1')$
27 $P(E|F_3')$ 28 $P(F_2'|E)$

The manufacturer of a new stuffed toy has three assembly lines, labeled A, B, and C. The probabilities that a toy produced by one of the assembly lines is flawed are .01, .04, and .03, respectively. The percentages of the entire production due to each of the assembly lines are 50%, 20%, and 30%, respectively. Determine the probability of each of the following events.

29 A randomly selected toy was made by Assembly Line B.
30 A randomly selected toy is faulty.
31 We get no faulty units when we choose 3 toys from each assembly line.
32 We get 1 or more faulty toys when we choose 2 units from Assembly Line A and 2 units from B.
33 A randomly chosen unit, that turned out to be faulty, was made by Assembly Line A.
34 A randomly chosen unit, that turned out to be faulty, was made by Assembly Line B.

The remaining problems pertain to optional Section 14.5. In problems 35 through 37, use the given transition matrix and the given initial state vector to compute the state vector after 2 transitions. Also, determine the steady state vector.

35 $\begin{bmatrix} .4 & .6 \\ .8 & .2 \end{bmatrix}$, (.6 .4)

36 $\begin{bmatrix} 1 & 0 \\ .3 & .7 \end{bmatrix}$, (.1 .9)

37 $\begin{bmatrix} .1 & .8 & .1 \\ .3 & .4 & .3 \\ .3 & .2 & .5 \end{bmatrix}$, (1 0 0)

A survey indicates that 10% of the people in Pleasantville play video games regularly. Each month 20% of those who do not play video games try them, and 60% of those who try video games continue to play them on a regular basis. Determine the percentages of people playing video games regularly at the end of each interval of time.

38 After 1 month
39 After 2 months
40 After many, many months

Social Science Applications

1 *Sociology*   About 20% of the home owners in the United States own condominiums, rather than detached single family dwellings. If 6 home owners are chosen randomly for an analysis of their life styles, what is the probability that all of them own condominiums? What is the probability that none of them own condominiums?

2 *Sociology*   The probability that a marriage will end in divorce is approximately .6. If 5 couples are chosen randomly, what is the probability that none of these marriages will end in divorce? What is the probability that exactly 3 will end in divorce?

3 *Sociology (Markov Processes)*   Each year 10% of the residents of Metropolis move to the suburbs, and only .1% move from the suburbs to the city. The number of city residents who move away from the area is the same as the number of people moving to the city from other areas. If the area's population is now evenly divided between the city and the suburbs,

what will be the division after 1 year? What will be the ultimate division of the population?

4   *Political Science*   In a certain community, 60% of the voters align themselves with a political party, and 40% identify themselves as independents. If 10 voters are selected randomly to complete a questionnaire, what is the probability that the sampling is truly representative, that is, there are exactly 4 independents in the sampling?

5   *Political Science*   A survey of a state legislature shows that the membership is 60% Democratic, 30% Republican, and 10% other. The probabilities that a bill for a balanced budget is supported by members chosen randomly from each of these 3 groups are .4, .9, and .5, respectively. If a legislator who voted for the bill is chosen to meet with the press, what is the probability that the chosen legislator is Republican?

6   *Urban Planning (Markov Processes)*   In Metropolis only 25% of the people use the city transportation system. As a result of advertising and other factors, each year 10% of those who have not used city transportation try it, and 80% of those who try it continue to use it. What percentage of the people will ultimately use the city transportation?

7   *Psychology*   A mouse is placed in a maze that has 8 forks; at each fork there is one correct way to go and one wrong way. If the mouse makes random decisions, what is the probability that the mouse goes through the maze without making a wrong turn?

8   *Health Care*   The probability that a person with cancer will get a positive outcome on cancer test X is .99, and the probability that a healthy person will get a positive outcome on test X is .08. If the probability that a randomly chosen person has cancer is .00001, what is the probability that a person with a positive outcome on test X has cancer?

9   *Genetics*   If each of two parents has curly hair and a recessive gene for straight hair, then the probability that the offspring has curly hair is .75. If these parents have 3 children, what is the probability that all 3 have curly hair?

CAREER PROFILE
Underwriter

If an explosion occurs at a nuclear power plant, how will the victims (or their surviving families) be compensated?

If a racing tire blows out during a speedway event, how will injuries to the driver and to spectators be covered?

Devising ways to provide insurance coverage to "share the risks" faced by individuals or companies, so that an unlikely accidental event (like those mentioned in the questions above) does not cause financial devastation to anyone, is one of the challenging tasks that faces an underwriter. Using the guidance provided by underwriters, insurance companies assume billions of dollars in risks each year by transferring the risk of loss from individual policy holders to the insurance company itself.

Mathematics training for undergraduates who are interested in a career in underwriting should include study of the basic principles of probability. An underwriter should know strategies for assessing the effects of uncertain future events in making present decisions. Courses such as finite mathematics, management science, and operations research often include discussion of probability and uncertainty and their role in decision making.

With the exception of topics in probability and statistics and their application to decision making, specific mathematical subjects are not usually required in a student's academic preparation for a career as an underwriter. However, the underwriter is expected to be an excellent problem solver, to be expert at asking pertinent questions, and to be adept at organizing and evaluating large quantities of information and at drawing sound conclusions. Challenging college courses in mathematics, science, and statistics can provide valuable opportunities for the prospective underwriter to develop these vital skills. Some background in accounting is useful, since underwriters need to be able to evaluate the financial condition of a company as they decide whether it is a good insurance risk.

Most underwriters specialize in one of the three major categories of insurance: life, property and liability, or health. Over 75% of the 76,000 insurance underwriters employed in 1980 worked in the area of property and liability. Most were located in the home or regional offices of their companies in a few large cities (New York, Chicago, San Francisco, Dallas, Philadelphia, and Hartford). Salaries for underwriters range from around $14,000 to over $30,000, depending on the individual's level of experience and responsibility.

Employment opportunities for underwriters probably will grow in the 1980s, but not necessarily within insurance companies. The high costs of insurance have led many large clients to consider "self-insurance" within their own organization or group. These enterprises may hire internal underwriters or create an increased demand for underwriters as consultants.

For beginning jobs in underwriting, most insurance companies seek trainees who are college graduates. Some prefer majors in business administration, but various liberal arts concentrations provide a good general background for underwriting; the critical factors are a good academic record and success in courses that demand strong organizing and reasoning skills.

The job duties for an underwriter-trainee begin with carefully supervised, routine tasks and gradually progress to complex evaluation situations with greater financial value. The process of deciding whether an insurer should assume a particular risk involves a sequence of information-gathering and decision-making steps. The underwriter must determine whether a given risk is a reasonable one to assume and, if so, at what rate. In addition to consideration of factual information, the underwriter must include assessment of certain intangible factors. The moral qualities of potential clients need to be evaluated: is the attempt to insure an honest one or might there be an attempt to defraud? Morale is another important factor, leading to questions such as, "Will insurance result in careless practices that lead to damage to insured items?"

Because of the many complexities of the profession, continuing education is required for an underwriter to advance. Independent study programs are available through the professional societies of underwriters in each specialty area. The demands for professional certification are rigorous. For example, certification as a Chartered Property Casualty Underwriter (CPCU) requires passing a series of ten difficult examinations. In 1982 only about 17,000 underwriters had this prestigious CPCU certification.

Sources of Additional Information
Information about career opportunities as an underwriter may be obtained from the home offices of large insurance companies and also from these sources:

Alliance of American Insurers, 20 N. Wacker Drive, Chicago, IL 60606.
Insurance Information Institute, 110 William St., New York, NY 10038.

Occupational Outlook Handbook. Bureau of Labor Statistics, U.S. Department of Labor, Washington, DC 20212. (Revised every two years, this handbook provides information about working conditions, level and places of employment, education and training requirements, advancement possibilities, job outlook, and earnings for about 250 occupations. Some of the information for this Career Profile was obtained from this source.)

Appendix

TABLE 1 Powers, Roots, Reciprocals

n	n^2	$\sqrt{n}$	$\sqrt{10n}$	n^3	$\sqrt[3]{n}$	$\sqrt[3]{10n}$	$\sqrt[3]{100n}$	$1/n$
1.0	1.0000	1.0000	3.1623	1.0000	1.0000	2.1544	4.6416	1.0000
1.1	1.2100	1.0488	3.3166	1.3310	1.0323	2.2240	4.7914	0.9091
1.2	1.4400	1.0954	3.4641	1.7280	1.0627	2.2894	4.9324	0.8333
1.3	1.6900	1.1402	3.6056	2.1970	1.0914	2.3513	5.0658	0.7692
1.4	1.9600	1.1832	3.7417	2.7440	1.1187	2.4101	5.1925	0.7143
1.5	2.2500	1.2247	3.8730	3.3750	1.1447	2.4662	5.3133	0.6667
1.6	2.5600	1.2649	4.0000	4.0960	1.1696	2.5198	5.4288	0.6250
1.7	2.8900	1.3038	4.1231	4.9130	1.1935	2.5713	5.5397	0.5882
1.8	3.2400	1.3416	4.2426	5.8320	1.2164	2.6207	5.6462	0.5556
1.9	3.6100	1.3784	4.3589	6.8590	1.2386	2.6684	5.7489	0.5263
2.0	4.0000	1.4142	4.4721	8.0000	1.2599	2.7144	5.8480	0.5000
2.1	4.4100	1.4491	4.5826	9.2610	1.2806	2.7589	5.9439	0.4762
2.2	4.8400	1.4832	4.6904	10.6480	1.3006	2.8020	6.0368	0.4545
2.3	5.2900	1.5166	4.7958	12.1670	1.3200	2.8439	6.1269	0.4348
2.4	5.7600	1.5492	4.8990	13.8240	1.3389	2.8845	6.2145	0.4167
2.5	6.2500	1.5811	5.0000	15.6250	1.3572	2.9240	6.2996	0.4000
2.6	6.7600	1.6125	5.0990	17.5760	1.3751	2.9625	6.3825	0.3846
2.7	7.2900	1.6432	5.1962	19.6830	1.3925	3.0000	6.4633	0.3704
2.8	7.8400	1.6733	5.2915	21.9520	1.4095	3.0366	6.5421	0.3571
2.9	8.4100	1.7029	5.3852	24.3890	1.4260	3.0723	6.6191	0.3448
3.0	9.0000	1.7321	5.4772	27.0000	1.4422	3.1072	6.6943	0.3333
3.1	9.6100	1.7607	5.5678	29.7910	1.4581	3.1414	6.7679	0.3226
3.2	10.2400	1.7889	5.6569	32.7680	1.4736	3.1748	6.8399	0.3125
3.3	10.8900	1.8166	5.7446	35.9370	1.4888	3.2075	6.9104	0.3030
3.4	11.5600	1.8439	5.8310	39.3040	1.5037	3.2396	6.9795	0.2941
3.5	12.2500	1.8708	5.9161	42.8750	1.5183	3.2711	7.0473	0.2857
3.6	12.9600	1.8974	6.0000	46.6560	1.5326	3.3019	7.1138	0.2778
3.7	13.6900	1.9235	6.0828	50.6530	1.5467	3.3322	7.1791	0.2703
3.8	14.4400	1.9494	6.1644	54.8720	1.5605	3.3620	7.2432	0.2632
3.9	15.2100	1.9748	6.2450	59.3190	1.5741	3.3912	7.3061	0.2564
4.0	16.0000	2.0000	6.3246	64.0000	1.5874	3.4200	7.3681	0.2500
4.1	16.8100	2.0248	6.4031	68.9210	1.6005	3.4482	7.4290	0.2439
4.2	17.6400	2.0494	6.4807	74.0880	1.6134	3.4760	7.4889	0.2381
4.3	18.4900	2.0736	6.5574	79.5070	1.6261	3.5034	7.5478	0.2326
4.4	19.3600	2.0976	6.6333	85.1840	1.6386	3.5303	7.6059	0.2273
4.5	20.2500	2.1213	6.7082	91.1250	1.6510	3.5569	7.6631	0.2222
4.6	21.1600	2.1448	6.7823	97.3360	1.6631	3.5830	7.7194	0.2174
4.7	22.0900	2.1679	6.8557	103.823	1.6751	3.6088	7.7750	0.2128
4.8	23.0400	2.1909	6.9282	110.592	1.6869	3.6342	7.8297	0.2083
4.9	24.0100	2.2136	7.0000	117.649	1.6985	3.6593	7.8837	0.2041

TABLE 1 *(Continued)*

n	n^2	$\sqrt{n}$	$\sqrt{10n}$	n^3	$\sqrt[3]{n}$	$\sqrt[3]{10n}$	$\sqrt[3]{100n}$	$1/n$
5.0	25.0000	2.2361	7.0711	125.000	1.7100	3.6840	7.9370	0.2000
5.1	26.0100	2.2583	7.1414	132.651	1.7213	3.7084	7.9896	0.1961
5.2	27.0400	2.2804	7.2111	140.608	1.7325	3.7325	8.0415	0.1923
5.3	28.0900	2.3022	7.2801	148.877	1.7435	3.7563	8.0927	0.1887
5.4	29.1600	2.3238	7.3485	157.464	1.7544	3.7798	8.1433	0.1852
5.5	30.2500	2.3452	7.4162	166.375	1.7652	3.8030	8.1932	0.1818
5.6	31.3600	2.3664	7.4833	175.616	1.7758	3.8259	8.2426	0.1786
5.7	32.4900	2.3875	7.5498	185.193	1.7863	3.8485	8.2913	0.1754
5.8	33.6400	2.4083	7.6158	195.112	1.7967	3.8709	8.3396	0.1724
5.9	34.8100	2.4290	7.6811	205.379	1.8070	3.8930	8.3872	0.1695
6.0	36.0000	2.4495	7.7460	216.000	1.8171	3.9149	8.4343	0.1667
6.1	37.2100	2.4698	7.8102	226.981	1.8272	3.9365	8.4809	0.1639
6.2	38.4400	2.4900	7.8740	238.328	1.8371	3.9579	8.5270	0.1613
6.3	39.6900	2.5100	7.9372	250.047	1.8469	3.9791	8.5726	0.1587
6.4	40.9600	2.5298	8.0000	262.144	1.8566	4.0000	8.6177	0.1563
6.5	42.2500	2.5495	8.0623	274.625	1.8663	4.0207	8.6624	0.1538
6.6	43.5600	2.5690	8.1240	287.496	1.8758	4.0412	8.7066	0.1515
6.7	44.8900	2.5884	8.1854	300.763	1.8852	4.0615	8.7503	0.1493
6.8	46.2400	2.6077	8.2462	314.432	1.8945	4.0817	8.7937	0.1471
6.9	47.6100	2.6268	8.3066	328.509	1.9038	4.1016	8.8366	0.1449
7.0	49.0000	2.6458	8.3666	343.000	1.9129	4.1213	8.8790	0.1429
7.1	50.4100	2.6646	8.4261	357.911	1.9220	4.1408	8.9211	0.1408
7.2	51.8400	2.6833	8.4853	373.248	1.9310	4.1602	8.9628	0.1389
7.3	53.2900	2.7019	8.5440	389.017	1.9399	4.1793	9.0041	0.1370
7.4	54.7600	2.7203	8.6023	405.224	1.9487	4.1983	9.0450	0.1351
7.5	56.2500	2.7386	8.6603	421.875	1.9574	4.2172	9.0856	0.1333
7.6	57.7600	2.7568	8.7178	438.976	1.9661	4.2358	9.1258	0.1316
7.7	59.2900	2.7749	8.7750	456.533	1.9747	4.2543	9.1657	0.1299
7.8	60.8400	2.7928	8.8318	474.552	1.9832	4.2727	9.2052	0.1282
7.9	62.4100	2.8107	8.8882	493.039	1.9916	4.2908	9.2443	0.1266
8.0	64.0000	2.8284	8.9443	512.000	2.0000	4.3089	9.2832	0.1250
8.1	65.6100	2.8460	9.0000	531.441	2.0083	4.3267	9.3217	0.1235
8.2	67.2400	2.8636	9.0554	551.368	2.0165	4.3445	9.3599	0.1220
8.3	68.8900	2.8810	9.1104	571.787	2.0247	4.3621	9.3978	0.1205
8.4	70.5600	2.8983	9.1652	592.704	2.0328	4.3795	9.4354	0.1190
8.5	72.2500	2.9155	9.2195	614.125	2.0408	4.3968	9.4727	0.1176
8.6	73.9600	2.9326	9.2736	636.056	2.0488	4.4140	9.5097	0.1163
8.7	75.6900	2.9496	9.3274	658.503	2.0567	4.4310	9.5464	0.1149
8.8	77.4400	2.9665	9.3808	681.472	2.0646	4.4480	9.5828	0.1136
8.9	79.2100	2.9833	9.4340	704.969	2.0723	4.4647	9.6190	0.1124
9.0	81.0000	3.0000	9.4868	729.000	2.0801	4.4814	9.6549	0.1111
9.1	82.8100	3.0166	9.5394	753.571	2.0878	4.4979	9.6905	0.1099
9.2	84.6400	3.0332	9.5917	778.688	2.0954	4.5144	9.7259	0.1087
9.3	86.4900	3.0496	9.6436	804.357	2.1029	4.5307	9.7610	0.1075
9.4	88.3600	3.0659	9.6954	830.584	2.1105	4.5468	9.7959	0.1064
9.5	90.2500	3.0822	9.7468	857.375	2.1179	4.5629	9.8305	0.1053
9.6	92.1600	3.0984	9.7980	884.736	2.1253	4.5789	9.8648	0.1042
9.7	94.0900	3.1145	9.8489	912.673	2.1327	4.5947	9.8990	0.1031
9.8	96.0400	3.1305	9.8995	941.192	2.1400	4.6104	9.9329	0.1020
9.9	98.0100	3.1464	9.9499	970.299	2.1472	4.6261	9.9666	0.1010
10.0	100.000	3.1623	10.000	1000.00	2.1544	4.6416	10.0000	0.1000

TABLE 2 Powers of e

x	e^x	e^{-x}	x	e^x	e^{-x}	x	e^x	e^{-x}
0.00	1.0000	1.00 000	0.50	1.6487	0.60 653	1.00	2.7183	0.36 788
0.01	1.0101	0.99 005	0.51	1.6653	0.60 050	1.01	2.7456	0.36 422
0.02	1.0202	0.98 020	0.52	1.6820	0.59 452	1.02	2.7732	0.36 059
0.03	1.0305	0.97 045	0.53	1.6989	0.58 860	1.03	2.8011	0.35 701
0.04	1.0408	0.96 079	0.54	1.7160	0.58 275	1.04	2.8292	0.35 345
0.05	1.0513	0.95 123	0.55	1.7333	0.57 695	1.05	2.8577	0.34 994
0.06	1.0618	0.94 176	0.56	1.7507	0.57 121	1.06	2.8864	0.34 646
0.07	1.0725	0.93 239	0.57	1.7683	0.56 553	1.07	2.9154	0.34 301
0.08	1.0833	0.92 312	0.58	1.7860	0.55 990	1.08	2.9447	0.33 960
0.09	1.0942	0.91 393	0.59	1.8040	0.55 433	1.09	2.9743	0.33 622
0.10	1.1052	0.90 484	0.60	1.8221	0.54 881	1.10	3.0042	0.33 287
0.11	1.1163	0.89 583	0.61	1.8404	0.54 335	1.11	3.0344	0.32 956
0.12	1.1275	0.88 692	0.62	1.8589	0.53 794	1.12	3.0649	0.32 628
0.13	1.1388	0.87 810	0.63	1.8776	0.53 259	1.13	3.0957	0.32 303
0.14	1.1503	0.86 936	0.64	1.8965	0.52 729	1.14	3.1268	0.31 982
0.15	1.1618	0.86 071	0.65	1.9155	0.52 205	1.15	3.1582	0.31 664
0.16	1.1735	0.85 214	0.66	1.9348	0.51 685	1.16	3.1899	0.31 349
0.17	1.1853	0.84 366	0.67	1.9542	0.51 171	1.17	3.2220	0.31 037
0.18	1.1972	0.83 527	0.68	1.9739	0.50 662	1.18	3.2544	0.30 728
0.19	1.2092	0.82 696	0.69	1.9937	0.50 158	1.19	3.2871	0.30 422
0.20	1.2214	0.81 873	0.70	2.0138	0.49 659	1.20	3.3201	0.30 119
0.21	1.2337	0.81 058	0.71	2.0340	0.49 164	1.21	3.3535	0.29 820
0.22	1.2461	0.80 252	0.72	2.0544	0.48 675	1.22	3.3872	0.29 523
0.23	1.2586	0.79 453	0.73	2.0751	0.48 191	1.23	3.4212	0.29 229
0.24	1.2712	0.78 663	0.74	2.0959	0.47 711	1.24	3.4556	0.28 938
0.25	1.2840	0.77 880	0.75	2.1170	0.47 237	1.25	3.4903	0.28 650
0.26	1.2969	0.77 105	0.76	2.1383	0.46 767	1.26	3.5254	0.28 365
0.27	1.3100	0.76 338	0.77	2.1598	0.46 301	1.27	3.5609	0.28 083
0.28	1.3231	0.75 578	0.78	2.1815	0.45 841	1.28	3.5966	0.27 804
0.29	1.3364	0.74 826	0.79	2.2034	0.45 384	1.29	3.6328	0.27 527
0.30	1.3499	0.74 082	0.80	2.2255	0.44 933	1.30	3.6693	0.27 253
0.31	1.3634	0.73 345	0.81	2.2479	0.44 486	1.31	3.7062	0.26 982
0.32	1.3771	0.72 615	0.82	2.2705	0.44 043	1.32	3.7434	0.26 714
0.33	1.3910	0.71 892	0.83	2.2933	0.43 605	1.33	3.7810	0.26 448
0.34	1.4049	0.71 177	0.84	2.3164	0.43 171	1.34	3.8190	0.26 185
0.35	1.4191	0.70 469	0.85	2.3396	0.42 741	1.35	3.8574	0.25 924
0.36	1.4333	0.69 768	0.86	2.3632	0.42 316	1.36	3.8962	0.25 666
0.37	1.4477	0.69 073	0.87	2.3869	0.41 895	1.37	3.9354	0.25 411
0.38	1.4623	0.68 386	0.88	2.4109	0.41 478	1.38	3.9749	0.25 158
0.39	1.4770	0.67 706	0.89	2.4351	0.41 066	1.39	4.0149	0.24 908
0.40	1.4918	0.67 032	0.90	2.4596	0.40 657	1.40	4.0552	0.24 660
0.41	1.5068	0.66 365	0.91	2.4843	0.40 252	1.41	4.0960	0.24 414
0.42	1.5220	0.65 705	0.92	2.5093	0.39 852	1.42	4.1371	0.24 171
0.43	1.5373	0.65 051	0.93	2.5345	0.39 455	1.43	4.1787	0.23 931
0.44	1.5527	0.64 404	0.94	2.5600	0.39 063	1.44	4.2207	0.23 693
0.45	1.5683	0.63 763	0.95	2.5857	0.38 674	1.45	4.2631	0.23 457
0.46	1.5841	0.63 128	0.96	2.6117	0.38 289	1.46	4.3060	0.23 224
0.47	1.6000	0.62 500	0.97	2.6379	0.37 908	1.47	4.3492	0.22 993
0.48	1.6161	0.61 878	0.98	2.6645	0.37 531	1.48	4.3939	0.22 764
0.49	1.6323	0.61 263	0.99	2.6912	0.37 158	1.49	4.4371	0.22 537
0.50	1.6487	0.60 653	1.00	2.7183	0.36 788	1.50	4.4817	0.22 313

TABLE 2 *(Continued)*

x	e^x	e^{-x}	x	e^x	e^{-x}	x	e^x	e^{-x}
1.50	4.4817	0.22 313	2.00	7.3891	0.13 534	2.50	12.182	0.082 085
1.51	4.5267	0.22 091	2.01	7.4633	0.13 399	2.51	12.305	0.081 268
1.52	4.5722	0.21 871	2.02	7.5383	0.13 266	2.52	12.429	0.080 460
1.53	4.6182	0.21 654	2.03	7.6141	0.13 134	2.53	12.554	0.079 659
1.54	4.6646	0.21 438	2.04	7.6906	0.13 003	2.54	12.680	0.078 866
1.55	4.7115	0.21 225	2.05	7.7679	0.12 873	2.55	12.807	0.078 082
1.56	4.7588	0.21 014	2.06	7.8460	0.12 745	2.56	12.936	0.077 305
1.57	4.8066	0.20 805	2.07	7.9248	0.12 619	2.57	13.066	0.076 536
1.58	4.8550	0.20 598	2.08	8.0045	0.12 493	2.56	13.197	0.075 774
1.59	4.9037	0.20 393	2.09	8.0849	0.12 369	2.59	13.330	0.075 020
1.60	4.9530	0.20 190	2.10	8.1662	0.12 246	2.60	13.464	0.074 274
1.61	5.0028	0.19 989	2.11	8.2482	0.12 124	2.61	13.599	0.073 535
1.62	5.0531	0.19 790	2.12	8.3311	0.12 003	2.62	13.736	0.072 803
1.63	5.1039	0.19 593	2.13	8.4149	0.11 884	2.63	13.874	0.072 078
1.64	5.1552	0.19 398	2.14	8.4994	0.11 765	2.64	14.013	0.071 361
1.65	5.2070	0.19 205	2.15	8.5849	0.11 648	2.65	14.154	0.070 651
1.66	5.2593	0.19 014	2.16	8.6711	0.11 533	2.66	14.296	0.069 948
1.67	5.3122	0.18 825	2.17	8.7583	0.11 418	2.67	14.440	0.069 252
1.68	5.3656	0.18 637	2.18	8.8463	0.11 304	2.68	14.585	0.068 563
1.69	5.4195	0.18 452	2.19	8.9352	0.11 192	2.69	14.732	0.067 881
1.70	5.4739	0.18 268	2.20	9.0250	0.11 080	2.70	14.880	0.067 206
1.71	5.5290	0.18 087	2.21	9.1157	0.10 970	2.71	15.029	0.066 537
1.72	5.5845	0.17 907	2.22	9.2073	0.10 861	2.72	15.180	0.065 875
1.73	5.6407	0.17 728	2.23	9.2999	0.10 753	2.73	15.333	0.065 219
1.74	5.6973	0.17 552	2.24	9.3933	0.10 646	2.74	15.487	0.064 570
1.75	5.7546	0.17 377	2.25	9.4877	0.10 540	2.75	15.643	0.063 928
1.76	5.8124	0.17 204	2.26	9.5831	0.10 435	2.76	15.800	0.063 292
1.77	5.8709	0.17 033	2.27	9.6794	0.10 331	2.77	15.959	0.062 662
1.78	5.9299	0.16 864	2.28	9.7767	0.10 228	2.78	16.119	0.062 039
1.79	5.9895	0.16 696	2.29	9.8749	0.10 127	2.79	16.281	0.061 421
1.80	6.0496	0.16 530	2.30	9.9742	0.10 026	2.80	16.445	0.060 810
1.81	6.1104	0.16 365	2.31	10.074	0.099 261	2.81	16.610	0.060 205
1.82	6.1719	0.16 203	2.32	10.176	0.098 274	2.82	16.777	0.059 606
1.83	6.2339	0.16 041	2.33	10.278	0.097 296	2.83	16.945	0.059 013
1.84	6.2965	0.15 882	2.34	10.381	0.096 328	2.84	17.116	0.058 426
1.85	6.3598	0.15 724	2.35	10.486	0.095 369	2.85	17.288	0.057 844
1.86	6.4237	0.15 567	2.36	10.591	0.094 420	2.86	17.462	0.057 269
1.87	6.4883	0.15 412	2.37	10.697	0.093 481	2.87	17.637	0.056 699
1.88	6.5535	0.15 259	2.38	10.805	0.092 551	2.88	17.814	0.056 135
1.89	6.6194	0.15 107	2.39	10.913	0.091 630	2.89	17.993	0.055 576
1.90	6.6859	0.14 957	2.40	11.023	0.090 718	2.90	18.174	0.055 023
1.91	6.7531	0.14 808	2.41	11.134	0.089 815	2.91	18.357	0.054 476
1.92	6.8210	0.14 661	2.42	11.246	0.088 922	2.92	18.541	0.053 934
1.93	6.8895	0.14 515	2.43	11.359	0.088 037	2.93	18.728	0.053 397
1.94	6.9588	0.14 370	2.44	11.473	0.087 161	2.94	18.916	0.052 866
1.95	7.0287	0.14 227	2.45	11.588	0.086 294	2.95	19.106	0.052 340
1.96	7.0993	0.14 086	2.46	11.705	0.085 435	2.96	19.298	0.051 819
1.97	7.1707	0.13 946	2.47	11.822	0.084 585	2.97	19.492	0.051 303
1.98	7.2427	0.13 807	2.48	11.941	0.083 743	2.98	19.688	0.050 793
1.99	7.3155	0.13 670	2.49	12.061	0.082 910	2.99	19.886	0.050 287
2.00	7.3891	0.13 534	2.50	12.182	0.082 085	3.00	20.086	0.049 787

TABLE 2 *(Continued)*

x	e^x	e^{-x}	x	e^x	e^{-x}	x	e^x	e^{-x}
3.00	20.086	0.049 787	3.50	33.115	0.030 197	4.00	54.598	0.018 316
3.01	20.287	0.049 292	3.51	33.448	0.029 897	4.01	55.147	0.018 133
3.02	20.491	0.048 801	3.52	33.784	0.029 599	4.02	55.701	0.017 953
3.03	20.697	0.048 316	3.53	34.124	0.029 305	4.03	56.261	0.017 774
3.04	20.905	0.047 835	3.54	34.467	0.029 013	4.04	56.826	0.017 597
3.05	21.115	0.047 359	3.55	34.813	0.028 725	4.05	57.397	0.017 422
3.06	21.328	0.046 888	3.56	35.163	0.028 439	4.06	57.974	0.017 249
3.07	21.542	0.046 421	3.57	35.517	0.028 156	4.07	58.557	0.017 077
3.08	21.758	0.045 959	3.58	35.874	0.027 876	4.08	59.145	0.016 907
3.09	21.977	0.045 502	3.59	36.234	0.027 598	4.09	59.740	0.016 739
3.10	22.198	0.045 049	3.60	36.598	0.027 324	4.10	60.340	0.016 573
3.11	22.421	0.044 601	3.61	36.966	0.027 052	4.11	60.947	0.016 408
3.12	22.646	0.044 157	3.62	37.338	0.026 783	4.12	61.559	0.016 245
3.13	22.874	0.043 718	3.63	37.713	0.026 516	4.13	62.178	0.016 083
3.14	23.104	0.043 283	3.64	38.092	0.026 252	4.14	62.803	0.015 923
3.15	23.336	0.042 852	3.65	38.475	0.025 991	4.15	63.434	0.015 764
3.16	23.571	0.042 426	3.66	38.861	0.025 733	4.16	64.072	0.015 608
3.17	23.807	0.042 004	3.67	39.252	0.025 476	4.17	64.715	0.015 452
3.18	24.047	0.041 586	3.68	39.646	0.025 223	4.18	65.366	0.015 299
3.19	24.288	0.041 172	3.69	40.045	0.024 972	4.19	66.023	0.015 146
3.20	24.533	0.040 762	3.70	40.447	0.024 724	4.20	66.686	0.014 996
3.21	24.779	0.040 357	3.71	40.854	0.024 478	4.21	67.357	0.014 846
3.22	25.028	0.039 955	3.72	41.264	0.024 234	4.22	68.033	0.014 699
3.23	25.280	0.039 557	3.73	41.679	0.023 993	4.23	68.717	0.014 552
3.24	25.534	0.039 164	3.74	42.098	0.023 754	4.24	69.408	0.014 408
3.25	25.790	0.038 774	3.75	42.521	0.023 518	4.25	70.105	0.014 264
3.26	26.050	0.038 388	3.76	42.948	0.023 284	4.26	70.810	0.014 122
3.27	26.311	0.038 006	3.77	43.380	0.023 052	4.27	71.522	0.013 982
3.28	26.576	0.037 628	3.78	43.816	0.022 823	4.28	72.240	0.013 843
3.29	26.843	0.037 254	3.79	44.256	0.022 596	4.29	72.966	0.013 705
3.30	27.113	0.036 883	3.80	44.701	0.022 371	4.30	73.700	0.013 569
3.31	27.385	0.036 516	3.81	45.150	0.022 148	4.31	74.440	0.013 434
3.32	27.660	0.036 153	3.82	45.604	0.021 928	4.32	75.189	0.013 300
3.33	27.938	0.035 793	3.83	46.063	0.021 710	4.33	75.944	0.013 168
3.34	28.219	0.035 437	3.84	46.525	0.021 494	4.34	76.708	0.013 037
3.35	28.503	0.035 084	3.85	46.993	0.021 280	4.35	77.478	0.012 907
3.36	28.789	0.034 735	3.86	47.465	0.021 068	4.36	78.257	0.012 778
3.37	29.079	0.034 390	3.87	47.942	0.020 858	4.37	79.044	0.012 651
3.38	29.371	0.034 047	3.88	48.424	0.020 651	4.38	79.838	0.012 525
3.39	29.666	0.033 709	3.89	48.911	0.020 445	4.39	80.640	0.012 401
3.40	29.984	0.033 373	3.90	49.402	0.020 242	4.40	81.451	0.012 277
3.41	30.265	0.033 041	3.91	49.899	0.020 041	4.41	82.269	0.012 155
3.42	30.569	0.032 712	3.92	50.400	0.019 841	4.42	83.096	0.012 034
3.43	30.877	0.032 387	3.93	50.907	0.019 644	4.43	83.931	0.011 914
3.44	31.187	0.032 065	3.94	51.419	0.019 448	4.44	84.775	0.011 796
3.45	31.500	0.031 746	3.95	51.935	0.019 255	4.45	85.627	0.011 679
3.46	31.817	0.031 430	3.96	52.457	0.019 063	4.46	86.488	0.011 562
3.47	32.137	0.031 117	3.97	52.985	0.018 873	4.47	87.357	0.011 447
3.48	32.460	0.030 807	3.98	53.517	0.018 686	4.48	88.235	0.011 333
3.49	32.786	0.030 501	3.99	54.055	0.018 500	4.49	89.121	0.011 221
3.50	33.115	0.030 197	4.00	54.598	0.018 316	4.50	90.017	0.011 109

TABLE 2 *(Continued)*

x	e^x	e^{-x}	x	e^x	e^{-x}	x	e^x	e^{-x}
4.50	90.017	0.011 109	5.00	148.41	0.006 7379	7.50	1,808.0	0.000 5531
4.51	90.922	0.010 998	5.05	156.02	0.006 4093	7.55	1,900.7	0.000 5261
4.52	91.836	0.010 889	5.10	164.02	0.006 0967	7.60	1,998.2	0.000 5005
4.53	92.759	0.010 781	5.15	172.43	0.005 7994	7.65	2,100.6	0.000 4760
4.54	93.691	0.010 673	5.20	181.27	0.005 5166	7.70	2,208.3	0.000 4528
4.55	94.632	0.010 567	5.25	190.57	0.005 2475	7.75	2,321.6	0.000 4307
4.56	95.583	0.010 462	5.30	200.34	0.004 9916	7.80	2,440.6	0.000 4097
4.57	96.544	0.010 358	5.35	210.61	0.004 7482	7.85	2,565.7	0.000 3898
4.58	97.514	0.010 255	5.40	221.41	0.004 5166	7.90	2,697.3	0.000 3707
4.59	98.494	0.010 153	5.45	232.76	0.004 2963	7.95	2,835.6	0.000 3527
4.60	99.484	0.010 052	5.50	244.69	0.004 0868	8.00	2,981.0	0.000 3355
4.61	100.48	0.009 9518	5.55	257.24	0.003 8875	8.05	3,133.8	0.000 3191
4.62	101.49	0.009 8528	5.60	270.43	0.003 6979	8.10	3,294.5	0.000 3035
4.63	102.51	0.009 7548	5.65	284.29	0.003 5175	8.15	3,463.4	0.000 2887
4.64	103.54	0.009 6577	5.70	298.87	0.003 3460	8.20	3,641.0	0.000 2747
4.65	104.58	0.009 5616	5.75	314.19	0.003 1828	8.25	3,827.6	0.000 2613
4.66	105.64	0.009 4665	5.80	330.30	0.003 0276	8.30	4,023.9	0.000 2485
4.67	106.70	0.009 3723	5.85	347.23	0.002 8799	8.35	4,230.2	0.000 2364
4.68	107.77	0.009 2790	5.90	365.04	0.002 7394	8.40	4,447.1	0.000 2249
4.69	108.85	0.009 1867	5.95	383.75	0.002 6058	8.45	4,675.1	0.000 2139
4.70	109.95	0.009 0953	6.00	403.43	0.002 4788	8.50	4,914.8	0.000 2035
4.71	111.05	0.009 0048	6.05	424.11	0.002 3579	8.55	5,166.8	0.000 1935
4.72	112.17	0.008 9152	6.10	445.86	0.002 2429	8.60	5,431.7	0.000 1841
4.73	113.30	0.008 8265	6.15	468.72	0.002 1335	8.65	5,710.1	0.000 1751
4.74	114.43	0.008 7386	6.20	492.75	0.002 2094	8.70	6,002.9	0.000 1666
4.75	115.58	0.008 6517	6.25	518.01	0.001 9305	8.75	6,310.7	0.000 1585
4.76	116.75	0.008 5656	6.30	544.57	0.001 8363	8.80	6,634.2	0.000 1507
4.77	117.92	0.008 4804	6.35	572.49	0.001 7467	8.85	6,974.4	0.000 1434
4.78	119.10	0.008 3960	6.40	601.85	0.001 6616	8.90	7,332.0	0.000 1364
4.79	120.30	0.008 3125	6.45	632.70	0.001 5805	8.95	7,707.9	0.000 1297
4.80	121.51	0.008 2297	6.50	665.14	0.001 5034	9.00	8,103.1	0.000 1234
4.81	122.73	0.008 1479	6.55	699.24	0.001 4301	9.05	8,518.5	0.000 1174
4.82	123.97	0.008 0668	6.60	735.10	0.001 3604	9.10	8,955.3	0.000 1117
4.83	125.21	0.007 9865	6.65	772.78	0.001 2940	9.15	9,414.4	0.000 1062
4.84	126.47	0.007 9071	6.70	812.41	0.001 2309	9.20	9,897.1	0.000 1010
4.85	127.74	0.007 8284	6.75	854.06	0.001 1709	9.25	10,405	0.000 0961
4.86	129.02	0.007 7505	6.80	897.85	0.001 1138	9.30	10,938	0.000 0914
4.87	130.32	0.007 6734	6.85	943.88	0.001 0595	9.35	11,499	0.000 0870
4.88	131.63	0.007 5970	6.90	992.27	0.001 0078	9.40	12,088	0.000 0827
4.89	132.95	0.007 5214	6.95	1,043.1	0.000 9586	9.45	12,708	0.000 0787
4.90	134.29	0.007 4466	7.00	1,096.6	0.000 9119	9.50	13,360	0.000 0749
4.91	135.64	0.007 3725	7.05	1,152.9	0.000 8674	9.55	14,045	0.000 0712
4.92	137.00	0.007 2991	7.10	1,212.0	0.000 8251	9.60	14,765	0.000 0677
4.93	138.38	0.007 2265	7.15	1,274.1	0.000 7849	9.65	15,522	0.000 0644
4.94	139.77	0.007 1546	7.20	1,339.4	0.000 7466	9.70	16,318	0.000 0613
4.95	141.17	0.007 0834	7.25	1,408.1	0.000 7102	9.75	17,154	0.000 0583
4.96	142.59	0.007 0129	7.30	1,480.3	0.000 6755	9.80	18,034	0.000 0555
4.97	144.03	0.006 9431	7.35	1,556.2	0.000 6426	9.85	18,958	0.000 0527
4.98	145.47	0.006 8741	7.40	1,636.0	0.000 6113	9.90	19,930	0.000 0502
4.99	146.94	0.006 8057	7.45	1,719.9	0.000 5814	9.95	20,952	0.000 0477
5.00	148.41	0.006 7379	7.50	1,808.0	0.000 5531	10.00	22,026	0.000 0454

TABLE 3 Common Logarithms

n	0	1	2	3	4	5	6	7	8	9
1.0	.0000	.0043	.0086	.0128	.0170	.0212	.0253	.0294	.0334	.0374
1.1	.0414	.0453	.0492	.0531	.0569	.0607	.0645	.0682	.0719	.0755
1.2	.0792	.0828	.0864	.0899	.0934	.0969	.1004	.1038	.1072	.1106
1.3	.1139	.1173	.1206	.1239	.1271	.1303	.1335	.1367	.1399	.1430
1.4	.1461	.1492	.1523	.1553	.1584	.1614	.1644	.1673	.1703	.1732
1.5	.1761	.1790	.1818	.1847	.1875	.1903	.1931	.1959	.1987	.2014
1.6	.2041	.2068	.2095	.2122	.2148	.2175	.2201	.2227	.2253	.2279
1.7	.2304	.2330	.2355	.2380	.2405	.2430	.2455	.2480	.2504	.2529
1.8	.2553	.2577	.2601	.2625	.2648	.2672	.2695	.2718	.2742	.2765
1.9	.2788	.2810	.2833	.2856	.2878	.2900	.2923	.2945	.2967	.2989
2.0	.3010	.3032	.3054	.3075	.3096	.3118	.3139	.3160	.3181	.3201
2.1	.3222	.3243	.3263	.3284	.3304	.3324	.3345	.3365	.3385	.3404
2.2	.3424	.3444	.3464	.3483	.3502	.3522	.3541	.3560	.3579	.3598
2.3	.3617	.3636	.3655	.3674	.3692	.3711	.3729	.3747	.3766	.3784
2.4	.3802	.3820	.3838	.3856	.3874	.3892	.3909	.3927	.3945	.3962
2.5	.3979	.3997	.4014	.4031	.4048	.4065	.4082	.4099	.4116	.4133
2.6	.4150	.4166	.4183	.4200	.4216	.4232	.4249	.4265	.4281	.4298
2.7	.4314	.4330	.4346	.4362	.4378	.4393	.4409	.4425	.4440	.4456
2.8	.4472	.4487	.4502	.4518	.4533	.4548	.4564	.4579	.4594	.4609
2.9	.4624	.4639	.4654	.4669	.4683	.4698	.4713	.4728	.4742	.4757
3.0	.4771	.4786	.4800	.4814	.4829	.4843	.4857	.4871	.4886	.4900
3.1	.4914	.4928	.4942	.4955	.4969	.4983	.4997	.5011	.5024	.5038
3.2	.5051	.5065	.5079	.5092	.5105	.5119	.5132	.5145	.5159	.5172
3.3	.5185	.5198	.5211	.5224	.5237	.5250	.5263	.5276	.5289	.5302
3.4	.5315	.5328	.5340	.5353	.5366	.5378	.5391	.5403	.5416	.5428
3.5	.5441	.5453	.5465	.5478	.5490	.5502	.5514	.5527	.5539	.5551
3.6	.5563	.5575	.5587	.5599	.5611	.5623	.5635	.5647	.5658	.5670
3.7	.5682	.5694	.5705	.5717	.5729	.5740	.5752	.5763	.5775	.5786
3.8	.5798	.5809	.5821	.5832	.5843	.5855	.5866	.5877	.5888	.5899
3.9	.5911	.5922	.5933	.5944	.5955	.5966	.5977	.5988	.5999	.6010
4.0	.6021	.6031	.6042	.6053	.6064	.6075	.6085	.6096	.6107	.6117
4.1	.6128	.6138	.6149	.6160	.6170	.6180	.6191	.6201	.6212	.6222
4.2	.6232	.6243	.6253	.6263	.6274	.6284	.6294	.6304	.6314	.6325
4.3	.6335	.6345	.6355	.6365	.6375	.6385	.6395	.6405	.6415	.6425
4.4	.6435	.6444	.6454	.6464	.6474	.6484	.6493	.6503	.6513	.6522
4.5	.6532	.6542	.6551	.6561	.6571	.6580	.6590	.6599	.6609	.6618
4.6	.6628	.6637	.6646	.6656	.6665	.6675	.6684	.6693	.6702	.6712
4.7	.6721	.6730	.6739	.6749	.6758	.6767	.6776	.6785	.6794	.6803
4.8	.6812	.6821	.6830	.6839	.6848	.6857	.6866	.6875	.6884	.6893
4.9	.6902	.6911	.6920	.6928	.6937	.6946	.6955	.6964	.6972	.6981
5.0	.6990	.6998	.7007	.7016	.7024	.7033	.7042	.7050	.7059	.7067
5.1	.7076	.7084	.7093	.7101	.7110	.7118	.7126	.7135	.7143	.7152
5.2	.7160	.7168	.7177	.7185	.7193	.7202	.7210	.7218	.7226	.7235
5.3	.7243	.7251	.7259	.7267	.7275	.7284	.7292	.7300	.7308	.7316
5.4	.7324	.7332	.7340	.7348	.7356	.7364	.7372	.7380	.7388	.7396
n	0	1	2	3	4	5	6	7	8	9

TABLE 3 *(Continued)*

n	0	1	2	3	4	5	6	7	8	9
5.5	.7404	.7412	.7419	.7427	.7435	.7443	.7451	.7459	.7466	.7474
5.6	.7482	.7490	.7497	.7505	.7513	.7520	.7528	.7536	.7543	.7551
5.7	.7559	.7566	.7574	.7582	.7589	.7597	.7604	.7612	.7619	.7627
5.8	.7634	.7642	.7649	.7657	.7664	.7672	.7679	.7686	.7694	.7701
5.9	.7709	.7716	.7723	.7731	.7738	.7745	.7752	.7760	.7767	.7774
6.0	.7782	.7789	.7796	.7803	.7810	.7818	.7825	.7832	.7839	.7846
6.1	.7853	.7860	.7868	.7875	.7882	.7889	.7896	.7903	.7910	.7917
6.2	.7924	.7931	.7938	.7945	.7952	.7959	.7966	.7973	.7980	.7987
6.3	.7993	.8000	.8007	.8014	.8021	.8028	.8035	.8041	.8048	.8055
6.4	.8062	.8069	.8075	.8082	.8089	.8096	.8102	.8109	.8116	.8122
6.5	.8129	.8136	.8142	.8149	.8156	.8162	.8169	.8176	.8182	.8189
6.6	.8195	.8202	.8209	.8215	.8222	.8228	.8235	.8241	.8248	.8254
6.7	.8261	.8267	.8274	.8280	.8287	.8293	.8299	.8306	.8312	.8319
6.8	.8325	.8331	.8338	.8344	.8351	.8357	.8363	.8370	.8376	.8382
6.9	.8388	.8395	.8401	.8407	.8414	.8420	.8426	.8432	.8439	.8445
7.0	.8451	.8457	.8463	.8470	.8476	.8482	.8488	.8494	.8500	.8506
7.1	.8513	.8519	.8525	.8531	.8537	.8543	.8549	.8555	.8561	.8567
7.2	.8573	.8579	.8585	.8591	.8597	.8603	.8609	.8615	.8621	.8627
7.3	.8633	.8639	.8645	.8651	.8657	.8663	.8669	.8675	.8681	.8686
7.4	.8692	.8698	.8704	.8710	.8716	.8722	.8727	.8733	.8739	.8745
7.5	.8751	.8756	.8762	.8768	.8774	.8779	.8785	.8791	.8797	.8802
7.6	.8808	.8814	.8820	.8825	.8831	.8837	.8842	.8848	.8854	.8859
7.7	.8865	.8871	.8876	.8882	.8887	.8893	.8899	.8904	.8910	.8915
7.8	.8921	.8927	.8932	.8938	.8943	.8949	.8954	.8960	.8965	.8971
7.9	.8976	.8982	.8987	.8993	.8998	.9004	.9009	.9015	.9020	.9025
8.0	.9031	.9036	.9042	.9047	.9053	.9058	.9063	.9069	.9074	.9079
8.1	.9085	.9090	.9096	.9101	.9106	.9112	.9117	.9122	.9128	.9133
8.2	.9138	.9143	.9149	.9154	.9159	.9165	.9170	.9175	.9180	.9186
8.3	.9191	.9196	.9201	.9206	.9212	.9217	.9222	.9227	.9232	.9238
8.4	.9243	.9248	.9253	.9258	.9263	.9269	.9274	.9279	.9284	.9289
8.5	.9294	.9299	.9304	.9309	.9315	.9320	.9325	.9330	.9335	.9340
8.6	.9345	.9350	.9355	.9360	.9365	.9370	.9375	.9380	.9385	.9390
8.7	.9395	.9400	.9405	.9410	.9415	.9420	.9425	.9430	.9435	.9440
8.8	.9445	.9450	.9455	.9460	.9465	.9469	.9474	.9479	.9484	.9489
8.9	.9494	.9499	.9504	.9509	.9513	.9518	.9523	.9528	.9533	.9538
9.0	.9542	.9547	.9552	.9557	.9562	.9566	.9571	.9576	.9581	.9586
9.1	.9590	.9595	.9600	.9605	.9609	.9614	.9619	.9624	.9628	.9633
9.2	.9638	.9643	.9647	.9652	.9657	.9661	.9666	.9671	.9675	.9680
9.3	.9685	.9689	.9694	.9699	.9703	.9708	.9713	.9717	.9722	.9727
9.4	.9731	.9736	.9741	.9745	.9750	.9754	.9759	.9763	.9768	.9773
9.5	.9777	.9782	.9786	.9791	.9795	.9800	.9805	.9809	.9814	.9818
9.6	.9823	.9827	.9832	.9836	.9841	.9845	.9850	.9854	.9859	.9863
9.7	.9868	.9872	.9877	.9881	.9886	.9890	.9894	.9899	.9903	.9908
9.8	.9912	.9917	.9921	.9926	.9930	.9934	.9939	.9943	.9948	.9952
9.9	.9956	.9961	.9965	.9969	.9974	.9978	.9983	.9987	.9991	.9996
n	0	1	2	3	4	5	6	7	8	9

TABLE 4 Natural Logarithms

n	0	1	2	3	4	5	6	7	8	9
1.0	0.0000	0.0100	0.0198	0.0296	0.0392	0.0488	0.0583	0.0677	0.0770	0.0862
1.1	0.0953	0.1044	0.1133	0.1222	0.1310	0.1398	0.1484	0.1570	0.1655	0.1740
1.2	0.1823	0.1906	0.1989	0.2070	0.2151	0.2231	0.2311	0.2390	0.2469	0.2546
1.3	0.2624	0.2700	0.2776	0.2852	0.2927	0.3001	0.3075	0.3148	0.3221	0.3293
1.4	0.3365	0.3436	0.3507	0.3577	0.3646	0.3716	0.3784	0.3853	0.3920	0.3988
1.5	0.4055	0.4121	0.4187	0.4253	0.4318	0.4383	0.4447	0.4511	0.4574	0.4637
1.6	0.4700	0.4762	0.4824	0.4886	0.4947	0.5008	0.5068	0.5128	0.5188	0.5247
1.7	0.5306	0.5365	0.5423	0.5481	0.5539	0.5596	0.5653	0.5710	0.5766	0.5822
1.8	0.5878	0.5933	0.5988	0.6043	0.6098	0.6152	0.6206	0.6259	0.6313	0.6366
1.9	0.6419	0.6471	0.6523	0.6575	0.6627	0.6678	0.6729	0.6780	0.6831	0.6881
2.0	0.6931	0.6981	0.7031	0.7080	0.7129	0.7178	0.7227	0.7275	0.7324	0.7372
2.1	0.7419	0.7467	0.7514	0.7561	0.7608	0.7655	0.7701	0.7747	0.7793	0.7839
2.2	0.7885	0.7930	0.7975	0.8020	0.8065	0.8109	0.8154	0.8198	0.8242	0.8286
2.3	0.8329	0.8372	0.8416	0.8459	0.8502	0.8544	0.8587	0.8629	0.8671	0.8713
2.4	0.8755	0.8796	0.8838	0.8879	0.8920	0.8961	0.9002	0.9042	0.9083	0.9123
2.5	0.9163	0.9203	0.9243	0.9282	0.9322	0.9361	0.9400	0.9439	0.9478	0.9517
2.6	0.9555	0.9594	0.9632	0.9670	0.9708	0.9746	0.9783	0.9821	0.9858	0.9895
2.7	0.9933	0.9969	1.0006	1.0043	1.0080	1.0116	1.0152	1.0188	1.0225	1.0260
2.8	1.0296	1.0332	1.0367	1.0403	1.0438	1.0473	1.0508	1.0543	1.0578	1.0613
2.9	1.0647	1.0682	1.0716	1.0750	1.0784	1.0818	1.0852	1.0886	1.0919	1.0953
3.0	1.0986	1.1019	1.1053	1.1086	1.1119	1.1151	1.1184	1.1217	1.1249	1.1282
3.1	1.1314	1.1346	1.1378	1.1410	1.1442	1.1474	1.1506	1.1537	1.1569	1.1600
3.2	1.1632	1.1663	1.1694	1.1725	1.1756	1.1787	1.1817	1.1848	1.1878	1.1909
3.3	1.1939	1.1969	1.2000	1.2030	1.2060	1.2090	1.2119	1.2149	1.2179	1.2208
3.4	1.2238	1.2267	1.2296	1.2326	1.2355	1.2384	1.2413	1.2442	1.2470	1.2499
3.5	1.2528	1.2556	1.2585	1.2613	1.2641	1.2669	1.2698	1.2726	1.2754	1.2782
3.6	1.2809	1.2837	1.2865	1.2892	1.2920	1.2947	1.2975	1.3002	1.3029	1.3056
3.7	1.3083	1.3110	1.3137	1.3164	1.3191	1.3218	1.3244	1.3271	1.3297	1.3324
3.8	1.3350	1.3376	1.3403	1.3429	1.3455	1.3481	1.3507	1.3533	1.3558	1.3584
3.9	1.3610	1.3635	1.3661	1.3686	1.3712	1.3737	1.3762	1.3788	1.3813	1.3838
4.0	1.3863	1.3888	1.3913	1.3938	1.3962	1.3987	1.4012	1.4036	1.4061	1.4085
4.1	1.4110	1.4134	1.4159	1.4183	1.4207	1.4231	1.4255	1.4279	1.4303	1.4327
4.2	1.4351	1.4375	1.4398	1.4422	1.4446	1.4469	1.4493	1.4516	1.4540	1.4563
4.3	1.4586	1.4609	1.4633	1.4656	1.4679	1.4702	1.4725	1.4748	1.4770	1.4793
4.4	1.4816	1.4839	1.4861	1.4884	1.4907	1.4929	1.4951	1.4974	1.4996	1.5019
4.5	1.5041	1.5063	1.5085	1.5107	1.5129	1.5151	1.5173	1.5195	1.5217	1.5239
4.6	1.5261	1.5282	1.5304	1.5326	1.5347	1.5369	1.5390	1.5412	1.5433	1.5454
4.7	1.5476	1.5497	1.5518	1.5539	1.5560	1.5581	1.5602	1.5623	1.5644	1.5665
4.8	1.5686	1.5707	1.5728	1.5748	1.5769	1.5790	1.5810	1.5831	1.5851	1.5872
4.9	1.5892	1.5913	1.5933	1.5953	1.5974	1.5994	1.6014	1.6034	1.6054	1.6074
5.0	1.6094	1.6114	1.6134	1.6154	1.6174	1.6194	1.6214	1.6233	1.6253	1.6273
5.1	1.6292	1.6312	1.6332	1.6351	1.6371	1.6390	1.6409	1.6429	1.6448	1.6467
5.2	1.6487	1.6506	1.6525	1.6544	1.6563	1.6582	1.6601	1.6620	1.6639	1.6658
5.3	1.6677	1.6696	1.6715	1.6734	1.6752	1.6771	1.6790	1.6808	1.6827	1.6845
5.4	1.6864	1.6882	1.6901	1.6919	1.6938	1.6956	1.6974	1.6993	1.7011	1.7029

TABLE 4 *(Continued)*

n	0	1	2	3	4	5	6	7	8	9
5.5	1.7047	1.7066	1.7084	1.7102	1.7120	1.7138	1.7156	1.7174	1.7192	1.7210
5.6	1.7228	1.7246	1.7263	1.7281	1.7299	1.7317	1.7334	1.7352	1.7370	1.7387
5.7	1.7405	1.7422	1.7440	1.7457	1.7475	1.7492	1.7509	1.7527	1.7544	1.7561
5.8	1.7579	1.7596	1.7613	1.7630	1.7647	1.7664	1.7681	1.7699	1.7716	1.7733
5.9	1.7750	1.7766	1.7783	1.7800	1.7817	1.7834	1.7851	1.7867	1.7884	1.7901
6.0	1.7918	1.7934	1.7951	1.7967	1.7984	1.8001	1.8017	1.8034	1.8050	1.8066
6.1	1.8083	1.8099	1.8116	1.8132	1.8148	1.8165	1.8181	1.8197	1.8213	1.8229
6.2	1.8245	1.8262	1.8278	1.8294	1.8310	1.8326	1.8342	1.8358	1.8374	1.8390
6.3	1.8405	1.8421	1.8437	1.8453	1.8469	1.8485	1.8500	1.8516	1.8532	1.8547
6.4	1.8563	1.8579	1.8594	1.8610	1.8625	1.8641	1.8656	1.8672	1.8687	1.8703
6.5	1.8718	1.8733	1.8749	1.8764	1.8779	1.8795	1.8810	1.8825	1.8840	1.8856
6.6	1.8871	1.8886	1.8901	1.8916	1.8931	1.8946	1.8961	1.8976	1.8991	1.9006
6.7	1.9021	1.9036	1.9051	1.9066	1.9081	1.9095	1.9110	1.9125	1.9140	1.9155
6.8	1.9169	1.9184	1.9199	1.9213	1.9228	1.9242	1.9257	1.9272	1.9286	1.9301
6.9	1.9315	1.9330	1.9344	1.9359	1.9373	1.9387	1.9402	1.9416	1.9430	1.9445
7.0	1.9459	1.9473	1.9488	1.9502	1.9516	1.9530	1.9544	1.9559	1.9573	1.9587
7.1	1.9601	1.9615	1.9629	1.9643	1.9657	1.9671	1.9685	1.9699	1.9713	1.9727
7.2	1.9741	1.9755	1.9769	1.9782	1.9796	1.9810	1.9824	1.9838	1.9851	1.9865
7.3	1.9879	1.9892	1.9906	1.9920	1.9933	1.9947	1.9961	1.9974	1.9988	2.0001
7.4	2.0015	2.0028	2.0042	2.0055	2.0069	2.0082	2.0096	2.0109	2.0122	2.0136
7.5	2.0149	2.0162	2.0176	2.0189	2.0202	2.0215	2.0229	2.0242	2.0255	2.0268
7.6	2.0281	2.0295	2.0308	2.0321	2.0334	2.0347	2.0360	2.0373	2.0386	2.0399
7.7	2.0412	2.0425	2.0438	2.0451	2.0464	2.0477	2.0490	2.0503	2.0516	2.0528
7.8	2.0541	2.0554	2.0567	2.0580	2.0592	2.0605	2.0618	2.0631	2.0643	2.0656
7.9	2.0669	2.0681	2.0694	2.0707	2.0719	2.0732	2.0744	2.0757	2.0769	2.0782
8.0	2.0794	2.0807	2.0819	2.0832	2.0844	2.0857	2.0869	2.0882	2.0894	2.0906
8.1	2.0919	2.0931	2.0943	2.0956	2.0968	2.0980	2.0992	2.1005	2.1017	2.1029
8.2	2.1041	2.1054	2.1066	2.1078	2.1090	2.1102	2.1114	2.1126	2.1138	2.1150
8.3	2.1163	2.1175	2.1187	2.1199	2.1211	2.1223	2.1235	2.1247	2.1258	2.1270
8.4	2.1282	2.1294	2.1306	2.1318	2.1330	2.1342	2.1353	2.1365	2.1377	2.1389
8.5	2.1401	2.1412	2.1424	2.1436	2.1448	2.1459	2.1471	2.1483	2.1494	2.1506
8.6	2.1518	2.1529	2.1541	2.1552	2.1564	2.1576	2.1587	2.1599	2.1610	2.1622
8.7	2.1633	2.1645	2.1656	2.1668	2.1679	2.1691	2.1702	2.1713	2.1725	2.1736
8.8	2.1748	2.1759	2.1770	2.1782	2.1793	2.1804	2.1815	2.1827	2.1838	2.1849
8.9	2.1861	2.1872	2.1883	2.1894	2.1905	2.1917	2.1928	2.1939	2.1950	2.1961
9.0	2.1972	2.1983	2.1994	2.2006	2.2017	2.2028	2.2039	2.2050	2.2061	2.2072
9.1	2.2083	2.2094	2.2105	2.2116	2.2127	2.2138	2.2148	2.2159	2.2170	2.2181
9.2	2.2192	2.2203	2.2214	2.2225	2.2235	2.2246	2.2257	2.2268	2.2279	2.2289
9.3	2.2300	2.2311	2.2322	2.2332	2.2343	2.2354	2.2364	2.2375	2.2386	2.2396
9.4	2.2407	2.2418	2.2428	2.2439	2.2450	2.2460	2.2471	2.2481	2.2492	2.2502
9.5	2.2513	2.2523	2.2534	2.2544	2.2555	2.2565	2.2576	2.2586	2.2597	2.2607
9.6	2.2618	2.2628	2.2638	2.2649	2.2659	2.2670	2.2680	2.2690	2.2701	2.2711
9.7	2.2721	2.2732	2.2742	2.2752	2.2762	2.2773	2.2783	2.2793	2.2803	2.2814
9.8	2.2824	2.2834	2.2844	2.2854	2.2865	2.2875	2.2885	2.2895	2.2905	2.2915
9.9	2.2925	2.2935	2.2946	2.2956	2.2966	2.2976	2.2986	2.2996	2.3006	2.3016

TABLE 5   Interest Rates

		$r = .005$		
n	$(1 + r)^n$	$(1 + r)^{-n}$	$a_{\overline{n}|r}$	$s_{\overline{n}|r}$
1	1.005000	0.995025	0.995025	1.000000
2	1.010025	0.990075	1.985099	2.005000
3	1.015075	0.985149	2.970248	3.015025
4	1.020151	0.980248	3.950496	4.030100
5	1.025251	0.975371	4.925866	5.050251
6	1.030378	0.970518	5.896384	6.075502
7	1.035529	0.965690	6.862074	7.105879
8	1.040707	0.960885	7.822959	8.141409
9	1.045911	0.956105	8.779064	9.182116
10	1.051140	0.951348	9.730412	10.228026
11	1.056396	0.946615	10.677027	11.279167
12	1.061678	0.941905	11.618932	12.335562
13	1.066986	0.937219	12.556151	13.397240
14	1.072321	0.932556	13.488708	14.464226
15	1.077683	0.927917	14.416625	15.536548
16	1.083071	0.923300	15.339925	16.614230
17	1.088487	0.918707	16.258632	17.697301
18	1.093929	0.914136	17.172768	18.785788
19	1.099399	0.909588	18.082356	19.879717
20	1.104896	0.905063	18.987419	20.979115
21	1.110420	0.900560	19.887979	22.084011
22	1.115972	0.896080	20.784059	23.194431
23	1.121552	0.891622	21.675681	24.310403
24	1.127160	0.887186	22.562866	25.431955
25	1.132796	0.882772	23.445638	26.559115
26	1.138460	0.878380	24.324018	27.691911
27	1.144152	0.874010	25.198028	28.830370
28	1.149873	0.869662	26.067689	29.974522
29	1.155622	0.865335	26.933024	31.124395
30	1.161400	0.861030	27.794054	32.280017
31	1.167207	0.856746	28.650800	33.441417
32	1.173043	0.852484	29.503284	34.608624
33	1.178908	0.848242	30.351526	35.781667
34	1.184803	0.844022	31.195548	36.960575
35	1.190727	0.839823	32.035371	38.145378
36	1.196681	0.835645	32.871016	39.336105
37	1.202664	0.831487	33.702504	40.532785
38	1.208677	0.827351	34.529854	41.735449
39	1.214721	0.823235	35.353089	42.944127
40	1.220794	0.819139	36.172228	44.158847
41	1.226898	0.815064	36.987291	45.379642
42	1.233033	0.811009	37.798300	46.606540
43	1.239198	0.806974	38.605274	47.839572
44	1.245394	0.802959	39.408232	49.078770
45	1.251621	0.798964	40.207196	50.324164
46	1.257879	0.794989	41.002185	51.575785
47	1.264168	0.791034	41.793219	52.833664
48	1.270489	0.787098	42.580318	54.097832
49	1.276842	0.783182	43.363500	55.368321
50	1.283226	0.779286	44.142786	56.645163

TABLE 5 *(Continued)*

		$r = .0075$		
n	$(1 + r)^n$	$(1 + r)^{-n}$	$a_{\overline{n}|r}$	$s_{\overline{n}|r}$
1	1.007500	0.992556	0.992556	1.000000
2	1.015056	0.985167	1.977723	2.007500
3	1.022669	0.977833	2.955556	3.022556
4	1.030339	0.970554	3.926110	4.045225
5	1.038067	0.963329	4.889440	5.075565
6	1.045852	0.956158	5.845598	6.113631
7	1.053696	0.949040	6.794638	7.159484
8	1.061599	0.941975	7.736613	8.213180
9	1.069561	0.934963	8.671576	9.274779
10	1.077583	0.928003	9.599580	10.344339
11	1.085664	0.921095	10.520675	11.421922
12	1.093807	0.914238	11.434913	12.507586
13	1.102010	0.907432	12.342345	13.601393
14	1.110276	0.900677	13.243022	14.703404
15	1.118603	0.893973	14.136995	15.813679
16	1.126992	0.887318	15.024313	16.932282
17	1.135445	0.880712	15.905025	18.059274
18	1.143960	0.874156	16.779181	19.194718
19	1.152540	0.867649	17.646830	20.338679
20	1.161184	0.861190	18.508020	21.491219
21	1.169893	0.854779	19.362799	22.652403
22	1.178667	0.848416	20.211215	23.822296
23	1.187507	0.842100	21.053315	25.000963
24	1.196414	0.835831	21.889146	26.188471
25	1.205387	0.829609	22.718755	27.384884
26	1.214427	0.823434	23.542189	28.590271
27	1.223535	0.817304	24.359493	29.804698
28	1.232712	0.811220	25.170713	31.028233
29	1.241957	0.805181	25.975893	32.260945
30	1.251272	0.799187	26.775080	33.502902
31	1.260656	0.793238	27.568318	34.754174
32	1.270111	0.787333	28.355650	36.014830
33	1.279637	0.781472	29.137122	37.284941
34	1.289234	0.775654	29.912776	38.564578
35	1.298904	0.769880	30.682656	39.853813
36	1.308645	0.764149	31.446805	41.152716
37	1.318460	0.758461	32.205266	42.461361
38	1.328349	0.752814	32.958080	43.779822
39	1.338311	0.747210	33.705290	45.108170
40	1.348349	0.741648	34.446938	46.446482
41	1.358461	0.736127	35.183065	47.794830
42	1.368650	0.730647	35.913713	49.153291
43	1.378915	0.725208	36.638921	50.521941
44	1.389256	0.719810	37.358730	51.900856
45	1.399676	0.714451	38.073181	53.290112
46	1.410173	0.709133	38.782314	54.689788
47	1.420750	0.703854	39.486168	56.099961
48	1.431405	0.698614	40.184782	57.520711
49	1.442141	0.693414	40.878195	58.952116
50	1.452957	0.688252	41.566447	60.394257

TABLE 5 *(Continued)*

			$r = .01$	
n	$(1 + r)^n$	$(1 + r)^{-n}$	$a_{\overline{n}|r}$	$s_{\overline{n}|r}$
1	1.010000	0.990099	0.990099	1.000000
2	1.020100	0.980296	1.970395	2.010000
3	1.030301	0.970590	2.940985	3.030100
4	1.040604	0.960980	3.901966	4.060401
5	1.051010	0.951466	4.853431	5.101005
6	1.061520	0.942045	5.795476	6.152015
7	1.072135	0.932718	6.728195	7.213535
8	1.082857	0.923483	7.651678	8.285671
9	1.093685	0.914340	8.566018	9.368527
10	1.104622	0.905287	9.471305	10.462213
11	1.115668	0.896324	10.367628	11.566835
12	1.126825	0.887449	11.255077	12.682503
13	1.138093	0.878663	12.133740	13.809328
14	1.149474	0.869963	13.003703	14.947421
15	1.160969	0.861349	13.865053	16.096896
16	1.172579	0.852821	14.717874	17.257864
17	1.184304	0.844377	15.562251	18.430443
18	1.196147	0.836017	16.398269	19.614748
19	1.208109	0.827740	17.226008	20.810895
20	1.220190	0.819544	18.045553	22.019004
21	1.232392	0.811430	18.856983	23.239194
22	1.244716	0.803396	19.660379	24.471586
23	1.257163	0.795442	20.455821	25.716302
24	1.269735	0.787566	21.243387	26.973465
25	1.282432	0.779768	22.023156	28.243200
26	1.295256	0.772048	22.795204	29.525631
27	1.308209	0.764404	23.559608	30.820888
28	1.321291	0.756836	24.316443	32.129097
29	1.334504	0.749342	25.065785	33.450388
30	1.347849	0.741923	25.807708	34.784892
31	1.361327	0.734577	26.542285	36.132740
32	1.374941	0.727304	27.269589	37.494068
33	1.388690	0.720103	27.989693	38.869009
34	1.402577	0.712973	28.702666	40.257699
35	1.416603	0.705914	29.408580	41.660276
36	1.430769	0.698925	30.107505	43.076878
37	1.445076	0.692005	30.799510	44.507647
38	1.459527	0.685153	31.484663	45.952724
39	1.474123	0.678370	32.163033	47.412251
40	1.488864	0.671653	32.834686	48.886373
41	1.503752	0.665003	33.499689	50.375237
42	1.518790	0.658419	34.158108	51.878989
43	1.533978	0.651900	34.810008	53.397779
44	1.549318	0.645445	35.455454	54.931757
45	1.564811	0.639055	36.094508	56.481075
46	1.580459	0.632728	36.727236	58.045885
47	1.596263	0.626463	37.353699	59.626344
48	1.612226	0.620260	37.973959	61.222608
49	1.628348	0.614119	38.588079	62.834834
50	1.644632	0.608039	39.196118	64.463182

TABLE 5 *(Continued)*

		$r = .0125$		
n	$(1 + r)^n$	$(1 + r)^{-n}$	$a_{\overline{n}|r}$	$s_{\overline{n}|r}$
1	1.012500	0.987654	0.987654	1.000000
2	1.025156	0.975461	1.963115	2.012500
3	1.037971	0.963418	2.926534	3.037656
4	1.050945	0.951524	3.878058	4.075627
5	1.064082	0.939777	4.817835	5.126572
6	1.077383	0.928175	5.746010	6.190654
7	1.090850	0.916716	6.662726	7.268038
8	1.104486	0.905398	7.568124	8.358888
9	1.118292	0.894221	8.462345	9.463374
10	1.132271	0.883181	9.345526	10.581666
11	1.146424	0.872277	10.217803	11.713937
12	1.160755	0.861509	11.079312	12.860361
13	1.175264	0.850873	11.930185	14.021116
14	1.189955	0.840368	12.770553	15.196380
15	1.204829	0.829993	13.600546	16.386335
16	1.219890	0.819746	14.420292	17.591164
17	1.235138	0.809626	15.229918	18.811053
18	1.250577	0.799631	16.029549	20.046192
19	1.266210	0.789759	16.819308	21.296769
20	1.282037	0.780009	17.599316	22.562979
21	1.298063	0.770379	18.369695	23.845016
22	1.314288	0.760868	19.130563	25.143078
23	1.330717	0.751475	19.882037	26.457367
24	1.347351	0.742197	20.624235	27.788084
25	1.364193	0.733034	21.357269	29.135435
26	1.381245	0.723984	22.081253	30.499628
27	1.398511	0.715046	22.796299	31.880873
28	1.415992	0.706219	23.502518	33.279384
29	1.433692	0.697500	24.200018	34.695377
30	1.451613	0.688889	24.888906	36.129069
31	1.469759	0.680384	25.569290	37.580682
32	1.488131	0.671984	26.241274	39.050441
33	1.506732	0.663688	26.904962	40.538571
34	1.525566	0.655494	27.560456	42.045303
35	1.544636	0.647402	28.207858	43.570870
36	1.563944	0.639409	28.847267	45.115505
37	1.583493	0.631515	29.478783	46.679449
38	1.603287	0.623719	30.102501	48.262942
39	1.623328	0.616019	30.718520	49.866229
40	1.643619	0.608413	31.326933	51.489557
41	1.664165	0.600902	31.927835	53.133177
42	1.684967	0.593484	32.521319	54.797341
43	1.706029	0.586157	33.107475	56.482308
44	1.727354	0.578920	33.686395	58.188337
45	1.748946	0.571773	34.258168	59.915691
46	1.770808	0.564714	34.822882	61.664637
47	1.792943	0.557742	35.380624	63.435445
48	1.815355	0.550856	35.931481	65.228388
49	1.838047	0.544056	36.475537	67.043743
50	1.861022	0.537339	37.012876	68.881790

TABLE 5 (Continued)

		$r = .015$		
n	$(1 + r)^n$	$(1 + r)^{-n}$	$a_{\overline{n}|r}$	$s_{\overline{n}|r}$
1	1.015000	0.985222	0.985222	1.000000
2	1.030225	0.970662	1.955883	2.015000
3	1.045678	0.956317	2.912200	3.045225
4	1.061364	0.942184	3.854385	4.090903
5	1.077284	0.928260	4.782645	5.152267
6	1.093443	0.914542	5.697187	6.229551
7	1.109845	0.901027	6.598214	7.322994
8	1.126493	0.887711	7.485925	8.432839
9	1.143390	0.874592	8.360517	9.559332
10	1.160541	0.861667	9.222185	10.702722
11	1.177949	0.848933	10.071118	11.863262
12	1.195618	0.836387	10.907505	13.041211
13	1.213552	0.824027	11.731532	14.236830
14	1.231756	0.811849	12.543382	15.450382
15	1.250232	0.799852	13.343233	16.682138
16	1.268986	0.788031	14.131264	17.932370
17	1.288020	0.776385	14.907649	19.201355
18	1.307341	0.764912	15.672561	20.489376
19	1.326951	0.753607	16.426168	21.796716
20	1.346855	0.742470	17.168639	23.123667
21	1.367058	0.731498	17.900137	24.470522
22	1.387564	0.720688	18.620824	25.837580
23	1.408377	0.710037	19.330861	27.225144
24	1.429503	0.699544	20.030405	28.633521
25	1.450945	0.689206	20.719611	30.063024
26	1.472710	0.679021	21.398632	31.513969
27	1.494800	0.668986	22.067617	32.986678
28	1.517222	0.659099	22.726717	34.481479
29	1.539981	0.649359	23.376076	35.998701
30	1.563080	0.639762	24.015838	37.538681
31	1.586526	0.630308	24.646146	39.101762
32	1.610324	0.620993	25.267139	40.688288
33	1.634479	0.611816	25.878954	42.298612
34	1.658996	0.602774	26.481728	43.933092
35	1.683881	0.593866	27.075595	45.592088
36	1.709140	0.585090	27.660684	47.275969
37	1.734777	0.576443	28.237127	48.985109
38	1.760798	0.567924	28.805052	50.719885
39	1.787210	0.559531	29.364583	52.480684
40	1.814018	0.551262	29.915845	54.267894
41	1.841229	0.543116	30.458961	56.081912
42	1.868847	0.535089	30.994050	57.923141
43	1.896880	0.527182	31.521232	59.791988
44	1.925333	0.519391	32.040622	61.688868
45	1.954213	0.511715	32.552337	63.614201
46	1.983526	0.504153	33.056490	65.568414
47	2.013279	0.496702	33.553192	67.551940
48	2.043478	0.489362	34.042554	69.565219
49	2.074130	0.482130	34.524683	71.608698
50	2.105242	0.475005	34.999688	73.682828

TABLE 5 *(Continued)*

		$r = .02$		
n	$(1 + r)^n$	$(1 + r)^{-n}$	$a_{\overline{n}|r}$	$s_{\overline{n}|r}$
1	1.020000	0.980392	0.980392	1.000000
2	1.040400	0.961169	1.941561	2.020000
3	1.061208	0.942322	2.883883	3.060400
4	1.082432	0.923845	3.807729	4.121608
5	1.104081	0.905731	4.713460	5.204040
6	1.126162	0.887971	5.601431	6.308121
7	1.148686	0.870560	6.471991	7.434283
8	1.171659	0.853490	7.325481	8.582969
9	1.195093	0.836755	8.162237	9.754628
10	1.218994	0.820348	8.982585	10.949721
11	1.243374	0.804263	9.786848	12.168715
12	1.268242	0.788493	10.575341	13.412090
13	1.293607	0.773033	11.348374	14.680332
14	1.319479	0.757875	12.106249	15.973938
15	1.345868	0.743015	12.849264	17.293417
16	1.372786	0.728446	13.577709	18.639285
17	1.400241	0.714163	14.291872	20.012071
18	1.428246	0.700159	14.992031	21.412312
19	1.456811	0.686431	15.678462	22.840559
20	1.485947	0.672971	16.351433	24.297370
21	1.515666	0.659776	17.011209	25.783317
22	1.545980	0.646839	17.658048	27.298984
23	1.576899	0.634156	18.292204	28.844963
24	1.608437	0.621721	18.913926	30.421862
25	1.640606	0.609531	19.523456	32.030300
26	1.673418	0.597579	20.121036	33.670906
27	1.706886	0.585862	20.706898	35.344324
28	1.741024	0.574375	21.281272	37.051210
29	1.775845	0.563112	21.844385	38.792235
30	1.811362	0.552071	22.396456	40.568079
31	1.847589	0.541246	22.937702	42.379441
32	1.884541	0.530633	23.468335	44.227030
33	1.922231	0.520229	23.988564	46.111570
34	1.960676	0.510028	24.498592	48.033802
35	1.999890	0.500028	24.998619	49.994478
36	2.039887	0.490223	25.488842	51.994367
37	2.080685	0.480611	25.969453	54.034255
38	2.122299	0.471187	26.440641	56.114940
39	2.164745	0.461948	26.902589	58.237238
40	2.208040	0.452890	27.355479	60.401983
41	2.252200	0.444010	27.799489	62.610023
42	2.297244	0.435304	28.234794	64.862223
43	2.343189	0.426769	28.661562	67.159468
44	2.390053	0.418401	29.079963	69.502657
45	2.437854	0.410197	29.490160	71.892710
46	2.486611	0.402154	29.892314	74.330564
47	2.536344	0.394268	30.286582	76.817176
48	2.587070	0.386538	30.673120	79.353519
49	2.638812	0.378958	31.052078	81.940590
50	2.691588	0.371528	31.423606	84.579401

TABLE 5 *(Continued)*

		$r = .025$		
n	$(1 + r)^n$	$(1 + r)^{-n}$	$a_{\overline{n}|r}$	$s_{\overline{n}|r}$
1	1.025000	0.975610	0.975610	1.000000
2	1.050625	0.951814	1.927424	2.025000
3	1.076891	0.928599	2.856024	3.075625
4	1.103813	0.905951	3.761974	4.152516
5	1.131408	0.883854	4.645828	5.256329
6	1.159693	0.862297	5.508125	6.387737
7	1.188686	0.841265	6.349391	7.547430
8	1.218403	0.820747	7.170137	8.736116
9	1.248863	0.800728	7.970866	9.954519
10	1.280085	0.781198	8.752064	11.203382
11	1.312087	0.762145	9.514209	12.483466
12	1.344889	0.743556	10.257765	13.795553
13	1.378511	0.725420	10.983185	15.140442
14	1.412974	0.707727	11.690912	16.518953
15	1.448298	0.690466	12.381378	17.931927
16	1.484506	0.673625	13.055003	19.380225
17	1.521618	0.657195	13.712198	20.864730
18	1.559659	0.641166	14.353364	22.386349
19	1.598650	0.625528	14.978891	23.946007
20	1.638616	0.610271	15.589162	25.544658
21	1.679582	0.595386	16.184549	27.183274
22	1.721571	0.580865	16.765413	28.862856
23	1.764611	0.566697	17.332110	30.584427
24	1.808726	0.552875	17.884986	32.349038
25	1.853944	0.539391	18.424376	34.157764
26	1.900293	0.526235	18.950611	36.011708
27	1.947800	0.513400	19.464011	37.912001
28	1.996495	0.500878	19.964889	39.859801
29	2.046407	0.488661	20.453550	41.856296
30	2.097568	0.476743	20.930293	43.902703
31	2.150007	0.465115	21.395407	46.000271
32	2.203757	0.453771	21.849178	48.150278
33	2.258851	0.442703	22.291881	50.354034
34	2.315322	0.431905	22.723786	52.612885
35	2.373205	0.421371	23.145157	54.928207
36	2.432535	0.411094	23.556251	57.301413
37	2.493349	0.401067	23.957318	59.733948
38	2.555682	0.391285	24.348603	62.227297
39	2.619574	0.381741	24.730344	64.782979
40	2.685064	0.372431	25.102775	67.402554
41	2.752190	0.363347	25.466122	70.087617
42	2.820995	0.354485	25.820607	72.839808
43	2.891520	0.345839	26.166446	75.660803
44	2.963808	0.337404	26.503849	78.552323
45	3.037903	0.329174	26.833024	81.516131
46	3.113851	0.321146	27.154170	84.554034
47	3.191697	0.313313	27.467483	87.667885
48	3.271490	0.305671	27.773154	90.859582
49	3.353277	0.298216	28.071369	94.131072
50	3.437109	0.290942	28.362312	97.484349

TABLE 5 *(Continued)*

		$r = .03$		
n	$(1 + r)^n$	$(1 + r)^{-n}$	$a_{\overline{n}|r}$	$s_{\overline{n}|r}$
1	1.030000	0.970874	0.970874	1.000000
2	1.060900	0.942596	1.913470	2.030000
3	1.092727	0.915142	2.828611	3.090900
4	1.125509	0.888487	3.717098	4.183627
5	1.159274	0.862609	4.579707	5.309136
6	1.194052	0.837484	5.417191	6.468410
7	1.229874	0.813092	6.230283	7.662462
8	1.266770	0.789409	7.019692	8.892336
9	1.304773	0.766417	7.786109	10.159106
10	1.343916	0.744094	8.530203	11.463879
11	1.384234	0.722421	9.252624	12.807796
12	1.425761	0.701380	9.954004	14.192030
13	1.468534	0.680951	10.634955	15.617790
14	1.512590	0.661118	11.296073	17.086324
15	1.557967	0.641862	11.937935	18.598914
16	1.604706	0.623167	12.561102	20.156881
17	1.652848	0.605016	13.166118	21.761588
18	1.702433	0.587395	13.753513	23.414435
19	1.753506	0.570286	14.323799	25.116868
20	1.806111	0.553676	14.877475	26.870374
21	1.860295	0.537549	15.415024	28.676486
22	1.916103	0.521893	15.936917	30.536780
23	1.973587	0.506692	16.443608	32.452884
24	2.032794	0.491934	16.935542	34.426470
25	2.093778	0.477606	17.413148	36.459264
26	2.156591	0.463695	17.876842	38.553042
27	2.221289	0.450189	18.327031	40.709634
28	2.287928	0.437077	18.764108	42.930923
29	2.356566	0.424346	19.188455	45.218850
30	2.427262	0.411987	19.600441	47.575416
31	2.500080	0.399987	20.000428	50.002678
32	2.575083	0.388337	20.388766	52.502759
33	2.652335	0.377026	20.765792	55.077841
34	2.731905	0.366045	21.131837	57.730177
35	2.813862	0.355383	21.487220	60.462082
36	2.898278	0.345032	21.832252	63.275944
37	2.985227	0.334983	22.167235	66.174223
38	3.074783	0.325226	22.492462	69.159449
39	3.167027	0.315754	22.808215	72.234233
40	3.262038	0.306557	23.114772	75.401260
41	3.359899	0.297628	23.412400	78.663298
42	3.460696	0.288959	23.701359	82.023196
43	3.564517	0.280543	23.981902	85.483892
44	3.671452	0.272372	24.254274	89.048409
45	3.781596	0.264439	24.518713	92.719861
46	3.895044	0.256737	24.775449	96.501457
47	4.011895	0.249259	25.024708	100.396501
48	4.132252	0.241999	25.266707	104.408396
49	4.256219	0.234950	25.501657	108.540648
50	4.383906	0.228107	25.729764	112.796867

TABLE 5 (Continued)

	$r = .035$			
n	$(1 + r)^n$	$(1 + r)^{-n}$	$a_{\overline{n}|r}$	$s_{\overline{n}|r}$
1	1.035000	0.966184	0.966184	1.000000
2	1.071225	0.933511	1.899694	2.035000
3	1.108718	0.901943	2.801637	3.106225
4	1.147523	0.871442	3.673079	4.214943
5	1.187686	0.841973	4.515052	5.362466
6	1.229255	0.813501	5.328553	6.550152
7	1.272279	0.785991	6.114544	7.779408
8	1.316809	0.759412	6.873956	9.051687
9	1.362897	0.733731	7.607687	10.368496
10	1.410599	0.708919	8.316605	11.731393
11	1.459970	0.684946	9.001551	13.141992
12	1.511069	0.661783	9.663334	14.601962
13	1.563956	0.639404	10.302738	16.113030
14	1.618695	0.617782	10.920520	17.676986
15	1.675349	0.596891	11.517411	19.295681
16	1.733986	0.576706	12.094117	20.971030
17	1.794676	0.557204	12.651321	22.705016
18	1.857489	0.538361	13.189682	24.499691
19	1.922501	0.520156	13.709837	26.357180
20	1.989789	0.502566	14.212403	28.279682
21	2.059431	0.485571	14.697974	30.269471
22	2.131512	0.469151	15.167125	32.328902
23	2.206114	0.453286	15.620410	34.460414
24	2.283328	0.437957	16.058368	36.666528
25	2.363245	0.423147	16.481515	38.949857
26	2.445959	0.408838	16.890352	41.313102
27	2.531567	0.395012	17.285365	43.759060
28	2.620172	0.381654	17.667019	46.290627
29	2.711878	0.368748	18.035767	48.910799
30	2.806794	0.356278	18.392045	51.622677
31	2.905031	0.344230	18.736276	54.429471
32	3.006708	0.332590	19.068865	57.334502
33	3.111942	0.321343	19.390208	60.341210
34	3.220860	0.310476	19.700684	63.453152
35	3.333590	0.299977	20.000661	66.674013
36	3.450266	0.289833	20.290494	70.007603
37	3.571025	0.280032	20.570525	73.457869
38	3.696011	0.270562	20.841087	77.028895
39	3.825372	0.261413	21.102500	80.724906
40	3.959260	0.252572	21.355072	84.550278
41	4.097834	0.244031	21.599104	88.509537
42	4.241258	0.235779	21.834883	92.607371
43	4.389702	0.227806	22.062689	96.848629
44	4.543342	0.220102	22.282791	101.238331
45	4.702359	0.212659	22.495450	105.781673
46	4.866941	0.205468	22.700918	110.484031
47	5.037284	0.198520	22.899438	115.350973
48	5.213589	0.191806	23.091244	120.388257
49	5.396065	0.185320	23.276564	125.601846
50	5.584927	0.179053	23.455618	130.997910

TABLE 5 *(Continued)*

$r = .04$				
n	$(1 + r)^n$	$(1 + r)^{-n}$	$a_{\overline{n}|r}$	$s_{\overline{n}|r}$
1	1.040000	0.961538	0.961538	1.000000
2	1.081600	0.924556	1.886095	2.040000
3	1.124864	0.888996	2.775091	3.121600
4	1.169859	0.854804	3.629895	4.246464
5	4.216653	0.821927	4.451822	5.416323
6	1.265319	0.790315	5.242137	6.632975
7	1.315932	0.759918	6.002055	7.898294
8	1.368569	0.730690	6.732745	9.214226
9	1.423312	0.702587	7.435332	10.582795
10	1.480244	0.675564	8.110896	12.006107
11	1.539454	0.649581	8.760477	13.486351
12	1.601032	0.624597	9.385074	15.025805
13	1.665074	0.600574	9.985648	16.626838
14	1.731676	0.577475	10.563123	18.291911
15	1.800944	0.555265	11.118387	20.023588
16	1.872981	0.533908	11.652296	21.824531
17	1.947900	0.513373	12.165669	23.697512
18	2.025817	0.493628	12.659297	25.645413
19	2.106849	0.474642	13.133939	27.671229
20	2.191123	0.456387	13.590326	29.778079
21	2.278768	0.438834	14.029160	31.969202
22	2.369919	0.421955	14.451115	34.247970
23	2.464716	0.405726	14.856842	36.617889
24	2.563304	0.390121	15.246963	39.082604
25	2.665836	0.375117	15.622080	41.645908
26	2.772470	0.360689	15.982769	44.311745
27	2.883369	0.346817	16.329586	47.084214
28	2.998703	0.333477	16.663063	49.967583
29	3.118651	0.320651	16.983715	52.966286
30	3.243398	0.308319	17.292033	56.084938
31	3.373133	0.296460	17.588494	59.328335
32	3.508059	0.285058	17.873551	62.701469
33	3.648381	0.274094	18.147646	66.209527
34	3.794316	0.263552	18.411198	69.857909
35	3.946089	0.253415	18.664613	73.652225
36	4.103933	0.243669	18.908282	77.598314
37	4.268090	0.234297	19.142579	81.702246
38	4.438813	0.225285	19.367864	85.970336
39	4.616366	0.216621	19.584485	90.409150
40	4.801021	0.208289	19.792774	95.025516
41	4.993061	0.200278	19.993052	99.826536
42	5.192784	0.192575	20.185627	104.819598
43	5.400495	0.185168	20.370795	110.012382
44	5.616515	0.178046	20.548841	115.412877
45	5.841176	0.171198	20.720040	121.029392
46	6.074823	0.164614	20.884654	126.870568
47	6.317816	0.158283	21.042936	132.945390
48	6.570528	0.152195	21.195131	139.263206
49	6.833349	0.146341	21.341472	145.833734
50	7.106683	0.140713	21.482185	152.667084

TABLE 5 *(Continued)*

			$r = .045$	
n	$(1 + r)^n$	$(1 + r)^{-n}$	$a_{\overline{n}|r}$	$s_{\overline{n}|r}$
1	1.045000	0.956938	0.956953	0.999996
2	1.092025	0.915730	1.872664	2.044996
3	1.141166	0.876297	2.748958	3.137016
4	1.192518	0.838562	3.587518	4.278180
5	1.246181	0.802451	4.389967	5.470695
6	1.302259	0.767896	5.157862	6.716874
7	1.360861	0.734829	5.892689	8.019130
8	1.422099	0.703186	6.595874	9.379988
9	1.486094	0.672905	7.268777	10.802084
10	1.552968	0.643928	7.912703	12.288173
11	1.622851	0.616199	8.528901	13.841137
12	1.695879	0.589665	9.118565	15.463986
13	1.772194	0.564272	9.682835	17.159860
14	1.851942	0.539974	10.222808	18.932049
15	1.935279	0.516721	10.739528	20.783986
16	2.022367	0.494470	11.233997	22.719262
17	2.113373	0.473177	11.707173	24.741624
18	2.208475	0.452801	12.159973	26.854993
19	2.307856	0.433303	12.593274	29.063459
20	2.411709	0.414644	13.007917	31.371308
21	2.520236	0.396788	13.404705	33.783013
22	2.633646	0.379702	13.784405	36.303240
23	2.752160	0.363351	14.147755	38.936881
24	2.876007	0.347704	14.495459	41.689037
25	3.005426	0.332731	14.828189	44.565032
26	3.140670	0.318403	15.146592	47.570452
27	3.282000	0.304692	15.451284	50.711119
28	3.429690	0.291572	15.742855	53.993110
29	3.584025	0.279016	16.021870	57.422788
30	3.745306	0.267001	16.288870	61.006807
31	3.913845	0.255503	16.544373	64.752103
32	4.089967	0.244501	16.788873	68.665940
33	4.274015	0.233972	17.022844	72.755889
34	4.466345	0.223897	17.246740	77.029898
35	4.667330	0.214255	17.460995	81.496230
36	4.877360	0.205029	17.666024	86.163555
37	5.096840	0.196200	17.862223	91.040899
38	5.326198	0.187751	18.049974	96.137727
39	5.565876	0.179666	18.229640	101.463914
40	5.816340	0.171929	18.401569	107.029779
41	6.078074	0.164526	18.566094	112.846102
42	6.351587	0.157441	18.723535	118.924165
43	6.637408	0.150661	18.874195	125.275731
44	6.936090	0.144173	19.018368	131.913124
45	7.248214	0.137965	19.156333	138.849198
46	7.574383	0.132024	19.288357	146.097398
47	7.915228	0.126339	19.414695	153.671734
48	8.271413	0.120898	19.535593	161.586956
49	8.643626	0.115692	19.651285	169.858364
50	9.032588	0.110710	19.761995	178.501960

TABLE 5 *(Continued)*

			$r = .05$	
n	$(1 + r)^n$	$(1 + r)^{-n}$	$a_{\overline{n}|r}$	$s_{\overline{n}|r}$
1	1.050000	0.952381	0.952381	1.000000
2	1.102500	0.907029	1.859410	2.050000
3	1.157625	0.863838	2.723248	3.152500
4	1.215506	0.822702	3.545951	4.310125
5	1.276282	0.783526	4.329477	5.525631
6	1.340096	0.746215	5.075692	6.801913
7	1.407100	0.710681	5.786373	8.142008
8	1.477455	0.676839	6.463213	9.549109
9	1.551328	0.644609	7.107822	11.026564
10	1.628895	0.613913	7.721735	12.577893
11	1.710339	0.584679	8.306414	14.206787
12	1.795856	0.556837	8.863252	15.917127
13	1.885649	0.530321	9.393573	17.712983
14	1.979932	0.505068	9.898641	19.598632
15	2.078928	0.481017	10.379658	21.578564
16	2.182875	0.458112	10.837770	23.657492
17	2.292018	0.436297	11.274066	25.840366
18	2.406619	0.415521	11.689587	28.132385
19	2.526950	0.395734	12.085321	30.539004
20	2.653298	0.376889	12.462210	33.065954
21	2.785963	0.358942	12.821153	35.719252
22	2.925261	0.341850	13.163003	38.505214
23	3.071524	0.325571	13.488574	41.430475
24	3.225100	0.310068	13.798642	44.501999
25	3.386355	0.295303	14.093945	47.727099
26	3.555673	0.281241	14.375185	51.113454
27	3.733456	0.267848	14.643034	54.669126
28	3.920129	0.255094	14.898127	58.402583
29	4.116136	0.242946	15.141074	62.322712
30	4.321942	0.231377	15.372451	66.438848
31	4.538039	0.220359	15.592811	70.760790
32	4.764941	0.209866	15.802677	75.298829
33	5.003189	0.199873	16.002549	80.063771
34	5.253348	0.190355	16.192904	85.066959
35	5.516015	0.181290	16.374194	90.320307
36	5.791816	0.172657	16.546852	95.836323
37	6.081407	0.164436	16.711287	101.628139
38	6.385477	0.156605	16.867893	107.709546
39	6.704751	0.149148	17.017041	114.095023
40	7.039989	0.142046	17.159086	120.799774
41	7.391988	0.135282	17.294368	127.839763
42	7.761588	0.128840	17.423208	135.231751
43	8.149667	0.122704	17.545912	142.993339
44	8.557150	0.116861	17.662773	151.143006
45	8.985008	0.111297	17.774070	159.700156
46	9.434258	0.105997	17.880066	168.685164
47	9.905971	0.100949	17.981016	178.119422
48	10.401270	0.096142	18.077158	188.025393
49	10.921333	0.091564	18.168722	198.426663
50	11.467400	0.087204	18.255925	209.347996

TABLE 5 *(Continued)*

			$r = .06$	
n	$(1 + r)^n$	$(1 + r)^{-n}$	$a_{\overline{n}|r}$	$s_{\overline{n}|r}$
1	1.060000	0.943396	0.943396	1.000000
2	1.123600	0.889996	1.833393	2.060000
3	1.191016	0.839619	2.673012	3.183600
4	1.262477	0.792094	3.465106	4.374616
5	1.338226	0.747258	4.212364	5.637093
6	1.418519	0.704961	4.917324	6.975319
7	1.503630	0.665057	5.582381	8.393838
8	1.593848	0.627412	6.209794	9.897468
9	1.689479	0.591898	6.801692	11.491316
10	1.790848	0.558395	7.360087	13.180795
11	1.898299	0.526788	7.886875	14.971643
12	2.012196	0.496969	8.383844	16.869941
13	2.132928	0.468839	8.852683	18.882138
14	2.260904	0.442301	9.294984	21.015066
15	2.396558	0.417265	9.712249	23.275970
16	2.540352	0.393646	10.105895	25.672528
17	2.692773	0.371364	10.477260	28.212880
18	2.854339	0.350344	10.827603	30.905653
19	3.025600	0.330513	11.158116	33.759992
20	3.207135	0.311805	11.469921	36.785591
21	3.399564	0.294155	11.764077	39.992727
22	3.603537	0.277505	12.041582	43.392290
23	3.819750	0.261797	12.303379	46.995828
24	4.048935	0.246979	12.550358	50.815577
25	4.291871	0.232999	12.783356	54.864512
26	4.549383	0.219810	13.003166	59.156383
27	4.822346	0.207368	13.210534	63.705766
28	5.111687	0.195630	13.406164	68.528112
29	5.418388	0.184557	13.590721	73.639798
30	5.743491	0.174110	13.764831	79.058186
31	6.088101	0.164255	13.929086	84.801677
32	6.453387	0.154957	14.084043	90.889778
33	6.840590	0.146186	14.230230	97.343165
34	7.251025	0.137912	14.368141	104.183755
35	7.686087	0.130105	14.498246	111.434780
36	8.147252	0.122741	14.620987	119.120867
37	8.636087	0.115793	14.736780	127.268119
38	9.154252	0.109239	14.846019	135.904206
39	9.703507	0.103056	14.949075	145.058458
40	10.285718	0.097222	15.046297	154.761966
41	10.902861	0.091719	15.138016	165.047684
42	11.557033	0.086527	15.224543	175.950545
43	12.250455	0.081630	15.306173	187.507577
44	12.985482	0.077009	15.383182	199.758032
45	13.764611	0.072650	15.455832	212.743514
46	14.590487	0.068538	15.524370	226.508125
47	15.465917	0.064658	15.589028	241.098612
48	16.393872	0.060998	15.650027	256.564529
49	17.377504	0.057546	15.707572	272.958401
50	18.420154	0.054288	15.761861	290.335905

TABLE 5 *(Continued)*

		$r = .07$		
n	$(1 + r)^n$	$(1 + r)^{-n}$	$a_{\overline{n}|r}$	$s_{\overline{n}|r}$
1	1.070000	0.934579	0.934579	1.000000
2	1.144900	0.873439	1.808018	2.070000
3	1.225043	0.816298	2.624316	3.214900
4	1.310796	0.762895	3.387211	4.439943
5	1.402552	0.712986	4.100197	5.750739
6	1.500730	0.666342	4.766540	7.153291
7	1.605781	0.622750	5.389289	8.654021
8	1.718186	0.582009	5.971299	10.259803
9	1.838459	0.543934	6.515232	11.977989
10	1.967151	0.508349	7.023582	13.816448
11	2.104852	0.475093	7.498674	15.783599
12	2.252192	0.444012	7.942686	17.888451
13	2.409845	0.414964	8.357651	20.140643
14	2.578534	0.387817	8.745468	22.550488
15	2.759032	0.362446	9.107914	25.129022
16	2.952164	0.338735	9.446649	27.888054
17	3.158815	0.316574	9.763223	30.840217
18	3.379932	0.295864	10.059087	33.999033
19	3.616528	0.276508	10.335595	37.378965
20	3.869684	0.258419	10.594014	40.995492
21	4.140562	0.241513	10.835527	44.865177
22	4.430402	0.225713	11.061240	49.005739
23	4.740530	0.210947	11.272187	53.436141
24	5.072367	0.197147	11.469334	58.176671
25	5.427433	0.184249	11.653583	63.249038
26	5.807353	0.172195	11.825779	68.676470
27	6.213868	0.160930	11.986709	74.483823
28	6.648838	0.150402	12.137111	80.697691
29	7.114257	0.140563	12.277674	87.346529
30	7.612255	0.131367	12.409041	94.460786
31	8.145113	0.122773	12.531814	102.073041
32	8.715271	0.114741	12.646555	110.218154
33	9.325340	0.107235	12.753790	118.933425
34	9.978114	0.100219	12.854009	128.258765
35	10.676581	0.093663	12.947672	138.236878
36	11.423942	0.087535	13.035208	148.913460
37	12.223618	0.081809	13.117017	160.337402
38	13.079271	0.076457	13.193473	172.561020
39	13.994820	0.071455	13.264928	185.640292
40	14.974458	0.066780	13.331709	199.635112
41	16.022670	0.062412	13.394120	214.609570
42	17.144257	0.058329	13.452449	230.632240
43	18.344355	0.054513	13.506962	247.776496
44	19.628460	0.050946	13.557908	266.120851
45	21.002452	0.047613	13.605522	285.749311
46	22.472623	0.044499	13.650020	306.751763
47	24.045707	0.041587	13.691608	329.224386
48	25.728907	0.038867	13.730474	353.270093
49	27.529930	0.036324	13.766799	378.999000
50	29.457025	0.033948	13.800746	406.528929

TABLE 5 *(Continued)*

			$r = .08$	
n	$(1 + r)^n$	$(1 + r)^{-n}$	$a_{\overline{n}|r}$	$s_{\overline{n}|r}$
1	1.080000	0.925926	0.925926	1.000000
2	1.166400	0.857339	1.783265	2.080000
3	1.259712	0.793832	2.577097	3.246400
4	1.360489	0.735030	3.312127	4.506112
5	1.469328	0.680583	3.992710	5.866601
6	1.586874	0.630170	4.622880	7.335929
7	1.713824	0.583490	5.206370	8.922803
8	1.850930	0.540269	5.746639	10.636628
9	1.999005	0.500249	6.246888	12.487558
10	2.158925	0.463193	6.710081	14.486562
11	2.331639	0.428883	7.138964	16.645487
12	2.518170	0.397114	7.536078	18.977126
13	2.719624	0.367698	7.903776	21.495297
14	2.937194	0.340461	8.244237	24.214920
15	3.172169	0.315242	8.559479	27.152114
16	3.425943	0.291890	8.851369	30.324283
17	3.700018	0.270269	9.121638	33.750226
18	3.996019	0.250249	9.371887	37.450244
19	4.315701	0.231712	9.603599	41.446263
20	4.660957	0.214548	9.818147	45.761964
21	5.033834	0.198656	10.016803	50.422921
22	5.436540	0.183941	10.200744	55.456755
23	5.871464	0.170315	10.371059	60.893296
24	6.341181	0.157699	10.528758	66.764759
25	6.848475	0.146018	10.674776	73.105940
26	7.396353	0.135202	10.809978	79.954415
27	7.988061	0.125187	10.935165	87.350768
28	8.627106	0.115914	11.051078	95.338830
29	9.317275	0.107328	11.158406	103.965936
30	10.062657	0.099377	11.257783	113.283211
31	10.867669	0.092016	11.349799	123.345868
32	11.737083	0.085200	11.434999	134.213537
33	12.676050	0.078889	11.513888	145.950620
34	13.690134	0.073045	11.586934	158.626670
35	14.785344	0.067635	11.654568	172.316804
36	15.968172	0.062625	11.717193	187.102148
37	17.245626	0.057986	11.775179	203.070320
38	18.625276	0.053690	11.828869	220.315945
39	20.115298	0.049713	11.878582	238.941221
40	21.724521	0.046031	11.924613	259.056519
41	23.462483	0.042621	11.967235	280.781040
42	25.339482	0.039464	12.006699	304.243523
43	27.366640	0.036541	12.043240	329.583005
44	29.555972	0.033834	12.077074	356.949646
45	31.920449	0.031328	12.108402	386.505617
46	34.474085	0.029007	12.137409	418.426067
47	37.232012	0.026859	12.164267	452.900152
48	40.210573	0.024869	12.189136	490.132164
49	43.427419	0.023027	12.212163	530.342737
50	46.901613	0.021321	12.233485	573.770156

TABLE 5 *(Continued)*

n	$(1+r)^n$	$(1+r)^{-n}$	$a_{\overline{n}|r}$	$s_{\overline{n}|r}$
		$r = .09$		
1	1.090000	0.917431	0.917432	1.000000
2	1.188100	0.841680	1.759112	2.090001
3	1.295029	0.772183	2.531296	3.278102
4	1.411582	0.708425	3.239721	4.573132
5	1.538624	0.649931	3.889653	5.984714
6	1.677101	0.596267	4.485920	7.523339
7	1.828040	0.547034	5.032955	9.200442
8	1.992563	0.501866	5.534821	11.028481
9	2.171894	0.460428	5.995249	13.021048
10	2.367365	0.422411	6.417660	15.192941
11	2.580428	0.387533	6.805193	17.560308
12	2.812666	0.355535	7.160727	20.140736
13	3.065806	0.326178	7.486906	22.953402
14	3.341729	0.299246	7.786153	26.019214
15	3.642485	0.274538	8.060690	29.360941
16	3.970308	0.251870	8.312560	33.003426
17	4.327636	0.231073	8.543633	36.973737
18	4.717124	0.211994	8.755627	41.301384
19	5.141665	0.194490	8.950117	46.018506
20	5.604415	0.178431	9.128548	51.160172
21	6.108813	0.163698	9.292245	56.764588
22	6.658606	0.150182	9.442427	62.873402
23	7.257881	0.137781	9.580208	69.532009
24	7.911091	0.126405	9.706613	76.789905
25	8.623090	0.115968	9.822581	84.701000
26	9.399168	0.106392	9.928974	93.324091
27	10.245092	0.097608	10.026581	102.723251
28	11.167153	0.089548	10.116130	112.968373
29	12.172196	0.082154	10.198284	124.135518
30	13.267694	0.075371	10.273655	136.307719
31	14.461787	0.069148	10.342803	149.575417
32	15.763349	0.063438	10.406241	164.037210
33	17.182053	0.058200	10.464442	179.800589
34	18.728436	0.053395	10.517836	196.982621
35	20.413996	0.048986	10.566822	215.711069
36	22.251263	0.044941	10.611764	236.125146
37	24.253870	0.041231	10.652994	258.376339
38	26.436726	0.037826	10.690821	282.630291
39	28.816021	0.034703	10.725523	309.066907
40	31.409473	0.031838	10.757361	337.883046
41	34.236324	0.029209	10.786570	369.292501
42	37.317596	0.026797	10.813367	403.528858
43	40.676178	0.024584	10.837951	440.846432
44	44.337036	0.022555	10.860506	481.522634
45	48.327370	0.020692	10.881198	525.859675
46	52.676830	0.018984	10.900182	574.187016
47	57.417747	0.017416	10.917598	626.863875
48	62.585365	0.015978	10.933576	684.281852
49	68.218048	0.014659	10.948235	746.867218
50	74.357674	0.013449	10.961683	815.085281

TABLE 5 *(Continued)*

		$r = .10$		
n	$(1 + r)^n$	$(1 + r)^{-n}$	$a_{\overline{n}|r}$	$s_{\overline{n}|r}$
1	1.100000	0.909091	0.909091	1.000000
2	1.210000	0.826446	1.735537	2.100000
3	1.331000	0.751315	2.486852	3.310000
4	1.464100	0.683013	3.169866	4.641000
5	1.610510	0.620921	3.790787	6.105100
6	1.771561	0.564474	4.355261	7.715609
7	1.948717	0.513158	4.868419	9.487170
8	2.143589	0.466507	5.334926	11.435886
9	2.357947	0.424098	5.759023	13.579474
10	2.593742	0.385543	6.144567	15.937422
11	2.853116	0.350494	6.495061	18.531163
12	3.138428	0.318631	6.813692	21.384283
13	3.452271	0.289664	7.103356	24.522708
14	3.797498	0.263331	7.366687	27.974978
15	4.177247	0.239392	7.606079	31.772471
16	4.594972	0.217629	7.823708	35.949722
17	5.054470	0.197845	8.021553	40.544697
18	5.559916	0.179859	8.201412	45.599161
19	6.115907	0.163508	8.364920	51.159073
20	6.727499	0.148644	8.513564	57.274987
21	7.400248	0.135131	8.648694	64.002477
22	8.140272	0.122846	8.771540	71.402723
23	8.954301	0.111678	8.883218	79.543011
24	9.849731	0.101526	8.984744	88.497316
25	10.834702	0.092296	9.077040	98.347018
26	11.918173	0.083905	9.160945	109.181731
27	13.109988	0.076278	9.237223	121.099885
28	14.420989	0.069343	9.306566	134.209893
29	15.863089	0.063039	9.369606	148.630889
30	17.449392	0.057309	9.426914	164.493927
31	19.194334	0.052099	9.479013	181.943344
32	21.113768	0.047362	9.526376	201.137681
33	23.225147	0.043057	9.569432	222.251477
34	25.547663	0.039143	9.608575	245.476633
35	28.102423	0.035584	9.644159	271.024233
36	30.912666	0.032349	9.676508	299.126670
37	34.003937	0.029408	9.705917	330.039375
38	37.404320	0.026735	9.732651	364.043206
39	41.144753	0.024304	9.756956	401.447534
40	45.259232	0.022095	9.779051	442.592326
41	49.785145	0.020086	9.799137	487.851459
42	54.763660	0.018260	9.817397	537.636616
43	60.240032	0.016600	9.833998	592.400335
44	66.264038	0.015091	9.849089	652.640395
45	72.890450	0.013719	9.862808	718.904511
46	80.179497	0.012472	9.875280	791.794985
47	88.197456	0.011338	9.886618	871.974583
48	97.017204	0.010307	9.896926	960.172064
49	106.718864	0.009370	9.906296	1057.188668
50	117.390755	0.008519	9.914815	1163.907573

TABLE 5 (Continued)

			$r = .11$	
n	$(1+r)^n$	$(1+r)^{-n}$	$a_{\overline{n}|r}$	$s_{\overline{n}|r}$
1	1.110000	0.900901	0.900900	0.999999
2	1.232100	0.811623	1.712523	2.109999
3	1.367631	0.731191	2.443714	3.342099
4	1.518070	0.658731	3.102445	4.709728
5	1.685058	0.593452	3.695895	6.227797
6	1.870414	0.534641	4.230537	7.912855
7	2.076159	0.481659	4.712194	9.783266
8	2.304537	0.433927	5.146121	11.859426
9	2.558036	0.390925	5.537046	14.163960
10	2.839419	0.352185	5.889230	16.721995
11	3.151756	0.317283	6.206514	19.561415
12	3.498449	0.285841	6.492355	22.713170
13	3.883278	0.257514	6.749869	26.211616
14	4.310438	0.231995	6.981864	30.094889
15	4.784585	0.209005	7.190868	34.405323
16	5.310890	0.188292	7.379161	39.189912
17	5.895087	0.169633	7.548793	44.500794
18	6.543547	0.152822	7.701615	50.395880
19	7.263336	0.137678	7.839293	56.939417
20	8.062303	0.124034	7.963327	64.202752
21	8.949157	0.111742	8.075070	72.265063
22	9.933562	0.100669	8.175738	81.214204
23	11.026253	0.090693	8.266431	91.147754
24	12.239142	0.081705	8.348136	102.174024
25	13.585445	0.073608	8.421744	114.413143
26	15.079844	0.066314	8.488058	127.998589
27	16.738626	0.059742	8.547800	143.078426
28	18.579872	0.053822	8.601621	159.817023
29	20.623661	0.048488	8.650109	178.396923
30	22.892262	0.043683	8.693792	199.020564
31	25.410406	0.039354	8.733146	221.912788
32	28.205553	0.035454	8.768600	247.323215
33	31.308159	0.031941	8.800541	275.528723
34	34.752052	0.028775	8.829316	306.836846
35	38.574768	0.025924	8.855239	341.588808
36	42.818001	0.023355	8.878594	380.163652
37	47.527973	0.021040	8.899634	422.981584
38	52.756039	0.018955	8.918590	470.509453
39	58.559196	0.017077	8.935666	523.265434
40	65.000717	0.015384	8.951051	581.824714
41	72.150803	0.013860	8.964911	646.825493
42	80.087410	0.012486	8.977397	718.976470
43	88.896980	0.011249	8.988646	799.063475
44	98.675659	0.010134	8.998780	887.960558
45	109.529999	0.009130	9.007910	986.636375
46	121.578239	0.008225	9.016135	1096.165838
47	134.951859	0.007410	9.023545	1217.744196
48	149.796600	0.006676	9.030221	1352.696397
49	166.274139	0.006014	9.036235	1502.492210
50	184.564316	0.005418	9.041653	1668.766545

TABLE 5 *(Continued)*

		$r = .12$		
n	$(1 + r)^n$	$(1 + r)^{-n}$	$a_{\overline{n}|r}$	$s_{\overline{n}|r}$
1	1.120000	0.892857	0.892856	0.999999
2	1.254400	0.797194	1.690050	2.119998
3	1.404928	0.711780	2.401830	3.374397
4	1.573519	0.635518	3.037347	4.779323
5	1.762341	0.567427	3.604774	6.352841
6	1.973822	0.506631	4.111405	8.115181
7	2.210680	0.452350	4.563754	10.088998
8	2.475961	0.403884	4.967637	12.299677
9	2.773076	0.360610	5.328247	14.775636
10	3.105845	0.321974	5.650221	17.548711
11	3.478546	0.287476	5.937697	20.654552
12	3.895972	0.256675	6.194372	24.133097
13	4.363487	0.229174	6.423546	28.029061
14	4.887105	0.204620	6.628166	32.392546
15	5.473557	0.182697	6.810862	37.279642
16	6.130383	0.163122	6.973984	42.753197
17	6.866029	0.145645	7.119629	48.883578
18	7.689952	0.130040	7.249668	55.749600
19	8.612744	0.116107	7.365775	63.439538
20	9.646275	0.103667	7.469442	72.052290
21	10.803825	0.092560	7.562002	81.698547
22	12.100284	0.082643	7.644644	92.502366
23	13.552314	0.073788	7.718432	104.602617
24	15.178594	0.065882	7.784315	118.154950
25	17.000017	0.058823	7.843138	133.333479
26	19.040022	0.052521	7.895659	150.333519
27	21.324820	0.046894	7.942553	169.373500
28	23.883801	0.041869	7.984422	190.698342
29	26.749851	0.037383	8.021805	214.582098
30	29.959826	0.033378	8.055183	241.331885
31	33.554996	0.029802	8.084985	271.291643
32	37.581600	0.026609	8.111594	304.846675
33	42.091396	0.023758	8.135352	342.428310
34	47.142353	0.021212	8.156564	384.519617
35	52.799442	0.018940	8.175504	431.662029
36	59.135361	0.016910	8.192414	484.461350
37	66.231590	0.015099	8.207512	543.596598
38	74.179359	0.013481	8.220993	609.828009
39	83.080864	0.012036	8.233030	684.007215
40	93.050613	0.010747	8.243777	767.088462
41	104.216660	0.009595	8.253372	860.138849
42	116.722633	0.008567	8.261939	964.355300
43	130.729324	0.007649	8.269589	1081.077727
44	146.416855	0.006830	8.276418	1211.807151
45	163.986832	0.006098	8.282516	1358.223628
46	183.665222	0.005445	8.287961	1522.210219
47	205.705002	0.004861	8.292822	1705.875053
48	230.389694	0.004340	8.297163	1911.580828
49	258.036407	0.003875	8.301038	2141.970110
50	289.000610	0.003460	8.304499	2400.005140

TABLE 5 *(Continued)*

		$r = .15$		
n	$(1 + r)^n$	$(1 + r)^{-n}$	$a_{\overline{n}|r}$	$s_{\overline{n}|r}$
1	1.150000	0.869565	0.869565	0.999999
2	1.322500	0.756144	1.625708	2.149998
3	1.520875	0.657516	2.283224	3.472497
4	1.749005	0.571754	2.854976	4.993369
5	2.011356	0.497177	3.352153	6.742373
6	2.313059	0.432328	3.784481	8.753727
7	2.660018	0.375937	4.160418	11.066785
8	3.059020	0.326902	4.487319	13.726799
9	3.517872	0.284263	4.771582	16.785816
10	4.045552	0.247185	5.018767	20.303682
11	4.652385	0.214944	5.233710	24.349232
12	5.350243	0.186907	5.420617	29.001618
13	6.152779	0.162528	5.583145	34.351858
14	7.075694	0.141329	5.724474	40.504628
15	8.137046	0.122895	5.847369	47.480307
16	9.357602	0.106865	5.954234	55.717349
17	10.761243	0.092926	6.047160	65.074954
18	12.375426	0.080805	6.127965	75.836177
19	14.231740	0.070265	6.198230	88.211602
20	16.366493	0.061100	6.259331	102.443290
21	18.821468	0.053131	6.312461	118.809792
22	21.644684	0.046201	6.358662	137.631229
23	24.891388	0.040175	6.398837	159.275923
24	28.625095	0.034934	6.433771	184.167307
25	32.918850	0.030378	6.464149	212.792338
26	37.856682	0.026415	6.490564	245.711218
27	43.535172	0.022970	6.513534	283.567816
28	50.065449	0.019974	6.533508	327.102999
29	57.575249	0.017369	6.550876	377.168333
30	66.211517	0.015103	6.565979	434.743459
31	76.143250	0.013133	6.579113	500.955008
32	87.564713	0.011420	6.590533	577.098096
33	100.699425	0.009931	6.600463	664.662846
34	115.804344	0.008635	6.609098	765.362312
35	133.174911	0.007509	6.616607	881.166096
36	153.151154	0.006529	6.623137	1014.341046
37	176.123840	0.005678	6.628815	1167.492295
38	202.542419	0.004937	6.633752	1343.616160
39	232.923645	0.004293	6.638045	1546.157668
40	267.862122	0.003733	6.641778	1779.080850
41	308.041534	0.003246	6.645025	2046.943609
42	354.247681	0.002823	6.647848	2354.984590
43	407.384979	0.002455	6.650302	2709.233256
44	468.492310	0.002135	6.652437	3116.615467
45	538.766357	0.001856	6.654293	3585.109130
46	619.581116	0.001614	6.655907	4123.874197
47	712.518066	0.001403	6.657310	4743.453882
48	819.396118	0.001220	6.658531	5455.974243
49	942.305298	0.001061	6.659592	6275.368793
50	1083.650757	0.000923	6.660515	7217.671874

TABLE 5 *(Continued)*

		$r = .18$		
n	$(1 + r)^n$	$(1 + r)^{-n}$	$a_{\overline{n}|r}$	$s_{\overline{n}|r}$
1	1.180000	0.847458	0.847457	0.999999
2	1.392400	0.718185	1.565641	2.179998
3	1.643031	0.608631	2.174271	3.572396
4	1.938777	0.515789	2.690060	5.215426
5	2.287756	0.437110	3.127169	7.154201
6	2.699552	0.370432	3.497601	9.441955
7	3.185471	0.313925	3.811526	12.141506
8	3.758855	0.266038	4.077564	15.326970
9	4.435448	0.225456	4.303020	19.085821
10	5.233828	0.191065	4.494085	23.521268
11	6.175916	0.161919	4.656004	28.755091
12	7.287580	0.137220	4.793224	34.930998
13	8.599342	0.116288	4.909511	42.218570
14	10.147224	0.098549	5.008061	50.817915
15	11.973722	0.083516	5.091577	60.965126
16	14.128988	0.070776	5.162353	72.938825
17	16.672203	0.059980	5.222333	87.067797
18	19.673199	0.050831	5.273164	103.739995
19	23.214373	0.043077	5.316240	123.413184
20	27.392956	0.036506	5.352746	146.627535
21	32.323673	0.030937	5.383683	174.020411
22	38.141941	0.026218	5.409901	206.344122
23	45.007488	0.022219	5.432119	244.486051
24	53.108810	0.018829	5.450949	289.493398
25	62.668411	0.015957	5.466906	342.602292
26	73.948692	0.013523	5.480429	405.270522
27	87.259445	0.011460	5.491889	479.219151
28	102.966171	0.009712	5.501601	566.478742
29	121.499985	0.008230	5.509831	669.444375
30	143.370010	0.006975	5.516806	790.944520
31	169.176544	0.005911	5.522717	934.314155
32	199.628296	0.005009	5.527726	1103.490557
33	235.561447	0.004245	5.531971	1303.119180
34	277.962372	0.003598	5.535569	1538.679878
35	327.995575	0.003049	5.538618	1816.642124
36	387.034729	0.002584	5.541202	2144.637431
37	456.700928	0.002190	5.543391	2531.671877
38	538.907043	0.001856	5.545247	2988.372530
39	635.909790	0.001573	5.546819	3527.276690
40	750.373962	0.001333	5.548152	4163.188773
41	885.441223	0.001129	5.549281	4913.562461
42	1044.819824	0.000957	5.550238	5798.999153
43	1232.888062	0.000811	5.551050	6843.822717
44	1454.807739	0.000687	5.551737	8076.709843
45	1716.671875	0.000583	5.552319	9531.510630
46	2025.673828	0.000494	5.552813	11248.188185
47	2390.293213	0.000418	5.553231	13273.851479
48	2820.545898	0.000355	5.553586	15664.144230
49	3328.243652	0.000300	5.553886	18484.687371
50	3927.329590	0.000255	5.554141	21812.942653

TABLE 5 *(Continued)*

		$r = .20$		
n	$(1 + r)^n$	$(1 + r)^{-n}$	$a_{\overline{n}|r}$	$s_{\overline{n}|r}$
1	1.200000	0.833333	0.833333	1.000000
2	1.440000	0.694444	1.527778	2.200000
3	1.728000	0.578704	2.106481	3.640000
4	2.073600	0.482253	2.588734	5.367998
5	2.488320	0.401878	2.990612	7.441598
6	2.985984	0.334898	3.325510	9.929918
7	3.583180	0.279082	3.604591	12.915899
8	4.299816	0.232568	3.837160	16.499079
9	5.159779	0.193807	4.030966	20.798893
10	6.191735	0.161506	4.192472	25.958675
11	7.430082	0.134588	4.327060	32.150410
12	8.916098	0.112157	4.439217	39.580489
13	10.699315	0.093464	4.532680	48.496576
14	12.839177	0.077887	4.610567	59.195887
15	15.407012	0.064906	4.675473	72.035062
16	18.488413	0.054088	4.729560	87.442066
17	22.186096	0.045073	4.774634	105.930483
18	26.623314	0.037561	4.812195	128.116572
19	31.947975	0.031301	4.843496	154.739879
20	38.337578	0.026084	4.869580	186.687893
21	46.005077	0.021737	4.891316	225.025392
22	55.206112	0.018114	4.909430	271.030566
23	66.247330	0.015095	4.924525	326.236656
24	79.496796	0.012579	4.937104	392.483987
25	95.396118	0.010483	4.947587	471.980601
26	114.475334	0.008736	4.956323	567.376684
27	137.370392	0.007280	4.963602	681.851974
28	164.844467	0.006066	4.969668	819.222354
29	197.813354	0.005055	4.974724	984.066794
30	237.376007	0.004213	4.978936	1181.880062
31	284.851288	0.003511	4.982447	1419.256471
32	341.821411	0.002926	4.985373	1704.107094
33	410.185822	0.002438	4.987811	2045.929153
34	492.222839	0.002032	4.989842	2456.114252
35	590.667542	0.001693	4.991535	2948.337773
36	708.800781	0.001411	4.992946	3539.003985
37	850.561218	0.001176	4.994122	4247.806186
38	1020.673035	0.000980	4.995101	5098.365287
39	1224.807983	0.000816	4.995918	6119.040054
40	1469.770020	0.000680	4.996598	7343.850262
41	1763.722168	0.000567	4.997165	8813.611037
42	2116.467285	0.000472	4.997638	10577.336662
43	2539.759766	0.000394	4.998031	12693.799112
44	3047.714600	0.000328	4.998360	15233.573339
45	3657.256104	0.000273	4.998633	18281.280926
46	4388.708496	0.000228	4.998861	21938.542971
47	5266.448242	0.000190	4.999051	26327.241799
48	6319.740234	0.000158	4.999209	31593.701878
49	7583.680176	0.000132	4.999341	37913.401726
50	9100.418945	0.000110	4.999451	45497.095744

TABLE 5 *(Continued)*

		$r = .24$		
n	$(1 + r)^n$	$(1 + r)^{-n}$	$a_{\overline{n}|r}$	$s_{\overline{n}|r}$
1	1.240000	0.806452	0.806451	1.000000
2	1.537600	0.650364	1.456816	2.239999
3	1.906624	0.524487	1.981303	3.777598
4	2.364213	0.422974	2.404276	5.684221
5	2.931624	0.341108	2.745384	8.048432
6	3.635213	0.275087	3.020471	10.980055
7	4.507664	0.221844	3.242315	14.615268
8	5.589503	0.178907	3.421222	19.122929
9	6.930983	0.144280	3.565501	24.712428
10	8.594417	0.116355	3.681856	31.643403
11	10.657076	0.093834	3.775690	40.237817
12	13.214774	0.075673	3.851363	50.894893
13	16.386320	0.061027	3.912390	64.109669
14	20.319035	0.049215	3.961605	80.495979
15	25.195601	0.039689	4.001294	100.815004
16	31.242540	0.032008	4.033302	126.010588
17	38.740742	0.025813	4.059114	157.253094
18	48.038517	0.020817	4.079931	195.993825
19	59.567753	0.016788	4.096718	244.032309
20	73.863998	0.013538	4.110257	303.600000
21	91.591347	0.010918	4.121175	377.463953
22	113.573257	0.008805	4.129980	469.055250
23	140.830826	0.007101	4.137080	582.628454
24	174.630264	0.005726	4.142807	723.459451
25	216.541489	0.004618	4.147425	898.089556
26	268.511505	0.003724	4.151149	1114.631296
27	332.954224	0.003003	4.154153	1383.142629
28	412.863159	0.002422	4.156575	1716.096535
29	511.950256	0.001953	4.158528	2128.959449
30	634.818237	0.001575	4.160103	2640.909381
31	787.174500	0.001270	4.161374	3275.727155
32	976.096252	0.001024	4.162398	4062.901143
33	1210.359131	0.000826	4.163224	5038.996491
34	1500.845215	0.000666	4.163891	6249.355202
35	1861.047852	0.000537	4.164428	7750.199555
36	2307.698975	0.000433	4.164861	9611.245942
37	2861.546387	0.000349	4.165211	11918.943544
38	3548.317139	0.000282	4.165492	14780.488408
39	4399.915527	0.000227	4.165720	18328.815107
40	5455.890625	0.000183	4.165903	22728.711446
41	6765.308105	0.000148	4.166051	28184.617736
42	8388.975586	0.000119	4.166170	34949.899056
43	10402.334961	0.000096	4.166266	43338.896639
44	12898.884766	0.000078	4.166344	53741.187725
45	15994.625977	0.000063	4.166406	66640.109725
46	19833.320313	0.000050	4.166457	82634.669816
47	24593.330078	0.000041	4.166497	102468.044283
48	30495.726563	0.000033	4.166530	127061.363517
49	37814.695313	0.000026	4.166557	157557.067324
50	46890.214844	0.000021	4.166578	195371.732883

TABLE 5 *(Continued)*

		$r = .25$		
n	$(1 + r)^n$	$(1 + r)^{-n}$	$a_{\overline{n}|r}$	$s_{\overline{n}|r}$
1	1.250000	0.800000	0.800000	1.000000
2	1.562500	0.640000	1.440000	2.250000
3	1.953125	0.512000	1.952000	3.812499
4	2.441406	0.409600	2.361600	5.765623
5	3.051757	0.327680	2.689280	8.207029
6	3.814696	0.262144	2.951424	11.258785
7	4.768370	0.209715	3.161139	15.073481
8	5.960462	0.167772	3.328911	19.841849
9	7.450578	0.134218	3.463129	25.802312
10	9.313221	0.107374	3.570503	33.252885
11	11.641526	0.085899	3.656403	42.566106
12	14.551905	0.068720	3.725122	54.207620
13	18.189882	0.054976	3.780098	68.759531
14	22.737352	0.043980	3.824078	86.949411
15	28.421690	0.035184	3.859263	109.686762
16	35.527111	0.028148	3.887410	138.108447
17	44.408886	0.022518	3.909928	173.635548
18	55.511105	0.018014	3.927942	218.044423
19	69.388878	0.014412	3.942354	273.555518
20	86.736092	0.011529	3.953883	342.944374
21	108.420105	0.009223	3.963107	429.680430
22	135.525131	0.007379	3.970485	538.100537
23	169.406342	0.005903	3.976388	673.625381
24	211.757919	0.004722	3.981111	843.031696
25	264.697479	0.003778	3.984888	1054.789941
26	330.871826	0.003022	3.987911	1319.487334
27	413.589752	0.002418	3.990329	1650.359046
28	516.987183	0.001934	3.992263	2063.948777
29	646.233948	0.001547	3.993810	2580.935849
30	807.792419	0.001238	3.995048	3227.169750
31	1009.740479	0.000990	3.996039	4034.962004
32	1262.175537	0.000792	3.996831	5044.702261
33	1577.719238	0.000634	3.997465	6306.877094
34	1972.149048	0.000507	3.997972	7884.596368
35	2465.186035	0.000406	3.998377	9856.744361
36	3081.482666	0.000325	3.998702	12321.930939
37	3851.853027	0.000260	3.998962	15403.412454
38	4814.815918	0.000208	3.999169	19255.264102
39	6018.519531	0.000166	3.999335	24070.078663
40	7523.149414	0.000133	3.999468	30088.598329
41	9403.935547	0.000106	3.999575	37611.743028
42	11754.919922	0.000085	3.999660	47015.680738
43	14693.648438	0.000068	3.999728	58770.595064
44	18367.060547	0.000054	3.999782	73464.243830
45	22958.808594	0.000044	3.999826	91831.236428
46	28698.509766	0.000035	3.999861	114790.041628
47	35873.132813	0.000028	3.999889	143488.534457
48	44841.417969	0.000022	3.999911	179361.675884
49	56051.804688	0.000018	3.999929	224203.223761
50	70064.750000	0.000014	3.999943	280255.006264

Answers

This appendix contains an answer to every problem appearing in the "Review Problems" for each chapter; it contains the answer to every odd-numbered problem in every other section of this text.

Chapter 1

Exercises 1.1

1. The number -115 is an integer and, hence, a rational number; it is not a counting number.
3. & 5. It is a rational number and not an integer.
7. & 9. It is an irrational number.
11. It is a counting number (hence, an integer and a rational number).
13., 15., & 17.

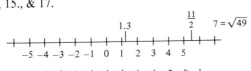

19.

21.

23.

25.

27.

29.

31. $|\sqrt{11}| = \sqrt{11}$

33. $\dfrac{5}{|-2|} = \dfrac{5}{2}$

35. $|3(5-4)| = |3(1)| = |3| = 3$

37. $|7| + |-3| = 7 + 3 = 10$
39. $|8| - |-5| = 8 - 5 = 3$
41. $6|5 - 2| = 6(3) = 18$
43. $\left|\dfrac{13 - (-2)}{5}\right| = \left|\dfrac{13 + 2}{5}\right| = \left|\dfrac{15}{5}\right| = |3| = 3$
45. $|-5||2| = 5(2) = 10$
47. $|2||9 - 14| = 2|-5| = 2(5) = 10$
49. The commutative property of addition.
51. The distributive property.
53. The associative property of addition.
55. The associative property of multiplication.
57. The additive inverse and the identity for addition.
59. The identity for addition.
61. The associative property for multiplication, the multiplicative inverse, and the identity for multiplication.
63. The distributive property.
65. The definition of division and the multiplicative inverse.
67. On the calculator, press the key 2 and then the key $\sqrt{x}$ to get 1.4142.
69. We write $p \geq \$2$ to say that "p must be at least \$2," and we write $p \leq \$5$ to say that "p cannot be more than \$5." Thus, we have $\$2 \leq p \leq \$5$.
71. We have the inequality $n \geq 0$, since the number of people must be nonnegative. We write $n \leq 100$ to say that the number "must not exceed 100."
$0 \leq n \leq 100$.
73. $n \leq 100{,}000$ and $n \geq 100$, $\quad 100 \leq n \leq 100{,}000$.

Exercises 1.2

1. $x = 3$
3. $x = 5/6$
5. $S = -7$
7. $S = 9$
9. $y = 3$
11. $y = 0$
13. $(\frac{1}{2} + \frac{1}{3})x = 1$
15. $x = 12/11$
$\quad \frac{5}{6}x = 1$
$\quad x = \frac{6}{5}$

17. $3(x + 1) + 5(x - 2) = -15$
$\quad 8x - 7 = -15$
$\quad x = -1$

19. $S = 50$
21. $S = 4$
23. $1 + 4 = 2S$
25. $x = -.5/4 = -1/8$
$\quad S = 5/2$

27. $x = 0$
29. $x = 7/4$
31. $x < 9$
33. $x \geq 12$
35. $x \geq 13$
37. $y > 12$
39. $y > 18$
41. $y \geq 3$
43. $w \geq 13/3$
45. $w < -120/11$
47. $w > -7/3$
49. $x \geq 2$
51. price $= p = \$25.50 - .1(\$25.50) = \$22.95$
53. $\$64.96$
55. price $= p = 1(\$20.60) + 1(\$20.60 - .05[\$20.60])$
$\quad + 1(\$20.60 - .1[\$20.60])$
$\quad + 1(\$20.60 - .15[\$20.60])$
$\quad + 1(\$20.60 - .2[\$20.60])$
$\quad p = \$20.60 + \$19.57 + \$18.54 + \$17.51$
$\quad + \$16.48$
$\quad p = \$92.70$

57. $x =$ amount put into investment A
$\$5000 - x =$ amount put into investment B

$.08x + .12(5000 - x) =$ return on A plus return
$\qquad\qquad\qquad\qquad$ on B
$\qquad\qquad\qquad\quad =$ desired return $= .11(5000)$

Thus, we must solve
$.08x + 600 - .12x = 550,$
$\quad$ and $\quad x = 1250.$

The amount put into A is $\$1250$, and the amount put into B is $(\$5000 - \$1250) = \$3,750.$

59. $x =$ amount put into A
$\$20,000 - x =$ amount put into B

$.22x + .15(20,000 - x) = 4015$

$\qquad\qquad x = 14,500$

Put $\$14,500$ into A and $(\$20,000 - \$14,500) = \$5500$ into B.

61. $r =$ rate of return on investment B

$.18(12,000) + r(8,000) = 3280$

$\qquad 8000r = 1120$

$\qquad\qquad r = .14 = 14\%$

63. $x =$ profit margin

$760,000x \geq 125,000$

$\quad x \geq .1645$ (rounded off to 4 decimal places)

$\quad$ or $\quad x \geq 16.45\%$

65. $x =$ total sales

$.12x \geq \$18,000$

$\quad x \geq \$150,000$

Exercises 1.3

1. $5x^2 + 5x + 1$
3. $-2\sqrt{2}\,x^2 + \sqrt{5}\,x - 8$
5. $13x^3 + 2x^2 + 6x + 2$
7. $3\sqrt{x} + 2x + 4$
9. $\sqrt[4]{x} - 6\sqrt[3]{x} + 4\sqrt{x}$
11. $1/x^2 + 8/x + 5 + 5x$
13. $5\sqrt{x} - 1/x - 4x$
15. $x + 1$
17. $5x^3 - 7x^2 + 2x + 1$
19. $12x^2 - 7x$
21. $x^3 + 8x^2 - 15x + 23$
23. $x^4 + 10x^2 + 36x$
25. $5x^2 - 5x - 20$
27. $-5x^3 + 11x^2 - 25x + 37$
29. $2x^2 + 3x + 2 + 3/x$
31. $x^{5/2} + 7x^{3/2} - x^{1/2} + 12x^2 + 83x - 19 + 1/x$
33. $8x^2 - 2x + 13$
35. x
37. $-10/(x^2 - 1)$
39. $(7x^2 - 32x + 15)/(x^2 - 25)$
41. $(8x - 7)/(x^2 - x - 2)$
43. $(x^2 - x + 22)/(x^2 + 2x - 8)$
45. $(11x + 1)/(x^2 - 3x - 10)$
47. $(-2x^3 - 2x^2 - 50x + 38)/(x^3 - 3x^2 - 13x + 15)$
49. $(3x^3 - 6x^2 - 4x + 125)/(x^3 + 2x^2 - 29x - 30)$
51. $4x + 3$
53. $(x^2 - 2x + 5)/(x - 2)$
55. $1/(x^3 + x^2 + x)$
57. $(9 + x)/(3x^3 - x + 5)$
59. $(x - 5)/2$
61. $(x^2 + 2x + 4)/3$
63. $2/(x - 6)$
65. $(x + 1)/(x - 1)$
67. $(x + 5)/(x^2 + 5x + 25)$

Exercises 1.4

1. $(x - 1)(x + 1) = 0, \quad x = 1, -1$
3. $x(x + 3) = 0, \quad x = 0, -3$
5. $y(3y - 4) = 0, \quad y = 0, 4/3$
7. $(x + 3)(x - 1) = 0, \quad x = -3, 1$
9. $(w - 7)(w + 1) = 0, \quad w = 7, -1$
11. $(p + 10)(p - 4) = 0, \quad p = -10, 4$
13. $(x + 2)(x + 1) = 0, \quad x = -2, -1$
15. $(y + 6)(y + 2) = 0, \quad y = -6, -2$
17. $(x + 10)(x + 5) = 0, \quad x = -10, -5$
19. $(t - 3)(t - 1) = 0, \quad t = 3, 1$
21. $(x - 8)(x - 5) = 0, \quad x = 8, 5$
23. $(p - 12)(p - 10) = 0, \quad p = 12, 10$
25. $(2x + 1)(x - 3) = 0, \quad x = -1/2, 3$
27. $(2x + 5)(x + 4) = 0, \quad x = -5/2, -4$

29. $(3r + 2)(2r - 1) = 0,\quad r = -2/3, 1/2$
31. $2(3x + 1)(3x - 6) = 0,\quad x = -1/3, 2$
33. No real roots.   35. $w = \pm\sqrt{7}$
37. $x = 1 \pm \sqrt{3}$ 39. $p = -3 \pm \sqrt{7}$
40. No real roots.   43. No real roots.
45. $r = -1/5 \pm \sqrt{10}/5$
47. We collect terms on the left and factor.

$(x + 4)(x - 2) = 0,\quad x = -4, 2$

49. $(x + 6)(x - 5) = 0,\quad x = -6, 5$
51. We simplify and apply the Quadratic Formula.

$x = (20 \pm 4\sqrt{37})/3$

53. We simplify and apply the Quadratic Formula.

$x = (19 \pm \sqrt{793})/6$

55. We simplify and apply the Quadratic Formula.

$x = (1 \pm \sqrt{37})/6$

57. We simplify and apply the Quadratic Formula.

$x = (-1 \pm \sqrt{41})/10$

59. We simplify and factor.

$(5x + 13)x = 0,\quad x = -13/5, 0$

61. We simplify and apply the Quadratic Formula.

$x = (11 \pm \sqrt{201})/4$

63. We factor.

$(x^2 + 9)(x - 3)(x + 3) = 0,\quad x = 3, -3$

65. $x(x - 1)^2 = 0,\quad x = 0, 1$
67. $(x^2 + 2)(x - 1)(x + 1) = 0,\quad x = 1, -1$
69. Squaring both sides, simplifying and using the Quadratic Formula, we get $x = (7 \pm \sqrt{17})/2$.
71. We simplify and apply the Quadratic Formula.

No real roots.

73. $(x + 3)/(x^2 + 2x + 1)$
75. $x - 1$ 77. $1/(x + 3)$
79. $(x - 5)/(x - 2)$ 81. $(x + 1)/(x + 6)$
83. After finding, the factorization $(x + 5)(x - 2) > 0$, we analyze the signs of the factors.

$x < -5\quad \text{or}\quad x > 2$

85. $x \le -6\quad \text{or}\quad x \ge 4$
87. $x < -5/4\quad \text{or}\quad x > 0$
89. All x.
91. $-2 < x < 7$
93. $-8 \le x \le 10$
95. $-7/3 < x < 0$
97. No solution.
99. $-3 - \sqrt{5} \le x \le -3 + \sqrt{5}$

Exercises 1.5

1. We must solve $.1x^2 + x + 50 = 250$.

$$.1x^2 + x - 200 = 0$$

$$x^2 + 10x - 2000 = 0$$

$$(x + 50)(x - 40) = 0$$

$$x = 40, -50$$

The answer $x = -50$ does not make sense. Thus, $x = 40$ is the only answer.

3. $x^2 + 25x = 350$

$$x^2 + 25x - 350 = 0$$

$$(x + 35)(x - 10) = 0$$

Only $x = 10$ makes sense.

5. If x is the width of the poster, then $(22 - x)$ is the length since (width) + (length) = 22. Then the area, which is stated to be 112, must be the product of the width and the length.

$$x(22 - x) = 112$$

$$-x^2 + 22x - 112 = 0$$

$$x^2 - 22x + 112 = 0$$

$$(x - 8)(x - 14) = 0$$

$$x = 8, 14$$

Note that $22 - 8 = 14$. The width of the poster can be 8 inches and the length 14 inches, or the width can be 14 inches and the length 8 inches.

7. Let $x =$ width, so $2x =$ length.

$$2x(x) = \text{area} = 288$$

$$x^2 = 144$$

$$x = 12,\quad \text{length} = 24$$

9. The length of the wall must be $560/10 = 56$. Thus, the sum of the lengths of the sides of the rectangular patio is 56. Let $x =$ width, so length = $(56 - 2x)/2$.

$$\frac{x(56 - 2x)}{2} = 192$$

$$x(28 - x) = 192$$

$$-x^2 + 28x = 192$$

$$0 = x^2 - 28x + 192$$

$$0 = (x - 16)(x - 12)$$

$$x = 16, 12$$

The dimensions of the patio are 16 feet by 12 feet.

11. Let $x =$ long leg, so $3x/4 =$ short leg. According to the Pythagorean theorem, we have

$$x^2 + \left(\frac{3x}{4}\right)^2 = 100^2$$

$$\tfrac{25}{16}x^2 = 10{,}000$$

$$x^2 = 16(400)$$

$$x = 4(20) = 80, \quad \text{short leg} = 60.$$

13. $\pi r^2 = 100\pi$

$r^2 = 100$

$r = 10$

15. $(10{,}000 + 500x)(20 - x) = 192{,}000$

$$200{,}000 - 500x^2 = 192{,}000$$

$$8{,}000 = 500x^2$$

$$x^2 = 16$$

$$x = 4$$

The monthly subscription rate should be reduced $4.

17. First, we note that a square with circumference of c has a side of $c/4$ and an area of $(c/4)^2 = c^2/16$. Second, we note that a rectangle with a long side which is twice the short side s has circumference $s + 2s + s + 2s = 6s$ and area $s(2s) = 2s^2$. Let x be the length of the piece of fence that is shaped into a square. Then the area of that square is $(x/4)^2 = x^2/16$, and $(100 - x)$ is left to shape the rectangle. From the above computations, we see that the short side of the rectangle is $(100 - x)/6$, and the area of the rectangle is $2(100 - x)^2/36 = (100 - x)^2/18$. Thus, we have

$$\frac{x^2}{16} + \frac{(100 - x)^2}{18} = 300$$

$$18x^2 + 16(100 - x)^2 = 18(16)(300)$$
$$= 86{,}400$$

$$18x^2 + 16(x^2 - 200x + 10{,}000) = 86{,}400$$

$$34x^2 - 3200x + 160{,}000 = 86{,}400$$

$$34x^2 - 3200x + 73{,}600 = 0$$

$$17x^2 - 1600x + 36{,}800 = 0$$

$$x = \frac{1600 \pm \sqrt{(1600)^2 - 4(17)(36{,}800)}}{34}$$

$$x = \frac{1600 \pm \sqrt{57{,}600}}{34} = \frac{1600 \pm 240}{34}$$

$$x = 40, \qquad \tfrac{920}{17} \approx 54.12$$

Either the square is 10 feet by 10 feet and the rectangle 10 feet by 20 feet or else the square is 230/17 feet by 230/17 feet and the rectangle is 130/17 feet by 260/17 feet.

19. If one number is x then the other is $8 - x$ and the product is $(8 - x)x$. Thus, we have

$$(8 - x)x = 15.75$$

$$0 = x^2 - 8x + 15.75$$

$$0 = 4x^2 - 32x + 63$$

$$0 = (2x - 9)(2x - 7)$$

$$x = 4.5, \, 3.5$$

21. $\qquad 175x - x^2 - 800 > 6700$

$$-x^2 + 175x - 7500 > 0$$

$$x^2 - 175x + 7500 < 0$$

$$(x - 75)(x - 100) < 0$$

One case is that $x - 75 < 0$ and $x - 100 > 0$, or $x < 75$ and $x > 100$, but this is clearly impossible.
    The other case is that $x - 75 > 0$ and $x - 100 < 0$ or $75 < x < 100$. The number of units manufactured should be between 75 and 100.

23. $x^2 + 40 \le 121$

$x^2 \le 81$

$x \le 9$

25. $x^2 + 4x + 60 \ge 200$

$x^2 + 4x - 140 \ge 0$

$(x + 14)(x - 10) \ge 0$

Case 1 $x + 14 \ge 0$ and $x - 10 \ge 0$

or $\qquad x \ge -14$ and $x \ge 10$

Since the first condition is automatically satisfied when the second condition holds, it suffices for the business to place 10 or more ads.

Case 2 $x + 14 \le 0$ and $x - 10 \le 0$

or $\qquad x \le -14$ and $x \le 10$

Since x is the number of customers, the first inequality cannot be satisfied and this cannot lead to an answer.

1. $2x + 3 = 7$ $\qquad$ or $\qquad$ $2x + 3 = -7$
$\qquad 2x = 4$ $\qquad\qquad\qquad$ $2x = -10$
$\qquad\quad x = 2$ $\qquad\qquad\qquad\quad$ $x = -5$

3. $9x + 4 = 0$

   $9x = -4$

   $x = -\frac{4}{9}$

5. There is no choice of x that makes the equation true.

7. $\frac{3}{x} + 4 = 5$ or $\frac{3}{x} + 4 = -5$

   $\frac{3}{x} = 1$ $\frac{3}{x} = -9$

   $\frac{1}{x} = \frac{1}{3}$ $\frac{1}{x} = -3$

   $x = 3$ $x = \frac{1}{-3}$

9. $|9x| = 3$

   $9x = 3$ or $9x = -3$

   $x = \frac{1}{3}$ or $x = -\frac{1}{3}$

11. $3|4x - 5| = 9$

   $|4x - 5| = 3$

   $4x - 5 = 3$ or $4x - 5 = -3$

   $4x = 8$ $4x = 2$

   $x = 2$ $x = \frac{1}{2}$

13. $|7x| + 5 = 19$

   $|7x| = 14$

   $7x = 14$ or $7x = -14$

   $x = 2$ $x = -2$

15. $2\left|\frac{3}{x} + 1\right| + 5 = 11$

   $2\left|\frac{3}{x} + 1\right| = 6$

   $\left|\frac{3}{x} + 1\right| = 3$

   $\frac{3}{x} + 1 = 3$ or $\frac{3}{x} + 1 = -3$

   $\frac{3}{x} = 2$ $\frac{3}{x} = -4$

   $x = \frac{3}{2}$ $x = -\frac{3}{4}$

17. $-20 \le 2x - 12 \le 20$

   $-10 \le x - 6 \le 10$

   $-4 \le x \le 16$

19. $-6 < 4 - 5x < 6$

   $-10 < -5x < 2$

   $2 > x > -\frac{2}{5}$

21. $-1 \le \frac{x}{2} + 3 \le 1$

   $-4 \le \frac{x}{2} \le -2$

   $-8 \le x \le -4$

23. $-9 < \frac{1}{x} + 7 < 9$

   $-16 < \frac{1}{x} < 2$

Since we do not know if x is positive or negative, these inequalities must be considered carefully.
Case 1 $x > 0$. In this case we have

   $-16x < 1 < 2x$ or

   $x > \frac{1}{-16}$ and $x > \frac{1}{2}$.

   Both of these inequalities are true when $x > 1/2$.
Case 2 $x < 0$. Here we have

   $-16x > 1 > 2x$ or

   $x < -1/16$ and $x < 1/2$.

   Both of these inequalities are true when $x < -1/16$.

25. $9x + 1 > 15$ or $9x + 1 < -15$

   $9x > 14$ $9x < -16$

   $x > \frac{14}{9}$ $x < -\frac{16}{9}$

27. $3 - 5x \ge 1$ or $3 - 5x \le -1$

   $-5x \ge -2$ $-5x \le -4$

   $x \le \frac{2}{5}$ $x \ge \frac{4}{5}$

29. Since $|5/x + 1| \ge 0 > -2$, the inequality is true for all $x \ne 0$.

31. The inequality is not true for any x.

33. $\frac{2}{x} + 1 \ge 3$ or $\frac{2}{x} + 1 \le -3$

   $\frac{2}{x} \ge 2$ $\frac{2}{x} \le -4$

   $\frac{1}{x} \ge 1$ $\frac{1}{x} \le -2$

For the first inequality to be true it must be that $x > 0$, so

   $1 \ge x > 0$.

For the second inequality to be true it must be that $x < 0$ and

   $1 \ge -2x$

   $-\frac{1}{2} \le x$.

The solution is all x such that either $1 \ge x > 0$ or $0 > x \ge -1/2$.

35. $5|10x + 3| > 20$
$|10x + 3| > 4$

$10x + 3 > 4$ or $10x + 3 < -4$

$10x > 1$ $10x < -7$

$x > \frac{1}{10}$ or $x < -\frac{7}{10}$

37. $4|-7x| + 5 \leq 19$
$4|-7x| \leq 14$
$|-7x| \leq \frac{7}{2}$

$\frac{-7}{2} \leq -7x \leq \frac{7}{2}$
$\frac{1}{2} \geq x \geq \frac{-1}{2}$

39. $\left|\dfrac{3}{x}\right| < 7$

$-7 < \dfrac{3}{x} < 7$

$\dfrac{-7}{3} < \dfrac{1}{x} < \dfrac{7}{3}$

Case 1 $x > 0$. In this case the inequality $-7/3 < 1/x$ is automatically satisfied.

$\dfrac{1}{x} < \dfrac{7}{3}$

$1 < \tfrac{7}{3}x$

$x > \tfrac{3}{7}$

Case 2 $x < 0$. In this case the inequality $1/x < 7/3$ is automatically satisfied.

$\dfrac{-7}{3} < \dfrac{1}{x}$

$\tfrac{-7}{3}x > 1$

$x < \tfrac{-3}{7}$

The solution is all x such that $x > 3/7$ or $x < -3/7$.

Review Problems: Chapter 1

1. -3.51 is a rational number but it is not an integer.
2. $22/7$ is a rational number but it is not an integer.
3. 111 is a counting number; thus, it is an integer and a rational number.
4. $\sqrt{10}$ is an irrational number.
5. -13 is an integer and, thus, a rational number, but it is not a counting number.

6.

7.

8.

9.

10.

11. The distributive property.
12. The commutative property of addition.
13. The associative property of addition.
14. The associative property of multiplication.
15. The identity for addition.

16. $9x = 17$
$x = \frac{17}{9}$

17. $2x - 2 + 3x + 6 = 6$
$5x = 2$
$x = \frac{2}{5}$

18. $-8x = 3 + 5x - 10$
$7 = 13x$
$x = \frac{7}{13}$

19. $x(\frac{1}{3} + \frac{4}{5}) = 1$
$x(\frac{17}{15}) = 1$
$x = \frac{15}{17}$

20. $5 + 3x = 2$
$3x = -3$
$x = -1$

21. $3 + 7 = 4(x - 2)$
$10 = 4x - 8$
$4x = 18$
$x = \frac{9}{2}$

22. $5 - 4(x + 1) = 6$
$5 - 4x - 4 = 6$
$-4x = 5$
$x = \frac{-5}{4}$

23. $5x < 27$
$x < \frac{27}{5}$

24. $-2x \geq 2$
$x \leq -1$

25. $\dfrac{-x}{3} \leq 2$
$x \geq -6$

26. $6x + 15 > 2x + 6$
$4x > -9$
$x > \frac{-9}{4}$

27. $3x + 18 < 30 + 2x$
$x < 12$

28. $2(2x + 1) + 60 > 5(x - 3)$
$4x + 2 + 60 > 5x - 15$
$77 > x$

29. $5x^2 - 3x + 4$

30. $6x^2 + 9x - 3$

31. $x - 2\sqrt{x}$

32. $3x^3 + 14x - 3 + 5/x$

33. $4x^2 + 4x + 3$

34. $-2x^2 - 5x - 18$

35. $28/(x^2 - 4)$

36. $(7x - 7)/(x^2 + x - 12)$

37. $(2x^2 + 5x + 5)/(x^2 - 1)$

38. $(x^3 + 7x^2 - 17x + 21)/(x^3 - 2x^2 - 5x + 6)$

39. $5x^2 - 6x + 10$ 40. $1/(9x^2 - 11x)$

41. $(x - 2)/(x^2 + 3x + 4)$ 42. $x - 4$

43. $(x + 2)/(x - 2)$

44. $(x - 5)(x + 5) = 0$
$x = 5, -5$

45. $(x + 1)^2 = 0$
$x = -1$

46. $(x+2)(x-1)=0$
    $x=-2, 1$
47. $x(7x-11)=0$
    $x=0, \frac{11}{7}$
48. $(x-4)(x-3)=0$
    $x=4, 3$
49. $(2x+1)(x-3)=0$
    $x=-\frac{1}{2}, 3$
50. $(4x-1)(2x-3)=0$
    $x=\frac{1}{4}, \frac{3}{2}$
51. $x=\dfrac{2\pm\sqrt{4+4}}{2}=1\pm\sqrt{2}\approx 2.41, -.41$
52. $x=\dfrac{4\pm\sqrt{16-20}}{2}$, no real roots
53. $x=\dfrac{\pm\sqrt{-36}}{2}$, no real roots
54. $x=\dfrac{5\pm\sqrt{25}}{2}=5, 0$
55. $x^2-x-2=0$
    $(x-2)(x+1)=0$
    $x=2, -1$
56. $x^2+x-12=0$
    $(x+4)(x-3)=0$
    $x=-4, 3$
57. $3(x+1)+5(x-2)=4(x+1)(x-2)$
    $8x-7=4(x^2-x-2)$
    $8x-7=4x^2-4x-8$
    $0=4x^2-12x-1$
    $x=\dfrac{12\pm\sqrt{144+16}}{8}$
    $=\dfrac{12\pm\sqrt{160}}{8}$
    $=\dfrac{12\pm4\sqrt{10}}{8}=\dfrac{3\pm\sqrt{10}}{2}$
    $x\approx 4.66, -1.66$
58. $3+2(x-1)=(x+1)(x-1)$
    $3+2x-2=x^2-1$
    $0=x^2-2x-2$
    $x=\dfrac{2\pm\sqrt{4+8}}{2}=\dfrac{2\pm\sqrt{12}}{2}$
    $=\dfrac{2\pm2\sqrt{3}}{2}=1\pm\sqrt{3}$
    $x\approx 2.73, -7.3$
59. $x(x^2-x-6)=0$
    $x(x-3)(x+2)=0$
    $x=0, 3, -2$

60. $(x^2+1)(x^2-1)=0$
    $(x^2+1)(x-1)(x+1)=0$
    $x=1, -1$
61. $x+2=(x-3)^2$
    $x+2=x^2-6x+9$
    $0=x^2-7x+7$
    $x=\dfrac{7\pm\sqrt{49-28}}{2}=\dfrac{7\pm\sqrt{21}}{2}\approx 5.79, 1.21$
62. $(x^2+4)(x^2-1)=0$
    $(x^2+4)(x-1)(x+1)=0$
    $x=1, -1$
63. $(x-6)(x-7)<0$
    Case 1 $x-6<0$ and $x-7>0$
               $x<6$ and $x>7$.
    This cannot happen.
    Case 2 $x-6>0$ and $x-7<0$
               $x>6$ and $x<7$.
    Solution: $6<x<7$.
64. $x^2-2x-15\le 0$
    $(x-5)(x+3)\le 0$
    Case 1 $x-5\le 0$ and $x+3\ge 0$
               $x\le 5$ and $x\ge -3$
    or, $-3\le x\le 5$.
    Case 2 $x-5\ge 0$ and $x+3\le 0$
               $x\ge 5$ and $x\le -3$.
    This cannot happen.
65. $x^2+2x-8>0$
    $(x+4)(x-2)>0$
    Case 1 $x+4>0$ and $x-2>0$
               $x>-4$ and $x>2$
    or simply, $x>2$.
    Case 2 $x+4<0$ and $x-2<0$
               $x<-4$ and $x<2$
    or simply, $x<-4$.
    Solution: either $x>2$ or $x<-4$.
66. $(x-3)(x+2)\ge 0$
    Case 1 $x-3\ge 0$ and $x+2\ge 0$
               $x\ge 3$ and $x\ge -2$
    or simply, $x\ge 3$.
    Case 2 $x-3\le 0$ and $x+2\le 0$
               $x\le 3$ and $x\le -2$
    or simply $x\le -2$.
    Solution: either $x\ge 3$ or $x\le -2$.
67. $7x+3=5$ or $7x+3=-5$
    $7x=2$ $7x=-8$
    $x=\frac{2}{7}$ $x=\frac{-8}{7}$
68. $8-3x=10$ or $8-3x=-10$
    $-3x=2$ $-3x=-18$
    $x=\dfrac{-2}{3}$ $x=6$

69. $1 + \dfrac{2}{x} = 4$ or $1 + \dfrac{2}{x} = -4$

$\dfrac{2}{x} = 3$ $\dfrac{2}{x} = -5$

$x = \tfrac{2}{3}$ $x = -\tfrac{2}{5}$

70. $5|x| - 2 = 3$ 71. $-12 \le 4x + 6 \le 12$

$5|x| = 5$ $-18 \le 4x \le 6$

$|x| = 1$

$x = 1, -1$ $\dfrac{-9}{2} \le x \le \dfrac{3}{2}$

72. $-6 < 8 - 2x < 6$

$-14 < -2x < -2$

$7 > x > 1$

73. $\dfrac{x}{3} + 5 > 8$ or $\dfrac{x}{3} + 5 < -8$

$\dfrac{x}{3} > 3$ $\dfrac{x}{3} < -13$

$x > 9$ or $x < -39$

74. $\dfrac{6}{x} + 5 \ge 9$ or $\dfrac{6}{x} + 5 \le -9$

$\dfrac{6}{x} \ge 4$ or $\dfrac{6}{x} \le -14$

For this to be   For this to be true, it must be
true, it must   that $x < 0$ and $\tfrac{3}{7} \le x$. Since
be that $x > 0$ that is impossible, this case
and $\tfrac{3}{2} \ge x > 0$.   does not lead to any solution.

75. $4|2 - 3x| \le 12$

$|2 - 3x| \le 3$

$-3 \le 2 - 3x \le 3$

$-5 \le -3x \le 1$

$\dfrac{5}{3} \ge x \ge \dfrac{-1}{3}$

76. $\$31 - .15(\$31) = \$26.35$
77. $100(\$5) + 100(\$4.50) + 200(\$4.00) = \$1750$
78. $.25x + .20(30,000 - x) = 6600$

$.05x + 6000 = 6600$

$.05x = 600$

$x = 12,000$

There is $12,000 invested in A and $18,000
invested in B.
79. $900,000x = 198,000$

$x = .22$ or 22%
80. $.3x^2 + 15x + 180 = 600$

$.3x^2 + 15x - 420 = 0$

$x^2 + 50x - 1400 = 0$

$(x + 70)(x - 20) = 0$

$x = 20$

The business can serve 20 customers.

81. $x =$ width of poster

$x(2x) = 162$

$x^2 = 81$

$x = 9,$ length $= 18$

82. $x =$ one leg of right triangle

$20^2 = x^2 + (48 - 20 - x)^2$

$400 = x^2 + x^2 - 56x + 784$

$0 = 2x^2 - 56x + 384$

$0 = x^2 - 28x + 192$

$0 = (x - 16)(x - 12)$

$x = 16, 12$

The two legs of the right triangle are 12 and 16.
83. $(100,000 + 1000x)(60 - x) = 5,500,000$

$6,000,000 - 40,000x - 1000x^2 = 5,500,000$

$0 = 1000x^2 + 40,000x - 500,000$

$0 = x^2 + 40x - 500$

$0 = (x + 50)(x - 10)$

$x = 10, -50$

The desired sum can be obtained by reducing the
subscription rate by $10.
84. $100x - x^2 - 600 > 1800$

$0 > x^2 - 100x + 2400$

$0 > (x - 60)(x - 40)$

Case 1 $x - 60 > 0$ and $x - 40 < 0$

$x > 60$ and $x < 40$

No x satisfies both of these.

Case 2 $x - 60 < 0$ and $x - 40 > 0$

$x < 60$ and $x > 40$

Answer: $60 > x > 40$.
85. Let x be one number, and note that $(15.5 - x)$ is
the other.

$x(15.5 - x) = 55$

$-x^2 + 15.5x = 55$

$0 = x^2 - 15.5x + 55$

$0 = 2x^2 - 31x + 110$

$0 = (2x - 11)(x - 10)$

$x = \tfrac{11}{2}, 10$

One number is 5.5 and the other is 10.

Social Science Applications: Chapter 1

1. mile > kilometer > meter
meter > yard > inch
inch > centimeter

3. 2.54 (length in inches) = length in centimeters
   .914 (length in yards) = length in meters
   1.610 (length in miles) = length in kilometers

5. Let x = the Celsius temperature.

   $\frac{9}{5}x + 32$ = equivalent Fahrenheit temperature

7. $\frac{11,720}{6,580} \approx 1.78 = 178\%$
9. $(.006)(\$100,000) = \$600$

Chapter 2

Exercises 2.1
1. $f(0) = 5$, $f(1) = 5$, $f(-1) = 5$, $f(\frac{1}{2}) = 5$
3. $f(0) = 2$, $f(1) = 5$, $f(-1) = -1$, $f(\frac{1}{2}) = 3.5$
5. $f(0) = 1$, $f(1) = -6$, $f(-1) = 8$, $f(\frac{1}{2}) = -2.5$
7. $f(0) = 1$, $f(1) = 2$, $f(-1) = 2$, $f(\frac{1}{2}) = 1.25$
9. $f(0) = 1$, $f(1) = 4$, $f(-1) = 2$, $f(\frac{1}{2}) = 2$
11. $f(0) = -2$, $f(1) = -2$, $f(-1) = 0$,
    $f(\frac{1}{2}) = -2.25$
13. $f(0) = 1$, $f(1) = .5$, $f(-1)$ is not defined,
    $f(\frac{1}{2}) = \frac{2}{3}$
15. $f(0) = -\frac{1}{3}$, $f(1) = -1$, $f(-1) = 0$, $f(\frac{1}{2}) = -\frac{3}{5}$
17. $f(0) = \sqrt{5}$, $f(1) = \sqrt{6}$, $f(-1) = 2$, $f(\frac{1}{2}) = \sqrt{5.5}$
19. $f(0) = 1$, $f(1) = 3$, $f(-1) = 1$, $f(\frac{1}{2}) = \frac{11}{8}$
21. All real x.
23. All t such that $t \geq 2$.
25. All s such that $-2 \leq s \leq 2$.
27. All x such that $x \neq 1$ and $x \neq -1$.
29. All real u.
31. It is a polynomial with degree 23.
33. It is not a polynomial.
35. It is a polynomial with degree 0.
37. It is a polynomial with degree 2.
39. This y is not a function of x.
41. This y is not a function of x.
43. This y is not a function of x.
45. This y is a function of x.
47. This y is not a function of x. When $x = 0$ then y can be any number.

49. $\dfrac{f(x+h) - f(x)}{h} = \dfrac{4(x+h) - 2 - 4x + 2}{h} = 4$

51. $\dfrac{f(x+h) - f(x)}{h}$

$= \dfrac{3(x+h)^2 - 6(x+h) + 7 - 3x^2 + 6x - 7}{h}$

$= \dfrac{3(x^2 + 2xh + h^2) - 6h - 3x^2}{h}$

$= 6x + 3h - 6$

53. $\dfrac{f(x+h) - f(x)}{h} = \dfrac{\sqrt{3(x+h)+1} - \sqrt{3x+1}}{h}$

$\cdot \dfrac{\sqrt{3(x+h)+1} + \sqrt{3x+1}}{\sqrt{3(x+h)+1} + \sqrt{3x+1}}$

$= \dfrac{3(x+h) + 1 - 3x - 1}{h(\sqrt{3(x+h)+1} + \sqrt{3x+1})}$

$= \dfrac{3}{\sqrt{3(x+h)+1} + \sqrt{3x+1}}$

55. $\dfrac{f(x+h) - f(x)}{h} = \dfrac{2(x+h)^3 + 1 - 2x^3 - 1}{h}$

$= \dfrac{2}{h}[(x+h)^3 - x^3]$

$= 2[(x+h)^2 + (x+h)x + x^2]$
$= 6x^2 + 6xh + 2h^2$

57. x = number of bricks sold

   $R(x) = .1x$, $\quad C(x) = .02x + 100,000$

   $P(x) = R(x) - C(x) = .08x - 100,000$

59. x^{23}, for example
61. Circumference $= C = 2\pi r$ where r = radius.

63. Area $= A = \sqrt{3}\, b^2/4$ where b is the length of one of the sides.

65.

x	1	2	3	4	5	6	7	8	9	10
x^9	1	512	19,683	262,144	1,953,125	10,077,696	40,353,607	134,217,728	387,420,489	1,000,000,000

Exercises 2.2
1–19. (odd numbers).

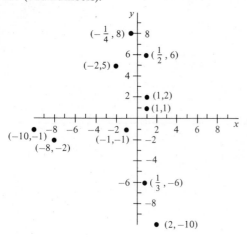

21. & 23.

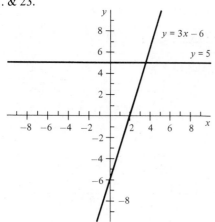

25. & 27.

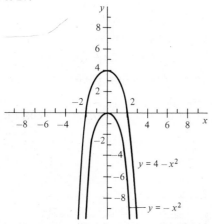

29.

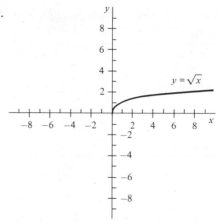

31. It is not a function.
33. It is not a function.
35. It is not a function.

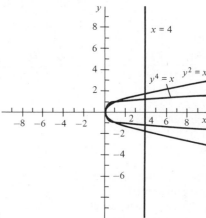

The axes show the graphs for problems 31, 33, and 35. 37. & 39. Each equation defines a function.

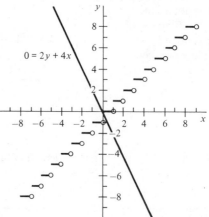

The axes show the graphs for problems 37 and 39.

41. It occurs twice since $(-1, 1)$ and $(1, 1)$ are on the graph.
43. It occurs once.
45. It occurs once.
47. It occurs once.
49. It does not occur as a y-value.
51. & 53. The range for each of these functions consists of all real numbers.

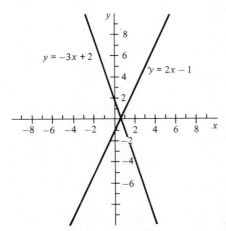

The axes show the graphs for problems 51 and 53.

55. The range consists of all numbers y such that $y \leq 1$.
57. The range consists of all numbers y such that $y \leq -3$.

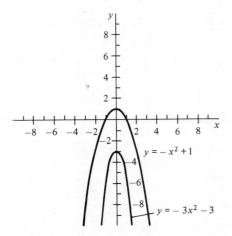

The axes show the graphs for problems 55 and 57.

59. The range consists of the two numbers -1 and 1.

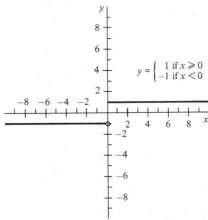

The axes show the graph for problem 59.

Exercises 2.3
1. $y = x - 1$
3. $y = (-3/2)x + 6$
5. $y = -x$
7. $y = (1/3)x$
9. $y = -(2/5)x$
11–15. (odd numbers).

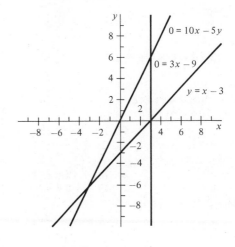

17. & 19.

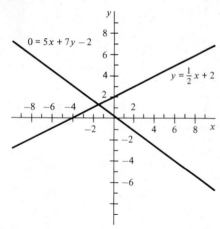

31. The graph for problem 31 is shown on the next axes.

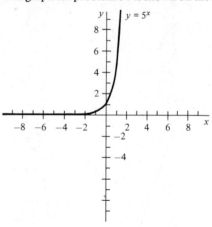

21. $y - 1 = x - 1$ 23. $y = 5x$
25. $y - 3 = 2x$
27. No point-slope equation exists; an equation is
    $x = -2$.
29. $y + 2 = -2x$ 31. $m = -1$
33. $m = -3$
35. This is a vertical line with no slope defined for it.
37. $m = -3$ 39. $m = 1/5$
41. $40   43. $20,000
45. $454,545.45

33. & 35. The next axes show the graphs for problems
33 and 35.

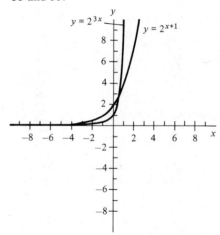

Exercises 2.4
1. $2^{3/2} = (2^{1/2})^3 \approx (1.414)^3 \approx 2.83$
3. $2^{3/6} = 2^{1/2} \approx 1.414$
5. $2^{-3/2} = (2^{1/2})^{-3} \approx (1.414)^{-3} \approx .35$
7. $2^{-5/2} = (2^{1/2})^{-5} \approx (1.414)^{-5} \approx .18$
9. $8^{1/6} = (2^3)^{1/6} = 2^{3/6} = 2^{1/2} \approx 1.414$
11. $3x = 1$, $x = \frac{1}{3}$ 13. $x - 2 = 2$, $x = 4$
15. $2x + 1 = -1$, $x = -1$ 17. $2x = 2$, $x = 1$
19. $3^{2x} = 3$, $2x = 1$, $x = \frac{1}{2}$
21. $e^x(1 + e^x)$ 23. $(e^x + 1)(e^x - 1)$
25. $(e^x - 3)(e^x + 2)$ 27. $(3 - 3^x)(3 + 3^x)$
29. $2^x(2^3 + 2^{-1}) = 8.5(2^x)$

37. The next axes show the graph for problem 37.

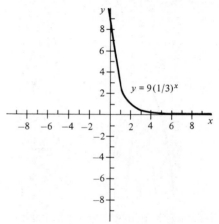

39. The next axes show the graph for problem 39.

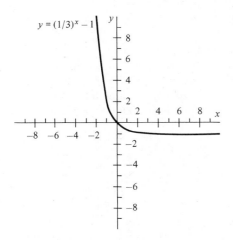

$y = (1/3)^x - 1$

41. According to Rule 1, we have

$$a^0 a^y = a^{0+y} = a^y,$$

so $$\frac{a^0 a^y}{a^y} = \frac{a^y}{a^y} = 1,$$

$$a^0 \cdot 1 = 1,$$

$$a^0 = 1.$$

43. Let t be the years that have passed since the average house cost $63,000 in 1980. Then $t/8$ is number of times that the price has doubled and the new price must be $p(t) = \$63,000(2^{t/8})$.

45. t = the number of years since the population was 100,000

$$p(t) = 100,000(2^{t/12}) = \text{population}$$

47. First year's depreciation = 2($90,000/45) = $4,000.

49.

x	2^x
1	2
2	4
3	8
4	16
5	32
6	64
7	128
8	256
9	512
10	1024

51.

x	$2^x/3^x$ (approximately)
1	.736
2	.541
3	.398
4	.293
5	.216
6	.159
7	.117
8	.086
9	.063
10	.047

Exercises 2.5

1. 2
3. 5
5. 4
7. 11
9. .2
11. e
13. 1.09861
15. 2.70805
17. .84510
19. -1.30103
21. $3x = \ln 8$, $x = (1/3)\ln 8$
23. $e^{8x} = e^{32}$, $8x = 32$, $x = 4$
25. $e^{2x} = e$, $2x = 1$, $x = 1/2$
27. $e^{3+x} = e^7$, $3 + x = 7$, $x = 4$
29. $x = 4$
31. $x = 2$
33. $10^{10x} = 1000$, $10x = 3$, $x = .3$
35. $10^{5x}/10^2 = e$, $10^{5x-2} = e$, $5x - 2 = \log e$, $5x = 2 + \log e$, $x = (1/5)(2 + \log e)$
37. $\log 10 + \log x = 9$, $\log x = 8$, $x = 10^{\log x} = 10^8 = 100,000,000$
39. $10^x = 10^2 \cdot 10^8 = 10^{10}$, $x = 10$
41. $\log(2/x)$
43. $\log[(x + 1)x(x + 2)]$
45. $\log(x^{1/2}) + \log[(x + 1)^{1/3}]$, $\log[x^{1/2}(x + 1)^{1/3}]$
47. $\log(x + 1)^{1/5} - \log x^2$, $\log[(x + 1)^{1/5}/x^2]$
49. $\log(x + 2)^3 - \log(.1)^2$, $\log[(x + 2)^3/.01]$, $\log[100(x + 2)^3]$
51. $500 = p = 10.46 \cdot 2^{t/4.5}$

$$47.80 = 2^{t/4.5}$$

$$\ln 47.80 = \frac{t}{4.5}\ln 2$$

$$\frac{t}{4.5} \approx 5.58$$

$t \approx 25.11$. It will reach $500 in 1999.

53. $1,000,000 = p(t) = 200,000 \, e^{.1t}$

$5 = e^{.1t}$

$\ln 5 = .1t$

$t = 10(\ln 5) \approx 16.09.$ It will happen in 1992.

55. $.5 = p(t) = e^{-.02t}$

$\ln .5 = -.02t$

$t = \dfrac{\ln .5}{-.02} \approx 34.66$

It will happen in 2014.

57.

x	1	2	3	4	5	6	7	8	9	10
log x	0	.301	.477	.602	.699	.778	.845	.903	.954	1

Exercises 2.6

1. Solve: $3b + 6m = 14$
   $6b + 14m = 35,$
   and find that $m = 7/2$ and $b = -7/3$. Thus,
   $y = \frac{7}{2}x - \frac{7}{3}.$

3. $y = 5x - 13$
5. $y = 2.25x + 9.08$
7. $y = -(13/10)x + 13$
9. $y = 5x - 11$
11. $y = (7/2)x - (5/2)$
13. $y = -5.2x + 15.6$
15. $y = 1.281x - 1.947$
17. $y = -7x + 51$
19. $y = (44/7)x - 20$

21. The least squares line for $(1, 25)$, $(2, 36)$, and $(3, 45)$ is $y = 10x + (46/3)$, and the sales for the next month are projected to be

$10(4) + \frac{46}{3} = 55 + \frac{1}{3} \approx 55.$

23. The least squares line for $(1, 10,000)$, $(2, 11,800)$, and $(3, 14,000)$ is $y = 2,000x + (23,800/3)$, and the projected worth of the portfolio is

$2000(4) + \frac{23,800}{3} \approx \$15,933.33.$

25. The least squares line for $(1, 300)$, $(2, 345)$, and $(3, 400)$ is $y = 50x + (745/3)$, and the projected electric bill is

$50(4) + \frac{745}{3} = \$448.33.$

27. The least squares line for $(1, 20,000)$, $(2, 22,000)$, $(3, 24,500)$, and $(4, 27,000)$ is $y = 2350x + 17,500$, and the projected value is

$2350(5) + 17,500 = \$29,250.$

29. The least squares line for $(1, 9,000)$, $(2, 9,100)$,

$(3, 9,220)$, and $(4, 9,310)$ is $y = 105x + 8,895$ and the projected profit is

$105(5) + 8,895 = \$9,420.$

Review Problems: Chapter 2
1. The domain consists of all numbers $x \geq -1$.
2. The domain consists of all real numbers.
3. The domain consists of all numbers $x \neq 3$.
4. The domain consists of all numbers $x \neq -2$ and $x \neq 1$.
5. no   6. no

7. yes   8. yes
9. yes   10. no
11. yes   12. no
13. The range consists of those numbers $y \leq 2$.

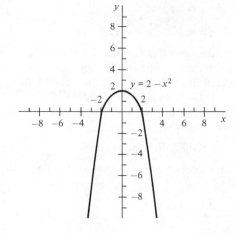

14. The range consists of all numbers y.

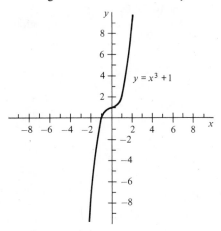

$y = x^3 + 1$

15. The range consists of all numbers y.

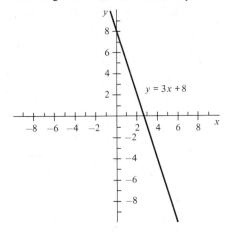

$y = 3x + 8$

16. The range consists of three numbers -2, 0, and 2.

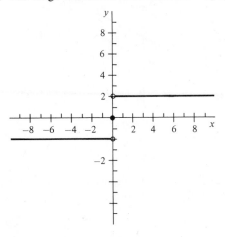

17. $y = 5$

18. $y = -x + 1$

19. $y = 2x - 3$

20–23. The graphs for problems 20 through 23 are shown on the next axes.

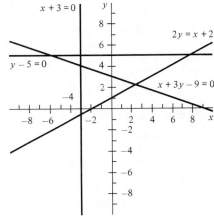

$x + 3 = 0$

$2y = x + 2$

$y - 5 = 0$

$x + 3y - 9 = 0$

24. $y = 9$
25. $y - 2 = 4(x + 2)$ or $y = 4x + 10$
26. $y - 9 = -9x$ or $y = -9x + 9$
27. $e^{5x+2} = e^{12}$, $5x + 2 = 12$, $x = 2$
28. $(2^{x-1})^3 = 1$, $2^{3x-3} = 2^0$, $3x - 3 = 0$, $x = 1$
29. $5^{3x+1+2-x} = 125$, $5^{2x+3} = 5^3$, $2x + 3 = 3$, $x = 0$
30. $10^{x^2-8} = 10^{-4}$, $x^2 - 8 = -4$, $x^2 = 4$, $x = \pm 2$
31. $5x = 10$, $x = 2$
32. $10^{-x} = 10^{-3}$, $x = 3$
33. $\log 100 + \log x = 2$, $2 + \log x = 2$, $x = 1$

34. $e^{\ln(x-2)} = 27$, $x^{-2} = 27$, $x^2 = 1/27$, $x = \dfrac{1}{3\sqrt{3}}$

35–36. The graphs for problems 35 and 36 are shown on the next axes.

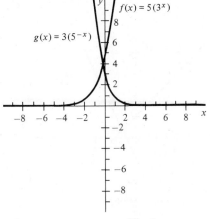

$f(x) = 5(3^x)$

$g(x) = 3(5^{-x})$

37. -3

38. 0

39. -1 40. 6
41. $y = (3/2)x + (1/6)$
42. $y = -(8/5)x + (23/10)$
43. $y = .1x + 4.3$
44. Costs $C = .05x$ dollars where $x =$ number of soft drinks sold.
45. Revenue $R = .25x$, profit $P = .20x$.
46. $p(x) = x^7$. This is one example; there are infinitely many possible examples.
47. First and third quadrants.
48. $24   49. $33,333.33
50. $120,000
51. The depreciation expense D for the tth year is

$$D(t) = (\tfrac{2}{40})100,000(1 - \tfrac{2}{40})^{t-1}.$$

Thus, the depreciation for the 10th year is $3151.25.

52. price $p(t) = \$100.00\ 2^{t/7}$

$t =$ years that have elapsed since now

53. $250 = 100\ 2^{t/7}$
    $2.5 = 2^{t/7}$

$$\ln 2.5 = \frac{t}{7} \ln 2$$

$$t = 7 \left(\frac{\ln 2.5}{\ln 2} \right) \approx 9.25$$

54. The least squares line for (1, 18,000), (2, 20,000), (3, 23,000), and (4, 24,000) is $y = 2100x + 16,000$, and the projected value of the portfolio is $2100(5) + 16,000 = \$26,500$.

Social Science Applications: Chapter 2
1. We need to find t_1 such that

$$.9P_0 = P(t_1) = P_0 e^{-.00012t_1}$$

or

$$.9 = e^{-.00012t_1}$$

We solve this equation.

$$\ln .9 = -.00012t_1$$

$$t_1 = -\ln \tfrac{9}{.00012} \approx 878$$

The fossils are 878 years old.

3. $n(8) = \dfrac{200,000}{1 + 1,999e^{-.8(8)}} \approx 46,280.73 \approx 46,281$

5. $n(24) = 1,000,000(1 - e^{-.03(24)}) \approx 513,247.74$
$\approx 513,248$

7. $y = .4(500) = 200$
The family spends $200 on "eating out."

Chapter 3

Exercises 3.1
1. $-2x + y = -2$ (I)
      $x + y = 2$ (II)
    add twice equation (II) to equation (I).

$3y = 2$

$y = \tfrac{2}{3}, \qquad x = \tfrac{4}{3}$

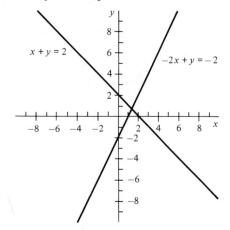

3. No solution.

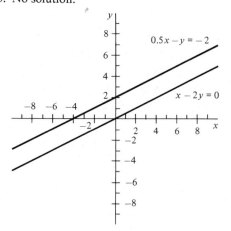

The graphs of the equations in problem 3 are shown on the axes above.

5. $y = 5x - 1$, x is arbitrary.

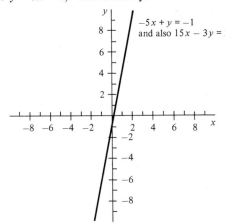

$-5x + y = -1$
and also $15x - 3y = 3$

The axes show the graphs of the equations in problem 5.

7. $x = 1/6$, $y = 1/18$
9. $x = (1 - 3y)/2$, y is arbitrary
11. No solution.
13. No solution.
15. $x = 39/35$, $y = 54/35$
17. $v = 7w - 3$, w is arbitrary
19. No solution.
21. $r = 1$, $s = 1$
23. No solution.

25. $y = \frac{12}{15}x + 2$ (I)

$4x - 5y = -10$ (II)

Substitute for y in equation (II) according to equation (I).

$4x - 5(\frac{12}{15}x + 2) = -10$

$-10 = -10$

This indicates that if equation (II) is solved for y then what we get is precisely equation (I). Thus, the general solution is $y = (4/5)x + 2$ for an arbitrary choice of x.

27. No solution.
29. $x = 1$, $y = 2$

31. $x =$ amount invested in business A
$y =$ amount invested in business B

$.12x + .20y = 24{,}000$ (I)

$.18x + .18y = 27{,}000$ (II)

Subtract 2 times equation (II) from 3 times equation (I).

$.24y = 18{,}000$

$y = \$75{,}000$, $x = \$75{,}000$

33. There are 100 light chairs and 80 heavy chairs made.
35. The market demand is 30 and the market price is $35.
37. There are 200 dimes and 120 quarters.
39. The number of wood screws is 50, and the number of sheet metal screws is 40.
41. $x \approx .96938$, $y \approx -.39535$
43. $x \approx -.01384$, $y \approx -.30041$

Exercises 3.2

1. $x_1 = -.5$, $x_2 = 1.5$, $x_3 = 2$
3. $x = 9/14$, $y = 1/2$, $z = -3/14$
5. $w = 2 + (y/5)$, $x = -3y/5$, y is arbitrary
7. $x_1 = (-4x_4 - 5)/3$, $x_2 = (-5x_4 + 11)/3$,
   $x_3 = (4x_4 - 7)/3$, x_4 is arbitrary
9. $w = 1 - x$, x is arbitrary, $y = -1$, $z = 2 - 2x$
11. No solution.
13. $x_1 = -1$, $x_2 = 4$, $x_3 = 3$
15. $r = .5$, $s = -1$, $t = -.5$
17. $r = (t + 1)/3$, $s = (5t + 2)/3$, t is arbitrary
19. $x_1 = 7/4$, $x_2 = 5/4$, $x_3 = -21/4$, $x_4 = -19/4$
21. No solution.

Exercises 3.3

1. The product AB exists and has order 2×3; the product BA does not exist.
3. Both products AB and BA exist, and each has order 3×3.
5. The products AB and BA exist and have orders 3×3 and 1×1, respectively.
7. The product AB exists and has order 3×2; the product BA does not exist.
9. The product AB exists and has order 1×4; the product BA does not exist.

11. $\begin{bmatrix} 3 & -2 \\ -1 & 12 \end{bmatrix}$ 13. $[5]$

15. There is no product.   17. $\begin{bmatrix} 7 & 3 \\ 5 & 2 \end{bmatrix}$

19. $\begin{bmatrix} .5 & .5 \\ .5 & .5 \end{bmatrix}$ 21. $\begin{bmatrix} 5 \\ 3 \\ 6 \end{bmatrix}$

23. $[-7]$ 25. $\begin{bmatrix} 56 & 29 & 11 \\ 7 & 8 & -8 \\ -19 & -11 & 47 \end{bmatrix}$

27. There is no product.   29. $\begin{bmatrix} 2 \\ 11 \end{bmatrix}$

31. $\begin{bmatrix} 3 & 2 \\ 1 & -5 \end{bmatrix} \begin{bmatrix} x_1 \\ x_2 \end{bmatrix} = \begin{bmatrix} 1 \\ 3 \end{bmatrix}$

33. $\begin{bmatrix} 2 & 1 & -1 \\ 1 & -3 & 1 \end{bmatrix} \begin{bmatrix} x \\ y \\ z \end{bmatrix} = \begin{bmatrix} 3 \\ -1 \end{bmatrix}$

35. $\begin{bmatrix} 1 & 3 \\ 1 & -1 \\ 3 & 4 \end{bmatrix} \begin{bmatrix} x_1 \\ x_2 \end{bmatrix} = \begin{bmatrix} 4 \\ 1 \\ 0 \end{bmatrix}$

37. $\begin{bmatrix} 1 & -3 & 1 \\ 1 & 1 & 5 \\ 1 & -1 & -1 \end{bmatrix} \begin{bmatrix} x_1 \\ x_2 \\ x_3 \end{bmatrix} = \begin{bmatrix} 4 \\ 4 \\ -3 \end{bmatrix}$

39. The numbers of units of the three products a, b, and c that are made are 35, 20, and 25, respectively.

41. The numbers of times that the breakfasts I, II, and III were ordered are 45, 25, and 40, respectively.

43. The total amounts in the investment portfolios of Adams, Brown, and Cain are $250,000, $20,000, and $100,000, respectively.

Exercises 3.4

1. The given matrices are inverses for each other.
3. The given matrices are inverses for each other.
5. The given matrices are not inverses for each other.
7. The given matrices are not inverses for each other.
9. The given matrices are inverses for each other.

11. $A^{-1} = \begin{bmatrix} \frac{1}{2} & 0 \\ 0 & \frac{1}{3} \end{bmatrix}$ 13. There is no inverse.

15. $A^{-1} = \begin{bmatrix} 1 & 0 \\ -1 & 1 \end{bmatrix}$ 17. There is no inverse.

19. $A^{-1} = \begin{bmatrix} 1 & 0 \\ 0 & 1 \end{bmatrix}$ 21. $A^{-1} = \begin{bmatrix} \frac{3}{7} & \frac{1}{7} \\ -\frac{1}{7} & \frac{2}{7} \end{bmatrix}$

23. $A^{-1} = \begin{bmatrix} 1 & 0 & 0 \\ 0 & \frac{1}{2} & 0 \\ 0 & 0 & \frac{1}{3} \end{bmatrix}$ 25. There is no inverse.

27. $A^{-1} = \begin{bmatrix} 1 & -\frac{2}{3} & \frac{1}{3} \\ 0 & \frac{1}{3} & -\frac{2}{3} \\ 0 & 0 & -1 \end{bmatrix}$

29. There is no inverse.

31. $\begin{bmatrix} x_1 \\ x_2 \end{bmatrix} = \begin{bmatrix} 1 & -\frac{2}{3} \\ 0 & \frac{1}{3} \end{bmatrix} \begin{bmatrix} 4 \\ 6 \end{bmatrix} = \begin{bmatrix} 0 \\ 2 \end{bmatrix}$

33. $\begin{bmatrix} x_1 \\ x_2 \end{bmatrix} = \begin{bmatrix} -\frac{3}{57} & \frac{8}{57} \\ \frac{9}{57} & -\frac{5}{57} \end{bmatrix} \begin{bmatrix} 20 \\ 30 \end{bmatrix} = \begin{bmatrix} \frac{60}{19} \\ \frac{10}{19} \end{bmatrix}$

35. $\begin{bmatrix} x_1 \\ x_2 \\ x_3 \end{bmatrix} = \begin{bmatrix} -\frac{1}{6} & \frac{1}{4} & \frac{1}{12} \\ \frac{1}{6} & -\frac{3}{4} & \frac{5}{12} \\ \frac{1}{12} & \frac{7}{8} & -\frac{13}{24} \end{bmatrix} \begin{bmatrix} 200 \\ 140 \\ 180 \end{bmatrix} = \begin{bmatrix} \frac{50}{3} \\ \frac{10}{3} \\ \frac{125}{3} \end{bmatrix}$

Review Problems: Chapter 3

1. $2x - 1 = -x + 1$
   $3x = 2$
   $x = \frac{2}{3}, \quad y = \frac{1}{3}$
   The graphs are shown below.

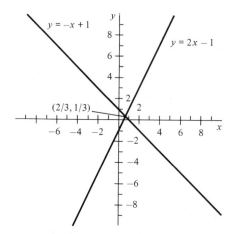

2. $\frac{3}{5}x + \frac{2}{5} = 5x - 6$
   $22x = 32$
   $x = \frac{16}{11}, \quad y = \frac{14}{11}$
   The graphs are shown below.

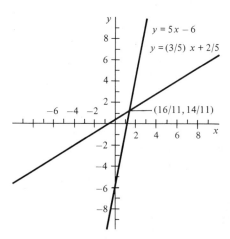

3. $5x - 7(-3x + 4) + 8 = 0$
   $26x - 20 = 0$
   $x = \frac{10}{13}, \quad y = \frac{22}{13}$

4. $3(2y - 1) - 9y = 11$
   $-3y = 14$
   $y = \frac{-14}{3}, \quad x = \frac{-31}{3}$

5. $\begin{bmatrix} 1 & 1 & | & 3 \\ 1 & -1 & | & 2 \end{bmatrix}$

$R_2 - R_1 \quad \begin{bmatrix} 1 & 1 & | & 3 \\ 0 & -2 & | & -1 \end{bmatrix}$

$x + y = 3$

$-2y = -1$

$y = \frac{1}{2}$

$x + \frac{1}{2} = 3$

$x = \frac{5}{2}$

6. $\begin{bmatrix} .5 & .1 & | & 4 \\ .1 & -.3 & | & -2 \end{bmatrix}$

$\begin{matrix} 10R_1 \\ 10R_2 \end{matrix} \quad \begin{bmatrix} 5 & 1 & | & 40 \\ 1 & -3 & | & -20 \end{bmatrix}$

$R_1 \leftrightarrow R_2 \quad \begin{bmatrix} 1 & -3 & | & -20 \\ 5 & 1 & | & 40 \end{bmatrix}$

$R_2 - 5R_1 \quad \begin{bmatrix} 1 & -3 & | & -20 \\ 0 & 16 & | & 140 \end{bmatrix}$

$x - 3y = -20$

$16y = 140$

$y = \frac{35}{4}$

$x - 3(\frac{35}{4}) = -20$

$x = \frac{25}{4}$

7. $v = -7/22, \quad w = -13/22$
8. $m = 78/23, \quad n = 6/23$
9. $x_1 = 8/3, \quad x_2 = 4, \quad x_3 = 5/3$
10. No solution.
11. $r = -1/3, \quad s = 29/9 - u, \quad t = 19/9 - u, \quad u$ is arbitrary
12. $w = 373/90, \quad x = -38/15, \quad y = 47/30, \quad z = -43/45$

13. There is no product.   14. $\begin{bmatrix} 4 & 20 & 28 & 36 \\ 6 & 30 & 42 & 54 \end{bmatrix}$

15. $[5]$ 16. $\begin{bmatrix} -2 & -2 \\ 7 & 4 \end{bmatrix}$

17. $\begin{bmatrix} 4 & 2 & 5 \\ 14 & 13 & -16 \end{bmatrix}$ 18. $\begin{bmatrix} -13 \\ -6 \\ 5 \end{bmatrix}$

19. $\begin{bmatrix} 1 & 0 & 0 & 0 \\ 0 & 1 & 0 & 0 \\ 0 & 0 & 1 & 0 \\ 0 & 0 & 0 & 1 \end{bmatrix}$

20. $\begin{bmatrix} 4 & -5 \\ 13 & 17 \end{bmatrix} \begin{bmatrix} x_1 \\ x_2 \end{bmatrix} = \begin{bmatrix} 9 \\ 23 \end{bmatrix}$

21. $\begin{bmatrix} 9 & -3 & 2 \\ 2 & 25 & -7 \\ 12 & -13 & 43 \end{bmatrix} \begin{bmatrix} x \\ y \\ z \end{bmatrix} = \begin{bmatrix} 1 \\ 35 \\ 81 \end{bmatrix}$

22. $A^{-1} = \begin{bmatrix} \frac{1}{33} & 0 \\ 0 & \frac{1}{22} \end{bmatrix}$

23. $A^{-1} = \begin{bmatrix} \frac{1}{18} & 0 & 0 \\ 0 & \frac{1}{5} & 0 \\ 0 & 0 & \frac{1}{43} \end{bmatrix}$

24. There is no inverse.

25. $A^{-1} = \begin{bmatrix} \frac{2}{13} & -\frac{3}{13} \\ -\frac{1}{13} & \frac{8}{13} \end{bmatrix}$

26. $A^{-1} = \begin{bmatrix} \frac{1}{2} & -\frac{1}{2} & \frac{1}{2} \\ \frac{1}{2} & \frac{1}{2} & -\frac{1}{2} \\ -\frac{1}{2} & \frac{1}{2} & \frac{1}{2} \end{bmatrix}$

27. $A^{-1} = \begin{bmatrix} \frac{7}{44} & \frac{9}{22} & \frac{1}{44} \\ \frac{9}{44} & -\frac{1}{22} & \frac{5}{44} \\ -\frac{5}{44} & \frac{3}{22} & \frac{7}{44} \end{bmatrix}$

28. $\begin{bmatrix} x \\ y \end{bmatrix} = \begin{bmatrix} -\frac{1}{9} & \frac{2}{9} \\ \frac{5}{18} & -\frac{1}{18} \end{bmatrix} \begin{bmatrix} 34 \\ 22 \end{bmatrix} = \begin{bmatrix} \frac{10}{9} \\ \frac{74}{9} \end{bmatrix}$

29. $\begin{bmatrix} x_1 \\ x_2 \\ x_3 \end{bmatrix} = \begin{bmatrix} \frac{9}{52} & -\frac{23}{52} & \frac{24}{52} \\ \frac{7}{52} & \frac{11}{52} & -\frac{16}{52} \\ -\frac{5}{52} & \frac{7}{52} & \frac{4}{52} \end{bmatrix} \begin{bmatrix} 100 \\ 50 \\ 200 \end{bmatrix}$

$= \begin{bmatrix} \frac{175}{2} \\ -\frac{75}{2} \\ \frac{25}{2} \end{bmatrix}$

30. The amount invested in A is $10,000 and the amount invested in B is $15,000.
31. The market demand and market price are 80/7 units and $68/7, respectively. Here $80/7 \approx 11.43$ and $68/7 \approx 9.71$.
32. The number of nickels is 264 and the number of quarters is 176.
33. There were 6 hamburgers ordered.

Social Science Applications: Chapter 3
1. Study hard.
3. 13 1 24 28 4 33 35 7 32 34 16 39
5. District 1 has 240,000 voters, and district 2 has 200,000 voters.
7. There were 560 copies of questionnaire A tabulated and 440 copies of questionnaire B.

9.

		(a)	(b)	(c)
		Percentages of Positive Responses by Categories of the Respondents		
		(a)	(b)	(c)
	(1)	.80	.70	.50
Questions	(2)	.90	.95	.70
	(3)	.85	.90	.80

Chapter 4

Exercises 4.1

1. $\begin{bmatrix} 4 & 4 \\ -4 & -2 \end{bmatrix}$

3. $\begin{bmatrix} -7 & 2 & 9 \\ 4 & -3 & 1 \end{bmatrix}$

5. $\begin{bmatrix} -3 \\ -8 \\ 11 \end{bmatrix}$

7. $\begin{bmatrix} 5 & -5 \\ 4 & 0 \end{bmatrix}$

9. The two matrices given cannot be added.

11. $\begin{bmatrix} 1 & 5 & 6 \\ 6 & 1 & 3 \\ 4 & 10 & 14 \end{bmatrix}$

13. $\begin{bmatrix} 8 & -34 & -20 \\ 15 & 22 & -18 \\ -15 & 24 & 28 \end{bmatrix}$

15. $\begin{bmatrix} \frac{3}{2} & 2 & 3 \\ \frac{11}{2} & 2 & \frac{1}{2} \\ 2 & 8 & \frac{21}{2} \end{bmatrix}$

17. $\begin{bmatrix} 4 & -8 & -4 \\ 9 & 8 & -6 \\ -3 & 12 & 14 \end{bmatrix}$

19. $\begin{bmatrix} -2 & 14 & 8 \\ -3 & 24 & 29 \\ 28 & -8 & 19 \end{bmatrix}$

21. $X = \frac{1}{2}(A + B)$

23. $X = A + B - C$

25. $X = (A + I)^{-1}(C - B)$

27. $X = \frac{1}{2}(B - C)(I + A)^{-1}$

29. $X = \frac{1}{2}(I + B)^{-1}(4A - C)$

31. $X = \begin{bmatrix} 5 & 4 \\ 2 & 3 \end{bmatrix}$

33. $X = \begin{bmatrix} 2 & 0 \\ 0 & 2 \end{bmatrix}$

35. $X = \begin{bmatrix} 3 & 5 \\ 3 & 1 \end{bmatrix}$

37. $X = \begin{bmatrix} 3 & 2 \\ 0 & -4 \end{bmatrix}$

39. $X = \begin{bmatrix} 3 & -2 \\ -\frac{5}{3} & 3 \end{bmatrix}$

41.
$$\begin{bmatrix} a_{11} & \cdots & a_{1n} \\ \vdots & & \vdots \\ a_{m1} & \cdots & a_{mn} \end{bmatrix} + \left(\begin{bmatrix} b_{11} & \cdots & b_{1n} \\ \vdots & & \vdots \\ b_{m1} & \cdots & b_{mn} \end{bmatrix} + \begin{bmatrix} c_{11} & \cdots & c_{1n} \\ \vdots & & \vdots \\ c_{m1} & \cdots & c_{mn} \end{bmatrix} \right)$$
$$= \begin{bmatrix} a_{11} & \cdots & a_{1n} \\ \vdots & & \vdots \\ a_{m1} & \cdots & a_{mn} \end{bmatrix} + \begin{bmatrix} b_{11} + c_{11} & \cdots & b_{1n} + c_{1n} \\ \vdots & & \vdots \\ b_{m1} + c_{m1} & \cdots & b_{mn} + c_{mn} \end{bmatrix}$$
$$= \begin{bmatrix} a_{11} + b_{11} + c_{11} & \cdots & a_{1n} + b_{1n} + c_{1n} \\ \vdots & & \vdots \\ a_{m1} + b_{m1} + c_{m1} & \cdots & a_{mn} + b_{mn} + c_{mn} \end{bmatrix}$$

$$= \begin{bmatrix} a_{11}+b_{11} & \cdots & a_{1n}+b_{1n} \\ \vdots & & \vdots \\ a_{m1}+b_{m1} & \cdots & a_{mn}+b_{mn} \end{bmatrix} + \begin{bmatrix} c_{11} & \cdots & c_{1n} \\ \vdots & & \vdots \\ c_{m1} & \cdots & c_{mn} \end{bmatrix}$$

$$\left(\begin{bmatrix} a_{11} & \cdots & a_{1n} \\ \vdots & & \vdots \\ a_{m1} & \cdots & a_{mn} \end{bmatrix} + \begin{bmatrix} b_{11} & \cdots & b_{1n} \\ \vdots & & \vdots \\ b_{m1} & \cdots & b_{mn} \end{bmatrix} \right) + \begin{bmatrix} c_{11} & \cdots & c_{1n} \\ \vdots & & \vdots \\ c_{m1} & \cdots & c_{mn} \end{bmatrix}$$

43. $$\begin{bmatrix} a_{11} & \cdots & a_{1n} \\ \vdots & & \vdots \\ a_{m1} & \cdots & a_{mn} \end{bmatrix} - \begin{bmatrix} b_{11} & \cdots & b_{1n} \\ \vdots & & \vdots \\ b_{m1} & \cdots & b_{mn} \end{bmatrix}$$

$$= \begin{bmatrix} a_{11}-b_{11} & \cdots & a_{1n}-b_{1n} \\ \vdots & & \vdots \\ a_{m1}-b_{m1} & \cdots & a_{mn}-b_{mn} \end{bmatrix} = \begin{bmatrix} a_{11} & \cdots & a_{1n} \\ \vdots & & \vdots \\ a_{m1} & \cdots & a_{mn} \end{bmatrix} + \begin{bmatrix} -b_{11} & \cdots & -b_{1n} \\ \vdots & & \vdots \\ -b_{m1} & \cdots & -b_{mn} \end{bmatrix}$$

$$= \begin{bmatrix} a_{11} & \cdots & a_{1n} \\ \vdots & & \vdots \\ a_{m1} & \cdots & a_{mn} \end{bmatrix} + (-1) \begin{bmatrix} b_{11} & \cdots & b_{1n} \\ \vdots & & \vdots \\ b_{m1} & \cdots & b_{mn} \end{bmatrix}$$

45. $$\left(\begin{bmatrix} b_{11} & b_{12} \\ b_{21} & b_{22} \end{bmatrix} + \begin{bmatrix} c_{11} & c_{12} \\ c_{21} & c_{22} \end{bmatrix} \right) \begin{bmatrix} a_{11} & a_{12} \\ a_{21} & a_{22} \end{bmatrix}$$

$$= \begin{bmatrix} b_{11}+c_{11} & b_{12}+c_{12} \\ b_{21}+c_{21} & b_{22}+c_{22} \end{bmatrix} \begin{bmatrix} a_{11} & a_{12} \\ a_{21} & a_{22} \end{bmatrix}$$

$$= \begin{bmatrix} a_{11}b_{11}+a_{11}c_{11}+a_{21}b_{12}+a_{21}c_{12} & a_{12}b_{11}+a_{12}c_{11}+a_{22}b_{12}+a_{22}c_{12} \\ a_{11}b_{21}+a_{11}c_{21}+a_{21}b_{22}+a_{21}c_{22} & a_{12}b_{21}+a_{12}c_{21}+a_{22}b_{22}+a_{22}c_{22} \end{bmatrix}$$

$$= \begin{bmatrix} a_{11}b_{11}+a_{21}b_{12} & a_{12}b_{11}+a_{22}b_{12} \\ a_{11}b_{21}+a_{21}b_{22} & a_{12}b_{21}+a_{22}b_{22} \end{bmatrix} + \begin{bmatrix} a_{11}c_{11}+a_{21}c_{12} & a_{12}c_{11}+a_{22}c_{12} \\ a_{11}c_{21}+a_{21}c_{22} & a_{12}c_{21}+a_{22}c_{22} \end{bmatrix}$$

$$= \begin{bmatrix} b_{11} & b_{12} \\ b_{21} & b_{22} \end{bmatrix} \begin{bmatrix} a_{11} & a_{12} \\ a_{21} & a_{22} \end{bmatrix} + \begin{bmatrix} c_{11} & c_{12} \\ c_{21} & c_{22} \end{bmatrix} \begin{bmatrix} a_{11} & a_{12} \\ a_{21} & a_{22} \end{bmatrix}$$

_____

Exercises 4.2

1. $A = \begin{bmatrix} .1 & .5 \\ .3 & .2 \end{bmatrix}$

3. $\frac{1}{.57} \begin{bmatrix} 70 \\ 69 \end{bmatrix} \approx \begin{bmatrix} 122.81 \\ 121.05 \end{bmatrix}$

5. $\frac{1}{.45} \begin{bmatrix} 470 \\ 575 \end{bmatrix} \approx \begin{bmatrix} 1044.44 \\ 1277.78 \end{bmatrix}$

7. $A = \begin{bmatrix} .1 & .3 & .25 \\ .3 & .1 & .25 \\ .2 & .4 & 0.00 \end{bmatrix}$

9. When $(I - A)^{-1}$ is computed to two decimal place accuracy, the answer is

$$\begin{bmatrix} 380 \\ 460.5 \\ 412.5 \end{bmatrix}.$$

11. When $(I - A)^{-1}$ is computed to two decimal place accuracy, the answer is

$$\begin{bmatrix} 1038 \\ 920 \\ 1533 \end{bmatrix}.$$

13.
$$X = XA + B$$
$$X - XA = B$$
$$X(I - A) = B$$
$$X = B(I - A)^{-1}$$

15. $AX - B = I$
$$AX = I + B$$
$$X = A^{-1}(I + B)$$

Exercises 4.3

1. 17
3. .22
5. 0
7. −10
9. 11
11. 0
13. 6
15. 0
17. 0
19. −3
21. 60
23. 0
25. Not invertible.
27. Invertible.
29. Not invertible.
31. Invertible.
33. Invertible.

Exercises 4.4

1. $x = 3/7, \quad y = 16/7$
3. Cramer's Rule does not apply because the determinant of the matrix of coefficients is 0.
5. $x = -2/11, \quad y = 5/11$
7. $r = 19/48, \quad s = -11/24$
9. Cramer's Rule does not apply because the determinant of the matrix of coefficients is 0.
11. $x_1 = 0, \quad x_2 = -1, \quad x_3 = 1$
13. $x = 35/8, \quad y = 25/4, \quad z = 5/8$
15. $x = -5/7$
17. $x = 0$
19. $x = -1/10$
21. $x = 0$
23. $x \approx 2.5923, \quad y \approx -.4765$
25. $x \approx .8142, \quad y \approx 3.2944$

Review Problems: Chapter 4

1. $\begin{bmatrix} 3 & 3 \\ 1 & 3 \end{bmatrix}$

2. $\begin{bmatrix} 6 \\ -5 \\ -3 \end{bmatrix}$

3. The indicated operation is not possible.

4. $\begin{bmatrix} -3 & -4 & 2 \\ 8 & -5 & 9 \\ 15 & 5 & -1 \end{bmatrix}$

5. [26]

6. $X = \begin{bmatrix} 0 & 2 \\ 3 & 1 \end{bmatrix}$

7. $X = \begin{bmatrix} -2 \\ -3 \end{bmatrix}$

8. $X = \begin{bmatrix} \frac{7}{3} & \frac{2}{3} \\ -\frac{2}{3} & \frac{2}{3} \end{bmatrix}$

9. $X = \begin{bmatrix} 4 & 2 \\ -2 & 1 \end{bmatrix}$

10. $A = \begin{bmatrix} \frac{1}{20} & \frac{7}{15} \\ \frac{9}{20} & \frac{2}{15} \end{bmatrix} \approx \begin{bmatrix} .05 & .47 \\ .45 & .13 \end{bmatrix}$

11. $\frac{1}{.615} \begin{bmatrix} 110.5 \\ 92.5 \end{bmatrix} \approx \begin{bmatrix} 179.67 \\ 150.41 \end{bmatrix}$

12. $\frac{1}{.615} \begin{bmatrix} 159.4 \\ 139 \end{bmatrix} \approx \begin{bmatrix} 259.19 \\ 226.02 \end{bmatrix}$

13. 0
14. 0
15. 8
16. 0
17. −12
18. 0
19. Invertible.
20. Not invertible.
21. Invertible.
22. $x = 50/23 \approx 2.1739, \quad y = -40/23 \approx -1.7391$
23. $r = 1, \quad s = 0$
24. Cramer's Rule does not apply because the determinant of the matrix of coefficients is 0.
25. $x = 1/3, \quad y = 1/3, \quad z = -1$
26. $x = 2/19 \approx .1053$
27. $x = 0$

Social Science Applications: Chapter 4

1. 1
3. Station P_3 can be linked to P_2 in two different ways using exactly 4 arcs, and that is the only pair of stations that can be linked in more than one way using exactly 4 arcs.
5. There are 500 employees who will get the larger raise and 300 who will get the smaller raise.
7. There were 40 trained mice among 100 mice that took the maze test.

Chapter 5

Exercises 5.1

1.

3.

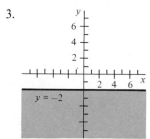

5.

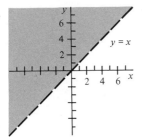

$y = x$

7.

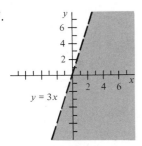

$y = 3x$

9.

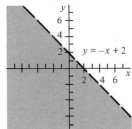

$y = -x + 2$

11.

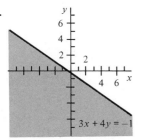

$3x + 4y = -1$

13.

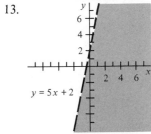

$y = 5x + 2$

15.

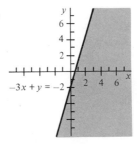

$-3x + y = -2$

17. $2x + 4y - 10 = 0$

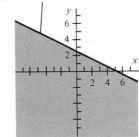

19.

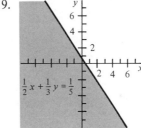

$\frac{1}{2}x + \frac{1}{3}y = \frac{1}{5}$

21.

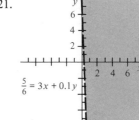

$\frac{5}{6} = 3x + 0.1y$

23.

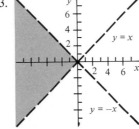

$y = x$

$y = -x$

25.

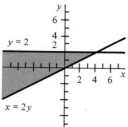

27.

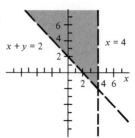

29.

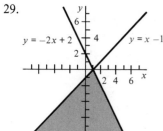

31.

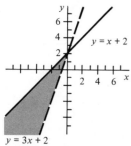

33.

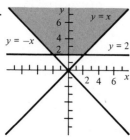

35. There are no points with coordinates solving all inequalities.

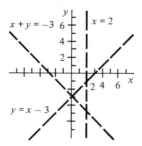

37.

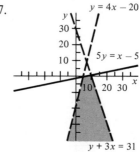

39. There are no points with coordinates solving all inequalities.

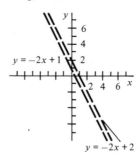

Exercises 5.2

1.

Vertex	Value
$(1, 0)$	2
$(0, 1)$	1
$(0, 0)$	0

The maximum is 2; it occurs at $(1, 0)$.

3.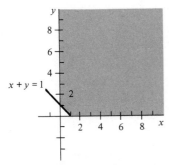

Vertex	Value
(1, 0)	2
(0, 1)	1

The minimum is 1; it occurs at (0, 1).

5.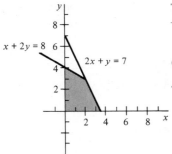

Vertex	Value
(2, 3)	12
(0, 4)	8
(3.5, 0)	10.5
(0, 0)	0

The maximum is 12; it occurs at (2, 3).

7.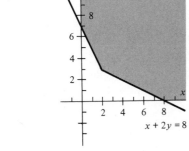

Vertex	Value
(0, 7)	14
(8, 0)	24
(2, 3)	12

The minimum is 12; it occurs at (2, 3).

9.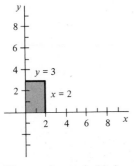

Vertex	Value
(0, 0)	0
(2, 0)	8
(0, 3)	12
(2, 3)	20

The maximum is 20; it occurs at (2, 3).

11.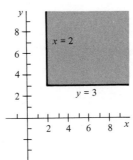

Vertex	Value
(2, 3)	20

The minimum is 20; it occurs at (2, 3).

13.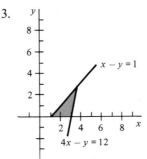

Vertex	Value
(1, 0)	3
(3, 0)	9
$\left(\frac{11}{3}, \frac{8}{3}\right)$	$\frac{73}{3}$

The maximum is $\frac{73}{3}$; it occurs at $\left(\frac{11}{3}, \frac{8}{3}\right)$.

15.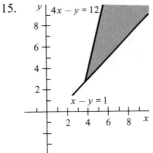

Vertex	Value
$\left(\frac{11}{3}, \frac{8}{3}\right)$	$\frac{73}{3}$

The minimum is $\frac{73}{3}$; it occurs at $\left(\frac{11}{3}, \frac{8}{3}\right)$.

17.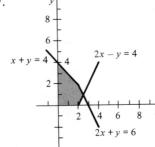

Vertex	Value
(0, 4)	16
(2, 0)	10
$\left(\frac{5}{2}, 1\right)$	$\frac{33}{2}$
(2, 2)	18

The maximum is 18; it occurs at (2, 2).

19.

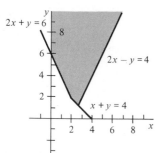

Vertex	Value
$(\frac{8}{3}, \frac{4}{3})$	$\frac{56}{3}$
$(2, 2)$	18
$(0, 6)$	24

The minimum is 18; it occurs at (2, 2).

21.

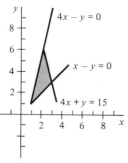

Vertex	Value
$(1, 1)$	2
$(3, 3)$	6
$(\frac{9}{4}, 6)$	$-\frac{51}{2}$

The minimum is $-\frac{51}{2}$; it occurs at $(\frac{9}{4}, 6)$. The maximum is 6; it occurs at (3, 3).

23.

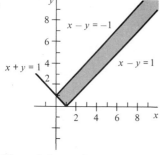

Vertex	Value
$(1, 0)$	1
$(0, 1)$	1

The minimum is 1; it occurs at (1, 0) and (0, 1). There is no maximum.

25.

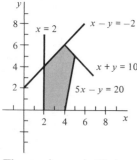

Vertex	Value
$(2, 0)$	4
$(2, 4)$	44
$(4, 0)$	8
$(5, 5)$	60
$(4, 6)$	68

The maximum is 68; it occurs at (4, 6). The minimum is 4; it occurs at (2, 0).

27. $x =$ amount in A bonds
$y =$ amount in BBB bonds
return $z = .15x + .16y$

$$x \geq y + 3000$$

$$x + y \leq 15,000$$

Vertex	Value
$(9000, 6000)$	$\$2310$
$(3000, 0)$	$\$ 450$
$(15,000\ 0)$	$\$2250$

The maximum return occurs when \$9000 is invested in A bonds and \$6000 is invested in BBB bonds.

29. Let x and y denote the number of bags of fertilizers no. 1 and no. 2, respectively; $z = 6x + 8y$.

$$5x + 10y \leq 100$$

$$15x + 10y \geq 120$$

$$5x + 30y \geq 120$$

Vertex	Value
$(2, 9)$	84
$(18, 1)$	116
$(6, 3)$	60

For minimum cost, the farmer should purchase 6 bags of fertilizer no. 1 and 3 bags of fertilizer no. 2.

31. $x =$ number of product A made
$y =$ number of product B made
profit $z = 16x + 14y$

$$6x + 3y \leq 90$$

$$3x + 6y \leq 75$$

$$5x + 4y \leq 80$$

Vertex	Value
$(0, \frac{25}{2})$	175
$(15, 0)$	240
$(\frac{40}{3}, \frac{10}{3})$	260
$(10, \frac{15}{2})$	265
$(0, 0)$	0

If it is possible to make a fraction of a unit, then the maximum profit is \$265, and it occurs when 10 and $\frac{15}{2}$ units of products A and B, respectively, are made. If it is not possible to make a fraction of a unit, then the maximum profit is \$258, and it occurs when 10 and 7 units of products A and B, respectively, are made.

33. $x =$ number of ounces of food A
    $y =$ number of ounces of food B
    cost $z = .10x + .11y$

    $12x + 8y \geq 120$

    $8x + 12y \geq 120$

    $10x + 10y \geq 110$

Vertex	Value
(0, 15)	$1.65
(15, 0)	$1.50
(6, 6)	$1.26

The minimum cost is $1.26 and it occurs when 6 ounces of each food are used.

35. $x =$ number of boxes carried for the toy company
    $y =$ number of boxes carried for the tool company
    revenue $z = x + .8y$

    $3x + y \leq 2100$

    $2x + 4y \leq 8000$

Vertex	Value
(0, 2000)	$1600
(0, 0)	$ 0
(700, 0)	$ 700
(40, 1980)	$1624

The maximum revenue is $1624; it occurs when the trucker hauls 40 boxes for the toy company and 1980 boxes for the tool company.

Exercises 5.3

1. $x + y + u = 1$

   $-2x - y + z = 0$

$$\begin{array}{ccccc} x & y & u & z & b \\ \left[\begin{array}{ccccc} \textcircled{1} & 1 & 1 & 0 & 1 \\ -2 & -1 & 0 & 1 & 0 \end{array}\right] \end{array}$$

$$\begin{array}{ccccc} x & y & u & z & b \\ \left[\begin{array}{ccccc} 1 & 1 & 1 & 0 & 1 \\ 0 & 1 & 2 & 1 & 2 \end{array}\right] \end{array}$$

$y = 0, \quad u = 0, \quad$ and $\quad x = 1$

The maximum is $z = 2 + 0 = 2$.

3. The maximum is $z = 12$, which occurs when $x = 2$, $y = 3$.

5. The maximum is $z = 20$, which occurs when $x = 2$, $y = 3$.

7. The maximum is $z = 18$, which occurs when $x = 2$, $y = 2$.

9. The maximum is $z = 6$, which occurs when $x = 2$, $y = 0$.

11. The maximum is $P = 8$, which occurs when $x = 0$, $y = 4$, and $z = 0$.

13. The maximum is $P = 16$, which occurs when $x = 0$, $y = 1/4$, $z = 21/4$.

15. The maximum is $P = 10$, which occurs when $x = 0$, $y = 2$, $z = 0$.

17. The maximum is $z = 7$, which occurs when $x_1 = 0$, $x_2 = .25$, $x_3 = 0$, $x_4 = 2.25$.

19. The maximum profit is $620, and it occurs when the numbers of simple, fancy, and extra fancy tables manufactured are 0, 10, and 20, respectively.

21. The maximum number of pounds of unsold meat that can be used is 1187.5, and that amount is used when the numbers of pounds of hot dogs, hamburgers, and sausage made are 875, 250, and 62.5, respectively.

23. The maximum revenue is $1652.80, and it occurs when the numbers of boxes of toys, tools, and crafts are 0, 1933, and 133, respectively.

Exercises 5.4

1. Maximize $z = 2v$ subject to the constraints $v \leq 5$ and $v \leq 10$.

3. Maximize $z = 4v_1 + 9v_2$ subject to the constraints

   $4v_1 + 2v_2 \leq 10$,

   $v_1 + 4v_2 \leq 8$, $\qquad v_1, v_2 \geq 0$.

5. Maximize $z = 10y_1 + 11y_2 + 49y_3$ subject to the constraints

   $5y_1 + y_2 + 2y_3 \leq 6$,

   $y_1 + 2y_2 + 10y_3 \leq 5$, $\qquad y_1, y_2, y_3 \geq 0$.

7. The minimum is $w = 8$, and it occurs when $x = 0$ and $y = 1$.

9. The minimum is $w = 4$, and it occurs when $x = 3$ and $y = 1$.

11. The minimum is $w = 2$, and it occurs at $x = 1$, $y = 1$.

13. The minimum is $w = 10$, and it occurs at $x_1 = 2$, $x_2 = 2$.

15. The minimum is $w = 3$, and it occurs at $x_1 = 1$, $x_2 = 2$, $x_3 = 0$.

17. The minimum is $w = 5$, and it occurs at $x_1 = 3$, $x_2 = 0$, $x_3 = 2$.

19. The minimum cost is $2.63, and it occurs when 7.5 ounces of each food is used.

21. The minimum cost is $910, and it results from purchasing 140 cans of fruit cocktail and 70 cans of peaches.

23. The minimum cost of an adequate diet is $4.40, and this cost is achieved by using 8 ounces of food A, none of food B, and 4 ounces of food C. It is also achieved by using none of food A, 4 ounces of food B, and 8 ounces of food C.

Exercises 5.5

1. The maximum is $z = 12$; it occurs at $x = 0$, $y = 4$.
3. The maximum is $z = 4$; it occurs at $x_1 = 0$, $x_2 = 4$ and at $x_1 = 4$, $x_2 = 0$.
5. The maximum is $z = 15$; it occurs at $x = 5$, $y = 0$.
7. The maximum is $z = 20$; it occurs at $x_1 = 5$, $x_2 = 0$, $x_3 = 0$.
9. The maximum is $z = 12$; it occurs at $x_1 = 1$, $x_2 = 0$, $x_3 = 0$.
11. The maximum is $z = 5$; it occurs at $x_1 = 5$, $x_2 = 0$, $x_3 = 0$.

Review Problems: Chapter 5

1.

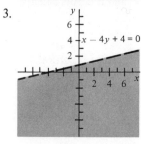

2.

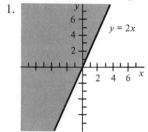

3.

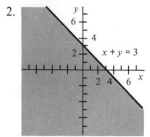

4.

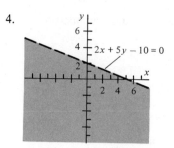

5.

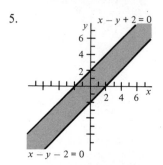

6.

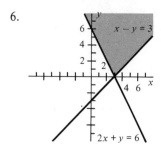

7.

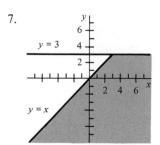

8.

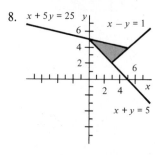

9.

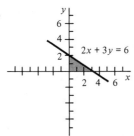

Vertex	Value
(0, 2)	2
(0, 0)	0
(3, 0)	3

The maximum is $z = 3$; it occurs at $x = 3$, $y = 0$.

13.

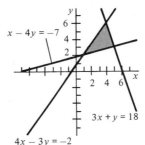

Vertex	Value
(4, 6)	46
(5, 3)	35
(1, 2)	14

The maximum is $z = 46$; it occurs at $x = 4$, $y = 6$.

10.

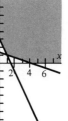

Vertex	Value
(0, 3)	9
(4, 0)	8
(1, 1)	5

The minimum is $z = 5$; it occurs at $x = 1$, $y = 1$.

11.

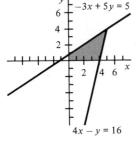

Vertex	Value
(5, 4)	31

The maximum is $z = 31$; it occurs at $x = 5$, $y = 4$.

12.

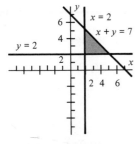

Vertex	Value
(2, 5)	7
(5, 2)	7
(2, 2)	4

The minimum is $z = 4$; it occurs at $x = 2$, $y = 2$.

14. The maximum is $z = 5$; it occurs at $x = 0$, $y = 5$.
15. The maximum is $z = 14$; it occurs at $x = 2$, $y = 2$.
16. The maximum is $z = 9$; it occurs at $x = 5$, $y = 4$.
17. The maximum is $z = 5/3 \approx 1.667$; it occurs at $x_1 = 0$, $x_2 = 5/6 \approx .833$, $x_3 = 5/6 \approx .833$.
18. The maximum is $z = 3$; it occurs at $x_1 = 0$, $x_2 = 3$, $x_3 = 0$.
19. The minimum is $w = 3$; it occurs at every point on the graph of the equation $x + y = 3$.
20. The minimum is $w = 5$; it occurs at $x = 1$, $y = 1$.
21. The minimum is $w = 4$; it occurs at $x = 2$, $y = 1$.
22. The minimum is $w = 5/3$; it occurs at $x_1 = 5/6$, $x_2 = 0$, $x_3 = 5/6$.
23. The minimum is $w = 6$; it occurs at $x_1 = 6$, $x_2 = 0$, $x_3 = 0$.
24. The maximum is $z = 10$; it occurs at $x = 0$, $y = 5$.
25. The maximum is $z = 31/8$; it occurs at $x = 3/8$, $y = 7/2$.
26. The maximum is $z = 20$; it occurs at $x_1 = 0$, $x_2 = 0$, $x_3 = 5$.
27. The linear programming method shows a maximum profit of $\$380/3 \approx \$126.67$ occurs when 20/3 economy models and 10/3 luxury models are made. Since 1/3 of a model cannot be made, make 7 economy models and 3 luxury models for a profit of $\$124$. (Note that these production numbers satisfy the constraints.)
28. There will be no meat left over when $800/3 \approx 266.67$ pounds of hot dogs and $1000/3 \approx 333.33$ pounds of hamburgers are made.
29. The diet that satisfies the requirements and costs the least consists of 4 ounces of each of foods A and B; it costs $\$.72$.
30. The greatest possible return is $\$4000$; it is achieved by investing $\$10,000$ in each kind of stock.
31. Costs are minimized by buying 25 cans of beef stew and 75 cans of chili. The minimal cost is $\$550$.

1. The most profitable mixture of sheep and cows is 40 cows and 50 sheep.
3. The maximum number of mice that can be trained is 45; this is achieved by training 30 from Group A and 15 from Group B.

Chapter 6

Exercises 6.1
1. Interest = $50, amount repaid = $1050.
3. Interest = $87.08, amount repaid = $1187.08.
5. Interest = $60, amount repaid = $960.
7. Interest = $2000, amount repaid = $7000.
9. Interest = $3150, amount repaid = $10,650.
11. Discount = $60, proceeds = $940.
13. Discount = $1620, proceeds = $7380.
15. Discount = $166.67, proceeds = $1833.33.
17. Discount = $640, proceeds = $7360.
19. Discount = $560, proceeds = $1440.
21. Effective interest rate = 21.95%.
23. Effective interest rate = 16.90%.
25. Effective interest rate = 6.82%.
27. $1666.67
29. 10%
31. 2 years

Exercises 6.2
1. Compound amount = $1061.52, compound interest = $61.52.
3. Compound amount = $1026.58, compound interest = $226.58.
5. Compound amount = $5125, compound interest = $125.
7. Compound amount = $93.05, compound interest = $92.05.
9. $797.19   11. $731.63
13. $97.06   15. $635.59
17. 12.68%   19. 10.25%
21. 24%
23. Approximately 6.12 years.
25. 19.56%   27. $355.89
29. $368.45   31. $983,840.87
33. $2581.24

Exercises 6.3
1. $17.45
3. $46,609.57
5. $52.59
7. At 11% inflation rate, the car would cost $64,498.49. At 8% inflation rate, the price would be $38,841.41 less.
9. At 12% inflation rate, the college education would cost $144,694.40; at 9% it would cost $60,628.23 less.
11. Approximately 8.04 years.
13. Approximately 10.24 years.
15. Approximately 3.98 years.

Exercises 6.4
1. The sum is approximately 1.97.
3. The sum is approximately .50.
5. The sum is approximately .33.
7. $S \approx \$2,697.35$ 9. $S \approx \$114,358.88$
11. $S \approx \$8496.77$ 13. $P \approx \$3889.65$
15. $P \approx \$1639.83$ 17. $P \approx \$8175.72$
19. John can afford to borrow approximately $3797.40.
21. The present value of the annuity is approximately $46,084.25; thus, Mary should take the $50,000 in cash.
23. Sue should pay approximately $4327.09.
25. She should pay him $3002.65.
27. The account contains $10,819.57.
29. $2145.58   31. $413.93
33. $18,293.07   35. $1013.12
37. $326,492.56

Exercises 6.5
1.

Payment	Principal Before Payment	Interest	Repayment of Principal	Principal After Payment
1	485.34	4.85	95.15	390.19
2	390.19	3.90	96.10	294.09
3	294.09	2.94	97.06	197.03
4	197.03	1.97	98.03	99.00
5	99.00	.99	99.01	−.01

3.

Payment	Principal Before Payment	Interest	Repayment of Principal	Principal After Payment
1	500.00	20.00	117.75	382.25
2	382.25	15.29	122.46	259.79
3	259.79	10.39	127.36	132.43
4	132.43	5.30	132.45	−.02

5. $83,926.09   7. $116,966.72
9. $22,400.02
11. Using the present value formula, we compute the
outstanding principal the instant after the 29th
payment is made; it is $4082.56. Thus, the
interest owed at the time of the 30th payment is
$40.83.
13. $41,335.35   15. $413.93

5. 1.25 years   6. 15%
7. $12,155.06   8. $1043.48
9. $10,482.60   10. 12.68%
11. $4971.77
12. Approximately 11.90 years.
13. $35,979.94   14. 5461
15. $37,700.63   16. $3144.35
17. $5650.22   18. $3797.40
19. $108,464.67

Review Problems: Chapter 6
1. $14,400.00   2. $3500.00
3. 21.95%   4. $4000

20.

Payment	Principal Before Payment	Interest	Repayment of Principal	Principal After Payment
1	90,000.00	12,600.00	26,165.83	63,834.17
2	63,834.17	8,936.78	29,829.05	34,005.12
3	34,005.12	4,760.72	34,005.11	.01

Chapter 7

Exercises 7.1
1. $y = 4x$
3. $y - 2 = -2(x - 1)$
5. $y - 1 = -.2(x + 1)$
7. $y - 1 = -.5(x - 1)$
9. $y - 1 = -10(x + 1)$
11. $y + 2 = (1/9)(x + 2)$
13. $y - 5 = 0$
15. $y + 3 = 0$
17. A point-slope equation is not possible.
19. B, E, F
21. B
23. F
25. E
27. $y = -x + 9$
29. $y = .2x + 1.4$
31. $y = 8x$
33. $y = x/9 + 49/9$
35. A slope-intercept equation is not possible.
37. A slope-intercept equation is not possible.
39. $y = -1$
41. $y = -4(x - 300)$
43. $y - 100 = 0$
45. $y - 150 = 0$

Exercises 7.2
1. $y - 1 = -2(x + 1)$, $f(-1.1) = 1.21$, $y = 1.2$
3. $y - 1 = -4(x + 1)$, $f(-1.1) = 1.4641$, $y = 1.4$
5. $y = 0$, $f(.1) = .0001$, $y = 0$

7. $y - 16 = 32(x - 2)$, $f(2.1) = 19.4481$, $y = 19.2$
9. $y + 1 = 3(x + 1)$, $f(-1.1) = -1.331$, $y = -1.3$
11. $y = 0$, $f(.1) = .001$, $y = 0$
13. $y - 8 = 12(x - 2)$, $f(2.1) = 9.261$, $y = 9.2$
15. $y - 1 = 3x$, $f(.1) = 1.3$, $y = 1.3$
17. $y - .5 = -.25(x - 2)$, $f(2.1) \approx .4762$, $y = .475$
19. -4.641, -4.06 21. 34.481, 32.24
23. 12.61, 12.06 25. -5, -4.08
27. It is less than 1.   29. It is less than 1.
31. It is more than 1.

Exercises 7.3
1. 10   3. $\sqrt{3}$
5. π 7. -1
9. 6.5   11. 3
13. 81   15. 0
17. 3.92   19. -7
21. 44   23. -6

25. $-1/5$

27. 7

29. $-4/3$

31. 0

33. 2

35. 2

37. -2

39. 21

41. 0

43. 0

45. 0

47. 0

49. The limit does not exist.

51. The limit does not exist.

53. 0

55. The limit does not exist.

57. 5

59. $\$12,500$

61. $\$41,666.67$

63. $12/5$

65. 0

Exercises 7.4

1. $f'(a) = \lim\limits_{h\to 0} \dfrac{5(a+h) - 5a}{h} = \lim\limits_{h\to 0} 5 = 5$

3. $f'(a) = \lim\limits_{h\to 0} \dfrac{7(a+h) + 9 - 7a - 9}{h} = \lim\limits_{h\to 0} 7 = 7$

5. $h'(a) = \lim\limits_{h\to 0} \dfrac{(a+h)^2 + 2(a+h) - 3 - a^2 - 2a + 3}{h}$

$= \lim\limits_{h\to 0} (2a + h + 2) = 2a + 2$

7. $g'(a) = \lim\limits_{h\to 0} \dfrac{(a+h)^2 - 9 - a^2 + 9}{h}$

$= \lim\limits_{h\to 0} (2a + h) = 2a$

9. $f'(a) = \lim\limits_{h\to 0} \dfrac{1}{h}\left(\dfrac{2}{a+h} - \dfrac{2}{a}\right) = \lim\limits_{h\to 0} \dfrac{-2}{a(a+h)} = \dfrac{-2}{a^2}$

11. $h'(a) = \lim\limits_{h\to 0} \dfrac{1}{h}\left(\dfrac{1}{(a+h)^2} - \dfrac{1}{a^2}\right)$

$= \lim\limits_{h\to 0} \dfrac{-2a - h}{a^2(a+h)^2} = \dfrac{-2}{a^3}$

13. $g'(a) = \lim\limits_{h\to 0} \dfrac{1}{h}\left[(a+h)^2 + \dfrac{1}{a+h} - a^2 - \dfrac{1}{a}\right]$

$= \lim\limits_{h\to 0}\left[2a + h - \dfrac{1}{a(a+h)}\right] = 2a - \dfrac{1}{a^2}$

15. $f'(a) = \lim\limits_{h\to 0} \dfrac{\sqrt{11} - \sqrt{11}}{h} = \lim\limits_{h\to 0} 0 = 0$

17. $h'(a) = \lim\limits_{h\to 0} \dfrac{\pi - \pi}{h} = \lim\limits_{h\to 0} 0 = 0$

19. $g'(a) = \lim\limits_{h\to 0}\left(\dfrac{\sqrt{a+h} - \sqrt{a}}{h} \dfrac{\sqrt{a+h} + \sqrt{a}}{\sqrt{a+h} + \sqrt{a}}\right)$

$= \lim\limits_{h\to 0} \dfrac{1}{\sqrt{a+h} + \sqrt{a}} = \dfrac{1}{2\sqrt{a}}$

21. $g'(a) = \lim\limits_{h\to 0} \dfrac{a + h + \sqrt{a+h} - a - \sqrt{a}}{h}$

$= \lim\limits_{h\to 0}\left[1 + \left(\dfrac{\sqrt{a+h} - \sqrt{a}}{h}\right)\left(\dfrac{\sqrt{a+h} + \sqrt{a}}{\sqrt{a+h} + \sqrt{a}}\right)\right]$

$= \lim\limits_{h\to 0}\left[1 + \dfrac{1}{\sqrt{a+h} + \sqrt{a}}\right] = 1 + \dfrac{1}{2\sqrt{a}}$

23. $h'(a) = \lim\limits_{h\to 0} \dfrac{(a+h)^3 + 6 - a^3 - 6}{h}$

$= \lim\limits_{h\to 0}\left[(a+h)^2 + (a+h)a + a^2\right] = 3a^2$

25. $g'(a) = \lim\limits_{h\to 0} \dfrac{(a+h)^3 - 5(a+h)^2 - a^3 + 5a^2}{h}$

$= \lim\limits_{h\to 0}\left[(a+h)^2 + (a+h)a + a^2 - 5(2a+h)\right]$

$= 3a^2 - 10a$

27. $f'(a) = \lim\limits_{h\to 0}\left[\dfrac{(a+h)^2 - 1 - a^2 + 1}{h}\right]$

$= \lim\limits_{h\to 0}(2a + h) = 2a$

29. $h'(a) = \lim\limits_{h\to 0} \dfrac{1}{h}\left[\dfrac{2(a+h)^3 - 3(a+h)}{a+h} - \dfrac{2a^3 - 3a}{a}\right],$

provided $a \neq 0$

$= \lim\limits_{h\to 0}\left[\dfrac{2(a+h)^2 - 3 - 2a^2 + 3}{h}\right]$

$= \lim\limits_{h\to 0}(4a + 2h) = 4a$

$h'(0)$ does not exist since $h(x)$ is not defined at $x = 0$.

31. $y - 4 = -4(x + 2)$

33. $y = 0$

35. $y - 4 = 4(x - 2)$

37. $y + 1 = 3(x + 1)$

39. $y - 1 = 3(x - 1)$

41. $y - 1/5 = (5 - x)/25$

43. $y + 1/5 = -(x + 5)/25$

45. $y - 16 = 32(x - 2)$

47. $y - 64 = 192(x - 2)$

49. $y - 256 = 1024(x - 2)$

51. The derivative of e^x at $x = 1$ is $e \approx 2.71828$.

53. The derivative of $\sqrt{x}$ at $x = 4$ is $1/4 = .25$.

Exercises 7.5

1. $.5$

3. $.04$

5. $.5x^{-1/2}$

7. $-2x + 1$

9. $-.02x + 1$

11. $3x^2$

13. $C'(10) = .16, \quad C(11) - C(10) = .168$

15. $C'(10) = 10, \quad C(11) - C(10) = 10.5$

17. $C'(10) = 40, \quad C(11) - C(10) = 41$

19. $C'(10) = 15, \quad C(11) - C(10) = 15$

21. $x = 5$

23. $x = 4$

25. $x = 10$

27. $x = 50$

29. $x = 2$

Exercises 7.6

1. There are no discontinuities.

3. There are no discontinuities.

5. $x = 0$ is an infinite discontinuity.

7. $x = 1$ is an infinite discontinuity.

9. There are no discontinuities.

11. $x = -2$ is a removable discontinuity.

13. There are no discontinuities.

15. $x = -1$ is a removable discontinuity.

17. $x = 2$ and $x = 3$ are both infinite discontinuities.

19. There are no discontinuities.

21. $x = 0$ is an infinite discontinuity.

23. $x = 0$ is a removable discontinuity.

25. There are no discontinuities.

27. $x = 0$ is a jump discontinuity.

29. There are no discontinuities.

31. $\lim\limits_{x \to 0} f(x) = \lim\limits_{x \to 0} 3 = 3 = f(0)$

33. $\lim\limits_{x \to 1} h(x) = \lim\limits_{x \to 1} 3(x - 1) = 3 \lim\limits_{x \to 1} (x - 1) = 3 \cdot 0$

$= 0 = h(1)$

35. $\lim\limits_{x \to 0} g(x) = \lim\limits_{x \to 0} (x^2 + x + 1)$

$= \lim\limits_{x \to 0} x^2 + \lim\limits_{x \to 0} x + \lim\limits_{x \to 0} 1$

$= 0^2 + 0 + 1 = g(0)$

37. $\lim\limits_{x \to 8} f(x) = \lim\limits_{x \to 8} x^{-1} = 8^{-1} = f(8)$

39. $\lim\limits_{x \to 1} h(x) = \lim\limits_{x \to 1} \dfrac{2x + 3}{3x - 2} = \dfrac{\lim\limits_{x \to 1} (2x + 3)}{\lim\limits_{x \to 1} (3x - 2)}$

$= \dfrac{2 \lim\limits_{x \to 1} x + 3}{3 \lim\limits_{x \to 1} x - 2} = \dfrac{2 \cdot 1 + 3}{3 \cdot 1 - 2} = h(1)$

41.

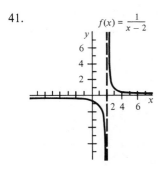

$f(x) = \dfrac{1}{x - 2}$

43.

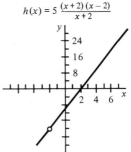

$h(x) = 5 \, \dfrac{(x + 2)(x - 2)}{x + 2}$

45.

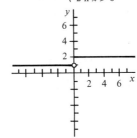

$g(x) = \begin{cases} 1 \text{ if } x < 0 \\ 2 \text{ if } x \geq 0 \end{cases}$

Review Problems: Chapter 7

1. $y - 1 = .2(x - 1)$

2. $y = -3x$

3. $y = 3$

4. It is not possible to write a point-slope equation.

5. $y - 4 = -2x$ 6. $y = -5x + 27$

7. $y = 2x/3 - 1/3$ 8. $y = x/3 - 3$

9. It is not possible to give a slope-intercept equation.

10. $y = 0$

11. $y = 0$, $f(.1) = -.0098$, 0

12. $y = 0$, $f(-.1) = -.0098$, 0

13. $y - 1 = 6(x - 1)$, $f(1.1) = 1.7182$, 1.6

14. $y - 1 = 6(x - 1)$, $f(.9) = .5022$, .4

15. Positive.   16. Positive.

17. Negative.   18. Negative.

19. -53 20. 0

21. -6 22. -4

23. $-.5$ 24. $23/9 \approx 2.556$

25. 2   26. 4

27. 0

28. This limit does not exist.

29. 0

30. 0

31. $f'(a) = \lim\limits_{h \to 0} \dfrac{-2(a + h) - 9 + 2a + 9}{h}$

$= \lim\limits_{h \to 0} - 2 = -2$

32. $g'(a) = \lim\limits_{h \to 0} \dfrac{(a+h)^2 + 3(a+h) + 12 - a^2 - 3a - 12}{h}$

$\qquad = \lim\limits_{h \to 0} (2a + h + 3) = 2a + 3$

33. $h'(a) = \lim\limits_{h \to 0} \dfrac{1}{h} \left[\dfrac{4}{(a+h)^2} - \dfrac{4}{a^2} \right]$

$\qquad = \lim\limits_{h \to 0} \dfrac{4}{h} \left[\dfrac{a^2 - (a+h)^2}{a^2(a+h)^2} \right]$

$\qquad = \lim\limits_{h \to 0} \dfrac{-8a - 4h}{a^2(a+h)^2} = \dfrac{-8}{a^3}$

34. $f'(a) = \lim\limits_{h \to 0} 2 \left(\dfrac{\sqrt{a+h} - \sqrt{a}}{h} \right) \left(\dfrac{\sqrt{a+h} + \sqrt{a}}{\sqrt{a+h} + \sqrt{a}} \right)$

$\qquad = \lim\limits_{h \to 0} \dfrac{2}{\sqrt{a+h} + \sqrt{a}} = \dfrac{1}{\sqrt{a}}$

35. $g'(a) = \lim\limits_{h \to 0} \dfrac{\sqrt{13} - \sqrt{13}}{h} = \lim\limits_{h \to 0} 0 = 0$

36. $y = 2x + 1$ 37. $y = 2s + 3$

38. $y = -16t + 12$ 39. $y = 100x - 99$

40. $y = 0$
41. $P'(5) = 0$, $P(6) - P(5) = -0.1$
42. $P'(5) = 10$, $P(6) - P(5) = 11$
43. $P'(5) = 5$, $P(6) - P(5) = 5$
44. $P'(5) = 9.25$, $P(6) - P(5) = 10.09$
45. $x = 25$
46. $x = 5$
47. There are no discontinuities.
48. There are no discontinuities.
49. $x = 0$ is a removable discontinuity.
50. $x = 0$ is an infinite discontinuity.
51. There are no discontinuities.
52. $x = 0$ is a removable discontinuity.
53. $y = x - 3$
54. \$66,666.67

Social Science Applications: Chapter 7

1. 1.3   3. 7.9
5. 3.1

Chapter 8

Exercises 8.1

1. $f'(x) = 11$
3. $h'(t) = 2t - 6$
5. $g'(x) = 9x^2 - 2x + 1$
7. $f'(s) = (4/3)s^{1/3}$
9. $h'(x) = -30x^{-6}$
11. $g'(r) = -r^{-2}$
13. $f'(x) = (15/2)x^{1/2}$
15. $h'(x) = 0$
17. $g'(s) = 102s^{101} + 83s^{82}$
19. $f'(t) = 42t^{41} - 13t^{-14} + (1/2)t^{-1/2}$
21. $h'(r) = 2r$ 23. $g'(x) = 3x^2 - 2x - 2$
25. $f'(x) = (3/2)x^{-1/2} + 3x^{1/2} - (1/2)x^{-3/2}$
27. $h'(t) = t$ 29. $g'(p) = 0$
31. $f'(x) = (2/3)x^{-1/3}$ 33. $h'(x) = 10x + 2$
35. $g'(x) = 2ex$ 37. $f'(x) = (13/3)x^{10/3}$
39. $h'(x) = 1$ 41. $-3/16 = -.1875$
43. -48 45. -3
47. 1   49. 1/6
51. 0   53. 0
55. -1 57. 0
59. $y = x$ 61. $y = (x/4) + 1$
63. $y = 0$ 65. $p/(p - 2)$
67. $p/(p - 10)$ 69. -2

Exercises 8.2

1. $f(g(y)) = 3y + \sqrt{y} - 1$, $g(f(x)) = \sqrt{3x^2 + x - 1}$
3. $f(g(y)) = -15y + 2$, $g(f(x)) = -15x - 6$
5. $f(g(y)) = (2/y^3) - (1/y)$, $g(f(x)) = 1/(2x^3 - x)$
7. $f(g(y)) = y$, $g(f(x)) = x$
9. $f(g(y)) = y^{220} - 1$, $g(f(x)) = (x^2 - 1)^{110}$
11. $f(g(y)) = 2e^{4y} + e^{2y} - 1$, $g(f(x)) = e^{2x^4 + x^2 - 1}$
13. $f(g(y)) = \ln(y^2 + 6)$, $g(f(x)) = (\ln x)^2 + 6$
15. $f(g(y)) = y$, $g(f(x)) = x$
17. $g(y) = y^{1/3}$, $u(x) = -3x + 5$
19. $g(y) = y^{95}$, $u(x) = x^2 - 13x + 15$
21. $g(y) = e^y$, $u(x) = 5 + 5x$ or $g(y) = e^{5y}$,
    $u(x) = 1 + x$
23. $g(y) = \ln y$, $u(x) = (x^4 + 2)/(x^2 + 1)$
25. $h'(s) = 770s(11s^2 - 13)^{34}$
27. $-516(w^2 - w - 2)^{-44}(2w - 1)$
29. $(x^3 - x)^{-2/5}3(3x^2 - 1)/5$
31. $5 + 58p(p^2 - 3)^{28}$
33. $23(s + 3)^{22}$,   provided $s \neq 3$
35. $90[(9s + 3)^{1/3} + 5]^{29}(9s + 3)^{-2/3}$
37. The marginal change in $p(x)$ when the dust level is reduced 20% is $1/1200 \approx .00083$. The marginal change in $p(x)$ when the dust level is reduced 50% is $1/7500 \approx .00013$.

Exercises 8.3

1. $f'(x) = 24x - 22$
3. $h'(x) = 3x^2 + 2x + 2$
5. $g'(s) = 4s^3 + 12s^2 - 90s + 2$
7. $f'(t) = 30t^2 - 42t - 25$
9. $h'(t) = (2t + 1)^{29}(-3t + 5)^{19}60(4 - 5t)$
11. $g'(r) = (3r - 1)^{14}(r + 1)^{22}(-4r + 2)^{29}4(-204r^2 - 25r + 41)$
13. $f'(p) = 15p^2 + 50 - (12 - p)^7(20 - 27p)$
15. $h'(p) = 27p^2 - 12p + 3$
17. $g'(w) = 26/(w + 7)^2$
19. $f'(x) = (-3x^2 + 4x + 5)/(x^2 + x + 1)^2$
21. $h'(x) = (8x^2 + 16x + 3)/(x + 1)^2$
23. $g'(z) = (15z^2 - 164z - 27)/(5z^2 + 9)^2$
25. $f'(x) = -(6x + 2)(3x^2 + 2x + 1)^{-2}$
27. $h'(x) = -(x + 5)^{-2} - (x + 1)^2(x - 2)^3(7x - 2)$
29. $g'(s) = (-4s^2 - 10s - 3)/(s + 1)^2(s - 2)^2$
31. $f'(x) = 5x(-3x^3 + 3x - 4)/(x^2 + 1)^2(3x - 2)^2$
33. $h'(x) = -2(33x^2 + 26x - 11)/(x - 3)^2(4x + 1)^2$
35. $g'(r) = (1 - r^2)/(r^2 + 1)^2$
37. $f'(x) = \sqrt{x^2 + 1} + x^2/\sqrt{x^2 + 1}$
39. $h'(x) = (-4x - 45)/3(x + 8)^{2/3}(2x + 3)^2$
41. $g'(x) = -.5x^{-3/2} + 15x^2 - 30x + 3$
43. $f'(p) = 4p^3 - 2p$
45. $h'(p) = -1/(p + 2)^{1/2}(3p + 4)^{3/2}$
47. $g'(x) = (12x + 5)/2\sqrt{(2x - 1)(3x + 4)}$
49. $f'(v) = .5\left(3v + \dfrac{5v}{v + 2}\right)^{-1/2}\left(3 + \dfrac{10}{(v + 2)^2}\right)$
51. $h'(v) = (v + 1)^4(2v - 1)^5(22v + 7)$
53. $g'(x) = (-8x^2 + 20x - 29)/2\sqrt{(2x - 3)(x + 7)}(5 - 4x)^{3/2}$
55. $f'(x) = 2x(1 - 2x)/(2x^2 + 4)^{1/2}(x^3 + 1)^{4/3}$
57. $h'(x) = 1.25x^{1/4} + .25x^{-3/4}$
59. $g'(x) = (115/42)x^{73/42}$
61. $A(x) = .1x + 40x^{-1}$, $A'(x) = .1 - 40x^{-2}$, $x = 20$
63. $A(x) = .01x + 36x^{-1}$, $A'(x) = .01 - 36x^{-2}$, $x = 60$

Exercises 8.4

1. $f'(x) = 2xe^x + x^2e^x$
3. $h'(x) = 3e^x(5x - 3)/(5x + 2)^2$
5. $g'(x) = \frac{1}{2}e^x - \frac{1}{2}e^{-x}$
7. $f'(x) = -2e^{-2x+1}$
9. $h'(x) = 42x^2 - e^{2x} - 2xe^{2x} + 27e^{3x}$
11. $g'(x) = 52e^x(e^x + 3)^{51}$
13. $f'(x) = 2(e^{2x} - e^{-2x})$
15. $h'(x) = e^{2x}(e^{2x} + 10)^{-1/2}$
17. $g'(x) = [e^x(-x^2 + 5x - 3) + 2x - 1]/(2e^x + 1)^{3/2}$
19. $f'(x) = 660e^{6x+92}$
21. Approximately $610.70.

23. With annual compounding, the account would be worth $404.56; with continuous compounding, the account would be worth $448.17, or $43.61 more.

Exercises 8.5

1. $f'(x) = x^{-1}$
3. $h'(x) = (2x + 5)/(x^2 + 5x + 2)$
5. $g'(x) = 18/(2x - 1) + 27/(3x + 2)$
7. $f'(x) = 8x/(4x^2 + 9)$
9. $h'(x) = (2x + 5)/(x^2 + 5x) - x^{-1}$
11. $g'(x) = 5/(5x + 1) - 2/(2x + 3)$
13. $f'(x) = 3e^{3x}$
15. $h'(x) = (2x - 10)e^{x^2-10x+2}$
17. $g'(x) = xe^{\sqrt{x^2+1}}/\sqrt{x^2 + 1}$
19. $f'(x) = e^{(x+2)(3x-1)}(6x + 5)$
21. $h'(x) = 1$
23. $g'(x) = 9e^{2x}(1 + 2x)$
25. $f'(x) = 12(e^{4x} + 3)^2e^{4x}$
27. $h'(x) = 5[\ln(x^2 + 3x + 1)]^4(2x + 3)/(x^2 + 3x + 1)$
29. $g'(x) = e^{3x}(3 \ln x + x^{-1})$
31. $f'(x) = 5^x \ln 5$
33. $h'(x) = 2(4^{2x+1})(\ln 4)$
35. $g'(x) = 7^{x^2}(2x)\ln 7$
37. $f'(x) = 1/x \ln 3$
39. $h'(x) = (2x + 5)/(x^2 + 5x - 2)\ln 2$
41. $g'(x) = 18/(3x + 2)$
43. $\dfrac{dz}{dx} = 7(18y + 3)$
45. $\dfrac{dz}{dx} = \dfrac{8e^{8y+2}(4x^3 + 2x)}{x^4 + x^2 + 1}$
47. $\dfrac{dz}{dx} = 10(x + 1)2^y \ln 2$
49. $\dfrac{dz}{dx} = \dfrac{x}{y(\ln 2)\sqrt{x^2 + 4}}$
51. .5   53. 350

Exercises 8.6

1. $y' = -x/y$
3. $y' = x/2y$
5. $y' = (3x^2 + 5y^2)/(1 - 10xy - 3y^2)$
7. $y' = (-16x^3y^4 - 1)/(16x^4y^3 + 2y)$
9. $y' = (y + 1)/[x - 1 + 4y^3(x + y)^2]$
11. $y' = -9y^{1/2}/x^{1/2}$
13. $y' = (-y^2 - xy \ln y)/(x^2 + xy \ln x)$
15. $y' = 1/2x(y + 1)$
17. Proceeding formally, we find that $y' = -x/y$, but examining the equation carefully we determine

that only $x = 0$, $y = 0$ makes the equation true. Since the formula for y' does not make sense for $x = 0$, $y = 0$, there is no derivative.

19. $y' = -y \ln y$
21. $y' = -y^2 e^{xy}/(xye^{xy} + 10[\ln y]^9)$
23. $\dfrac{dp}{dq} = \dfrac{-1}{2p}$
25. $\dfrac{dp}{dq} = -1$
27. $\dfrac{dp}{dq} = \dfrac{-p}{q}$

Review Problems: Chapter 8

1. $f'(s) = 60s^2 - 2s + 5$
2. $g'(t) = 1.5t^{1/2} + 20t^3 + 2t^{-2}$
3. $h'(r) = -(11/2)r^{-3/2}$
4. $f'(p) = 660p^{109} + .5p^{-1/2}$
5. $g'(w) = 180w(3w^2 + 1)^{29}$
6. $h'(u) = .5(3u^2 + 2u + 1)/(u^3 + u^2 + u)^{1/2}$
7. $f'(x) = 4x + 1$
8. $g'(s) = (2s + 5)(3s - 7) + 2(s - 3)(3s - 7)$
    $+ 3(s - 3)(2s + 5)$
    $= 2(9s^2 - 17s - 19)$
9. $h'(t) = 20(t + 6)^{19}(2t - 7)^{39}(6t + 17)$
10. $f'(r) = 8(r + 9)^{11}(2r - 4)^{23}(5r + 1)^7(55r^2 + 319r - 132)$
11. $g'(p) = -106/(5p - 6)^2$
12. $h'(w) = (-5w^2 - 4w - 36)/(w^2 + 3w - 6)^2$
13. $f'(z) = 4(-3z^2 + 4z + 6)/(4z^2 + 8)^2$
14. $g'(x) = -(4x + 1)^{-2} + 3(7x + 2)^{15}$
    $+ 105(3x - 5)(7x + 2)^{14}$
    $= -(4x + 1)^{-2} + (7x + 2)^{14}(336x - 519)$
15. $h'(v) = 23/2(8v + 1)^{1/2}(v + 3)^{3/2}$
16. $f'(r) = [(6r + 19)(2r + 5)(r - 9)$
    $- (4r - 13)(r + 7)(3r$
    $- 2)]/(2r + 5)^2(r - 9)^2$
17. $g'(s) = (s + 8)s/(s^2 + 1)^{3/2}(s^3 - 8)^{2/3}$
18. $h'(t) = 5e^{5t}$
19. $f'(p) = 2pe^{-2p}(1 - p)$
20. $g'(u) = -80(10 - e^u)^{79}e^u$
21. $h'(x) = [(2x + 3)(e^x + e^{-x}) - (e^x - e^{-x})(x^2 + 3x + 1)]/(e^x + e^{-x})^2$
22. $f'(u) = e^u + ue^u - 3u^2 + 18u$
23. $g'(w) = 1.5e^{3w}/\sqrt{e^{3w} + 1}$
24. $h'(s) = 5(6s - 1)e^{3s^2 - s + 11}$
25. $f'(v) = 1/v$
26. $g'(x) = 1/x \ln 2$
27. $h'(t) = 9/(9t + 5)$
28. $f(x) = 3 \ln(2x + 1) + 3 \ln(x - 6)$,
    $f'(x) = 6/(2x + 1) + 3/(x - 6)$
29. $g(r) = \ln(7r + 1) - \ln(r + 3)$,
    $g'(r) = 7/(7r + 1) - 1/(r + 3)$

30. $h'(x) = 5^x \ln 5$
31. $f'(s) = 3s^2 e^{s+1} + s^3 e^{s+1} - 8se^{4s^2}$
32. $g'(w) = e^w \ln w + w^{-1}e^w$
33. $h'(r) = 100r(\ln(5r^2 + 9))^9/(5r^2 + 9)$
34. $f'(x) = (2x + 3)/(x^2 + 3x + 1)\ln 5$
35. $g'(z) = 2(3^{2z+5})\ln 3$
36. $y - 1 = x$
37. $y = x - 1$
38. $y + 5 = 3x$
39. $y - 5 = 5(\ln 5)(x - 1)$
40. $\dfrac{dz}{dx} = -28e^{7y}(6x - 1)$
41. $\dfrac{dz}{dx} = \dfrac{52ye^{2x}}{y^2 + 1}$
42. $\dfrac{dz}{dx} = 6(2y^3 + 3y)$
43. $\dfrac{dz}{dx} = \dfrac{18x + 7}{2\sqrt{y + 2}}$
44. $y' = -3x/4y$
45. $y' = (y - 5x^4y^5)/(5x^5y^4 - x)$
46. $y' = -x/y$
47. Proceeding formally, we obtain $y' = -2x/y$. However, a close examination of the equation shows that the only solution is $x = 0$, $y = 0$. Thus, there is no derivative.
48. $y' = y/[x - (x - y)^2y]$
49. $\dfrac{dp}{dq} = \dfrac{3p}{(3p - 6)}$
50. When the noise reduction is 40%, the marginal change is $1/80 = .0125$. When the noise reduction is 70%, the marginal change is $1/140 \approx .0071$.
51. $\dfrac{ds}{dx} = \dfrac{ds}{dp}\dfrac{dp}{dx} = (.4)(7) = 2.8$
52. $\$1000e^{5(.16)} \approx \$2225.54$

Social Science Applications: Chapter 8

1. When 11 items are being memorized, the person is memorizing items at a rate of $f'(11) = 29/6 \approx 4.83$. When 18 items are being memorized, the person is memorizing items at a rate of $f'(18) = 25/4 = 6.25$.
3. $x'(0) = 1998$
   $x'(5) = 1998(10^6)e^{-10}/(999e^{-10} + 1)^2 \approx 83,008.67$
5. The rate of decay of carbon 14, as a percentage of the amount present, is $-.00012$.
7. The rate of decrease of sulfur dioxide 1 mile from the smokestack is $y'(1) = -.1$, and 10 miles from the smokestack it is $y'(10) = -.0001$.

Chapter 9

Exercises 9.1
1. None.
3. $x = 0$
5. $x = 2$
7. $x = 2$
9. $x = 1, -2$
11. None.
13. $x = .5, -.5$
15. None.
17. $x = 0$
19. None.
21. $x = e^{-1}$
23. $x = 1, -3, -7/9$
25. $x = 0$, since the derivative does not exist there.
27. None.
29. $(-4, -28)$ is a relative minimum.
31. None.
33. $(0, 10)$ is a relative maximum; $(5, -115)$ is a relative minimum.
35. None.
37. $(1 - 1/\sqrt{3}, -1.86) \approx (.42, -1.86)$ is a relative maximum;
$(1 + 1/\sqrt{3}, 25.86) \approx (1.58, 25.86)$ is a relative minimum.
39. $(0, 2)$ is a relative minimum.
41. $(e^{-1}, -5/e) \approx (.37, -1.84)$ is a relative minimum.
43. $(-1, 0)$ is a relative maximum; $(.2, -8.40)$ is a relative minimum.
45. $(0, 3)$ is a relative minimum.
47. It is increasing for all x.
49. It is decreasing for all $x \neq 0$.
51. For $x < 2$, $g(x)$ is decreasing; for $x > 2$, $g(x)$ is increasing.
53. It is increasing for all x.
55. It is decreasing for all $x \neq 1$.
57. We are only concerned with $x > 0$, since $\ln x$ is only defined for such x; $g(x)$ is increasing for all $x > 0$.
59. It is increasing for $x < -2$ and for $x > -.8$; it is decreasing for $-2 < x < -.8$.
61. It is decreasing for $x < -1$ and increasing for $x > -1$.
63. It is decreasing for $x < 0$ and increasing for $x > 0$.

Exercises 9.2
1. $f''(s) = 120s - 2$
3. $h''(r) = (44/9)r^{-7/3}$
5. $g''(w) = 180(3w^2 + 1)^{28}(177w^2 + 1)$
7. $f''(x) = 4$
9. $h''(t) = 20(t + 6)^{18}(2t - 7)^{38}(708t^2 + 4012t + 5443)$
11. $g''(x) = 1060(5x - 6)^{-3}$
13. $g''(x) = 12(4x + 1)^{-5/2} + 60(7x + 9)^{13}(588x - 763)$
15. $h''(t) = 25e^{5t}$

17. $g''(u) = -800e^u(10 - e^u)^{78}(1 - 8e^u)$
19. $g''(w) = 4.5e^{3w}(e^{3w} + 1)^{-1.5}(.5e^{3w} + 1)$
21. $f''(x) = -x^{-2}$
23. $h''(t) = -81/(9t + 5)^2$
25. $g''(r) = -49/(7r + 1)^2 + 1/(r + 3)^2$
27. $g''(w) = e^w \ln w + 2e^w/w - e^w/w^2$
29. $f''(x) = -(2x^2 + 6x + 7)/(x^2 + 3x + 1)^2 \ln 5$
31. $(-1, -6)$ is a relative minimum.
33. $(0, 10)$ is a relative maximum.
35. $(-2, 28)$ is a relative maximum; $(1, 1)$ is a relative minimum.
37. $(2, -21)$ is a relative minimum.
39. $(-.2, .00032)$ is a relative maximum; $(0, 0)$ is a relative minimum.
41. None.
43. None.
45. None.
47. $(0, 2)$ is a relative minimum.
49. $(-1, -1/e)$ is a relative minimum.
51. $(-.2, 8.40)$ is a relative maximum; $(1, 0)$ is a relative minimum.
53. $(0, 0)$ is an inflection point.
55. None.
57. $(-.5, 14.5)$ is an inflection point.
59. $(0, -5)$ and $(1.33, -14.48)$ are inflection points.
61. None.
63. $(-2, -2/e^2)$ is an inflection point.
65. None.
67. $(-2, 0)$, $(-.93, 4.56)$, and $(.53, 3.58)$ are inflection points.

Exercises 9.3
1.

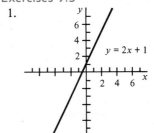

Exercises 9.3 (*continued*)

3. $y = x^2 - 3x - 4$

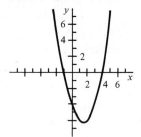

5. $y = -x^2 - 5x - 6$

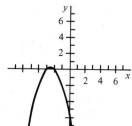

7. $y = x^2 - 4x + 4$

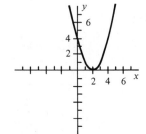

9. $y = x^3 + 6x^2 + 12x + 8$

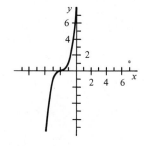

11. $y = -x^3 - 9x^2 - 27x - 27$

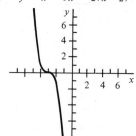

13. $y = 2x^3 - 3x^2 - 12x + 1$

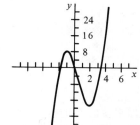

15. $y = -x^3 + 27x - 10$

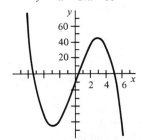

17. $y = 3x^4 - 4x^3 - 72x^2 + 30$

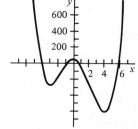

19. $y = 3x^5 - 5x^3 + 1$

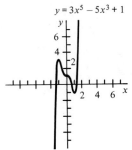

21. $y = x^4 - 2x^2 + 1$

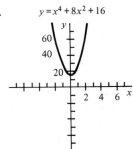

23. $y = x^4 + 8x^2 + 16$

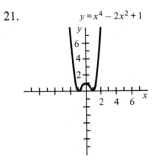

25.

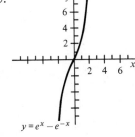

$y = e^x - e^{-x}$

27. $y = (x+3)^2 (x-2)^2$

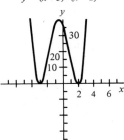

29. $y = (x+1)^3 (x-2)^2$

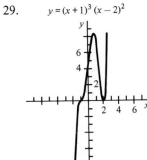

31.

$y = xe^x$

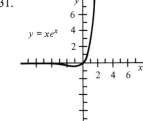

33.

$y = e^{x^2}$

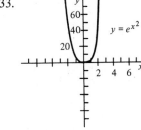

Exercises 9.4

1.
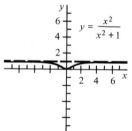

3.
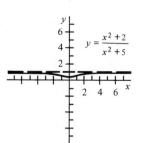

5.
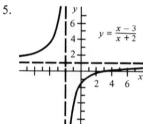

7.
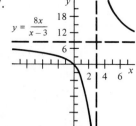

9.
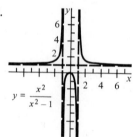

11.
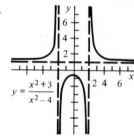

13.
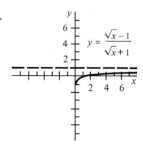

15.
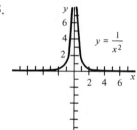

17.
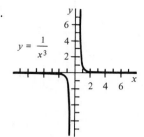

19.
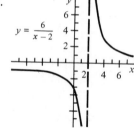

21.

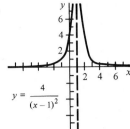

$$y = \frac{4}{(x-1)^2}$$

23.

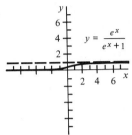

$$y = \frac{e^x}{e^x + 1}$$

25.

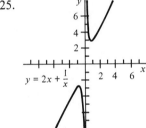

$$y = 2x + \frac{1}{x}$$

Exercises 9.5

1. Since the only critical value is $x = 50$ and it is a relative maximum, the maximum profit occurs when $x = 50$.

3. Since the only critical value is $x = 1.25$ and it is a relative maximum, the fee that results in the maximum revenue is $\$15 - \$1.25 = \$13.75$.

5. Since the only critical value for average cost is $x = 20$ and it is a relative minimum, the value of x that results in the minimum average cost is $x = 20$.

7. Since the only critical value for profit is $x = 120$ and it is a relative maximum, the value of x that results in the maximum profit is 120.

9. The only critical value for the net revenue is $x = 2.07$, and it is a relative maximum. The maximum net revenue results from spending $\$1000(2^{2.07}) = \$4198.87$ in advertising.

11. The minimum costs result from ordering 200 units at one time.

13. The largest possible product is $xy = 25$, and it results from choosing $x = 5$ and $y = 5$.

15. The absolute minimum value is $c - (b^2/4a)$, which results from choosing $x = -b/2a$.

Exercises 9.6

1. The least expensive patio has brick walls on two 7-foot sides and one 8-foot side. Thus, one 8-foot side has a concrete block wall, and the total cost is $\$560$.

3. The largest area that the printed matter can cover is 128 square inches. The maximum occurs when the poster is 10 inches by 20 inches.

5. The dimensions of the largest rectangular storage yard possible are 49 feet by 98 feet. The warehouse wall is used as one 98-foot side.

7. The largest rectangular package with square bottom has dimensions of 18 inches by 18 inches by 36 inches.

9. The largest package in the shape of a right circular cylinder has a height of 24 inches, and the radius of the base is $24/\pi$ inches.

11. The container with a maximum volume has a height of 4 inches, and the radius of the base is 2 inches.

13. The function that gives the sum of the areas of the circle and the square has only one critical value, and it is a relative minimum. Thus, the maximum area occurs at an endpoint; in particular, the maximum area occurs when all of the fence is used to make a circle.

Review Problems: Chapter 9

1. None.
2. $(3, 12)$ is a relative maximum.
3. None.
4. $(-2, 84)$ is a relative maximum; $(3, -41)$ is a relative minimum.
5. $(-1, -2)$ is a relative maximum; $(3, 6)$ is a relative minimum.
6. $(0, -1)$ is a relative maximum.
7. $(-1, 0)$ is a relative maximum; $(-.2, -3456/3125) \approx (-.2, -1.11)$ is a relative minimum.
8. None.
9. The function is increasing for $x < -2$ and $x > 2$; it is decreasing for $-2 < x < 0$ and $0 < x < 2$.
10. The function is increasing for $x < -2$ and $x > 3$; it is decreasing for $-2 < x < 3$.

11. The function is increasing for $x > 1$ and decreasing for $0 < x < 1$.
12. The function is increasing for $x < -1$ and $x > -.2$; it is decreasing for $-1 < x < -.2$.
13. $f''(x) = (10/3x^{1/3}) + 6 + 4x^{-3}$
14. $g''(x) = 4(x^2 + 4)^{-3/2}$
15. $h''(x) = 4(2x + 1)^6(x - 3)^{10}(380x^2 - 684x + 249)$
16. $f''(x) = 2(2x^2 + 1)e^{x^2+3}$
17. $g''(x) = -46(x + 4)^{-3}$
18. $h''(x) = -4(x + 7)^{-2}$
19. $f''(x) = -2/x^2 \ln 3$
20. $g''(x) = 4^x(\ln 4)^2$
21. $(-1, 1)$ is the only inflection point
22. $(2.5, 27.5)$ is the only inflection point
23. $(0, -2)$ is the only inflection point
24. The inflection points are $(-1.631, -.005)$, $(-1.227, -.004)$, and $(-1, 0)$.

25. $y = x^2 - 4x + 3$

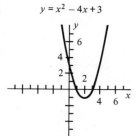

26. $y = 2x^3 - 15x^2 - 36x$

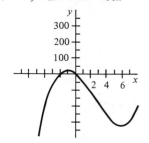

27. $y = x^3 + 3x^2 + x$

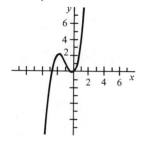

28. $y = (x - 2)^2 (x + 1)^2$

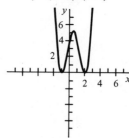

29. $y = x^5/5 + 2x^3/3 + x$

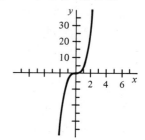

30. $y = x^4 - 18x^2 + 10$

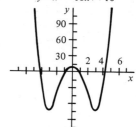

31. $y = \dfrac{x + 1}{x - 1}$

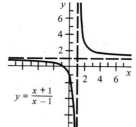

32. $y = 3e^{2x}$

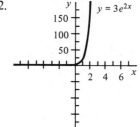

33.

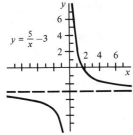

$y = \frac{5}{x} - 3$

34.

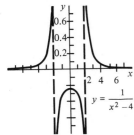

$y = \frac{1}{x^2 - 4}$

35. The maximum profit, which is 100, occurs when $x = 10$.
36. The maximum yield per acre, which is 800 pounds, results from planting 40 trees per acre.
37. The maximum revenue, which is $112,500, results from reducing the rent to $300 per month.
38. The minimum average cost, which is $50.03, occurs when $x = 100$.

39. In order for the area covered by the printed matter to be the maximum, which is 64 square inches, the page should be 10 inches by 10 inches.
40. The minimum cost of the fences is $528; it results from the rectangular area being 12 feet by 22 feet and the fence that divides the rectangular area being 12 feet.
41. For maximum area the yard should be a square with each side 20 feet.
42. The dimensions of the base of the box should be 6 inches by 12 inches; provided the height is 8 inches, the box will use the least amount of material.
43. In order for the can to use the least amount of material the radius of the base and the height should both equal 2 feet.
44. The rectangle below the semicircle should have a height of $12/(4 + \pi) \approx 1.68$ feet and a width of $24/(4 + \pi) \approx 3.36$ feet. Thus, the radius of the semicircle is $12/(4 + \pi) \approx 1.68$ feet.

Social Science Applications: Chapter 9
1. The maximum rate of learning occurs when $x = 2$; then the maximum rate is $3\sqrt{2} \approx 4.24$.
3. The average speed that minimizes the total costs of the trip is $(360)^{2/3} \approx 50.61$. At that speed the trip takes 1.98 hours and costs $1069.20.
5. The maximum population, which was 3,500,000, occurred in 1974 (for $t = 10$).
7. Since the derivative of $f(x)$ is $f'(x) = -.1x^{-3}$, which is negative for all positive values of x, the function is decreasing for all nonnegative x. There is no minimum value.

Chapter 10

Exercises 10.1
1. $5x + C$
3. $ex + C$
5. $1.5x^2 - 5x + C$
7. $x^3/3 - x^2/2 + x + C$
9. $x^4/4 + 5x^3/3 - 3x^2 + 3x + C$
11. $x^5 + x^3 - 8x + C$
13. $.75x^{4/3} + C$
15. $2x^{7/2}/7 + C$
17. $-5/x^2 + C$
19. $-9/(2x^2) + C$
21. $4/\sqrt{x} + x^2/2 + C$
23. $6x^{1/2} + C$
25. $x^3/3 - x + C$
27. $2x^3/3 + 3x^2/2 + C$
29. $x^2/2 + \ln x + C$
31. $x^2/2 - x - 2\ln x + C$
33. $x^{\pi+1}/(\pi + 1) + C$
35. $x^{5.2}/5.2 + 1/.2x^{.6} + C$
37. $7e^x + C$
39. $.5e^{2x} - e^{-x} + 10x + C$
41. $-2e^{-3x} + 8x + C$
43. $e^{5+x} - x + C$
45. $.5e^{2x} - e^x + C$
47. $x^3/3 - e^{3x}/3 + 11x + C$

49. $e^{4+x} + C$
51. $2x^{3/2}/3 - .3x^{10/3} + 2e^{2x} + C$
53. $e^{6x}/6 + e^{3x}/3 + C$
55. $.5e^{2x} - x + C$
57. $e^{3x}/3 + 7/2e^{2x} + C$
59. $e^{12x}/12 - e^{6x}/3 + x + 2x^{3/2}/3 + C$
61. $C(x) = .3x^2 - .3x + 220$
63. $C(x) = x^3 - .2x^2 - .6x + 400$
65. $R(x) = 5x$
67. $R(x) = x^2 - x$
69. $C(x) = .3x^2 + 2x + 50$
71. $R(x) = 2x + 2x^{3/2}/3$

Exercises 10.2
1. $2(x + 5)^{3/2}/3 + C$
3. $(2x + 1)^{11}/22 + C$

5. $2(5x + 1)^{1/2}/5 + C$
7. $-9/44(4x + 3)^{11} + C$
9. $(4x^2 + 9)^{3/2}/12 + C$
11. $(x^2 + 7)^{12}/24 + C$
13. $(4x^2 + 3)^{1/2}/4 + C$
15. $-3/26(x^2 + 4)^{13} + C$
17. $x^5/5 + 2x^3/3 + x + C$
19. $x^8/8 - 4x^5/5 + 2x^2 + C$
21. $x^3 + x^2 + C$
23. $(x^2 + 4x)^{1/2} + C$
25. $(3x^2 + 6x - 8)^6/36 + C$
27. $2(x^2 + x + 1)^{1/2} + C$
29. $-1/6(5x^2 + 2x + 1)^3 + C$

31. $(2\sqrt{x} + 1)^6/6 + C$
32. $2x + 2\sqrt{x} + C$
35. $x^2/2 + 2x^{3/2} + C$
37. $3(x^{1/3} + 9)^2/2 + C$
39. $(x^{1/3} + 2)^6/2 + C$
41. $x^2/3 + x/2 + (7/6)\ln x + C$
43. $x^2/2 + 6/x + C$
45. $3x^2/2 - 3x + C$
47. $4x^{3/2}/3 - 4x + C$
49. $(x^{16} + 53)^{11}/176 + C$
51. $R(x) = (4x - 3)^{11}/44 - 1/44$
53. $P(x) = -1/16(x^2 + 1)^8$

Exercises 10.3

1. $e^{2x}/2 + C$
3. $-4/5e^{5x} + C$
5. $3e^{x^2}/2 + C$
7. $e^{4x^2}/8 + C$
9. $e^{2x^3}/6 + C$
11. $2e^{x^3} + C$
13. $x^2/2 + C$
15. $e^{7x}/7 + e^{3x^2}/6 + 5x + C$
17. $x^{e+1}/(e + 1) + C$
19. $e^{6x}/6 - 3/e^x + C$
21. $e^{(x^2+1)/2} + C$
23. $x^3/3 + x^2 + C$
25. $.1 \ln(5x^2 + 2) + C$
27. $(-5/8)\ln(9 - 4x^2) + C$
29. $(1/9)\ln(9x - 1) + C$
31. $(-2/5)\ln(6 - 5x) + C$
33. $.5 \ln(x^2 + 4x + 3) + C$
35. $(1/3)\ln(x^3 - 3x) + C$
37. $2 \ln x - 5x^{-1} + C$
39. $6x^{3/2} + 12x + 8\sqrt{x} + C$
41. $.5(\ln x)^2 + C$
43. $(1/3)(\ln x)^3 + C$
45. $.5[\ln(2x)]^2 + C$
47. 110.52
49. 6.93 years
51. Let $A(t)$ denote the pounds of uranium-234 t thousands of years from now. Then we have

$$A(t) = 10e^{-.0028t}$$

53. Let $N(t)$ be the number of million repeat customers after t weeks of the advertising campaign. Solving the differential equation

$$N'(t) = k(100 - N(t)),$$

we get

$$N(t) = 100 - Ce^{-kt}.$$

Using the condition that $N(0) = 0$, we find that $C = 100$, and using the equation $N(10) = 10$, we determine that $k = .0105$. Thus, we have

$$N(t) = 100 - 100e^{-.0105t}.$$

Exercises 10.4

1.
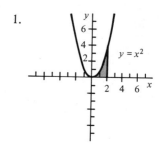

3.
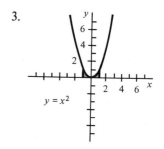

5.
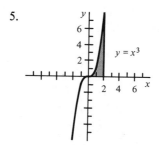

7.
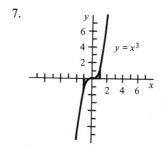

9.

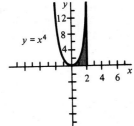

11.

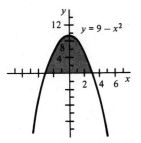

13.

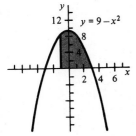

15.

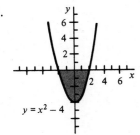

17.

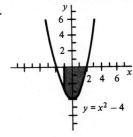

19.

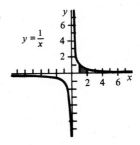

21.

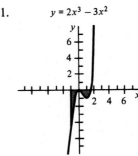

23.

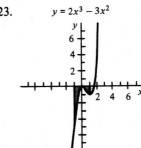

25.

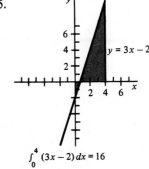

$$\int_0^4 (3x - 2)\, dx = 16$$

27.

$$\int_2^8 (0.5x+4)\, dx = 39$$

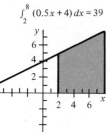

29.

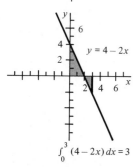

$y = 4 - 2x$

$$\int_0^3 (4-2x)\, dx = 3$$

31.

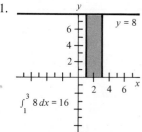

$y = 8$

$$\int_1^3 8\, dx = 16$$

33.

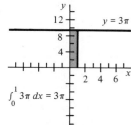

$y = 3\pi$

$$\int_0^1 3\pi\, dx = 3\pi$$

35.

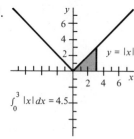

$y = |x|$

$$\int_0^3 |x|\, dx = 4.5$$

37.

$$\int_0^{100} 0.5x\, dx = 2500 = R(100)$$
Only $x \geqslant 0$ makes sense.

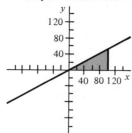

39.

$$\int_0^{100} (0.3x-2)\, dx = 1300 = P(100)$$

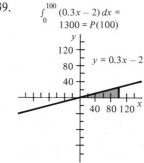

$y = 0.3x - 2$

Exercises 10.5

1. 5/6
3. 89/12
5. 48
7. 12
9. 15/4
11. 6
13. $9/2 - 2 \ln 4$
15. $.5e^2 - 1/e + 3.5$
17. $e^{12}/6 + e^6/3 - .5$
19. $.5e^2 - .5/e^4 - 3$
21. 38/3
23. $98/12 \approx 8.17$
25. $(3/26)(1/4^{13} - 1/5^{13}) \approx .0000000016$
27. $-8/5$
29. $7448/3 \approx 2482.67$
31. 94.5
33. $.5 \ln(25/9) \approx .511$
35. $(\ln 5.5)/3 \approx .568$
37. $15/4 + 2 \ln 4 \approx 6.523$
39. 0
41. $64/3 \approx 21.33$
43. 21
45. 1024
47. 36
49. $32/3 \approx 10.67$
51. $\ln 5 \approx 1.61$
53. $e - 1 \approx 1.72$
55. 1
57. 118/3
59. 64/3
61. 4.25
63. 1
65. $e^2 - 3 + 1/e \approx 4.76$
67. $16{,}000/3 \approx 5333.33$
69. $40{,}000/3 \approx 13{,}333.33$
71. $233/6 \approx 38.83$
73. 22
75. 105
77. $14{,}990/3 \approx 4996.67$

Exercises 10.6

1. 22/3
3. 3

5. $1.5 + 4 \ln 4 \approx 7.05$

7. $32/3$

9. $125/6 \approx 20.83$

11. $125/6 \approx 20.83$

13. $64/3$

15. $15/8 - \ln 4 \approx .49$

17. 2

19. $14/3$

21. $e^2 - 1 \approx 6.39$

23. $e^2 - 3 \approx 4.39$

25. $e^2/2 + 5/2 - 2e \approx .76$

27. $2(4x - x^3/3 + x^2/2)|_{.5(1-\sqrt{17})}^{0} \approx 7.52$

29. 9

31. $CS = 2,\ PS = 1$

33. $CS = 2,\ PS = 2$

35. $CS = \ln(1 + \sqrt{17}) + (\sqrt{17} - 9)/8 - \ln 2 \approx .331$

   $PS = (9 - \sqrt{17})/16 \approx .305$

Exercises 10.7

1. $\displaystyle\sum_{k=1}^{9} k^2$

3. $\displaystyle\sum_{k=1}^{12} k(k+1)$

5. $\displaystyle\sum_{k=1}^{40} 3k$

7. $\displaystyle\sum_{k=1}^{18} (2k+3)$

9. $\displaystyle\sum_{k=1}^{45} (k+4)$ or $\displaystyle\sum_{k=5}^{49} k$

11. 2.25

13. 6

15. $1/e^2 + 1/e + 1 + e \approx 4.22$

17. -14

19. $e^3 + e^2 + e + 1 \approx 31.19$

21. 13

23. 98

25. $4 + .5(e^{.5} + e + e^{1.5} + e^2) \approx 12.12$

27. $13/2$

29. 14

Review Problems: Chapter 10

1. $3e^2 x + C$
2. $(e + 9)x + C$
3. $1.5x^2 - 5x + C$
4. $x^3/30 - 3x^2/20 + 3x/5 + C$
5. $x^4/4 - 3x^{4/3}/4 + C$
6. $8 \ln x + 4x^{13/4}/13 + C$
7. $x^{e+3}/(e + 3) + C$
8. $15x^2/2 + 2x^3 + C$
9. $.5x^2 + x - 2 \ln x + C$
10. $9e^x + 3ex + C$
11. $5e^{3x}/3 + e^{-x} + C$
12. $(2x + 3)^{3/2}/3 + C$
13. $(5x + 1)^{12}/60 + C$
14. $(4x^2 + 25)^{3/2}/12 + C$
15. $3(x^2 - 9)^{2/3}/4 + C$
16. $x^5/5 + 4x^3/3 + 4x + C$
17. $4(x^2 + 9x)^{1/2} + C$
18. $5(x^2 - x + 6)^6/6 + C$
19. $x + 15x^{2/3}/2 + C$
20. $(\sqrt{x} + 5)^{10}/5 + C$
21. $x + 10\sqrt{x} + C$
22. $x^2/2 + 6/x + C$
23. $(x^{19} + 20)^2/38 + C$
24. $5e^{2x} + C$
25. $e^{4x^2}/8 + C$
26. $.25x^4 + C$
27. $.2e^{2.5x^2+15} + C$
28. $(1/8)\ln(4x^2 + 9) + C$
29. $1.5 \ln(x^2 - 5) + C$
30. $(\ln x)^2/10 + C$
31. $(2\sqrt{x} - 1)^3/3 + C$
32. $2 \ln x + C$
33. $(11/3)\ln(3x + 5) + C$
34. $(\ln 5x)^3/3 + C$
35. $f(x) = x^4/4 - x^3/3 + x^2/2 - 3x + 10$
36. $g(x) = 2x^{3/2}/3 + 3x + 6$
37. $-9/4$
38. $16/3$
39. $3e^2/2 + 2/e - 7/2$
40. $.5 \ln 1.8$
41. $8/3$
42. $500/3$
43. 5
44. $27/2$
45. $125/6$
46. $9/2$
47. $22/3$
48. $.75 - \ln 2 \approx .0569$
49. $13/6 \approx 2.167$
50. $149/3 \approx 16.333$
51. 18.5
52. $CS = 2,\ PS = 4$
53. $CS = (\ln 3) - 2/3 \approx .432,\ PS = 1/3$
54. $125/8 \approx 15.63$
55. $225/16 \approx 14.06$
56. 15
57. $(\ln 2)/.18 \approx 3.85$
58. $100e^{-(\ln 2/24,800)} \approx 99.9972$
59. $539/32 \approx 16.84$
60. $1089/64 \approx 17.02$
61. $33/2 = 16.5$

Social Science Applications: Chapter 10

1. 3466.67

3. $P'(t) = -.00012\, P(t)$

$$\frac{d}{dt} \ln P(t) = \frac{P'(t)}{P(t)} = -.00012$$

$$\ln P(t) = -.00012t + C$$

$$P(t) = e^{C - .00012t} = e^C e^{-.00012t}$$

$$P_0 = P(0) = e^C$$

$$P(t) = P_0 e^{-.00012t}$$

5. $\qquad n'(t) = .03(1,000,000 - n(t))$

$$\frac{n'(t)}{1,000,000 - n(t)} = .03$$

$$\frac{d}{dt} \ln(1,000,000 - n(t)) = \frac{-n'(t)}{1,000,000 - n(t)} = -.03$$

$$\ln(1,000,000 - n) = -.03t + C$$

$$n(t) = 1,000,000 - e^C e^{-.03t}$$

$$0 = n(0) = 1,000,000 - e^C$$

$$e^C = 1,000,000$$

$$n(t) = 1,000,000(1 - e^{-.03t})$$

7. $\qquad H'(t) = .0369 H(t)$

$$\frac{H'(t)}{H(t)} = .0369$$

$$\ln H(t) = .0369t + C$$

$$H(t) = e^C e^{.0369t}$$

$$1,999,316 = H(0) = e^C$$

$$H(t) = 1,999,316 e^{.0369t}$$

Chapter 11

Exercises 11.1

1. $(e^{3x}/3)(x - 1/3) + C$
3. $(5e^{6t-4}/6)(t - 1/6) + C$
5. $3e^x(x^2 - 2x + 2) + C$
7. $e^{t+1}(t^2 - 4t + 5) + C$
9. $e^{x^2}/2 + C$
11. $(3x^2/2)(\ln(5x) - 1/2) + C$
13. $(2t^{3/2}/3)(\ln(3t) - 2/3) + C$
15. $(x^3/3 + 4x)\ln x - x^3/9 - 4x + C$
17. $t \ln(9t) - t + C$
19. $2(x + 5)^{3/2}(3x/5 - 2) + C$
21. $3x^2(x + 2)^{5/2}/5 - 18x(x + 2)^{8/3}/40 + 54(x + 2)^{11/3}/440 + C$
23. $7e^x(x^3 - 3x^2 + 6x - 6) + C$
25. $9(e^2 + 1)/4 \approx 18.88$
27. $12.5 \ln 20 - .5 \ln 4 - 6 \approx 30.75$
29. $506/15 \approx 33.73$

Exercises 11.2

1. $\displaystyle \int \frac{4}{1 - x^2}\, dx = \int \frac{2}{1 - x}\, dx + \int \frac{2}{1 + x}\, dx$

$\displaystyle = \ln\left[\left(\frac{1 + x}{1 - x}\right)^2\right] + C$

3. $\displaystyle \int \frac{7x + 6}{x^2 + x - 6}\, dx = \int \frac{3}{x + 3}\, dx + \int \frac{4}{x - 2}\, dx$

$\displaystyle = \ln[(x + 3)^3(x - 2)^4] + C$

5. $\displaystyle \int \frac{9x - 8}{3x^2 - 10x - 8}\, dx = \int \frac{3}{3x + 2}\, dx + \int \frac{2}{x - 4}\, dx$

$\displaystyle = \ln[(3x + 2)(x - 4)^2] + C$

7. $\ln\sqrt{x^2 - 4x + 13} + C$

9. $\displaystyle \int \frac{16x - 10}{x^3 - 3x^2 - 10x}\, dx = \int \frac{1}{x}\, dx + \int \frac{2}{x - 5}\, dx$

$\displaystyle - \int \frac{3}{x + 2}\, dx$

$\displaystyle = \ln\left[\frac{x(x - 5)^2}{(x + 2)^3}\right] + C$

11. $\displaystyle \int \frac{7x^2 + 11x - 6}{x^3 + 6x^2}\, dx = \int \frac{2}{x}\, dx - \int \frac{1}{x^2}\, dx$

$\displaystyle + \int \frac{5}{x + 6}\, dx$

$\displaystyle = \ln[x^2(x + 6)^5] + \frac{1}{x} + C$

13. $\displaystyle \int \frac{3x^2 - 12x - 11}{x^3 - 6x^2 - 11x - 40}\, dx = \int \frac{1}{x - 8}\, dx + \int \frac{2x + 2}{x^2 + 2x + 5}\, dx = \ln[(x - 8)(x^2 + 2x + 5)] + C$

15. $\displaystyle \int \frac{x^2 + 9}{x^3 + 6x}\, dx = \frac{3}{2}\int \frac{1}{x}\, dx - \frac{1}{2}\int \frac{x}{x^2 + 6}\, dx = \frac{3}{2}\ln x - \frac{1}{4}\ln(x^2 + 6) + C$

17. $\displaystyle \int \frac{4x^2 - 14x + 15}{x^3 - 4x^2 + 5}\, dx = \int \frac{3}{x + 1}\, dx + \frac{4(10 + 3\sqrt{5})}{(5 + 7\sqrt{5})}\int \frac{1}{2x - 5 - \sqrt{5}}\, dx + \frac{4(10 - 3\sqrt{5})}{(5 - 7\sqrt{5})}\int \frac{1}{2x - 5 + \sqrt{5}}\, dx$

$\displaystyle = 3\ln(x + 1) + \frac{2(10 + 3\sqrt{5})}{(5 + 7\sqrt{5})}\ln(2x - 5 - \sqrt{5}) + \frac{2(10 - 3\sqrt{5})}{(5 - 7\sqrt{5})}\ln(2x - 5 + \sqrt{5}) + C$

$\displaystyle \approx 3\ln(x + 1) + 1.62\ln(2x - 7.24) - .62\ln(2x - 2.76) + C$

19. $\displaystyle \int \frac{4x^2 - 14x + 14}{x^3 - 5x^2 + 9x - 5}\, dx = \frac{8}{3}\int \frac{1}{x + 1}\, dx - \frac{1}{2}\int \frac{1}{x - 1}\, dx + \frac{11}{6}\int \frac{1}{x - 5}\, dx$

$\displaystyle = \frac{8}{3}\ln(x + 1) - \frac{1}{2}\ln(x - 1) + \frac{11}{6}\ln(x - 5) + C$

Exercises 11.3

1. $(x^2 + 5)^{3/2}/3 + C$
3. $3(5x + 2)^{4/3}/20 + C$
5. $(2x + 6)^{51}/102 + C$
7. $(x^2 + 2)^{21}/42 + C$
9. $\ln\sqrt{x^2 + 10} + C$
11. $\ln\sqrt[5]{5x - 1} + C$
13. $-1/22(x^2 + 6)^{11} + C$
15. $\ln\sqrt[3]{x^3 + 3x^2 + 6x + 1} + C$
17. $.5e^{x^2} + C$
19. $-.2e^{-5x} + C$
21. $2e^{\sqrt{x}} + C$
23. $.5(\ln x)^2 + C$
25. $\ln(\ln x) + C$
27. $\sqrt{x^2 + 1} + 1/\sqrt{x^2 + 1} + C$
29. $2(x + 3)^{7/2}/7 - 12(x + 3)^{5/2}/5 + 6(x + 3)^{3/2} + C$
31. $3\ln(\sqrt[3]{x} - 1) + C$
33. $2x + 2x^{3/2}/3 + C$
35. $(x^2 + 1)^{7/2}/7 - 2(x^2 + 1)^{5/2}/5 + (x^2 + 1)^{3/2}/3 + C$
37. $2\sqrt{x + 1} + 5\ln(\sqrt{x + 1} - 1) - 5\ln(\sqrt{x + 1} + 1) + C$
39. $2\sqrt{x} - 4\ln(\sqrt{x} + 2) + C$
41. $-x + 12\sqrt{x} - 36\ln(\sqrt{x} + 3) + C$
43. $\ln[x/(x - 2\sqrt{x} + 1)] + C$
45. $\ln(\sqrt[3]{x^2} - \sqrt[3]{x} + 1) - \ln(\sqrt[3]{x} + 1) + C$
47. $2\ln\left(\sqrt{\dfrac{x + 3}{x - 1}} + 1\right) - 2\ln\left(\sqrt{\dfrac{x + 3}{x - 1}} - 1\right)$
$+ 2/\left(\sqrt{\dfrac{x + 3}{x - 1}} - 1\right) + 2/\left(\sqrt{\dfrac{x + 3}{x - 1}} + 1\right) + C$

Exercises 11.4

1. $-.75\ln\left(\dfrac{2 + \sqrt{4 - x^2}}{x}\right) + C$, by formula (3).
3. $.4\sqrt{(7 + 5x)^3} + C$, by formula (7).
5. $-3\sqrt{.4}\ln\left(\dfrac{\sqrt{5} + \sqrt{5 - 3x^2}}{\sqrt{3}\,x}\right) + C$, by formula (3).
7. $(\sqrt{19}/3)\sqrt{5 + 6x} + C$, by formula (5).
9. $\ln(x + \sqrt{x^2 + 16}) - .25\ln\left(\dfrac{4 + \sqrt{x^2 + 16}}{x}\right) + C$, by formulas (1) and (2).
11. $(4/3)(x - 18)\sqrt{9 + x} + 6\sqrt{9 + x} + C$, by formulas (5) and (6).
13. $\ln(x + 3 + \sqrt{x^2 + 6x + 5}) + C$, by formula (1).
15. $.5\ln(2x + 1 + \sqrt{4x^2 + 4x}) + C$, by formula (1).

Exercises 11.5

1. 1.104
3. 1.117
5. .325
7. $8/3 \approx 2.667$
9. .916
11. .191
13. 1.048
15. 1.611

Exercises 11.6

1. 1
3. It does not exist.
5. 1
7. It does not exist.
9. 1/6
11. $(\ln 4)/3$
13. 4
15. 2
17. It does not exist.
19. 9
21. It does not exist.
23. It does not exist.
25. Consumers' surplus $= 1$

Review Problems: Chapter 11

1. $(2/9)e^{9x} + C$
2. $.5e^{2x}(x - .5) + C$
3. $x\ln(4x) - x + C$
4. $.25(\ln x)^4 + C$
5. $(3x^2/4)(2\ln x - 1) + C$
6. $-1/(2 + x) + C$
7. $(-.2/e^{5x})(x^2 + .4x + .08) + C$
8. $-x - 4\sqrt{x} - 4\ln(1 - \sqrt{x}) + C$
9. $e^{7x^2}/14 + C$
10. $x - 4\ln(x + 5) + C$
11. $(2/5)(x + 2)^{5/2} - (4/3)(x + 2)^{3/2} + C$
12. $(1/3)(2x + 5)^{3/2} + C$
13. $2.5\ln(x^2 + 3) + C$
14. $(1/5)(x^2 + 1)^{5/2} - (1/3)(x^2 + 1)^{3/2} + C$
15. $x - \ln(e^x + 1) + C$
16. $2\ln(x - 1) - \ln\sqrt{x^2 + 1} + C$
17. $\ln(x^3 - x^2 + x - 1) + C$
18. $2\ln(\sqrt{x} + 1) + C$
19. $(1/9)\ln(x - 4) - (1/9)\ln(x + 5) + C$
20. $(2/3)\ln x + (1/3)\ln(x + 3) + C$
21. $\ln(x^3 + 2x^2) - 1/x + C$
22. 4
23. It does not exist.
24. 1
25. $-1/56$
26. 2
27. It does not exist.
28. It does not exist.
29. .261
30. .511
31. 1.793

Social Science Applications: Chapter 11

1. 9.94%
3. 94.72%
5. 68.31%

Chapter 12

Exercises 12.1

1.

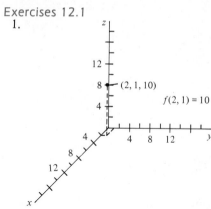

3.

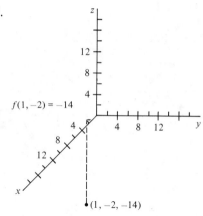

5.

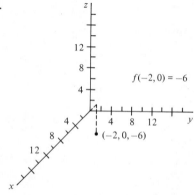

7.

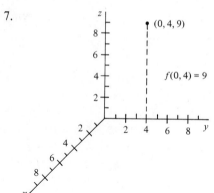

9.

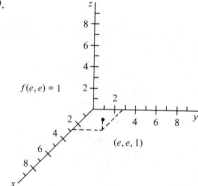

11.

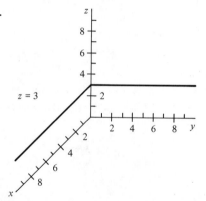

13. $z = 5x$

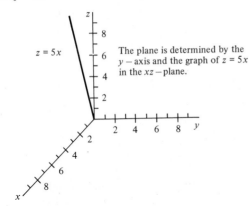

The plane is determined by the y – axis and the graph of $z = 5x$ in the xz – plane.

15.

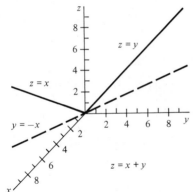

17.

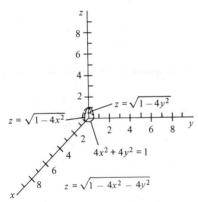

19.

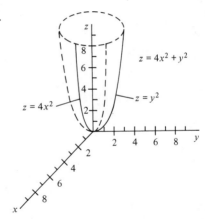

Exercises 12.2

1. $f_x(x, y) = 10$, $f_y(x, y) = -22$
3. $f_x(x, y) = 2x$, $f_y(x, y) = 0$
5. $f_x(x, y) = 2xy^3 + 4y^2 - 3x^2$,
   $f_y(x, y) = 3x^2y^2 + 8xy$
7. $f_x(x, y) = x/\sqrt{x^2 + 4y^2}$, $f_y(x, y) = 4y/\sqrt{x^2 + 4y^2}$
9. $f_x(x, y) = 2x/(1 + y^2)$,
   $f_y(x, y) = -2y(3 + x^2)/(1 + y^2)^2$
11. $f_x(x, y) = (3x^2y^3 + 4xy - 2x^4)/(y + x^3)^2$,
    $f_y(x, y) = (x^3y^2 + 2x^6y - 2x^2)/(y + x^3)^2$
13. $f_x(x, y) = 1/2\sqrt{xy + x}$, $f_y(x, y) = -\sqrt{x}/2(y + 1)^{3/2}$
15. $f_x(x, y) = 2xe^{x^2+y^2}$, $f_y(x, y) = 2ye^{x^2+y^2}$
17. $f_x(x, y) = 10x/(x^2 + y^4)$, $f_y(x, y) = 20y^3/(x^2 + y^4)$
19. $f_x(x, y) = 9e^x \ln y$, $f_y(x, y) = 9e^x/y$
21. $v_x = yz$, $v_y = xz$, $v_z = xy$
23. $w_x = ze^{x+y}$, $w_y = ze^{x+y}$, $w_z = e^{x+y}$
25. $u_w = xy$, $u_x = e^{x/y}\sqrt{z^2 + 1}/y - wy$,
    $u_y = -xe^{x/y}\sqrt{z^2 + 1}/y^2 - wx$, $u_z = ze^{x/y}/\sqrt{z^2 + 1}$
27. $f_x(x, y) = 0$, $f_y(x, y) = 0$, $f_{xx}(x, y) = 0$,
    $f_{yy}(x, y) = 0$, $f_{xy}(x, y) = 0$, $f_{yx}(x, y) = 0$
29. $f_x(x, y) = 23$, $f_y(x, y) = -6$, $f_{xx}(x, y) = 0$,
    $f_{yy}(x, y) = 0$, $f_{xy}(x, y) = 0$, $f_{yx}(x, y) = 0$
31. $f_x(x, y) = 3x^2y^5 + 6x^5y^4$,
    $f_y(x, y) = 5x^3y^4 + 4x^6y^3$,
    $f_{xx}(x, y) = 6xy^5 + 30x^4y^4$,
    $f_{yy}(x, y) = 20x^3y^3 + 12x^6y^2$,
    $f_{xy}(x, y) = 15x^2y^4 + 24x^5y^3$,
    $f_{yx}(x, y) = 15x^2y^4 + 24x^5y^3$
33. $f_x(x, y) = -2xy/(x^2 + 1)^2$, $f_y(x, y) = 1/(x^2 + 1)$,
    $f_{xx}(x, y) = (6x^2y - 2y)/(x^2 + 1)^3$, $f_{yy}(x, y) = 0$,
    $f_{xy}(x, y) = -2x/(x^2 + 1)^2$,
    $f_{yx}(x, y) = -2x/(x^2 + 1)^2$
35. $f_x(x, y) = ye^{xy}$, $f_y(x, y) = xe^{xy}$, $f_{xx}(x, y) = y^2e^{xy}$,
    $f_{yy}(x, y) = x^2e^{xy}$, $f_{xy}(x, y) = e^{xy}(xy + 1)$,
    $f_{yx}(x, y) = e^{xy}(xy + 1)$

37. $R_x = 10,\ R_y = 2y + 3$

39. $P_x = 9,\ P_y = y + 3$

41. The coordinates of the curve satisfy both of the equations $z = \sqrt{25 - x^2 - y^2}$ and $y = 3$. Thus, the coordinates satisfy the equation

$$z = \sqrt{25 - x^2 - 9} \quad \text{or} \quad z = \sqrt{16 - x^2}.$$

$$\frac{dz}{dx} = -x/\sqrt{16 - x^2}, \qquad \frac{dz}{dx}\bigg|_{(3,3,\sqrt{7})}$$

$$= -3/\sqrt{7} \approx -1.13$$

The slope of the tangent line is $-3/\sqrt{7} \approx -1.13$.

43. The coordinates of the curve satisfy both of the equations $z = x^2 + y^2$ and $x = 1$. Thus, the coordinates satisfy the equation

$$z = 1 + y^2.$$

$$\frac{dz}{dy} = 2y, \qquad \frac{dz}{dy}\bigg|_{(1,3,10)} = 6.$$

Exercises 12.3

1. The only critical value is $(1/3, -3)$.

3. The only critical value is $(0, 0)$.

5. The only critical value is $(-5/8, -1/8)$.

7. The only critical value is $(\sqrt[3]{4/3}, -\sqrt[3]{9/2}) \approx (1.10, -1.65)$.

9. The only critical values are $(\sqrt{3}, -3) \approx (1.73, -3)$ and $(-\sqrt{3}, -3) \approx (-1.73, -3)$.

11. The only critical value is $(0, 0)$; $(0, 0, 0)$ is a saddle point.

13. The only critical value is $(-1, -.4)$; $(-1, -.4, 12.2)$ is a saddle point.

15. The only critical value is $(-\sqrt[3]{18}, -\sqrt[3]{18/3}) \approx (-2.62, -.87)$; $(-\sqrt[3]{18}, -\sqrt[3]{18/3}, -\sqrt[3]{324}) \approx (-2.62, -.87, -6.87)$ is a relative maximum.

17. The critical values are $([-1 + \sqrt{37}]/6, [-1 + \sqrt{37}]/6) \approx (.85, .85)$ and $([-1 - \sqrt{37}]/6, [-1 - \sqrt{37}]/6) \approx (-1.18, -1.18)$. The first critical value gives a saddle point, and the second one gives a relative minimum.

19. One critical value is $(1, 3)$; $(1, 3, -132)$ is a relative minimum. Another critical value is $(1, -4)$; $(1, -4, 211)$ is a saddle point. Another critical value is $(-2, 3)$; $(-2, 3, -105)$ is a saddle point. The only other critical value is $(-2, -4)$; $(-2, -4, 238)$ is a relative maximum.

21. The box with minimum surface area has a base that is 10 inches by 10 inches and a height of 5 inches.

23. The box with the maximum volume is a cube with each edge equal to one foot.

25. For maximum revenue each of the two products should sell for $1000; then, the revenue is $9,000,000.

Exercises 12.4

1. $(3/2, 1, 3/2)$

3. $(1/\sqrt{2}, 1/\sqrt{2}, \sqrt{2})$

5. $(\sqrt{5}/3, -1/3, 2/3)$ and $(-\sqrt{5}/3, -1/3, 2/3)$

7. $(-3\sqrt{3}/2, 3/2, -9\sqrt{3}/4)$

9. $(1, 1, -1, -5)$

11. $(1/3, 2/3, 2/3, 3)$

13. $(0, 3, 0, 9)$ and $(0, -3, 0, 9)$

15. For maximum productivity there should be $40/13 \approx 3.08$ units of capital and $60/13 \approx 4.62$ units of labor. Then, production is $18(40/13)^{5/3}(60/13)^{1/2} \approx 251.71$.

17. The box with maximum volume has a base that is 5 inches by 5 inches and a height of 2.5 inches.

19. The point on the parabola that is closest to $(1/2, 2)$ is $(1, 1)$; its distance to $(1/2, 2)$ is $\sqrt{5}/2 \approx 1.12$.

Review Problems: Chapter 12

1.

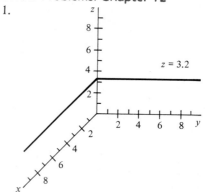

2.

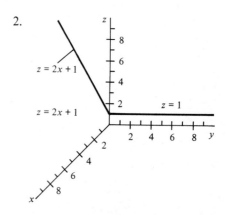

3.

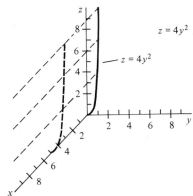

$z = 4y^2$

$z = 4y^2$

4.

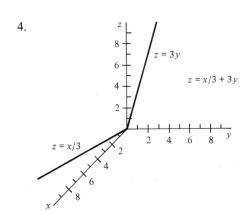

$z = 3y$

$z = x/3 + 3y$

$z = x/3$

5.

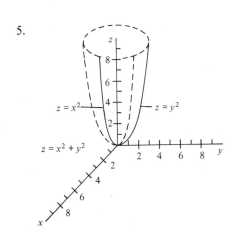

$z = x^2$

$z = y^2$

$z = x^2 + y^2$

6. $f_x(x, y) = .8x + .1, \quad f_y(x, y) = -2y + .5$

7. $g_x(x, y) = 2x - 3y, \quad g_y(x, y) = -3x + 10y$

8. $f_x(x, y) = 2x/(3x^2 + y^2)^{2/3},$
   $f_y(x, y) = 2y/3(3x^2 + y^2)^{2/3}$
9. $g_x(x, y) = -2x(y^2 + 5y - 6)/(x^2 + 4)^2,$
   $g_y(x, y) = (2y + 5)/(x^2 + 4)$
10. $f_x(x, y) = x/\sqrt{(x^2 + 1)(y^2 + 9)},$
    $f_y(x, y) = -y\sqrt{x^2 + 1}/\sqrt{(y^2 + 9)^3}$
11. $g_x(x, y) = e^{x^2+y}(1 + 2x^2), \quad g_y(x, y) = xe^{x^2+y}$
12. $f_x(x, y) = 10xy^2/(5x^2y^2 + 1),$
    $f_y(x, y) = -2/(5x^2y^3 + y)$
13. $g_x(x, y) = e^y/(x + 1), \quad g_y(x, y) = e^y \ln(x + 1)$
14. $f_x(x, y) = 0, \quad f_y(x, y) = 0, \quad f_{xx}(x, y) = 0,$
    $f_{yy}(x, y) = 0, \quad f_{xy}(x, y) = 0, \quad f_{yx}(x, y) = 0$
15. $g_x(x, y) = 2xy + y^2 - y, \quad g_{xx}(x, y) = 2y,$
    $g_{xy}(x, y) = 2x + 2y - 1,$
    $g_y(x, y) = x^2 + 2xy - x, \quad g_{yy}(x, y) = 2x,$
    $g_{yx}(x, y) = 2x + 2y - 1$
16. $h_x(x, y) = 2xe^{x^2+y^2}, \quad h_y(x, y) = 2ye^{x^2+y^2},$
    $h_{xx}(x, y) = 2e^{x^2+y^2}(1 + 2x^2),$
    $h_{xy}(x, y) = 4xye^{x^2+y^2},$
    $h_{yy}(x, y) = 2e^{x^2+y^2}(1 + 2y^2), \quad h_{yx}(x, y) = 4xye^{x^2+y^2}$
17. The saddle point $(0, 0, 0)$ occurs at the only critical value $(0, 0)$.
18. The relative minimum $(0, 0, 0)$ occurs at the only critical value $(0, 0)$.
19. There are no critical values.
20. The saddle point $(1/16, 8, -1/2)$ occurs at the only critical value $(1/16, 8)$.
21. The relative minimum $(2, 2, -13)$ occurs at the critical value $(2, 2)$. The saddle point $(-1, -1, 14)$ occurs at the only other critical value.
22. $(1/2, 1/2, -5/4)$
23. $(\sqrt{2}, 1/\sqrt{2}, 1)$ and $(-\sqrt{2}, -1/\sqrt{2}, 1)$
24. $(-4/3), -2/3, -4/3, -6)$
25. $(0, 0, 2, 4)$
26. The least expensive box has a base that is 3 inches by 3 inches and a height of 5 inches. The cost of the box is $2.70.
27. Maximum revenue occurs when $x = 100$ and $y = 200$; it is $1,100,000.
28. Maximum production occurs when $x = 6$ and $y = 5$; it is $60 \cdot 6^{6/5} \cdot 5^{4/5} \approx 1866.85$.
29. The point on the circle closest to $(4, 2)$ is $(2 + \sqrt{2}, \sqrt{2})$; its distance to $(4, 2)$ is $2\sqrt{3 - 2\sqrt{2}} \approx .83$.
30. The largest possible product is 1; it occurs when $x = 1, y = 1, z = 1$.

Social Science Applications: Chapter 12
1. 237.6   3. 3.96
5. 10 inches   7. 1
9. 1004   11. 40

Chapter 13

Exercises 13.1

1. {0}
3. {0, 3, 6, 9, . . .} = {3*n*: *n* = 0, 1, 2, 3, . . .}
5. {*A, B, C, D, E*}
7. ∅
9. ∅, {0}, {5}, {0, 5}
11. ∅, {4}, {5}, {6}, {*e*}, {4, 5}, {4, 6}, {4, *e*}, {5, 6},
    {5, *e*}, {6, *e*}, {4, 5, 6}, {4, 5, *e*}, {4, 6, *e*}, {5, 6, *e*},
    {4, 5, 6, *e*}
13. {1, 2, 3, 4, 5, 6, 7, 8}
15. {1, 2, 3, 5, 6, 7}
17. {1, 5}
19. ∅
21. {1, 3, 5, 7}
23. {2, 4, 6, 8}
25. ∅
27. {1, 3, 4, 5, 7, 8}
29. {1, 2, 3, 4, 5, 6, 7, 8}
31. {1, 2, 3, 4, 5, 6, 7, 8}
33. {all college instructors who are good at their jobs}
35. {all college students who are good at their jobs}
37. {all people who are either college students or lazy}
39. {all lazy college students who are good at their jobs}

41. $A \cap B \cap C$ 43. $A' \cap B' \cap C$

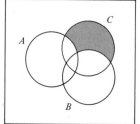

45. $A' \cap B' \cap C'$ 47. $(A' \cup B) \cap C$

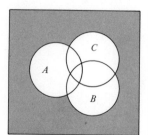

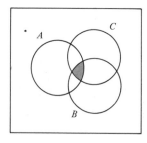

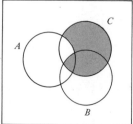

49. $A' \cup B' \cup C'$

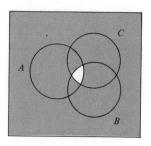

51. We give an explicit description of *E*.

$$E = \{1973, 1974, 1978, 1979, 1980\}$$

$$B \cap D = \varnothing$$

$$B \cap E = \{1974, 1979, 1980\}$$

(Some economists have noted that surges in the money supply are often followed by surges in inflation about two years later.)

Exercises 13.2

1. It has 4 elements.   3. It has 4 elements.
5. It is infinite.   7. It is infinite.
9. It has 2 elements.   11. It has no elements.
13. It is infinite.   15. 5
17. 5   19. 7
21. 22   23. 18
25. 4   27. 12
29. 11   31. 23
33. a. 50, b. 50, c. 40, d. 70, e. 50, f. 30
35. a. 350, b. 820, c. 880, d. 80, e. 130, f. 110

Exercises 13.3

1. 1   3. 5040
5. 8   7. 20
9. 360   11. 604,800
13. 6   15. 60
17. 336   19. 15
21. 260,000   23. 10,000
25. 210   27. 11,880
29. 60,466,176   31. 15
33. 2,176,782,336

Exercises 13.4

1. 1   3. 3
5. 28   7. 28

9. 1   11. 4
13. 16   15. 64
17. $x^6 + 6x^5y + 15x^4y^2 + 20x^3y^3$
19. $x^8 + 8x^7y + 28x^6y^2 + 56x^5y^3$
21. 126   23. 120
25. 10   27. 133,784,560
29. 792   31. 15
33. $C(3, 1)C(6, 3)C(4, 1) = 240$

Review Problems: Chapter 13
1. $\{k^2: k = 0, 1, \ldots, 10\} =$
   $\{0, 1, 4, 9, 16, 25, 36, 49, 64, 81, 100\}$
2. $\{3, 9, x, 4, y\}$, $\varnothing$, $\{3\}$, $\{a\}$, $\{x\}$, $\{4\}$, $\{y\}$, $\{3, a\}$,
   $\{3, x\}$, $\{3, 4\}$, $\{3, y\}$, $\{a, x\}$, $\{a, 4\}$, $\{a, y\}$, $\{x, 4\}$,
   $\{x, y\}$, $\{4, y\}$, $\{x, 4, y\}$, $\{a, 4, y\}$, $\{a, x, y\}$, $\{a, x, 4\}$,
   $\{3, 4, y\}$, $\{3, x, y\}$, $\{3, x, 4\}$, $\{3, a, y\}$, $\{3, a, 4\}$,
   $\{3, a, x\}$, $\{a, x, 4, y\}$, $\{3, x, 4, y\}$, $\{3, a, 4, y\}$,
   $\{3, a, x, y\}$, $\{3, a, x, 4\}$
3. $\{4, 5, 6\}$ 4. $\varnothing$
5. $\{1, 2, 7, 8\}$ 6. $\{1, 2, 7, 8\}$
7. $\{1, 2, 3, 4, 5, 6\}$ 8. $A \cap B \cap C'$

9. 1   10. 120
11. 720   12. 20
13. 604,800   14. 1
15. 5   16. 56
17. 60   18. 32
19. $20x^3y^3 + 15x^2y^4 + 6xy^5 + y^6$
20. 12   21. 0
22. 6   23. a. 50, b. 20, c. 15
24. 20   25. 17,576,000
26. 177,100   27. 133,784,560
28. 3,628,800   29. 10,272,278,170
30. 22,308

Social Science Applications: Chapter 13
1. There are 100 ways to complete EUN8; thus, there
   are 100 cars that might be the getaway car.
3. $C(100, 2)C(435, 2) = 467,255,250$
5. $C(25, 10) = 3,268,760$

Chapter 14

Exercises 14.1
1. $\{HH, HT, TH, TT\}$ 3. $\{1, 2, 3, 4, 5, 6\}$
5. $\{(1, 1), (1, 2), (1, 3), (1, 4), (1, 5), (1, 6), (2, 1),$
   $(2, 2), (2, 3), (2, 4), (2, 5), (2, 6), (3, 1), (3, 2),$
   $(3, 3), (3, 4), (3, 5), (3, 6), (4, 1), (4, 2), (4, 3),$
   $(4, 4), (4, 5), (4, 6), (5, 1), (5, 2), (5, 3), (5, 4),$
   $(5, 5), (5, 6), (6, 1), (6, 2), (6, 3), (6, 4), (6, 5), (6, 6)\}$
7. $\{Ja, F, Mar, Ap, May, Jun, Jul, Au, S, O, N, D\}$
9. 3/8   11. 7/8
13. 3/8   15. 1/36
17. 1/9   19. 1/6
21. 1/4   23. 11/16
25. 5/16   27. 1/17
29. 1/221   31. 4/17
33. .95   35. .95
37. .20   39. 1 to 1

41. 1 to 51   43. 1 to 7
45. Two.

Exercises 14.2
1. .5   3. 1/18
5. 1/13   7. $33/33,320 \approx .00099$
9. .5   11. .25
13. $33/66,640 \approx .00050$ 15. $45/1176 \approx .03827$
17. $33/16,660 \approx .00198$ 19. $143/16,660 \approx .00858$
21. $1/46,410 \approx .00002$ 23. 1/6
25. 5/6   27. 5/36
29. 1/32   31. 1/32
33. .5   35. .5
37. .6   39. .6
41. .4

43. The probability (to five decimal places) that 1 or more pairs of the randomly drawn slacks has flaws is .42691. If 5 pairs are style A and 5 pairs are style C then the probability is .26767.
45. The probability (to five decimal places) that 1 or more of the transistors is bad is .46443.
47. .06   49. .576

Exercises 14.3
1. $8/9 \approx .88889$ 3. .28
5. .50   7. $1/3 \approx .33333$
9. .40   11. $4/23 \approx .17391$
13. $10/23 \approx .43478$ 15. .54
17. $14/23 \approx .60870$ 19. $252/259 \approx .97297$
21. $1/117 \approx .00855$ 23. $27/41 \approx .65854$
25. The probability that the faulty unit was produced at Plant A is .375; the probability that it was produced at Plant B is .250; the probability that it was produced at Plant C is .375.
27. $1/1501 \approx .00067$

Exercises 14.4
1. $35/128 \approx .27344$ 3. $7/64 \approx .10938$
5. $1/256 \approx .00391$ 7. $1/32 = .03125$
9. $125/3888 \approx .03215$ 11. $3125/7776 \approx .40188$
13. $1/7776 \approx .00013$ 15. .001536
17. .000064   19. .08192
21. The probability that exactly 1 faulty transistor is chosen is approximately .00597006.
23. The probability that no faulty transistors are chosen is .99401498.
25. The probability of exactly 1 sale is approximately .16986931.
27. The probability of no sales is approximately .81537270.
29. .14880348   31. .43046721

Exercises 14.5
1. After two transitions the state vector is (.386 .614). The steady state vector is $(5/13\ 8/13) \approx$ (.3846 .6154).
3. After two transitions the state vector is (.712 .288); the steady state vector is (1 0).
5. After two transitions the state vector is (.1 .9); every vector is a steady state vector.
7. After two transitions the state vector is (.267 .291 .442); the steady state vector is $(5/19\ 11/38\ 17/38) \approx$ (.263 .289 .447).
9. After two transitions the state vector is (.120 .676 .204); the steady state vector is (0 1 0).

11. The matrix is a transition matrix.
13. The matrix is not a transition matrix, since the sum of the entries in some row is not equal to 1.
15. The matrix is a transition matrix.
17. The matrix is not a transition matrix, since the sum of the entries in some row is not equal to 1.
19. The vector is a state vector.
21. The vector is not a state vector, since not all of the entries are nonnegative.
23. The vector is not a state vector, since the sum of the entries is not equal to 1.
25. The vector is a state vector.
27. The vector is not a state vector, since the sum of the entries is not equal to 1.
29. 31%   31. 51.16%
33. 70.5%   35. 71.03625%

Review Problems: Chapter 14
1. 32   2. 1/32
3. 1/32   4. 5/16
5. 5/32   6. 1/2
7. 31/32   8. 1/2
9. 1/1296   10. 1/1296
11. 25/216   12. 5/324
13. 5/1296   14. 35/1296
15. 1/6   16. 1296
17. $1/108,290 \approx .00000923$
18. $1/649,740 \approx .00000154$
19. $33/66,640 \approx .00049520$
20. $143/39,984 \approx .00357643$
21. $3/16 = .1875$ 22. $9/460 \approx .019565$
23. .8   24. .52
25. $2/13 \approx .15385$ 26. $8/15 \approx .53333$
27. $7/15 \approx .46666$ 28. $11/13 \approx .84615$
29. .200   30. .022
31. $(.99)^3(.96)^3(.97)^3 \approx .7835$
32. $1 - (.99)^2(.96)^2 \approx .0967$
33. $5/22 \approx .2273$ 34. $4/11 \approx .3636$
35. After two transitions the state vector is (.576 .424); the steady state vector $(4/7\ 3/7) \approx$ (.5714 .4286).
36. After two transitions the state vector is (.559 .441); the steady state vector is (1 0).
37. After two transitions the state vector is (.28 .42 .30); the steady state vector is $(1/4\ 7/16\ 5/16) \approx$ (.2500 .4375 .3125).
38. After 1 month 24% of the people play video games.
39. After 2 months 29.6% of the people play video games.
40. After many many months approximately 1/3 of the people play video games, since the steady state for the transition matrix is (1/3 2/3).

Social Science Applications: Chapter 14

1. The probability that all of them own condominiums is .000064, and the probability that none of them owns a condominium is .262144.

3. After 1 year, 45.05% of the people live in the city. Ultimately the fraction of the population living in the city will be $1/101 \approx .0099$.

5. The probability that the chosen legislator is Republican is $27/56 \approx .4821$.

7. $1/256 \approx .0039$

9. $(.75)^3 = .421875$

Photo Credit List

Chapter 1

Chapter Opener: © Paul Sequeira/Photo Researchers
Career Profile: Arthur Glauberman/Photo Researchers

Chapter 2

Chapter Opener: © Paolo Koch/Photo Researchers
Career Profile: © Sherry Suris, Photo Researchers

Chapter 3

Chapter Opener: F.B. Grunzweig/Photo Researchers
Career Profile: © The Photo Works/Photo Researchers

Chapter 4

Chapter Opener: Margot Granitsas/Photo Researchers
Career Profile: Art Stein/Photo Researchers

Chapter 5

Chapter Opener: © Paolo Koch/Photo Researchers
Career Profile: The Photo Works/Photo Researchers

Chapter 6

Chapter Opener: Russ Kinne/Photo Researchers
Career Profile: © Paul Sequeira/Photo Researchers

Chapter 7

Chapter Opener: Arthur Treas/Photo Researchers
Career Profile: Robert A. Isaacs, 1980/Photo Researchers

Chapter 8

Chapter Opener: Steve Niedorf 1983/The Image Bank
Career Profile: © M.E. Warren, Photo Researchers

Chapter 9

Chapter Opener: Robert A. Isaacs 1982/Photo Researchers
Career Profile: A.G. Schoenfeld 1983/Photo Researchers

Chapter 10

Chapter Opener: The Image Bank
Career Profile: The Photo Works/Photo Researchers

Chapter 11

Chapter Opener: Larry Dale Gordon, 1981/The Image Bank
Career Profile: James Foote/Photo Researchers

Chapter 12

Chapter Opener: © Tom McHugh/Photo Researchers
Career Profile: Peter Angelo Simon 1976/Photo Researchers

Chapter 13

Chapter Opener: Steve Kagan 1980/Photo Researchers
Career Profile: Steven L. Feldman/Photo Researchers

Chapter 14

Chapter Opener: © Thomas S. England 1981/Photo Researchers
Career Profile: © 1978 Guy Gillette/Photo Researchers

Index